Silicon Nanoparticles

Silicon Nanoparticles

Synthesis, Uptake, and Applications

Edited by

Archana Singh
Department of Plant Molecular Biology
University of Delhi South Campus
New Delhi, Delhi, India

Aryadeep Roychoudhury
Discipline of Life Sciences, School of Sciences
Indira Gandhi National Open University
Maidan Garhi, New Delhi, Delhi, India

Indrakant K. Singh
Department of Zoology, Deshbandhu College
University of Delhi
New Delhi, Delhi, India

Dr. B. R. Ambedkar Centre for Biomedical Research, &
Delhi School of Public Health, Institute of Eminence
University of Delhi, Delhi, India

WILEY

Published by John Wiley & Sons, Inc., Hoboken, New Jersey.

Library of Congress Cataloging-in-Publication Data Applied for:

Hardback ISBN: 9781394261468

Cover Design: Wiley

Set in 9.5/12.5pt STIXTwoText by Straive, Chennai, India

This book is dedicated to the Divine Almighty for the blessings and family members for their endless encouragement and support.

Contents

List of Contributors

Alan Mendoza Alcantara
Carrera en Ingeniería en Innovación
Agrícola Sustentable
Tecnológico Nacional de México
Zamora
Mexico

Anupam Awasthi
Department of Botany
Multanimal Modi College, Modinagar
Ghaziabad
Uttar Pradesh
India

Debjyoti Bandhu Banerjee
Department of Botany
Ramakrishna Mission Vivekananda
Centenary College
Kolkata
West Bengal
India

Reeddhi Banerjee
Department of Biotechnology
St. Xavier's College (Autonomous)
Kolkata
West Bengal
India

Anirban Bhar
Department of Botany
Ramakrishna Mission Vivekananda
Centenary College
Kolkata
West Bengal
India

Nikhil Bhosale
Department of Pharmaceutical
Analysis
SVKM's Dr. Bhanuben Nanavati
College of Pharmacy
Mumbai
Maharashtra
India

Braulio Edgar Herrera Cabrera
Campus Puebla
Colegio de Postgraduados
Puebla
Mexico

Hebert Jair Barrales Cureño
Carrera en Ingeniería en Innovación
Agrícola Sustentable
Tecnológico Nacional de México
Zamora
Mexico

Ramesh Chandra Deka
Department of Molecular Biology and Biotechnology
Cotton University
Guwahati
Assam
India

and

Department of Chemical Sciences
Tezpur University
Tezpur
Assam
India

Bharat Engleng
Department of Molecular Biology and Biotechnology
Cotton University
Guwahati
Assam
India

P. Ganseh
Department of Botany
Yogi Vemana University
Kadapa
Andhra Pradesh
India

Rafael Salgado Garciglia
Instituto de Investigaciones Químico Biológicas
Universidad Michoacana de San Nicolás de Hidalgo
Morelia
México

Sagnik Goswami
Department of Botany
Ramakrishna Mission Vivekananda Centenary College
Kolkata
West Bengal
India

Nancy Sanjay Gupta
Molecular Biology Research Lab
Department of Zoology
Deshbandhu College
University of Delhi
New Delhi
Delhi
India

Victor Manuel Ocaño Higuera
Departamento de Ciencias Químico Biológicas
Universidad de Sonora
Hermosillo
Mexico

Himanshu
Department of Horticulture
School of Agriculture
Lovely Professional University
Phagwara
Punjab
India

Eeshan Kalita
Department of Molecular Biology and Biotechnology
Cotton University
Guwahati
Assam
India

Sumit Kumar Kashyap
Department of Molecular and Cellular Engineering
Jacob Institute of Biotechnology and Bioengineering
Sam Higginbottom University of Agriculture
Technology and Sciences
Prayagraj
Uttar Pradesh
India

Bharti Kaushik
Department of Botany
Multanimal Modi College, Modinagar
Ghaziabad
Uttar Pradesh
India

S. Khader Basha
Department of Botany
Yogi Vemana University
Kadapa
Andhra Pradesh
India

Tabassum Khan
Department of Pharmaceutical Chemistry
SVKM's Dr. Bhanuben Nanavati College of Pharmacy
Mumbai
Maharashtra
India

Arun Kumar
Department of Horticulture
School of Agriculture
Lovely Professional University
Phagwara
Punjab
India

Naveen Kumar
Department of Plant Molecular Biology
University of Delhi South Campus
Dhaula Kuan
New Delhi
Delhi
India

and

Department of Botany
Shivaji College
University of Delhi
New Delhi
Delhi
India

Prasann Kumar
Department of Agronomy
School of Agriculture
Lovely Professional University
Phagwara
Punjab
India

Maya Madhavan
Department of Biochemistry
Government College Kariavattom
Thiruvananthapuram
Kerala
India

Kalpak Majumdar
Department of Botany
Ramakrishna Mission Vivekananda
Centenary College
Kolkata
West Bengal
India

Arun Kumar Maurya
Department of Botany
Multanimal Modi College, Modinagar
Ghaziabad
Uttar Pradesh
India

Upasana Medhi
Department of Molecular Biology and
Biotechnology
Cotton University
Guwahati
Assam
India

Francisco Javier Verduzco Miramón
Carrera en Ingeniería en Innovación
Agrícola Sustentable
Tecnológico Nacional de México
Zamora
Mexico

Disha Mittal
Department of Life Sciences
School of Biosciences and Technology
Galgotias University
Greater Noida
Uttar Pradesh
India

Venkataraja V. Mokshith
Department of Molecular Biology and
Biotechnology
S.V Agricultural College
Acharya NG Ranga Agricultural
University (ANGRAU)
Tirupati
Andhra Pradesh
India

Sabeena Mustafa
Department of AI and Bioinformatics
King Abdullah International Medical
Research Center (KAIMRC)
King Saud Bin Abdulaziz University
for Health Sciences
King Abdulaziz Medical City
Ministry of National Guard Health
Affairs (MNGHA)
Riyadh
Kingdom of Saudi Arabia

Shrishti Naidu
Department of Plant Molecular
Biology
University of Delhi South Campus
New Delhi
Delhi
India

S. Naseem
Department of Botany
Yogi Vemana University
Kadapa
Andhra Pradesh
India

Shreyosi Pahari
Department of Biotechnology
St. Xavier's College (Autonomous)
Kolkata
West Bengal
India

K. Pallavi
Department of Botany
Yogi Vemana University
Kadapa
Andhra Pradesh
India

Subhransu Panda
KIIT-Technology Business Incubator
(KIIT-TBI)
KIIT University
Bhubaneswar
Odisha
India

Ruby Panwar
Department of Botany
Gargi College
University of Delhi
New Delhi
Delhi
India

Sampa Paul
Department of Molecular Biology and
Biotechnology
Cotton University
Guwahati
Assam
India

Aman Prakash
Department of Marine Biology
Leon H. Charney School of Marine
Sciences
University of Haifa
Haifa
Israel

Adeeb Rahman
Plant RNAi Biology Group
International Centre for Genetic
Engineering and Biotechnology
New Delhi
Delhi
India

Vaishali Raikwar
Department of Physics
Ramniranjan Jhunjhunwala College of
Arts
Science and Commerce (Autonomous)
Mumbai
Maharashtra
India

Ambika Rajendran
Division of Genetics
ICAR-Indian Agricultural Research
Institute (IARI)
Pusa Campus
New Delhi
Delhi
India

Dhandapani Raju
Division of Plant Physiology
ICAR-Indian Agricultural Research
Institute
Pusa Campus
New Delhi
Delhi
India

Ayyagari Ramlal
Plant Biotechnology Laboratory
School of Biological Sciences
Universiti Sains Malaysia (USM)
Penang
Malaysia

and

Division of Genetics
ICAR-Indian Agricultural Research Institute (IARI)
Pusa Campus
New Delhi
Delhi
India

Shashi Ranjan
Amrita Centre for Nanosciences and Molecular Medicine
Amrita Vishwa Vidyapeetham
Kochi
Kerala
India

Sujata Rathi
Department of Botany
Multanimal Modi College, Modinagar
Ghaziabad
Uttar Pradesh
India

Aryadeep Roychoudhury
Discipline of Life Sciences
School of Sciences
Indira Gandhi National Open University
Maidan Garhi
New Delhi
Delhi
India

Anjana Rustagi
Department of Botany
Gargi College
University of Delhi
New Delhi
Delhi
India

Sayan Sadhukhan
Department of Botany
Ramakrishna Mission Vivekananda Centenary College
Kolkata
West Bengal
India

Yogendra Prasad Saksena
Department of Botany
Multanimal Modi College, Modinagar
Ghaziabad
Uttar Pradesh
India

P. Shaik Sha Valli Khan
Department of Botany
Yogi Vemana University
Kadapa
Andhra Pradesh
India

Aarti Sharma
Department of Life Sciences
School of Biosciences and Technology
Galgotias University
Greater Noida
Uttar Pradesh
India

Aparna Sharma
Acharya Narendra Dev College
University of Delhi
New Delhi
Delhi
India

Sheetal
Department of Botany
Multanimal Modi College, Modinagar
Ghaziabad
Uttar Pradesh
India

Shubham Kumar Shrivastav
Molecular Biology Research Lab
Department of Zoology
Deshbandhu College
University of Delhi
New Delhi
Delhi
India

R.S. Silpa
Thejas
Neyyattinkara
Kerala
India

Ankur Singh
KIIT-Technology Business Incubator
(KIIT-TBI)
KIIT University
Bhubaneswar
Odisha
India

Archana Singh
Department of Plant Molecular
Biology
University of Delhi South Campus
New Delhi
Delhi
India

Indrakant K. Singh
Department of Zoology
Deshbandhu College
University of Delhi
New Delhi
Delhi
India

and

Molecular Biology Research Lab
Department of Zoology
Deshbandhu College
University of Delhi
New Delhi
Delhi
India

and

Delhi School of Public Health
Institute of Eminence
University of Delhi
New Delhi
Delhi
India

and

Dr. B. R. Ambedkar Centre for
Biomedical Research, & Delhi School
of Public Health
Institute of Eminence
University of Delhi
Delhi
India

Jatinder Singh
Department of Horticulture
School of Agriculture
Lovely Professional University
Phagwara
Punjab
India

Joginder Singh
Department of Botany
Nagaland University
Kohima
Nagaland
India

Sreeramanan Subramaniam
Plant Biotechnology Laboratory
School of Biological Sciences
Universiti Sains Malaysia (USM)
Penang
Malaysia

and

Centre for Chemical Biology (CCB)
Universiti Sains Malaysia (USM)
Penang
Malaysia

and

Department of Biology
Faculty of Science and Technology
Universitas Airlangga
Surabaya
Indonesia

Sumali Sucharita
KIIT-Technology Business Incubator
(KIIT-TBI)
KIIT University
Bhubaneswar
Odisha
India

V. Suneetha
Department of Botany
Yogi Vemana University
Kadapa
Andhra Pradesh
India

D. Suresh
Department of Botany
Yogi Vemana University
Kadapa
Andhra Pradesh
India

Seena Thomas
Department of Biostatistics
King Abdullah International Medical
Research Center (KAIMRC)
King Saud Bin Abdulaziz University
for Health Sciences
King Abdulaziz Medical City
Ministry of National Guard Health
Affairs (MNGHA)
Riyadh
Kingdom of Saudi Arabia

O.N. Tiwari
Centre for Conservation and Utilization of Blue Green Algae
Division of Microbiology
Indian Agricultural Research Institute (ICAR)
New Delhi
India

Luis Germán López Valdez
Laboratorio de Productos Naturales
Área de Química
Universidad Autónoma Chapingo
Texcoco
México

Prabhavathi Venkat Raman
Department of Botany
Shivaji College
University of Delhi
New Delhi
Delhi
India

Vikanksha
Department of Horticulture
School of Agriculture
Lovely Professional University
Phagwara
Punjab
India

Ameen Pareesa Zidhiq
M. Sc. Biotechnology
Vellore Institute of Technology (VIT)
Vellore
Tamil Nadu
India

Preface

In the current scenario of global warming and climate change, human beings and plants are going to experience more incidences of environmental stresses, which will affect human health and restrain crop production. Crises such as reduced fertilizer efficacy, dwindling water content, heat stress, drought, soil fertility degradation, rising pathogen and pest incidence, the socio-economic problems and food insecurity all pose substantial hurdles to sustainable agriculture. Likewise, health issues, including skin diseases, insomnia, psychological disorders, depression, high blood pressure and lethal diseases such as cancers may be caused due to environmental stresses. Nanotechnology is gaining precedence over other means to deal with the issue of sustainability. One of nature's most beneficial/abundant metalloids, silicon (Si), shows ameliorative effects against environmental challenges and is a promising element for fertilizer delivery and for drug delivery as nanovehicles. Silicon/silica nanoparticles (Si/SiO_2 NPs) have gained special attention due to their significant chemical and optoelectronic capabilities. Their mesoporous nature, easy availability, and least biological toxicity have made them highly attractive to researchers. Si/SiO_2 NPs can be synthesized by chemical, physical and biological methods. To enhance agricultural production, Si/SiO_2 NPs are supplied to plants by foliar, soil or seed priming. Upon uptake and translocation, they reach their destined cells and cause optimum growth, development and tolerance against environmental stresses as well as pest attack and pathogen infection. Si/SiO_2 NPs have been recognized as a versatile element for improved, safe and more efficient cancer treatment. Using Si/SiO_2 NPs as a supplement can be an eco-friendly and cost-effective option for better health management and sustainable agriculture, as they facilitate the delivery of drugs and nutrients and assist plants in mitigating abiotic and biotic stresses.

This book has 21 chapters that aim to present an overview of the methods of formulation of Si/SiO_2 NPs, their application, uptake and translocation; emphasize the role of Si/SiO_2 NPs in drug delivery, treatment of diseases such as dermatitis, infectious diseases and cancers as well as how application of Si/SiO_2 NPs can boost the growth and development of plants and their conventional advantage as fertilizers with special emphasis on their mitigating effects towards abiotic and biotic stresses. The usage of Si/SiO_2 NPs is one of the most suitable approaches in the present scenario since it provides an environment-friendly, economic, and successful system. This book will provide the most recent information on basic and the most advanced insights on synthesis, uptake of Si/SiO_2 NPs and their applications in the field of human health and agriculture. All the chapters are highlighted with well-drawn, clear figures, flowcharts and tables to make them impressive to the readers. Every chapter also highlights the future prospective of research. The content of the book is designed in such a manner that it will attract the attention of both beginners, such as students of undergraduate and postgraduate level, research scholars and young scientists who are newly introduced to the subject, as well as renowned scientists dealing with advanced research. The chapters of this book have been contributed by leading scientists and educationists from different disciplines like nanotechnology, agriculture, genetic engineering, biomedical sciences and biotechnology.

Silica nanoparticles have emerged as an efficient and eco-friendly solution to environmental stresses and human diseases. Moreover, they can be tailored by several stimuli-sensitive gatekeepers for releasing their cargos and can serve as nanofertilizer as well as an agent for gene delivery and drug administration. Considering their importance, the current book has been designed using a streamlined approach and straightforward language to make the concepts easy to follow. The chapters have been authored by academicians, scientists and experts in the field, and this book is an amalgamation of current knowledge on silica nanoparticles and their applications, along with figures and tables that complete the whole content in a more comprehensive way. The contents have been structured logically to deliver fundamental as well as updated knowledge on silica NPs. It will be a valuable resource to lay a firm foundation for research scholars and young scientists. This book has a competitive edge over others, as every effort has been made to include the use of unambiguous and lucid language, rich pedagogy and labelled illustrations to comprehend the concepts better. The book will convey an overview of our current knowledge on traditional, green and microbial synthesis of Si NPs and their applications in human health and agriculture. This title is devised for researchers, professionals and students in the field of agriculture, health science, nanotechnology, biotechnology, microbiology, biochemistry and genetic engineering.

We, the editors, would like to thank all the contributors for their considerable energy, time and effort that they spent in making this book an advanced source of knowledge for understanding the synthesis, uptake and applications of Si NPs in agriculture and health.

We are optimistic that this book will be effective in broadcasting the latest knowledge about the synthesis, uptake and applications of Si NPs in agriculture and health.

Acknowledgements

We, the two editors Dr. Archana Singh and Dr. Indrakant K. Singh, are grateful to our parents for their constant support and blessings. We would like to express our heartfelt appreciation to our little angels, Saumya and Kimaya, for creating such a beautiful ambience, enabling us to work tirelessly and providing unwavering emotional support. We are extremely thankful to Prof. Yogesh Singh, Vice Chancellor, University of Delhi; Prof. Shree Prakash Singh, Director, University of Delhi South Campus; and Prof. Sunit K. Singh, Director, Dr. B. R. Ambedkar Centre for Biomedical Research, University of Delhi, for providing overall support for our research and academic pursuits. We would like to convey our gratitude to our Ph.D. mentors Prof. Praveen K. Verma and Prof. A. K. Singh. We also acknowledge our research scholars for their diligent efforts while working on this project. We, the IoE Fellows, acknowledge Delhi School of Climate Change and Sustainability and Delhi School of Public Health, Institution of Eminence, University of Delhi, for providing overall support to our research and academic pursuits.

The other Editor Dr. Aryadeep Roychoudhury would also like to acknowledge the immense support of all his family members.

Last but not the least, our sincere thanks to all the authors and publisher.

1

An Introduction to Nanoparticles: Synthesis, Applications and Ethical Issues

Aryadeep Roychoudhury[1], Shreyosi Pahari[2] and Reeddhi Banerjee[2]

[1]*Discipline of Life Sciences, School of Sciences, Indira Gandhi National Open University, Maidan Garhi, New Delhi, Delhi, India*
[2]*Department of Biotechnology, St. Xavier's College (Autonomous), Kolkata, West Bengal, India*

1.1 Introduction

Nanotechnology is a branch of science that involves the use of matter at the atomic, molecular and supramolecular scale, generally ranging from 1 to 100 nm. It deals with designing, producing, characterizing and using devices, structures and systems by controlling their shape and size at the nanoscale.

The concept of manipulating atoms and molecules was first proposed by physicist Richard Feynman in his famous lecture 'There's Plenty of Room at the Bottom' in 1959 (Bayda et al. 2019). The term 'nanotechnology' was first coined by Norio Taniguchi in 1974 to describe precision engineering of materials at the nanoscale. The foundation for further research on nanotechnology was laid through the development of scanning tunnelling microscopy and atomic force microscopy (AFM) (Bayda et al. 2019). In 1986, three eminent scientists, Harry Kroto, Richard Smalley and Robert Curl, discovered a new class of carbon, viz., fullerene, that opened new possibilities in the field of nanomaterials. Significant advancements like the synthesis of carbon nanotubes and the development of nanolithography techniques were prominent during the 1900s. In 2000, several developments in research in the field of nanotechnology were coordinated by the National Nanotechnology Initiative launched in the United States (Bayda et al. 2019; Zhang et al. 2023).

Nanotechnology started gaining attention in agriculture as researchers began to explore the exploding potential of nanoparticles in improving plant growth, enhancing nutrient uptake and protecting the plants from pests and diseases. In 2003, the United States Department of Agriculture established the Nanoscale Science and Engineering for Agriculture and Food States (NanoSAFENET) to

Silicon Nanoparticles: Synthesis, Uptake, and Applications, First Edition.
Edited by Archana Singh, Aryadeep Roychoudhury, and Indrakant K. Singh.

promote further research in nanotechnology applications for food and agriculture. The importance of nanotechnology in forestry and related agricultural sectors was discussed by the International Conference on Nanotechnology for the Forest Products Industry in 2007. In 2015, the International Union of Pure and Applied Chemistry organized a conference on 'Nanotechnology in Sustainable Agriculture' to discuss the role of nanotechnology in dealing with agricultural challenges. Recent years have witnessed the advancement of nanotechnology with a special focus on nanosensors used for monitoring the health, nanopesticides and nano-fertilizers for targeted pest control and efficient nutrient delivery, respectively, and nanomaterials for water purification in agricultural systems. These enhancements throw light on the ability of nanotechnology to revolutionize agriculture in various aspects (Zhang et al. 2023).

Nanotechnology is evolving at a very fast pace and its benefits are traced in every domain like medicine, electronics, energy, etc. It has had a strong impact on different industries and society as a whole in all these years.

Nanotechnology has numerous applications in agriculture that affect crop production, soil management, sustainability and pest control (Fraceto et al. 2016). Nanoparticles are extensively used to enhance the nutrient uptake by plants, reduce nutrient leaching and increase crop yield while reducing environmental impacts. Soil remediation is done with the help of nano-clays and nanoscale zero-valent iron to improve soil structure and water retention capacity and remove pollutants (Haris et al. 2023). Nanopesticides and nanoherbicides target specific weeds and pests and help mitigate environmental contamination. Nanosensors can monitor soil moisture, enable optimal water management practices and check plant health in real time (Haris et al. 2023). Nanoparticles like nanosilica and nanochitosan are growth promoters that stimulate development, seed germination and overall plant growth (Fraceto et al. 2016)

1.2 Methods of Synthesis of Nanoparticles

With the use of different synthetic processes described below, a wide variety of nanoparticles can be synthesized. Nanostructures, which lie in the range of 1–100 nm, can be generated from atoms or by diminishing the size of microparticles to nanoparticles.

1.2.1 Co-precipitation

This reaction involves the process of nucleation, growth, coarsening and/or agglomeration. Some of its characteristics are as follows:

- Products require high supersaturated conditions to be formed and they are generally insoluble in nature.

- The nucleation step is highly crucial as it forms a large number of small particles.
- The size, morphology and properties of particles are largely affected by the secondary processes like Ostwald ripening and aggregation.

The chemical reaction that takes place is as follows:

$$xAy + (aq) + yBx - (aq) \rightarrow AxBy\ (s)$$

Some of the advantages are as follows:

- Particle size and composition can be easily controlled.
- Rapid preparation with energy efficiency and low temperature is required.
- Organic solvents are not required.

Some of the disadvantages are as follows:

- Trace amounts of impurity can get precipitated with the product.
- It is a time-consuming method which cannot be applied to charged particles.
- Reactants should not have varying rates of precipitation.

1.2.2 Hydrothermal Technique

This technique was first used by British geologist Roderick Murchison which describes the role of water in changing the earth's crust to form rocks and minerals at elevated temperature and pressure. This technique has been exploited in preparing solids like microporous crystals, chemical sensing oxides, superionic conductors, magnetic materials, etc.

The synthesis of single crystals based on hydrothermal techniques depends on the solubility of minerals in hot water under pressure. The growth of the crystals occurs in an autoclave consisting of nutrient supply dissolved in water. The growth chamber is maintained in such a way that a temperature gradient is created when the hotter end dissolves the nutrients and the cooler end allows the seed to undergo further growth.

Some of the advantages are as follows:

i) Almost all materials can be made soluble in a suitable solvent by applying heat and pressure (close to the critical point).
ii) Substances having low melting point and high vapour pressure can be obtained.
iii) The size, shape and crystallinity of the final product can be controlled precisely by adjusting the reaction time, temperature, type of solvent and surfactant.

The disadvantages of the hydrothermal technique are as follows:

i) The cost of an autoclave is expensive.
ii) The reaction process cannot be observed, a phenomenon known as the 'black box'.
iii) There could be certain safety issues.

1.2.3 Inert Gas Condensation

This method is the most frequently and widely used one in synthesizing metal nanoparticles. An ultra-ultrahigh vacuum chamber, filled with helium or argon gas at high pressure, is used to evaporate the metals. Eventually, the metal atoms collide with the gas and lose their kinetic energy, thereby condensing into small particles. These then undergo Brownian coagulation and coalescence to form nanoparticles.

This was the first technique used to make nanocrystalline metals and alloys. Clusters aggregate by the homogeneous nucleation in the gaseous phase and then grow by coalescence. The size of the particles depends on the residence time of the particles in the growth phase, the type of inert gas and the vapour pressure of the evaporating material. The size of the nanoparticle is directly proportional to the gas and vapour pressure and mass of the inert gas. The particles are collected with the help of a rotating cylindrical device cooled with liquid nitrogen.

The smaller ones (2–50 nm) are derived from the gas flow and deposited as a powder on the surface of the collection device. These are then scraped off from the surface in the form of a metallic plate.

Advantages:

i) The particle size can be controlled by adjusting the flow rate of the inert gas.

Disadvantages:

i) The quantity is limited to the laboratory scale. It is time-consuming and hence industrial-level production is restricted.

1.2.4 Sputtering

Sputtering is the process of ejecting atoms from the surface of a material through bombardment of energetic particles. Bombarding ions drive off atoms from the target. These atoms then travel and strike a substrate, where they get deposited to form a layer (Yu et al. 2008). The ions are supplied by the glow discharge phenomenon due to an applied electric field between two electrodes in a gas at low pressure. After reaching a certain voltage, the gas breaks down to conduct electricity. This ionized form of gas is called plasma, whose ions are accelerated by a large electric field towards a target. After hitting the target, these ions are ejected into the plasma, where they are carried further and deposited on the substrate surface. This is known as the DC sputtering. The sputtered gas used is an inert one to avoid any chemical reaction between the sputtering gas and the sputtered atoms (Yu et al. 2008). When these ions hit the target, the electric charge is balanced and they get transformed back into atoms within the process. This balancing results in a positive charge on the surface when the target is an insulator. The sputtering process halts when the incoming ions are repelled at high charge density. To resume the process, the polarity needs to be switched to draw sufficient electrons from the plasma and remove the surface charge.

This technique offers numerous process variables that can be adjusted to customize and enhance material properties. The use of solid targets allows precise control over material type and composition, eliminating the need for complex process chemistry. Sputtering enables the deposition of films with the same composition as the target source, making it a versatile process which can be applied to various types of material. However, sputtering targets can be costly; most incident energy on the target is converted to heat, requiring heat management; and in reactive sputter deposition, gas composition control is crucial to avoid target contamination.

1.2.5 Microemulsion

This method is considered as an excellent technique for preparing inorganic nanoparticles, although the precise mechanism of nanoparticle formation within microemulsions is not completely understood. When the microemulsion components like reactants are mixed, rapid reactant exchange occurs during collisions between water droplets in the microemulsion. This exchange leads to a precipitation reaction within nanodroplets, followed by nucleation, growth and coagulation of primary particles, resulting in the formation of final nanoparticles that are either surrounded by water or stabilized by surfactants.

The advantages of this method are as follows:

i) **Thermodynamic Stability**: Microemulsions offer improved stability and longer shelf life due to their inherent thermodynamic stability.
ii) **Ease of Preparation**: Microemulsions form spontaneously at room temperature and are simple to produce.
iii) **Nanoparticle Characteristics and Small Droplet Size**: They produce nanometre-sized inorganic particles with minimal agglomeration along with a crystalline structure and a high specific surface area.

The disadvantages are as follows:

1) **Surfactant Usage**: The formation of microemulsion requires large amounts of surfactants, which can be toxic at high concentrations.
2) **Limited Solubility**: They have limited capacity to solubilize substances with high melting points.

1.2.6 Microwave-Assisted Synthesis of Nanoparticles

Microwave-assisted synthesis of nanoparticles is used in various fields like nanotechnology and biochemical processes. This yields a higher amount of product and lesser by-products. The reactors offer precise control over reaction mixing and can tolerate high temperature and pressure. This allows better separation between nucleation and growth phases and also selectively activates nanomaterial precursors, enhancing scalability and providing selectivity in heating either the solvent or the precursor molecules during the preparation stage.

1.2.7 Laser Ablation

This involves the removal of material from a solid or liquid surface by exposing it to a laser beam. When the laser flux is low, the material heats up and either evaporates or sublimates due to absorbed laser energy. At higher flux levels, the material is usually transformed into plasma. Laser ablation refers to material removal using a pulsed laser, although it is feasible to ablate material with a continuous wave laser beam if the intensity of the laser is sufficiently high.

1.2.8 Sol–Gel Method

This is used to create solid materials from small molecules. During the process, the solution undergoes transition gradually into a gel-like system, comprising both liquid and solid phases, with morphologies ranging from discrete particles to continuous polymer networks. Precipitation allows for the formation of ultrafine and uniform ceramic powders. The raw materials should be uniformly mixed in the liquid phase, followed by hydrolysis and polycondensation reactions to form a stable and clear solution system. With time, the colloidal particles aggregate to form gel with 3D structures, which are then dried, producing materials with microstructures and sometimes nanostructures (Yu et al. 2008).

The advantages of this method are as follows:

i) The final products are homogeneous in nature.
ii) Controllable degree of porosity and higher purity of products.
iii) Enables synthesis at low temperature and allows for uniform dispersion of small quantities of dopants in the final product.

The disadvantages are as follows:

i) Relatively longer reaction time.
ii) The use of organic solvents may be harmful to the human body.

1.2.9 Ultrasound

This has become a crucial tool in nanoparticle synthesis due to its ability to induce ultrasonic cavitation in liquids. This leads to various physical and chemical effects like high temperature, pressure and cooling rates, creating an environment conducive to chemical reactions under extreme conditions.

Ultrasound offers several advantages in nanoparticle synthesis:

i) Nanoparticles can be synthesized at ambient temperature.
ii) It is a quick and straightforward method, allowing the synthesis of nanoparticles with varying morphologies like nanobelts, nanorings, core-in-shell structures and bimetallic nanoparticles.

There are several disadvantages to consider:

i) Heat-sensitive materials may not withstand acoustic cavitation.
ii) It is energy-intensive and scaling up the production may be challenging.
iii) The energy required for sonochemical synthesis is high, resulting in a relatively low yield per unit of energy supplied.

1.2.10 Spark Discharge

This is a promising approach for cost-effective industrial-scale nanofabrication of advanced materials. Electric sparks or abrupt electrical discharges occur when a high electric field generates an ionized and conductive channel through typically insulating media like air mixtures. The distinctive features of this method are as follows:

i) Ability to synthesize extremely small primary nanoparticles with diameters less than 10 nm.
ii) Applicable for various materials like metals, oxides, semiconductors and carbon-based materials.
iii) The high chemical purity of the synthesized nanoparticles with low impurity content.
iv) Relative simplicity in implementation.

A limitation to this method is its low production rate which restricts its application at the industrial scale (Hachem et al. 2022).

1.2.11 Template Synthesis

This is a straightforward and environmentally sustainable approach for producing nanoparticles. It is a highly promising method for creating uniform inorganic nanoparticles. The template serves dual purpose by enabling precise structure replication with high reproducibility and functioning as a framework to organize various device functions like active components and interfaces (Hachem et al. 2022).

1.2.12 Biological Synthesis

It represents a sustainable approach that bridges nanotechnology and biotechnology. Despite their stability, biological nanoparticles often lack monodispersity (uniformity in size) and exhibit a slow synthesis rate. This is due to differences in the concentration of synthesized macromolecules or nucleation components over time, leading to prolonged nucleation periods and resulting in polydisperse nanoparticles with reduced synthesis rates.

To overcome these challenges, several optimization strategies are used like refining microbial cultivation methods and extraction techniques. Photobiological methods can also be used to delve into the biochemical, cellular and molecular mechanisms that govern the synthesis of biological nanoparticles to the synthesis rate and properties of the nanoparticles.

1.3 Application of Nanoparticles in Crop Improvement

1.3.1 Application of Nanotechnology in Pests and Plant Disease Management

Nowadays, pesticides and fertilizers are being increasingly used to control pests and diseases. This adversely affects the human body, pollinating insects and even domestic animals. The biological control methods are not affordable; hence, chemicals at the nanoscale are used to combat this problem. The target area of these chemicals is the part of the plant that is attacked by the pest (Haris et al. 2023). Nanotechnology has the best application in agricultural sciences and mitigates environmental pollution using nanoparticles and nanocapsules in synthesizing fertilizers and pesticides. Pesticide or drug delivery has become much easier with the implementation of polymeric nanoparticles, iron oxide and gold nanoparticles (Hachem et al. 2022).

The main challenge with disease management is detecting the exact stage of prevention. The pesticides are often added as a precaution, which leads to increased toxicity. Among the various diseases, viral diseases are the most dangerous ones, as the spread of the disease needs to be prevented with the use of vectors (Fraceto et al. 2016). Viral disease can only be controlled if detection occurs at a particular stage, like viral DNA replication or the initial viral protein synthesized (Haris et al. 2023). Even in animal systems, nanotechnology-based viral diagnostics like multiplexed diagnostic kits have been used to detect the exact strain of the virus (He et al. 2019).

1.3.2 Potential of Nanotechnology in Sustainable Agriculture

Pests in the field are controlled using pesticides. This has posed a serious problem to humans and pollinating species and has also entered the water bodies and soil. Nanoscale compounds are viable and sustainable alternatives in this regard. Polymers, nanopesticides, FeONPs and AuNPs are easy to produce and can be used as medicine carriers or pesticides. They are used to enhance crop productivity, mitigate plant diseases and boost agricultural output (Fraceto et al. 2016; Haris et al. 2023).

1.3.2.1 Nanopesticides

Biopesticides are known to control insects and pests more effectively than synthetic ones. The non-toxic and environment friendly pesticides have superior properties like antibacterial, anti-fungal and anti-pathogenic characteristics (Haris et al. 2023). These nanomaterials are known to improve plant development, fertilization and overall agricultural yield (He et al. 2019).

1.3.2.2 Nano-insecticides

Large amounts of insecticides are dispersed in the environment without achieving successful results. This raises the cost of production and affects the environment as well (Haris et al. 2023). The formulation of nano-insecticide and nano-encapsulation has protected the plants better against pest invasion and crop loss (He et al. 2019). The nano-encapsulation process coats the active ingredients of pesticides with nano-sized materials. Nano-formulation and nano-encapsulation of insecticides serve significant properties required for sustainable agricultural practices such as increased solubility, crystallinity, thermal stability, biodegradability, permeability, etc. (He et al. 2019; Mohammad et al. 2022).

1.3.2.3 Nano-antimicrobial Agents

Nanomaterials like AgNPs are used as antimicrobial agents in food packaging. The application of Ag as a disinfectant in commercial water has been approved by the Food and Drug Administration. Other nanoparticles like TiO_2 also show antimicrobial properties (Mohammad et al. 2022; Mahmoud Nasrollahzadeh et al. 2019).

1.3.2.4 Nano-fungicides and Nano-fertilizers

Sometimes, nanoparticles are reported to inhibit the growth of beneficial soil microbes. Nano-fertilizers containing Si, Fe, TiO_2, Zn and Au ensure the efficient nutrient uptake by the plants (Fraceto et al. 2016; Haris et al. 2023). However, this efficacy also depends on a number of other factors like plant species, size, susceptibility, composition, as well as concentration and chemical characteristics of the nanocompounds (Haleem et al. 2023).

1.4 Applications of Nanoparticles on Human Health

Scientists regard nanotechnology as the future of science since unique properties of nanoparticles attract the attention of physicians, doctors and biological

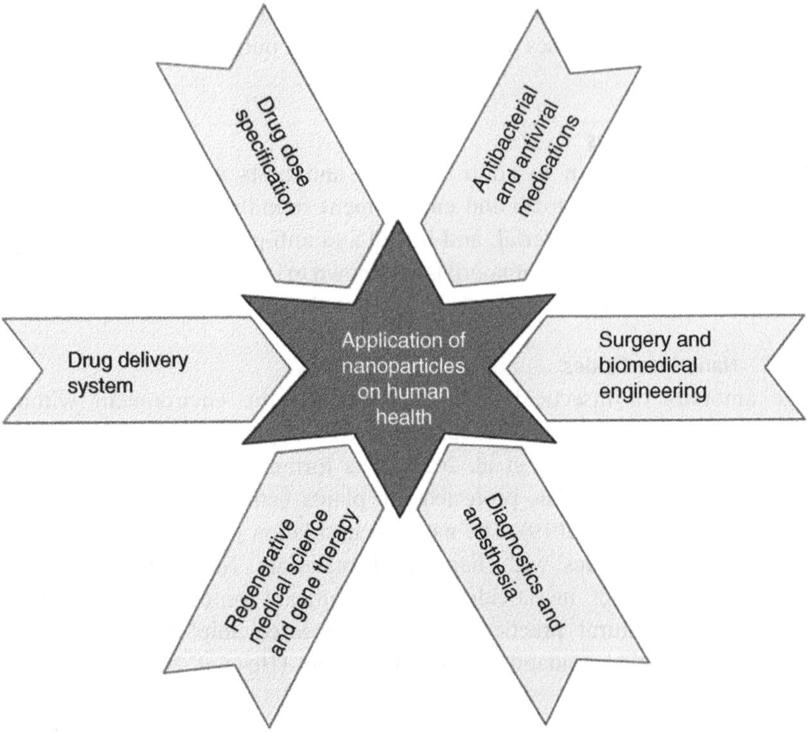

Figure 1.1 Applications of nanoparticles in human health.

researchers (Figure 1.1). This field is now applying quantum phenomena at the nanoscale to other fields of biology and medicine, including medicine, biomedical science, bioengineering, food technology, biochemistry and biophysics (Caldorera-Moore and Peppas 2009; Malik et al. 2023).

1.4.1 Application of Nanoscience on Drug Dose Specification

The development of better therapeutic medications with higher efficacy and lower toxicity has been made possible by nanoscience, completely changing the pharmaceutical sector (Sheth et al. 2012). Nanoparticles can enhance the solubility, stability and bioavailability of drugs, hence improving their pharmacokinetics. Additionally, in order to lessen adverse effects and increase their efficacy, they might target particular tissues and cells (Malik et al. 2023). Precise criteria for drug dosage and administration are necessary due to the distinct particle size and physicochemical features (Kateb et al. 2011; Tan et al. 2023). Several variables, including size, shape, surface characteristics and mode of administration, affect the quantity of nanoparticles (Tan et al. 2023). Because of the potential for variations in absorption and biodistribution, oral administration necessitates greater doses to produce the same effect as intravenous treatment. The development of self-nano-emulsifying drug delivery systems (SNEDDSs) is a response to the limited oral bioavailability of recently identified therapeutic compounds. Surfactant, co-surfactant and oil phase make up these systems. It is essential to optimize formulations using statistical design or phase diagrams. The solubilization, stability and absorption of SNEDDSs are improved, which increases bioavailability. They can be changed into solid dose forms to improve patient compliance and stability (Buya et al. 2020; Hu et al. 2018). DNA nanotechnology is a promising field that uses basic but effective design strategies to create self-assembling nanostructures that improve therapeutic targeting and reduce toxicity. It has been successfully loaded onto potent anticancer therapeutic compounds such as doxorubicin and CpG oligonucleotides, increasing cell absorption efficiency (Shen et al. 2021). Researchers are developing nanosponge-like particles to absorb harmful medication dosages in blood, decrease the negative effects of overdosing and cure fluid imbalances (Iravani and Varma 2022). These nano-assemblies, when combined with squalene, are superior anticancer compounds for treating human cancer cells, while additional development beyond in vitro experiments is still continuing (Dong et al. 2019; Levit et al. 2020).

1.4.2 Impact of Nanotechnology on Drug Delivery Systems

Nanotechnology can be strategically exploited to produce drug delivery systems, hence expanding drug markets. This method is especially useful for medications that have not reached clinical trials due to weak biopharmacological characteristics (Schek et al. 2004). This strategy has the potential to improve drug performance, effectiveness, safety, patient adherence and healthcare costs by prolonging product lives and improving drug effectiveness (Hughes 2005; Sahoo and Labhasetwar 2003).

1.4.2.1 Significance of Nanocarriers

i) Liposomes are self-assembling artificial vesicles made of amphiphilic phospholipids. They exhibit desirable biological features such as biocompatibility, biodegradability, drug separation and drug entrapment for both hydrophilic and hydrophobic compounds. These features are easily modified by adding agents or changing the surface chemistry of the lipid membrane. Liposomes are clinically recognized nanosystems for drug administration that have been shown to reduce systemic effects and toxicity while also slowing drug clearance (Blume and Cevc 1990). Modified liposomes offer remarkable pharmacokinetic properties when delivering deoxyribonucleic acid (DNA), antisense oligonucleotides, siRNA, proteins and chemotherapeutic drugs. Doxorubicin, an anticancer medicine, is widely used though it is extremely hazardous, harming tumour tissue, heart and kidney (Papahadjopoulos et al. 1991). The discovery of doxorubicin-encapsulated liposomes resulted in an approved nanomedical drug delivery system that reduced distribution to the heart and renal system while enhancing tumour tissue accumulation via the enhanced permeability and retention effect (Abraham et al. 2005; Berry et al. 1998; Forssen and Tökes 1979; Safra et al. 2000).

ii) Polymeric nanoparticles (NPs) are colloidal particles ranging in size from 10 to 100 nm that can be spherical, branching or core-shell in shape. They can be made from biodegradable synthetic polymers or natural polymers (Dinarvand et al. 2011). Advances in polymer research and engineering have resulted in the creation of smart polymers that can modify their properties in response to environmental stimuli (Alcalá-Alcalá et al. 2015). These polymers may be tuned to certain stimuli, enabling more precise and programmable drug delivery. Polymer sources are versatile and easy to combine, allowing us to control sensitivity within a limited range.

iii) Nanotechnology provides an excellent opportunity to convert pharmaceuticals with low water solubility and dissolution rate into water-soluble dispersions. By transforming medications into nano-sized versions, they can act as their own carriers (Bunjes 2010; Merisko-Liversidge and Liversidge 2008). The most practicable technique is to reduce drug particle size to nanometres and stabilize the drug surface using non-ionic surfactants or polymeric macromolecules (Kipp 2004; Merisko-Liversidge et al. 2003). This enhances the surface area of the drug, resulting in improved solubility and dissolution. Once nano-sized, the medicine can be manufactured in a variety of dosage forms, including oral, nasal and injectable. This method has advantages over association colloids (Baba et al. 2007).

1.4.3 Implications of Nanotechnology in Gene Therapy

Gene therapy is a promising method for treating genetic diseases, but its use has its challenges. Nanomedicine, a nonviral approach, offers a new prospect. Nanocarriers that respond to stimuli, such as internal or external cues, can effectively deliver therapeutic nucleic acids. Recent studies have explored various stimulus-responsive techniques for gene therapy (Yu et al. 2021). Nanobiotechnology in gene therapy uses nanoparticle-based delivery methods to transport therapeutic genes to target cells, enhancing efficiency and safety by shielding DNA molecules from degradation and enhancing their ability to permeate the cell membrane (DeLuca et al. 2020; Hu et al. 2019).

Nano-assemblies are used in gene therapy to attach biodegradable and non-biodegradable particles, allowing for DNA binding and access across cellular surfaces (Yang et al. 2022). These technologies are also being developed for intravenous medication injections, which will lead to advances in nanogenetic therapy for a variety of ailments, including cancer, genetic abnormalities and infections (George et al. 2019) (Noreen et al. 2022). Polymeric materials are being investigated as gene delivery vectors, with synthetic designs aimed at controlling charge density, masking toxicity and optimizing structure. Pi-conjugated structures provide simple fluorescence tracking and DNA/RNA complexation. New designs and new delivery techniques will lead to better therapeutic outcomes. Combining physical approaches such as electroporation and sonoporation with polyplex transfection improves targeting and minimizes off-target damage (Salameh et al. 2020). Non-viral administration of plasmid DNA or messenger RNA is a safe and promising treatment for serious medical conditions. Carriers composed of cationic polymers or lipids resist breakdown and escape host clearance. Ligands promote fast absorption, but carrier toxicity reduces adverse responses and cell function (Uchida and Kataoka 2019). Green nanotechnology-driven medication delivery systems, including nanometal particles, polymers and biological materials are transforming the sector by using eco-friendly chemical processes or natural biomaterials like plant extracts and microorganisms (Jahangirian et al. 2017). The growing demand for green nanotechnology-driven drug delivery systems has led to the development of various delivery devices like inorganic nanoparticles, quantum dots, organic polymeric nanoparticles, mesoporous silica nanoparticles, dendrimers, nanostructured lipid carriers and solid lipid nanoparticles (Kanwar et al. 2019).

1.4.4 Importance of Nanotechnology in Regenerative Medical Science

Nanotechnology is being used in bone regeneration technologies to generate precise, targeted materials that promote bone growth such as nanoparticles that carry pharmaceuticals or chemicals precisely to afflicted areas (Pardi et al. 2015). A research work investigated the use of lipid nanoparticles (LNPs) to deliver the reporter molecule, green fluorescent protein (GFP), to various tissues of zebrafish embryos. Injections of LNP-packaged *gfp* mRNA led to GFP expression in neuronal, vascular, cardiac and skeletal muscle tissues, revealing that LNP-packaged *gfp* mRNA can be transported and taken up by cells in various organs without affecting the survival of the embryo. The study concluded that LNPs can be used as an mRNA delivery platform in zebrafish, allowing for the investigation of therapeutic effects of LNP-packaged candidate mRNAs in a variety of disease models (Patton et al. 2018). Biomineralization research aims to lower the particle size of bone components for incorporation into collagen fibres, transforming osteology and bone tissue engineering. This research is also developing artificial joints and nanoscale collagen-mimicking coatings for knees and hips, enhancing bone repair and regeneration results, such as faster healing times and increased strength (Pardi et al. 2015; Yang and Webster 2009). Orthopaedists confront difficulties in treating big bone fractures because bone regeneration is a complex process involving osteoblasts, osteoclasts and immune cells (Wang et al. 2021). Osteoimmunology investigates this mechanism, which has resulted in the creation of biomaterials for enhanced bone regeneration. Nanoparticle-based drug delivery systems allow for controlled therapeutic release and targeting of certain cell types (Wen et al. 2023).

The study created a reticuloendothelial system-specific blocking system based on a 'don't-eat-us' technique. A CD47-derived peptide ligand was loaded onto liposomes, which then got adsorbed onto hepatic phagocyte membranes, providing a long-lasting mask. This reduced the interactions between phagocytes and injected nanoparticles. This method was tested on a cryptococcal meningitis model, demonstrating improved brain accumulation and medication delivery (Tang et al. 2019).

1.4.5 Diagnostics and Nanotechnology

Nanotechnology presents a promising tool for in vivo imaging and cancer cell detection, which facilitates rapid, reliable, economically viable and well-regulated cancer management. It enables simultaneous diagnosis and therapy with nano-theragnostic particles, allowing for early identification and specific cell elimination (Dessale et al. 2022). Nanoparticles exhibit unique biophysical properties that improve contrast in biomedical imaging and allow for tissue-specific diagnosis. Variations in size or composition can change their optical, magnetic or electrical properties, facilitating multiplexing. One of the fundamental applications includes nanoparticle-based diagnostic imaging, where nanoparticles can be bound to specific biomarkers to improve imaging modalities such as magnetic resonance imaging, computerized tomography scans and positron emission tomography scans, making them more sensitive, accurate and specific (Singh and Amiji 2022).

Biosensors, such as electrochemical nano-biosensors, have recently acquired popularity due to their sensitive and selective detection of oncomarkers. A proficient research work showed the prevalence, pathobiology and most often used electrochemical nano-biosensors for point-of-care detection of several breast cancer oncomarkers (Mao et al. 2022; Sadeghi et al. 2023).

Nanomaterials have features such as elasticity, hardness, tensile strength and magnetic and optical qualities, making them valuable in healthcare. Computational models, used in 'executable biology', enable computer science and pharmacokinetic simulations. They are frequently utilized in pharmaceutical research. Future research should centre on multiscale simulations and collaboration between the pharmaceutical industry and computer scientists (Singh and Amiji 2022; Rajput et al. 2023).

1.4.6 Application of Nanotechnology in Antibacterial and Antiviral Medications

The combination of pharmacology with nanotechnology has resulted in the development of novel antimicrobial drugs. TiO_2 nanoparticles have photocatalytic intrinsic capabilities and are stable, non-toxic and affordable with safe physicochemical features (Amiri et al. 2022). Some research works showed that silver nanoparticles (AgNPs) have unique physicochemical and antibacterial properties, making them beneficial in a variety of applications, including wound care management. The mechanisms underlying antiviral and antibacterial capabilities of AgNPs as well as their applications and limitations such as human and environmental cytotoxicity have been well studied (Salleh et al. 2020).

1.4.7 Role of Nanotechnology in Surgery and Biomedical Engineering

Conventional biomedical devices have difficulties such as short service life, excessive equipment size and safety risks. Nanogenerators (NGs), which gather micro/nanomechanical and thermal energy from humans, are an excellent solution for self-powered biomedical equipment. The integration of NGs and biomedicine has accelerated the development of self-powered medical equipment (Wang et al. 2022). Replacing implantable medical devices (IMDs) is critical for their continued operation, and NGs are gaining popularity as viable power solutions due to their compact size and safety. NGs create electrical potential as a result of body movement, allowing IMDs to monitor and detect events instantly. Ultrasound-driven NGs have the ability to noninvasively power IMDs that require more energy (Sun et al. 2021; Yoon and Kim 2020). Research methodologies in nanotechnology development have been focused on applications such as nano-networks and communication, biosensors, nanoprobes, drug delivery systems and nano-implants in order to treat a wide range of medical conditions, including chronic pain, wound healing and heart disease (Buniyamin et al. 2022). In the pharmaceutical sector, nanotechnology is utilized to develop medications with nano-sized active components for the treatment of diseases such as human immunodeficiency virus and cancer. These medications can help to rebuild lost tissue as well as monitor, diagnose and treat illnesses. Smaller drug delivery systems allow medication to be placed in previously inaccessible locations, which is critical for disease therapy and diagnosis. Target treatment is one of the most recent drug delivery innovations (Malik et al. 2023; Mazumder et al. 2020). Triboelectric nanogenerators (TENGs), being versatile, efficient, environment-friendly, biodegradable and partially biocompatible, are perfect for harvesting power from live organisms. Recent advances in TENG technology provide unique clinical healthcare solutions, allowing continuous power supply for wearable and implanted medical devices that are light in weight, thin, biocompatible and soft tissue conformable (Liu et al. 2023; Quan et al. 2022).

1.4.8 Importance of Nanotechnology in Anaesthesia

Anaesthetics are crucial medications used in clinical surgery to relieve pain and encourage relaxation. Potential toxic consequences are a concern, as in the case of the blood–brain barrier, which makes general anaesthetics difficult to penetrate. Anaesthetic nanomedicines loaded onto nanocarriers could address these challenges, improve anaesthesia effectiveness, lessen toxic effects and improve the efficacy of general anaesthetics that cross the blood–brain barrier (Zhang et al. 2023). This nanoscale anaesthetic activity contributes to the intended result, which is swiftly achieved through an equal distribution of anaesthetic in the projected organ, such as the dental surface. The sensitivity action can also be adjusted for a specific tooth that requires surgical intervention. Following surgery, nanorobots are controlled by nanocomputers to restore tooth sensitivity to normal levels (Bhansali et al. 2021; Malik et al. 2023).

1.5 Risks of Using Nanoparticles

The extensive use of manufactured nanoparticles in a variety of products leads to potential exposure throughout their lifetime. Inhalation, cutaneous and oral exposure are all possible exposure routes. In rats, nanoparticles can accumulate in the respiratory system, causing oxidative stress-related inflammatory reactions as well as tumour-related consequences. According to certain findings, uptake occurs in the brain via the olfactory epithelium. Although nanoparticles may enter the systemic circulation following inhalation, there is conflicting evidence on the amount of translocation (Borm et al. 2006).

Nanoparticles, which are used in food and pharmaceutical components, have been shown to be absorbed in the gastrointestinal tract after oral administration. However, only a few distinct nanoparticles have been studied in test systems, making it difficult to generalize results to different materials. Air pollution research has implicated combustion-derived nanoparticles with health impacts (Warheit et al. 2007). Nanomaterials released from various sources into the environment, both intentionally and unintentionally, have the potential to pollute soil and spread to water bodies. The most significant environmental release occurs from spills. In general, nanoparticles can be released in closed cycles via leaks or during the transfer of an intermediate or final product to further handling phases. Such nanoparticle exposure is mostly relevant to workers because all of these processes occur in workplace circumstances. The public can only be exposed to nanoparticles through the product or through their release into the environment via stack or diffusive emissions.

Deposition of inhaled particles takes place throughout the respiratory organs, including pharyngeal, nasal, tracheobronchial and alveolar regions, depending on the particle dimensions. Ambient particulate air pollution has been associated with cardiovascular morbidity and mortality, but the link between nanoparticle exposure and health outcomes is less clear (De Hartog et al. 2003; Lanki et al. 2008; Pekkanen et al. 2002). According to research, tiny particles are more closely related to cardio-respiratory symptoms than nanoparticles, and the number of nanoparticles is more strongly correlated with health impacts than mass (Peters et al. 1997).

Nanoparticles can deposit in the respiratory system and have nanostructure-influenced toxicity due to high surface area, surface activity, unusual morphology, small diameters or degradation. Particles from degradation may pose a risk if they exhibit nanostructure-dependent biological activity. Epidemiological studies have found that the ultrafine particle fraction in the atmosphere has a specific purpose, but its dosimetry and biokinetic destiny are not entirely understood. Cardiovascular effects show that ultrafine particles (UFPs) are transported more efficiently from the respiratory epithelium to circulation and target organs, generating negative effects on heart function, blood coagulation and central nervous system (CNS) function. Toxicological studies seek to establish the biological plausibility of health consequences and to discover causal pathways for pathophysiological changes and chronic disease (Kreyling et al. 2006). A pilot investigation was conducted to evaluate if ultrafine elemental carbon particles translocate to the liver and other organs following inhalation. The results showed that the translocation was effective within one day. However, the study took into account input from UFPs in the gastrointestinal system, as respiratory tract deposits may not account for the entire hepatic (13)C burden (Oberdörster et al. 2002). A research study reveals that airborne solid UFPs (<100 nm) can target the CNS by depositing in the olfactory mucosa of the nasopharyngeal area of the respiratory tract and then translocating via the olfactory nerve. During nasal breathing, more than half of the inhaled UFPs can end up in the nasopharyngeal region. Approximately 20% of UFPs in the rat olfactory mucosa can enter the CNS through the olfactory bulb (Oberdörster et al. 2004). Nanoscale particles (NSP) or small particles of the same chemistry are more physiologically active than bigger particles because they have a larger surface area per mass. This action has the potential to be both inflammatory and pro-oxidant, as well as antioxidant. Early investigations revealed inconsistent results regarding the toxicity of NSPs to environmentally significant species. NSPs that penetrate the skin spread through the lymphatic system. The presence of mitochondrial distribution and an oxidative stress response, following NSP endocytosis, suggests that more investigation is needed into their interactions with subcellular structures (Oberdörster et al. 2005). One study investigated the relationship between dosimetry parameters and harmful pulmonary consequences from inhaled particles. It used a model that incorporates TiO_2 particles of varying sizes (20 nm diameter, ultrafine; 250 nm diameter, fine) with the same crystalline structure. The findings indicated a link between the persistence of effects in the alveolar and interstitial spaces and the persistence of particles in the corresponding compartment (Oberdörster et al. 1994). Another research study indicated that lung disease influences the deposition of UFPs in the respiratory tract, especially in people with obstructive or restrictive lung disease. Particle deposition was dramatically enhanced in obstructive lung disease patients, with plausible causes including increased transit time, aberrant expiratory collapse and flow perturbations caused by lower airway calibre (Anderson et al. 1990).

Metal nanoparticles have been shown to produce brain dysfunction in animals and exacerbate brain disease caused by whole-body heat. Normal rats treated with nanoparticles demonstrated modest cognitive impairment and cellular changes. In comparison to naive mice, exposure to whole-body hyperthermia caused substantial cognitive and motor defects, worsening of blood–brain barrier disruption, oedema development and brain pathology. These data indicate that nanoparticles worsen brain disease and cognitive failure in hyperthermia (Sharma and Sharma 2007).

In one study, it has been found that carbon nanoparticles increased heart rate but only caused low-grade pulmonary inflammation (Schulz et al. 2005). However, it is a contradiction to previous reports that blood hypercoagulability was linked to heart rate (Donaldson et al. 2001; Nemmar et al. 2002). It has also been found that inhaling diesel soot in humans impairs the function of endothelial cells in the forearm, leading to impaired vasomotor and secretory responses. This results in a systemic inflammatory response and altered cardiac control, but no evidence of endothelial dysfunction (Park et al. 2008).

Table 1.1 summarizes the types of nanoparticles, their sources and risks associated with their application.

1.6 Ethical Issues of Nanoparticle Application

Due to the rapid advancement of nanomedical technology, it is critical to evaluate the legal, social and ethical implications of its larger therapeutic applications while being aware of potential hazards. In the absence of scientific certainty on

Table 1.1 Types of nanoparticles, their sources and their probable risks.

Type of nanoparticles and their sources	Probable risks	References
Manufactured nanoparticles in a variety of products	Oxidative stress-related inflammatory reactions as well as tumour-related consequences	Borm et al. (2006)
This NP exposure to workers occur in workplace circumstances	Deposition of inhaled particle takes place throughout the respiratory organs including pharyngeal, nasal, tracheobronchial and alveolar regions. Associated to cardiovascular morbidity and mortality	De Hartog et al. (2003), Pekkanen et al. (2002), Lanki et al. (2008)
Ultrafine elemental carbon particles	Translocate to the liver and other organs following inhalation Cardiovascular effects show that ultrafine particles are transported more efficiently from the respiratory epithelium to circulation and target organs, generating negative effects on heart function, blood coagulation and central nervous system (CNS) function Can target the central nervous system by depositing in the olfactory mucosa of the nasopharyngeal area of the respiratory tract and then translocating via the olfactory nerve Systemic inflammatory response and altered cardiac control, but no evidence of endothelial dysfunction	Kreyling et al. (2006), Oberdörster et al. (2002), Oberdörster et al. (2004), Donaldson et al. (2001)
Nano scale particles (NSP)	Can be both inflammatory and pro-oxidant, as well as antioxidant Can penetrate the skin and spread through lymphatic systems	Oberdörster et al. (2005)
TiO_2 particles of varying sizes (20 nm diameter, ultrafine; 250 nm diameter, fine) with crystalline structure	Obstructive lung disease, including increased transit time, aberrant expiratory collapse and flow perturbations caused by lower airway calibre	Oberdörster et al. (1994), Anderson et al. (1990)
Metal nanoparticles	Brain dysfunction in normal animals and exacerbated brain disease caused by whole-body heat Substantial cognitive and motor deficits, worsening of blood–brain barrier disruption, oedema development and brain pathology	Sharma and Sharma (2007)

the potential health impacts of occupational nanoparticle exposure, there is a need for advice in decision-making about potential risks and their management. Identifying the ethical problems involved may be beneficial to decision-makers, notably employers, workers, investors and health authorities. Because the purpose of occupational safety and health is to avoid sickness among workers, the situations with the most ethical implications for workers have been recognized. The Centres for Nanotechnology in Society at Arizona State University and the University of California, Santa Barbara, conducted a study on the legal and ethical concerns surrounding nanotechnology. Despite increased interest in nanotechnology ethics since the early 2000s, a few recent studies have thoroughly examined ethical and legal issues concerning nanomedical technology research and development such as reviews on contemporary ethical challenges in nanomedicine and nanomedical technology but did not address legal issues related to rising nanotechnological practices (Allon et al. 2017; Yasri and Wiwanitkit 2017). Nanomedicine is ushering in a new age in medical treatment, with an emphasis on monitoring, managing and optimizing human biological systems through the use of man-made nanodevices and nanomaterials. This interdisciplinary topic poses complicated ethical and societal issues, including privacy, secrecy, dangers, disease definition and augmentation (Bawa and Johnson 2007).

Because of its heightened reactivity, nanomedical technology poses a serious risk to human health. Nanoparticles can induce DNA damage, lung fibrosis and enhanced toxicity in the human body (Karlsson et al. 2008; Könczöl et al. 2011; Lu et al. 2014). Discharging nanomaterials from the body may reduce these hazards, but it may also have an impact on the environment, potentially resulting in inadvertent exposure and harmful consequences (Atalla et al. 2016). Countries have established legal regulations for nanomaterial research and development; however, international cooperation and global regulatory standards could unify research approaches and provide equal protection against harm from nanomaterials and nanotechnology, ensuring global equality (Marchant et al. 2010; Nel et al. 2011; Sophia et al. 2023). Marchant et al. (2010) developed a framework for international cooperation that includes common goals and safety measures for nano-research and development. These rules provide proper courses of action and substantive safeguards against potential harm in nanomedical technology (Marchant et al. 2010; Sophia et al. 2023).

The challenge lies in conveying the risks associated with nanotechnology to the research subjects and to society as a whole. Due to ethical and legal requirements, investigators must inform subjects of the purpose of the study, the procedures, the advantages and disadvantages, the risks associated with the study, the alternatives to the study, the confidentiality of the study and other relevant information (Resnik and Tinkle 2007). The most crucial ethical considerations for workers in nanomaterial-related occupations include hazard identification, risk acceptance, controls, medical screening involvement and adequate investment in toxicological and exposure control studies. Ethical considerations include hazard and risk assessment, no malefice (do no harm), autonomy, justice (fair distribution of hazards), privacy (in processing of medical information) and respect for persons (Schulte and Salamanca-Buentello 2007). Biases and values have an impact on scientific knowledge, which guides ethical decision-making in occupational safety and health. No scientific theory is completely objective, but some may be more objective than others. Ethical choices are influenced by the portrayal of nanotechnology, possible advantages and the risks and dangers involved (Kantrowitz 1995; Schrader-Frechette 2002). Assessments of risk in nanotechnology are based on personal opinions and heavily influenced by politics, resulting in biased conclusions. Due to the diverse nature of the field, not a single scenario can accurately represent all risks and controls. Moral dilemmas pertain to the field of expertise and manner of creation. Researchers propose the necessity of different assessments to encompass ethical and political values in nanotechnology policy (Schrader-Frechette 2002). Employers and employees rely on scientists to analyse dangerous and risky details, which can push them to move past their scholarly studies. The intersection of science and ethics is intricate; yet, scientists must fulfil moral responsibilities to the community. Using suitable qualifiers, avoiding generalizations and contemplating the implications of their work are crucial for ensuring ethical responsibilities in nanotechnology research, even in cases where all scientific details are not fully understood (Pimple 2002; Schrader-Frechette 2002; Weil 2002). Understanding the scientific data on the dangers of nanomaterials is essential for conveying the risks and hazards faced by workers. Employers must understand and convey hazard and risk information as part of their risk management responsibilities. Employers will concentrate on determining the specific preventive measures to implement in order to ensure a safe and healthy work environment (Schulte and Salamanca-Buentello 2007).

Acknowledgement of chance includes assessing the certainty, seriousness, well-being impacts, intentional nature, options and remuneration (Fischoff 1994). The choice of work and working conditions is not free, but financial and social conditions altogether impact their determination, hazard resistance and support in chance administration (Jensen 2002; Lynn 1997; Shearn 2005). Justice in the decision-making of workers entails considering whether workers face more hazards than the general public and whether incentives should be exchanged for increased risk from nanoparticle exposure. Society strives to offer a safe and healthy workplace for all employees (Schrader-Frechette 2002). The problem may be mitigated if nanoparticle restrictions decrease the risk level of workers, as society strives to create a safe and healthy workplace for all workers (Schulte and Salamanca-Buentello 2007).

The ethical challenge in nanoparticle control is whether adequate controls are in place to prevent injury and disease, as worker exposures may raise the risk of harm. Although the risk caused by nanoparticles is not well understood, preliminary data suggests that engineered nanomaterials should be treated with the same level of care (Hett 2004; Stern and McNeil 2008). UFP exposure is determined by their potential toxicity and the extent of exposure. Given the limited evidence and growing concern, recognizing the manufactured nanoparticles as occupational hazards and developing interim preventive measures based on risk is advocated. Certain manufactured nanoparticles may be more harmful per unit mass than bigger particles containing the same substances. Workers may not be at a significantly increased risk of lung disease if existing controls for fine or UFPs are used. Animal studies indicate that workers, exposed to ultrafine TiO_2 at 0.1 mg/m^3 for 45 years, have a risk of lung cancer of less than 1/1000, potentially approaching nil (Kantrowitz 1995). The scope of this investigation extends to additional nanoparticles and requires acceptable animal or human data (Oberdörster et al. 2005). The current creation of nanoparticles poses insufficient concerns; therefore, risks are likely to be mitigated by adopting controls for recognized industrial UFPs. This conclusion is supported by a general risk assessment based on the surface area and adherence to classical physics and aerodynamic principles. Future assessments may differ depending on bio-persistence, structure, surface activity and translocation data (Schulte and Salamanca-Buentello 2007).

Medical screening refers to detection of early disease stages in asymptomatic persons using various tests. It varies in the workplace depending on professional conditions and employer duties (Halperin et al. 1986). Ethical concerns arise over the reasoning, voluntary nature, actions for positive testing and access to test information for those who test positive (Ashford et al. 1990; Schulte 1986). Medical screening for nanomaterials is often not suggested due to uncertain toxicity and worker danger. There is insufficient evidence to support routine screening for diseases such as lung cancer. However, for non-malignant respiratory consequences in nanotechnology processes, screening should be part of a complete risk management programme that includes respiratory, cardiovascular, neurologic and other organ system hazards (Oberdörster et al. 2005; Radomski et al. 2005; Schulte and Salamanca-Buentello 2007). If harmful consequences are identified, medical examination may be necessary (Ashford et al. 1990).

The issue over nanotechnology revolves around enduring eminent adverse effects to some while maximizing societal benefits (Radomski et al. 2005). This utilitarian viewpoint may excuse injury to one person, while opposing the ethical ideal of respect for persons, which emphasizes individual liberties and the golden rule (Harris 2003; Schulte and Salamanca-Buentello 2007). This idea in the workplace compels companies and governments to provide a safe and healthy environment (Harris 2003). The right of employers to property and gain from nanotechnology may clash with the rights of the workers, necessitating a logical decision. Conducting site-specific danger evaluations and implementing relevant controls is consistent with the principles of respect for humans, autonomy, beneficence and nonmaleficence. However, the level of control mechanisms required may be an important consideration. The current generation of modified nanoparticles can be controlled using existing technologies; the question is how much to invest in workplace applications (Sophia et al. 2023).

1.7 Conclusion and Future Prospective

Nanotechnology has substantially enhanced crop productivity and agricultural techniques, with the goal of meeting global food demand while also protecting the environment. The development of man-made nanomaterials has transformed the global agricultural environment, allowing for improved input management and conservation (Shang et al. 2019). However, there are still knowledge gaps regarding the absorption capacity, acceptable limit and ecotoxicity of certain nanomaterials (Kim et al. 2018; Li et al. 2016). More research is needed to understand how altered agricultural inputs behave and interact with biomacromolecules in living systems and surroundings.

Nanomedicine is crucial for personalized medicine, enabling precise disease identification and monitoring. It uses nanoscale materials for sensitive sensors and biomarkers, allowing for precise mapping of diseases. Nanomedicine can efficiently attack cells while reducing side effects and harm to healthy cells. Current products include nano-encapsulated doxorubicin. Future challenges include drug loading and the development of metallic nanoparticles for diagnosis and treatment (Malik et al. 2023). Nanotechnology is driving the healthcare revolution by emphasizing preventive population health management, minimizing adverse effects and increasing therapeutic efficacy. It is especially effective in cancer detection, treatment and gene therapy. Nanomedicine encompasses vaccinations, pharmaceutical administration, wearable technology, diagnostics and antimicrobial items. Combining traditional anticancer medications with nanoscale technologies has enormous promise for therapeutic discovery (Malik et al. 2023). The use of nano-based goods in healthcare services confronts hurdles due to their toxicity, which can cause allergenic reactions. Toxicity is influenced by factors such as morphology, size, dosage and treatment time (Kanwar et al. 2012). Standardization and reproducibility in nanomaterial investigations are also critical. Controlling their activities in delicate locations is difficult, and the high prices, contaminants and environmental consequences pose serious hazards (Jamkhande et al. 2019).

Ethical issues about nanotechnology in the workplace stem from a lack of understanding of nanomaterial dangers and potential worker risks. An interim review is required to ensure that employees may exercise their autonomy. Employers must conform to the ideals of autonomy, beneficence, nonmaleficence, fairness, privacy and respect for people by correctly describing hazards, taking precautions, communicating with employees and reducing risks so that they are reasonable and acceptable to workers (Stern and McNeil 2008).

References

Abraham, S.A., Waterhouse, D.N., Mayer, L.D. et al. (2005). The liposomal formulation of doxorubicin. *Methods in Enzymology* 391: 71–97.

Alcalá-Alcalá, S., Benítez-Cardoza, C.G., Lima-Muñoz, E.J. et al. (2015). Evaluation of a combined drug-delivery system for proteins assembled with polymeric nanoparticles and porous microspheres; characterization and protein integrity studies. *International Journal of Pharmaceutics* 489 (1–2): 139–147. https://doi.org/10.1016/j.ijpharm.2015.04.074.

Allon, I., Ben-Yehudah, A., Dekel, R. et al. (2017). Ethical issues in nanomedicine: tempest in a teapot. *Medicine, Health Care, and Philosophy* 20: 3–11. https://doi.org/10.1007/s11019-016-9720-7.

Amiri, M.R., Alavi, M., Taran, M., and Kahrizi, D. (2022). Antibacterial, antifungal, antiviral, and photocatalytic activities of TiO_2 nanoparticles, nanocomposites, and bio-nanocomposites: recent advances and challenges. *Journal of Public Health Research* 11 (2): https://doi.org/10.1177/22799036221104151.

Anderson, P.J., Wilson, J.D., and Hiller, F.C. (1990). Respiratory tract deposition of ultrafine particles in subjects with obstructive or restrictive lung disease. *Chest* 97 (5): 1115–1120. https://doi.org/10.1378/chest.97.5.1115.

Ashford, N.A., Spadafor, C.J., Hattis, D.B., and Caldart, C.C. (1990). *Monitoring the Worker for Exposure and Disease: Scientific Legal and Ethical Considerations in the Use of Biomarkers*. Baltimore, MD: The Johns Hopkins University Press.

Atalla, K., Chaudhary, A., Eshaghian-Wilner, M.M. et al. (2016). Ethical, privacy, and intellectual property issues in nanomedicine. In: *Wireless Computing in Medicine* (ed. M.M. Eshaghian-Wilner), 567–600. New Jersey: Wiley. https://doi.org/10.1002/9781118993620.ch20.

Baba, K., Pudavar, H.E., Roy, I. et al. (2007). New method for delivering a hydrophobic drug for photodynamic therapy using pure nanocrystal form of the drug. *Molecular Pharmaceutics* 4 (2): 289–297. https://doi.org/10.1021/mp060117f.

Bawa, R. and Johnson, S. (2007). The ethical dimensions of nanomedicine. *The Medical Clinics of North America* 91 (5): 881–887. https://doi.org/10.1016/j.mcna.2007.05.007.

Bayda, S., Adeel, M., Tuccinardi, T. et al. (2019). The history of nanoscience and nanotechnology: from chemical-physical applications to nanomedicine. *Molecules (Basel, Switzerland)* 25 (1): 112. https://doi.org/10.3390/molecules25010112.

Berry, G., Billingham, M., Alderman, E. et al. (1998). The use of cardiac biopsy to demonstrate reduced cardiotoxicity in AIDS Kaposi's sarcoma patients treated with pegylated liposomal doxorubicin. *Annals of Oncology* 9 (7): 711–716.

Bhansali, D., Teng, S.L., Lee, C.S. et al. (2021). Nanotechnology for pain management: current and future therapeutic interventions. *Nano Today* 39: 101223. https://doi.org/10.1016/j.nantod.2021.101223.

Blume, G. and Cevc, G. (1990). Liposomes for the sustained drug release in vivo. *Biochimica et Biophysica Acta – Biomembranes* 1029 (1): 91–97.

Borm, P.J., Robbins, D., Haubold, S. et al. (2006). The potential risks of nanomaterials: a review carried out for ECETOC. *Particle and Fibre Toxicology* 3: 11. https://doi.org/10.1186/1743-8977-3-11.

Buniyamin, I., Akhir, R.M., Asli, N.A. et al. (2022). Nanotechnology applications in biomedical systems. *Current Nanomaterials* 7 (3): 167–180.

Bunjes, H. (2010). Lipid nanoparticles for the delivery of poorly water-soluble drugs. *The Journal of Pharmacy and Pharmacology* 62 (11): 1637–1645. https://doi.org/10.1111/j.2042-7158.2010.01024.x.

Buya, A.B., Beloqui, A., Memvanga, P.B., and Préat, V. (2020). Self-nano-emulsifying drug-delivery systems: from the development to the current applications and challenges in oral drug delivery. *Pharmaceutics* 12 (12): 1194. https://doi.org/10.3390/pharmaceutics12121194.

Caldorera-Moore, M. and Peppas, N.A. (2009). Micro-and nanotechnologies for intelligent and responsive biomaterial-based medical systems. *Advanced Drug Delivery Reviews* 61 (15): 1391–1401.

De Hartog, J.J., Hoek, G., Peters, A. et al. (2003). Effects of fine and ultrafine particles on cardiorespiratory symptoms in elderly subjects with coronary heart disease: the ULTRA study. *American Journal of Epidemiology* 157 (7): 613–623.

DeLuca, M., Shi, Z., Castro, C.E., and Arya, G. (2020). Dynamic DNA nanotechnology: toward functional nanoscale devices. *Nanoscale Horizons* 5 (2): 182–201.

Dessale, M., Mengistu, G., and Mengist, H.M. (2022). Nanotechnology: a promising approach for cancer diagnosis, therapeutics and theragnosis. *International Journal of Nanomedicine* 17: 3735–3749. https://doi.org/10.2147/IJN.S378074.

Dinarvand, R., Sepehri, N., Manoochehri, S. et al. (2011). Polylactide-co-glycolide nanoparticles for controlled delivery of anticancer agents. *International Journal of Nanomedicine* 6: 877–895. https://doi.org/10.2147/IJN.S18905.

Donaldson, K., Stone, V., Seaton, A., and MacNee, W. (2001). Ambient particle inhalation and the cardiovascular system: potential mechanisms. *Environmental Health Perspectives* 109 (suppl 4): 523–527.

Dong, P., Rakesh, K.P., Manukumar, H.M. et al. (2019). Innovative nano-carriers in anticancer drug delivery-a comprehensive review. *Bioorganic Chemistry* 85: 325–336. https://doi.org/10.1016/j.bioorg.2019.01.019.

Fischoff, B. (1994). Acceptable risk: a conceptual proposal. Risk 5.1. http://www.piercelaw.edu/risk/vol5/winter/Fischhof.htm.

Forssen, E.A. and Tökes, Z.A. (1979). In vitro and in vivo studies with adriamycin liposomes. *Biochemical and Biophysical Research Communications* 91 (4): 1295–1301.

Fraceto Leonardo F., Grillo R., de Medeiros G. A., Scognamiglio V., Rea G., and Bartolucci C. (2016). Nanotechnology in agriculture: Which innovation potential does it have? *Frontiers in Environmental Science*, 4. https://doi.org/10.3389/fenvs.2016.00020.

George, A., Shah, P.A., and Shrivastav, P.S. (2019). Natural biodegradable polymers based nano-formulations for drug delivery: a review. *International Journal of Pharmaceutics* 561: 244–264. https://doi.org/10.1016/j.ijpharm.2019.03.011.

Hachem, K., Ansari, M.J., Saleh, R.O. et al. (2022). Methods of chemical synthesis in the synthesis of nanomaterial and nanoparticles by the chemical deposition method: a review. *BioNanoScience* 12: 1032–1057. https://doi.org/10.1007/s12668-022-00996-w.

Haleem, A., Javaid, M., Singh, R.P. et al. (2023). Applications of nanotechnology in medical field: a brief review. *Global Health Journal* 7 (2): 70–77. https://doi.org/10.1016/j.glohj.2023.02.008.

Halperin, W.E., Ratcliffe, J., Frazier, T.M. et al. (1986). Medical screening in the workplace: proposed principles. *Journal of Occupational Medicine: Official Publication of the Industrial Medical Association* 28 (8): 547–552. https://doi.org/10.1097/00043764-198608000-00004.

Haris, M., Hussain, T., Mohamed, H.I. et al. (2023). Nanotechnology a new frontier of nano-farming in agricultural and food production and its development. *Science of the Total Environment* 857 (Part 3): 159639. https://doi.org/10.1016/j.scitotenv.2022.159639.

Harris, C.E. (2003). State of the art in engineering ethics: Methodologies for case studies in engineering ethics. In: *Emerging Technologies and Ethical Issues in Engineering: Papers from a workshop*, 79–93. Washington, DC: National Academy of Sciences.

He, X., Deng, H., and Hwang, H.-m. (2019). The current application of nanotechnology in food and agriculture. *Journal of Food and Drug Analysis* 27 (1): 1–21, ISSN 1021-9498. 10.1016/j.jfda.2018.12.002.

Hett, A. (2004). *Nanotechnology: Small Matter, Many Unknowns*. Zurich: Swiss Reinsurance Company. http://www.swissre.com/INTERNET/pwsfilpr.nsf/vwFilebyIDKEYLu/ULR-68AKZF/$FILE/Publ04_nano_en.pdf.

Hu, Q., Wang, S., Wang, L. et al. (2018). DNA nanostructure-based systems for intelligent delivery of therapeutic oligonucleotides. *Advanced Healthcare Materials* 7 (20): e1701153. https://doi.org/10.1002/adhm.201701153.

Hu, Q., Li, H., Wang, L. et al. (2019). DNA nanotechnology-enabled drug delivery systems. *Chemical Reviews* 119 (10): 6459–6506. https://doi.org/10.1021/acs.chemrev.7b00663.

Hughes, G.A. (2005). Nanostructure-mediated drug delivery. *Nanomedicine* 1: 23–30.

Iravani, S. and Varma, R.S. (2022). Nanosponges for drug delivery and cancer therapy: recent advances. *Nanomaterials (Basel, Switzerland)* 12 (14): 2440. https://doi.org/10.3390/nano12142440.

Jahangirian, H., Lemraski, E.G., Webster, T.J. et al. (2017). A review of drug delivery systems based on nanotechnology and green chemistry: green nanomedicine. *International Journal of Nanomedicine* 12: 2957–2978. https://doi.org/10.2147/IJN.S127683.

Jamkhande, P.G., Ghule, N.W., Bamer, A.H., and Kalaskar, M.G. (2019). Metal nanoparticles synthesis: an overview on methods of preparation, advantages and disadvantages, and applications. *Journal of Drug Delivery Science and Technology* 53: 101174.

Jensen, P.L. (2002). Assessing assessment: the Danish experience of worker participation in risk assessment. *Economic and Industrial Democracy* 23 (2): 201–227.

Kantrowitz, A. (1995). The separation of facts and values. Risk 6:105. http:///www.piercelaw.edu/Risk/vol6/spring/Kantro2.htm.

Kanwar, J.R., Sun, X., Punj, V. et al. (2012). Nanoparticles in the treatment and diagnosis of neurological disorders: untamed dragon with fire power to heal. *Nanomedicine: Nanotechnology, Biology, and Medicine* 8 (4): 399–414. https://doi.org/10.1016/j.nano.2011.08.006.

Kanwar, R., Rathee, J., Salunke, D.B., and Mehta, S.K. (2019). Green nanotechnology-driven drug delivery assemblies. *ACS Omega* 4 (5): 8804–8815. https://doi.org/10.1021/acsomega.9b00304.

Karlsson, H.L., Cronholm, P., Gustafsson, J., and Möller, L. (2008). Copper oxide nanoparticles are highly toxic: a comparison between metal oxide nanoparticles and carbon nanotubes. *Chemical Research in Toxicology* 21 (9): 1726–1732. https://doi.org/10.1021/tx800064j.

Kateb, B., Chiu, K., Black, K.L. et al. (2011). Nanoplatforms for constructing new approaches to cancer treatment, imaging, and drug delivery: what should be the policy? *NeuroImage* 54 (Suppl 1): S106–S124. https://doi.org/10.1016/j.neuroimage.2010.01.105.

Kim, D.Y., Kadam, A., Shinde, S. et al. (2018). Recent developments in nanotechnology transforming the agricultural sector: a transition replete with opportunities. *Journal of the Science of Food and Agriculture* 98 (3): 849–864. https://doi.org/10.1002/jsfa.8749.

Kipp, J.E. (2004). The role of solid nanoparticle technology in the parenteral delivery of poorly water-soluble drugs. *International Journal of Pharmaceutics* 284 (1–2): 109–122. https://doi.org/10.1016/j.ijpharm.2004.07.019.

Könczöl, M., Ebeling, S., Goldenberg, E. et al. (2011). Cytotoxicity and genotoxicity of size-fractionated iron oxide (magnetite) in A549 human lung epithelial cells: role of ROS, JNK, and NF-κB. *Chemical Research in Toxicology* 24 (9): 1460–1475. https://doi.org/10.1021/tx200051s.

Kreyling, W.G., Semmler-Behnke, M., and Möller, W. (2006). Ultrafine particle-lung interactions: does size matter? *Journal of Aerosol Medicine: The Official Journal of the International Society for Aerosols in Medicine* 19 (1): 74–83. https://doi.org/10.1089/jam.2006.19.74.

Lanki, T., Hoek, G., Timonen, K.L. et al. (2008). Hourly variation in fine particle exposure is associated with transiently increased risk of ST segment depression. *Occupational and Environmental Medicine* 65 (11): 782–786. https://doi.org/10.1136/oem.2007.037531.

Levit, S.L., Gade, N.R., Roper, T.D. et al. (2020). Self-assembly of pH-labile polymer nanoparticles for paclitaxel prodrug delivery: formulation, characterization, and evaluation. *International Journal of Molecular Sciences* 21 (23): 9292. https://doi.org/10.3390/ijms21239292.

Li, M., Ahammed, G.J., Li, C. et al. (2016). Brassinosteroid ameliorates zinc oxide nanoparticles-induced oxidative stress by improving antioxidant potential and redox homeostasis in tomato seedling. *Frontiers in Plant Science* 7: 615. https://doi.org/10.3389/fpls.2016.00615.

Liu, S., Tong, W., Gao, C. et al. (2023). Environmentally-friendly natural materials for triboelectric nanogenerators: a review. *Journal of Materials Chemistry A* 11: 9270.

Lu, X., Zhu, T., Chen, C., and Liu, Y. (2014). Right or left: the role of nanoparticles in pulmonary diseases. *International Journal of Molecular Sciences* 15 (10): 17577–17600. https://doi.org/10.3390/ijms151017577.

Lynn, F.M. (1997). Public participation in risk management decisions: the right to define, the right to know, and the right to act. *Risk* 1: 95. http://www.piercelaw.edu/risk/vol1/spring/lynn.htm.

Mahmoud Nasrollahzadeh, S., Sajadi, M., Sajjadi, M., and Issaabadi, Z. (2019). Chapter 4 – applications of nanotechnology in daily life. In: *Interface Science and Technology*, vol. 28 (ed. S. Mahmoud Nasrollahzadeh, M. Sajadi, M. Sajjadi, et al.), 113–143. Elsevier. https://doi.org/10.1016/B978-0-12-813586-0.00004-3.

Malik, S., Muhammad, K., and Waheed, Y. (2023). Emerging applications of nanotechnology in healthcare and medicine. *Molecules (Basel, Switzerland)* 28 (18): 6624. https://doi.org/10.3390/molecules28186624.

Mao, H., Yu, L., Tu, M. et al. (2022). Recent advances on the metal-organic frameworks-based biosensing methods for cancer biomarkers detection. *Critical Reviews in Analytical Chemistry* 54: 1–17.

Marchant, G.E., Sylvester, D.J., Abbott, K.W., and Danforth, T.L. (2010). International harmonization of regulation of nanomedicine. *Studies in Ethics, Law, and Technology* 3 (3): 1–16. https://doi.org/10.2202/1941-6008.1120.

Mazumder, S., Biswas, G.R., and Majee, S.B. (2020). Applications of nanorobots in medical techniques. *IJPSR* 11: 3150.

Merisko-Liversidge, E.M. and Liversidge, G.G. (2008). Drug nanoparticles: formulating poorly water-soluble compounds. *Toxicologic Pathology* 36 (1): 43–48. https://doi.org/10.1177/0192623307310946.

Merisko-Liversidge, E., Liversidge, G.G., and Cooper, E.R. (2003). Nanosizing: a formulation approach for poorly-water-soluble compounds. *European Journal of Pharmaceutical Sciences: Official Journal of the European Federation for Pharmaceutical Sciences* 18 (2): 113–120. https://doi.org/10.1016/s0928-0987(02)00251-8.

Mohammad, Z.H., Ahmad, F., Ibrahim, S.A. et al. (2022). Application of nanotechnology in different aspects of the food industry. *Discover Food* 2: 12. https://doi.org/10.1007/s44187-022-00013-9.

Nel, A., Grainger, D., and Alvarez, P.J. (2011). Badesha. Nanotechnology environmental, health, and safety issues. In: *Nanotechnology Research Directions for Societal Needs in 2020* (ed. M.C. Roco), 159–220. (New York: Springer. https://doi.org/10.1007/978-94-007-1168-6_5.

Nemmar, A., Hoet, P.H., Vanquickenborne, B. et al. (2002). Passage of inhaled particles into the blood circulation in humans. *Circulation* 105 (4): 411–414. https://doi.org/10.1161/hc0402.104118.

Noreen, S., Ma, J.X., Saeed, M. et al. (2022). Natural polysaccharide-based biodegradable polymeric platforms for transdermal drug delivery system: a critical analysis. *Drug Delivery and Translational Research* 12 (11): 2649–2666. https://doi.org/10.1007/s13346-022-01152-3.

Oberdörster, G., Ferin, J., and Lehnert, B.E. (1994). Correlation between particle size, in vivo particle persistence, and lung injury. *Environmental Health Perspectives* 102 (Suppl 5): 173–179. https://doi.org/10.1289/ehp.102-1567252.

Oberdörster, G., Sharp, Z., Atudorei, V. et al. (2002). Extrapulmonary translocation of ultrafine carbon particles following whole-body inhalation exposure of rats. *Journal of Toxicology and Environmental Health. Part A* 65 (20): 1531–1543. https://doi.org/10.1080/00984100290071658.

Oberdörster, G., Sharp, Z., Atudorei, V. et al. (2004). Translocation of inhaled ultrafine particles to the brain. *Inhalation Toxicology* 16 (6–7): 437–445. https://doi.org/10.1080/08958370490439597.

Oberdörster, G., Oberdörster, E., and Oberdörster, J. (2005). Nanotoxicology: an emerging discipline evolving from studies of ultrafine particles. *Environmental Health Perspectives* 113 (7): 823–839. https://doi.org/10.1289/ehp.7339.

Papahadjopoulos, D., Allen, T.M., Gabizon, A. et al. (1991). Sterically stabilized liposomes: improvements in pharmacokinetics and antitumor therapeutic efficacy. *Proceedings of the National Academy of Sciences* 88: 11460–11464.

Pardi, N., Tuyishime, S., Muramatsu, H. et al. (2015). Expression kinetics of nucleoside-modified mRNA delivered in lipid nanoparticles to mice by various routes. *Journal of Controlled Release: Official Journal of the Controlled Release Society* 217: 345–351. https://doi.org/10.1016/j.jconrel.2015.08.007.

Park, B., Donaldson, K., Duffin, R. et al. (2008). Hazard and risk assessment of a nanoparticulate cerium oxide-based diesel fuel additive – a case study. *Inhalation Toxicology* 20 (6): 547–566.

Patton, C., Farr, G.H. 3rd, An, D. et al. (2018). Lipid nanoparticle packaging is an effective and nontoxic mRNA delivery platform in embryonic zebrafish. *Zebrafish* 15 (3): 217–227. https://doi.org/10.1089/zeb.2017.1511.

Pekkanen, J., Peters, A., Hoek, G. et al. (2002). Particulate air pollution and risk of ST-segment depression during repeated submaximal exercise tests among subjects with coronary heart disease: the exposure and risk assessment for fine and ultrafine particles in ambient air (ULTRA) study. *Circulation* 106 (8): 933–938. https://doi.org/10.1161/01.cir.0000027561.41736.3c.

Peters, A., Wichmann, H.E., Tuch, T. et al. (1997). Respiratory effects are associated with the number of ultrafine particles. *American Journal of Respiratory and Critical Care Medicine* 155 (4): 1376–1383. https://doi.org/10.1164/ajrccm.155.4.9105082.

Pimple, K.D. (2002). Six domains of research ethics. A heuristic framework for the responsible conduct of research. *Science and Engineering Ethics* 8 (2): 191–205. https://doi.org/10.1007/s11948-002-0018-1.

Quan, Y., Wu, X., Zhu, S. et al. (2022). Triboelectric nanogenerators for clinical diagnosis and therapy: a report of recent progress. *Medicine in Novel Technology and Devices* 16: 100195.

Radomski, A., Jurasz, P., Alonso-Escolano, D. et al. (2005). Nanoparticle-induced platelet aggregation and vascular thrombosis. *British Journal of Pharmacology* 146 (6): 882–893. https://doi.org/10.1038/sj.bjp.0706386.

Rajput, A., Shevalkar, G., Pardeshi, K., and Prashant, P. (2023). Computational nanoscience and technology, *OpenNano*, Volume 12, 100147, 2352-9520, https://doi.org/10.1016/j.onano.2023.100147.

Resnik, D.B. and Tinkle, S.S. (2007). Ethics in nanomedicine. *Nanomedicine* 2 (3): 345–350. https://doi.org/10.2217/17435889.2.3.345.

Sadeghi, M., Sadeghi, S., Naghib, S.M., and Garshasbi, H.R. (2023). A comprehensive review on electrochemical nano biosensors for precise detection of blood-based oncomarkers in breast cancer. *Biosensors* 13 (4): 481. https://doi.org/10.3390/bios13040481.

Safra, T., Muggia, F., Jeffers, S. et al. (2000). Pegylated liposomal doxorubicin (Doxil): reduced clinical cardiotoxicity in patients reaching or exceeding cumulative doses of 500 mg/m2. *Annals of Oncology* 11 (8): 1029–1033.

Sahoo, S.K. and Labhasetwar, V. (2003). Nanotech approaches to drug delivery and imaging. *Drug Discovery Today* 8 (24): 1112–1120.

Salameh, J.W., Zhou, L., Ward, S.M. et al. (2020). Polymer-mediated gene therapy: recent advances and merging of delivery techniques. *Wiley Interdisciplinary Reviews. Nanomedicine and Nanobiotechnology* 12 (2): e1598. https://doi.org/10.1002/wnan.1598.

Salleh, A., Naomi, R., Utami, N.D. et al. (2020). The potential of silver nanoparticles for antiviral and antibacterial applications: a mechanism of action. *Nanomaterials (Basel, Switzerland)* 10 (8): 1566. https://doi.org/10.3390/nano10081566.

Schek, R.M., Hollister, S.J., and Krebsbach, P.H. (2004). Delivery and protection of adenoviruses using biocompatible hydrogels for localized gene therapy. *Molecular Therapy* 9 (1): 130–138.

Schrader-Frechette, K. (2002). *Environmental Justice: Creating Equality, Reclaiming Democracy*. New York: Oxford University Press.

Schulte, P.A. (1986). Problems in notification and screening of workers at high risk of disease. *Journal of Occupational Medicine* 28: 951–957.

Schulte, P.A. and Salamanca-Buentello, F. (2007). Ethical and scientific issues of nanotechnology in the workplace. *Environmental Health Perspectives* 115 (1): 5–12. https://doi.org/10.1289/ehp.9456.

Schulz, H., Harder, V., Ibald-Mulli, A. et al. (2005). Cardiovascular effects of fine and ultrafine particles. *Journal of Aerosol Medicine* 18 (1): 1–22.

Shang, Y., Hasan, M.K., Ahammed, G.J. et al. (2019). Applications of nanotechnology in plant growth and crop protection: a review. *Molecules (Basel, Switzerland)* 24 (14): 2558. https://doi.org/10.3390/molecules24142558.

Sharma, H.S. and Sharma, A. (2007). Nanoparticles aggravate heat stress induced cognitive deficits, blood-brain barrier disruption, edema formation and brain pathology. *Progress in Brain Research* 162: 245–273. https://doi.org/10.1016/S0079-6123(06)62013-X.

Shearn, P. (2005). Workforce participation in occupational health and safety in non-unionized workplaces. HSE Report no. HSL/2005/41. London: Health and Safety Executive.

Shen, L., Wang, P., and Ke, Y. (2021). DNA nanotechnology-based biosensors and therapeutics. *Advanced Healthcare Materials* 10 (15): e2002205. https://doi.org/10.1002/adhm.202002205.

Sheth, P., Sandhu, H., Singhal, D. et al. (2012). Nanoparticles in the pharmaceutical industry and the use of supercritical fluid technologies for nanoparticle production. *Current Drug Delivery* 9 (3): 269–284. https://doi.org/10.2174/156720112800389052.

Singh, A. and Amiji, M.M. (2022). Application of nanotechnology in medical diagnosis and imaging. *Current Opinion in Biotechnology* 74: 241–246. https://doi.org/10.1016/j.copbio.2021.12.011.

Sophia, W., Ho, L.I., Sumin, K. et al. (2023). Ethical and legal challenges in nanomedical innovations: a scoping review. *Frontiers in Genetics* 14: https://doi.org/10.3389/fgene.2023.1163392.

Stern, S.T. and McNeil, S.E. (2008). Nanotechnology safety concerns revisited. *Toxicological Sciences: An Official Journal of the Society of Toxicology* 101 (1): 4–21. https://doi.org/10.1093/toxsci/kfm169.

Sun, M., Li, Z., Yang, C. et al. (2021). Nanogenerator-based devices for biomedical applications. *Nano Energy* 89: 106461.

Tan, P., Chen, X., Zhang, H. et al. (2023). Artificial intelligence aids in development of nanomedicines for cancer management. *Seminars in Cancer Biology* 89: 61–75. https://doi.org/10.1016/j.semcancer.2023.01.005.

Tang, Y., Wang, X., Li, J. et al. (2019). Overcoming the reticuloendothelial system barrier to drug delivery with a "don't-eat-us" strategy. *ACS Nano* 13 (11): 13015–13026. https://doi.org/10.1021/acsnano.9b05679.

Uchida, S. and Kataoka, K. (2019). Design concepts of polyplex micelles for in vivo therapeutic delivery of plasmid DNA and messenger RNA. *Journal of Biomedical Materials Research. Part A* 107 (5): 978–990. https://doi.org/10.1002/jbm.a.36614.

Wang, Z., Agrawal, P., and Zhang, Y.S. (2021). Nanotechnologies and nanomaterials in 3D (Bio) printing toward bone regeneration. *Advanced NanoBiomed Research* 1 (11): 2100035.

Wang, W., Pang, J., Su, J. et al. (2022). Applications of nanogenerators for biomedical engineering and healthcare systems. *InfoMat* 4 (2): e12262. https://doi.org/10.1002/inf2.12262.

Warheit, D.B., Borm, P.J., Hennes, C., and Lademann, J. (2007). Testing strategies to establish the safety of nanomaterials: conclusions of an ECETOC workshop. *Inhalation Toxicology* 19 (8): 631–643. https://doi.org/10.1080/08958370701353080.

Weil, V. (2002). Making sense of scientists' responsibilities at the interface of science and society. Commentary on "Six Domains of Research Ethics". *Science and Engineering Ethics* 8 (2): 223–227. https://doi.org/10.1007/s11948-002-0023-4.

Wen, J., Cai, D., Gao, W. et al. (2023). Osteoimmunomodulatory nanoparticles for bone regeneration. *Nanomaterials (Basel, Switzerland)* 13 (4): 692. https://doi.org/10.3390/nano13040692.

Yang, L. and Webster, T.J. (2009). Nanotechnology controlled drug delivery for treating bone diseases. *Expert Opinion on Drug Delivery* 6 (8): 851–864. https://doi.org/10.1517/17425240903044935.

Yang, C., Lin, Z.I., Chen, J.A. et al. (2022). Organic/inorganic self-assembled hybrid nano-architectures for cancer therapy applications. *Macromolecular Bioscience* 22 (2): 2100349.

Yasri, S. and Wiwanitkit, V. (2017). Important ethical issues for nanomedicine. *Journal of Research in Medical Sciences : The Official Journal of Isfahan University of Medical Sciences* 22: 138. https://doi.org/10.4103/jrms.JRMS_856_17.

Yoon, H.J. and Kim, S.W. (2020). Nanogenerators to power implantable medical systems. *Joule* 4 (7): 1398–1407.

Yu, C.-H., Tam, K., and Tsang, E.S.C. (2008). Chapter 5 chemical methods for preparation of nanoparticles in solution. In: *Handbook of Metal Physics*, vol. 5 (ed. J.A. Blackman), 113–141. Elsevier. https://doi.org/10.1016/S1570-002X(08)00205-X.

Yu, C., Li, L., Hu, P. et al. (2021). Recent advances in stimulus-responsive nanocarriers for gene therapy. *Advanced Science (Weinheim, Baden-Wurttemberg, Germany)* 8 (14): 2100540. https://doi.org/10.1002/advs.202100540.

Zhang, P., Lynch, I., Handy, R.D., and White, J.C. (2023). 1 – a brief history of nanotechnology in agriculture and current status. In: *Nano-Enabled Sustainable and Precision Agriculture* (ed. P. Zhang, I. Lynch, R.D. Handy, and J.C. White), 3–14. Academic Press. https://doi.org/10.1016/B978-0-323-91233-4.00002-8.

Zhang, S., Wang, Y., Zhang, S. et al. (2023). Emerging anesthetic nanomedicines: current state and challenges. *International Journal of Nanomedicine* 18: 3913–3935. https://doi.org/10.2147/IJN.S417855.

2

An Overview of the Synthesis and Delivery of Nanoparticles in Biological Organisms

Nikhil Bhosale[1] *and Tabassum Khan*[2]

[1]*Department of Pharmaceutical Analysis, SVKM's Dr. Bhanuben Nanavati College of Pharmacy, Mumbai, Maharashtra, India*
[2]*Department of Pharmaceutical Chemistry, SVKM's Dr. Bhanuben Nanavati College of Pharmacy, Mumbai, Maharashtra, India*

2.1 Introduction

Nanoparticles (NPs) have garnered significant interest due to their unique physicochemical properties and applications in the biological and agricultural domains. In agriculture, NPs may enhance crop yield, nutrient absorption and insect management. They also provide choices for therapeutic treatments in biological research, targeted drug administration and diagnostic imaging. Understanding the synthesis and delivery of NPs in living organisms is essential to their full use in a wide range of applications (Blanco et al. 2015).

NPs are particles with sizes ranging from one to one hundred nanometres. Their unique physicochemical properties set them apart from bulk materials. With their large surface area-to-volume ratios, changeable surface chemistry and quantum size effects, NPs are versatile and have a wide range of applications in biological systems (Kah et al. 2018).

Nanoparticle creation is an amalgamation of several procedures that are used to generate a variety of materials, such as metals, semiconductors, polymers and lipids. These techniques (chemical, physical and biological methods) are coupled widely for the production of NPs. Every approach possesses several phenomenal features, such as the production of a certain chemical structure, particle-size and shape control. Physically comprised techniques encompass vaporizing, condensing or irradiation, while the chemical methods constitute splitting of metal salts in a solution. Biomimetic synthesis involves natural agents like plants, microbes or enzymes to decrease the effects of these particles in an effort to produce NPs without polluting the environment (Nel et al. 2006).

Silicon Nanoparticles: Synthesis, Uptake, and Applications, First Edition.
Edited by Archana Singh, Aryadeep Roychoudhury, and Indrakant K. Singh.

It is crucial for the NPs to be successfully used in biological systems to realize those market expectations and succeed at microlaser systems, imaging, etc., in targeted medication administration. Different routes of administration, such as intravenous injection, intranasal and oral drug delivery, as well as the use of convection-enhanced delivery strategies, are all aimed at improving the distribution, cellular intake and effectiveness of NPs in vivo. These modified treatments, based on specific biological targets and range of delivery alternatives, incorporate a range of delivery approaches, including but not limited to intravenous injection, oral consumption, inhalation, topical application and implantation (Rai et al. 2015).

Even if the use of NPs possesses the prospect of novelty and inventiveness in environmental sciences, medicine and agriculture, the fact that problems stem from toxicity, safety and environmental impact has to be considered. To provide the safest solutions to the possible problems arising during the development and application of NPs in technologies, risk assessment frameworks, standardized methods of characterization and toxicity tests and legal stipulations should be established (Khan et al. 2019).

At the end of the look-back, it is evaluated that the complexity of the capacity of the biological organisms to disperse and synthesize nanomaterial advanced technologies creates innovation and the starting point of a green environment, clearing agriculture and health care. The main contribution of nanotechnology is in the advancement of NP production, their transport system and the way they interact with biological materials. Through these, the scientists can effectively use nanotechnology to tackle global issues that need long-term solutions. Finally, nanotechnology can elevate the living standard of the entire human population.

2.2 Methods of Synthesis of NPs

Generally speaking, NP refers to inorganic materials rather than individual molecules. The reason that NPs and ultrafine particles (UFPs) are defined similarly is because in the 1970s and 1980s, Granqvist and Buhrman from the United States were conducting the first comprehensive basic investigations with NPs, which were referred to as UFPs in Japan. The term NP had gained popularity in the 1990s, nevertheless, prior to the establishment of the US National Nanotechnology Initiative. The size-related characteristics of bulk materials or tiny particles might differ dramatically from those of NPs. Craftspeople used NPs as early as the fourth century in Rome, in the dichroic glass Lycurgus cup, and in the ninth century in Mesopotamia, to give pots a shimmering appearance. Antique and Renaissance ceramics frequently have a noticeable metallic sheen that is either copper- or gold-coloured. The translucent glazing surface has a metallic coating put to it, which is the reason for its brilliance. If the film has not been affected by other weathering processes or air oxidation, the brilliance can still be seen. The film itself, which has uniformly distributed copper and silver NPs in the glassy ceramic glaze matrix, is the source of the shine. The artists produced these NPs by coating the previously glazed pottery with vinegar, ochre, clay, salts and oxides of copper and silver. Despite being linked to contemporary research, NPs have a lengthy past (Rane et al. 2018).

NPs are defined by the worldwide federation of national standard bodies, the International Organization for Standardization (ISO), as nano objects with all external dimensions in the nanoscale, where the length of the longest and shortest axes of nano objects does not differ significantly. Though nanoscale is basically ranged from 1 to 100 nm (Vert et al. 2012), NPs can be categorized by three size ranges: larger than 500 nm, between 100 and 500 nm and between 1 and 100 nm (Vert et al. 2012). With respect to the size and size distribution, NPs may exhibit size-related intensive properties. There are diverse methods used to synthesize NPs as indicated in Figure 2.1.

2.2.1 Co-precipitation Technique

Co-precipitation reactions occur when nucleation, growth, coarsening and/or agglomeration activities take place all at once. Co-precipitation reactions include the following features:

- In most cases, the products are insoluble species that were created in a supersaturated environment.
- Nucleation is an important stage in which a lot of tiny particles will develop.
- Secondary processes that have a significant impact on the product size, shape and quality include Ostwald ripening and aggregation.

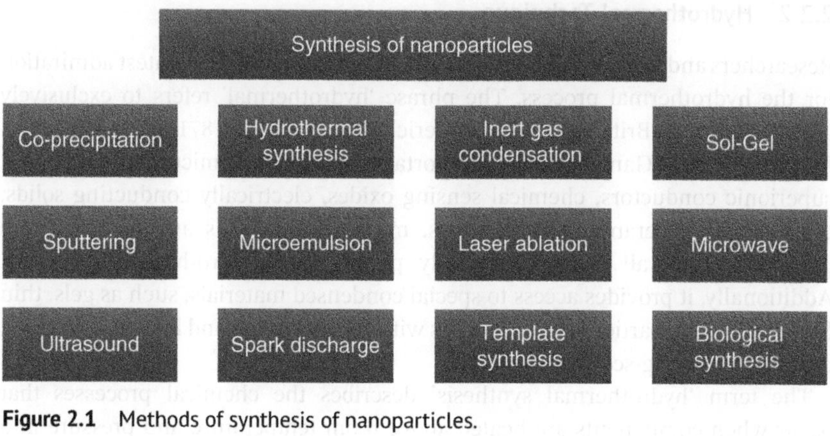

Figure 2.1 Methods of synthesis of nanoparticles.

- The following chemical reactions often produce the supersaturation conditions required to cause precipitation. $Xay^{+}{}_{(aq)} + YBx^{-}{}_{(aq)}\ AxBy_{(s)}$

Methods of co-precipitation synthesis:

- Metals generated from aqueous solutions by electrochemical reduction, reduction from non-aqueous solutions and breakdown of metallorganic precursors.
- Both aqueous and non-aqueous solutions produced oxides.
- Metal chalconides produced by molecular precursor processes.

Advantages and disadvantages of the co-precipitation method (Sud et al. 2022):

Advantages:

- Easy and quick to prepare.
- Particle size and composition can be easily controlled, and there are several ways to change the particle surface state and overall homogeneity.
- Low body temperature.
- Energy-conserving.
- Eliminates the need for an organic solvent.

Disadvantages:

- Not relevant to species without charges.
- It takes time and there is a chance that trace contaminants will precipitate with the result.
- Issues with repeatability from batch to batch.
- If the precipitation rates of the reactants are significantly varied, this approach is not effective.

2.2.2 Hydrothermal Technique

Researchers and technicians from many fields have shown the greatest admiration for the hydrothermal process. The phrase 'hydrothermal' refers to exclusively geological origin. British geologist Roderick Murray (1792–1871) coined the term for the first time (Gan et al. 2020). Important solids such as microporous crystals, superionic conductors, chemical sensing oxides, electrically conducting solids, complex oxide ceramic and fluorides, magnetic materials and luminescence phosphors have all been successfully prepared by hydrothermal synthesis. Additionally, it provides access to special condensed materials, such as gels, thin films, nanoscale particles and materials with distinct chiral and helical structures, especially stacking-sequence material.

The term 'hydrothermal synthesis' describes the chemical processes that occur when components are heated above room temperature and pressure in a sealed solution. The solubility of minerals in heated water under high pressure is a prerequisite for the hydrothermal synthesis of single crystals. The crystal development process is carried out in a device that consists of an autoclave, a steel pressure vessel that holds water and nutrients. The growing chamber is kept at two different temperatures, with the hotter end dissolving the nutrients and the colder end encouraging seeds to develop more.

Advantages of solvothermal and hydrothermal synthesis (Gan et al. 2020):

- The majority of the material may be heated to near-critical temperature and then crushed until it dissolves in the appropriate solvent.
- The use of hydrothermal/solvothermal synthesis can result in a significant increase in the chemical activity of the reactant, the ability to substitute solid-state synthesis and the development of materials that would not be possible through solid-state reaction.
- It is possible to manufacture unique compounds of metastable state and other specialized condensed states, as well as products of intermediate state, metastable state and specific phase with ease.
- The size of the final product, form distribution and crystallinity may be easily and precisely controlled by adjusting several factors, including reaction temperature, time, solvent type, surfactant type and precursor type.

Drawbacks of hydrothermal/solvothermal synthesis:

- The requirement for expensive autoclaves.
- Safety concerns throughout the course of the response.
- The reaction process is an unobservable 'black box.'

2.2.3 Inert Gas Condensation Technique

Inert gas condensation (IGC) is the most useful technique for creating metal NPs. In IGC, metals are evaporated at extremely high pressure inside an ultrahigh vacuum chamber that is filled with argon or helium gas. Collisions with the gas cause the evaporated metal atoms to lose their kinetic energy and condense into tiny particles. Following Brownian coagulation and coalescence, these particles eventually expand to create nanocrystals.

The initial method for creating nanocrystalline metals and alloys was gas condensation. This method was developed first for the synthesis of inorganic or metallic materials. By homogeneous nucleation in the gas phase, clusters are created close to the source and expand by coalescence and integration of gas-phase atoms. The residence duration of the particles in the growth regime is a significant factor that determines the cluster or particle size. Other factors that may have an impact include the gas pressure, the kind of inert gas (He, Ar or Kr), and the evaporation rate and vapour pressure of the evaporating substance. The average particle size of the NPs rises with increasing gas pressure, vapour pressure and inert gas mass. For the purpose of collecting particles, a revolving cylindrical device chilled by liquid nitrogen is employed. The 2–50 nm NPs are separated from the gas flow and placed as a low-density, agglomeration-free powder on the surface of the collecting device (https://shellzero.wordpress.com/2012/05/14/inert-gas-condensation-method). Following that, a scraper in the shape of a metal plate is used to remove the NPs from the surface of the cylinder. These techniques, however, are limited to systems intended for gas flow, i.e., those in which a dynamic vacuum is produced by a combination of continuous pumping and gas entry via a mass flow controller.

Advantages:

- The enhanced control over particle size is a key benefit over convectional gas movement. Particle sizes are decreased by 80% and standard deviations by 18%, depending on the He-gas flow rate.

Disadvantages:

- Although a wide range of metallic and ceramic materials have been prepared using the gas condensation process, including its modifications, the amount produced has thus far only been on a laboratory scale. The process is really sluggish.
- Metal concentration is less than 1 g/day, whereas oxide concentrations for simple oxides, such ZrO_2 or CeO_2, can reach 20 g/day. These amounts are enough for evaluating materials, but not for large-scale manufacturing (https://shellzero.wordpress.com/2012/05/14/inert-gas-condensation-method).

2.2.4 Sol-Gel Technique

Sol-gel is a technique that turns tiny molecules into solid materials. The sol (or solution) in this chemical process progressively transforms into a gel-like diphasic system with a liquid phase and a solid phase, with morphologies ranging from discrete particles to continuous polymer networks. Precipitation is a method for creating homogeneous and ultrafine ceramic powders. For use in dentistry and biomedicine, these single- and multi-component composition powders may be made with nanoscale particle sizes (Jain et al. 2012).

The following is the standard procedure for creating metal NPs:

- In the liquid phase, evenly combine the raw ingredients.
- Conduct polycondensation and hydrolysis activities to create a transparent and stable sol system.
- After sols age, colloidal particles gradually agglomerate to create gels with a three-dimensional network architecture.

- The preparation of materials containing microstructures or even nanostructures might include a number of steps, including drying and sintering (Jain et al. 2012).

Advantages:

- The starting components are at the molecular level, and the produced final products are uniform.
- Products have a comparatively greater level of purity, a controlled porosity level, ease of preparation in a range of sizes and convenience of use.
- Permits low-temperature synthesis.
- Enables precise control over the chemical makeup of the final product, making it ideal for the preparation of multi-component products.
- Allows dopants to be added in little amounts to the sol and evenly distribute themselves throughout the finished product.

Disadvantages:

- Comparatively slower in response time.
- The human body may be harmed by organic solvents.

2.2.5 Sputtering Technique

Sputtering is the process of ejecting atoms from a substance (the target) by subjecting it to an intense particle bombardment. In the momentum transfer process known as sputtering, ions are used to force off atoms from a cathode or target. Travelling until they come into contact with a substrate, spewed atoms deposit to create the appropriate layer. The most straightforward source of ions for sputtering is the well-known glow discharge phenomenon, which results from an applied electric field between two electrodes in a low-pressure gas. Once a minimal voltage is attained, the gas disintegrates to carry electricity. We refer to this type of ionized gas as plasma. A strong electric field accelerates the ions of the plasma as they approach the target. Atoms (or molecules) are propelled from the surface of the target into the plasma when the ions strike it, where they are transported away and eventually deposit on the substrate. Direct current (DC) sputtering' is the term for this kind of sputtering. The sputtering gas is typically an inert gas, such as argon, to prevent any chemical interaction between the sputtered atoms and the gas. To ensure that the deposited layer is a chemical compound, a reactive gas is intentionally introduced to argon in some applications, such as the deposition of oxides and nitrides. 'Reactive sputtering' is the term for this kind of sputtering (Patankar et al. 2012).

The electrical charge of the plasma ions is neutralized when they hit the target, and they return to the process as atoms. The neutralizing procedure leaves the target surface positively charged if the target is an insulator. There is a chance that this charge will get so high that bombardment ions will be rejected, stopping the sputtering process. In order to remove surface charge, more electrons from the plasma must be drawn in by reversing the polarity. This cyclical polarity reversal is automatically accomplished by putting the target assembly under a radio-frequency (RF) voltage. As a result, this kind of sputtering is called 'RF sputtering' (Patankar et al. 2012).

It is typical for the sputtering source to have some magnetic confinement using a magnetron source in order to boost the efficiency of the sputtering process. The plasma may be run at a greater density because of the ability of the magnetic field to spiral electrons, increasing their likelihood of experiencing an ionizing collision. 'Magnetron sputtering' is the term for this kind of sputtering, which may be combined with either DC or RF sputtering.

Depending on the desired qualities of the NPs to be synthesized, several modifications may be made to the ion sputtering synthesis process by adjusting the magnetic field or RF. Elements like pure metals and elemental semiconductors, alloys and compounds including oxides, nitrides, sulfides and carbides are among the materials that can be sputtered.

Advantages:

- Films with the same composition as the target source can be deposited using sputtering.
- It is a flexible method that works with nearly any kind of material.
- Its numerous process variables allow for the customization and modification of the qualities of the materials.
- The kind and content of the material may be controlled by using solid targets, which eliminates the need for complex process chemistry (Patankar et al. 2012).

Disadvantages:

- Sputtering targets are frequently costly, and sputtering rates are modest when compared to thermal evaporation rates.
- The majority of the energy that strikes the object turns into heat, which has to be expelled.
- To avoid poisoning the sputtering target, the gas composition in reactive sputter deposition needs to be carefully regulated.

2.2.6 Microemulsion Technique

Although the microemulsion approach is one of the best ways to manufacture inorganic NPs, little is known about the process underlying the production of NPs in the microemulsion. Nonetheless, some scientists have proposed a method for the creation of NPs in microemulsions. Reactant exchange occurs when the components of the microemulsion, including the reactants, are combined and the water droplets in the microemulsion collide. Reactant exchange proceeds too quickly, causing a precipitation reaction in the nanodroplets. This is followed by primary particle nucleation, growth and coagulation, producing final NPs stabilized by surfactants and/or surrounded by water (http://shodhganga.inflibnet.ac.in/bitstream/10603/11272/11/11_chapter%203.pdf).

Advantages:

- **Preparedness**: Compared to liposomes and macroemulsions, which need high pressure and homogenization during preparation, microemulsions develop spontaneously at ambient temperature and are simpler to make.
- **Thermodynamic Stability**: Microemulsions, which are composed of tiny droplets with a large interfacial area, are thermodynamically stable, and this stability contributes to the increased stability and shelf life of the formulation. These properties make them easier to utilize in the creation of NPs.
- Inorganic particles that are nanometre in size and exhibit little aggregation.
- The creation of NPs with a high specific surface area and crystalline structure; strict control over particle size and content.

Disadvantages:

- Large quantities of surfactant and/or cosurfactant are needed for microemulsion production, and these substances are often unpleasant.
- External variables that affect the stability of microemulsions include pH and temperature.
- Limited ability to dissolve materials with high melting points.

2.2.7 Laser Ablation Technique

The technique of using a laser beam to remove material from a solid (or occasionally liquid) surface is known as laser ablation (Figure 2.2). When the laser flux is low, the absorbed laser energy heats the material, causing it to either evaporate or sublimate. Usually, the material turns into plasma at high laser flux. Typically, laser ablation in the Synthesis of Inorganic Nanomaterials refers to the removal of material using a pulsed laser; but, if the laser intensity is high enough, material can also be ablated using a continuous wave laser beam (Penide et al. 2019).

2.2.8 Microwave-Assisted Synthesis

Microwave-assisted synthesis finds widespread use in fields spanning from nanotechnology to biological processes. Chemical reactions provide better yields and less by-products than conventional convection heating techniques. They are also frequently quicker. In addition to withstanding high temperature and pressure and exhibiting repeatability from reaction to reaction, microwave reactors offer

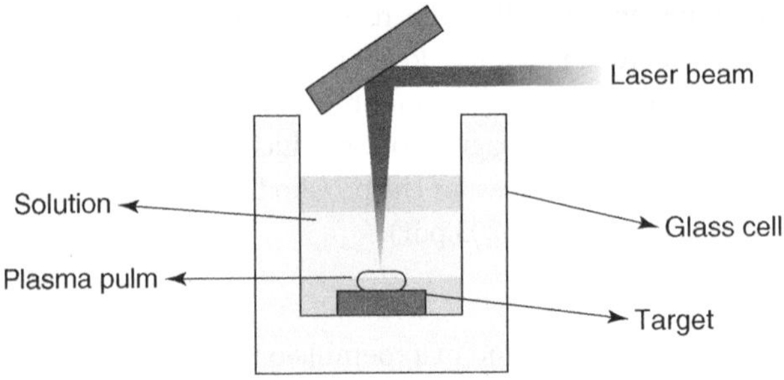

Figure 2.2 Preparation of nanoparticles by laser ablation.

superior control over reaction mixing. When the reaction is started from room temperature, microwave-assisted approaches offer better engineering control over the separation of the nucleation and growth stages of nanomaterial production. Scalability depends on the potential for selectivity in the activation of the precursor materials for nanomaterials, which may be achieved using microwave-assisted heating. For the creation of nanomaterials, microwave synthesis offers the ability to selectively heat the precursor molecules or the solvent (Lyell 1911).

2.2.9 Ultrasound Technique

The use of ultrasound in the creation of NPs has grown in importance. Ultrasonic cavitation is the result of ultrasonic radiation applied to liquids. Extreme temperature, high pressure and rapid cooling rates are just a few of the physical and chemical phenomena produced by ultrasonic cavitation, which creates a special environment for chemical reactions. A good technique for creating NPs with regulated morphologies is ultrasound (Sandhya et al. 2021).

Advantages:

- The process of creating NPs occurs at room temperature.
- When ultrasonography is used, a very high response rate is obtained.
- Consequently, the time needed to create NPs is reduced. It is an easy and quick procedure.

Disadvantages:

- Acoustic cavitation is an energy-intensive process that cannot be handled by heat-sensitive materials.
- Scaling up is also challenging.
- Even though sonochemical synthesis produces NPs at room temperature, it requires a significant amount of energy, which means that the yield per unit energy given is poor.

2.2.10 Spark Discharge Technique

Spark discharge NP synthesis is a viable approach for low-cost industrial nanofabrication of innovative materials. An electric spark is a sudden electrical discharge that occurs when an electrically conductive channel is created through an insulating medium (often air or other gases or gas combinations) by an ionized electric field that is high enough (Efimov et al. 2016).

The unique characteristics of this approach include:

- The ability to create extremely tiny primary NPs from metals, oxides, semiconductors and carbon compounds that have a diameter of less than 10 nm.
- Relatively easy technique implementation; high chemical purity of the produced NPs with low impurity concentration.
- One constraint has been the low pace of output (Efimov et al. 2016).

2.2.11 Template Synthesis Technique

The development of a simple, eco-friendly process for the synthesis of NPs is becoming more and more relevant, especially with the growing emphasis on green chemistry. One of the most promising techniques for creating monodispersed inorganic NPs is reportedly template synthesis, which uses the uniform void spaces of porous materials as hosts to confine the created NPs as guests. The template serves two purposes. Firstly, it serves as a skeleton to arrange the many functionalities of a device, the active components and the various interfaces while enabling the optimal repeatability of the structure (Pantidos and Horsfall 2014).

2.2.12 Biological Synthesis Technique

The ability of the microorganisms to biosynthesize NPs is an environmentally benign technology. Metal oxides like titanium oxide and zinc oxide, as well as a variety of prokaryotes and eukaryotes, are employed in the production of metallic NPs, including those of silver, gold, platinum, zirconium, palladium, iron and cadmium. These microorganisms consist of algae, fungus, bacteria and actinomycetes. Depending on where the NPs are located, the creation of the particles might be either extracellular or intracellular (Li et al. 2011).

Intracellular synthesis of NPs by fungi

Using this technique, ions are transported into microbial cells where they combine with enzymes to produce NPs. The size of the inside-formed NPs is less than that of the extracellularly reduced NPs. The ability of the particles to nucleate inside of the organisms is likely connected to the size restriction.

Extracellular synthesis of NPs by fungi

Since extracellular production of NPs lacks the need for extracellular components to be present within the cell, it has a wider range of applications than intracellular synthesis. Due in large part to their massive secretory components, which are involved in the reduction and capping of NPs, fungi are known to create NPs extracellularly.

Microbes for the production of NPs

Inorganic minerals are produced either extracellularly or intracellularly by both unicellular and multicellular organisms. The quest for novel materials makes use of the capacity of the microorganisms to regulate the creation of metallic NPs, such as bacteria and fungus. Fungal bioaccumulation of metals and their tolerance have made them the focus of investigations on the biological production of metallic NPs.

2.3 Methods of Delivery of NPs to Plants

Plant absorption, or the delivery of materials into plants, is an essential process for the growth and development of plants. For their life and development, plants need a variety of vital nutrients, water and gases. The two main mechanisms that allow materials to be taken up by plants are the absorption of water and nutrients. Because it is involved in several physiological processes and acts as a solvent for nutrients, water is vital to plants (Dontsova et al. 2019).

To fully use NPs in biotechnology and agriculture, it is essential to comprehend how they enter plants. It is possible for the NPs to be designed in the way to act as a carrier in case when certain molecules, compounds or genetic materials are to be transferred to certain areas of plants (Sembada and Lenggoro 2024). Some of the applications using carriers are shown in Figure 2.3.

The precise and effective use of NPs can benefit from their regulated distribution. Delivery of nutrients or fertilizer, genetic material and pesticides or herbicides can all be accomplished with NPs. Essential nutrients, such as micronutrients like iron, zinc and other trace elements, may be incorporated into plant cells using NPs. Micronutrient NP composites were effectively produced by Tombuloglu et al. and used in *Hordeum vulgare* L. (Handy et al. 2008). They gave an example of how these micronutrients may be successfully incorporated into plant tissue. The number of elements in the root and leaf tissues was greatly enhanced by the conveyance of these NPs. Specifically, the contents of Fe, Zn and Cu were raised to approximately 5, 3 and 18 (Hong et al. 2021) times higher than the control, respectively. This method, referred to as 'NP-mediated nutrient delivery', aims to improve the absorption and use of nutrients. NPs, due to their small size, can be taken up by plant roots or leaves. They may enter plant cells through various mechanisms, such as diffusion, endocytosis or direct uptake through ion channels and transporters.

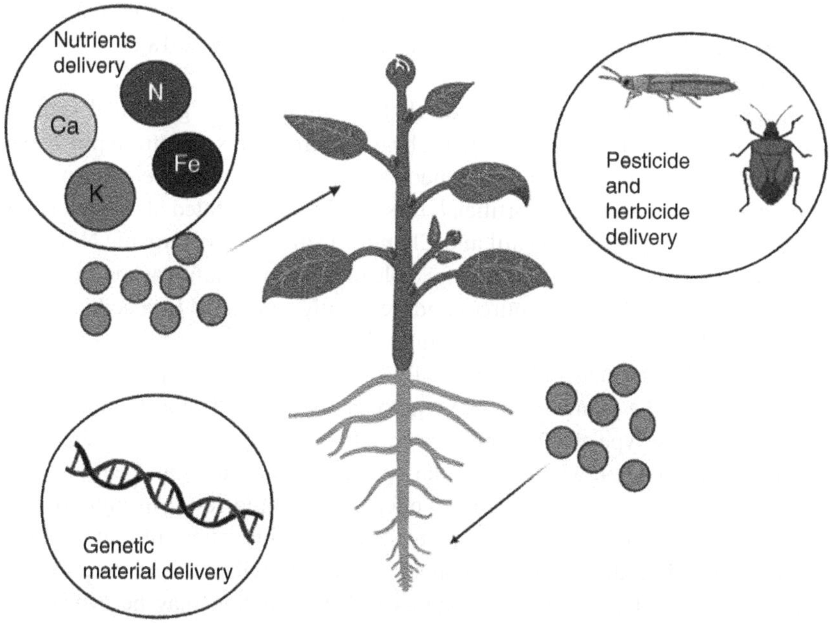

Figure 2.3 Nanoparticles used as carriers for different materials.

The contained nutrients are released by the NPs once they are within the plant cells. The regulated release process guarantees that the plant receives the nutrients at a pace that corresponds with its requirements. Through enhanced nutrient absorption and use, nutrient delivery mediated by NPs can minimize the environmental effects of excess nutrient runoff and leaching while also reducing the requirement for conventional fertilizers (Jat et al. 2020).

2.3.1 Different Types of NPs for Transportation

A diverse variety of NPs are reported to be used for transportation in plants (Table 2.1).

1) **Commercial NPs**
   Commercial NPs, also called engineered NPs, are particles like metal NPs, metal oxide NPs, carbon nanotubes, quantum dots, nanocomposites and polymeric NPs that are purposefully designed, synthesized and produced with particular characteristics for a range of industrial, technological, agricultural or consumer applications (Barhoum et al. 2022). These NPs, which usually have a size range of 1 to 100 nm, are produced via engineering techniques to get desired qualities at the nanoscale. The deliberate modification of surface

characteristics, composition, size and form sets manufactured NPs apart from their naturally occurring counterparts (Jeevanandam et al. 2018).

2) **Biogenic NPs**
   Biogenic NPs are defined as those that are produced or generated spontaneously, without the help of humans, by living organisms or natural processes. Biogenic NPs are created spontaneously by biological and environmental processes, in contrast to artificial NPs, which are created deliberately by humans for certain uses (Patil and Chandrasekaran 2020). Organic NPs such as those derived from bacteria, fungi, volcano, clay or wildfire ashes are found in huge variety of natural sources and frequently connected to specific species or natural phenomena (Jeevanandam et al. 2018).

   Those NPs formed either naturally or as a man-made product can be the cause for the plants in many ways to be restrained. The influence always depends on the type of NPs, their concentration, as well as how much exposure time is required and which particular species of plants are involved. The NPs may manifest a range of phytotoxic effects that would result in injured cells and tissue as well as a disrupted plant structure. This might ultimately cause a corresponding drop in plant biomass, slower growth rate and deteriorated plant health. The root morphology as well as function may be evolved in the case of exposure to some NPs. This includes adjusting the surface area, height and the root hair structure of the plant. That adjustment is responsible for increasing the uptake of water and nutrients by these plants. An active reaction with oxygen occurs as a result of nanospheres. These nanospheres generate reactive oxygen species (ROS) in biological systems such as plants (Rastogi et al. 2017). Elevated ROS levels may lead to a number of effects on plant health with possible impairment of the cellular membranes, proteins and DNA (Nair and Chung 2014; Rastogi et al. 2017). Some of these NPs could be toxic to any seed or spore within its reach and thus cause a threat to the future plant life. These fears have also sprung up relative to the possibility of the bioaccumulation of NPs in the plant tissues, which might have negative consequences on the higher order species such as birds, animals, and so on, in the chain of food production.

2.3.2 Methods of NP Transport into Plants

There are two primary methods for NP transport in plants (Table 2.2):

1) **Assisted Delivery**
   In the context of transferring NPs to the plant, the term 'assisted delivery' refers to the application of external force or power to aid in the transfer of NPs into the plant tissues (Dontsova et al. 2019). This external force facilitates delivery by frequently getting through obstacles that would prevent passive transport.

Table 2.1 Types of nanoparticles used for transport in plants.

Classes	Nanoparticles	Sources	Target site	References
Metal	Au	*Oryza sativa, Lolium perenne, Raphanus sativus, Cucurbita mixta*	Roots	Zhu et al. (2012)
	Co	Tomato	Roots	Li et al. (2021)
	Ni	Tomato	Roots	Li et al. (2021)
	SiO_2	*Cucumis sativus*	Leaves	Song et al. (2013)
	TiO_2	*Lycopersicum esculentum*	Roots	Wang et al. (2011)
	TiO_2	*Arabidopsis thaliana*	Roots	El-Temsah and Joner (2012)
	Ag	*Linum usitatissimum, Lolium perenne, Hordeum vulgare*	Roots	Lin and Xing (2007)
	Al	*Brassica napus, Raphanus sativus, Lolium perenne, Lactuca sativa, Zea mays, Cucumis sativus*	Roots	Koelmel et al. (2013)
	Au	*Oryza sativa*	Roots	Lizzi et al. (2021)
	ZnO	*Brassica napus, Raphanus sativus, Lolium perenne, Lactuca sativa, Zea mays, Cucumis sativus*	Roots	Koelmel et al. (2013)
	Zn	*Brassica napus, Raphanus sativus, Lolium perenne, Lactuca sativa, Zea mays, Cucumis sativus*	Roots	Koelmel et al. (2013)
Carbon based	CeO_2	*Holcus lanatus, Diplotaxis tenuifolia*	Roots	Zhang et al. (2011)
	CeO_2	*Cucumis sativus*	Roots	Demirer et al. (2020)
	Multi-walled carbon nanotubes (MWCNT)	*Brassica napus, Raphanus sativus, Lolium perenne, Lactuca sativa, Zea mays, Cucumis sativus*	Roots	Koelmel et al. (2013)

(Continued)

Table 2.1 (Continued)

Classes	Nanoparticles	Sources	Target site	References
	Single-walled carbon nanotubes (SWCNT)	*Nicotiana benthamiana*	Leaves	Zhao et al. (2021)
	Carbon	*Zea mays*	Roots	Tripathi et al. (2011)
	Carbon nanotubes (CNT)	*Cicer arietinum*	Roots	Modlitbová et al. (2018)
Semiconductor	3-mercaptopropionic acid (MPA) quantum dots (QDs)	*Lemna minor*	Leaves	Modlitbová et al. (2018)
	Cd-based QDs	*Allium cepa*	Roots	Modlitbová et al. (2018)
	CdSe/CdZnS QDs	*Populus deltoides*	Roots	Wang et al. (2014)
Magnetic	Superparamagnetic iron oxide (SPION)	*Glycine max*	Roots	Divya et al. (2019)
Polymeric	Chitosan	*Oryza sativa*	Roots	Divya et al. (2019)
	Thiamine loaded chitosan	*Cicer arietinum*	Roots	Muthukrishnan et al. (2019)

Assisted delivery techniques for introducing NPs into plants in vitro:

- **Biolistic**
  A biolistic (gene gun) is a device that propels NPs with the use of an external force, such as pressurized gas or helium (Rajkumari et al. 2021; Roychoudhury 2020). In order to achieve biolistic gene delivery in *Nicotiana tabacum* L., Ag NPs were used as gene carriers instead of Au microcarriers. Their findings demonstrated that the transformation efficiency of Ag NPs as carriers was much greater than that of Au microparticles.
- **Electroporation**
  Through the process of electroporation, plant cells are exposed to external electric fields, which cause transient holes to form in their membranes. This technique facilitates NP penetration by using an external electrical power source (Silva et al. 2010).
- **Sonoporation**
  Plant cell membranes are temporarily perforated by ultrasonic vibrations during the sonoporation process, which permits NPs to enter the cells (Das et al. 2015).

Table 2.2 Different assisted delivery techniques.

Method of delivery	NPs	Source	Site of delivery	References
Magnetic field	Carbon-coated iron	*Cucurbita pepo*	Roots	Ozyigit (2020)
	Fe_2O_3	*Solanum lycopersicum*	Roots	Shankramma et al. (2016)
	Fe_3O_4	*Hordeum vulgare*	Roots	Tombuloglu et al. (2019)
Microinjection	Magnesium oxide nanoparticles (MgNPs)	*Brassica napus*	Cells	Hao et al. (2013)
	MgNPs	*Brassica napus*	Cells	Liu et al. (2009)
Biolistic (gene gun)	Ag	*Nicotiana tabacum*	Leaf explants	Rajkumari et al. (2021)
	Au	*Nicotiana benthamiana*	Leaf explants	Makhotenko et al. (2018)
	Au-MSN	*Nicotiana tabacum,* Teosinte	Leaf explants	Martin-Ortigosa et al. (2012)
	Fe	*Nicotiana benthamiana*	Leaf explants	Makhotenko et al. (2018)
Electroporation	Conjugated polymer nanoparticles	Tobacco	Protoplasts	Silva et al. (2010)
Sonoporation	Hyperbranched Polyami-doamine (hPAMAM-G2)	*Medicago sativa*	Cells	Amani et al. (2018)
	PEI-MSN	*Nicotiana tabacum*	Cells	Zolghadrnasab et al. (2021)

- **Microinjection**
  NPs may be injected into certain plant cells using a combination of external manipulation and microinjection. When transferring NPs in vitro – that is, outside of living entity and usually in a controlled laboratory environment – the majority of aided delivery techniques are employed (Hao et al. 2013; Liu et al. 2009).

- **Magnetic Field**
  External magnetic fields may be used to direct magnetic NPs to particular plant tissues, which helps with the targeted administration of these NPs (Li et al. 2016; Shankramma et al. 2016; Tombuloglu et al. 2019; Ozyigit 2020).

2) **Passive Delivery**
   When discussing the transport of NPs in plants, the term 'passive delivery' usually refers to the transportation of particles across organic pathways without the aid of active processes or external pressure. It includes all of the ways that NPs can get inside a plant, usually through roots and leaves (Shankramma et al. 2016). The routes of passive delivery:
   - **Root uptake**
     Water and other dissolved materials are naturally absorbed by plants, and when they do so, NPs may enter the plant through their roots. This is a typical technique for in vivo NP delivery to plants.
   - **Foliar uptake**
     Certain NPs, such as those found in foliar fertilizers and pesticides, can be absorbed by plants through the cuticle of their leaves or through the stomata, which are tiny holes on the surfaces of their leaves. These two techniques, which make use of the in vivo NP transport pathway, are demonstrated above.

2.3.3 Transport Pathway of NP in Leaves

NPs can travel a great distance within a plant via the plastid or extracellular pathways once they reach mesophyll cells through the stomata and cuticle. Transport across the extracellular space is often referred to as the extracellular route. Depending on their particle size and surface charge, NPs can move across the extracellular space (such as the cell wall, a longitudinal channel between cell walls, intermediate lamella and xylem) after passing through the cell wall. The protoplast route is mostly carried by plasmodesmata, an intercellular channel with a diameter of two to twenty nanometres. Following their passage via the plasmodesmata, NPs gather in the cytoplasm and go to the Casparian strip and plant endothelium layer (Wang et al. 2012).

According to some research, the plastid pathway in plants is often used to transport NPs, smaller than 50 nm, but the apoplast system is typically used to transport NPs larger than 200 nm. The vascular system-phloem transport channels subsequently facilitate the transfer of NPs from leaf to root. The enormous diameter of sieve tube facilitates the easy passage of particles with a diameter of around 0.405 μm through the phloem.

2.3.4 Distribution of NPs Absorbed via Leaves

Numerous studies have demonstrated the ability of NPs sprayed through leaves to reach many areas of plants, including stems, roots, flowers, fruits and even the soil in the rhizosphere. CuO NPs were sprayed on maize leaves, and deposits of NPs were seen at the leaf epidermis. CuO NPs were transported by plasmodesmata between cells after entry. In a different investigation, ICP-MS was used to establish that gold NPs were absorbed and accumulated in the watermelon leaf cells during a 48-hour spraying period. Watermelon stems and roots were also found to contain gold, suggesting that phloem carried the NPs from leaves to roots. Additionally, after CeO_2 NPs acted on cucumber leaves for 72 hours, transmission electron microscopy revealed the presence of CeO_2 NPs in cucumber root slices, offering concrete proof of the transfer of CeO_2 NPs from leaves to cucumber stems and roots. When azoline oxalate-loaded porous hollow silica NPs (Azo@PHSNs) were sprayed on tomato, on the second and fourth days, the amount of Si present in various plant sections was identified and measured. The findings demonstrated that the order of Si concentration in the tomato tissues was stem > root > young leaf > mature leaf, suggesting that the effectiveness of PHSN delivery to the different tomato tissues varied. The transportation of NPs from stem to root was simpler than that between separate leaves (Wang et al. 2023).

2.4 Methods of Delivery of NPs in Animals

Administering NPs is not just a part of a recovery procedure but an act of ingenuity to attain maximal curative result, lowered toxicity and appropriated targeting. Many novel approaches have been developed for animal care and biomedical science that are all unique in the sense that the benefit is determined by the delivery system, which can either be very effective or can bring about some drawbacks (Geiser 2010). Different nanocarriers are reported to be used to deliver medicines to animals some of which are shown in Figure 2.4.

- **Intravenous Injection**
  Through the controversial intravenous administration procedure, a dogmatic pathway is formed for NP entry in the animal system causing a rapid circulation and distribution to the desired tissues. This choice offers maximum absorption and specific timing of the drug taken. Nevertheless, the NPs elicit inflammation, toxicity and other immunogenic effects (Jain et al. 2016).
- **Oral Administration**
  Animal feed can be enriched with NPs and they may be ingested in the form of oral suspension or tablets. Many prefer this way as it does not need another

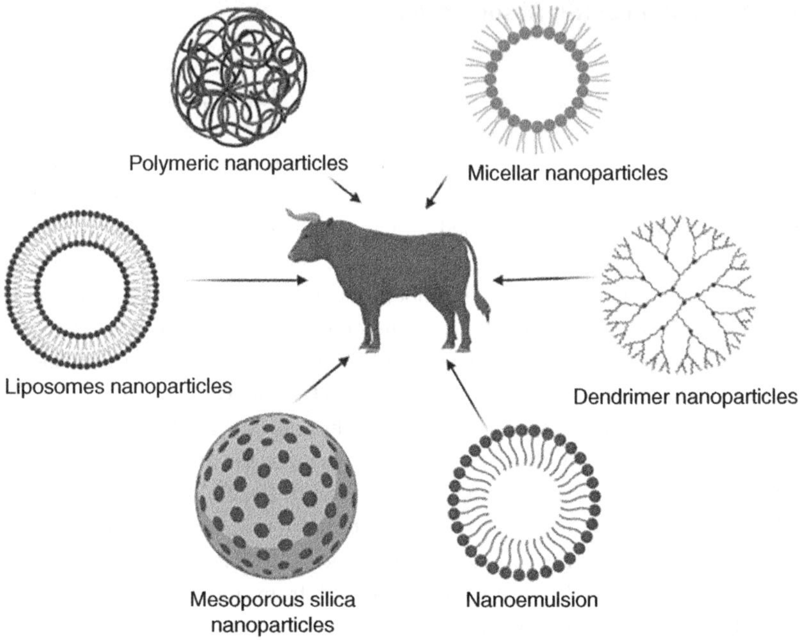

Figure 2.4 Nanocarriers used to deliver medicines in animals (veterinary medications).

route. It is thus simple and can be used for all patients, including animals seeking treatment (Li et al. 2016).

- **Inhalation**
  Inhale delivery refers to the intake of NPs in the form of an aerosol or dispersion or powder, which are intended to be targeted in case of respiratory disease and for systemic delivery to the lungs. This route directly connects the lungs and the first cycle of metabolism. While these NPs can provide a targeted dosage, the crucial factors such as the size of particles, deposition in the lungs and potentially the toxicity need to be considered (Li et al. 2016).
- **Topical Application**
  Topical NP-carrying formulations, adhered to the skin or mucous membrane for compliance with local or systemic use, are applied. The strategic path can be applied for dermatological conditions, wound healing and transdermal drug administration. Some of the factors that the development of NPs for topical skin delivery has to overcome are the skin barrier penetration and irritation potential (Niska et al. 2013).
- **Implantation or Surgical Placement**
  Putting NPs or surgery implantation means embedding them in implants, devices or tissues and releasing them or delivering them to a point target.

This technique leads both to struck off and also to time release kinetics. While surgical methods involve risks and may limit application to particular cases, they can result in precise outcomes and are effective in a wide range of situations (Peer et al. 2007).

- **Targeted Delivery Systems**
  NPs have a broad range of clinical applications as they can be modified to use various targeting systems such as ligand-conjugated materials or NP-drug conjugates in order to deliver therapeutics to specific organs or cells. Tackle ligands, e.g., antibodies or peptides, which contribute to selective adhesion and events of internalization, increase treatment effectiveness and minimize off-target effects (Suh et al. 2014).
- **Carrier-Mediated Delivery**
  There are several carrier-mediated delivery systems that are utilized, including liposomes, micelles and polymeric NPs which act as a vehicle to attach drugs or therapeutic agents to overcome issues of stability, solubility and bioavailability. These nano-sized delivery systems perform the functions of shielding the payload from degradation and releasing it at the precise time and hence prolong the circulation path and enhance efficacy. The selection of an ideal delivery model may be based on a combination of several factors like the specification of the NP, the target tissue, the anticipated therapeutic effect and the animal species. Integration of method optimization, delivery parameters, animal models and clinical settings is the key factor that should be considered in the process of preliminary and early clinical trials (Torchilin 2014).

2.5 Impact of NPs on Biological Organisms

Filings of NPs possess unique physicochemical properties so that they can be used in many applications, but the interactions of NPs with living systems are a reason for concerns about their impacts on health and environment. Knowledge of NP interactions with biological organisms is important in the evaluation of their safety and optimal usage in biomedical, agricultural and environmental events (Borm et al. 2006).

- **Toxicity and Cellular Responses**
  NPs are capable of inducing numerous toxic effects in biota via ROS production, inducing inflammation and disruption of cellular functions. With regard to the toxicity of NPs, size, relation, surface chemistry and surface charge all should be considered. The characteristic property of an NP for cellular entry due to its small size allows further interaction with intracellular components, leading to cell proliferation (Dobrovolskaia and McNeil 2007).

- **Bioaccumulation and Biodistribution**
  NPs are prone to bioaccumulation in various forms in the human body by a process of ingestion, inhalation or even dermal exposure, leading systemic toxicity and contamination of the environment. It is after absorption that NPs express their tendency to end up in a variety of other organs and tissues with some risk of long-term health effects and ecological hazards. The study of the biodistribution kinetics of NPs is the foundation for the safety assessment on exposure risks as well as on the prediction of their environmental fate (Maurer-Jones et al. 2013).
- **Immunological Responses**
  NPs interfere with inflammatory mechanism, cells activation and cytokine production. Immunomodulatory effects of NPs could result in allergic reactions, hypersensitivity responses or immune-mediated toxicity. Both positive and negative immunomodulatory effects of NPs have been found. Learning the requirements of NP-immune system can allow assessing the reactions of these components in the development of therapies and vaccines (Nel et al. 2006).
- **Ecotoxicity and Environmental Impact**
  Human activities have introduced treatment or disposal fugitive NPs into the environment purposefully or unintentionally, and the NP particles have shown the capacity to cause long-term problems for ecological systems and biodiversity. NPs may decrease growth potential, pollute resources and damage the natural life cycle of organisms. One of the common findings is that the aquatic living organisms like fishes, amphibians and invertebrates are more likely to get exposed to NP, which can potentially end up disrupting the ecosystem in a broader sense (Stone et al. 2010).
- **Regulatory Considerations and Risk Assessment**
  Creating a risk assessment framework for NPs is necessary to take into account exposure routes, dose–response relationships and how people or the environment might potentially be impacted by their presence. Regulatory departments, which are being represented in every country, are developing guidelines and standards for NP characterization, toxicity testing and risk management to guarantee that the development and applying nanotechnology will be carried out attentively. Acquiring the necessary knowledge about the dynamic interplay between NPs and biological entities is fundamentally crucial in order to prevent risks, optimize NP design and make the most of it as a tool for beneficial applications (Stone et al. 2010).

2.6 Conclusion and Future Prospective

Though NPs are used in agriculture and clinical areas, questions related to human health and the environment must be considered as well (Kumar et al. 2020). The recent advancement in NP producing techniques, which include chemical, physical and biological approaches, made the production of NP with particular dimensions, forms and surface chemistry much easier. Substantial business potential for NP has emerged; it is now extensively used in agriculture for increasing crop productivity and nutrient absorption and in healthcare for disease targeting (e.g., drug delivery and imaging) (Hua et al. 2018). Special approaches for establishing interactions, distribution and reaching optimal therapeutic effectiveness within biological systems while reducing side effects is vital when delivering NPs to organisms. There are a variety of delivery methods like topical treatments, oral medications, intravenous injections and inhalation, which offer a high degree of adaptability in addressing various sites of action in different animals and organs (Nel et al. 2006). Application of NPs could be considered beneficial, yet the matter still concerns more research and regulatory process because of the concern about their toxicity, bioaccumulation and impact on the environment. If risk assessment mechanisms are to be established and applied throughout the process of safe development and use of nanotechnology, NPs, with certainty, can prove to be beneficial.

The future research direction in nanotechnology includes influencing the development of green synthesis that is biocompatible and sustainable, minimizing harmful substances and promoting eco-safe industrial processes with the help of these methods. Molecular and cellular mechanisms can be unravelled in order to get the overall picture of NP interactions with the cells and detect the main routes leading to cytotoxic consequences (Fadeel et al. 2013). By designing multipurpose conveyers with a combination of medicinal chemicals, imaging probes and ligand targets, we can better attack the cancer cells and control the accuracy of the treatment. Determining NP-related protocols along with toxicity testing and risk assessment would become simpler for products that incorporate such particles and consequently, they might receive regulatory approval and proceed for sale. Thus, there is a hope for problem solving with regard to the environment, agriculture and healthcare (Dreaden et al. 2012).

References

Amani, A., Zare, N., Asadi, A., and Asghari-Zakaria, R. (2018). Ultrasound-enhanced gene delivery to alfalfa cells by hPAMAM dendrimer nanoparticles. *Turkish Journal of Biology* 42: 63–75. https://doi.org/10.3906/biy-1706-6.

Barhoum, A., García-Betancourt, M.L., Jeevanandam, J. et al. (2022). Review on natural, incidental, bioinspired, and engineered nanomaterials: history, definitions, classifications, synthesis, properties, market, toxicities, risks, and regulations. *Nanomaterials* 12: 177.

Blanco, E., Shen, H., and Ferrari, M. (2015). Principles of nanoparticle design for overcoming biological barriers to drug delivery. *Nature Biotechnology* 33 (9): 941–951. https://doi.org/10.1038/nbt.3330.

Borm, P.J., Robbins, D., Haubold, S. et al. (2006). The potential risks of nanomaterials: a review carried out for ECETOC. *Particle and Fibre Toxicology* 3 (1): 11.

Das, A.K., Gupta, P., and Chakraborty, D. (2015). Physical methods of gene transfer: kinetics of gene delivery into cells: a Review. *Agricultural Reviews* 36: 61–66.

Demirer, G.S., Zhang, H., Goh, N.S. et al. (2020). Carbon nanocarriers deliver siRNA to intact plant cells for efficient gene knockdown. *Science Advances* 6 (26): https://doi.org/10.1126/sciadv.aaz0495.

Divya, K., Vijayan, S., Nair, S.J., and Jisha, M.S. (2019). Optimization of chitosan nanoparticle synthesis and its potential application as germination elicitor of Oryza sativa L. *International Journal of Biological Macromolecules* 124: 1053–1059. https://doi.org/10.1016/j.ijbiomac.2018.11.185.

Dobrovolskaia, M.A. and McNeil, S.E. (2007). Immunological properties of engineered nanomaterials. *Nature Nanotechnology* 2 (8): 469–478.

Dontsova, T.A., Nahirniak, S.V., and Astrelin, I.M. (2019). Metaloxide nanomaterials and nanocomposites of ecological purpose. *Journal of Nanomaterials* 2019: 5942194.

Dreaden, E.C., Alkilany, A.M., Huang, X. et al. (2012). The golden age: gold nanoparticles for biomedicine. *Chemical Society Reviews* 41 (7): 2740–2779.

Efimov, A.A., Lizunova, A.A., Volkov, I.A. et al. (2016). A new approach to the high-yield synthesis of nanoparticles by spark discharge. *Journal of Physics Conference Series* 741: 012035. https://doi.org/10.1088/1742-6596/741/1/012035.

El-Temsah, Y.S. and Joner, E.J. (2012). Impact of Fe and Ag nanoparticles on seed germination and differences in bioavailability during exposure in aqueous suspension and soil. *Environmental Toxicology* 27 (1): 42–49. https://doi.org/10.1002/tox.20610.

Fadeel, B., Farcal, L., Hardy, B. et al. (2013). Advanced tools for the safety assessment of nanomaterials. *Nature Nanotechnology* 8 (12): 871–874.

Gan, Y.X., Jayatissa, A.H., Yu, Z. et al. (2020, 8917013). Hydrothermal synthesis of nanomaterials. *Journal of Nanomaterials* 1–3.

Geiser, M. (2010). Update on macrophage clearance of inhaled micro- and nanoparticles. *Journal of Aerosol Medicine and Pulmonary Drug Delivery* 23 (4): 207–217.

Handy, R.D., Von der Kammer, F., Lead, J.R. et al. (2008). The ecotoxicology and chemistry of manufactured nanoparticles. *Ecotoxicology* 17: 287.

Hao, Y., Yang, X., Shi, Y. et al. (2013). Magnetic gold nanoparticles as a vehicle for fluorescein isothiocyanate and DNA delivery into plant cells. *Botany* 91 (7): 457–466. https://doi.org/10.1139/cjb-2012-0281.

Vert, M., Yosihura, D., Hellwich, K.H. et al. (2012). Terminology for biorelated polymers and applications (IUPAC recommendations 2012). *Pure and Applied Chemistry* 84 (2): 377–410. https://doi.org/10.1351/PAC-REC-10-12-04.

Hong, J., Wang, C., Wagner, D.C. et al. (2021). Foliar application of nanoparticles: mechanisms of absorption, transfer, and multiple impacts. *Environmental Science. Nano* 8: 1196–1210.

Hua, S., de Matos, M.B.C., Metselaar, J.M., and Storm, G. (2018). Current trends and challenges in the clinical translation of nanoparticulate nanomedicines: pathways for translational development and commercialization. *Frontiers in Pharmacology* 9: 790.

Jain, R., Chaurasia, S.K., and Kalga, A. (2012). Report on sol gel method for nanoparticle synthesis.

Jain, A., Jain, S.K., Ganesh, N., and Barve, J. (2016). Nanocarriers for oral drug delivery: a review. *Current Drug Delivery* 13 (6): 867–880.

Jat, S.K., Bhattacharya, J., and Sharma, M.K. (2020). Nanomaterial based gene delivery: a promising method for plant genome engineering. *Journal of Materials Chemistry B* 8: 4165–4175.

Jeevanandam, J., Barhoum, A., Chan, Y.S. et al. (2018). Review on nanoparticles and nanostructured materials: history, sources, toxicity and regulations. *Beilstein Journal of Nanotechnology* 9: 1050–1074.

Kah, M., Kookana, R.S., Gogos, A., and Bucheli, T.D. (2018). A critical evaluation of nanopesticides and nanofertilizers against their conventional analogues. *Nature Nanotechnology* 13 (8): 677–684. https://doi.org/10.1038/s41565-018-0131-1.

Khan, I., Saeed, K., and Khan, I. (2019). Nanoparticles: properties, applications and toxicities. *Arabian Journal of Chemistry* 12 (7): 908–931. https://doi.org/10.1016/j.arabjc.2017.05.011.

Koelmel, J., Leland, T., Wang, H. et al. (2013). Investigation of gold nanoparticles uptake and their tissue level distribution in rice plants by laser ablation-inductively coupled-mass spectrometry. *Environmental Pollution* 174: 222–228. https://doi.org/10.1016/j.envpol.2012.11.026.

Kumar, A., Sharma, G., Chhabra, R. et al. (2020). Intranasal delivery of nanoparticles: a game changer in the drug repurposing scenario. *AAPS PharmSciTech* 21 (6): 228.

Li, X., Xu, H., Chen, Z.-S., and Chen, G. (2011). Biosynthesis of nanoparticles by microorganisms and their applications. *Journal of Nanomaterials* 2011: 1–16. https://doi.org/10.1155/2011/270974.

Li, J., Hu, J., Ma, C. et al. (2016). Uptake, translocation and physiological effects of magnetic iron oxide (γ-Fe_2O_3) nanoparticles in corn (Zea mays L.). *Chemosphere* 159: 326–334.

Li, J., Zou, Y., Ye, D. et al. (2016). Delivery of nanoparticles for treatment of brain tumor. *Current Drug Metabolism* 17 (8): 745–754.

Li, S. et al. (2021). Comparative physiological and metabolomic analyses revealed that foliar spraying with zinc oxide and silica nanoparticles modulates metabolite profiles in cucumber (*Cucumis sativus* L.). *Food and Energy Security* 10 (1): https://doi.org/10.1002/fes3.269.

Lin, D. and Xing, B. (2007). Phytotoxicity of nanoparticles: inhibition of seed germination and root growth. *Environmental Pollution* 150 (2): 243–250. https://doi.org/10.1016/j.envpol.2007.01.016.

Liu, Q., Chen, B., Wang, Q. et al. (2009). Carbon nanotubes as molecular transporters for walled plant cells. *Nano Letters* 9 (3): 1007–1010.

Lizzi, D., Mattiello, A., Adamiano, A. et al. (2021). Influence of cerium oxide nanoparticles on two terrestrial wild plant species. *Plants* 10 (2): 335. https://doi.org/10.3390/plants10020335.

Lyell, C. (1911). *The Student's Lyell: The Principles and Methods of Geology, as Applied to the Investigation of the Past History of the Earth and Its Inhabitants*. J. Murray.

Makhotenko, A.V., Snigir, E.A., Kalinina, N.O. et al. (2018). Data on a delivery of biomolecules into Nicothiana benthamiana leaves using different nanoparticles. *Data in Brief* 16: 1034–1037. https://doi.org/10.1016/j.dib.2017.12.031.

Martin-Ortigosa, S., Valenstein, J.S., Lin, V.S.-Y. et al. (2012). Gold functionalized mesoporous silica nanoparticle mediated protein and DNA codelivery to plant cells via the biolistic method. *Advanced Functional Materials* 22 (17): 3576–3582. https://doi.org/10.1002/adfm.201200359.

Maurer-Jones, M.A., Gunsolus, I.L., Murphy, C.J., and Haynes, C.L. (2013). Toxicity of engineered nanoparticles in the environment. *Analytical Chemistry* 85 (6): 3036–3049.

Modlitbová, P. et al. (2018). Comparative investigation of toxicity and bioaccumulation of Cd-based quantum dots and Cd salt in freshwater plant Lemna minor L. *Ecotoxicology and Environmental Safety* 147: 334–341. https://doi.org/10.1016/j.ecoenv.2017.08.053.

Modlitbová, P. et al. (2018). Short-term assessment of cadmium toxicity and uptake from different types of Cd-based quantum dots in the model plant Allium cepa L. *Ecotoxicology and Environmental Safety* 153: 23–31. https://doi.org/10.1016/j.ecoenv.2018.01.044.

Muthukrishnan, S., Murugan, I., and Selvaraj, M. (2019). Chitosan nanoparticles loaded with thiamine stimulate growth and enhances protection against wilt disease in Chickpea. *Carbohydrate Polymers* 212: 169–177. https://doi.org/10.1016/j.carbpol.2019.02.037.

Nair, P.M.G. and Chung, I.M. (2014). Physiological and molecular level effects of silver nanoparticles exposure in rice (Oryza sativa L.) seedlings. *Chemosphere* 112: 105–113.

Nel, A., Xia, T., Mädler, L., and Li, N. (2006). Toxic potential of materials at the nanolevel. *Science* 311 (5761): 622–627. https://doi.org/10.1126/science.1114397.

Niska, K., Zielinska, E., Radomski, M.W., and Inkielewicz-Stepniak, I. (2013). Metal nanoparticles in dermatology and cosmetology: interactions with human skin cells. *Chemical Society Reviews* 42 (21): 8844–8866.

Ozyigit, I.I. (2020). Gene transfer to plants by electroporation: methods and applications. *Molecular Biology Reports* 47 (4): 3195–3210. https://doi.org/10.1007/s11033-020-05343-4.

Pantidos, N. and Horsfall, L.E. (2014). Biological synthesis of metallic nanoparticles by bacteria, fungi and plants. *Journal of Nanomedicine & Nanotechnology* 5 (5): 1.

Patankar, S., Magdum, D., Patil, P., and Pakhale, S. (2012). Report on synthesis of nanoparticles using ion sputtering method.

Patil, S. and Chandrasekaran, R. (2020). Biogenic nanoparticles: a comprehensive perspective in synthesis, characterization, application and its challenges. *Journal, Genetic Engineering & Biotechnology* 18: 67.

Peer, D., Karp, J.M., Hong, S. et al. (2007). Nanocarriers as an emerging platform for cancer therapy. *Nature Nanotechnology* 2 (12): 751–760.

Penide, J. et al. (2019). Laser surface blasting of granite stones using a laser scanning system. *Coatings* 9 (2): 131. https://doi.org/10.3390/coatings9020131.

Rai, M., Ingle, A.P., Birla, S. et al. (2015). Strategic role of selected noble metal nanoparticles in medicine. *Critical Reviews in Microbiology* 1–24. https://doi.org/10.3109/1040841X.2015.1018131.

Rajkumari, N., Alex, S., Soni, K.B. et al. (2021). Silver nanoparticles for biolistic transformation in Nicotiana tabacum L. *3 Biotech* 11 (12): 497. https://doi.org/10.1007/s13205-021-03043-9.

Rane, A.V., Kanny, K., Abitha, V.K., and Thomas, S. (2018). Methods for synthesis of nanoparticles and fabrication of nanocomposites. In: *Synthesis of Inorganic Nanomaterials* (ed. S. Mohan, B. Oluwatobi, S. Oluwafemi, et al.), 121–139. Elsevier. https://doi.org/10.1016/B978-0-08-101975-7.00005-1.

Rastogi, A., Zivcak, M., Sytar, O. et al. (2017). Impact of metal and metal oxide nanoparticles on plant: a critical review. *Frontiers in Chemistry* 5: 78.

Roychoudhury, A. (2020). Nanobiolistics: new generation transfection system for animals and plants. *Journal of Molecular and Cellular Biology Forecast* 3: 1020.

Sandhya, M., Ramasamy, D., Sudhakar, K. et al. (2021). Ultrasonication an intensifying tool for preparation of stable nanofluids and study the time influence on distinct properties of graphene nanofluids – a systematic overview. *Ultrasonics Sonochemistry* 73: 105479. https://doi.org/10.1016/j.ultsonch.2021.105479.

Sembada, A.A. and Lenggoro, I.W. (2024). Transport of nanoparticles into plants and their detection methods. *Nanomaterials* 14 (2): 131. https://doi.org/10.3390/nano14020131.

Shankramma, K., Yallappa, S., Shivanna, M.B., and Manjanna, J. (2016). Fe_2O_3 magnetic nanoparticles to enhance S. lycopersicum (tomato) plant growth and their biomineralization. *Applied Nanoscience* 6 (7): 983–990. https://doi.org/10.1007/s13204-015-0510-y.

Silva, A.T., Nguyen, A., Ye, C. et al. (2010). Conjugated polymer nanoparticles for effective siRNA delivery to tobacco BY-2 protoplasts. *BMC Plant Biology* 10 (1): 291. https://doi.org/10.1186/1471-2229-10-291.

Song, U. et al. (2013). Functional analyses of nanoparticle toxicity: a comparative study of the effects of TiO_2 and Ag on tomatoes (Lycopersicon esculentum). *Ecotoxicology and Environmental Safety* 93: 60–67. https://doi.org/10.1016/j.ecoenv.2013.03.033.

Stone, V., Nowack, B., Baun, A. et al. (2010). Nanomaterials for environmental studies: classification, reference material issues, and strategies for physico-chemical characterisation. *Science of the Total Environment* 408 (7): 1745–1754.

Sud, D., Singla, A.K., and Gupta, M.K. (2022). *Nanoparticles in Manufacturing Procesess*. CRC Press.

'Suh, W.H., Suslick, K.S., Stucky, G.D. et al. (2014). Nanotechnology, nanotoxicology, and neuroscience. *Progress in Neurobiology* 123: 103–138.

Tombuloglu, H., Slimani, Y., Tombuloglu, G. et al. (2019). Uptake and translocation of magnetite (Fe_3O_4) nanoparticles and its impact on photosynthetic genes in barley (Hordeum vulgare L.). *Chemosphere* 226: 110–122. https://doi.org/10.1016/j.chemosphere.2019.03.075.

Torchilin, V.P. (2014). Multifunctional, stimuli-sensitive nanoparticulate systems for drug delivery. *Nature Reviews Drug Discovery* 13 (11): 813–827.

Tripathi, S., Sonkar, S.K., and Sarkar, S. (2011). Growth stimulation of gram (Cicer arietinum) plant by water soluble carbon nanotubes. *Nanoscale* 3 (3): 1176. https://doi.org/10.1039/c0nr00722f.

Wang, S., Kurepa, J., and Smalle, J.A. (2011). Ultra-small TiO_2 nanoparticles disrupt microtubular networks in *Arabidopsis thaliana*. *Plant, Cell & Environment* 34 (5): 811–820. https://doi.org/10.1111/j.1365-3040.2011.02284.x.

Wang, Z., Xie, X., Zhao, J. et al. (2012). Xylem-and phloem-based transport of CuO nanoparticles in maize (Zea mays L.). *Environmental Science & Technology* 46: 4434–4441.

Wang, J., Yang, Y., Zhu, H. et al. (2014). Uptake, translocation, and transformation of quantum dots with cationic versus anionic coatings by *populus deltoides × nigra* cuttings. *Environmental Science & Technology* 48 (12): 6754–6762. https://doi.org/10.1021/es501425r.

Wang, X., Xie, H., Wang, P., and Yin, H. (2023). Nanoparticles in plants: uptake, transport and physiological activity in leaf and root. *Materials* 16 (8): 3097. https://doi.org/10.3390/ma16083097.

Zhang, Z. et al. (2011). Uptake and distribution of ceria nanoparticles in cucumber plants. *Metallomics* 3 (8): 816. https://doi.org/10.1039/c1mt00049g.

Zhao, F. et al. (2021). Use of carbon nanoparticles to improve soil fertility, crop growth and nutrient uptake by corn (Zea mays L.). *Nanomaterials* 11 (10): 2717. https://doi.org/10.3390/nano11102717.

Zhu, Z.-J. et al. (2012). Effect of surface charge on the uptake and distribution of gold nanoparticles in four plant species. *Environmental Science & Technology* 46 (22): 12391–12398. https://doi.org/10.1021/es301977w.

Zolghadrnasab, M., Mousavi, A., Farmany, A., and Arpanaei, A. (2021). Ultrasound-mediated gene delivery into suspended plant cells using polyethyleneimine-coated mesoporous silica nanoparticles. *Ultrasonics Sonochemistry* 73: 105507. https://doi.org/10.1016/j.ultsonch.2021.105507.

Section B

Source and Synthesis of Si/SiO_2 NPs

3

Traditional Methods of Synthesis and Characterization of SiNPs for Biomedical Applications and Sustainable Agriculture

Ruby Panwar and Anjana Rustagi

Department of Botany, Gargi College, University of Delhi, New Delhi, Delhi, India

3.1 Introduction

Despite their importance to life on earth, plants are continually challenged by environmental changes that have a detrimental influence on their growth. Plants, being sedentary organisms, have developed a variety of defence systems to deal with these challenges (Saxena et al. 2016). However, the growing consequences of human activity and climate change exacerbate these challenges, emphasizing the critical need for cutting-edge approaches to overcome plant development barriers (Jalil and Ansari 2019). Global agriculture has long contended with environmental difficulties (biotic and abiotic stress), which reduce crop production and output. Environmental stresses can affect agricultural output by up to 40–60% when biotic stress is present and 50–70% when abiotic stress is prevalent (Tiwari et al. 2020). Abiotic challenges include salt, drought, temperature and chemical pollutants, whereas biotic pressures include phytopathogen infections such as bacteria, viruses, fungi and nematodes, which limit contemporary agriculture. Environmental restrictions limit plants' capacity to grow and develop by influencing their morphological, physiological and molecular processes (Shalaby et al. 2022). Innovative strategies for increasing agricultural yields in difficult situations have been devised, utilizing standard fertilizers and stress-tolerant varieties.

The widespread use of agricultural pesticides to meet rising food demand and climate change has broken the delicate balance between crop yield and the environment. The use of these agricultural compounds begins with the delivery of refined nutrients to boost crop production and growth. Nonetheless, a large proportion of these chemicals seep from the field and end up in water sources, causing eutrophication. Continuous use of these agricultural pesticides increases abiotic and biotic stresses, negatively impacting the environment, soil, human health and microbial ecology (Verma et al. 2022).

Silicon Nanoparticles: Synthesis, Uptake, and Applications, First Edition.
Edited by Archana Singh, Aryadeep Roychoudhury, and Indrakant K. Singh.

Nanotechnology is a well-established scientific and technological field with a considerable impact on the agriculture business. It is an area of study that focuses on the use of nanoparticles (NPs) to create new high-quality nano-based products that may be utilized in a range of industries, including health, manufacturing, electronics, wastewater treatment and agriculture. Nanotechnology, like biotechnology, contributes significantly to agricultural improvement (Tubana et al. 2016). The capacity of nanotechnology to identify and manage plant infections has proven useful in the agricultural business, resulting in enhanced animal output (Adebisi et al. 2018). NPs, which are also extremely effective fertilizers for agricultural production, can be utilized to manage pesticides and plant diseases (Adebisi et al. 2020). The development of phytonanotechnology refers to the application of sophisticated nanotechnological approaches to improve plant development features. This trend is also observed in silicon nanoparticles (SiNPs), which are more stable than elemental silicon and have a high surface-to-volume ratio, excellent consumption quality, easy surface moderation, customizable pore structure, increased reactivity and a safe delivery system (Shanmugam et al. 2020). As a result, SiNPs have emerged as an effective tool for increasing plant tolerance to biotic and abiotic stresses (Tripathi et al. 2017). The use of SiNPs promotes stress tolerance by activating various plant defence mechanisms (De Sousa et al. 2019; El-Shetehy et al. 2021).

The distinct properties of SiNPs have piqued global interest in agriculture for increasing plant growth and stress resistance, particularly when compared to bulk silicon. Silicon is the second most common element in the lithosphere, accounting for 28% of the crust's mass after oxygen (Bansal et al. 2022). It is commonly found in the terrestrial environment in two distinct forms: liquid and solid. The solid phase is composed of crystalline (primary and secondary silicates) and amorphous (biogenic and non-biogenic) forms, and it is sparingly soluble and biogeochemically inactive (Souri et al. 2021). These forms are the primary means by which both natural and human activities transport silicon (Si) into soil and water bodies (Guntzer et al. 2012). In addition to organic and inorganic complexes, the liquid phase contains a combination of monosilicic and polysilicic acids (Tubana et al. 2016). The most soluble form of silicon is monosilicic acid ($H_4SiO_4/Si(OH)_4$) that plants may use. The concentration ranges (0.1–0.6 mM) (Imtiaz et al. 2016; Tubaña and Heckman 2015), and the roots absorb it via different silicon transporters (Tubana et al. 2016).

Si is generally recognized as a plant beneficial element because of its growth-promoting properties on plants, particularly those cultivated under harsh conditions. With the gradual integration of agriculture and nanotechnology, the use of NPs in crop production has been shown to be an effective agronomic approach to meeting rising global food demand (Agathokleous et al. 2019; Sharma et al. 2022; Wang et al. 2020). In nano-enabled agriculture, SiNPs stand out among the other NPs for their remarkable benefits and suitability in enhancing plant stress tolerance and guaranteeing consistent crop output (Dhakate et al. 2022; Rastogi et al. 2019). Silicon has been demonstrated to boost agricultural productivity by improving nutrient absorption and plant tolerance to a variety of environmental stresses (both biotic and abiotic) (Sarkar et al. 2022; Shalaby et al. 2022; Shi et al. 2022; Sun et al. 2022; Zhou et al. 2019; Zhu et al. 2020). Silica deposition on the epidermis fortifies the plant's cell wall, creating a physical barrier that protects against environmental stressors, microbial attack and mechanical damage (Ahammed and Yang 2021; Reynolds et al. 2016; Sarkar et al. 2022).

Nano silicon may be more effective than elemental silicon in reducing specific stressors due to its smaller size and improved penetration (Suriyaprabha et al. 2014). SiNPs have allegedly improved plant tolerance to biotic and abiotic stress relative to other nano fertilizers, minimizing the adverse consequences (Khan et al. 2020). In addition, it strengthens plants' defences against several heavy metal elements (Jia-Wen et al. 2013). Nanotechnology enhances agricultural crops' quality and output rate at a low cost without harming the environment or living things. Since SiNPs are manufactured using greener ideas and will lessen the bad impacts of chemical fungicides, their main premise is to eliminate harmful environmental inputs and excessive fertilizer expenses. SiNPs offer promise for use in agriculture because of their executive properties, which include their vast surface area and compact size, which guarantee realistic dispersion in plant tissues (Bhat et al. 2021). Researchers have investigated the use of NPs as soil stabilizers. Before being moved and deposited in shoot sections, a SiNP is first polymerized in root tissues (Debona et al. 2017). However, the confirmed mechanism has to be investigated (Coutris et al. 2012). SiNPs were also applied to plants directly as insecticides, herbicides and fertilizers (Malik et al. 2021). SiNPs also serve as transporters for the movement of molecules such as proteins, nucleotides and other chemicals in flora, and they might be used successfully in farming to improve soil water holding capacity (Bapat et al. 2020). SiNPs have a wide range of uses since they are inexpensive to make on a large scale, hydrophobic, have a large surface area and pore volume and are biocompatible. SiNPs, for example, have been used to solve agricultural difficulties due to their exceptional adsorption power and non-toxicity.

SiNPs have lately been used to increase agricultural output, improve plant growth and enhance disease resistance (Goswami et al. 2024). However, certain research works have discovered that SiNPs can cause phytotoxicity, which varies based on particle size, surface area, concentration, pH and plant type. SiNP toxicity is mostly influenced by size and surface area. Smaller SiNPs have a greater surface area and may easily permeate cellular compartments, resulting in higher toxicity than their larger counterparts. For example, a study on SiO_2 NPs (10–20 nm) found that they had a negative impact on *Scenedesmus obliquus* growth and photosynthetic pigments when compared to bulk SiO_2

(Wei et al. 2010). Slomberg and Schoenfisch (2012) discovered that larger SiNPs (50 and 200 nm) induced poisoning symptoms in *Arabidopsis thaliana*, such as decreased growth and chlorosis, independent of NPs. These phytotoxic effects were associated with pH levels and mineral nutrient adsorption on SiNP surfaces. The concentration of SiNPs also determines their damaging effects, which vary according to the plant's tolerance. High concentrations (more than 2000 mg/l) of SiNPs were shown to inhibit root growth and reduce leaf number in *Arabidopsis thaliana* seedlings. Similarly, concentrations of SiO_2 NPs larger than 200 mg/l decreased fresh and dry mass, photosynthetic pigments, lipid peroxidation and antioxidant activity in wheat seedlings. In other studies, significant doses of SiNPs reduced plant biomass and photosynthesis while increasing the formation of malondialdehyde (MDA), a marker of membrane damage, and proline (a stress injury signal) in cherry tomatoes and sugar beetroot. Recent toxicity studies on *Allium cepa* using SiNPs of different sizes at various doses found substantial decreases in root advancement, germinating ratio and rate of cell division as nanoparticle concentration rose. These data indicate that if SiNPs are not used in the appropriate size and concentration ranges, they may have phytotoxic effects. However, the exact toxicity mechanisms of SiNPs in plants remain unknown, necessitating more experimental research to investigate their interactions at the transcriptional and post-transcriptional levels.

3.2 Synthesis of Silica Nanoparticles

Chemical, physical and green synthesis approaches are being used to produce nanoscale metals by many researchers. When it comes to managing the shape, size, structure and performance of NPs, synthesis approaches are crucial. The optical, electrical and physiochemical characteristics of silica NPs are also impacted. Sometimes, when silica NPs are coated, their characteristics are preserved after they precipitate out of solutions. There are two types of silica particle synthesis methods: top-down and bottom-up. There are three different ways to create silica NPs: physical, chemical and green synthesis.

3.3 Physical Methods

Physical techniques rely on the application of thermal energy, radiation energy and mechanical pressure forces to create nanomaterials. This leads to the materials' condensation, dissolution, evaporation, or abrasion. Physical approaches to creating nanomaterials are different from chemical ones in that they do not harm the environment, create uniform NPs and involve no use of chemical solvents for their production (Kumari et al. 2023). Typically, top-down physical processes that fragment bulk materials are used to create nanomaterials (Baig et al. 2021). The following are some physical techniques for creating nanomaterials and NPs: (Anwar 2018; Ijaz et al. 2020; Irshad et al. 2021; Krishnia et al. 2022; Salem and Fouda 2021; Thakur et al. 2022; Yun et al. 2009).

3.3.1 Method of Physical Crushing

In order to crush solid material particles to the desired size, mechanical tension is first exerted from outside the object being crushed. It is important to consider the potential for fire and explosion to occur during the crushing of the raw material since this method would result in less pure and irregularly distributed silica NPs when used in industrial manufacturing.

3.3.2 The Process of Mechanical Ball Milling

With this method, reactants are mechanically crushed into the required size using balls. It is commonly used to break down large materials into small particles. In industry, continuous production is quite simple, but the ball milling process usually takes a long time, which makes it inappropriate for laboratory studies requiring the manufacture of silica.

3.4 Ion Sputtering Technique

This method produces the glow discharge phenomenon by ionizing argon or other inert gases with an applied electric field. The ionization-generated particles then attack the target, sputtering out and landing on it. Ion beam sputtering offers the ability to independently control ion energy, flux, species and angle of incidence across a wide range.

3.4.1 Laser Ablation or Pulsed Laser Deposition

To create nanomaterials, a powerful laser beam is used. Laser ablation (LA) or pulsed laser deposition (PLD) are the names given to these techniques. By exposing the original material or primary material to a high-intensity laser beam, particles are evaporated or removed from its solid surface, resulting in the creation of NPs (Ismail et al. 2020).

3.4.2 Molecular Condensation or Inert Gas Condensation

Numerous nanomaterials, such as metallic materials, alloys, semiconductors, metal NPs and their oxide NPs, are produced using the inert gas condensation (IGC) approach (Kumari et al. 2023). By compressing metal materials with an inert or inactive gas, such as helium, argon or neon, inside a pressure chamber container, the molecular or inert gas condensation process causes the metal material's particles to evaporate. After that, liquid nitrogen is used to chill the metal vapours. The atom rapidly loses energy when metal vapours cool, forming NPs that range in size from 2 to 100 nm. For instance, an inert gas was used to evaporate copper NPs and liquid nitrogen was used to chill the vapours (Sp et al. 2020).

3.4.3 Pyrolysis by Laser Light

Laser pyrolysis is used to create magnetic NPs. This method uses a high-power laser from carbon dioxide to stimulate and then quench the interacting gaseous or liquid components. Reactants are also sprayed with a salt solution, where they evaporate and form particles that are dried and heated to a higher temperature to become concentrated particles (Jing et al. 2014). An inert photosensitizer, sulphur hexafluoride, absorbs a high-energy infrared carbon dioxide laser. When the reactants achieve saturation and condense during the evaporation stage, nanomaterials are created. Molybdenum sulphide NPs, silicon carbide NPs, SiNPs, composite NPs, ceramic NPs, ternary nanocomposites and metal NPs may also be produced using this technique (Kim et al. 2017).

3.4.4 Utilizing Electrospinning

One of the most significant physical developments in the creation of nanomaterials is coaxial electrospinning. It is an easy and efficient method for creating nanofibers from polymers, and it takes a top-down approach (Farré and Barceló 2012). A high-voltage electrical source, a blunt injection needle, a solution tank, a pump and a collector make up the electro-spindle's primary components (Zhou et al. 2019).

3.4.5 Method of Pulsed Wire Discharge

It is one of the most used physical procedures for producing metallic NPs. This method involves a pulsing electric current flowing over two electrodes to vaporize a metal wire. The surrounding atmosphere then cools the vapours to create NPs (Boddolla and Thodeti 2018; Sp et al. 2020).

3.5 Arc Discharge Technique

There are two types of arc discharge methods: the Electric Arc Discharge in Aerosol Dielectrics method in air insulators, which requires a gaseous medium, and the Electric Arc Discharge with the Liquid Dielectric method, which requires a liquid medium (Altammar 2023).

3.5.1 The Electric Arc Discharge in Aerosol Dielectrics

With this technique, electrodes are used to produce the electric arc at atmospheric pressure while keeping a crucial spacing between them. A plasma column forms as a result of the gas ionizing and the current flowing between the electrodes expanding when the electric potential between them hits the threshold. After the electrodes disperse, the metal vaporizes and crystallizes into elementary particles, which are tiny particles with a diameter of a few nanometres. Aerosols are particles that combine with the gas. This results in the formation of NPs (Singh et al. 2019).

3.5.2 The Electric Arc Discharge with Liquid Dielectric

The fundamental idea behind the arc discharge technique in an insulating liquid media is to harness the electrical energy between the electrodes to generate heat energy through a sequence of electrical discharges. In this process, the electrodes may get as hot as several thousand degrees Celsius. The electrode material melts and then evaporates as a result. After condensing in an insulating liquid media, the resultant vapours form NPs. Both gold and silver NPs may be made using this technique (Tien et al. 2008).

3.6 Chemical Techniques

Chemical techniques are used to generate silica NPs. Sol–gel, microemulsion, electrolysis, chemical vapour deposition, precipitation, hydrothermal reaction, solvent evaporation and other methods are some of these methods (Anwar 2018; De Souza et al. 2019; Jamkhande et al. 2019; Sreelekha et al. 2021; Yaqoob et al. 2020; Ying et al. 2022; Zhao and Friedrich 2017).

3.6.1 Sol–Gel Method

The method of creating solid silica NPs from tiny molecules is known as the 'bottom-up' method of nanoparticle creation. The monomer is transformed into a condensed solution (Sol), which serves as a forerunner for network polymers (gels) or discrete particles, usually metal alkoxide precursors (Rahman and Padavettan 2012). This process can yield silica NPs with consistent sizes and good product purity.

3.6.2 Microemulsion Method

This approach is a simple and widely used controlled nanoparticle production process that was developed in recent years. Microemulsion, also known as reversed-phase microemulsion, is commonly used to create silica NPs, with the 'water core' acting as a reaction site for silica nanoparticle formation (also known as a 'micro-reactor'). The product can only develop in a confined region since the surfactant and cosurfactant are attached to the NPs, allowing for nanoparticle size control. The distribution of silica NPs generated by this approach is uniform, high purification is easy to accomplish, the product is simple to create, and the acquired product monodispersity is good.

3.6.3 Electrolysis Method

Aqueous solution electrolysis and molten salt electrolysis are the two methods often used in the electrolysis process. By using direct current to propel non-spontaneous chemical processes, electrolysis makes it possible to separate elements from natural resources like ores. High-purity metal ultrafine particles that are difficult or impossible to produce with other traditional technologies may be produced using this technology.

3.7 Chemical Vapour Deposition

In this method, silica NPs are produced by chemical reactions on a heated substrate in chemical vapour deposition. This process is widely used to create silica NPs and is thought to be simple, providing a narrow particle size range and facile product acquisition.

3.8 Precipitation of Chemicals

The arrested precipitation approach is used in this strategy to manage size. The fundamental strategy has been to create and investigate the nanomaterial in situ, that is, inside the same liquid media, avoiding the physical alterations and small crystallite aggregation. By using non-aqueous solvents for synthesis at lower temperatures, double-layer repulsion of crystallites regulated thermal coagulation and Oswald ripening. A reaction between the component materials in an appropriate solvent was required for the synthesis. Before the precipitation reaction, dopamine is added to the parent solution. To keep the generated particles apart, surfactant is utilized. Centrifugation is used to separate the resulting nanocrystals, which are then cleaned and vacuum dried. For the purpose of potentially polymerizing the surfactant capping layer on the surface, the dried material was also exposed to ultra-violet (UV) curing (Konrad et al. 2001; Rostislav 1994; Sharma et al. 2009).

3.8.1 Hydrothermal Reaction Method

The method uses high temperature and pressure to interact with source ingredients in an aqueous solution to create silica NPs. This process has benefits including high purity, simple size control and optimum product dispersion.

3.8.2 Solvent Evaporation Method

A supersaturated condition is produced by the evaporation of the solvent at high temperature in the solvent evaporation method, which is a clear growth approach that provides enough driving force for the formation of silica NPs. This approach produces tiny, evenly distributed silica NPs; however, the synthesis process is intricate and challenging to carry out. Both physical and chemical techniques, each with unique advantages and disadvantages, can be used to create silica NPs. Numerous sizes of NPs may be produced physically using top-down and bottom-up techniques. But, although they need specific tools and materials, chemical processes like hydrothermal synthesis, sol–gel and precipitation provide more regulated sizes, forms and compositions. In the end, the technique selection is determined by the required silica nanoparticle qualities.

3.8.3 Characterization

SiNPs are often stored in powder or colloidal form after fabrication and are characterized by their shape, size, mass, crystallinity and other characteristics. UV-visible spectroscopy is used to study NPs of various metals ranging in size from 2 to 100 nm. The wavelengths of 300–800 nm are often used. This method determines the generation and stability of NPs in aqueous solutions (Chanda 2014; Mohammadlou et al. 2016). Scanning electron microscopy (SEM) is generally utilized for NPs greater than 100 nm, whereas Transmission electron microscopy (TEM) is favoured for NPs less than 100 nm. Although SEM and TEM are frequently used to evaluate the form and size of NPs, their examination capabilities are restricted. Dynamic light scattering (DLS) is often used to analyse particle size distributions (Jain et al. 2021; Wang et al. 2012).

Small-angle X-ray scattering (SAXS) or differential centrifugal sedimentation (DCS) can be used to calculate particle mass from known size and density measurements (Minelli et al. 2018). Capillary or solid microchannel resonators can also be used to measure the mass of a single nanoparticle (Ko et al. 2019, 2021).

SiNPs produced by reduction techniques may not be entirely reduced to pure silicon and may contain impurities. X-ray photoelectron spectroscopy (XPS) may be used to study the elemental composition of nanoparticle clusters (Thiessen et al. 2019). Another important factor is the crystallinity of the NPs; amorphous silicon is preferable for battery anodes, but crystalline silicon is better for optical applications (Choi et al. 2021; Ulvestad et al. 2020). This may be assessed using high-resolution TEM (HRTEM), X-ray diffraction (XRD) and Raman spectroscopy (Doğan and van de Sanden 2013; Elangovan et al. 2020; Thiessen et al. 2019). The atomic structure of materials is determined using XRD. It is suitable for both qualitative and quantitative analyses. It is used to determine the crystal structure, calculate the size of crystalline NPs and confirm the NPs (Chanda 2014; Mohammadlou et al. 2016). Raman spectroscopy also reveals crystallinity, which causes shifts and widening of the Full width at half maximum (FWHM) when particle size decreases due to phonon vibrations trapped inside the NPs (Tan et al. 2011).

SiNPs exhibit Mie resonance of electric and magnetic dipoles in the visible light spectrum (Evlyukhin et al. 2012; Fu et al. 2013). To examine this, single NPs are put on a glass substrate, and their Mie scattering properties are captured in dark field images. This effect is best visible in NPs with diameters ranging from 100 to 200 nm, and varying sizes enable Mie resonators to produce vivid, broad-spectrum colours on a dark backdrop to red blue green (RGB) (Chaâbani et al. 2019). Fourier-transform infrared spectroscopy (FTIR) provides information on bond strength, functional group placement and molecular structure interactions (Hülser et al. 2011). When SiNPs are lowered to nanoscale sizes, quantum confinement provides adjustable photoluminescence that is proportional to nanoparticle size (Priolo et al. 2014). SiNP colloids were found to exhibit photoluminescence quantum yields of up to 60% (Jurbergs et al. 2006). Efforts to fine-tune photoluminescence by nanoparticle size alterations are ongoing, with a blue shift seen when nanoparticle size decreases due to enhanced quantum confinement (Wen et al. 2015). Small-angle neutron scattering (SANS) gives particle size information and may use scattering data to estimate the interior surface area per volume unit, making it useful for characterizing materials with changing surfaces or porous structures (Lai et al. 2019).

3.9 Conclusion

As demonstrated by the earlier research mentioned above, SiNPs' ability to enhance plant development and resistance has been thoroughly investigated at the morphological and physiological levels. Nevertheless, further research is necessary to completely grasp the molecular processes by which SiNPs affect the protein composition as well as the expression of genes linked to stress and growth. In previous decades, there has been a significant advancement in our understanding of silicon and its nanoscale form in agriculture (Pavlovic et al. 2021). Effects of silicon on crop improvement and field conditions, however, are still in their infancy. There is a potential for SiNPs in agricultural nanotechnology. It is now crucial to deepen our knowledge of how NPs interact with plants in order to make use of the potential advantages of SiNPs. It is now crucial to deepen our knowledge of how NPs interact with plants and related microorganisms in order to fully use the potential advantages of SiNPs.

Agronomists throughout the world have been more interested in SiNPs in recent decades than in other metalloid NPs because of their many features that enhance plant resistance and growth. Because of their high surface area-to-volume ratio, which allows for more surface contact, and their capacity to undergo surface changes. Their tunable pore shape improves adsorption capacity, while their high reactivity allows for interactions with a wide range of biological and chemical agents. Furthermore, SiNPs provide a reliable and effective delivery mechanism for certain agricultural applications. In recent years, SiNPs' morphological and physiological functions in enhancing the development and yield of both exhausted and non-exhausted plants have been closely examined. But the chemical mechanism by which SiNPs affect host plant growth and stress-related gene expression, signalling pathways and important elements of the relation process (between these NPs and plants) remains unanswered and requires further investigation. Furthermore, previous studies have primarily looked at how NPs affect above-ground components. The subterranean component, especially the related microbial consortia, is similarly crucial for sustainable agriculture in promoting plant growth and development. Therefore, future studies must focus on how SiNPs affect the phytomicrobiome. Considerable advancements have been achieved in the green synthesis of SiO_2NPs (using microbial substrate, vegetal and agricultural deteriorate) as an economical, non-toxic and ecologically friendly alternative to chemical synthesis. However, this technique has been limited on a laboratory scale due to low repeatability, yield and varying morphology (size, shape and surface) under the inventive circumstances. The development of biologic NPs with regulated characteristics that may be used in agriculture in a variety of climates may be facilitated by machine learning-assisted research on SiNP manufacturing. Plants are likely to experience harmful consequences due to the morphophysiological attributes of these NPs, especially their tiny size and high concentration. The phytotoxic effects of SiO_2NPs were described in certain research without a systematic justification. Toxicological investigations of silicon-based nanopesticides and fertilizers and their effects on both baited and non-baited organisms are therefore another area of future study that must be prioritized. A deeper comprehension of the transcriptional and post-transcriptional mechanisms of the relation between these NPs and the plants will reveal several undiscovered facts regarding SiNPs for putting this method into practice with encouraging results in the field. It is anticipated that the analysis of SiNP manufacturing and uses in enhancing plant development and stress tolerance would pave the way for future agri-nanotechnology research.

3.10 Future Prospective

The capacity to mass-produce SiNPs at an economical cost and with good quality is crucial to their future. Large-scale manufacture is now feasible because of improvements in synthesis techniques, although maintaining constant quality control is still difficult. For both commercial and scientific applications, it is essential to guarantee consistency in particle size, shape and characteristics. It is evident that the standard will be defined by a production method that effectively accomplishes this homogeneity. Cutting production expenses is also crucial. The price of SiNPs is anticipated to drop as research advances and new manufacturing methods are developed, increasing their accessibility for a range of sectors.

Real-time characterization of SiNPs during synthesis is made possible by the application of in situ measuring methods, which yield important insights that were previously unattainable. These methods offer immediate feedback, making it easier to modify the production process to achieve the required qualities. Furthermore, the drive for high-speed and high-throughput metrologies guarantees

the rapid characterization of large quantities of SiNPs, enabling quality control in situations involving mass manufacturing.

References

Adebisi, J., Agunsoye, J., Bello, S. et al. (2018). Extraction of silica from cassava periderm using modified sol-gel method. *Nigerian Journal of Technological Development* 15 (2): 57–65.

Adebisi, J.A., Agunsoye, J.O., Bello, S.A. et al. (2020). Green production of silica nanoparticles from maize stalk. *Particulate Science and Technology* 38 (6): 667–675.

Agathokleous, E., Feng, Z., Iavicoli, I., and Calabrese, E.J. (2019). The two faces of nanomaterials: a quantification of hormesis in algae and plants. *Environment International* 131: 105044.

Ahammed, G.J. and Yang, Y. (2021). Mechanisms of silicon-induced fungal disease resistance in plants. *Plant Physiology and Biochemistry* 165: 200–206.

Altammar, K.A. (2023). A review on nanoparticles: characteristics, synthesis, applications, and challenges. *Frontiers in Microbiology* 14: 1155622.

Anwar, S.H. (2018). A brief review on nanoparticles: types of platforms, biological synthesis and applications. *Research & Reviews: Journal of Material Sciences* 6: 109–116.

Baig, N., Kammakakam, I., and Falath, W. (2021). Nanomaterials: a review of synthesis methods, properties, recent progress, and challenges. *Materials Advances* 2 (6): 1821–1871.

Bansal, K., Hooda, V., Verma, N. et al. (2022). Stress alleviation and crop improvement using silicon nanoparticles in agriculture: a review. *Silicon* 14: 10173–10186.

Bapat, G., Zinjarde, S., and Tamhane, V. (2020). Evaluation of silica nanoparticle mediated delivery of protease inhibitor in tomato plants and its effect on insect pest Helicoverpa armigera. *Colloids and Surfaces B: Biointerfaces* 193: 111079.

Bhat, J.A., Rajora, N., Raturi, G. et al. (2021). Silicon nanoparticles (SiNPs) in sustainable agriculture: major emphasis on the practicality, efficacy and concerns. *Nanoscale Advances* 3: 4019–4028.

Boddolla, S. and Thodeti, S. (2018). A review on characterization techniques of nanomaterials. *International Journal of Engineering, Science and Mathematics* 7 (1): 169–175.

Chaâbani, W., Proust, J., Movsesyan, A. et al. (2019). Large-scale and low-cost fabrication of silicon mie resonators. *ACS Nano* 13 (4): 4199–4208.

Chanda, S. (2014). *Silver Nanoparticles (Medicinal Plants Mediated): A New Generation of Antimicrobials to Combat Microbial Pathogens – A Review. Microbial Pathogens and Strategies for Combating Them: Science Technology and Education*, 1314–1323. Badajoz, Spain: FORMATEX Research Center.

Choi, D., Kim, J.H., Kwon, D.C. et al. (2021). Crystalline silicon nanoparticle formation by tailored plasma irradiation: self-structurization, nucleation and growth acceleration, and size control. *Nanoscale* 13 (23): 10356–10364.

Coutris, C., Joner, E.J., and Oughton, D.H. (2012). Aging and soil organic matter content affect the fate of silver nanoparticles in soil. *Science of the Total Environment* 420: 327–333.

De Souza, C.D., Nogueira, B.R., and Rostelato, M.E.C. (2019). Review of the methodologies used in the synthesis gold nanoparticles by chemical reduction. *Journal of Alloys and Compounds* 798: 714–740.

Debona, D., Rodrigues, F.A., and Datnoff, L.E. (2017). Silicon's role in abiotic and biotic plant stresses. *Annual Review of Phytopathology* 55: 85–107.

Dhakate, P., Kandhol, N., Raturi, G. et al. (2022). Silicon nanoforms in crop improvement and stress management. *Chemosphere* 305: 135165.

Doğan, İ. and van de Sanden, M.C.M. (2013). Direct characterization of nanocrystal size distribution using raman spectroscopy. *Journal of Applied Physics* 114 (13): 134310. https://doi.org/10.1063/1.4824178.

Elangovan, H., Sengupta, S., Narayanan, R., and Chattopadhyay, K. (2020). Silicon nanoparticles with UV range photoluminescence synthesized through cryomilling induced phase transformation and etching. *Journal of Materials Science* 56 (2): 1515–1526.

El-Shetehy, M., Moradi, A., Maceroni, M. et al. (2021). Silica nanoparticles enhance disease resistance in Arabidopsis plants. *Nature Nanotechnology* 16: 344–353.

Evlyukhin, A.B., Novikov, S.M., Zywietz, U. et al. (2012). Demonstration of magnetic dipole resonances of dielectric nanospheres in the visible region. *Nano Letters* 12 (7): 3749–3755.

Farré, M. and Barceló, D. (2012). Introduction to the analysis and risk of nanomaterials in environmental and food samples. In: *Comprehensive Analytical Chemistry*, vol. 59 (ed. M. Farre and D. Barcelo), 1–32. London: Elsevier.

Fu, Y.H., Kuznetsov, A.I., Miroshnichenko, A.E. et al. (2013). Directional visible light scattering by silicon nanoparticles. *Nature Communications*. https://doi.org/10.1038/ncomms2538.

Goswami, S., Gupta, P., Mahanta, S.P. et al. (2024). Tuning exchange bias in silica-coated Co-CoO core-shell nanostructure. *Journal of Nanoparticle Research* 26 (3): 56.

Guntzer, F., Keller, C., and Meunier, J.D. (2012). Benefits of plant silicon for crops: a review. *Agronomy for Sustainable Development* 32: 201–213.

Hülser, T., Schnurre, S.M., Wiggers, H., and Schulz, C. (2011). Gas-phase synthesis of nanoscale silicon as an economical route towards sustainable energy technology. *Kona Powder and Particle Journal* 29: 191–207.

Ijaz, I., Gilani, E., Nazir, A., and Bukhari, A. (2020). Detail review on chemical, physical and green synthesis, classification, characterizations and applications of nanoparticles. *Green Chemistry Letters and Reviews* 13 (3): 223–245.

Imtiaz, M., Rizwan, M.S., Mushtaq, M.A. et al. (2016). Silicon occurrence, uptake, transport and mechanisms of heavy metals, minerals and salinity enhanced tolerance in plants with future prospects: a review. *Journal of Environmental Management* 183: 521–529.

Irshad, M.A., Nawaz, R., Zia ur Rehman, M. et al. (2021). Synthesis, characterization and advanced sustainable applications of titanium dioxide nanoparticles: a review. *Ecotoxicology and Environmental Safety* 212: 111978.

Ismail, R.A., Mohsin, M.H., Ali, A.K. et al. (2020). Preparation and characterization of carbon nanotubes by pulsed laser ablation in water for optoelectronic application. *Physica E* 119: 113997.

Jain, N., Jain, P., Rajput, D., and Patil, U.K. (2021). Green synthesized plant-based silver nanoparticles: therapeutic prospective for anticancer and antiviral activity. *Micro and Nano Systems Letters*. https://doi.org/10.1186/s40486-021-00131-6.

Jalil, S.U. and Ansari, M.I. (2019). Nanoparticles and abiotic stress tolerance in plants: synthesis, action, and signaling mechanisms. *Plant Signaling Molecules* 549–561.

Jamkhande, P.G., Ghule, N.W., Bamer, A.H. et al. (2019). Metal nanoparticles synthesis: an overview on methods of preparation, advantages and disadvantages, and applications. *Journal of Drug Delivery Science and Technology* 53: 101174.

Jia-Wen, W.U., Yu, S.H.I., Yong-Xing, Z.H.U. et al. (2013). Mechanisms of enhanced heavy metal tolerance in plants by silicon: a review. *Pedosphere* 23 (6): 815–825.

Jing, Y., Wang, H., Chen, X. et al. (2014). Pulsed laser deposited Ag nanoparticles on nickel hydroxide nanosheet arrays for highly sensitive surface-enhanced Raman scattering spectroscopy. *Applied Surface Science* 316: 66–71.

Jurbergs, D., Rogojina, E., Mangolini, L., and Kortshagen, U. (2006). Silicon nanocrystals with ensemble quantum yields exceeding 60%. *Applied Physics Letters*. https://doi.org/10.1063/1.2210788.

Khan, Z.S., Rizwan, M., Hafeez, M. et al. (2020). Effects of silicon nanoparticles on growth and physiology of wheat in cadmium contaminated soil under different soil moisture levels. *Environmental Science andPollution Research* 27: 4958–4968.

Kim, M., Osone, S., Kim, T. et al. (2017). Synthesis of nanoparticles by laser ablation: a review. *Kona Powder and Particle Journal* 34: 80–90.

Ko, J., Lee, D., Lee, B.J. et al. (2019). Micropipette resonator enabling targeted aspiration and mass measurement of single particles and cells. *ACS Sensors* 4 (12): 3275–3282.

Ko, J., Jeong, J., Son, S., and Lee, J. (2021). Cellular and biomolecular detection based on suspended microchannel resonators. *Biomedical Engineering Letters* 11 (4): 367–382.

Konrad, A., Herr, U., Tidecks, R., and Samwer, F. (2001). Luminescence of bulk and nanocrystalline cubic yttria. *Journal of Applied Physics* 90 (7): 3516–3523.

Krishnia, L., Thakur, P., and Thakur, A. (2022). Synthesis of nanoparticles by physical route. In: *Synthesis and Applications of Nanoparticles* (ed. A. Thakur, P. Thakur, and S.M.P. Khurana), 45–59. Springer.

Kumari, S., Raturi, S., Kulshrestha, S. et al. (2023, 2023). A comprehensive review on various techniques used for synthesizing nanoparticles. *Journal of Materials Research and Technology* 27: 1739–1763.

Lai, S.Y., Knudsen, K.D., Sejersted, B.T. et al. (2019). Silicon nanoparticle ensembles for lithium-ion batteries elucidated by small-angle neutron scattering. *ACS Applied Energy Materials* 2 (5): 3220–3227.

Malik, M.A., Wani, A.H., Mir, S.H. et al. (2021). Elucidating the role of silicon in drought stress tolerance in plants. *Plant Physiology and Biochemistry* 165: 187–195.

Minelli, C., Sikora, A., Garcia-Diez, R. et al. (2018). Measuring the size and density of nanoparticles by centrifugal sedimentation and flotation. *Analytical Methods* 10 (15): 1725–1732.

Mohammadlou, M., Maghsoudi, H., and Jafarizadeh-Malmiri, H.J.I.F.R.J. (2016). A review on green silver nanoparticles based on plants: synthesis, potential applications and eco-friendly approach. *International Food Research Journal* 23 (2): 446–463c.

Pavlovic, J., Kostic, L., Bosnic, P. et al. (2021). Interactions of silicon with essential and beneficial elements in plants. *Frontiers in Plant Science* 12: 697592.

Priolo, F., Gregorkiewicz, T., Galli, M., and Krauss, T.F. (2014). Silicon nanostructures for photonics and photovoltaics. *Nature Nanotechnology* 9 (1): 19–32.

Rahman, I.A. and Padavettan, V. (2012). Synthesis of silica nanoparticles by sol–gel: size-dependent properties, surface modification, and applications in silica-polymer nanocomposites—a review. *Journal of Nanomaterials* 1: 132424.

Rastogi, A., Tripathi, D.K., Yadav, S. et al. (2019). Application of silicon nanoparticles in agriculture. *3 Biotech* 9: 1–11.

Reynolds, O.L., Padula, M.P., Zeng, R., and Gurr, G.M. (2016). Silicon: potential to promote direct and indirect effects on plant defense against arthropod pests in agriculture. *Frontiers in Plant Science* 7: 744.

Rostislav, A.A. (1994). The synthesis and properties of nanocrystalline refractory compounds. *Russian Chemical Reviews* 63: 411–427.

Salem, S.S. and Fouda, A. (2021). Green synthesis of metallic nanoparticles and their prospective biotechnological applications: an overview. *Biological Trace Element Research* 199: 344–370.

Sarkar, M.M., Mathur, P., and Roy, S. (2022). Silicon and nano-silicon: new frontiers of biostimulants for plant growth and stress amelioration. In: *Silicon and Nano-Silicon in Environmental Stress Management and Crop Quality Improvement*, 17–36. Academic Press.

Saxena, R., Tomar, R.S., and Kumar, M. (2016). Exploring nanobiotechnology to mitigate abiotic stress in crop plants. *Journal of Pharmaceutical Sciences and Research* 8: 974.

Shalaby, T.A., Bayoumi, Y., Eid, Y. et al. (2022). Can nanofertilizers mitigate multiple environmental stresses for higher crop productivity? *Sustainability* 14: 3480.

Shanmugam, S., Krishnaswamy, S., Chandrababu, R. et al. (2020). Optimal immobilization of Trichoderma asperellum laccase on polymer coated Fe_3O_4@ SiO_2 nanoparticles for enhanced biohydrogen production from delignified lignocellulosic biomass. *Fuel* 273: 117777.

Sharma, A.B., Sharma, M., and Pandey, R.K. (2009). Synthesis, properties and potential applications of semiconductor quantum particles. *Asian Journal of Chemistry* 21 (10): S033–S038.

Sharma, A., Sagar, A., Rana, J., and Rani, R. (2022). Green synthesis of silver nanoparticles and its antibacterial activity using fungus *Talaromyces purpureogenus* isolated from *Taxus baccata* linn. *Micro and Nano Systems Letters*. https://doi.org/10.1186/s40486-022-00144-9.

Shi, Y., Guo, S., Zhao, X. et al. (2022). Comparative physiological and transcriptomics analysis revealed crucial mechanisms of silicon-mediated tolerance to iron deficiency in tomato. *Frontiers in Plant Science* 13.

Singh, P.K., Kumar, P., and Das, A.K. (2019). Unconventional physical methods for synthesis of metal and non-metal nanoparticles: a review. *Proceedings of the National Academy of Sciences, India Section A: Physical Sciences* 89: 199–221.

Slomberg, D.L. and Schoenfisch, M.H. (2012). Silica nanoparticle phytotoxicity to Arabidopsis thaliana. *Environmental Science and Technology* 46: 10247–10254.

Souri, Z., Khanna, K., Karimi, N., and Ahmad, P. (2021). Silicon and plants: current knowledge and future prospects. *Journal of Plant Growth Regulation* 40: 906–925.

Sp, V., Udayabhanu, U., and Lalithamba, H.S. (2020). Plant-mediated green synthesis of Ag nanoparticles using *Rauvolfia tetraphylla* (L) flower extracts: characterization, biological activities, and screening of the catalytic activity in formylation reaction. *Scientia Iranica* 27 (6): 3353–3366.

Sreelekha, E., George, B., Shyam, A. et al. (2021). A comparative study on the synthesis, characterization, and antioxidant activity of green and chemically synthesized silver nanoparticles. *BioNanoScience* 11: 489–496.

Sun, S., Yang, Z., Song, Z. et al. (2022). Silicon enhances plant resistance to Fusarium wilt by promoting antioxidant potential and photosynthetic capacity in cucumber (Cucumis sativus L.). *Frontiers in Plant Science* 13.

Suriyaprabha, R., Karunakaran, G., Kavitha, K. et al. (2014). Application of silica nanoparticles in maize to enhance fungal resistance. *IET Nanobiotechnology* 8: 133–137.

Tan, D., Ma, Z., Xu, B. et al. (2011). Surface passivated silicon nanocrystals with stable luminescence synthesized by femtosecond laser ablation in solution. *Physical Chemistry* 13 (45): 20255.

Thakur, A., Thakur, P., and Khurana, S.P. (2022). *Synthesis and Applications of Nanoparticles*. Springer Nature.

Thiessen, A.N., Ha, M., Hooper, R.W. et al. (2019). Silicon nanoparticles: are they crystalline from the core to the surface? *Chemistry of Materials* 31 (3): 678–688.

Tien, D.C., Tseng, K.H., Liao, C.Y. et al. (2008). Discovery of ionic silver in silver nanoparticle suspension fabricated by arc discharge method. *Journal of Alloys and Compounds* 463 (1–2): 408–411.

Tiwari, R.K., Lal, M.K., Naga, K.C. et al. (2020). Emerging roles of melatonin in mitigating abiotic and biotic stresses of horticultural crops. *Scientia Horticulturae* 272: 109592.

Tripathi, D.K., Singh, S., Singh, V.P. et al. (2017). Silicon nanoparticles more effectively alleviated UV-B stress than silicon in wheat (Triticum aestivum) seedlings. *Plant Physiology and Biochemistry* 110: 70–81.

Tubaña, B.S. and Heckman, J.R. (2015). Silicon in soils and plants. *Silicon and Plant Diseases* 7–51.

Tubana, B.S., Babu, T., and Datnoff, L.E. (2016). A review of silicon in soils and plants and its role in US agriculture: history and future perspectives. *Soil Science* 181: 393–411.

Ulvestad, A., Reksten, A.H., Andersen, H.F. et al. (2020). Crystallinity of silicon nanoparticles: direct influence on the electrochemical performance of lithium ion battery anodes. *ChemElectroChem* 7 (21): 4349–4353.

Verma, K.K., Song, X.P., Joshi, A. et al. (2022). Recent trends in nano-fertilizers for sustainable agriculture under climate change for global food security. *Nanomaterials* 12: 173.

Wang, Q., Bao, Y., Zhang, X. et al. (2012). Uptake and toxicity studies of poly-acrylic acid functionalized silicon nanoparticles in cultured mammalian cells. *Advanced Healthcare Materials* 1 (2): 189–198.

Wang, L., Hasanzadeh Kafshgari, M., and Meunier, M. (2020). Optical properties and applications of plasmonic-metal nanoparticles. *Advanced Functional Materials* 30 (51): 2005400.

Wei, C., Zhang, Y., Guo, J. et al. (2010). Effects of silica nanoparticles on growth and photosynthetic pigment contents of Scenedesmus obliquus. *Journal of Environmental Sciences* 22: 155–160.

Wen, X., Zhang, P., Smith, T.A. et al. (2015). Tunability limit of photoluminescence in colloidal silicon nanocrystals. *Scientific Reports*. https://doi.org/10.1038/srep12469.

Yaqoob, A.A., Umar, K., and Ibrahim, M.N.M. (2020). Silver nanoparticles: various methods of synthesis, size affecting factors and their potential applications – a review. *Applied Nanoscience* 10: 1369–1378.

Ying, S., Guan, Z., Ofoegbu, P.C. et al. (2022). Green synthesis of nanoparticles: current developments and limitations. *Environmental Technology and Innovation* 26: 102336.

Yun, K.M., Suryamas, A.B., Hirakawa, C. et al. (2009). A new physical route to produce monodispersed microsphere nanoparticle–polymer composites. *Langmuir* 25 (18): 11038–11042.

Zhao, J. and Friedrich, B. (2017). *Synthesis of Gold Nanoparticles via the Chemical Reduction Methods*. Oak Park, IL: Shaker.

Zhou, H., Shi, Z., Wan, X. et al. (2019). The relationships between process parameters and polymeric nanofibers fabricated using a modified coaxial electrospinning. *Nanomaterials* 9 (6): 843.

Zhu, Y., Jiang, X., Zhang, J. et al. (2020). Silicon confers cucumber resistance to salinity stress through regulation of proline and cytokinins. *Plant Physiology and Biochemistry* 156: 209–220.

4

Green Synthesis of Silicon Nanoparticles and Their Application in Agriculture

P. Shaik Sha Valli Khan[1], *V. Suneetha*[1], *S. Naseem*[1], *S. Khader Basha*[1], *P. Ganseh*[1], *K. Pallavi*[1], *D. Suresh*[1] *and O.N. Tiwari*[2]

[1] *Department of Botany, Yogi Vemana University, Kadapa, Andhra Pradesh, India*
[2] *Centre for Conservation and Utilization of Blue Green Algae, Division of Microbiology, Indian Agricultural Research Institute (ICAR), New Delhi, India*

4.1 Introduction

Agriculture is crucial for growing economies, providing food and improving the overall quality of life. However, it faces challenges such as climate change; soil pollution from harmful pollutants like pesticides, heavy metals and fertilizers; and increased food demand due to the growing world population (Liu et al. 2023). The UN projections suggest that the world population will increase to 9.0 billion, necessitating a 50% increase in food production to meet this demand (Okeke et al. 2023). At present, biotic and abiotic stressors such as drought, extreme climate, salt, diseases and pests cause crop loss worldwide. The influence of abiotic stresses on crop production causes an annual loss of crop yield of 51–82% (Oshunsanya et al. 2019). Farmers often use pesticides and fertilizers to increase crop yield, posing a significant threat to the environment. Approximately, 70% of India's workforce relies on agriculture for their livelihood, and a significant portion of them are subsistence farmers. They produce barely enough food in each year to feed their families which makes them extremely vulnerable to the effects of stressful conditions on crop yields. Therefore, eco-friendly technologies should be developed to help plants overcome environmental stresses and ensure optimal crop yield and agricultural sustainability.

Silicon (Si) is the second most abundant element in the earth's crust (28%), found in quartz, feldspar, mica and clay minerals. Most Si exists in an unavailable form of silicon dioxide. Si occurs in various forms of amorphous silica in phytoliths and silica-rich plants in biological systems. The biogenic share of silicon is about 1–3% of the total Si pool in the soil. Plants uptake Si in the form of silicic acid

Silicon Nanoparticles: Synthesis, Uptake, and Applications, First Edition.
Edited by Archana Singh, Aryadeep Roychoudhury, and Indrakant K. Singh.

or monosilicic acid [$Si(OH)_4$ or H_4SiO_4] (Ma et al. 2001). Its concentration varies from 0.1% to 10% Si on a dry weight basis and has been observed in different plant species (Hodson et al. 2005). Despite its abundant accumulation in plants, Si is considered as a quasi-essential element since most plant species complete their life cycle in its absence (Arnon and Stout 1939). Si has been shown to improve plant growth, biomass, seed yield and quality, structural rigidity and physiological processes such as photosynthesis, respiration, translocation, ion uptake, stomatal behaviour, water relations, mineral nutrition and resistance to biotic and abiotic stresses (Bhat et al. 2019; Singh and Roychoudhury 2020; Zargar et al. 2019). In this context, using nanotechnology in Si research has emerged as a necessary alternative approach to increase the efficiency of using Si in crop improvement.

Nanotechnology, a rapidly emerging field, permitted the synthesis of a wide class of materials with one dimension or more in the nanoscale range from 1 to 100 nm. These synthesized chemical materials in the nanoscale range are called nanomaterials (NMs). The size is the difference between bulk materials and NMs. The bulk materials are >100 nm in all dimensions and visible through a simple microscope, whereas NMs <100 nm in at least one dimension and cannot be seen by the naked eye. Physical properties of NMs are dependent on their shape and size, whereas physical properties of bulk materials are independent of size. The unique physicochemical properties (optical, mechanical, electrical and chemical properties), large surface area-to-weight ratios and small sizes that increase the solubility and transportation speed underlie the importance of NMs. They have been classified as lipid-based, metallic, ceramic, semiconductor, polymeric and carbon NPs. NMs show unique properties that allow excellent performance, and they have been applied in different fields such as agriculture (Alsaiari et al. 2023).

Silicon/silica nanoparticles (Si/SiO_2 NPs) have gained special attention due to their optoelectronic and significant chemical capabilities. Its mesoporous nature, easy availability and low biological toxicity have made SiNPs attractive to researchers (Naidu et al. 2023). Silicon nanoparticles (SiNPs) are small, with an average diameter of 7 nm and a wide absorption surface area (400 m^2), making them easily penetrate the cell wall barrier and enter the plant system or get easily absorbed and transported by transporters (Bhat et al. 2021). In comparison with bulk materials, SiO_2 NPs have a high surface-to-volume ratio, thus improving their reactivity and biochemical activity (Saw et al. 2023). SiNPs are commercially synthesized by chemical, physical and biological methods. SiNPs can be supplied to plants by foliar, soil, or seed priming (Banerjee et al. 2021). Upon uptake and translocation, SiNPs reach their destined cells and cause optimum growth, development and tolerance against environmental stresses. Silicon strengthens the physical barrier of plants through deposition on to plant cell walls. SiNPs can address various agricultural issues, such as weeds, pathogenicity, pest infestation and crop loss (Islam et al. 2021). Using SiNPs as a supplement can be an eco-friendly and cost-effective option for sustainable agriculture (Naidu et al. 2023).

This chapter explores the different routes of synthesis of Si/SiO_2 NPs, especially collective information about green synthesis, using biological resources (bacteria, actinomyces, yeast, fungi, microalgae and plant residues), their characterization, uptake, translocation and usage in agriculture. It highlights the role of SiNPs in promoting plant growth and development, its effects against abiotic stresses and its potential as a pesticide, insecticide, herbicide and fertilizer for crop protection and nutrition (Roychoudhury 2020a). The chapter also discusses toxicity effects and biosafety concerns related to their use in agriculture. It also highlights and discusses certain prospects and challenges.

4.2 Different Methods for the Synthesis of SiNPs

SiNPs could also be classified as porous or nonporous in structure. SiNPs are being commercially synthesized from metals and metal oxides using different physical and chemical methods. Stöber et al. synthesized SiO_2 NPs in 1962 using tetraethyl orthosilicate (TEOS) as the precursor, ethyl alcohol and water as solvents and ammonia as an alkaline catalyst. Since then, various methods have been used to synthesize SiNPs, including sputtering, lithography, laser ablation, mechanical milling, sonication, milling, UV irradiation, reverse micro-emulsion, atomic condensation, spray pyrolysis, aerosol process, chemical vapour condensation bioreduction and fame synthesis (Liberman et al. 2014). However, these methods have disadvantages, such as high costs, high-temperature furnaces, slower growth rates and toxic reagents. Fame synthesis has limitations in particle size, morphology and phase composition. Laser ablation has limitations in wavelength, pulse duration, fluency and the use of surfactants. Reverse microemulsion is costly and difficult to remove surfactants from final products (Babu et al. 2015). The structure of SiNPs significantly impacts their properties, such as particle size, shape, pore size and dissolution rate, which affects their use and interaction with living organisms (Bose et al. 2018).

Physical and chemical techniques are divided into two main approaches: top-down and bottom-up. Top-down involves synthesizing nanoparticles from bulk, which are then broken down into smaller particles using physical methods like size reduction techniques. Despite its simplicity, top-up methods are less widely used due to surface structure imperfections. The bottom-up method uses a standard route to synthesize SiNPs on an atomic or molecular scale. Green synthesis is a bottom-up approach, which uses biopolymers, biomolecules and bioresources and is becoming more popular due to their lower time-consuming, cost-effective, high-yielding, non-toxic and environmentally friendly nature (Verma et al. 2021).

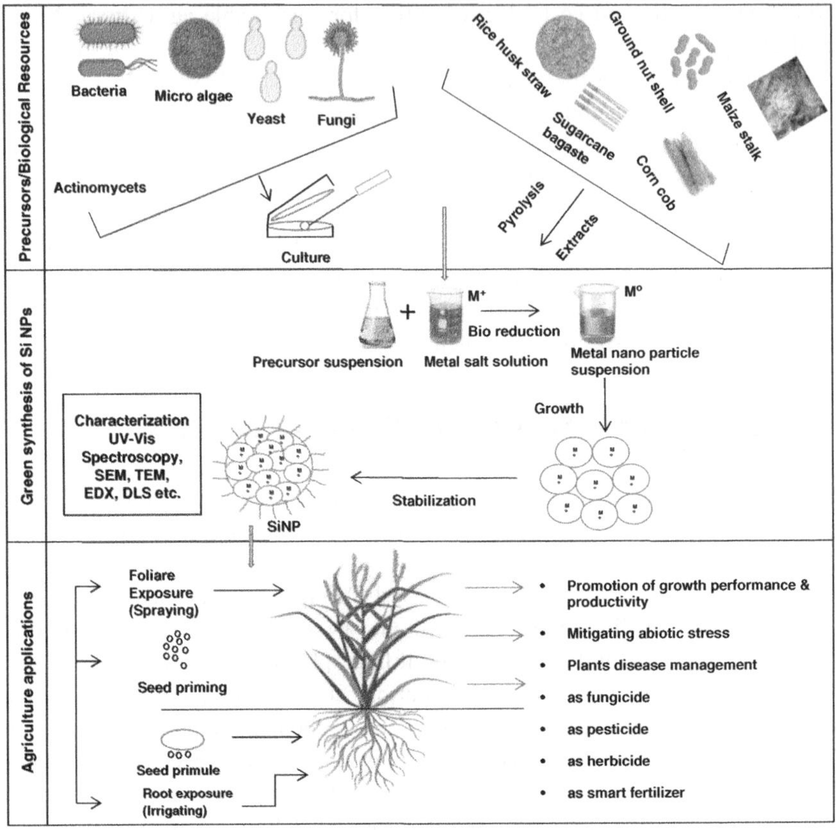

Figure 4.1 Green synthesis of SiNPs from various biological resources.

4.2.1 Biological/Green Synthesis of SiNPs

The synthesis of nanoparticles using biological resources like plants, bacteria, actinomyces, yeast, fungi, and algae has led to the development of 'green nanotechnology,' which combines biological principles with physical and chemical procedures to create eco-friendly, nano-sized particles with specific functions. Green methods offer several advantages and can overcome some drawbacks of other synthesis techniques (Paul and Roychoudhury 2021; Roychoudhury 2020b). SiNPs produced by green methods are of similar quality to chemical ones and can be used in various technical applications, including medicine, pharmaceuticals, electricals, paints and cosmetics (Bhat et al. 2021). Figure 4.1 illustrates the various biological resources available for the green synthesis of SiO_2 NPs (Table 4.1).

Table 4.1 Comparative analysis of green synthesis methods for SiNPs.

Biological source	Possible mechanisms	Advantages	Disadvantages
Bacteria	The bacterial cell utilizes specific reducing enzymes, such as NADH-dependent reductase or nitrate dependent reductase, to reduce metal ions	This method allows for the production of highly pure SiNPs, which can be utilized in biomedical applications due to their purity	The quantitative yields and production costs for this approach have not been reported yet
Actinomycetes	Actinomycetes use extracellular or intracellular approaches to synthesize metal and metal oxide nanoparticles	Biosilification is an economically beneficial, energy-saving, and deco-friendly method for obtaining highly pure SiNPs	The method has the potential to synthesize SiNPs on a large scale, but it requires addressing scale-up issues
Algae	Polysaccharides, with their hydroxyl groups and other functionalities, play crucial roles in the reduction and stabilization of nanoparticles	Microalgae-based synthesis has wide range of potential applications	Nanosilica precursors occur at low concentrations and required considerable energy input
Yeast	Membrane-bound and cytosolic oxido-reductases and quinones are essential enzymes involved in the production	This method allows for the production of highly pure SiNPs, which can be utilized in biomedical applications due to their purity	The quantitative yields and production costs for such approaches have not been reported yet
Fungi	The process of biometric mineralization involves reducing enzymes either intracellularly or extracellularly	Biosilification is an economically beneficial, energy-saving, and deco-friendly method for obtaining highly pure SiNPs	The method has the potential to synthesize SiNPs on a large scale, but it requires addressing scale-up issues

(Continued)

Table 4.1 (Continued)

Biological source	Possible mechanisms	Advantages	Disadvantages
Plants	Secondary metabolites serve as reducing and stabilizing agents	The green method ensures the production of high-quality, pure, and safe SiNPs due to the abundance of natural resources used	The yield of SNPs obtained using this technique may be affected, and their stability must be compared to those obtained using chemical methods
Agricultural wastes/plant residues	The cellular enzymes of different plant tissues require complicated steps to attract the synthesized NMs by cracking the cell wall of the plant residue	The availability of raw materials at the end of each harvesting season can be low or free, resulting in superior quality of SiNPs	The papers mainly report laboratory-scale synthesis of SiNPs, suggesting potential for large-scale synthesis, but need to address scale-up issues

4.2.1.1 Synthesis of SiNPs by Microorganisms

Microorganisms like bacteria, actinomyces, fungi, yeasts and microalgae can synthesize metallic nanoparticles, either intra- or extra-cellularly. These microorganisms can produce organic matter and transport it to the outside of their cells. Extracellular synthesis mechanisms are particularly interesting as they eliminate costly and complex downstream processing steps. Control over size and shape can be achieved by manipulating key parameters that control the growth and other cellular activities of microorganisms. Microorganisms offer an eco-friendly, cost-effective and non-toxic method for synthesizing SiNPs, requiring minimal energy and offering a wide range of sizes, shapes, compositions and physicochemical properties. However, the main drawbacks include complicated steps like microbial sampling, isolation, culturing and storage. Table 4.2 summarizes the microbial synthesis of SiNPs using different types of organisms.

4.2.1.2 Synthesis of SiNPs by Bacteria

Bacteria can synthesize various nanomaterials using two main strategies: extracellular and intracellular. The extracellular method is more advantageous as it eliminates complex steps for extracting nanomaterials. Bacteria species are chosen based on cost, eco-friendliness and large-scale synthesis. Bacteria are favoured due to their ease of handling and genetic manipulation, reduced time, better control over size and shape, and industrial scaling. However, there

Table 4.2 Details of major studies demonstrating green synthesis of SiNPs from microorganisms.

Substrate	Characterization	Reference
Bacteria		
Bacterial culture	UV-Vis NMR spectroscopy, TEM, SEM, EDX	Manivasagan et al. (2023)
		Vetchinkina et al. (2017)
	XRF, TEM, FTIR	Singh et al. (2008)
BKHI bacteria		Show et al. (2015)
Bacillus girulens MTCC 6811	UV-Vis spectroscopy, FTIR, PSA, FESED, TEM	Yadav et al. (2023)
Actinomycetes		
Actinomycetes	SEM, EXD	Patel (2015)
Streptomyces	UV-VIS spectroscopy, PSA, TEM, SAED	Mohanraj et al. (2022)
Trichoderma roatroviride *Streptomyces griseus* *Pseudomonos fluoresum*	FTRI, UV-VIS spectroscopy, EDX, XRD, TEM	Kaur et al. (2023)
Yeast		
Saccharomyces cerevisiae	FTIR, XRD, SEM, DRS, TGA	Manivasagan et al. (2016)
		Shah et al. (2015)
		Zamani et al. (2020)
Fungi		
Fusarium culmorum	SEM	Zhao et al. (2022)
Fusarium oxysporum	FTIR, GC-MS	Hachicha et al. (2022)
		Khanna et al. (2019)
		Chaudhary et al. (2020)
White rot fungus	PCR	Pavithra et al. (2020)
Aspergillus parasiticus	SEM, STEM, EDX, FTIR	Caliskan et al. (2022)
		Zielonka et al. (2018)
Algae		
Scytonema guyanense	FTIR, SEM, HRTEM, EDS, XRD	Nath et al. (2016)

are limitations in the shapes and sizes of bacteria-synthesized nanoparticles compared to conventional methods (Singh et al. 2016). Silicon nanospheres with sizes ranging from 5 to 250 nm were synthesized using bacterial culture liquids and *Actinobacter* species (Vetchinkina et al. 2017). The bacterial reductase and oxidase enzymes lead to the synthesis of SiO_2 NPs (Singh et al. 2008). A green protocol was developed to produce SiNPs using thermophilic bacteria BKH1, inorganic magnesium trisilicate precursor and organic tetraethyl orthosilicate precursor (Show et al. 2015). Bacteria have extensive potential for SiNP synthesis due to their replication rate and simple cultivation techniques.

4.2.1.3 Synthesis of SiNPs by Actinomycetes

Actinomycetes are Gram-positive and anaerobic, bacteria with unique characteristics resembling fungi and bacteria. They are considered superior for their saprophytic behaviour and production of numerous secreted secondary metabolites. Actinomycetes can synthesize metal and metal oxide nanoparticles using extracellular or intracellular approaches. Intracellular nanostructures are formed when metal ions attach to the carboxylate group of an enzyme in the cell wall. Actinomycetes have protein, enhancing nanoparticle synthesis efficiency and making it easier to scale up. They are also simple to extract by filtration, unlike other methods that require methodical tools (Alsaiari et al. 2023). SiO_2 nanocomposites are synthesized from rice husk and wheat bran using actinomycetes, which biotransform amorphous silica into crystalline silica within 24 hours in the size range of 232–250 nm (Kaur et al. 2023). Actinomycetes are the least explored and are an efficient candidate for SiNP synthesis.

4.2.2 Synthesis of SiNPs by Yeast

Yeasts are single-cell eukaryotes that reproduce through budding and have around 1500 species from Ascomycetes such as *Saccharomyces* and *Candida* or members from Basidiomycetes such as *Filobasidiella* and *Rhodotorula* (Bhattacharya and Gupta 2005). They can grow easily on low-cost media and can adapt to harsh environmental conditions like temperature and pH changes. Yeasts can absorb and accumulate toxic metal ions due to their wide surfaces. They can detoxify hazardous metals through methods like biosorption, extracellular sequestration, chelation and bioprecipitation. Studies have shown that yeasts can successfully synthesize SiNPs, with some investigations showing their applicability in oil recovery (Alsaiari et al. 2023). The baker's yeast *Saccharomyces cerevisiae* is used for the biochemical synthesis of SiO_2 NPs, which can be used for enhanced oil recovery (Zamani et al. 2020).

4.2.3 Synthesis of SiNPs by Fungi

Fungi, eukaryotic organisms found in various habitats, are decomposer organisms with an estimated population of 5.1 million (Blackwell 2011). They digest extracellular food and release enzymes to hydrolyse complex compositions into simpler molecules, which are used as energy. Fungi have gained attention in nanobiotechnology for their tolerance and metal bioaccumulation capabilities. Their easy scale-up and secretion of extracellular enzymes make large enzyme production feasible, making fungal-mediated nanoparticle synthesis a green approach. Fungi are fast-growing organisms with high wall-binding and intracellular metal uptake capabilities, making them ideal for the green synthesis of SiNPs. However, there are drawbacks, such as determining the best fungus for producing desired nanoparticles, determining growth parameters, sterile conditions and time required for growth (Moghaddam et al. 2015). Research shows that various species of fungi can be used for green synthesis, with *Fusarium*, *Aspergillus* and *Penicillium* showing potential for extracellular bioproduction and *Verticillium* sp. for intracellular nanoparticle synthesis. Numerous reports exist on synthesizing SiNPs using fungi (Table 4.2).

SiO_2 NPs were obtained from corn cob husk bioconversion using *Fusarium culmorum* (Pieła et al. 2020). Bansal et al. (2005a) developed a novel method for producing SiO_2 NPs using *Fusarium oxysporum* fungus. The two-step process involves extracellular bioleaching of silica from sand grains in the form of silicic acid by organic acids produced by fungal biomass. The silicate complex is hydrolysed by specific proteins or enzymes in the fungus to silica. Both steps depend on cultivation conditions and culture medium compositions. This bioleaching method is rapid and takes place within one day of the biomass reaction with sand (Bansal et al. 2005a). Bansal et al. (2005b) developed a method to synthesize SiO_2 NPs from rice husk or rice ash using *F. oxysporum* and anionic complex of SiF_6^{-2}. A new way to synthesize nanosilica introduced from rice husk (RH) or RH ash using *F. oxysporum* was developed by Bansal et al. (2007); these fungal species are capable of rapid biotransformation of the naturally occurring biosilica from RH into crystalline SiO_2 NPs (Bansal et al. 2007). SiO_2 NPs were produced from agro-waste rice husk by fungal-mediated biotransformation using *Aspergillus parasiticus* species, which generates SiNPs of size 3–400 nm (Zielonka et al. (2018).

4.2.4 Synthesis of SiNPs by Algae

Microalgae, single-celled aquatic microorganisms, undergo photosynthesis and form colonies without cell differentiation. They grow in various environment, influenced by temperature, light intensity and nutrient concentration. Algal extracts, including microalgae and cyanobacteria, can be used as a living cell factory for efficient green nanoparticle synthesis due to their unique characteristics, including no toxic by-products, minimal energy input and the presence of biomolecules acting as reducing and capping agents (Khanna et al. 2019). Microalgae-based synthesis of the MtNPs, known as 'phyco-nanotechnology', is an emerging field with a wide range of potential applications (Chaudhary et al. 2020). Diatoms and brown algae can design and control nanostructure synthesis, possessing a silica exoskeleton called frustules, composed of SiO_2 NPs (50–100 nm) assembled in a highly organized structure. Frustules can be formed within hours from naturally occurring precursors at low concentrations at ambient temperature (Roychoudhury et al. 2022).

4.2.5 Synthesis of SiNPs from Plant Residues/Agriculture Residues

Plant-mediated synthesis is a cost-effective and straightforward method for synthesizing SiNPs from plant tissues like leaves, roots, fruits and husks. This process can be classified into intracellular, extracellular and phytochemical synthesis. The extracellular method yields higher NMs due to the high number of phytochemicals in the plant extract. The intracellular method requires complex steps to attract synthesized NMs by cracking the cell wall of the plant. Thus, plant-mediated synthesis is a cost-effective and eco-friendly method for synthesizing NMs (Adinarayana et al. 2020). Green synthesis of SiNPs using plants has been reported in various studies. One such method involves laser ablation, resulting in an environmentally friendly procedure for the synthesis of SiNPs from sugarbeet bagasse (San et al. 2014). Another method uses *Azadirachta indica* leaf extract as a chelating agent (Jabeen et al. 2017). *Cynodon dactylon* plant is used as a source of SiNPs production, and the produced SiNPs are applied in antimicrobial studies (Babu et al. 2018). Bamboo leaves are used to synthesize SiO_2, with 99.9% purity. Bamboo leaf ash is used to enhance the separation properties of SiO_2. *Bambusa vulgaris* leaf ash is used for sol-gel production, with environmental applications (Durairaj et al. 2019; Sethy et al. 2019). *Thuja orientalis* leaf extract is used for biofilm formation. Silica is extracted from banana stems using acid-leaching. *Equisetum arvense* extract is pyrolysed using a microwave oven to generate SiNPs with fluorescent properties (Adinarayana et al. 2020). These methods demonstrate the potential of plants in the green synthesis of SiNPs.

Agricultural waste is utilized as a precursor for synthesizing SiNPs due to its abundant availability at the end of harvesting seasons. The silica extraction from agricultural wastes requires a pre-treatment. Acid leaching is a cost-effective pre-treatment method for removing metallic impurities and breaking down complex compounds, but high impurity content like iron oxide hinders the process of obtaining high-purity silica. Therefore, acid leaching is necessary to retain the biogenic structure while removing the impurities (September et al. 2023). Rice hull, a waste product from agriculture, is being utilized to produce high-quality SiO_2 NPs. Biochemical and thermochemical conversion pathways are used to convert biomass into bioproducts. The former involves breaking down biomass into fuels, while the latter uses heat to break down the complex chemical structure into power, materials and chemicals. The thermochemical conversion path is ideal for extracting silica. Silica microparticles and nanoparticles were synthesized using rice hull for the first time. Rice straw is used as a source of silica and synthesis of highly pure amorphous silica nano-discs . Sticky, red and brown rice husk ash is used to synthesize biogenerated SiNPs (Sankar et al. 2016). Similarly, several researchers synthesized SiNPs from rice husk (Bose et al. 2018; Chen et al. 2013; Wang et al. 2012).

Bagasse production, generated millions of tonnes annually, produces 279 million metric tonnes of waste. It is used as a fuel source in sugar mills, producing ash and SCBA. Sugarcane bagasse ash is used to synthesize SiNPs and nano-silicon through magnesio-thermic reactions, highlighting the significant waste produced in the sugar industry (Falk et al. 2019). The production of SiNPs was reported from sugarcane bagasse (Mohd et al. 2017). This green approach helps in agricultural waste utilization by economic route. SiNPs were synthesized from agricultural waste like rice husk, bamboo leaves, sugarcane bagasse and groundnut shells and recovered 52–78% silica from respective wastes. Tef straw is an agricultural residue available in high amounts in Ethiopia. Wassie and Srivastava (2017) reported the synthesis of nano silica by using Tef straw using heat and acid treatment. This method involves acid treatment, calcination at 700° C, modified sol-gel and post-filtration treatments.

Sorghum bicolor, a type of agricultural residue, can be used as a precursor for synthesizing SiO_2 NPs (SiNPs), which have various applications in food industry (Adinarayana et al. 2020). Sorghum residues contain a high amount of silica, making them suitable for synthesis. Balamurugan and Saravanan (2012) reported the use of *Sorghum vulgare* seed heads for nanosilica synthesis. *Carex riparia*, a weed that produces large amounts of agro-waste, was used to synthesize SiNPs, providing a new method for obtaining useful nanomaterials from waste. Agunsoye et al. (2018) explored the use of Cassava periderm as a new source of silica. Pal et al. synthesized SiO_2 NPs from coconut shells. All these examples prove the worth of various plants in the green synthesis of SiNPs. Therefore, by looking at all the above examples from the literature, it can be concluded that agricultural waste emerges as a good source for the synthesis of SiNPs.

4.3 Characterization of SiNP/SiO_2

Characterization techniques are used to study the physical, chemical and other properties of SiNPs, including topography, roughness, surface chemistry, stability, distribution, surface area, shape, size and composition. These techniques depend on the type of application and can be applied to various materials, including materials like SiNPs. Several instrumental approaches such as transmission electron microscope (TEM), scanning electron microscope (SEM), field emission scanning electron microscope (FESEM), X-ray crystallography (XRD), dynamic light scattering (DLS), UV-Vis spectrophotometry, fourier transform infrared (FTIR) spectroscopy and ultrasonication can be used to characterize NPs. UV-visible spectroscopy is generally used to confirm the synthesis and stability of SiNPs (Goswami et al. 2022). Fourier transform infrared (FTIR) spectroscopy is used to measure the properties of SiNPs such as chemical concentration, surface chemistry, surface functional groups and atomic arrangement. Transmission electron microscopy (TEM), scanning electron microscopy (SEM) and atomic force microscopy (AFM) can be used to visualize the position, size and morphology of SiNPs. X-ray powder diffraction (XRD) is used to determine the crystallographic structure. The elemental composition of SiNPs is usually examined by energy-dispersive X-ray spectroscopy (EDX). Dynamic light scattering (DLS) method is mainly used to evaluate the size as well as surface charge of SiNPs (Tables 4.2 and 4.3).

4.4 Application of Si/SiO_2 NPs in Agriculture

Agriculture is facing challenges like shrinking farmland, limited water availability, low organic soil and diminishing resources due to population growth. Researchers are exploring nanotechnology as an alternative technology to improve productivity and variety in crops. SiO_2 NPs have physiological properties that enable them to penetrate plants and manipulate their metabolic actions (Rea et al. 2022). SiNPs are used as fertilizers, insecticides, herbicides and potential agents against heavy metal impacts, salt and drought stress. As biocompatible, biodegradable and non-toxic agents, they are ideal for sustainable agricultural practices (Goswami et al. 2022). As a result, SiNPs have the potential to enhance crops for long-term farming (Rastogi et al. 2019). SiNPs have been extensively utilized in promoting sustainable agriculture due to their diverse functionalities, according to recent reports by Dhakate et al. (2022). Thus, this section includes studies on SiNPs and discusses their effects on numerous factors of agriculture.

4.4.1 Modes of Application of Si/SiO_2 NPs

Si/SiO_2 NPs are commonly administered to plants through foliar, soil and seed priming methods (Figure 1). In foliar application, SiNPs are sprayed on shoot tissues, which are transported through the cuticle or stomata. The transport path of solutes via the cuticle has two routes: lipophilic for non-polar solutes via diffusion and penetration and hydrophilic for polar solutes via water pores. Hydrophilic substances are transported via stomatal pores. After entering leaves, nanomaterials can translocate back to roots via the phloem. Following the uptake of SiNPs, a binary film is formed at the epidermal cell wall which imparts a structural color

Table 4.3 Details of major studies demonstrating green synthesis of SiNPs from plants, plant residues and agri-wastes.

Substrate	Nanoparticle shape/size	Characterization	Reference
Plants			
Cynodon dactylon	Amorphous, spherical SiNPs; 7–80 nm	UV-Vis spectroscopy, FT-IR, DLS, XRD, SEM, TEM, EDX	Babu et al. (2018)
Bambusa vulgaris leaves	Crystalline, porous SiNPs around 200 nm	SEM, TEM, XRD	Kumar et al. (2016) Vaibhav et al. (2015)
Azadirachta leaves	Cetyltrimethyl ammonium bromide 200 nm	SEM, EDX, XRD, FTIR	Jabeen et al. (2017)
Thuja oreintalis	Crystalline, 11.1868 nm	XRD, AFM, FTIR	
Banana stem	40–70 nm	SEM, XRD	Yusaidi et al. (2019)
Cassava periderm	Higher silica yield and lower particle sizes	SEM, EDX, XRD, FT-IR, Raman, TEM and PSA	Adebisi et al. (2018)
Citrus lemon	Nanoparticles, average size 20–57 nm	TEM, UV-Vis absorption and PL, FTIR	Tiwari et al. (2019)
Agri-wastes			
Sugarcane bagasse	Amorphous, spherical SiNPs; around 30 nm	FT-IR, XRD, SEM	Mohd et al. (2017)
Sugar beet bagasse	Spheroid-shaped SiNP 38–190 nm	SEM-EDS, TEM, DLS, FT-IR, Raman Spetroscopy	San et al. (2014)
Rice husk	Biogenic, porous SiNPs 25–30 nm	SEM-XRD, TEM	Chen et al. (2013) Wang et al. (2012)
Rice straw	Amorphous, 25–55 nm	EDX, FTIR, FESEM, TEM	Singh and Patel (2022)
Rice hull	Clustered spherical shape	SEM, TEM, XRD, FTIR	Sankar et al. (2016)
Maize stalk	Amorphous 30 nm	SEM, XRD, EDX, TEM, FTIR, Raman Spectroscopy	

(Continued)

Table 4.3 (Continued)

Substrate	Nanoparticle shape/size	Characterization	Reference
Corn cob stalk	Spherical shape 40–70 nm	SEM, STEM, EDX, FT-IR	Okoronkwo et al. (2013); Pieła et al. (2020)
Teff straw	40–70 nm	XRF, XRD, FTIR, TEM	Wassie and Srivastva (2017); 44
Ground nut shell	High-energy nanoparticles	TEM, XRF, FTIR, XRD	Yaro et al. (2017)
Wheat husk	40–70 nm	SEM, EDX, XRD, FTIR, ESI, MS, NMR	Patel et al. (2018)
Sorghum bicolor	BSNs had spherical morphology, diameters of 30–90 nm, amorphous nature	XRD, FTIR, SEM, TEM	Athinarayanan et al. (2020)
Sedge *Carex riparia*	Pore sizes 4.1 nm, surface areas 1174 m^2/g, total pore volume 0.98 cm^3/g	XRD, XRF, SEM, TEM, TGA	Ghorbani et al. (2013)

to the plant (Strout et al. 2013). Si/SiO_2 NPs can be introduced directly to soil or nutrient solutions, with efficiency depending on soil factors like pH, texture and salt concentration. Foliar application is less efficient due to reduced absorption by root tissues (Suriyaprabha et al. 2014). Root secretions release organic acids, phenolics, amino acids, secondary metabolites, proteins and mucilages, which combine with NPs to form complexes that attach to proteins or transporters, passing via ion channels, aquaporins or endocytosis (Rastogi et al. 2019; Wang et al. 2020).

Si/SiO_2 NPs can be applied through seed priming, which provides controlled hydration and stimulates specific physiological states in plants before seed sowing. This method has been shown to boost crop and vegetable growth, including wheat and tomato (Hussain et al. 2019; Liu et al. 2014). However, many researchers believe that seed priming is more critical than other methods of NP administration (Rizwan et al. 2019; Sundaria et al. 2019). Seed priming maintains a constant moisture level until the germination process begins and stimulates a specific physiological state in plants before seed planting (Jisha et al. 2013). Despite the availability of these three modes of SiNP application, there is limited information on their effectiveness. Therefore, efforts are needed to determine the relative efficacy of these methods in different plant species to fully utilize the potential of SiNPs application (Bhat et al. 2021).

4.4.2 Uptake and Translocation of Si/SiO_2 NPs by Plants

Plant species vary significantly in their Si content, ranging from 1.0 to 100 g/kg. They can be classified into Si accumulators (tissue contents above 10 g/kg), excluders (less than 5 g/kg), and intermediates (between these limits) (Ma and Takahashi 2002). Shoot Si concentrations tend to decline in certain plant orders, such as liverworts, horsetails, club mosses, mosses, angiosperms, gymnosperms and ferns (Hodson et al. 2005). Si accumulation is generally greater in monocots than in dicots but is largely restricted to certain monocot clades namely Poaceae, Cyperaceae and Commelinaceae (Ma and Takahashi 2002). Si concentrations in crop plants generally increase in the order of legumes, fruit crops, vegetables, grasses and grain crops. Significant differences in Si accumulation between genotypes of the same species and significant variations in the Si content of plant parts can be attributed to distinct mechanisms involved in the uptake and transport of Si among different plant species (Ma and Takahashi 2002).

Silicon uptake in plants can be active, passive or rejective for accumulator, intermediate and excluder plants, respectively. Accumulators have efficient transporters and specialized silica cells for increased uptake and accumulation. Active uptake in Si-accumulator plants is mediated by plasma membrane transporters (Ma and Hill 2006), identified for several species (Liang et al. 2008). In rice, wheat, maize, barley and pumpkin (*Cucurbita moschata*), it is mediated by influx (Lsi1) and efflux (Lsi2) protein transporters (Yamaji et al. 2012). Silicic acid is transported from external solution into root exodermal cells by Lsi1 and released into the apoplast of the cortex by Lsi2. Non-accumulators usually lack silica cells, and many dicot families have lost Si-transporter AQPs (Deshmukh and Bélanger 2016).

4.4.3 SiNPs Promote Plant Growth and Development

Plants are exposed to various environmental stresses, such as biotic (caused by viral and bacterial pathogens or fungi and herbivores) and abiotic stresses (saline, high and low temperature, flooding, UV, wind, drought, metal toxicity, light and mineral deficiency or excess). Studies show that SiNPs have significant potential to positively influence plant growth, especially under biotic and abiotic conditions. These effects are mainly due to their positive effects on photosynthesis, antioxidant defense systems, ROS scavenging and oxidative damage. Additionally, SiNPs have been found to have beneficial effects on nutrient homeostasis and carbon metabolism. This highlights the potential of SiNPs in enhancing plant stress tolerance (Naidu et al. 2023; Wang et al. 2022).

Si/SiO_2 NPs interact with plants, altering their morphological and physiological characteristics, enhancing crop yield and quality (Suriyaprabha et al. 2014). They increase germination percentage, vigor index, seedling growth, lateral root development, plant biomass, chlorophyll content and photosynthetic activity. They also boost nutrient bioavailability and uptake in the rhizosphere, resulting in seedlings with enhanced fresh and dry weight. Si/SiO_2 NPs also enhance xylem wall lignification, cell wall thickening, protein concentration, enzyme activity and promote blooming (Liu and Lal 2015). Silicon application has also been widely reported to increase yield of a wide range of crops due to increased leaf erectness and a more upright leaf position favours light interception and increased photosynthesis. Si application also strengthens cell wall and provides mechanical support, particularly in monocots, by enhancing suberization, lignification and silicification (Guerriero et al. 2016).

The effect of SiNPs varies among plant species, making it crucial to understand the role and mechanism involved in SiNP-mediated regulation of plant growth. Si/SiO_2 NPs have been found to harm plants due to changes in the pH of the growth medium. The introduction of Si/SiO_2 NPs primarily causes these effects. When pH fluctuations caused by Si/SiO_2 NPs are removed, the harmful impact on *Arabidopsis thaliana* (Slomberg and Schoenfisch 2012) is eliminated. Critical consideration of factors such as size, concentration, surface area, physicochemical properties, plant species, plant age and NP stability is necessary to maximize the benefits of SiNPs in agriculture (Rastogi et al. 2017).

4.4.4 Si/SiO_2 NPs in the Mitigation of Abiotic Stresses

Plant growth and productivity are limited by extreme environmental conditions. SiNPs have been recognized as a regulator for plant protection against abiotic and biotic stresses. SiNPs form a dual layer on the epidermal cell wall, reducing the harmful impact of stress. There are two major processes related to biotic and abiotic stress resistance: (i) a physico-mechanical mechanism produced by SiNPs and (ii) a biochemical mechanism controlling metabolic activity. SiNPs can provide resistance to plants, viz., SiNPs form a dual layer on the epidermal cell wall and help to reduce the deleterious impact of stress. Exogenous application of SiO_2-NPs can alleviate abiotic stress damage by increasing osmotic regulatory substances, reducing electrolyte leakage, malondialdehyde content, scavenging toxic effects of ROS, and decreasing excessive uptake of Na^+. This leads to increased leaf water content, stomatal conductance, chlorophyll content, PSII activity and photosynthetic enzyme activity, ultimately improving photosynthesis and crop growth (Wang et al. 2022). The interaction between plants and NPs is determined by parameters such as dimension, morphology, mode of treatment and physicochemical characteristics (Rastogi et al. 2017). Recent research suggests that SiNPs can directly react with flora and manipulate their morphology and physiology, enhancing plant growth and yield (Rastogi et al. 2019).

4.4.5 Radiation Stress

Studies show that treating rice seedlings with SiNPs can alleviate toxic effects like UV radiation, reactive anions, pesticides and microplastics (Okeke et al. 2023). The growth rate of rice seedlings is reduced when irradiated with gamma rays; SiNPs enhance UV resistance due to their defensive deposition on the plant cell wall. Similar results were observed in wheat seedlings, modulating antioxidant defense systems and decreasing harmful UV ray impacts like low fresh weight and chlorophyll decrease (Tripathi et al. 2015, 2017).

4.4.6 Heavy Metal Stress

Heavy metals, such as chromium, cadmium, nickel, arsenic and lead, can cause plant growth retardation and decreased productivity. The toxic concentrations of heavy metals trigger physiological and metabolic changes, with visual symptoms, varying depending on the action site of the metal. Common responses include reduced growth, loss of turgor, reduced photosynthetic rate and leaf chlorosis/necrosis (Hall, 2002). These effects are influenced by ultrastructural, biochemical and molecular changes caused by the presence of metals. For instance, heavy metals can damage chloroplast ultrastructure, inhibit enzymes involved in C assimilation and replace Mg in chlorophyll molecules (Hossain et al. 2012).

Plant tolerance to metal toxicity is regulated by physiological and molecular mechanisms, preventing toxic concentration buildup in cells, especially in shoots. The two main strategies are avoidance, where plants restrict metal uptake/translocation and tolerance, where plants survive high metal concentrations through intercellular sequestration and vacuolar compartmentation (Viehweger 2014). Avoidance in Al tolerance involves precipitation or complexation of metals in the root environment before uptake (Barcelo and Poschenrieder 2002). This can involve increased rhizosphere pH or ion excretion, while root exudation of organic acid anions and polyphenols can cause metal complexation, making them less available to the plant. Additionally, the plant may accumulate metals in the apoplast instead of intracellular localization (Hossain et al. 2012). Metal tolerance in plants involves transport processes at the plasma membrane, reducing metal uptake or stimulating efflux pumping (Manara 2012). Plants can regulate metal distribution by restricting translocation to shoots. Metals can be sequestered in the cytosol by high-affinity ligands like amino acids, organic acids, phytochelatins and metallothioneins. Another strategy is sequestration in vacuoles, which involves active transport across the tonoplast (Hall 2002). Stress protein production and activation of antioxidant defense systems (both enzymatic and nonenzymatic systems) are also important aspects of metal tolerance (Manara 2012).

SiNPs have been found to alleviate metal toxicity in plants by reducing total metal uptake and translocation to shoots (Wu et al. 2015). An alleviative function of SiNPs on metal toxicity such as zinc (Zn), cadmium (Cd), iron (Fe), aluminium (Al) and manganese (Mn) has been studied. Different mechanisms have been studied, with Si-containing minerals like sodium silicates and alkaline silica affecting rhizospheric pH, reducing heavy metal concentration in the soil. Si accumulation in the endodermis also improves manganese accumulation in the leaf and enhances antioxidant enzyme control in heavy metal-treated plants. These effects are influenced by the plant variety (Adrees et al. 2015).

SiO_2 NPs stabilize chloroplasts, maintain photosystem integrity and increase pigment content, reducing toxic aluminium concentration. The defensive effect of silica on aluminium toxicity varies with plant species. In pea plants treated with Cr (chromium), SiNPs provided defense by removing oxidative stress; the activities of enzymes such as SOD (superoxide dismutase) and APXs (ascorbate peroxidase) were improved extensively due to the presence of SiNPs, whereas CAT (catalase), GSR (Glutathione reductase), and DHAR (dehydroascorbate reductase) were less repressed by Cr in the presence of SiNPs (Tripathi et al. 2015). Chen et al. found that rice plants bioaccumulate heavy metals due to exposure to Cd, leading to oxidative stress and poor productivity. However, treatment with SiNPs at an optimum concentration of 25 mM significantly decreased Cd accumulation in mature rice plants. Thind et al. (2021) reported that wheat plants treated with SiNPs showed increased growth rates, biomass, photosynthetic pigment levels and upregulation of non-enzymatic antioxidant and antioxidant enzyme production. El-Saadony et al. (2021) and Nweze et al. (2022) also reported similar findings regarding improved nutrient uptake in plants treated with SiNPs.

4.4.7 Salt Stress

Soil salinization is a global issue affecting resources and ecological environment. By 2050, 50% of agricultural land is expected to be covered by saline soil due to climate change and irrigation water use (Chhipa 2017). Salinity causes reduced plant growth due to osmotic stress and ion toxicity (particularly those of Na^+ and Cl^-). The high concentration of salt results in osmotic stress, limiting water availability to plants. This leads to reduced cell expansion in root tips and young leaves, as well as stomatal closure (Eckhard et al. 2012). High concentrations of accumulated Na^+ and Cl^- can be phytotoxic, causing necrosis in older leaves, decreased leaf area, reduced photosynthetic capacity, and decreased growth and yields. Na^+ uptake leads to nutrient imbalances, particularly reduced uptake and accumulation of K^+ and Ca^{2+} (Munns and Tester 2008). Salt stress damages plants through osmotic stress, oxidative stress, nutritional disorders, reduced enzyme activity, altered photosynthetic and metabolic processes and genotoxicity (Rastogi et al. 2019). Over-accumulation of reactive oxygen species negatively impacts cell membrane integrity, ion transport, hormone balance, respiration, oxidation/reduction, DNA/protein synthesis, and ultimately affects crop growth, development and yield (Liang et al. 2023; Zhao et al. 2021).

SiNPs help alleviate NaCl stress by preventing NaCl uptake and reducing the uptake of ions from the soil. In maize, SiNP treatment reduces ion transport (Rangaraj et al. 2014). A study found that *Pisum sativum* grown in high salt concentration experienced reduced growth, yield and water retention, along with distorted antioxidant enzyme levels, and increased Na^+ ion concentration in the shoots and leaves (Ismail et al. 2022). SiNP treatment reversed toxic effects in soybean, enhancing vegetative growth, fresh and dry weight, and yield (Ismail et al. 2022). High NaCl concentration (10 dS/m) caused oxidative stress in soybean, leading to increased expression of antioxidant enzymes and lipid peroxidation (Farhangi-Abriz and Torabian 2018). Applying SiNPs to stressed plants restores redox balance, alleviates lipid peroxidation and removes high Na^+ ion concentration. It also improves the expression of antioxidant molecules and enzymes (Farhangi-Abriz and Torabian 2018). Similar findings have been reported by Karimian et al. (2021).

4.4.8 Drought Stress

The addition of Si can enhance drought tolerance in various crops. Water stress is a major abiotic stress encountered by plants, and crop yields are often restricted by a water shortage. Water limitation affects plant processes from cellular to whole-plant levels, resulting in reduced photosynthesis, senescence, nutrient uptake, growth and yield. Plants close stomata to prevent transpirational water loss, reducing CO_2 flux and photosynthetic rate. Reductions in enzyme activities in C assimilation and ATP synthesis also contribute to reduced photosynthesis. Water stress typically leads to the accumulation of stress proteins, osmotic solutes and reactive oxygen species in plant tissues (Farooq et al. 2009). Plants respond to drought stress in a multitude of ways and the effectiveness of such responses determines their drought tolerance (Bartels and Sunkar 2005).

SiNPs in plant tissues can reduce water stress by decreasing transpiration and improving light interception by keeping the leaf blade erect. They form a thick silicated layer on leaf surfaces, reducing cuticular transpiration. After treatment, SiNPs regulate gas transfer and water losses, improving survival chances for plants in low water or drought conditions (Alsaeedi et al. 2019). The application also reduces oxidative tension in plant tissues, leading to significant improvements in vegetative growth and plant yield, as reported in banana (Mahmoud et al. 2020; Malik et al. 2021). SiNPs induce resistance in plants against drought conditions by reducing stomatal conductance and also modifying the cell wall properties. SiNP enclosing absorbents can slowly discharge nutrients and store large amounts of water, helping the flora resist drought and saline conditions without harming the environment (Rajput et al. 2021). Therefore, Si application could improve water status under drought conditions.

4.4.9 SiNPs for Plant Disease Management

Surface-modified hydrophobic SiNPs enhance plant tolerance to various pathogens, including mammalian, arthropod, molluscan herbivores, fungal, bacterial pathogens, viruses, and nematodes, thereby protecting crops. A list of a range of foliar diseases, soil-borne diseases, seed-borne fungal pathogens and bacterial diseases that have been controlled by SiNPs application has been complied by Haynes (2017) and Goswami et al. (2022). Plants use constitutive and inducible defense mechanisms to combat pathogens. Constitutive defenses involve physical barriers like cuticle layers and cell wall, as well as antimicrobial-effecting organic molecules. Inducible defense strategies are activated once a pathogen penetrates the cell wall. Pathogen-derived effector proteins trigger a signal transduction cascade, while biotrophic pathogens activate a hormone pathway based on salicylic acid, jasmonic acid and ethylene, activating defense genes and producing defense chemicals (Haynes 2017).

The physical defense induced by Si deposition in plant parts, primarily phytoliths, was the first theory for studying stress tolerance to pests. The deposition of phytoliths enhances plant immunity and physical resilience and serves as a barrier against pathogen penetration. Si also increases the stiffness and mechanical strength of pectin and hemicellulose in the cell wall (Tripathi et al. 2021). Silicon deposition can damage insect-feeding mouthparts, decrease plant digestibility and harm herbivores, especially in highly attacked plants, and may induce more plant tissue silicification. SiNPs interact with plant cell wall physically to create a second layer of cuticle that prevents pathogen penetration and lowers the frequency of diseases in plants, including *Pyricularia oryzae* Cavara, *Bipolaris sorokiniana* Shoemaker, *Pyricularia grisea* Sacc. and *Rhizocotnia solani* Kühn, among others (Asgari-Targhi et al. 2018). SiNPs could decrease the bacterial load by eightfold. It was also reported that Si/SiO_2 NPs do not affect bacterial growth directly; rather they enrich plant resistance against invaders (Kandhol et al. 2022; Rastogi et al. 2019).

SiNPs enhance plant–pathogen interactions by activating defense-related enzymes, promoting antimicrobial compound synthesis, controlling signaling pathways, and triggering defense-related gene expression. They improve plant biochemical resistance by activating defensive enzymes and antimicrobial molecules like phenolics, flavonoids and pathogenesis-related proteins. The defensive enzymes responsible for the synthesis of biologically active antimicrobial compounds include polyphenol oxidase, β-1,3 glucanase, peroxidase and phenylalanine ammonia-lyase (PAL), superoxide dismutase, ascorbate peroxidase, glutathione reductase, catalase, lipoxygenase, and chitin (Waewthongrak et al. 2015). SiNPs regulate signalling pathways for salicylic acid, jasmonic acid and ethylene, thereby making nano biopesticides more environment friendly than conventional pesticides due to their less toxic effects (El-Shetehy et al. 2021).

4.4.10 SiNPs as Fungicide

Si formulation can address fungal infection issues by reducing spore degradation, potentially leading to new applications in plant–pathogen interaction and management. SiNPs exhibit antifungal activity, reducing plant diseases caused by fungal pathogens. They have been reported to reduce sheath blight, brown spots and grain discolouration in rice and inhibit early blight of tomatoes caused by *Alternaria solani* and black root rot in tomato (Debona et al. 2017). The application of SiNPs effectively suppressed rice blast disease, powdery mildew, cucumber leaf spot and *Fusarium* wilt. The volatile compounds encapsulated with SiNP showed long-term activity against *Aspergillus niger*, compared to pure compounds. The effects of these compounds were compared with pure compounds (Janatova et al. 2015). Mesosporous SiNPs effectively control rice blast and brown spot fungal diseases, with mesoporous Si serving as a podium for antifungal compounds. Mesoporous Si reduces blast disease severity and enhances antifungal activity against *Candida albicans* due to its high porosity (Tang and Cheng 2013). It is a useful and nontoxic substitute to fungicides to control early blight of tomato caused by *A. solani* (Derbalah et al. 2018).

4.4.11 SiNPs as an Insecticide

Si significantly impacts plant resistance to various insect-feeding guilds, including folivores, borers and phloem or xylem feeders. A list of insect herbivores, other arthropod pests (e.g., mites) and nematodes causing damage to particular crops has been reported (Reynolds et al. 2016). Plants have both constitutive and inducible defense strategies against insect pests. Constitutive defenses include physical barriers like cuticles, thorns and trichomes, as well as insecticidal metabolites. Inducible defenses are triggered by the detection/recognition of elicitor molecules in insect oral secretions and compounds in oviposition fluids, which are transduced through pathways like Ca^{2+} influxes, kinase cascades, reactive oxygen species and defense hormone pathways (War et al. 2012).

SiNPs can be used as field pesticides or nanocarriers to release commercial pesticides. SiO_2 nanoparticles are lethal to *Callosobruchus maculates* (Rouhani et al. 2013). Si increases plant resistance to insect attack by forming a silica barrier in the epidermis of plant tissue, causing mechanical resistance to insect feeding and plant penetration, and reduced digestibility/palatability due to increased hardness and abrasiveness (Massey and Hartley 2009). SiNPs can cause pest death through constriction of the spiracle and trachea, adsorption, and damage to the cuticular wax layer. Physical adsorption breaches the protective lipid hydrophilic membrane, leading to the death of the targeted organisms. Desiccation causes the absorption and destruction of SiNPs (Ayoub et al. 2017; Elnahal et al. 2022). These NPs might also damage the digestive system of pest herbivores that consume the SiNP-treated plants (Thabet et al. 2021).

4.4.12 SiNPs as Herbicides

Market-based herbicides inhibit weed growth but not underground plant parts like rhizomes and tubers, causing crop growth interference and yield loss to the tune of 10% of crop production. A nano-herbicide formulation is available to combat weed plants, as Si accumulation increases plant tissue, reducing palatability and digestibility for herbivores. Size of NP allows for better penetration through the cuticle, allowing active ingredients to be released onto the target site, reducing the impact on crop production, specifically in soybean (against the cotton bollworm *Helicoverpa armigera* by attracting the predatory beetle *Dicranolaius bellulus*) and grapevine (against *Helicoverpa punctigera* by attracting the predatory beetle *Dicranolaius bellulus*) (against the light brown apple moth *Epiphyas postvittana* by attracting *D. bellulus*) (Johnson et al. 2020; Kvedaras et al. 2010).

4.4.13 SiNPs as Fertilizers

Fertilizers are essential for plant growth and development. Excessive use of fertilizers, particularly those containing silicon (Si), can lead to environmental issues such as water pollution and soil degradation, while deficiency can result in reduced crop yield and increased susceptibility to biotic and abiotic stresses (Zargar et al. 2019). SiNPs are a promising fertilizer delivery system due to their small size, high surface area and ability to penetrate plant cell (Goswami et al. 2022). The development of SiNP delivery systems for fertilizers has been achieved through various methods including coating, encapsulation and functionalization. For example, Wanyika et al. (2012) developed a mesoporous SiNP-based fertilizer delivery system by coating urea with SiNPs. The study found that urea coated with SiNP had a slower release rate and lasted five times longer than uncoated urea, suggesting the potential of SiNP delivery systems for improving fertilizer efficiency. Janmohammadi et al. (2016) found that combining organic fertilizers with SiNP foliar spray significantly improved safflower production, indicating its potential as an effective agricultural strategy. SiNPs enhance fertilizer delivery efficiency, leading to increased crop yield and reduced environmental impact.

4.5 Toxic Effects and Safety Concerns

The use of SiNPs in agriculture has gained attention, but their toxicity to plants remains controversial. The interaction between plant cells and nanomaterials affects gene expression and growth. Silica, abundant in human diets and commercial products, is degraded in the biosphere. Therefore, SiNPs pose a low risk to the environment (Dekkers et al. 2011). SiNP applications to plants are considered safe because they are naturally and abundantly found in soil and plant parts such as the leaf epidermis and root endodermis. Besides, they show excellent biocompatibility, less toxicity and are 'Generally Recognized As Safe' (GRAS). However, the potential toxicity of SiNPs in plants depends on the conditions of application, as well as their physical and chemical properties (Croissant et al. 2017).

The size of nanoparticles significantly influences their toxicity, making them relatively non-toxic. However, studies have shown that the application of SiNPs in agroecosystems can have adverse effects, influenced by their physicochemical properties like shape, charge, size, dispersity and particle size. Nano-sized particles tend to be more toxic than their counterparts with the same chemical composition (Fadeel and Garcia-Bennett 2010). Studies have reported risks associated with silicon-based nanoparticles, with the size of NPs potentially influencing their toxicity due to their large surface area and easy penetration into cells. The toxic effect of SiNPs is likely related to the concentration applied, with different plants having varying degrees of tolerance. Slomberg and Schoenfisch (2012) found that Arabidopsis growth was not affected at 1000 ppm, while Lee et al. (2010) found that high concentrations may be toxic to root growth. A similar result was found in sugar beet, where 2 mM SiNPs increased MDA content under drought stress (Lee et al. 2010).

SiNPs have been found to reduce carbon and nitrogen content in soil microbial biomass, alter bacterial organization, and decrease bacteria that can use organic nitrogen (Simonin and Richaume 2015). They also decrease amylolytic bacteria, oligotrophs, and cellulolytic and microscopic fungi. Additionally, soil enzymes like peroxidase, polyphenol oxidase and urease have been observed to decrease with the application of SiNPs, according to Lebedev et al. (2019). Furthermore, soil enzymes have been recorded to decrease on the application of SiNPs. Soil enzyme activities such as peroxidase, polyphenol oxidase and urease were decreased in the findings by Lebedev et al. (2019). Studies show that SiNPs negatively affect earthworm protein metabolism, a major pathway for other pollutants to exacerbate their toxicity, as reported in studies on megafauna (Lebedev et al. 2019; Zhang et al. 2017). Di Marzio et al. (2018) also noted some cytotoxicity and genotoxicity in earthworms. The impacts of these changes could potentially cause the death of microfauna and megafauna, leading to a decline in soil organism populations. This could result in reduced degradation of organic matter, decreased nutrient turnover, deformation of soil structure, modification of hydraulics and temperature conductivity (Maggi and Tang 2021).

Nanoparticles can cause phytotoxicity in plants, as they adhere to plant roots, affecting the transportation of water and minerals within the plant (Bhat et al. 2021). Silicon-based nanoparticles, applied at 100 mg/ml concentrations caused DNA damage in *Allium cepa* root meristem and decreased *Bt*-cotton plant growth and biomass (Bhat et al. 2021). Silicon-based nanoparticles below 20 nm inhibited rice seedling germination, while those above 20 nm showed positive effect for other growth parameters (Asgari-Targhi et al. 2018). The presence of silicon-based nanoparticles in soil increases the stability of more hazardous nanoparticle suspensions (TiO_2, CuO and ZnO) in soil (de Oliveira et al. 2017). The toxicity of SiNPs depends on their concentration, causing complications at different concentrations. To avoid these issues, it is crucial to understand the role of SiNPs in plant growth, considering factors like size, shape, charge, stability, surface area and concentration. Currently, few studies have documented the relationship between plant health and size, dosage and accumulation of SiNPs in plants. Thus, the toxicity of SiNPs in plants needs further study.

4.6 Conclusions and Future Directions

SiNPs can be synthesized using physical, chemical and green synthesis methods. Green synthesis methods use natural resources like plants, minerals, microorganisms and waste products from agriculture and industries to prepare various shapes of SiO_2 NPs. Green synthesis eliminates the use of unsafe chemicals and reduces environmental risks. All the examples of green synthesis of SiNPs discussed above show that variously shaped SiO_2 NPs can be prepared using simple biological and eco-friendly methods using green resources. These methods do not produce hazardous by-products and provide better particle shape, size and morphology control. Green synthesis technology has gained attention due to its economic prospects. While green synthesis from plant residues is easier than microorganisms, microorganisms are more cost-effective. Microorganism-assisted synthesis yields high-purity nanoparticles without complex purification steps. This eliminates hazardous organic solvents used for purification, making it a more environment friendly alternative to chemical methods.

The green synthesis of SiNPs has made significant progress, but several issues remain to be addressed. The demand for SiNPs is expected to increase, necessitating large-scale production. Large-scale production is primarily done through chemical routes due to high yield and industrial feasibility. However, there is still room for improvement in green synthesis methods for large-scale production. Innovations and further research are needed to upgrade current green methods and processes. The changing attitude of international community towards sustainable development, improving environmental conditions, and minimizing harmful artificial waste presents a promising future for green synthesis of SiNPs and their application in various fields, including agriculture. Given the various challenges posed by population growth and global climate change, using SiNPs in agriculture significantly helps overcome the damage caused by the excessive use of pesticides and chemical fertilizers for increasing crop production. SiNPs can potentially improve growth and development, especially under adverse environmental stresses. Studies have shown that SiNPs are more efficient than bulk-Si as a fertilizer for plants, and they show similar performance for enhancing plant growth under stress conditions in both accumulators and low/non-accumulators. However, the mechanism and other aspects related to the positive effect of SiNPs on plant growth must be understood. The wide surface area of SiNPs makes them ideal carriers for pesticides and fertilizers, leading to concepts like nanopesticides or nanofertilizers. They facilitate site-targeted, controlled delivery of nucleic acids and nutrients, providing increased crop protection. However, the promise of significant use of nanotechnology can only be achieved if the ecotoxicity of these nanomaterials is fully assessed and adequately managed. Researchers are encouraged to adopt a multidisciplinary approach to fill gaps related to SiNPs-mediated improved stress tolerance, which could lead to sustainable agriculture.

References

Adebisi, J.A., Agunsoye, J.O., Bello, S.A. et al. (2018). Extraction of silica from cassava periderm using modified sol-gel method. *Nigerian Journal of Technological Development* 15: 57–65. https://doi.org/10.4314/njtd.v15i2.4.

Adinarayana, T.V.S., Mishra, A., Singhal, I., and Reddy, D.R.K. (2020). Facile green synthesis of silicon nanoparticles from *Equisetum arvense* for fluorescence based detection of Fe (III) ions. *Nanoscale Advances* 2: 4125–4132.

Adrees, M., Ali, S., Rizwan, M. et al. (2015). The effect of excess copper on growth and physiology of important food crops: a review. *Environmental Science and Pollution Research* 22: 8148–8162.

Agunsoye, J.O., Adebisi, J.A., Bello, S.A., et al., 2018. Synthesis of silicon nanoparticles from cassava periderm by reduction method. *Materials Science and Technology* Conference, Columbus, Ohio, USA.

Alsaeedi, A., El-Ramady, H., Alshaal, T. et al. (2019). Silica nanoparticles boost growth and productivity of cucumber under water deficit and salinity stresses by balancing nutrients uptake. *Plant Physiology and Biochemistry* 139: 1–10.

Alsaiari, N.S., Alzahrani, F.M., Amari, A. et al. (2023). Plant and microbial approaches as green methods for the synthesis of nanomaterials: synthesis, applications, and future perspectives. *Molecules* 28: 463.

Arnon, D.I. and Stout, P.R. (1939). The essentiality of certain elements in minute quantity for plants with special reference to copper. *Plant Physiology* 14: 371.

Asgari-Targhi, G., Iranbakhsh, A., and Ardebili, Z.O. (2018). Potential benefits and phytotoxicity of bulk and nano-chitosan on the growth, morphogenesis, physiology, and micropropagation of *Capsicum annuum*. *Plant Physiology and Biochemistry* 127: 393–402.

Athinarayanan, J., Jaafari, S.A.A.H., Periasamy, V.S. et al. (2020). Fabrication of biogenic silica nanostructures from *Sorghum bicolor* leaves for food industry applications. *Silicon* 12: 2829–2836.

Ayoub, H.A., Khairy, M., Rashwan, F.A., and Abdel-Hafez, H.F. (2017). Synthesis and characterization of silica nanostructures for cotton leaf worm control. *Journal of Nanostructure in Chemistry* 7: 91–100.

Babu, K., Maurya, N.K., Mandal, A., and Saxena, V.K. (2015). Synthesis and characterization of sodium methyl ester sulfonate for chemically-enhanced oil recovery. *Brazilian Journal of Chemical Engineering* 32: 795–803.

Babu, R.H., Yugandhar, P., and Savithramma, N. (2018). Synthesis, characterization and antimicrobial studies of bio silica nanoparticles prepared from *Cynodon dactylon* L.: a green approach. *Bulletin of Materials Science* 41: 1–8.

Balamurugan, M. and Saravanan, S. (2012). Producing nanosilica from *Sorghum vulgare* seed heads. *Powder Technology* 224: 345–350.

Banerjee, A., Singh, A., Sudarshan, M., and Roychoudhury, A. (2021). Silicon nanoparticle-pulsing mitigates fluoride stress in rice by fine tuning the ionomic and metabolomic balance and refining agronomic traits. *Chemosphere* 262: 127826.

Bansal, V., Sanyal, A., Rautaray, D. et al. (2005a). Bioleaching of sand by the fungus *Fusarium oxysporum* as a means of producing extracellular silica nanoparticles. *Advanced Materials* 17: 889–892.

Bansal, V., Rautaray, D., Bharde, A. et al. (2005b). Fungus-mediated biosynthesis of silica and titania particles. *Journal of Materials Chemistry* 15 (26): 2583–2589.

Bansal, V., Syed, A., Bhargava, S.K. et al. (2007). Zirconia enrichment in zircon sand by selective fungus-mediated bioleaching of silica. *Langmuir* 23: 4993–4998.

Barcelo, J. and Poschenrieder, C. (2002). Fast root growth responses, root exudates, and internal detoxification as clues to the mechanisms of aluminium toxicity and resistance: a review. *Environmental and Experimental Botany* 48: 75–92.

Bartels, D. and Sunkar, R. (2005). Drought and salt tolerance in plants. *Critical Reviews in Plant Sciences* 24: 23–58.

Bhat, U.H., Sami, F., Siddiqui, H. et al. (2021). Nitric oxide alleviates zinc oxide nanoparticles-induced phytotoxicity in *Brassica juncea*. *Russian Journal of Plant Physiology* 68: 559–568.

Bhat, J.A., Shivaraj, S.M., Singh, P. et al. (2019). Role of silicon in mitigation of heavy metal stresses in crop plants. *Plants* 8: 71.

Bhattacharya, D. and Gupta, R.K. (2005). Nanotechnology and potential of microorganisms. *Critical Reviews in Biotechnology* 25 (4): 199–204.

Blackwell, M. (2011). The Fungi: 1, 2, 3 … 5.1 million species. *American Journal of Botany* 98: 426–438.

Bose, S., Ke, D., Sahasrabudhe, H., and Bandyopadhyay, A. (2018). Additive manufacturing of biomaterials. *Progress in Materials Science* 93: 45–111.

Caliskan, G., Mutaf, T., Agba, H.C., and Elibol, M. (2022). Green synthesis and characterization of titanium nanoparticles using microalga, *Phaeodactylum tricornutum*. *Geomicrobiology Journal* 39: 83–96. https://doi.org/10.1080/01490451.2021.2008549.

Chaudhary, R., Nawaz, K., Khan, A.K. et al. (2020). An overview of the algae-mediated biosynthesis of nanoparticles and their biomedical applications. *Biomolecules* 10: 1498.

Chen, H., Wang, W., Martin, J.C. et al. (2013). Extraction of lignocellulose and synthesis of porous silica nanoparticles from rice husks: a comprehensive utilization of rice husk biomass. *ACS Sustainable Chemistry and Engineering* 1: 254–259. https://doi.org/10.1021/sc300115r.

Chhipa, H. (2017). Nanofertilizers and nanopesticides for agriculture. *Environmental Chemistry Letters* 15: 15–22.

Croissant, J.G., Fatieiev, Y., and Khashab, N.M. (2017). Degradability and clearance of silicon, organosilica, silsesquioxane, silica mixed oxide, and mesoporous silica nanoparticles. *Advanced Materials* 29: 1604634.

Debona, D., Rodrigues, F.A., and Datnoff, L.E. (2017). Silicon's role in abiotic and biotic plant stresses. *Annual Review of Phytopathology* 55: 85–107.

Dekkers, S., Krystek, P., Peters, R.J. et al. (2011). Presence and risks of nanosilica in food products. *Nanotoxicology* 5: 393–405.

Derbalah, A., Shenashen, M., Hamza, A. et al. (2018). Antifungal activity of fabricated mesoporous silica nanoparticles against early blight of tomato. *Egyptian Journal of Basic and Applied Sciences* 5: 145–150.

Deshmukh, R. and Bélanger, R.R. (2016). Molecular evolution of aquaporins and silicon influx in plants. *Functional Ecology* 30: 1277–1285.

Dhakate, P., Kandhol, N., Raturi, G. et al. (2022). Silicon nanoforms in crop improvement and stress management. *Chemosphere* 305: 135165.

Di Marzio, W.D., Curieses, S.P., Alberdi, J.L., and Sáenz, M.E. (2018). Cyto and genotoxicity of positive and negative coated Silica nano particles on Celomocytes of earthworms Eisenia fetida (Oligocheta, Annelida). *Journal of Advanced in Environmental Studies* 2 (1): 74–81.

Durairaj, K., Senthilkumar, P., Velmurugan, P. et al. (2019). Sol-gel mediated synthesis of silica nanoparticle from *Bambusa vulgaris* leaves and its environmental applications: kinetics and isotherms studies. *Journal of Sol-Gel Science and Technology* 90: 653–664.

Eckhard, G., Horst, W.J., and Neuman, E. (2012). Adaptation of plants to adverse chemical soil conditions. In: *Marschner's Mineral Nutrition of Higher Plants* (ed. P. Marschner), 409–472. Amsterdam: Elsevier.

Elnahal, A.S., El-Saadony, M.T., Saad, A.M. et al. (2022). The use of microbial inoculants for biological control, plant growth promotion, and sustainable agriculture: a review. *European Journal of Plant Pathology* 162: 759–792.

El-Saadony, M.T., Saad, A.M., Taha, T.F. et al. (2021). Selenium nanoparticles from *Lactobacillus paracasei* HM1 capable of antagonizing animal pathogenic fungi as a new source from human breast milk. *Saudi Journal of Biological Sciences* 28: 6782–6794.

El-Shetehy, M., Moradi, A., Maceroni, M. et al. (2021). Silica nanoparticles enhance disease resistance in Arabidopsis plants. *Nature Nanotechnology* 16: 344–353.

Fadeel, B. and Garcia-Bennett, A.E. (2010). Better safe than sorry: understanding the toxicological properties of inorganic nanoparticles manufactured for biomedical applications. *Advanced Drug Delivery Reviews* 62: 362–374.

Falk, G., Shinhe, G.P., Teixeira, L.B. et al. (2019). Synthesis of ilica nanoparticles from sugarcane bagasse ash and nano-silicon via magnesiothermic reactions. *Ceramics International* 45: 21618–21624. https://doi.org/10.1016/j.ceramint.2019.07.157.

Farhangi-Abriz, S. and Torabian, S. (2018). Nano-silicon alters antioxidant activities of soybean seedlings under salt toxicity. *Protoplasma* 255: 953–962.

Farooq, M., Wahid, A., Kobayashi, N.S.M.A. et al. (2009). Plant drought stress: effects, mechanisms and management. In: *Sustainable Agriculture* (ed. E. Lichtfouse et al.), 153–188. Dordrecht: Springer.

Ghorbani, F., Younesi, H., Mehraban, Z. et al. (2013). Preparation and characterization of highly pure silica from sedge as agricultural waste and its utilization in the synthesis of mesoporous silica MCM-41. *Journal of the Taiwan Institute of Chemical Engineers* 44: 821–828. https://doi.org/10.1016/j.jtice.2013.01.019.

Goswami, P., Mathur, J., and Srivastava, N. (2022). Silica nanoparticles as novel sustainable approach for plant growth and crop protection. *Heliyon* 8: e09908.

Guerriero, G., Hausman, J.F., Strauss, J. et al. (2016). Lignocellulosic biomass: biosynthesis, degradation, and industrial utilization. *Engineering in Life Sciences* 16: 1–16.

Hachicha, R., Elleuch, F., Ben Hlima, H. et al. (2022). Biomolecules from microalgae and cyanobacteria: applications and market survey. *Applied Sciences* 12: 1924.

Hall, J.Á. (2002). Cellular mechanisms for heavy metal detoxification and tolerance. *Journal of Experimental Botany* 53: 1–11.

Haynes, R.J. (2017). Significance and role of Si in crop production. *Advances in Agronomy* 146: 83–166.

Hodson, M.J., White, P.J., Mead, A., and Broadley, M.R. (2005). Phylogenetic variation in the silicon composition of plants. *Annals of Botany* 96: 1027–1046.

Hossain, M.A., Piyatida, P., da Silva, J.A.T., and Fujita, M. (2012). Molecular mechanism of heavy metal toxicity and tolerance in plants: central role of glutathione in detoxification of reactive oxygen species and methylglyoxal and in heavy metal chelation. *Journal of Botany* .

Hussain, A., Rizwan, M., Ali, Q., and Ali, S. (2019). Seed priming with silicon nanoparticles improved the biomass and yield while reduced the oxidative stress and cadmium concentration in wheat grains. *Environmental Science and Pollution Research* 26: 7579–7588.

Islam, A.M., Nasir, M., Akter Mou, M. et al. (2021). Preliminary reports on comparative weed competitiveness of Bangladeshi monsoon and winter rice varieties under puddled transplanted conditions. *Sustainability* 13: 5091.

Ismail, L.M., Soliman, M.I., Abd El-Aziz, M.H., and Abdel-Aziz, H.M. (2022). Impact of silica ions and nano silica on growth and productivity of pea plants under salinity stress. *Plants* 11: 494.

Jabeen, N., Maqbool, Q., Sajjad, S. et al. (2017). Biosynthesis and characterisation of nano-silica as potential system for carrying streptomycin at nano-scale drug delivery. *IET Nanobiotechnology* 11: 557–561.

Janatova, A., Bernardos, A., Smid, J. et al. (2015). Long-term antifungal activity of volatile essential oil components released from mesoporous silica materials. *Industrial Crops and Products* 67: 216–220.

Janmohammadi, M., Amanzadeh, T., Sabaghnia, N., and Dashti, S. (2016). Impact of foliar application of nano micronutrient fertilizers and titanium dioxide nanoparticles on the growth and yield components of barley under supplemental irrigation. *Acta Agriculturae Slovenica* 107: 265–276.

Jisha, K.C., Vijayakumari, K., and Puthur, J.T. (2013). Seed priming for abiotic stress tolerance: an overview. *Acta Physiologiae Plantarum* 35: 1381–1396.

Johnson, S.N., Rowe, R.C., and Hall, C.R. (2020). Silicon is an inducible and effective herbivore defence against *Helicoverpa punctigera* (Lepidoptera: Noctuidae) in soybean. *Bulletin of Entomological Research* 110 (3): 417–422.

Kandhol, N., Jain, M., and Tripathi, D.K. (2022). Nanoparticles as potential hallmarks of drought stress tolerance in plants. *Physiologia Plantarum* 174: e13665.

Karimian, S., Shekoohiyan, S., and Moussavi, G. (2021). Health and ecological risk assessment and simulation of heavy metal-contaminated soil of Tehran landfill. *RSC Advances* 11: 8080–8095.

Kaur, D., Pandey, O.P., and Reddy, M.S. (2023). Production of silica nanoparticles from rice husk and rice straw by fungal-mediated biotransformation using *Pleurotus florida*. *Biomass Conversion and Biorefinery* 15: 803–812.

Khanna, P., Kaur, A., and Goyal, D. (2019). Algae-based metallic nanoparticles: synthesis, characterization and applications. *Journal of Microbiological Methods* 163: 105656. https://doi.org/10.1016/j.mimet.2019.105656.

Kumar, V., Tiwari, P., Krishnia, L. et al. (2016). Green route synthesis of silicon/silicon oxide from bamboo. *Advanced Materials Letters* 7: 271–276. https://doi.org/10.5185/amlett.2016.6151.

Kvedaras, O.L., An, M., Choi, Y.S., and Gurr, G.M. (2010). Silicon enhances natural enemy attraction and biological control through induced plant defences. *Bulletin of Entomological Research* 100 (3): 367–371.

Lebedev, S.V., Gavrish, I.A., Galaktionova, L.V. et al. (2019). Assessment of the toxicity of silicon nanooxide in relation to various components of the agroecosystem under the conditions of the model experiment. *Environmental Geochemistry and Health* 41: 769–782.

Lee, S.K., Sohn, E.Y., Hamayun, M. et al. (2010). Effect of silicon on growth and salinity stress of soybean plant grown under hydroponic system. *Agroforestry Systems* 80: 333–340.

Liang, Y., Liu, H., Fu, Y. et al. (2023). Regulatory effects of silicon nanoparticles on the growth and photosynthesis of cotton seedlings under salt and low-temperature dual stress. *BMC Plant Biology* 23: 504.

Liang, S., Wang, K., Chen, D. et al. (2008). Shear enhanced interfacial interaction between carbon nanotubes and polyethylene and formation of nanohybrid shish–kebabs. *Polymer* 49: 4925–4929.

Liberman, A., Mendez, N., Trogler, W.C., and Kummel, A.C. (2014). Synthesis and surface functionalization of silica nanoparticles for nanomedicine. *Surface Science Reports* 69: 132–158.

Liu, R. and Lal, R. (2015). Potentials of engineered nanoparticles as fertilizers for increasing agronomic productions. *Science of the Total Environment* 514: 131–139.

Liu, X., Liu, C., Pan, G., and Clarke, N. (2023). Biochar-based technology in food production, climate change mitigation, and sustainable agricultural soil management: Post Terra Preta Era. In: *Innovation for Environmentally-friendly Food Production and Food Safety in China*, 93–112. Singapore: Springer Nature Singapore.

Liu, C., Wei, L., Zhang, S. et al. (2014). Effects of nanoscale silica sol foliar application on arsenic uptake, distribution and oxidative damage defense in rice (Oryza sativa L.) under arsenic stress. *RSC Advances* 4 (100): 57227–57234.

Ma, M. and Hill, R.M. (2006). Superhydrophobic surfaces. *Current Opinion in Colloid and Interface Science* 11: 193–202.

Ma, J.F., Miyake, Y., and Takahashi, E. (2001). Silicon as a beneficial element for crop plants. *Studies in Plant Science* 8: 17–39.

Ma, J.F. and Takahashi, E. (2002). *Soil, Fertilizer, and Plant Silicon Research in Japan*. Elsevier.

Mahmoud, L.M., Dutt, M., Shalan, A.M. et al. (2020). Silicon nanoparticles mitigate oxidative stress of *in vitro*-derived banana (*Musa acuminata* 'Grand Nain') under simulated water deficit or salinity stress. *South African Journal of Botany* 132: 155–163.

Malik, M.A., Wani, A.H., Mir, S.H. et al. (2021). Elucidating the role of silicon in drought stress tolerance in plants. *Plant Physiology and Biochemistry* 165: 187–195.

Manara, A. (2012). Plant responses to heavy metal toxicity. In: *Plants and Heavy Metals* (ed. A. Furini), 27–53. Dordrecht: Springer.

Manivasagan, P., Khan, F., Dhatchayeny, D.R. et al. (2023). Antibody-conjugated and streptomycin-chitosan oligosaccharide-modified gold nanoshells for synergistic chemo-photothermal therapy of drug-resistant bacterial infection. *Journal of Advanced Research* 48: 87–104. https://doi.org/10.1016/j.jare.2022.08.009.

Manivasagan, P., Nam, S.Y., and Oh, J. (2016). Marine microorganisms as potential biofactories for synthesis of metallic nanoparticles. *Critical Reviews in Microbiology* 42: 1007–1019. https://doi.org/10.3109/1040841X.2015.1137860.

Massey, F.P. and Hartley, S.E. (2009). Physical defences wear you down: progressive and irreversible impacts of silica on insect herbivores. *Journal of Animal Ecology* 78: 281–291.

Moghaddam, F.M., Foroushani, B.K., and Rezvani, H.R. (2015). Nickel ferrite nanoparticles: an efficient and reusable nanocatalyst for a neat, one-pot and four-component synthesis of pyrroles. *RSC Advances* 5: 18092–18096.

Mohanraj, R., Gnanamangai, B.M., Poornima, S. et al. (2022). Decolourisation efficiency of immobilized silica nanoparticles synthesized by actinomycetes. *Materials Today Proceedings* 48: 129–135. https://doi.org/10.1016/j.matpr.2020.04.139.

Mohd, N.K., Wee, N.N.A.N. and Azmi, A.A., (2017). Green synthesis of silica nanoparticles using sugarcane bagasse. *AIP conference proceedings: 1885*, No. 1. AIP Publishing. https://doi.org/10.1063/1.5002317.

Munns, R. and Tester, M. (2008). Mechanisms of salinity tolerance. *Annual Review of Plant Biology* 59: 651–681.

Naidu, S., Pandey, J., Mishra, L.C. et al. (2023). Silicon nanoparticles: synthesis, uptake and their role in mitigation of biotic stress. *Ecotoxicology and Environmental Safety* 255: 114783.

Nath, A., Das, A., Deb, S. et al. (2016). Green synthesis of novel antioxidant luminescent silica nanoparticle embedded carbon nanocomposites from a blue-green alga. *Green Processing and Synthesis* 5: 189–194. https://doi.org/10.1063/1.5002317.

Nweze, E.J., Ubani, C.S., Okeke, E.S. et al. (2022). Health risk assessment of heavy metals associated with Terminalia catappa fruit consumption obtained from an automobile workshop cluster in Nsukka, Nigeria. *Science and Technology* 22 (2): https://doi.org/10.4314/EJESM.V3I3.63962.

Okeke, E.S., Nweze, E.J., Ezike, T.C. et al. (2023). Silicon-based nanoparticles for mitigating the effect of potentially toxic elements (PTEs) and plant stress in agroecosystems: a sustainable pathway towards food security. *The Science of the Total Environment* 165446.

Okoronkwo, E.A., Imoisili, P.E., and Olusunle, S.O.O. (2013). Extraction and characterization of amorphous silica from corn cob ash by sol-gel method. *Chemistry and Materials Research* 3: 68–72.

Oliveira, E.M.D., Garcia-Rojas, E.E., Valadão, I.C.R.P. et al. (2017). Effects of the silica nanoparticles (NPSiO $_2$) on the stabilization and transport of hazardous nanoparticle suspensions into landfill soil columns. *REM-International Engineering Journal* 70: 317–323.

Oshunsanya, S.O., Nwosu, N.J., and Li, Y. (2019). Abiotic stress in agricultural crops under climatic conditions. In: *Sustainable Agriculture, Forest and Environmental Management* (ed. M.K. Jhariya et al.), 71–100. Springer Nature, Singapore Pte Ltd.

Patel, R.K. (2015). Biomimetic sequestration of CO, using carbonic anhydrase from calcite encrust forming marine actinomycetes. *Science International* 3: 48–57.

Patel, K.G., Misra, N.M., Vekariya, R.H., and Shettigar, R.R. (2018). One-pot multicomponent synthesis in aqueous medium of 1, 4-dihydropyrano [2, 3-c] pyrazole-5-carbonitrile and derivatives using a green and reusable nano-SiO_2 catalyst from agricultural waste. *Research on Chemical Intermediates* 44: 289–304.

Paul, A. and Roychoudhury, A. (2021). Go green to protect plants: repurposing the antimicrobial activity of biosynthesized silver nanoparticles to combat phytopathogens. *Nanotechnology for Environmental Engineering* 6: 10.

Pavithra, R., Kannan, R., Ramamoorthy, V. et al. (2020). Exploration of major fungal pathogens associated with water hyacinth (*Eichhornia crassipes* (C. Mart) Solms) and evaluation of their potential as mycoherbicides by proving pathogenicity. *International Journal of Current Microbiology and Applied Sciences* 9: 1897–1903. https://doi.org/10.20546/ijcmas.20209.8.217.

Pieła, A., Żymańczyk-Duda, E., Brzezińska-Rodak, M. et al. (2020). Biogenic synthesis of silica nanoparticles from corn cobs husks. Dependence of the productivity on the method of raw material processing. *Bioorganic Chemistry* 99: 103773. https://doi.org/10.1016/j.bioorg.2020.103773.

Rajput, V.D., Minkina, T., Kumari, A. et al. (2021). Coping with the challenges of abiotic stress in plants: new dimensions in the field application of nanoparticles. *Plants* 10: 1221.

Rangaraj, S., Gopalu, K., Muthusamy, P. et al. (2014). Augmented biocontrol action of silica nanoparticles and Pseudomonas fluorescens bioformulant in maize (Zea mays L.). *RSC Advances* 4: 8461–8465.

Rastogi, A., Tripathi, D.K., Yadav, S. et al. (2019). Application of silicon nanoparticles in agriculture. 3 Biotech, 9: 1–11.

Rastogi, A., Zivcak, M., Sytar, O. et al. (2017). Impact of metal and metal oxide nanoparticles on plant: a critical review. *Frontiers in Chemistry* 5: 78.

Rea, R.S., Islam, M.R., Rahman, M.M. et al. (2022). Growth, nutrient accumulation, and drought tolerance in crop plants with silicon application: a review. *Sustainability* 14: 4525.

Reynolds, D., Duray, G.Z., Omar, R. et al. (2016). A leadless intracardiac transcatheter pacing system. *New England Journal of Medicine* 374: 533–541.

Rizwan, M., Ali, S., Ali, B. et al. (2019). Zinc and iron oxide nanoparticles improved the plant growth and reduced the oxidative stress and cadmium concentration in wheat. *Chemosphere* 214: 269–277.

Rouhani, M., Samih, M.A., and Kalantari, S. (2013). Insecticidal effect of silica and silver nanoparticles on the cowpea seed beetle, Callosobruchus maculatus F. (Col.: Bruchidae). *Journal of Entomological Research* 4: 297–305.

Roychoudhury, A. (2020a). Silicon-nanoparticles in crop improvement and agriculture. *International Journal on Recent Advancement in Biotechnology & Nanotechnology* 3: 54–65.

Roychoudhury, A. (2020b). Yeast-mediated green synthesis of nanoparticles for biological applications. *Indian Journal of Pharmaceutical and Biological Research* 8 (3): 26–31.

Roychoudhury, P., Bose, R., Dąbek, P., and Witkowski, A. (2022). Photonic nano-/microstructured diatom based biosilica in metal modification and removal–a review. *Materials* 15: 6597.

San, N.O., Kurşungöz, C., Tümtaş, Y. et al. (2014). Novel one-step synthesis of silica nanoparticles from sugar beet bagasse by laser ablation and their effects on the growth of freshwater algae culture. *Particuology* 17: 29–35. https://doi.org/10.1016/j.partic.2013.11.003.

Sankar, S., Sharma, S.K., Kaur, N. et al. (2016). Biogenerated silica nanoparticles synthesized from sticky, red, and brown rice husk ashes by a chemical method. *Ceramics International* 42: 4875–4885. https://doi.org/10.1016/j.ceramint.2015.11.172.

Saw, G., Nagdev, P., Jeer, M., and Murali-Baskaran, R.K. (2023). Silica nanoparticles mediated insect pest management. *Pesticide Biochemistry and Physiology* 105524.

September, L.A., Kheswa, N., Seroka, N.S., and Khotseng, L. (2023). Green synthesis of silica and silicon from agricultural residue sugarcane bagasse ash–a mini review. *RSC Advances* 13: 1370–1380.

Sethy, N.K., Arif, Z., Mishra, P.K., and Kumar, P. (2019). Synthesis of SiO_2 nanoparticle from bamboo leaf and its incorporation in PDMS membrane to enhance its separation properties. *Journal of Polymer Engineering* 39: 679–687.

Shah, M., Fawcett, D., Sharma, S. et al. (2015). Green synthesis of metallic nanoparticles via biological entities. *Materials* 8: 7278–7308.

Show, S., Tamang, A., Chowdhury, T. et al. (2015). Bacterial (BKH1) assisted silica nanoparticles from silica rich substrates: a facile and green approach for biotechnological applications. *Colloids and Surfaces B: Biointerfaces* 126: 245–250.

Simonin, M. and Richaume, A. (2015). Impact of engineered nanoparticles on the activity, abundance, and diversity of soil microbial communities: a review. *Environmental Science and Pollution Research* 22: 13710–13723.

Singh, S., Bhatta, U.M., Satyam, P.V. et al. (2008). Bacterial synthesis of silicon/silica nanocomposites. *Journal of Materials Chemistry* 18: 2601–2606.

Singh, R. and Patel, M. (2022). Effective utilization of rice straw in value-added by-products: a systematic review of state of art and future perspectives. *Biomass and Bioenergy* 159: 106411. https://doi.org/10.1016/j.biombioe.2022.126411.

Singh, A. and Roychoudhury, A. (2020). Silicon-regulated antioxidant and osmolyte defense and methylglyoxal detoxification functions co-ordinately in attenuating fluoride toxicity and conferring protection to rice seedlings. *Plant Physiology and Biochemistry* 154: 758–769.

Singh, P., Singh, H., Kim, Y.J. et al. (2016). Extracellular synthesis of silver and gold nanoparticles by *Sporosarcina koreensis* DC4 and their biological applications. *Enzyme and Microbial Technology* 86: 75–83.

Slomberg, D.L. and Schoenfisch, M.H. (2012). Silica nanoparticle phytotoxicity to Arabidopsis thaliana. *Environmental Science and Technology* 46: 10247–10254.

Strout, G., Russell, S.D., Pulsifer, D.P. et al. (2013). Silica nanoparticles aid in structural leaf coloration in the Malaysian tropical rainforest understorey herb *Mapania caudata*. *Annals of Botany* 112: 1141–1148.

Sundaria, N., Singh, M., Upreti, P. et al. (2019). Seed priming with iron oxide nanoparticles triggers iron acquisition and biofortification in wheat (*Triticum aestivum* L.) grains. *Journal of Plant Growth Regulation* 38: 122–131.

Suriyaprabha, R., Karunakaran, G., Kavitha, K. et al. (2014). Application of silica nanoparticles in maize to enhance fungal resistance. *IET Nanobiotechnology* 8: 133–137.

Tang, L. and Cheng, J. (2013). Nonporous silica nanoparticles for nanomedicine application. *Nano Today* 8: 290–312.

Tang, F.H. and Maggi, F. (2021). Pesticide mixtures in soil: a global outlook. *Environmental Research Letters* 16: 044051.

Thabet, A.F., Boraei, H.A., Galal, O.A. et al. (2021). Silica nanoparticles as pesticide against insects of different feeding types and their non-target attraction of predators. *Scientific Reports* 11: 14484.

Thind, S., Hussain, I., Ali, S. et al. (2021). Silicon application modulates growth, physio-chemicals, and antioxidants in wheat (*Triticum aestivum* L.) exposed to different cadmium regimes. *Dose-Response* 19: 15593258211014646.

Tiwari, A., Sherpa, Y.L., Pathak, A.P. et al. (2019). One-pot green synthesis of highly luminescent silicon nanoparticles using Citrus limon (L.) and their applications in luminescent cell imaging and antimicrobial efficacy. *Materials Today Communications* 19: 62–67. https://doi.org/10.1016/j.mtcomm.2018.12.005.

Tripathi, O.P., Charlton-Perez, A., Sigmond, M., and Vitart, F. (2015). Enhanced long-range forecast skill in boreal winter following stratospheric strong vortex conditions. *Environmental Research Letters* 10: 104007.

Tripathi, D.K., Singh, S., Singh, S. et al. (2017). An overview on manufactured nanoparticles in plants: uptake, translocation, accumulation and phytotoxicity. *Plant Physiology and Biochemistry* 110: 2–12.

Tripathi, D.K., Vishwakarma, K., Singh, V.P. et al. (2021). Silicon crosstalk with reactive oxygen species, phytohormones and other signaling molecules. *Journal of Hazardous Materials* 408: 124820.

Vaibhav, V., Vijayalakshmi, U., and Roopan, S.M. (2015). Agricultural waste as a source for the production of silica nanoparticles. *Spectrochimica Acta Part A: Molecular and Biomolecular Spectroscopy* 139: 515–520. https://doi.org/10.1016/j.saa.2014.12.083.

Verma, S., Chakdar, H., Kumar, M. et al. (2021). Microorganisms as a sustainable alternative to traditional biofortification of iron and zinc: status and prospect to combat hidden hunger. *Journal of Soil Science and Plant Nutrition* 21: 1700–1717.

Vetchinkina, E.P., Loshchinina, E.A., Vodolazov, I.R. et al. (2017). Biosynthesis of nanoparticles of metals and metalloids by basidiomycetes. Preparation of gold nanoparticles by using purified fungal phenol oxidases. *Applied Microbiology and Biotechnology* 101: 1047–1062.

Viehweger, K. (2014). How plants cope with heavy metals. *Botanical Studies* 55: 1–12.

Waewthongrak, W., Pisuchpen, S., and Leelasuphakul, W. (2015). Effect of *Bacillus subtilis* and chitosan applications on green mold (*Penicilium digitatum* Sacc.) decay in citrus fruit. *Postharvest Biology and Technology* 99: 44–49.

Wang, W., Martin, J.C., Fan, X. et al. (2012). Silica nanoparticles and frameworks from rice husk biomass. *ACS Applied Materials and Interfaces* 4: 977–981.

Wang, Q., Ou, E.L., Wang, P.C. et al. (2022). *Bacillus amyloliquefaciens* GB03 augmented tall fescue growth by regulating phytohormone and nutrient homeostasis under nitrogen deficiency. *Frontiers in Plant Science* 13: 979883.

Wang, X., Qiu, Y., Wang, M. et al. (2020). Endocytosis and organelle targeting of nanomedicines in cancer therapy. *International Journal of Nanomedicine* 9447–9467.

Wanyika, H., Gatebe, E., Kioni, P. et al. (2012). Mesoporous silica nanoparticles carrier for urea: potential applications in agrochemical delivery systems. *Journal of Nanoscience and Nanotechnology* 12: 2221–2228.

War, A.R., Paulraj, M.G., Ahmad, T. et al. (2012). Mechanisms of plant defense against insect herbivores. *Plant Signaling & Behavior* 7: 1306–1320.

Wassie, A.B. and Srivastava, V.C. (2017). Synthesis and characterization of nano-silica from teff straw. *Journal of Nano Research* 46: 64–72.

Wu, J., Guo, J., Hu, Y., and Gong, H. (2015). Distinct physiological responses of tomato and cucumber plants in silicon-mediated alleviation of cadmium stress. *Frontiers in Plant Science* 6: 141391.

Yadav, V.K., Amari, A., Mahdhi, N. et al. (2023). A novel and economical approach for the synthesis of short rod-shaped mesoporous silica nanoparticles from coal fly ash waste by *Bacillus circulans* MTCC 6811. *World Journal of Microbiology and Biotechnology* 39: 289.

Yamaji, N., Chiba, Y., Mitani-Ueno, N., and Feng Ma, J. (2012). Functional characterization of a silicon transporter gene implicated in silicon distribution in barley. *Plant Physiology* 160: 1491–1497.

Yaro, S.A., Olajide, O.S., Asuke, F., and Popoola, A.P.I. (2017). Synthesis of groundnut shell nanoparticles: characterization and particle size determination. *The International Journal of Advanced Manufacturing Technology* 91: 1111–1116.

Yusaidi, N., Abdullah, S., and Zarib, N. (2019). Structural analysis of silica extract from banana stems via acid leaching under different reaction time. *International Journal of Engineering Advance Technology* 9: 5858–5862. https://doi.org/10.35940/ijeat.A3017.109119.

Zamani, H., Jafari, A., Mousavi, S.M., and Darezereshki, E. (2020). Biosynthesis of silica nanoparticle using *Saccharomyces cervisiae* and its application on enhanced oil recovery. *Journal of Petroleum Science and Engineering* 190: 107002. https://doi.org/10.1016/j.petrol.2020.107002.

Zargar, S.M., Mahajan, R., Bhat, J.A. et al. (2019). Role of silicon in plant stress tolerance: opportunities to achieve a sustainable cropping system. *3 Biotech* 9: 73.

Zhang, L., Duan, X., He, N. et al. (2017). Exposure to lethal levels of benzo [a] pyrene or cadmium trigger distinct protein expression patterns in earthworms (Eisenia fetida). *Science of the Total Environment* 595: 733–742.

Zhao, X., Chen, Y., Li, H., and Lu, J. (2021). Influence of seed coating with copper, iron and zinc nanoparticles on growth and yield of tomato. *IET Nanobiotechnology* 15: 674–679.

Zhao, L., Su, P., Hou, B. et al. (2022). The black necrotic lesion enhanced *Fusarium graminearum* resistance in wheat. *Frontiers in Plant Science* 13: 926621. https://doi.org/10.3389/fpls.2022.926621.

Zielonka, A., Żymańczyk-Duda, E., Brzezińska-Rodak, M. et al. (2018). Nanosilica synthesis mediated by *Aspergillus parasiticus* strain. *Fungal Biology* 122: 333–344. https://doi.org/10.1016/j.funbio.2018.02.0045.

5

'From Lab to Life': An Overview of Synthesis and Delivery of Nanoparticles in Biological Organisms

Debjyoti Bandhu Banerjee, Kalpak Majumdar, Sagnik Goswami, Sayan Sadhukhan and Anirban Bhar

Department of Botany, Ramakrishna Mission Vivekananda Centenary College, Kolkata, West Bengal, India

5.1 Introduction

Nanoparticles (NPs) are characterized by their small size, ranging from 10 to 100 nm (Mohanraj and Chen 2006). The term 'nano' is derived from the Greek word meaning 'dwarf', it is 10^{-9} or one billionths of a meter. NPs possess a wide array of chemical natures, for example, metal (copper, gold, silver and zinc) or metal oxide, polymer and organic. Moreover, there are a number of shapes of NPs; they may be spherical, cylindrical or tubular (Pal et al. 2019). NPs have extensive applications in agriculture, drug delivery, biosensing, tissue engineering and diagnosis (Solanki et al. 2015). Metal nanoparticles (MNPs) have significant catalytic properties that are useful in clinical diagnosis, biosensing, sewage water management and food safety (Sajid et al. 2015). The synthesis and application of NPs are not unique to contemporary study; there is evidence of their use dating back thousands of years. Viruses, proteins, polysaccharides (organic); iron oxyhydroxides and aluminosilicates (inorganic) are naturally obtained NPs, synthesized for natural activities, for example, weathering, volcanic eruption, wildfire, etc. (Heiligtag and Niederberger 2013).

NPs can be produced through different synthesis methods, for example, the constructive bottom-up method and the destructive top-down method. Extensive development of instrumentation led to better characterization and application of NPs. Photochemical reduction, laser ablation and lithography technologies have been used in the synthesis of NPs. These techniques are not cost-effective and produce several hazardous chemicals. Therefore, the synthesis of NPs with naturally occurring substances like microorganisms, plant extracts and biodegradable polymers, as reducing and capping agents, opens up a new approach known

Silicon Nanoparticles: Synthesis, Uptake, and Applications, First Edition.
Edited by Archana Singh, Aryadeep Roychoudhury, and Indrakant K. Singh.

as green synthesis of NPs (Chakraborty et al. 2022). Nanotechnology is one of the most promising approaches for a clean and sustainable future (Ealia and Saravanakumar 2017; Kharissova et al. 2013). The term 'nanotechnology' was first used by Professor Norio Taniguchi in 1974 and popularized by Eric Drexler (Ferdous and Nemmar 2020). Nanotechnology has immense importance in transforming biological systems. Moreover, it is extensively used in agriculture, pathogen diagnostics and packaging equipment. Nanobiotechnology is a modern approach to combat different biotic and abiotic stresses in plants with the help of NPs (Khan and Upadhyaya 2019). Furthermore, it is also used in bioremediation, wastewater treatment and other biomedical and biotechnological fields (Manna and Bandyopadhyay 2019; Roychoudhury and Bhowmik 2023).

NPs have immense importance in drug delivery for their several positive attributes, such as surface properties useful in active as well as passive drug targeting, promoting drug-therapeutic efficiency, site-specific targeting, etc. (Mohanraj and Chen 2006). Nanomedicine is an emerging branch that deals with the potential of NPs in medicine and diagnostics. NPs increase the therapeutic index of drugs; for this reason, 20 types of NPs are primarily used in clinical applications (Sajid et al. 2015).

Small particles of 200nm or less than that, may penetrate cells easily through different mechanisms such as caveolae-mediated endocytosis, clathrin-mediated endocytosis or by passive mechanisms. Upon internalization, NPs interact with cellular organelles, cytoskeletal elements or nuclei (Gehr 2018). NPs may enter via different ways, for example, oral, dermal and respiratory tract. Human exposure to different NPs may induce pulmonary and cardiac disease (Roy et al. 2014). NPs may act as a 'magic bullet' to deliver herbicides and pesticides to plants; surprisingly, there are several reports demonstrating stress conditions arising from NPs in plants, thus altering plant growth and development (Siddiqui et al. 2015).

In this chapter, we have discussed different methods of synthesis of NPs and their mode of delivery to plant and animal systems. Moreover, the impact of NPs, depending on their composition, size and other properties, has also been analysed.

5.2 Methods of Synthesis of NPs

NPs may be synthesized using numerous techniques. Different synthesis methods are developed or modified to improve the mechanical, physical and chemical properties of NPs, lowering production costs (Ealia and Saravanakumar 2017; Rane et al. 2018). In the top-down approach, the bulk counterpart is used to start the synthesis, followed by a gradual leak out bit by bit to produce the NPs; the bottom-up approach creates a variety of NPs by assembling or condensing atoms and molecules. Top-down processes like photolithography, electron beam lithography, milling procedures, anodization and ion and plasma etching are frequently used for large-scale production of NPs, whereas sol–gel processing, laser pyrolysis, chemical vapour deposition (CVD), plasma or flame spraying synthesis and bio-assisted synthesis are included under the bottom-up approach (Dhand et al. 2015). The breakdown process of large bulk matters may include physical or chemical means. It is the simplest process for the formation of NPs, but the irregularities in the shape of the particles are a major drawback of this process (Tripathy 2023). Chemical synthesis processes have numerous negative impacts since they include harmful chemicals adsorbed onto their surface. The biological method of NPs synthesis with bacteria, enzymes, fungi, plants and plant extracts came out as an alternative environment-friendly approach (Hasan 2015; Roychoudhury 2020).

5.2.1 Physical Methods of Synthesis of NPs

Physical method is a top-down approach to NPs synthesis that often involves the application of mechanical pressure, high-energy radiation, thermal energy or electrical energy to change the state of the material. Since the physical method involves a top-down approach, a huge amount of waste is produced in this method, which makes it disadvantageous. Thermal evaporation–condensation, high-energy ball milling, laser ablation and sputtering are some of the most predominantly used physical methods that help in the synthesis of NPs (Dhand et al. 2015).

5.2.1.1 Thermal Evaporation–Condensation

Thermal evaporation–condensation is the most primitive process that was used to synthesize nanocrystalline metals and alloys. In this endothermic method, chemical decomposition was achieved by thermal evaporation, for example, electron beam evaporation device, Joule heated refractory crucibles, at an atmospheric pressure of 1–50 mbar. The specific temperature at which the chemical decomposition occurs is termed as the decomposition temperature (Ijaz et al. 2020; Rajput 2015).

5.2.1.2 High-Energy Ball Milling

It is the simplest and cheapest process for the synthesis of NPs. This process generally includes the breakdown of the materials into a fine-ground powder. Then, this fine powder is subjected to high-energy ball milling to synthesize NPs. For example, 60 μm Cu powder when subjected to milling at a milling speed of 450 rpm for 0–60 hours, CuO NPs of size 11 nm are produced (Ijaz et al. 2020).

5.2.1.3 Laser Ablation

It is also one of the simplest methods to synthesize NPs. Generally, laser ablation is used to produce NPs from different kinds of solvents (Ealia and Saravanakumar 2017). The basic principle of laser ablation is that a solid target material in the ambient media (gas or liquid) will vaporize when the laser beam is focused on its surface because the temperature of the irradiated spot will rise quickly. The atoms and clusters that collided to form the evaporating species and the surrounding molecules cause the electron state to be excited, which is followed by light emission, the creation of electrons and ions, and the formation of a laser-induced plasma plume (Kim et al. 2017). Laser ablations have several advantages. It is a highly reproducible method and devoid of any stabilizing agent or chemical; therefore, it is also regarded as a 'green way'. In addition to creating metal NPs, this approach may be used to create organic, magnetic and semiconducting NPs in a variety of forms and sizes (Naser et al. 2019).

5.2.2 Chemical Synthesis Method

5.2.2.1 Microemulsion Process

Microemulsions are thermodynamically stable and isotropic dispersions of two immiscible liquids where a surfactant stabilizes the microdomains in combination with a co-surfactant (Lawrence and Rees 2012). The microemulsion process is abundantly used to synthesize inorganic NPs (Cele 2020). The formation of NPs through the microemulsion method is widely considered as an ideal technique, although the exact mechanism of NPs formation within microemulsions remains elusive. Despite this, some researchers have proposed a mechanism for synthesizing NPs in microemulsions. When the microemulsion material containing reactants is agitated, reactant exchange takes place during collisions between water droplets in the microemulsion. This rapid exchange of reactants leads to a precipitation reaction of the nano-droplets, followed by nucleation and growth of primary particles, ultimately resulting in the formation of NPs surrounded by water and stabilized by surfactants (Rane et al. 2018). There is great potential for micro-emulsions in drug targeting and drug delivery systems (Arredondo-Ochoa and Silva-Martínez 2022).

5.2.3 Biological Synthesis of NPs

Synthesis of NPs through biological means has come out as a more environmentally friendly, rapid and pharmacologically active method than the physical and chemical methods of synthesis (Singh et al. 2016). Green synthesis of NPs can be classified as a 'bottom-up' approach where natural product extracts replace the expensive chemical reducing agents; it deserves more attention as biogenic reduction is sustainable, lowers chemical contamination and is budget-friendly (Hussain et al. 2016; Roychoudhury et al. 2023). Metallic NPs synthesized by biological methods are primarily used in the field of biomedicine such as bioimaging, cancer treatment, sensor construction and medical diagnosis for their antimicrobial, anti-metastatic and antioxidant properties (Nadaroglu et al. 2017).

Microorganisms such as bacteria, fungi, leaf extracts, flowers, roots and fruits are used in the green synthesis of metal NPs as they contain proteins and polyphenols to reduce metal ions. Silver, gold, zinc oxide, palladium and iron NPs can be easily synthesized through green synthesis. Leaf extracts of *Cassia auriculata*, *Cinnamomum zeylanicum* and *Pogestemon benghalensis* are used in making gold NPs that reduce chloroauric acid to AuNPs. On the other hand, silver NPs are produced by mixing plant extracts with silver nitrate solutions; the leaf powder of *Tephrosia purpurea* is used in producing silver NPs (Ying et al. 2022). Enzymes are also used in green synthesis due to their well-defined structure and purity. Au NPs with anti-fungal activity in response to human pathogenic fungi were developed with the purified product of the sulphite reductase enzyme. Agrochemical waste such as coir of coconut, rice and wheat bran and corn is also used in green synthesis as these contain flavonoids, phenolics and proteins as reducing agents. Furthermore, vitamins, for example, Vitamin B2 and ascorbic acid, are also reported to be used in NP synthesis as reducing and capping agents (Gour and Jain 2019).

5.3 Methods of Delivery of NPs for Animals

Drug delivery is necessary because humans have experienced different diseases throughout their life cycle (Tiwari et al. 2012). Delivery of drugs is generally carried out in two ways, i.e., oral and injection. In case of injections, the drug is directly delivered into the body either through the veins (intravenous) or muscles (intramuscular), or skin (subcutaneous and intradermal) (Langer 1990). The method of the conventional drug delivery system could not be sustained due to several drawbacks (Tiwari et al. 2012), including poor biodistribution, i.e., the transfer of chemicals from one part to another becomes poor, transitory effectiveness and ineffective selectivity, giving rise to other side effects (Wilczewska et al. 2012). It was observed that some drugs worked in their target organ very efficiently in vitro but could not withstand the high concentration of acids and enzymes inside the gastrointestinal tract, leading to ineffective methods of delivering the drugs (Rizvi and Saleh 2018).

Nanotechnology played an important role in overcoming all these challenges. Since NPs have a diminutive size, they can easily penetrate the target tissue without hampering other tissues and start their therapeutic action, showing minimized side effects (Wilczewska et al. 2012). Due to all these dynamic physicochemical properties of NPs, they have been used to address all the odds against the conventional methods of drug delivery (Naahidi et al. 2013).

Nanotechnology offers various methods for the delivery of drugs using the distinctive properties of NPs. Liposomes are round-shaped or, more specifically, spherical vesicles composed of lipid bilayers, as described in 1965 (Allen and Cullis 2013). When drugs are encapsulated within the liposomes, the toxic effects of the drugs can be reduced, which helps in the minimization of off-target side effects (Sharma 1997). The surface of liposomes can be modified with ligands that help them deliver to specific cells or tissues in animals. In various diseases in animals including cancer, anti-bacterial and anti-fungal disorders, inflammatory conditions, and most prominently in veterinary medicine, the contributions of liposomes are much broader (Goyal et al. 2005).

Dendrimers are highly branched macromolecules with a tree-like appearance and a well-defined structure. Poly(amidoamine) (PAMAM) dendrimers are commonly used for drug delivery, which was first reported by Tomalia in 1985 (Chauhan 2018). It consists of a central core molecule, branching units and multiple layers of repeating units known as dendrons (Caminade et al. 2005). The core molecule generally consists of ammonia or ethylene diamine, depending on the properties of the dendrimer. The arms-like structures from the core molecule are the branching units that are generally repeating in nature, forming the dendritic arms. The outermost layer of the dendrimers contains the surface functional groups, which help in the binding or attachment of drugs and other molecules (Pedziwiatr-Werbicka et al. 2019). This region is used for the application of drugs in animals. Dendrimers can solubilize hydrophobic drugs by encapsulating them within their hydrophobic interior, improving solubility and bioavailability (Smith et al. 2021). Dendrimers are also used as a shield of protection from enzyme degradation or harsh bodily conditions. A major function of dendrimers is releasing the drug in a controlled manner and often sustaining drug release over time (Chauhan 2018). PAMAM dendrimers are used to encapsulate the famous anticancer drug cisplatin, which helps in slower release, higher accumulation over the solid tumours, and is less toxic in comparison to free cisplatin (Svenson and Tomalia 2012). The most promising feature of dendrimers is in vivo behaviour, which helps to synthesize anti-viral drugs, repair tissue scaffolds, target the carriers of chemotherapeutics and develop optical oxygen sensors (Lee et al. 2005).

Solid-liquid NPs (SLNs), introduced in 1991, are colloidal in nature and are made up of lipids that are solid not only at room temperature but also at body temperature (Hou et al. 2003). They are generally spherical in shape having a lipid core. SLN structure consists of a lipid matrix at the core, which can be composed of one or more solid lipids such as triglycerides, fatty acids or waxes. They are solid at physiological temperature and thus provide stability to the NP structure (Mehnert and Mäder 2012). Surrounding the central lipid core, a layer of surfactant molecules is present. Surfactants are generally amphiphilic compounds that have both hydrophilic and hydrophobic regions, which enhance colloidal stability (Naseri et al. 2015). According to the desired properties, the surface can be modified with various functional groups or coating materials. Surface modification can be done by incorporation of polymers, PEGylation (attachment of polyethylene glycol chains) or by other means (Arana et al. 2019). Therapeutic drugs can be incorporated into the lipid core matrix during SLN preparation. The lipophilic nature of the lipid core helps in the encapsulation of the hydrophobic drugs, increasing drug stability (Baek and Cho 2017). Nanocrystals are minute, grained, polycrystalline solids with a dimensional range from 1 to 100 nm (Suryanarayana and Froes 1992). They display some unique properties like quantum confinement effects and a large surface-area-to-volume ratio due to their small size (Delley and Steigmeier 1993). These make them useful in various applications including electronics, catalysis and biomedical imaging. Depending on the material they are made of the structure of nanocrystals varies but generally possesses a crystalline lattice structure (Smith and Nie 2010). In case of animal drug delivery applications, drugs are either encapsulated within or stuck to the surface of the nanocrystals. The minute size of the nanocrystals gives them the advantage to penetrate all the barriers efficiently, such as the cell membrane and blood–brain barrier, to hit and act on the target effectively (Misra and Pathak 2022). Moreover, due to the encapsulation of the drugs within the nanocrystals, they are spared from degradation and clearance by the immune system of the body. There are several methods to synthesize drug-loaded nanocrystals including

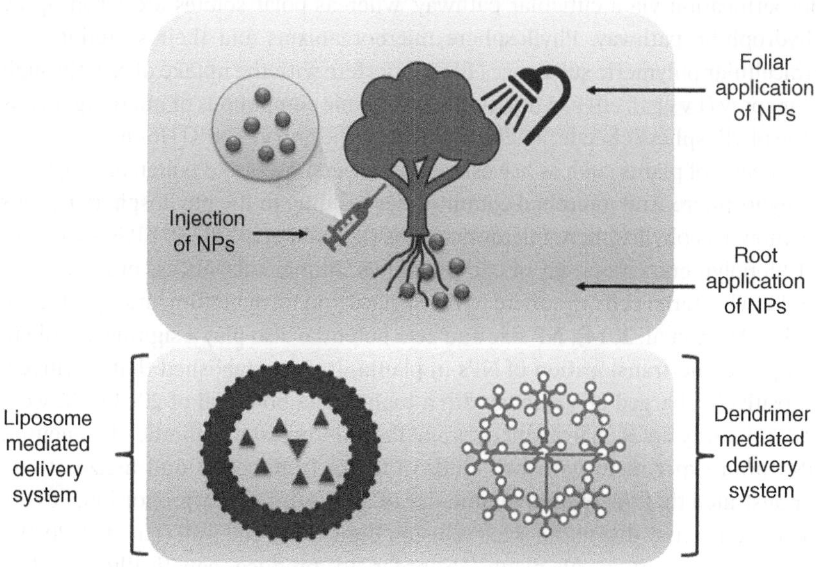

Figure 5.1 Methods of delivery of NPs in plants and animals.

some bottom-up synthesis techniques like sol–gel methods and precipitation, and top-down approaches like milling (Sinha et al. 2013). Nanocrystals are in high demand and explored in the field of drug delivery systems. Due to their high surface area-to-volume ratio, they are used as efficient drug carriers (Gao et al. 2008) (Figure 5.1).

5.4 Methods of Delivery of NPs for Plants

For sustainable agriculture and overcoming increasing food insecurity throughout the world, NPs are immensely used to utilize nutrients, mitigate plant disease and yield improvements. Foliar application, root application and trunk injection are the three most popular methods of NP delivery to plants. Among these, foliar application and root application are useful for delivering nanofertilizers and nanopesticides (Su et al. 2019).

Foliar applications of NPs have several positive attributes, for example, they lower the synthesis of reactive oxygen species (ROS), escalate the shelf life of agricultural produce, lower abiotic stress-mediated damage and promote better assimilation and absorption of foliar fertilizer. Besides, it is also cost-effective and soil health-friendly. Stomata, water pores or cracks, wound sites, stigma and trichomes are some sites through which plants uptake NP following foliar application (Hong et al. 2021). Nonpolar solutes are taken up by plants through diffusion and permeation via a cuticular pathway, whereas polar solutes are taken up by a hydrophilic pathway. Phyllosphere microorganisms and their secretion, i.e., extracellular polymeric substance (EPS), interfere with the uptake of NPs through foliar spray (Lv et al. 2019). Organic and inorganic compounds of microorganisms in the phyllosphere chelate, acidify and reduce the entry of NPs (Hong et al. 2021). Aerial parts of plants such as leaves, stems, flowers, fruits and pollen are regarded as phyllosphere, and microbial communities residing in the phyllosphere regions are known as phyllosphere microorganisms (Sohrabi et al. 2023). EPSs are chemical biopolymers, composed of carbohydrates, humic substances, proteins, etc., that help bacterial cells to adhere to the surface and form biofilm and a protective barrier (More et al. 2014). NP size and zeta potential also play a significant role in the uptake and translocation of NPs in planta. It was established that negatively or positively charged NPs, along with a higher zeta potential of 20–30 mV, were predominantly uptaken by plant cells and the chloroplast membrane, than neutral NPs. In an experiment with two kinds of leaves from cotton and maize, it was demonstrated that the hydrodynamic size of NPs is negatively proportional to the delivery efficiency due to the size exclusion limit of the cell wall (Hu et al. 2020).

The transformation of plants in order to produce genetically modified organisms to combat the reduction of crop yield due to climate change and other environmental factors is quite difficult. Nanoparticle-mediated delivery of biomolecules can overcome the hindrance effect of the additional physical barrier, the plant cell wall. Carbon nanotubes and quantum dots are directly uptaken in plants (Cunningham et al. 2018). Single-walled carbon nanotubes (SWCNTs) and multi-walled carbon nanotubes (MWCNTs) are the two main categories of carbon nanotubes. SWCNTs are comprised of graphene encompassed in a cylindrical nanostructure of diameter 0.7–3.0 nm, whereas MWCNTs consist of multiple SWCNTs with a diameter of 220 nm. SWCNTs are efficient as a gene delivery system in plants because of their biocompatibility, surface-area-to-volume ratio and tensile strength. Therefore, SWNCTs are used to deliver large quantities of DNA, small interfering RNA (siRNA) and plasmid DNA to plants (Lv et al. 2020). Quantum dot (QD) are semiconducting material nanocrystals are sufficiently tiny to display quantum mechanical characteristics. It demonstrates robust optical and electrical features that depend on size. QDs can have a diameter of 10–50 atoms and a volume of as much as 100,000 atoms (Dhand et al. 2015). Plant growth regulators like gibberellic acid (GA), salicylic acid (SA), soil fertilizers, plasmid DNA, siRNA, proteins, pesticides and herbicides can be delivered through a chitosan-mediated delivery system (Mujtaba et al. 2020). Chitosan, a chemical mucopolysaccharide, is produced from the deacetylation of chitin. Chitin is found in the exoskeleton layer of crustacean animals. Chitosan NPs are extensively used in drug delivery because of their low toxicity, better stability and easier preparation methods (Nagpal et al. 2010) (Figure 5.1).

5.5 Impact of NPs on Biological System

Entry of NPs depends on their charge, surface characteristics, hydrophobicity and hydrophilicity. However, the entry of NPs through plants and the cell wall of some microorganisms is quite difficult due to their pore size. Oxide NPs mostly enter by binding with the inner side of the vesicle, whereas polymeric or a few derivatized metal NPs enter by floating freely in the cytosol, in the case of mammalian cells (Stark 2011). Upon internalization, NPs undergo intracellular trafficking, affecting their distribution within cells and tissues. Modern imaging techniques such as super-resolution microscopy and live cell imaging have provided valuable insights into the dynamics of NPs uptake and distribution (Albanese et al. 2012; Nel et al. 2006). Nanotechnology has achieved exponential growth in the last few years in nanomedicine, molecular diagnostics, energy and electronics. Besides, excessive use of NPs also imposes cytotoxic effects on biological systems that also need to be addressed (Auría-Soro et al. 2019).

5.5.1 Impact of NPs on Plants

The impact of NPs on plants may be positive as well as negative, depending on their concentration. The effects of engineered NPs (ENPs) depend upon composition, size, concentration and physical and chemical properties (Siddiqui et al. 2015). Man-made NPs or ENPs are 100 nm or less in size, produced by different types of nanomaterials, mainly used in nanomedicine due to their several positive attributes such as specificity, functionality, sensitivity and efficiency (Albalawi et al. 2021). ENPs may be classified into inorganic NPs consisting of metal and metal oxide NPs, organic NPs composed of biologically compatible and polymeric NPs, NPs derived from carbon such as carbon nanotubes, carbon nanofibres, etc., and lastly organic–inorganic hybrid NPs (Sajid et al. 2015). NPs can alter the physiology of crop plants by modulating the antioxidant and ROS metabolism, light absorbance and photosynthetic efficiency and lowering the uptake of toxic heavy metals. Moreover, they can also mitigate different abiotic and biotic stresses. NPs can upregulate antioxidant enzymes that lower ROS accumulation in cells during abiotic stress conditions, thus preventing the oxidation of carbohydrates, proteins, lipids and DNA (Khan and Upadhyaya 2019). NPs have immense importance in controlling phytopathogenic organisms to combat biotic stress in crop plants. Carbon-based, metallic and polymeric NPs are primarily used in mitigating pathogens. Ag, CuO, Cu and TiO_2-based NPs are used to control bacterial pathogens; among them, AgNPs are utilized efficiently against a wide range of pathogens, viz., bacteria, fungi and viruses (Shoala 2018).

NPs that are released into the environment are taken up by plants. It was reported that the concentration of both carbon as well as engineered metal-based NPs beyond the threshold level negatively impacts plants by modulating the physiological, genetic and morphoanatomical parameters of plants (Kumar et al. 2018). Besides the use of AgNPs in agriculture for their antibacterial and anti-fungal properties, the indiscriminate use of these NPs imposes potential risks to the ecosystem. Silver ions are leached from AgNPs and are the most toxic materials to aquatic organisms, following mercury. Investigations on the effects of AgNPs revealed that they hinder seed germination, growth of roots, lower leaf area and biomass, destabilize thylakoid membrane structure, reduce chlorophyll content and promote excess ROS production (Yan and Chen 2019). An interesting study with $AgNO_3$ and AgNPs on *Brassica* sp. seedlings revealed that $AgNO_3$ imparts more reduction in fresh weight, root and shoot length, total protein content and photosynthetic parameters than AgNPs in comparison with the control condition (Upadhyaya and Roychoudhury 2024; Vishwakarma et al. 2017). NPs like ZnO, TiO_2, CeO, and AgNPs accumulate in the cell surface and organelles, thus promoting oxidative stress by inducing oxidative stress signaling. AgNPs interact with the vacuolar and endoplasmic reticulum (ER) proteins, indicating that these are the target sites for AgNPs (Hossain et al. 2015).

5.5.2 Impact of NPs on Animal Systems and Human Health

NP toxicity in humans mainly depends on its properties and some other factors such as shape, size, mobility, reactivity, surface chemistry and charge and the way of interaction with the human body. Occupational exposure to NPs mainly occurs through research-scale synthesis and industrial manufacturing of NP-based products. The human body primarily interacts with NPs through inhalation into the respiratory system, ingestion into the digestive system and infiltration via skin nodes (Sajid et al. 2015). The respiratory system is prone to being affected by occupational exposure to NPs. Systemic distribution throughout the body and accumulation in the lungs of NPs mainly depend on the size. Upon exposure of NP in lungs, it is surrounded by biomolecules like albumin and surfactant proteins that increase the size of particles and determine the fate of NPs, i.e., whether they would be phagocytosed by alveolar macrophages or enter into the systemic circulation (Pietroiusti et al. 2018). Larger NPs with 5–30 μm accumulate in the nasopharyngeal region, while 1–5 μm particles accumulate in the tracheobronchial region. Nanosized particles are mainly deposited in the alveolar region and can overcome the blood–air–tissue barrier, thereby entering the bloodstream. Moreover, small-sized inhaled ultrafine particles (UFPs) accumulate in the olfactory mucosa and are transported to the central nervous system (CNS), thus inducing neurotoxicity (Gupta and Xie 2018).

On exposure, AgNPs have the ability to cause oxidative stress and inflammation. Moreover, they can penetrate through a number of biological barriers and reach the bloodstream, also inducing several pathophysiological effects in different organs. in vitro studies with AgNPs with different types of cell lines, for example, alveolar epithelial cells (A549), bronchial epithelial cells (BEAS-2B), hepatocytes (C3A, HepG2), skin keratinocytes (HaCaT), human epidermal keratinocytes (HEKs), embryonic kidney cells (HEK293T) and porcine kidney cells (Pk 15) have demonstrated several adverse effects on the above cell lines. As an illustration, a high concentration of AgNPs in A549 cells induces cell shrinkage and cell death, and exposure to C3A cells with AgNPs induces cytotoxicity, oxidative stress and an increase in inflammatory proteins (Ferdous and Nemmar 2020). NPs induce oxidative stress by activating nuclear factor $\kappa\beta$ (NF$\kappa\beta$) through regulating the transcription of different pro-inflammatory genes, tumour necrosis factor α (TNFα), interleukin (IL)-1, IL-6 and IL-8. Oxidative stress leads to damaged DNA and apoptosis (Brohi et al. 2017). ENMs have been used extensively in aerospace, airplanes, architecture, biomedicine, computer memory and computer products due to their unique physical and chemical properties. Dispersion of ENMs in air and water imposes a significant risk to the gastrointestinal and respiratory systems of human beings (Mu et al. 2014). NPs also impose toxicity on the reproductive system and have deleterious effects on stages of pregnancy and alterations of reproductive cycles. Long-term exposure to TiO_2 NPs alters gene expression associated with estrogen, progesterone synthesis and metabolism in adult mice; furthermore, they also induce alterations of gene expression associated with inflammatory and immune responses and oxidative stress. The male reproductive system is also affected by the toxicity of NPs, such as the reduction of sperm production by altering the expression of genes associated with spermatogenesis (Brohi et al. 2017). A study investigated that TiO_2 NPs, deposited in the spleen of mice at maximum concentration, induce liver fibrosis and necrosis of liver cells. Short exposure to TiO_2 alters heart rate and blood pressure, and long exposure to it induces heart arrhythmia (Baranowska-Wójcik et al. 2020).

Surface runoff from soil can introduce NPs into aquatic systems indirectly or directly through industrial discharges or wastewater treatment effluents. When NPs dissolve, potentially hazardous substances may be released into the surrounding environment (Gupta and Xie 2018). Aquatic animals are also negatively affected by exposure to NPs. Ingestion and entry through gills are the two main ways of NP entry to aquatic animals. ENPs with nickel, copper, silver, aluminium, etc., impose hazardous effects on algae, zebrafish and dolphins. For example, AgNPs induce apoptotic damage in zebrafish liver cells, increasing malondialdehyde as a result of lipid peroxidation (Sajid et al. 2015).

5.6 Conclusion and Future Prospective

In summary, the creation and transport of NPs in biological organisms is an emerging field that has great promise for advancing both practical applications and scientific understanding. Using inventive methodologies and cross-disciplinary cooperation, researchers have achieved noteworthy progress in the production of NPs with precise attributes customized for particular medicinal objectives and agricultural applications. Furthermore, the creation of sophisticated delivery methods has made it easier for these NPs to be released within biological systems in a targeted and regulated manner. This has maximized their efficacy in plants and animals and reduced off-target effects.

The creation and transport of NPs have a promising future. Further developments in material science, biotechnology and nanotechnology will probably result in the production of NPs with even higher functionality, accuracy and biocompatibility. These developments will make it possible to create cutting-edge therapeutic strategies for the treatment of a variety of illnesses such as cancer, infectious diseases and neurological conditions. Additionally, the synthesis and application of NPs in agriculture present viable paths forward for tackling urgent issues including crop protection, food security and sustainable farming methods. NPs can be engineered to boost nutrient uptake, lessen the effects of abiotic stressors and increase agricultural yields through creative research and development. In addition, sophisticated delivery mechanisms provide the precise and regulated dispersal of these NPs, enhancing their effectiveness while reducing their ecological footprint. In future, more research into the application of nanotechnology in agriculture has the potential to transform agricultural methods, advance environmental sustainability and boost global food supply in a world that is changing quickly.

Acknowledgement

The authors are extremely thankful to the editors for inviting for chapter submission and giving the opportunity to contribute.

Glossary

Alloy Alloy is a substance produced either through combination of more than one metal or by mixing non-metallic elements with metals. Brass an alloy composed of zinc and copper.

Antioxidants Antioxidants are natural or man-made substances that protect cells from damage due to free radicals, thus delaying or preventing oxidative damage of cell.

Biosensing Biosensing may be defined as the method of detecting a target molecule with similar mechanisms of living system, for example, immune system. Detection, sensitivity and specificity are important parameters of biosensing. Proteins are used as a biomarker to detect a specific type of disease in medical biosensing.

Bioremediation Bioremediation is the use of microorganisms to degrade the environmental contaminants into less toxic forms. Microorganisms can be specifically designed for bioremediation using genetic engineering techniques. It is environment friendly, less expensive and useful for the destruction of a wide variety of contaminants.

Colloids Colloid or colloidal solution may be defined as the mixture of microscopically dispersed insoluble particles ranging from 1 to 1000 nm in diameter, suspended on another substance. Aerosol, emulsion, sol and foam are different types of colloids.

Endocytosis Endocytosis is an energy-dependent cellular process through which cells internalize extracellular materials or substances by engulfing them by plasma membrane invagination.

Hazardous Chemicals Hazardous chemicals are referred to those substances or mixtures that impose deleterious effects on environment and human health. These chemicals may act as an irritant, corrosive, flammable or explosive etc. Cleaning and disinfecting agents, mercury and pesticides are some common hazardous chemicals in health sector.

Nanotechnology Nanotechnology is a branch of science and engineering that deals with producing, designing and manipulating of matter near the atomic scale. Medicine, consumer products, energy sector undergoes scientific advancement through this technology.

Nanobiotechnology Nanobiotechnology is an emerging field of science that deals with physicochemical and biological properties of nanoscale structures to investigate biological systems and develop new medical technologies. It is useful in the medical and agricultural sectors. For example, anti-fungal properties of metal nanoparticles are useful in both humans and plants.

Oxidative Stress It is a stress condition caused due to imbalance between ROS synthesis and accumulation. Oxidative stress imparts damage to cells and tissues.

Reducing Agent Reducing agent or reductant is an electron donor that donates its electron to reduce a reactant in oxidation reduction chemical reaction. Hydrogen, formic acid, oxalic acid and alkali metals are common reducing agents.

Reactive Oxygen Species (ROS) It is an unstable, highly reactive molecule composed of radical oxygen, produced as a toxic by-product of cellular metabolism.

Sewage Management Sewage management refers to a process of collecting, treating and disposing wastewater generated from household, industries and commercial sector in favour of public health and environment.

Volcanic Eruption It is a natural phenomenon of eruption of molten rock, rock fragments, and hot gases through the vent of volcano on the earth surface.

Zeta Potential Zeta potential may be defined as the physical properties that are expressed as the difference of electric potential between the surface fluid layer of dispersed phase and dispersion medium.

References

Albalawi, F., Hussein, M.Z., Fakurazi, S., and Masarudin, M.J. (2021). Engineered nanomaterials: the challenges and opportunities for nanomedicines. *International Journal of Nanomedicine* 16 (January): 161–184. https://doi.org/10.2147/IJN.S288236.

Albanese, A., Tang, P.S., and Chan, W.C.W. (2012). The effect of nanoparticle size, shape, and surface chemistry on biological systems. *Annual Review of Biomedical Engineering* 14. Annual Reviews: 1–16.

Allen, T.M. and Cullis, P.R. (2013). Liposomal drug delivery systems: from concept to clinical applications. *Advanced Drug Delivery Reviews* 65 (1): 36–48. https://doi.org/10.1016/j.addr.2012.09.037.

Arana, L., Bayón-Cordero, L., Sarasola, L. et al. (2019). Solid lipid nanoparticles surface modification modulates cell internalization and improves chemotoxic treatment in an oral carcinoma cell line. *Nanomaterials* 9 (3): 464. https://doi.org/10.3390/nano9030464.

Arredondo-Ochoa, T. and Silva-Martínez, G.A. (2022). Microemulsion based nanostructures for drug delivery. *Frontiers in Nanotechnology* 3 (January): 753947. https://doi.org/10.3389/fnano.2021.753947.

Auría-Soro, C., Nesma, T., Juanes-Velasco, P. et al. (2019). Interactions of nanoparticles and biosystems: microenvironment of nanoparticles and biomolecules in nanomedicine. *Nanomaterials* 9 (10): 1365. https://doi.org/10.3390/nano9101365.

Baek, J.-S. and Cho, C.-W. (2017). Surface modification of solid lipid nanoparticles for oral delivery of curcumin: improvement of bioavailability through enhanced cellular uptake, and lymphatic uptake. *European Journal of Pharmaceutics and Biopharmaceutics* 117 (August): 132–140. https://doi.org/10.1016/j.ejpb.2017.04.013.

Baranowska-Wójcik, E., Szwajgier, D., Oleszczuk, P., and Winiarska-Mieczan, A. (2020). Effects of titanium dioxide nanoparticles exposure on human health—a review. *Biological Trace Element Research* 193 (1): 118–129. https://doi.org/10.1007/s12011-019-01706-6.

Brohi, R.D., Wang, L., Talpur, H.S. et al. (2017). Toxicity of nanoparticles on the reproductive system in animal models: a review. *Frontiers in Pharmacology* 8 (September): 606. https://doi.org/10.3389/fphar.2017.00606.

Caminade, A., Laurent, R., and Majoral, J. (2005). Characterization of dendrimers. *Advanced Drug Delivery Reviews* 57 (15): 2130–2146. https://doi.org/10.1016/j.addr.2005.09.011.

Cele, T. (2020). Preparation of nanoparticles. In: *Engineered Nanomaterials – Health and Safety* (ed. S.M. Avramescu, K. Akhtar, I. Fierascu, et al.). IntechOpen. https://doi.org/10.5772/intechopen.90771.

Chakraborty, S., Singh, A., and Roychoudhury, A. (2022). Biogenic nanoparticles and generation of abiotic stress-resilient plants: a new approach for sustainable agriculture. *Plant Stress* 6: 100117.

Chauhan, A. (2018). Dendrimers for drug delivery. *Molecules* 23 (4): 938. https://doi.org/10.3390/molecules23040938.

Cunningham, F.J., Goh, N.S., Demirer, G.S. et al. (2018). Nanoparticle-mediated delivery towards advancing plant genetic engineering. *Trends in Biotechnology* 36 (9): 882–897. https://doi.org/10.1016/j.tibtech.2018.03.009.

Delley, B. and Steigmeier, E.F. (1993). Quantum confinement in Si nanocrystals. *Physical Review B* 47 (3): 1397–1400. https://doi.org/10.1103/PhysRevB.47.1397.

Dhand, C., Dwivedi, N., Loh, X.J. et al. (2015). Methods and strategies for the synthesis of diverse nanoparticles and their applications: a comprehensive overview. *RSC Advances* 5 (127): 105003–105037. https://doi.org/10.1039/C5RA19388E.

Ealia, S.A.M. and Saravanakumar, M.P. (2017). A review on the classification, characterisation, synthesis of nanoparticles and their application. *IOP Conference Series: Materials Science and Engineering* 263 (November): 032019. https://doi.org/10.1088/1757-899X/263/3/032019.

Ferdous, Z. and Nemmar, A. (2020). Health impact of silver nanoparticles: a review of the biodistribution and toxicity following various routes of exposure. *International Journal of Molecular Sciences* 21 (7): 2375. https://doi.org/10.3390/ijms21072375.

Gao, L., Zhang, D., and Chen, M. (2008). Drug nanocrystals for the formulation of poorly soluble drugs and its application as a potential drug delivery system. *Journal of Nanoparticle Research* 10 (5): 845–862. https://doi.org/10.1007/s11051-008-9357-4.

Gehr, P. (2018). Interaction of nanoparticles with biological systems. *Colloids and Surfaces B: Biointerfaces* 172 (December): 395–399. https://doi.org/10.1016/j.colsurfb.2018.08.023.

Gour, A. and Jain, N.K. (2019). Advances in green synthesis of nanoparticles. *Artificial Cells, Nanomedicine, and Biotechnology* 47 (1): 844–851. https://doi.org/10.1080/21691401.2019.1577878.

Goyal, P., Goyal, K., Kumar, S.G.V. et al. (2005). Liposomal drug delivery systems-clinical applications. *Acta Pharmaceutica* 55 (1): 1–25.

Gupta, R. and Xie, H. (2018). Nanoparticles in daily life: applications, toxicity and regulations. *Journal of Environmental Pathology, Toxicology and Oncology* 37 (3): 209–230.

Hasan, S. (2015). A review on nanoparticles: their synthesis and types. *Research Journal of Recent Sciences* 2277: 2502.

Heiligtag, F.J. and Niederberger, M. (2013). The fascinating world of nanoparticle research. *Materials Today* 16 (7–8): 262–271. https://doi.org/10.1016/j.mattod.2013.07.004.

Hong, J., Wang, C., Wagner, D.C. et al. (2021). Foliar application of nanoparticles: mechanisms of absorption, transfer, and multiple impacts. *Environmental Science: Nano* 8 (5): 1196–1210. https://doi.org/10.1039/D0EN01129K.

Hossain, Z., Mustafa, G., and Komatsu, S. (2015). Plant responses to nanoparticle stress. *International Journal of Molecular Sciences* 16 (11): 26644–26653. https://doi.org/10.3390/ijms161125980.

Hou, D.Z., Xie, C.S., Huang, K.J., and Zhu, C.H. (2003). The production and characteristics of solid lipid nanoparticles (SLNs). *Biomaterials* 24 (10): 1781–1785. https://doi.org/10.1016/S0142-9612(02)00578-1.

Hu, P., An, J., Faulkner, M.M. et al. (2020). Nanoparticle charge and size control foliar delivery efficiency to plant cells and organelles. *ACS Nano* 14 (7): 7970–7986. https://doi.org/10.1021/acsnano.9b09178.

Hussain, I., Singh, N.B., Singh, A. et al. (2016). Green synthesis of nanoparticles and its potential application. *Biotechnology Letters* 38 (4): 545–560. https://doi.org/10.1007/s10529-015-2026-7.

Ijaz, I., Gilani, E., Nazir, A., and Bukhari, A. (2020). Detail review on chemical, physical and green synthesis, classification, characterizations and applications of nanoparticles. *Green Chemistry Letters and Reviews* 13 (3): 223–245.

Khan, Z. and Upadhyaya, H. (2019). Impact of nanoparticles on abiotic stress responses in plants: an overview. In: *Nanomaterials in Plants, Algae and Microorganisms* (ed. D.K. Tripathi, P. Ahmad, S. Sharma, et al.), 305–322. Elsevier.

Kharissova, O.V., Rasika Dias, H.V., Kharisov, B.I. et al. (2013). The greener synthesis of nanoparticles. *Trends in Biotechnology* 31 (4): 240–248. https://doi.org/10.1016/j.tibtech.2013.01.003.

Kim, M., Osone, S., Kim, T. et al. (2017). Synthesis of nanoparticles by laser ablation: a review. *Kona Powder and Particle Journal* 34: 80–90. https://doi.org/10.14356/kona.2017009.

Kumar, V., Sharma, M., Khare, T., and Wani, S.H. (2018). Impact of nanoparticles on oxidative stress and responsive antioxidative defense in plants. In: *Nanomaterials in Plants, Algae, and Microorganisms* (ed. D.K. Tripathi, P. Ahmad, S. Sharma, et al.), 393–406. Elsevier https://doi.org/10.1016/B978-0-12-811487-2.00017-7.

Langer, R. (1990). New methods of drug delivery. *Science* 249 (4976): 1527–1533. https://doi.org/10.1126/science.2218494.

Lawrence, M.J. and Rees, G.D. (2012). Microemulsion-based media as novel drug delivery systems. *Advanced Drug Delivery Reviews* 64 (December): 175–193. https://doi.org/10.1016/j.addr.2012.09.018.

Lee, C.C., MacKay, J.A., Fréchet, J.M.J., and Szoka, F.C. (2005). Designing dendrimers for biological applications. *Nature Biotechnology* 23 (12): 1517–1526. https://doi.org/10.1038/nbt1171.

Lv, J., Christie, P., and Zhang, S. (2019). Uptake, translocation, and transformation of metal-based nanoparticles in plants: recent advances and methodological challenges. *Environmental Science: Nano* 6 (1): 41–59. https://doi.org/10.1039/C8EN00645H.

Lv, Z., Jiang, R., Chen, J., and Chen, W. (2020). Nanoparticle-mediated gene transformation strategies for plant genetic engineering. *The Plant Journal* 104 (4): 880–891. https://doi.org/10.1111/tpj.14973.

Manna, I. and Bandyopadhyay, M. (2019). A review on the biotechnological aspects of utilizing engineered nanoparticles as delivery systems in plants. *Plant Gene* 17 (March): 100167. https://doi.org/10.1016/j.plgene.2018.100167.

Mehnert, W. and Mäder, K. (2012). Solid lipid nanoparticles. *Advanced Drug Delivery Reviews* 64 (December): 83–101. https://doi.org/10.1016/j.addr.2012.09.021.

Misra, S.K. and Pathak, K. (2022). Drug nanocrystals as drug delivery systems. In: *Industrial Applications of Nanocrystals* (ed. S. Mallakpour and C.M. Hussain), 153–178. Elsevier. https://doi.org/10.1016/B978-0-12-824024-3.00026-9.

Mohanraj, V.J. and Chen, Y.J.T.J.O.P.R. (2006). Nanoparticles – a review. *Tropical Journal of Pharmaceutical Research* 5 (1): 561–573.

More, T.T., Yadav, J.S.S., Yan, S. et al. (2014). Extracellular polymeric substances of bacteria and their potential environmental applications. *Journal of Environmental Management* 144 (November): 1–25. https://doi.org/10.1016/j.jenvman.2014.05.010.

Mu, Q., Jiang, G., Chen, L. et al. (2014). Chemical basis of interactions between engineered nanoparticles and biological systems. *Chemical Reviews* 114 (15): 7740–7781. https://doi.org/10.1021/cr400295a.

Mujtaba, M., Khawar, K.M., Camara, M.C. et al. (2020). Chitosan-based delivery systems for plants: a brief overview of recent advances and future directions. *International Journal of Biological Macromolecules* 154 (July): 683–697. https://doi.org/10.1016/j.ijbiomac.2020.03.128.

Naahidi, S., Jafari, M., Edalat, F. et al. (2013). Biocompatibility of engineered nanoparticles for drug delivery. *Journal of Controlled Release* 166 (2): 182–194. https://doi.org/10.1016/j.jconrel.2012.12.013.

Nadaroglu, H., Azize, A.G., and Selvi, I. (2017). Synthesis of nanoparticles by green synthesis method. *International Journal of Innovative Research and Reviews* 1 (1): 6–9.

Nagpal, K., Singh, S.K., and Mishra, D.N. (2010). Chitosan nanoparticles: a promising system in novel drug delivery. *Chemical and Pharmaceutical Bulletin* 58 (11): 1423–1430.

Naser, H., Alghoul, M.A., Hossain, M.K. et al. (2019). The role of laser ablation technique parameters in synthesis of nanoparticles from different target types. *Journal of Nanoparticle Research* 21 (11): 249. https://doi.org/10.1007/s11051-019-4690-3.

Naseri, N., Valizadeh, H., and Zakeri-Milani, P. (2015). Solid lipid nanoparticles and nanostructured lipid carriers: structure, preparation and application. *Advanced Pharmaceutical Bulletin* 5 (3): 305–313. https://doi.org/10.15171/apb.2015.043.

Nel, A., Xia, T., Madler, L., and Li, N. (2006). Toxic potential of materials at the nanolevel. *Science* 311 (5761): 622–627.

Pal, G., Rai, P., and Pandey, A. (2019). Green synthesis of nanoparticles: a greener approach for a cleaner future. In: *Green Synthesis, Characterization and Applications of Nanoparticles* (ed. A.K. Shukla and S. Iravani), 1–26. Elsevier.

Pedziwiatr-Werbicka, E., Milowska, K., Dzmitruk, V. et al. (2019). Dendrimers and hyperbranched structures for biomedical applications. *European Polymer Journal* 119 (October): 61–73. https://doi.org/10.1016/j.eurpolymj.2019.07.013.

Pietroiusti, A., Stockmann-Juvala, H., Lucaroni, F., and Savolainen, K. (2018). Nanomaterial exposure, toxicity, and impact on human health. *WIREs Nanomedicine and Nanobiotechnology* 10 (5): e1513. https://doi.org/10.1002/wnan.1513.

Rajput, N. (2015). Methods of preparation of nanoparticle – a review. *International Journal of Advances in Engineering & Technology* 7 (6): 1806.

Rane, A.V., Kanny, K., Abitha, V.K., and Thomas, S. (2018). Methods for synthesis of nanoparticles and fabrication of nanocomposites. In: *Synthesis of Inorganic Nanomaterials* (ed. S.M. Bhagyaraj, O.S. Oluwafemi, N. Kalarikkal, and S. Thomas), 121–139. Elsevier. https://doi.org/10.1016/B978-0-08-101975-7.00005-1.

Rizvi, S.A.A. and Saleh, A.M. (2018). Applications of nanoparticle systems in drug delivery technology. *Saudi Pharmaceutical Journal* 26 (1): 64–70. https://doi.org/10.1016/j.jsps.2017.10.012.

Roy, R., Kumar, S., Tripathi, A. et al. (2014). Interactive threats of nanoparticles to the biological system. *Immunology Letters* 158 (1–2): 79–87.

Roychoudhury, A. (2020). Yeast-mediated green synthesis of nanoparticles for biological applications. *Indian Journal of Pharmaceutical and Biological Research* 8 (3): 26–31.

Roychoudhury, A. and Bhowmik, R. (2023). Nanoremediation of heavy metals in agricultural soil. In: *Agricultural and Environmental Nanotechnology* (ed. F. Fernandez-Luqueno and J.K. Patra), 433–450. Springer Nature Singapore Pte Ltd.

Roychoudhury, A., Sarkar, S., and Chakraborty, S. (2023). Fungal-mediated synthesis of gold and titanium nanoparticles and their application in agriculture. In: *Myconanotechnology and Application of Nanoparticles in Biology* (ed. M. Kuddus, I.Z. Ahmad, and C.M. Hussain), 79–92. Academic Press (Elsevier).

Sajid, M., Ilyas, M., Basheer, C. et al. (2015). Impact of nanoparticles on human and environment: review of toxicity factors, exposures, control strategies, and future prospects. *Environmental Science and Pollution Research* 22 (6): 4122–4143. https://doi.org/10.1007/s11356-014-3994-1.

Sharma, A. (1997). Liposomes in drug delivery: progress and limitations. *International Journal of Pharmaceutics* 154 (2): 123–140. https://doi.org/10.1016/S0378-5173(97)00135-X.

Shoala, T. (2018). Positive impacts of nanoparticles in plant resistance against different stimuli. In: *Nanobiotechnology Applications in Plant Protection*, Nanotechnology in the Life Sciences (ed. K.A. Abd-Elsalam and R. Prasad), 267–279. Cham: Springer International Publishing. https://doi.org/10.1007/978-3-319-91161-8_10.

Siddiqui, M.H., Al-Whaibi, M.H., Firoz, M., and Al-Khaishany, M.Y. (2015). Role of nanoparticles in plants. In: *Nanotechnology and Plant Sciences* (ed. M.H. Siddiqui, M.H. Al-Whaibi, and F. Mohammad), 19–35. Cham: Springer International Publishing. https://doi.org/10.1007/978-3-319-14502-0_2.

Singh, P., Kim, Y.-J., Zhang, D., and Yang, D.-C. (2016). Biological synthesis of nanoparticles from plants and microorganisms. *Trends in Biotechnology* 34 (7): 588–599. https://doi.org/10.1016/j.tibtech.2016.02.006.

Sinha, B., Müller, R.H., and Möschwitzer, J.P. (2013). Bottom-up approaches for preparing drug nanocrystals: formulations and factors affecting particle size. *International Journal of Pharmaceutics* 453 (1): 126–141. https://doi.org/10.1016/j.ijpharm.2013.01.019.

Smith, A.M. and Nie, S. (2010). Semiconductor nanocrystals: structure, properties, and band gap engineering. *Accounts of Chemical Research* 43 (2): 190–200. https://doi.org/10.1021/ar9001069.

Smith, R.J., Gorman, C., and Menegatti, S. (2021). Synthesis, structure, and function of internally functionalized dendrimers. *Journal of Polymer Science* 59 (1): 10–28. https://doi.org/10.1002/pol.20200721.

Sohrabi, R., Paasch, B.C., Liber, J.A., and He, S.Y. (2023). Phyllosphere microbiome. *Annual Review of Plant Biology* 74 (1): 539–568. https://doi.org/10.1146/annurev-arplant-102820-032704.

Solanki, P., Bhargava, A., Chhipa, H. et al. (2015). Nano-fertilizers and their smart delivery system. In: *Nanotechnologies in Food and Agriculture* (ed. M. Rai, C. Ribeiro, L. Mattoso, and N. Duran), 81–101. Cham: Springer International Publishing. https://doi.org/10.1007/978-3-319-14024-7_4.

Stark, W.J. (2011). Nanoparticles in biological systems. *Angewandte Chemie International Edition* 50 (6): 1242–1258. https://doi.org/10.1002/anie.200906684.

Su, Y., Ashworth, V., Kim, C. et al. (2019). Delivery, uptake, fate, and transport of engineered nanoparticles in plants: a critical review and data analysis. *Environmental Science: Nano* 6 (8): 2311–2331. https://doi.org/10.1039/C9EN00461K.

Suryanarayana, C. and Froes, F.H. (1992). The structure and mechanical properties of metallic nanocrystals. *Metallurgical Transactions A* 23 (4): 1071–1081. https://doi.org/10.1007/BF02665039.

Svenson, S. and Tomalia, D.A. (2012). Dendrimers in biomedical applications—reflections on the field. *Advanced Drug Delivery Reviews* 64 (December): 102–115. https://doi.org/10.1016/j.addr.2012.09.030.

Tiwari, G., Tiwari, R., SaurabhK Bannerjee, L. et al. (2012). Drug delivery systems: an updated review. *International Journal of Pharmaceutical Investigation* 2 (1): 2. https://doi.org/10.4103/2230-973X.96920.

Tripathy, S. (2023). Top-down and bottom-up approaches for synthesis of nanoparticles. In: *Materials Research Foundations*, 1e, vol. 145 (ed. J.N. Cruz), 92–130. Materials Research Forum LLC. https://doi.org/10.21741/9781644902370-4.

Upadhyaya, G. and Roychoudhury, A. (2024). Potential of silver and zinc nanoparticles in mediating abiotic stress tolerance in crop plants. In: *Metal and Metal-Oxide Based Nanomaterials* (ed. R. Bachheti, A. Bachheti, and A. Husen), 157–177. Springer Nature Singapore.

Vishwakarma, K., Shweta, N.U., Singh, J. et al. (2017). Differential phytotoxic impact of plant mediated silver nanoparticles (AgNPs) and silver nitrate ($AgNO_3$) on *Brassica* Sp. *Frontiers in Plant Science* 8 (October): 1501. https://doi.org/10.3389/fpls.2017.01501.

Wilczewska, A.Z., Niemirowicz, K., Markiewicz, K.H., and Car, H. (2012). Nanoparticles as drug delivery systems. *Pharmacological Reports* 64 (5): 1020–1037. https://doi.org/10.1016/S1734-1140(12)70901-5.

Yan, A. and Chen, Z. (2019). Impacts of silver nanoparticles on plants: a focus on the phytotoxicity and underlying mechanism. *International Journal of Molecular Sciences* 20 (5): 1003. https://doi.org/10.3390/ijms20051003.

Ying, S., Guan, Z., Ofoegbu, P.C. et al. (2022). Green synthesis of nanoparticles: current developments and limitations. *Environmental Technology & Innovation* 26 (May): 102336. https://doi.org/10.1016/j.eti.2022.102336.

6

Modes of Applications of Nanoparticles in Plants and Animals

Braulio Edgar Herrera Cabrera[1], Rafael Salgado Garciglia[2], Victor Manuel Ocaño Higuera[3], Luis Germán López Valdez[4], Alan Mendoza Alcantara[5], Francisco Javier Verduzco Miramón[5] and Hebert Jair Barrales Cureño[5]

[1]*Campus Puebla, Colegio de Postgraduados, Puebla, Mexico*
[2]*Instituto de Investigaciones Químico Biológicas, Universidad Michoacana de San Nicolás de Hidalgo, Morelia, México*
[3]*Departamento de Ciencias Químico Biológicas, Universidad de Sonora, Hermosillo, Mexico*
[4]*Laboratorio de Productos Naturales, Área de Química, Universidad Autónoma Chapingo, Texcoco, México*
[5]*Carrera en Ingeniería en Innovación Agrícola Sustentable, Tecnológico Nacional de México, Zamora, Mexico*

6.1 Introduction

Currently, one of the main problems in the area of human health is the low efficiency of drug delivery systems (DDSs), causing serious side effects caused by their use (Singh and Roychoudhury 2022). The nanometer scale is ideal in terms of delivery vehicles, since there are larger materials that are more likely to cause acute immune responses, since their sizes are within the range of bacteria and the efficiency of cellular uptake decreases. In recent years, the use of nanoparticles (NPs) in both biomedical research and industry has increased exponentially (Roychoudhury and Singh 2024). Scientific advances are based on the use of nanoparticles as transport vehicles, information systems, actuation surfaces, catalytic applications, new materials or biomedicines (Llinás 2014). Recently, advances in drug delivery has facilitated the choice of specific tissues. With the advent of nanotechnology, these tissues become specific organelles within individual cells. Nanomedicines are products that are transformed into a billion-dollar industry due to the inherent ability of their components to overcome solubility and stability issues, direct drug delivery to specific targets, as well as diagnose in vivo through imaging. NPs-SiO_2 are excellent marker agents for real-time imaging and are ideal systems for the design of non-toxic carriers for controlled DDSs. These types of nanoparticles are often an exceptional choice for the release of

Silicon Nanoparticles: Synthesis, Uptake, and Applications, First Edition.
Edited by Archana Singh, Aryadeep Roychoudhury, and Indrakant K. Singh.

substances of pharmacological, cosmetic or nutritional interest among inorganic nanoparticles, mainly due to their chemical versatility, biocompatibility and high stability (Aravind et al. 2012). Currently, mesoporous silica nanoparticles (MSNs) are making an impact due to their potential properties for biomedical applications. Due to their large surface area and mesoporous structure, MSNs as DDSs show many significant advantages compared to other traditional drug nanocarriers (Tang et al. 2012). In the last decade, MSNs have attracted strong attention due to their potential properties for biomedical applications. Due to their mesoporous structure and large surface area, MSNs DDSs show significant advantages compared to other traditional drug nanocarriers (Tang et al. 2012). The advantages of MSNs are mainly related to uniform mesopores, easy surface functionalization and significant biocompatibility, which have been very useful and functionalized in the last two decades for biomedical applications.

6.2 Fabrication and Characteristics of Si/SiO_2 Nanoparticles

The large surface area and porous compartments provide a natural platform to construct a multifunctional theranostic agent. The unique topology functionalizes MSNs three distinct domains which are the silica structure, the nanochannels/nanopores and the outermost surface of the nanoparticles (Wu et al. 2011). Tang et al. (2012) observed that inorganic nanomaterials have unique physicochemical properties and special structures. Among inorganic materials, MSNs possess a mesoporous structure, a pore volume that makes them outstanding compared to other porous materials and a specific surface area. These properties give them unique benefits for encapsulating a wide variety of therapeutic agents with applications in medicine. Importantly, the fabrication of MSNs is simple, scalable, inexpensive and controllable.

MSNs offer a wide variety of applications due to their different properties; therefore, for their biomedical application, precise control over the size and shape of the nanoparticle is very important. These nanoparticles have excellent textural and structural properties, being able to host various and multiple drugs that will later be released on specific tissues in a controlled manner. Some laboratory tests have allowed the evaluation of mesoporous Si nanoparticles, which have shown excellent reactivity on their surface and, therefore, can be functionalized with molecules of different sizes, shapes and functions with the ability to respond to a stimulus causing a change in their properties, and, as a consequence, the controlled release of the drug occurs (Ambrogio et al. 2011; Valle-Regí 2022).

Particle size affects the cellular uptake pathway and efficiency by influencing particle adhesion and cell interaction. Vallhov et al. (2007) reported that SiO_2 particle size is an important factor affecting the response of human dendritic cells. Particle size is considered to be one of the most important properties of nanomaterials. The effect of nanoparticle size on their biocompatibility has also been evaluated. Tang et al. (2012) reported that particle size is one of the most important parameters for mediating the biocompatibility of nanomaterials. Lu et al. (2009) observed that for biomedical applications, the size of MSNs is important because it can affect the efficiency and cellular uptake pathway by influencing particle adhesion and cellular interaction. The diameter of MSNs greatly affects their biological effects, such as their capacity for cellular uptake and inducing hemolysis. Hemolysis is the destruction of red blood cells; the importance of this not occurring lies in the fact that regeneration of red blood cells from precursor cells is inadequate to replenish the destroyed cells (Silverstein and Hopper 2014).

6.3 Design of Silica Nanoparticles

There are different methods to control the diameters of silica nanoparticles. Particle size is controlled through the addition of suitable additive agents such as alcohols, amines, bases or inorganic salts. These agents alter the hydrolysis and condensation of the silica precursor. They act as catalysts, resulting in smaller particle sizes (Narayan et al. 2018; Roychoudhury 2020); the above is because the nucleation mechanisms and growth of MSNs are strongly affected by such factors. There are other factors that affect the size of MSNs, one of the most notorious being temperature; for example, Yu et al. (2012) reported the production of colloidal MSNs at different temperature and found that the diameter of MSNs increased proportionally with increasing reaction temperature.

Stöber's method is a method of obtaining SiO_2 nanoparticles (Yamamoto and Kuroda 2016). In 1968, the first publication on the use of silica particles for a biomedical application was reported (Figure 6.1).

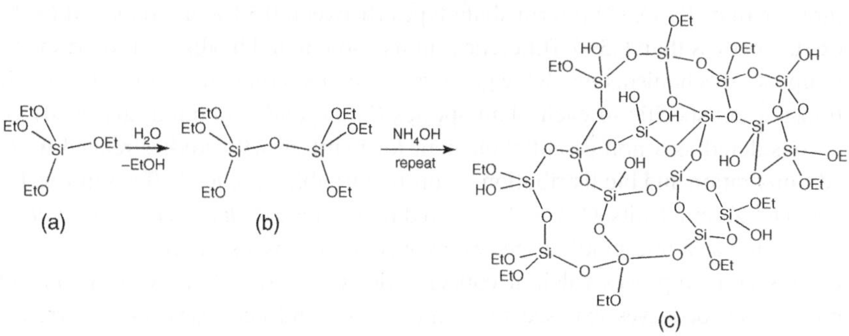

Figure 6.1 Simplified representation of the hydrolysis and condensation of tetraethyl orthosilicate (TEOS) in the Stöber method.

The synthesis method is facile, involving the condensation of tetraethyl orthosilicate (TEOS) in ethanol. Stöber's method allows obtaining an optimal control over the size, particle size distribution and an ideal spherical morphology for the resulting silica particles (Rossi et al. 2005).

Stöber's method uses TEOS as the precursor material; TEOS is dissolved in an organic solvent such as ethanol. Subsequently, a controlled hydrolysis is carried out to produce the silica precursor. The precursor is condensed with ammonium hydroxide (NH_4OH), which synthesizes the Si—O—Si bond, producing water as a by-product, to finally obtain the silica structure. The size of the silicon dioxide nanoparticles is controlled taking into account the variants of their synthesis, the most significant being the temperature. The higher the temperature, the smaller the silica nanoparticles will be obtained, and conversely, the lower the temperature in the preparation process, the larger nanoparticles will be produced.

6.4 Delivery of SiO_2 in Plants

At present, anthropogenic activities cause a series of environmental problems that pose a risk to the development of agriculture. Current problems constitute the contamination of water and soil with heavy metals, which due to their characteristics (toxicity, bioavailability, bioaccessibility, persistence and high solubility) cause various problems in living organisms (Ruíz-Huerta et al. 2017). The use of bio-nanotechnology in agronomy and environmental protection has grown exponentially in recent years due to the physicochemical properties of nanomaterials (Cui et al. 2017). Silicon (Si) is the second most abundant element in the earth's crust. Silicon is a normal component in plants, ranging from 0.1% to 10.0%. The function of Si involves strengthening the cell walls as well as reducing metal accumulation in plants (Roychoudhury 2020). According to Si concentration, plant species are classified into three groups: (1) Si accumulators (greater than 1.0%); (2) intermediate types (between 0.5% and 1.0%), and (3) Si excluders (less than 0.5%). Differences in its content and biodistribution respond to uptake mechanisms involving genotypic responses that are regulated through transporters specific to each plant species (Singh and Roychoudhury 2020). Si is considered as a non-essential element for plant growth; studies related to its administration and biodistribution report remarkable beneficial effects in vegetables, cereals and fruits. Cao et al. observed that tomato (*Solanum lycopersicum*) is considered as a Si-excluding species because it absorbs less than 0.5%. Si uptake by this crop improves calcium concentrations in leaves and fruits, alters the metabolism of leaves exposed to salinity stress, and decreases the occurrence of blossom end rot, increasing the total number of fruits and productivity. Si offers plant resistance to pests and diseases, increases in yield and availability of phosphorus in the soil, drought tolerance, decreased toxicity to aluminum, iron and manganese by forming silicates and tolerance to salinity. Silicon is absorbed by the plant in the form of orthosilicic acid [$Si(OH)_4$] and metasilicic acid (H_2SiO_3). Si is found in soil solution within the range of 0.1–0.6 mol/m^3. Once adsorbed, it is transformed to silicon oxide (SiO_2) in order to be transported into the plant. This molecule accumulates and changes its composition to a colloidal gel that is absorbed under the epidermis and crystallizes on the surface.

On the other hand, silicon is assimilated in three ways: active, passive and selective. The active form involves energy expenditure as in corn, barley, rice and sugar cane crops. Passive assimilation refers to assimilation by the amount present in the soil solution, for example in strawberry, melon, cucumber and soybean. Finally, the selective form is based on its absorption depending on the phenological stage of the crop, such as tomato, where higher concentrations of silicon are absorbed in the flowering stage compared to the vegetative growth stage. Silicon applications are common in Japan and Korea as they increase rice production; in countries such as India, Brazil, South Africa and Australia, it is used as a fertilizer in sugarcane production (Raya and Aguirre 2012).

6.5 Pest and Disease Resistance in Plants

Regarding pests and diseases, Si application reduces by 50% the number of lesions caused by *Frankliniella schultzei* and by 56.2% the incidence of *Ralstonia solanacearum*. In rice (*Oryza sativa*), Si concentrations range from 0.1 to 0.6 millimoles (mM). Guntamukkala and Pusarla observed that Si uptake occurs via lateral roots and its translocation helps the plant to keep its leaves firm, minimize transpiration, and increase plant resistance against fungal diseases and insect pests. Borda et al. observed that in oats (*Avena sativa*), Si helps to stimulate crop growth and productivity by increasing the availability of different elements such as P, Ca, K, etc. Likewise, Si can help reduce salt stress in herbaceous plants such as carnation (*Dianthus caryophyllus*), as well as increase pest resistance in maize plants to prevent crop death by *Rhopalosiphum maydis*. Silicon is an element with multiple benefits; it improves soil conditions and nutrient uptake in plants (Zargar et al. 2019). It increases Ca and K uptake in maize leaves under abiotic stress (Kaya et al. 2006). Minimal concentrations of silicon enhance K uptake both edaphologically and hydroponically by activating H^+-ATPase (Mali and Aery 2008). The mechanism of silicon uptake varies among different plant species and depends on the presence of specific transporters of the element (Kaur and Greger 2019).

Silicon-induced tolerance to heavy metal stress is generated through mechanisms including decreased metal ion concentration in the substrate or soil, co-precipitation of toxic metals, chelation (metal ions exist in solution in a highly hydrated form, surrounded by water molecules and upon replacement of water molecules by a chelating agent, a complex ring structure is formed), stimulation of antioxidants and structural alterations in plants (Bhat et al. 2019; Samanta and Roychoudhury 2021).

Silicon nanoparticles reduce the toxic effect of As on plants (Tripathi et al. 2015). For example, application of Si nanoparticles increases the photosynthetic rate even in the presence of cadmium, facilitating nutrient transport through the xylem (Gao et al. 2018) and increase water and nutrient use efficiency (Alsaeedi et al. 2019).

González-Moscoso et al. (2021) reported that tomato plants were affected in their growth by both As and SiO_2 nanoparticles. As decreased root and aerial part biomass at low concentrations, while they were stimulated at high concentrations. SiO_2 nanoparticles did not develop toxicity symptoms such as chlorosis or necrosis due to As application. SiO_2 nanoparticles positively influenced the uptake and concentration of Zn and Cu in roots and Fe in leaves. In addition, the uptake of macronutrients in roots and leaves was increased with SiO_2 nanoparticles.

6.6 Delivery in Animals

Nanoscale carriers for targeted and release drug delivery are a challenge for pharmacotherapy because nanoparticles improve the pharmacokinetics of loaded therapeutic drugs based on their physical properties. Tang et al. (2017) designed 5-ASA–loaded silicon dioxide nanoparticles using microemulsion method, validating the therapeutic effects in mouse model of ulcerative colitis. The 5-ASA-SiO_2 nanoparticles are a selective drug delivery system targeting the inflamed colon and can greatly enhance the therapeutic efficacy in ulcerative colitis.

MSNs were first proposed in 2001 as a possible drug delivery system, and since then, devices based on MSNs have been developed that generate controlled release. The surface of the nanoparticles can be functionalized using organic components such as gates or ligands and hydrophilic polymers, allowing the nanoparticles to be targeted to specific tissues or cell types. This type of devices without premature release of the cargo are particularly useful when the cargo is toxic, as occurring in case of antitumor drugs, in which the release is directed to a specific tissue or cell type. The coating of nanoparticles with organic materials facilitates anchoring to ligands for specific cellular recognition, thus improving their biocompatibility (Argyo et al. 2013; Slowing et al. 2008). Regarding the biocompatibility of MSNs, viability and proliferation studies have been performed with different mammalian cells showing that MSN internalization is not affected up to concentrations of 100 μg/ml per 10^5 cells even after seven cell cycles. In addition, the cells maintain normal cell morphology, mitochondrial activity, membrane integrity and growth rates. Regarding long-term biocompatibility and biodegradation, there are model animals in which silica implants have been performed, and no derived toxic effects have been observed within 42 days, leading to believe that MSNs would have at least the same biocompatibility as other silica materials. in vivo tests with MSNs show satisfactory results in terms of biodistribution and excretion, cytotoxicity and biodegradation. They have recently been recruited for clinical trials in the form of 'cornell dots', this being the first clinical trial with this type of materials and a major advance toward their clinical acceptance (Na et al. 2012). The synthesis of MSNs was first performed in 1992, and numerous studies have been conducted. Their synthesis is based on the use of silica precursors such as TEOS, sodium silicate, or tetramethylammonium silicate in the presence of cationic surfactants under basic conditions. The surfactants form micelles and, successively, supramicellar structures that act as molds and lead to pore formation when combined with the silica precursors. TEOS is usually used in combination with the cationic surfactant cetyltrimethylammonium bromide (CTAB) at pH 11. Finally, the surfactant is removed from inside the pores, which can be done by three methods: treatment with ammonium nitrate, refluxing in alcohol acidified with hydrochloric acid (HCl) or calcination. The removal of the surfactant from inside the pores allows the mesoporous nanoparticles to be obtained (Benezra et al. 2011; Wang et al. 2015).

6.7 Fluorescence Imaging

There are several strategies to conjugate a variety of dyes into the pores or on the surfaces of MSNs to form fluorescence MSNs (FMSNs) (Pratiwi et al. 2018). Using different synthesis techniques, fluorescent nanoparticles with controlled shapes and sizes have been synthesized by using $SiCl_4$, and these nanoparticles exhibited fluorescence near 2.95 eV (420 nm), originated from optically active defects of silanone and dioxasilyrano and were subsequently applied for in vitro bioimaging of lung epithelial carcinoma cells (A549) where a result without cytotoxicity was given (Chang et al. 2019).

Functionalization of nanoparticles with other elements also shows favorable results towards fluorescence. Araichimani et al. (2020) investigated nanoparticles integrated with europium which demonstrated strong red fluorescence imaging ability attributed to 5 D 0→7 F 1 and 5 D 0→7 F 2 transitions of europium ions in the silica host matrix. Similarly, Du et al. (2016) functionalized silsesquioxane-structured nanoparticles with bridged dual fluorophore in HeLa cell culture, where strong green and red dual fluorescence emission was exhibited in the cells under single wavelength excitation, and synchronous or asynchronous release of different drugs, suggesting its potential as an agent in bioimaging and as a fluorescently tractable drug carrier.

6.8 Phototherapy and MSN-Based Nanocarriers

Phototherapy is a technique using ultraviolet light to treat skin-related diseases. Recently, MSNs have been designed as nanocarriers for two types of phototherapy, viz., photothermal therapy (PTT) and photodynamic therapy (PDT) or their combination, which have been widely used for cancer treatments using photosensitizers and photothermal agents, respectively (Li et al. 2019). Depending on the types of cargo, MSN-based nanocarriers can be used for various purposes. For example, fluorescent dyes or photosensitizers have been physically trapped inside nanochannels or chemically attached to the surface of MSN-based nanocarriers for bioimaging and PDT in cancer treatment (Ma et al. 2017).

A strategy to prepare a multifunctional MSN-based PDT/PTT system was developed by Lv et al. (2017), using strong upconversion strong luminescent GdOF:Ln (UCL) GdOF:Ln nanoparticles as cores and coating them with mesoporous silica. The markedly enhanced red emission of Yb/Er/Mn co-doped on GdOF could efficiently transfer energy to the conjugated PDT agent (ZnPc) which in turn produced high singlet oxygen ($^1O_2$). The system demonstrated significantly improved therapeutic efficacy against tumor growth both in vitro and in vivo. Doping of NPs has also been used for phototherapy. de Souza Oliveira et al. (2016) doped nanoparticles with pink bengal (RB) and 9,10-anthraquinone-2-carboxylic acid (OCAq) for application in PDT against lung cancer cell lines; the result was that after 15 minutes, there was a cell survival of 20–30% for both materials and that the concentration of OCAq was half that of the RB nanomaterial.

6.9 Positron Emission Tomography

Positron emission tomography (PET) is based on obtaining tomographic images of the 3D distribution in the organism of some positron emitting radiopharmaceuticals, which represent certain biochemical processes in vivo. One of the limitations of this technique is the penetration of light into the tissue. To overcome this limitation, the pores of MSNs have been used to load them with radionuclides such as zirconium, arsenic, copper, technetium, titanium, fluorine, or carbon. The effect of nanoparticles in this technique is to increase both tissue selectivity and resolution due to higher overall signals. Nanoparticles with metal cores have also been used such as gold core spherical silica nanoparticles (AuMSS); the gold core has the ability to enhance image contrast in structural imaging modalities. This was studied by Kobayashi et al. (2010) where they prepared a colloidal solution of silica core gold nanoparticles with a gold core of 17 nm and a total particle size of 136 nm to be used as a contrast agent for computed tomography. In their work, the authors demonstrated that the attenuation capacity of AuMSS nanoparticles was almost seven times greater than that of Iopamiron®, a commercial iodine-based X-ray contrast agent.

6.10 Biosensors

Biosensors have also been used for the identification of vital protein constituents such as L-tyrosine which establishes a positive nitrogen balance. Rahimi-Mohseni et al. (2018) designed a new and simple method for the electrochemical determination of L-tyrosine using 3-mercaptopropyl trimethoxysilane functionalized nanoparticles where the biosensor showed a linear response for L-tyrosine in the wide concentration range of 0.05–600 μM and a low detection limit of about 0.02 μM due to the co-catalytic effect of the enzyme and the nanoparticles. Additionally, nanoparticles have also been used to amplify biomarker signals. Jimenez et al. (2019) used MSNs to amplify the detection of carcinoembryonic antigen (CEA), a relevant cancer biomarker by being loaded with the redox probe methylene blue and coated with an avidin/immunobiotin stimulus response gate-like assembly; such signal amplification can also be developed by fluorescence intensification.

Yun et al. (2019) prepared a metal-enhanced fluorescence (MEF) biosensor platform on fibrous substrates using a silica layer that acted as a spacer between silver nanoparticles and fluorescent molecules to optimize the MEF effect. Regarding pathogenic bacteria, their detection is of critical importance for health and safety issues in clinical diagnosis, food industry, water quality and environment, as well as fight against possible bioterrorism. Mathelié-Guinlet et al. (2019) presented results on the design of an electrochemical biosensor, in which the presence of nanoparticles, functionalized with specific polyclonal antibodies, is used to target and bind to *Escherichia coli* for detection and quantification.

6.11 Conclusion

Nanomedicines could generate individualized medicine. Future innovations in the field of nanomedicines could generate multifunctional entities, capable of simultaneously diagnosing, treating and monitoring personalized treatment. SiO_2 nanoparticles present potential applications including bioimaging, in which silica nanoparticles contribute to image resolution, tissue selectivity and enhancement of enhancement. For biosensors, the high porosity of silica provides extensive areas to immobilize sensing agents. With respect to phototherapy silica nanoparticles are used to enhance light reception. Fluorescence imaging and traceable fluorescent agents are loaded into the pores of the nanoparticles. Finally, tissue engineering is another area where silica is used to promote osteogenesis in scaffolds created from nanoparticles. The use of silica composites is emerging as an ideal icon to be used for prolonged drug delivery in animals along with multiple applications in the future.

References

Alsaeedi, A., El-Ramady, H., Alshaal, T. et al. (2019). Silica nanoparticles boost growth and productivity of cucumber under water deficit and salinity stresses by balancing nutrients uptake. *Plant Physiology and Biochemistry* 139: 1–10. https://doi.org/10.1016/j.plaphy.2019.03.008.

Ambrogio, M.W., Thomas, C.R., Zhao, Y. et al. (2011). Mechanized silica nanoparticles: a new frontier in theranostic nanomedicine. *Accounts of Chemical Research* 44 (10): 903–913. https://doi.org/10.1021/ar200018x.

Araichimani, P., Prabu, K., Suresh-Kumar, G. et al. (2020). Rare-earth ions integrated silica nanoparticles derived from rice husk via microwave-assisted combustion method for bioimaging applications. *Ceramics International* 46 (11): 18366–18372. https://doi.org/10.1016/j.ceramint.2020.04.125.

Aravind, A., Veeranarayanan, S., Poulose, A.C. et al. (2012). Aptamer-functionalized silica nanoparticles for targeted cancer therapy. *BioNanoScience* 2 (1): 1–8.

Argyo, C., Weiss, V., Bräuchle, C., and Bein, T. (2013). Multifunctional mesoporous silica nanoparticles as a universal platform for drug delivery. *Chemistry of Materials* 26 (1): 435–451.

Benezra, M., Penate-Medina, O., Zanzonico, P.B. et al. (2011). Multimodal silica nanoparticles are effective cancer-targeted probes in a model of human melanoma. *The Journal of Clinical Investigation* 121 (7): 2768–2780.

Bhat, J.A., Shivaraj, S.M., Singh, P. et al. (2019). Role of silicon in mitigation of heavy metal stresses in crop plants. *Plants* 8 (3): 71. https://doi.org/10.3390/plants8030071.

Chang, H., Yoo, J., Kim, H. et al. (2019). Controlled growth of fluorescent silica nanoparticles using two-phase orthogonal solvents for bioimaging. *Journal of Luminescence* 214: 1–8. https://doi.org/10.1016/j.jlumin.2019.116529.

Cui, J., Liu, T., Li, F. et al. (2017). Silica nanoparticles alleviate cadmium toxicity in rice cells: mechanisms and size effects. *Environmental Pollution* 228: 363–369. https://doi.org/10.1016/j.envpol.2017.05.014.

Du, X., Li, X., Xiong, L. et al. (2016). Biomaterials mesoporous silica nanoparticles with organo-bridged silsesquioxane framework as innovative platforms for bioimaging and therapeutic agent delivery. *Biomaterials* 91: 90–127. https://doi.org/10.1016/j.biomaterials.2016.03.019.

Gao, M., Zhou, J., Liu, H. et al. (2018). Foliar spraying with silicon and selenium reduces cadmium uptake and mitigates cadmium toxicity in rice. *Science of the Total Environment* 631–632: 1100–1108. https://doi.org/10.1016/j.scitotenv.2018.03.047.

González-Moscoso, M., Martínez-Villegas, N.V., Meza-Figueroa, D. et al. (2021). Las Nanopartículas de SiO_2 Mejoran la Absorción de Nutrientes en Plantas de Tomate Desarrolladas en Presencia de Arsénico. *Revista Bio Ciencias* 8: e1084. Epub 04 de octubre de 2021.

Jimenez-Falcao, S., Parra-Nieto, J., and Pérez-Cuadrado, H. (2019). Avidin-gated mesoporous silica nanoparticles for signal amplification in electrochemical biosensor. *Electrochemistry Communications* 108: 106556. https://doi.org/10.1016/j.elecom.2019.106556.

Kaur, H. and Greger, M. (2019). A review on Si uptake and transport system. *Plants* 8 (4): 81. https://doi.org/10.3390/plants8040081.

Kaya, C., Tuna, L., and Higgs, D. (2006). Effect of silicon on plant growth and mineral nutrition of maize grown under waterstress conditions. *Journal of Plant Nutrition* 29 (8): 1469–1480. https://doi.org/10.1080/01904160600837238.

Kobayashi, H., Ogawa, M., Alford, P.L. et al. (2010). New strategies for fluorescent probe design in medical diagnostic imaging. *Chemical Reviews* 110 (5): 2620–2640. https://doi.org/10.1021/cr900263j.

Li, T., Shi, S., Goel, S. et al. (2019). Recent advancements in mesoporous silica nanoparticles towards therapeutic applications for cancer. *Acta Biomaterialia* 89: 1–13. https://doi.org/10.1016/j.actbio.2019.02.031.

Llinás, M.C.-G. (2014). Nanopartículas de sílice: preparación y aplicaciones en biomedicina. *Afinidad* 71 (565): 20–31.

Lu, F., Wu, S.H., Hung, Y., and Mou, C.Y. (2009). Size effect on cell uptake in well-suspended, uniform mesoporous silica nanoparticles. *Small* 5 (12): 1408–1413.

Lv, Y., Cao, Y., Li, P. et al. (2017). Ultrasound-triggered destruction of folate-functionalized mesoporous silica nanoparticle-loaded microbubble for targeted tumor therapy. *Advanced Healthcare Materials* 6 (18): 1–10. https://doi.org/10.1002/adhm.201700354.

Ma, X., Feng, H., Liang, C. et al. (2017). Mesoporous silica as micro/nano-carrier: from passive to active cargo delivery, a mini review. *Journal of Materials Science and Technology* 33 (10): 1067–1074. https://doi.org/10.1016/j.jmst.2017.06.007.

Mali, M. and Aery, N.C. (2008). Influence of silicon on growth, relative water contents and uptake of silicon, calcium and potassium in wheat grown in nutrient solution. *Journal of Plant Nutrition* 31 (11): 1867–1876. https://doi.org/10.1080/01904160802402666.

Mathelié-Guinlet, M., Cohen-Bouhacina, T., Gammoudi, I. et al. (2019). Silica nanoparticles-assisted electrochemical biosensor for the rapid, sensitive and specific detection of *Escherichia coli*. *Sensors and Actuators B: Chemical* 292: 314–320. https://doi.org/10.1016/j.snb.2019.03.144.

Na, H.K., Kim, M.H., Park, K. et al. (2012). Efficient functional delivery of siRNA using mesoporous silica nanoparticles with ultralarge pores. *Small* 8 (11): 1752–1761.

Narayan, R., Nayak, U.Y., Raichur, A.M., and Garg, S. (2018). Mesoporous silica nanoparticles: a comprehensive review on synthesis and recent advances. *Pharmaceutics* 10 (3): 118.

Pratiwi, F.W., Kuo, C.W., and Wu, S. (2018). Mesoporous silica-based nanomaterials and biomedical applications Part A, 43, cap. 6, The Bioimaging Applications of Mesoporous Silica Nanoparticles. [online]. Taipei. https://www.sciencedirect.com/science/article/pii/S1874604718300076 (fecha de consulta 10 de abril 2020).

Rahimi-Mohseni, M., Raoof, J.B., Ojani, R. et al. (2018). Development of a new paper-based nano-biosensor using the co-catalytic effect of tyrosinase from banana peel tissue (*Musa Cavendish*) and functionalized silica nanoparticles for voltammetric determination of L-tyrosine. *International Journal of Biological Macromolecules* 113: 648–654. https://doi.org/10.1016/j.ijbiomac.2018.02.060.

Raya, J.C. and Aguirre, C. (2012). *El papel del silicio en los organismos y ecosistemas*. México: Instituto Tecnológico de Roque, 5 p.

Rossi, L.M., Shi, L., Quina, F.H., and Rosenzweig, Z. (2005). Stöber synthesis of monodispersed luminescent silica nanoparticles for bioanalytical assays. *Langmuir* 21 (10): 4277–4280.

Roychoudhury, A. (2020). Silicon-nanoparticles in crop improvement and agriculture. *International Journal on Recent Advancement in Biotechnology & Nanotechnology* 3: 54–65.

Roychoudhury, A. and Singh, A. (2024). Applications of nanoparticles as anti-microbial agents. In: *The Nanotechnology Driven Agriculture* (ed. S. Roy and A. Hossain), 264–276. CRC Press (Taylor & Francis Group).

Ruíz-Huerta, E.A., De la Garza Varela, A., Gómez-Bernal, J.M. et al. (2017). Arsenic contamination in irrigation water, agricultural soil and maize crop from an abandoned smelter site in Matehuala, Mexico. *Journal of Hazardous Materials* 339: 330–339. https://doi.org/10.1016/j.jhazmat.2017.06.0417.

Samanta, S. and Roychoudhury, A. (2021). Recent trend in nanoparticle research in regulating arsenic bioaccumulation and mitigating arsenic toxicity in plant species. *Journal of Plant Biochemistry and Biotechnology* 30 (4): 793–812.

Silverstein, D. and Hopper, K. (2014). *Small Animal Critical Care Medicine-E-Book*. Elsevier Health Sciences.

Singh, A. and Roychoudhury, A. (2020). Silicon-regulated antioxidant and osmolyte defense and methylglyoxal detoxification functions co-ordinately in attenuating fluoride toxicity and conferring protection to rice seedlings. *Plant Physiology and Biochemistry* 154: 758–769.

Singh, A. and Roychoudhury, A. (2022). Role of nanoparticles in remediation of environmental contaminants. In: *Hazardous and Trace Materials in Soil and Plants* (ed. M. Naeem, T. Aftab, A.A. Ansari, et al.), 327–340. Academic Press (Elsevier).

Slowing, I.I., Vivero-Escoto, J.L., Wu, C.-W., and Lin, V.S.Y. (2008). Mesoporous silica nanoparticles as controlled release drug delivery and gene transfection carriers. *Advanced Drug Delivery Reviews* 60 (11): 1278–1288.

de Souza Oliveira, R.C., Corrêa, R.J., Teixeira, R.S.P. et al. (2016). Silica nanoparticles doped with anthraquinone for lung cancer phototherapy. *Journal of Photochemistry and Photobiology B: Biology* 165: 1–9. https://doi.org/10.1016/j.jphotobiol.2016.10.008.

Tang, F., Li, L., and Chen, D. (2012). Mesoporous silica nanoparticles: synthesis, biocompatibility and drug delivery. *Advanced Materials* 24 (12): 1504–1534.

Tang, H., Xiang, D., Wang, F. et al. (2017). 5-ASA-loaded SiO_2 nanoparticles-a novel drug delivery system targeting therapy on ulcerative colitis mice. *Molecular Medicine Reports* 15: 1117–1122.

Tripathi, D.K., Singh, V.P., Prasad, S.M. et al. (2015). Silicon nanoparticles (SiNp) alleviate chromium (VI) phytotoxicity in *Pisum sativum* (L.) seedlings. *Plant Physiology and Biochemistry* 96: 189–198. https://doi.org/10.1016/j.plaphy.2015.07.026.

Vallet-Regí, M. (2022). Our contributions to applications of mesoporous silica nanoparticles. *Acta Biomaterialia* 137: 44–52. https://doi.org/10.1016/j.actbio.2021.10.01.

Vallhov, H., Gabrielsson, S., Strømme, M. et al. (2007). Mesoporous silica particles induce size dependent effects on human dendritic cells. *Nano Letters* 7 (12): 3576–3582.

Wang, Y., Zhao, Q., Han, N. et al. (2015). Mesoporous silica nanoparticles in drug delivery and biomedical applications. *Nanomedicine: Nanotechnology, Biology and Medicine* 11 (2): 313–327.

Wu, S.H., Hung, Y., and Mou, C.Y. (2011). Mesoporous silica nanoparticles as nanocarriers. *Chemical Communications* 47 (36): 9972–9985.

Yamamoto, E. and Kuroda, K. (2016). Colloidal mesoporous silica nanoparticles. *Bulletin of the Chemical Society of Japan* 89 (5): 501–539.

Yu, M., Zhou, L., Zhang, J. et al. (2012). A simple approach to prepare monodisperse mesoporous silica nanospheres with adjustable sizes. *Journal of Colloid and Interface Science* 376 (1): 67–75.

Yun, B.J., Kwon, J.E., Lee, K., and Koh, W.G. (2019). Highly sensitive metal-enhanced fluorescence biosensor prepared on electrospun fibers decorated with silica-coated silver nanoparticles. *Sensors and Actuators B: Chemical* 284: 140–147. https://doi.org/10.1016/j.snb.2018.12.096.

Zargar, S.M., Mahajan, R., Bhat, J.A. et al. (2019). Role of silicon in plant stress tolerance: opportunities to achieve a sustainable cropping system. *3 Biotech* 9: 73. https://doi.org/10.1007/s13205-0191613-z.

7

Technologies to Identify and Characterize Si/SiO_2 Nanoparticles

Vaishali Raikwar

Department of Physics, Ramniranjan Jhunjhunwala College of Arts, Science and Commerce (Autonomous), Mumbai, Maharashtra, India

7.1 Introduction

Nanomaterials are emerging as a class of materials that surpass their bulk counterparts in any application area. The nanoparticles show many unique and improved properties that can be advantageous in myriad applications. These properties include physical, chemical, optical, electrical and mechanical properties (Gkika et al. 2023; Murty et al. 2013; Saleh 2020). Many synthesis approaches are applied in the preparation of nanomaterials that can be broadly classified as the top-down approach and bottom-up approach (Abid et al. 2022; Kolahalam et al. 2019). The characterization techniques are to be carefully chosen as it is a way to analyse the synthesized materials. If the method is not suitable, it may wrongly interpret the results and hence the analysis may go in the wrong direction. Hence, it is important to know thoroughly about the different characterization techniques available for analysing the sample and interpret the data obtained through them. Various data software available with the characterization and testing tools/instruments should also be used in a systematic and optimum way to interpret the resultant data, graphs, plots and images. Many a times, a particular technique is not sufficient to analyse a sample in the nano regime. In such cases, a comprehensive approach that combines multiple analytical techniques should be used.

Metal oxides play an important role in chemistry, physics and material science (Henrich and Cox 1994; Vorokhta et al. 2018). They are widely used in the fabrication of sensors, microelectronic circuits, as catalysts, on a surface as a coating, etc. (Manjakkal et al. 2018; Rim et al. 2015). Silica (SiO_2) is an extensively studied metal oxide nanoparticle over the past two decades (Rafigh and Heydarinasab 2017). Silicon (Si) is the semiconductor material that is an

Silicon Nanoparticles: Synthesis, Uptake, and Applications, First Edition.
Edited by Archana Singh, Aryadeep Roychoudhury, and Indrakant K. Singh.

integral part of any solid-state device. The conductivity of intrinsic Si can be increased by doping it with an impurity ion material. The versatility of synthesis methods and availability of modern characterization tools with advancements in the field of science and technology have contributed to positive growth in research on these nanomaterials. They are widely studied for their applications as a dielectric, photocatalyst, in pharmaceuticals, in photovoltaics, in light emitting diodes (LEDs) and in semiconductor thin film devices (Bley et al. 1996; Greish et al. 2012; Kustov et al. 2018; Morozova et al. 2020). It is therefore the need of the hour to know more about their characterization techniques in detail.

This chapter gives an overview of various characterization tools available to analyse Si/SiO_2 nanoparticles. Sometimes the characterization tools for a specific study become exclusive and limited to that study. In general, there are certain testing parameters and tools needed to analyse a sample at the nanoscale regime, and those are the minimum requirements for any nanoscale material study. Herein, we described those tools in detail and their applications in various fields. For nanoscale materials, the size, shape, assembly of atoms, surface morphology, crystal structure and functional groups are important aspects. To study these properties and parameters, researchers have to use characterization tools such as X-Ray diffraction techniques, Fourier transform infrared spectroscopy (FTIR), nuclear magnetic resonance (NMR) technique, low-energy ion scattering (LEIS), scanning electron microscopy (SEM), UV-Vis spectroscopy, photoluminescence (PL) spectroscopy, the mass spectrometry (MS) and secondary ion mass spectrometry (SIMS) (Sato and Swihart 2006; Song and Chen 2014). Figure 7.1 gives

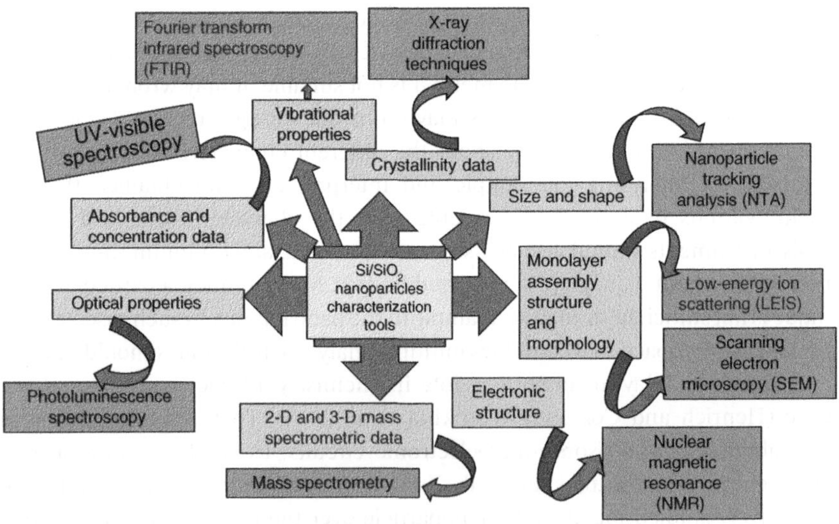

Figure 7.1 Characterization tools to study Si/SiO_2 nanoparticles.

the broad outline of the characterization tools required to study the silicon nanoparticles (SiNP).

7.2 X-Ray-Based Techniques

7.2.1 X-Ray Diffraction Measurement

X-rays are the electromagnetic radiation having photon energies of 100 eV–100 keV, of which a very short wavelength X-rays typically from a few angstroms to 0.1 angstrom (1–120 keV) are needed for diffraction applications. As the wavelength of X-rays is comparable to the size of atoms, they are ideally suitable for investigating the structural arrangement of atoms and molecules in a wide range of materials. They can be penetrated deep into the materials and provide information about the bulk structure (Kargina et al. 2019; Kim and Lee 2023).

When X-rays interact with electrons in atoms and collide, some photons from the incident beam are deflected away from the original. Elastic scattering takes place if the wavelength of scattered X-rays does not change. These diffracted X-rays are important as they carry information about the electron distribution in materials. The waves from different atoms get diffracted at different angles and can interfere with each other, and the resultant intensity distribution is strongly modulated by this interaction. Since the crystalline structure has a periodic arrangement of atoms, the diffracted waves will have sharp interference maxima that match the distribution of atoms. Therefore, measuring the diffraction pattern helps us to determine the distribution of atoms in the crystal lattice of a material. When X-rays are incident on a crystalline solid, they are scattered and interfere constructively to produce a diffracted beam. According to W. L. Bragg, the constructive interference condition is given by $n\lambda = 2d\sin\theta$. This equation is called the Bragg equation, where d is the interplanar spacing, n represents the order of diffraction, λ is the wavelength of the X-ray, and θ signifies the scattering angle. The d spacing can be calculated from the diffraction pattern as it is the distance between similar atomic planes in a crystal and is measured in angstroms (Azaroff 1974; Cullity 1978; Connolly 2007). The XRD peak broadening is attributed to various effects such as instrumental configuration, size of the crystallite and lattice strain. The peak properties can be analysed by different methods, including simplified integral breadth methods, double Voigt methods and Fourier methods. By using the X-ray diffraction pattern, one can easily determine crystallite size from XRD peak broadening using a simplified expression obtained by (Scherrer 1918; Patterson 1939) as given by

$$T = \frac{0.9\,\lambda}{B\cos\theta}$$

where T is the size of the crystallite of the scattering particle, in nm, λ denotes the X-ray wavelength in nm and B is the full width at half maximum (FWHM) that can be taken from one of the fundamental peaks scattered at an angle of θ, in radians. The Scherrer equation does not take into consideration the strain factor while calculating the crystallite size. According to the Scherrer equation, the broadening due to small crystallite size is given by

$$\beta_c = \frac{K\lambda}{d\cos\theta} \tag{7.1}$$

where β_c is the broadening due to small crystallite size, K is the constant whose value depends on particle shape and is usually taken as less than or equal to one, d is the crystallite size in nm, λ is the wavelength of the incident X-ray beam in nm, and θ is Bragg's angle.

According to Wilson, the broadening due to lattice strain is given by

$$\beta_s = 4\eta\tan\theta \tag{7.2}$$

where β_s is the peak broadening due to lattice strain, θ is Bragg's angle and η is the strain distribution within the material.

From Eqs (7.1) and (7.2), the total peak broadening β_r is given by

$$\beta_r = \frac{K\lambda}{d\cos\theta} + 4\eta\tan\theta \tag{7.3}$$

Multiplying both sides of equation (7.3) by $\cos\theta$

$$\beta_r\cos\theta = \frac{K\lambda}{d} + 4\eta\sin\theta \tag{7.4}$$

$\beta_r\cos\theta$ is plotted on the Y-axis and $4\sin\theta$ is plotted on the X-axis, which gives slope η and Y intercept $K\lambda/d$. From the W–H plot (Williamson and Hall 1953; Wilson 1949), the Y-intercept can be obtained, from which the crystallite size is calculated. The grain size that is determined from the W–H formula is slightly greater than that calculated using Scherrer's formula because in the W–H formula, the strain is also considered.

The actual photograph of the XRD machine is shown in Figure 7.2a, and the sample holder chamber is shown in Figure 7.2b. The schematic diagram of the XRD measurement system is shown in Figure 7.2c. The diffraction pattern acts as a fingerprint of that crystalline substance. Therefore, the crystalline phase of a material can also be identified by examining the diffraction pattern. The width of the diffraction lines is closely related to the size, size distribution, defects and strain in crystals.

7.2.2 X-Ray Photoelectron Spectroscopy

X-Ray photoelectron spectroscopy (XPS) is a quantitative analysis of a sample that is based on the photoelectric effect. It can identify elements present in the sample.

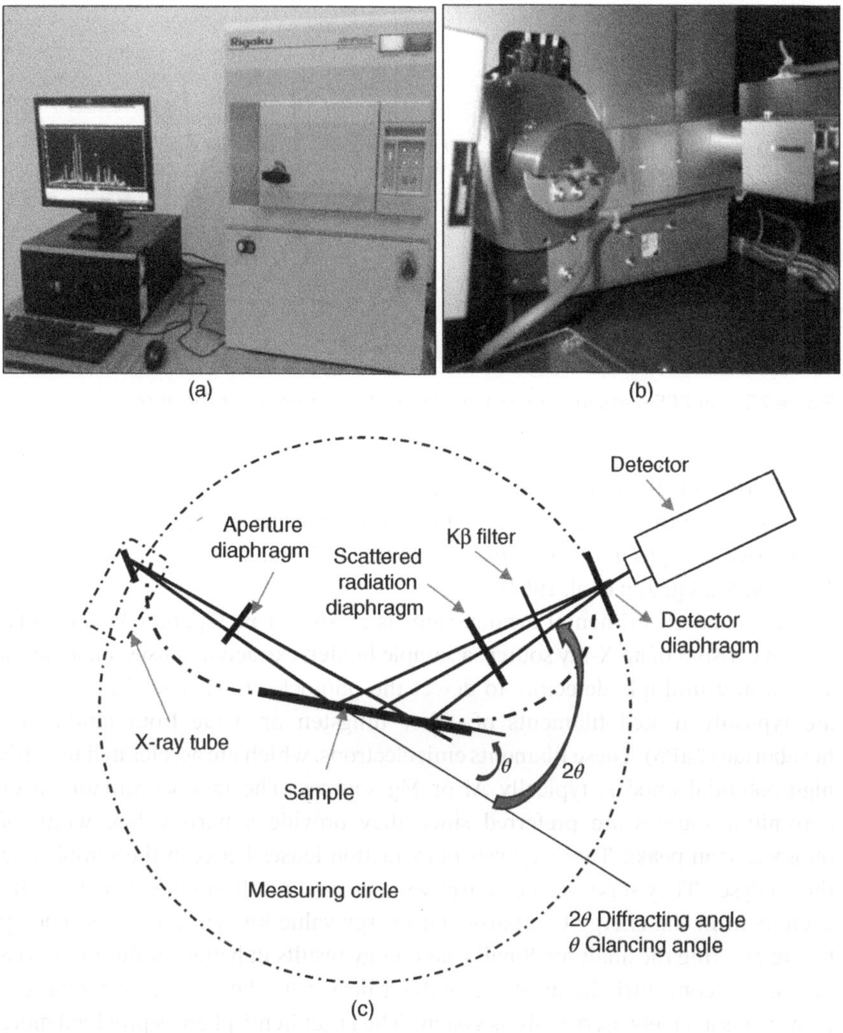

Figure 7.2 XRD measurement system. (a) actual photograph, (b) sample holder chamber, (c) the schematic diagram.

It is a surface-sensitive quantitative spectroscopic technique. The X-rays bombard the surface of the material, and the kinetic energy of the ejecting electrons is measured. The chemical state information of all the elements can be revealed by XPS, making it a powerful tool for detecting the elemental composition of Si/SiO_2 structure or composites. It was first discovered by Heinrich Hertz in 1887. Albert Einstein more formally described the effect in 1905 and was awarded the

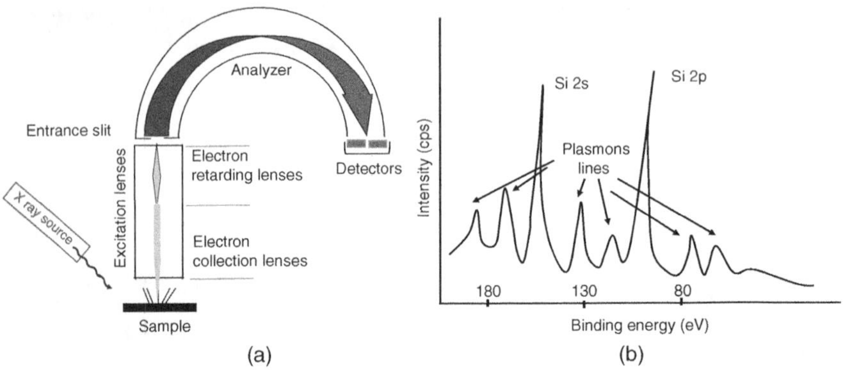

Figure 7.3 (a) XPS measurement system, (b) photoelectron peaks for silicon.

Nobel Prize for the photoelectric effect in 1921. More rigorous work in this field to convert it into a technological application was done by Kai Siegbahn at the University of Uppsala in Sweden during the 1950s and 1960s (Nordling et al. 1957, 1958; Siegbahn et al. 1967).

The typical experimental arrangement is as shown in Figure 7.3a. The XPS system consists of an X-ray source, a sample holder, extraction lenses, an analysis system and multiple detectors to detect the photoelectrons. The X-ray sources are typically heated filaments of either tungsten or made from lanthanum hexaboride (LaB6). These filaments emit electrons, which are accelerated towards high-potential anodes, typically Al or Mg sources. The monochromatic Al or chromium sources are preferred since they provide a narrow line width of photoelectron peaks. There are sets of extraction lenses between the sample and the analyser. They serve various purposes such as the collection of electrons and decrement in the energy to a particular energy value known as the pass energy before entering the analyser. Small pass energy results in better resolution of XPS peaks. The concentric hemisphere analyser design has better energy resolution. Hence, it is a widely used analysis system. The outer hemisphere is provided more negative voltage than the inner one. The varying voltages to these hemispheres are provided to allow specific energy photoelectrons to travel through the analyser chamber. The larger diameter analyser provides better resolution. The detectors are generally electron multipliers. Often multiple detectors are arranged to collect more electrons. The efficiency of detectors will affect the resulting peaks (Watts and Wolstenholme 2019). The whole instrument needs an ultra-high vacuum environment. This is because while travelling through the analyser, the photoelectron must not scatter with air molecules, and surface contamination must also be avoided (Aziz and Ismail 2017; Korin et al. 2017; Stevie and Donley 2020; Van der Heide 2011).

When the sample gets irradiated with soft X-rays, which have energies less than 6 keV, the emitted photoelectrons acquire the X-ray energy. This can be expressed by the following equation: $h\nu = BE + KE + \varphi_{spec}$ where ν is the frequency of X-ray, BE is the binding energy of electrons, KE is the kinetic energy of the emitted electrons and φ_{spec} is the spectrometer work function. The photoelectron peaks are generated and represented by the element and the orbital from which they originate. For example, 'Si 2s' denotes that electrons are emitted from the 2S orbital of the Si atom (Figure 7.3b).

7.3 Fourier Transform Infrared Spectroscopy

FTIR is the preferred method of infrared spectroscopy. In this kind of spectroscopy, Infra-red (IR) radiation is allowed to pass through a sample, some of which is absorbed by the sample and some of it is transmitted. The resultant is called a molecular fingerprint of the sample as it represents absorption and transmission by the molecules. Each molecular structure produces its unique infrared spectrum. That is the reason why infrared spectroscopy can be useful for several types of analysis. The other dispersive instruments have many drawbacks that can be easily overcome by FTIR spectrometry (Kravets et al. 2005).

The interferometer is an integral part of the FTIR instrument. A unique type of signal that comprises the whole infrared frequency band 'encoded' into it is generated by the interferometer. This is considered to be a fast technique wherein the signal can be measured within a second. A beam splitter splits the beam into two light beams. One beam gets reflected off a flat fixed mirror. Another beam gets reflected by a flat movable mirror. These two reflected beams are recombined when they meet back at the beam splitter. As one path is of a fixed length and the other is changing because of the mirror movement, the signal that is coming out of the interferometer is the resultant of the interference of these two beams 'interfering' with each other. The interferogram has this unique property that each data point (that was plotted) has information about every infrared frequency that originates from the excitation source of infrared light. In FTIR, all the incoming frequencies are measured simultaneously. This makes the device faster. As the frequency spectrum is expected to be analysed, a technique of 'decoding' is required that measures the individual frequencies. This mathematical technique is called as the Fourier transformation. This part of the transformation is computed by a computer, which can be further extracted using the FTIR instrument software for analysis of the sample (Petreanu et al. 2023; Zou et al. 2004).

The instrumental block diagram is as shown in Figure 7.4.

Source: A glowing black-body source is used as an Infrared energy source. This IR beam is then allowed to pass through an aperture to control the intensity of the IR light passing through the sample.

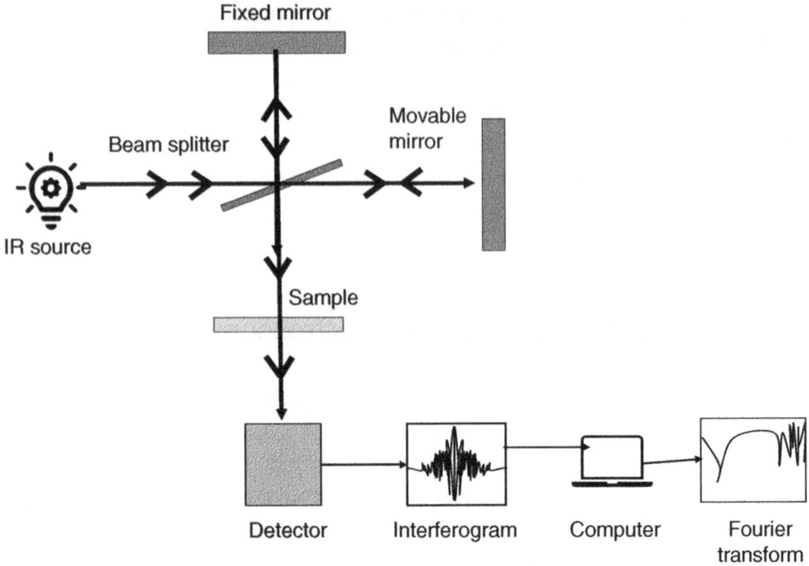

Figure 7.4 Schematic diagram of Fourier transform infrared spectroscopy.

Interferometer: The Infrared beam is then allowed to enter the interferometer in which its 'spectral encoding' takes place. The resulting 'interferogram signal' then exits the interferometer.

Sample: The interferogram signal thus enters the sample chamber in which there are reflecting surfaces and lenses that transmit or reflect the beam, depending on what type of analysis is to be accomplished. In this chamber, the specific frequency absorption of the incident IR light by the sample takes place.

Detector: The detector is used for the final measurement, which is specially designed to measure the specific 'interferogram signal.'

Computer: The digitization of the signal and the Fourier transformation take place in the computer section of the instrument. The interferogram consists of a series of peaks that represent a specific wave number plotted on the *X* axis and transmittance or reflectance on the *Y* axis.

7.4 Nuclear Magnetic Resonance Spectroscopy

NMR spectroscopy uses the magnetic field of certain nuclei to determine chemical and physical properties of molecules or atoms to which they belong. This analysis tool is designed to provide detailed information about the reaction state, the structure, the "chemical environment of molecules", etc. This technique is also used to determine the content and purity of the sample by material scientists and for quality control by industry personnel. The basic principle includes the influence of an external magnetic field that forces protons into two possible orientations, which have different energies. The NMR spectrum is the result of energy emitted or energy absorbed due to a transition between these energy levels. In general, the nuclear magnetic fields are randomly oriented. When they are placed in an external magnetic field, the spins can be oriented either in the direction or opposite to the direction of the applied field. There is a small difference in the energies of these two types of spins. When an external applied electromagnetic radiation in the form of radio waves causes the spin to flip, resonance phenomena occur with the magnetic field. The NMR instrument has the simple principle of resonance, wherein nuclei of a given type will resonate at various energies depending on their chemical and electronic environment. The position of peaks and pattern of the NMR signals gives important information about the "chemical environment of the nuclei".

The nuclei will precess with the Larmor frequency along the direction of the magnetic field B. The Larmor frequency depends on the external magnetic field B and its gyromagnetic ratio of a certain nucleus. It can be expressed as $\nu = \frac{\gamma B}{2\pi}$. The energy difference ΔE between two energy states is equal to hν. Therefore, $\Delta E = \frac{h\gamma B}{2\pi}$. When the applied radiofrequency becomes equal to the Larmor frequency of nuclei under the influence of an external magnetic field B, NMR phenomena occur (Casabianca 2020; Jackman and Sternhell 2013; Nishi 1971). Recently, the quantitative NMR (qNMR) technique was used by Cheng R et al. for quantifying the purity and the chemical composition of silica nanoparticles. They focused on the spin-lattice relaxation and quantified the relaxation delay. Thus, using a simple technique of NMR, they studied the porous silica structure (Cheng et al. 2021). Figure 7.5 shows the schematic of resonance phenomena and the resulting NMR spectrum.

7.5 Low-energy Ion Scattering Spectroscopy

The technique that uses the energy of backscattering electrons is called as ion scattering spectroscopy (ISS). This can be broadly classified as low-energy ion scattering spectroscopy (LEIS) that has primary energies from 100 eV to 10 keV, medium-energy ion scattering spectroscopy (MEIS) that comprises energies between 100 and 200 keV, and high-energy ion scattering spectroscopy (HEIS) in a range from 1 to several MeV. The LEIS is better known for its ion scattering spectroscopy, while the HEIS spectroscopy is known as Rutherford backscattering spectroscopy (Ebnesajjad 2011). The typical experimental setup for LEIS is shown in Figure 7.6. An ion gun is mounted on the top from which the ions of noble gas can be directed on the surface of the sample perpendicularly. The ion beam can

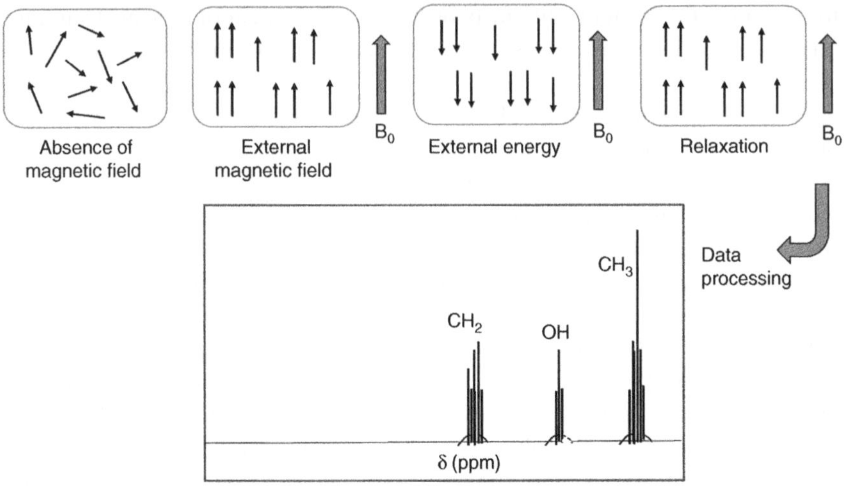

Figure 7.5 The resonance phenomena and NMR spectroscopy pattern.

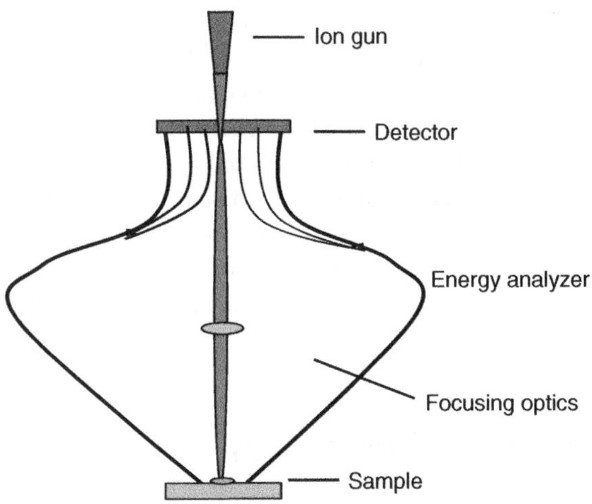

Figure 7.6 Low energy ion scattering (LEIS) set up.

be focused using optical elements and deflection plates. The beam can typically be scanned over an area of $1\,mm^2 \times 1\,mm^2$ to $2\,mm^2 \times 2\,mm^2$. The spot size is of the order of $10\,\mu m$. During the analysis of Si nanoparticles, the sample is analysed in an ensemble form in a particular analysis area. An energy analyser is mounted in an ultrahigh vacuum chamber. This analyser accepts all ions scattered into the cone, which was scanned at a fixed scattering angle θ and 360-degree azimuthal movement. This enables the collection of as many ions as possible for a well-defined energy spectrograph. The ions are transmitted with a specific kinetic energy onto the sample (Zheng and Chen 2023). Thus, LEIS investigates the elemental composition of the outermost atomic layer of a nanomaterial, and it has a high sensitivity and a capability of quantitative analysis of silicon or silica nanoparticles coated on various catalysts, sensing elements, or biomaterials (Aono and Souda 1985). It is a powerful technique that studies surface chemistry and surface-related phenomena such as adhesion, contamination, wetting and thin film growth or monolayer surface assembly. The high sensitivity of the LEIS technique makes it a popular technique for the study of thin film materials.

7.6 Scanning Electron Microscopy

The SEM technique is very useful for determining the surface morphology of Si/SiO_2 nanoparticles (Prabha et al. 2021). Electrons are emitted from a field emission electron source, which are then accelerated with the application of a "high electrical field gradient". The primary electron beam is then focused and deflected by electronic lenses to produce a narrow scan beam that bombards the sample. The whole scanning is done in a "high vacuum column". Due to bombardment by primary electrons, secondary electrons are emitted from each spot on the sample. The structure of the surface of the sample is related to the diffraction angle and velocity of these secondary electrons. The secondary electrons are detected by the detector, and it then produces an electronic signal that is further amplified and transformed into a video scan image that is visualized on a monitor (Huseynov et al. 2016). The digital image can be saved, and analysis can be done. Figure 7.7a depicts the schematic diagram of the experimental set-up of SEM.

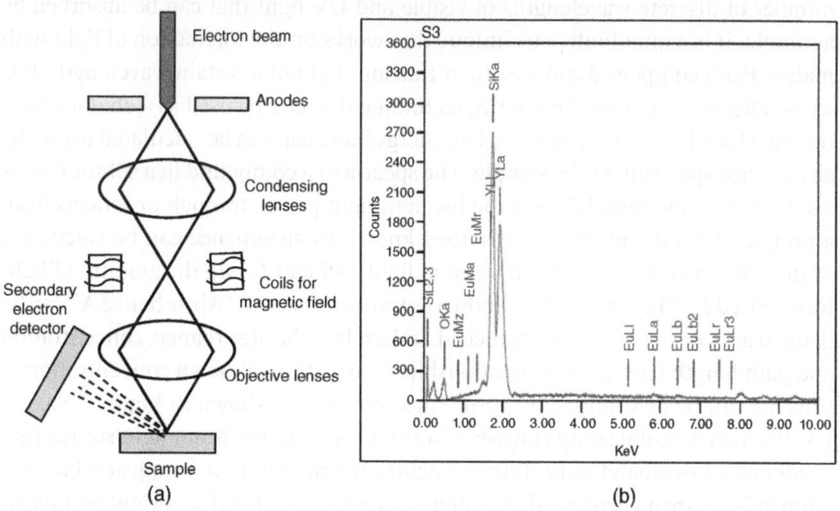

Figure 7.7 (a) Schematic diagram of the SEM imaging technique, (b) EDX elemental composition pattern.

"Energy-dispersive X-ray spectroscopy" (EDS, EDAX, or XEDS) is an illustrative methodology utilized for the essential examination or chemical characterization of a sample. It depends on the examination of the interaction of many sources of excitation by X-rays and the sample. EDAX systems work in association with SEM and TEM, where the imaging ability of the instrument recognizes the illustration of captured data. The data created by EDX examination include a spectrum that shows peaks comparing the components making up the veritable composition of the test being analysed (Mbakaan et al. 2018; Sharififard and Rezvanpanah 2021).

To strengthen the surge of characteristic X-rays from an illustration, a high-energy column of charged particles, such as electrons or protons, or a narrow beam of X-rays, is bombarded into the sample being inspected. When the atom is at rest, it has a ground state that contains unexcited electrons. The external trigger like an X-ray energy beam may energize an electron in an inside shell, propelling it from the shell, making an electron crevice where the electron was. An electron from an outside, higher-energy shell at that point fills the place, and the contrast in imperativeness between the higher-energy shell and the lower-energy shell may be released inside the frame of an X-ray. Figure 7.7b shows the EDX spectroscopy showing "the elemental composition of a sample".

7.7 UV-Visible Spectroscopy

UV-Visible (UV-Vis) spectroscopy is an analytical technique that measures the number of discrete wavelengths of visible and UV light that can be absorbed by a sample. It is a quantitative technique that works on the interaction of light with matter. Each compound can absorb or transmit light of a certain wavelength. It is a cost-effective and non-destructive technique that can be used as a characterization tool for silicon nanoparticles. The energy band gap can be calculated using the absorbance spectrum of the sample. The spectra of a compound in a solution or as a solid can be measured. When the incident light passes through or reflects from a sample, the amount of light absorbed known as absorbance can be calculated as the difference between the amount of light incident I_0 and the amount of light transmitted I_T. The ratio I_T/I_0 is known as transmittance T. "Absorbance A" can be defined as $A = -\log T$. As per the Beer-Lambert law, the absorbance, concentration and path length have a linear relationship. Hence, the unknown concentration of solvents can be determined by absorbance spectra. As shown in Figure 7.8a, the UV-Vis instrumental setup consists of a single or a double-beam light source that generates a broadband light. Either a deuterium arc lamp or a tungsten halogen lamp or both are incorporated. A source selector is used to select a single source or a combined broadband can be used for spectrum recording. However, a xenon light source is the ultimate choice where high diffusion reflectance spectra are to be

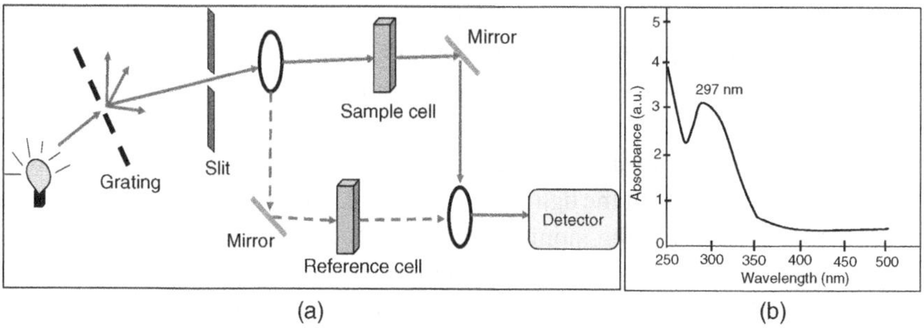

Figure 7.8 (a) Schematic diagram of a UV-Vis spectrophotometer, (b) UV absorbance spectrum of silica nanoparticles.

recorded. A dispersion device is used to select a particular band of wavelength for a particular sample. "Holographic gratings" are used as dispersion devices in most UV-Vis spectrophotometers. Aluminium coating is used as a reflectance source that reflects the light at different angles. The monochromator has an entrance slit, a dispersion device and an exit slit. A photomultiplier tube is generally used as a detector. Either concave mirrors or lenses can be used to focus the light through the instrument. Figure 7.8b shows the UV absorbance spectrum of silica nanoparticles showing an absorbance peak around 295 nm. Various parameters can be determined using this spectrum. It is possible to determine the absorbance or transmittance through a liquid or solid over a range of UV or visible wavelengths. It can also determine the concentration of a particular molecule in a solution (Gaabour 2019; Ghorbanpour et al. 2017). One can characterize the reflectance properties of a surface of a sample or study chemical reactions or biological processes. It is a versatile instrument that is available with various types of measurements by combining different accessories and sample holders (Presentato et al. 2020).

7.8 Others

7.8.1 Photoluminescence Spectroscopy

PL property is an important parameter that is needed for luminescent silicon materials (Kravitz et al. 2010; Valenta et al. 2002). It can be recorded using a fluorescence spectrophotometer (Campos et al. 2016). The block diagram of a general "fluorescence spectrophotometer" is shown in Figure 7.9a. The excitation source (1) is a light source for providing excitation light to irradiate a specimen; a xenon lamp is usually used. The light from the white light source enters a spectroscope (2), which is towards the excitation side. While measuring an excitation spectrum, this excitation side of the spectroscope (2) is moved to successively change the wavelength used to irradiate the specimen (this is called scanning or wavelength scanning). In contrast, when measuring a fluorescence spectrum, a specific wavelength of the excitation source is selected and fixed in the excitation side spectroscope. The excitation side spectroscope emits the light, which then travels towards the specimen in order to excite it. Along the way, a half-mirror is used to split the light beam (diagonal line in the figure above) so that a portion also travels to the monitor detector (3). The intensity of the excitation light which irradiates the specimen is monitored by detector monitors. Generally, a photoelectric tube (phototube), a photodiode, a photomultiplier tube (also called a photomultiplier or photomul), etc., are used. When the excitation light reaches the specimen, the latter is excited to emit fluorescent light. The emitted fluorescent light then enters a spectroscope (4), which is on the fluorescent light side. When measuring the excitation spectrum, a specific wavelength is selected and fixed in the fluorescent light side spectroscope (4). When measuring a fluorescence spectrum, the fluorescent light side spectroscope is moved to measure the wavelengths of the emitted fluorescent light. The light beam leaving the spectroscope thus enters the fluorescence detector (5). Generally, a photomultiplier tube is used as a detector. This detector converts the fluorescent light into an analogue electric signal, which needs to be converted into a digital signal; that task is done by an A/D conversion circuit (6). The wavelength scanning and digital signal processing are done on a computer. It uses 150 W xenon lamps with "ozone self-dissociation function". "Large stigmatic concave grating having 900 lines/mm is used as monochromators on both excitation and emission sides with eagle mounting (F: 3). Blaze wavelengths are 300 nm on excitation side and 400 nm on emission side". Photomultiplier R372 F is used for measurements, and photoelectric tube R 518 is used for monitoring (Moreno and Cox 2010).

The following performance parameters of the system are considered during measurement.

Sensitivity: This is also called as signal/noise (S/N) ratio. This index indicates how the data can be plotted with a smooth curve having minimum noise. The greater the sensitivity, the better the performance of the system.

Slit: It is also known as bandpass. It is the width of the slit that is placed at the inlet/outlet of a spectroscope. If the slit width is large, data with relatively lower noise can be obtained since the intensity of light increases. If the slit is narrow, the instrument can record even minute peaks of the spectrum.

Response: If a large value of response is chosen, the PL spectrum becomes smooth; however, the shape can be changed. The optimal response is an important parameter during the measurement of excitation or emission spectrum. Depending on the scanning speed, it should be decided.

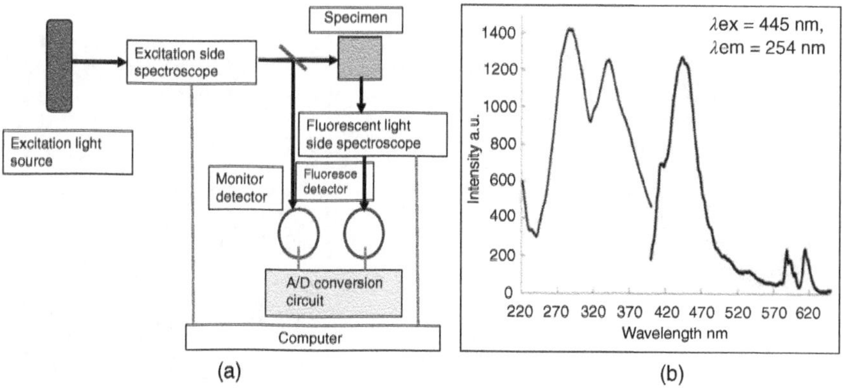

Figure 7.9 (a) Block diagram of a fluorescence spectrophotometer and (b) PL Ex and Em spectrum.

The sample chamber has a round sample holder that can be used to hold a powder sample, a quartz disc (transparent window through which light can fall on the sample), and a threaded cap. The quartz disc is to be fixed into the sample holder, and the powder sample is spread on it. There is a threaded cap that can be held tightly to hold the powder sample. When the metal frame is fixed on the sample cell, the front protrusion of the cell can be fitted into the metal frame aperture. To record the PL excitation (EX) spectra, initially, we have to set the emission wavelength at zero order, keeping the other parameters as specified in the user manual. On the other hand, the emission (EM) spectra can be scanned after finding the excitation wavelengths (Figure 7.9b). The superposition of direct scattered light may be taken with the EX-spectrum; that is done by selecting a particular band in the emission spectra for scanning the EX. The same procedure is followed exactly for identifying the correct EX peak positions, and EM can be recorded for each EX-band separately (Larsson et al. 2003; Kanemitsu and Okamoto 1997).

7.8.2 Nanoparticle Tracking Analysis

"Nanoparticle Tracking Analysis" (NTA) is a system for sizing particles from about 30 to 1000 nm, where the lower detection limit is size-dependent on the refractive index of the nanoparticles. This technique is the combination of laser light scattering microscopy and a charge-coupled device (CCD) camera, which enables the visualization and recording of nanoparticles in solution. The corresponding software then identifies and tracks individual nanoparticles that move by Brownian motion. The movement of a particle is then related to a particle size according to

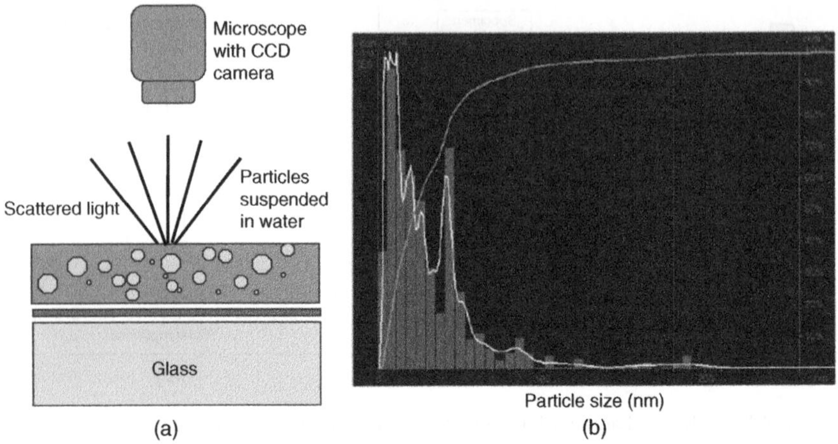

Figure 7.10 (a) NTA working principle. (b) Particle size distribution of nanoparticles.

the formula, which is derived from the Stokes–Einstein Equation as follows:

$$D_t = \frac{2TK_\beta}{3\pi\eta r_h}$$

where K_β is the Boltzmann constant and Dt is a particle diffusion coefficient at a room temperature T in a medium of viscosity η with a hydrodynamic radius of r_h. Thus, NTA converts the distances moved by the particles into a particle size and plots accumulated results in real time as a "particle size distribution profile" (Comfort et al. 2021; Dragovic et al. 2011; Hole et al. 2013). Figure 7.10a shows the NTA instrument that uses a laser light source to illuminate nano-scale particles. The particles suspended in a liquid are loaded into a sample holder. The sample chamber is then illuminated by a laser beam of approximately 50 μm wide. The particle in Brownian motion in a solution that comes in the path of laser light gets scattered and detected by a microscopic objective and a digital camera. The image analysis NTA software suite allows the user to automatically track and size nanoparticles on an individual basis (Figure 7.10b). Results are displayed as a frequency size distribution graph and output to a spreadsheet. Video clips of images can also be retained (Filipe et al. 2010).

Advantages:

- Real-time dynamic nano-particle visualization
- Particle-by-particle analysis
- Particle counting and sizing
- Particle size distributions are displayed as histograms and output to a spreadsheet
- Full reporting and batch processing facilities

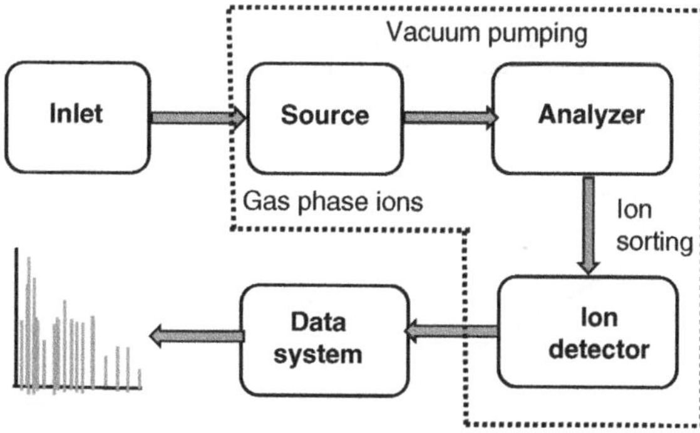

Figure 7.11 Mass spectrometry schematic diagram.

7.8.3 Mass Spectrometry

MS is another elemental composition determination analytical tool. It is highly sensitive and capable of determining small compounds in a sample, making it very useful for analysing traces of compounds in a complex mixture. It reveals the detailed molecular structure of a compound (Rogers et al. 2022; Wen et al. 2007). It allows the user to record high-resolution mass spectrograph, which can identify and perform quantitative analysis of a sample. A large number of samples can be processed at a time, making it a versatile device for mass spectrometry. It measures the mass-to-charge ratio and abundance of gas phase ions in the compound (Li et al. 2023). In this technique, electrons are allowed to fall on the sample, and the sample then gets converted to rapidly moving positive ions. These charged particles are then separated according to their masses. The relative abundance of ions is then plotted against the ratio of mass/charge (m/e). For most ions, the charge is one and hence the m/e ratio is simply the molecular mass of the ion. The ions pass through magnetic and electric fields to finally reach the detector, which detects them, and the signals are recorded as a mass spectrum. Ions get deflected by the magnetic field due to differences in their masses. The lighter masses get deflected more. The positive charge also affects the deflection. The greater the positive charge of the ion, the more it is deflected. The beam of ions, when passed through the mass analyser, is detected by the detector, the basis of which is the m/e ratio (Bolea-Fernandez et al. 2017). Figure 7.11 shows the schematic of a mass spectrometer.

7.8.4 Secondary Ion Mass Spectrometry

SIMS is a non-destructive technique that requires less sample (Gulin et al. 2019). It has high sensitivity meaning that it can analyse low-concentration samples down to the ppb level. Hence, they can be used to determine the trace constituents in nonconducting surfaces like SiO_2 in the semiconductor industry

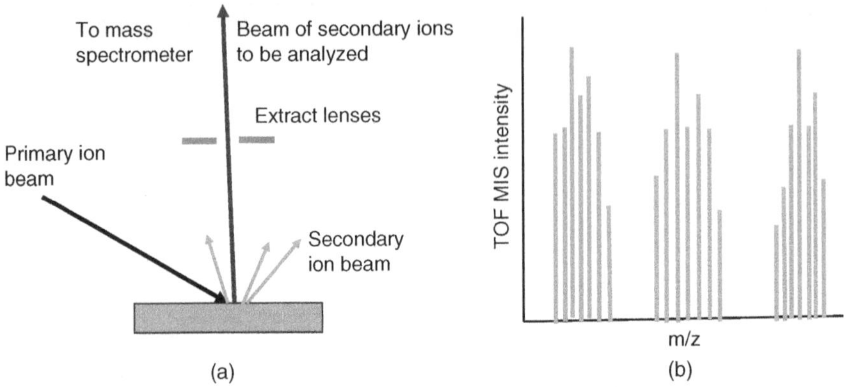

Figure 7.12 (a) Working principle of SIMS. (b) SIMS spectrograph.

(Perego et al. 2004). High sensitivity also facilitates the depth profiling and the molecular and elemental abundances. The instrument consists of a primary ion source either made from Cs for positive ions or oxygen for negative ions. This ion beam is bombarded on a sample, which in turn generates secondary ions, which are then transferred to the detector (Figure 7.12a). Most of the removed atoms and molecules from the sample surface are neutral, but some of them are ionized, which are then focused, accelerated and detected by a detector. The primary beams of positive ions favour the generation of negative secondary ions, and negative primary beams favour the production of positive ions (Benninghoven et al. 1987; Laeter and John 2001). The SIMS spectra are generally plotted with signal intensity or ion counts on the Y axis and the ratio m/z on the X axis. The height of the signal is directly proportional to the number of ions detected by the detector, and the ratio m/z is the mass of a particular ion if it is singly charged. Depth profiles can also be created by sputtering the sample with a primary ion beam. The Cs or Ga is generally used as a primary ion source. The flight path can be either circular or linear, wherein reflectors can be used. Using SIMS elemental and chemical mapping at the submicron level is done with great accuracy. It has a high mass resolution and high sensitivity to trace elements of the order of ppm or ppb, making it a powerful technique for checking the purity of the sample in material science.

7.9 Conclusion and Future Prospective

There are numerous characterization techniques through which the Si/SiO_2 nanoparticles can be studied. The structural properties can be determined using X-ray techniques that include X-ray diffraction and X-ray photoelectron spectroscopy. The optical properties can be studied using UV-vis spectroscopy and a photoluminescence spectrophotometer. The surface morphology of the nanosized materials can be analysed using SEM images, and elemental composition can be conveniently done using either EDX or MS. For a small trace of any material of the order of ppm or ppb, SIMS can be used. FTIR helps the material scientists and the researchers detect the type of bond and functional groups present in the sample. The reaction rate and chemical structure can be effectively studied using NMR spectroscopy. The important aspect about the characterization tools is their resolution and accuracy. The advancement in technology, the miniaturization of instruments and the multimodal operations make these characterization tools perform better day by day. The clear images, vast data and the facility of computer software to process this data are all leading to the increased use of these tools. Centrally funded research facilitation centres are helping research students and remote area research aspirants to use these characterization tools in their research areas. Though the lack of analysis guidelines and literature at a large scale restricts some of these tools, the instrumentation facilities at different research institutes are always a good source of helping hands for smaller institutes and researchers in remote areas.

Glossary

EDS, EDAX, or XEDS Energy-dispersive X-ray spectroscopy used for elemental analysis of a sample.

FTIR Fourier Transform Infrared, which is a molecular fingerprint of the sample.

LEIS Spectroscopy Low energy ion scattering spectroscopy that studies surface chemistry and surface-related phenomena such as adhesion, contamination, wetting, and thin film growth or monolayer surface assembly.

MS Mass Spectrometry is an analytical tool for elemental composition determination.

NMR Spectroscopy The technique that is designed to provide detailed information about the reaction state, the structure, the chemical environment of molecules, etc.

NTA Nanoparticle tracking analysis is an analytical tool that gives particle size distribution.

PL Spectroscopy Photoluminescence spectroscopy determines the excitation and emission properties of a sample.

SEM Scanning electron microscopy that produces an image of a surface to reveal surface morphology.

SIMS The secondary ion mass spectrometry is a non-destructive technique used to determine mass of the order of ppm or ppb.

UV-Vis Spectroscopy This technique is used to determine the UV or visible absorption wavelength and energy band gap from absorbance spectra.

XPS The quantitative analysis technique of a sample based on the photoelectric effect.

References

Abid, N., Khan, A.M., Shujait, S. et al. (2022). Synthesis of nanomaterials using various top-down and bottom-up approaches, influencing factors, advantages, and disadvantages: a review. *Advances in Colloid and Interface Science* 300: 102597.

Aono, M. and Souda, R. (1985). Quantitative surface atomic structure analysis by low-energy ion scattering spectroscopy (ISS). *Japanese Journal of Applied Physics* 24 (10R): 1249.

Azaroff, L.V. (1974). *X-Ray Diffraction*. McGraw Hill Company.

Aziz, M. and Ismail, A.F. (2017). X-ray photoelectron spectroscopy (XPS). In: *Membrane Characterization* (ed. N. Hilal, A.F. Ismail, T. Matsurra, and D. Oatley-Radcliffe), 81–93. Elsevier.

Benninghoven, A., Rudenauer, F.G., and Werner, H.W. (1987). *Secondary Ion Mass Spectrometry: Basic Concepts, Instrumental Aspects, Applications, and Trends*, 1227. New York: Wiley.

Bley, R.A., Kauzlarich, S.M., Davis, J.E., and Lee, H.W. (1996). Characterization of silicon nanoparticles prepared from porous silicon. *Chemistry of Materials* 8 (8): 1881–1888.

Bolea-Fernandez, E., Leite, D., Rua-Ibarz, A. et al. (2017). Characterization of SiO_2 nanoparticles by single particle-inductively coupled plasma-tandem mass spectrometry (SP-ICP-MS/MS). *Journal of Analytical Atomic Spectrometry* 32 (11): 2140–2152.

Campos, B.B., Gelde, L., Algarra, M. et al. (2016). Characterization of cellulose membranes modified with luminescent silicon quantum dots nanoparticles. *Carbohydrate Polymers* 151: 939–946.

Casabianca, L.B. (2020). Solid-state nuclear magnetic resonance studies of nanoparticles. *Solid State Nuclear Magnetic Resonance* 107: 101664.

Cheng, R., Wang, S., Moslova, K. et al. (2021). Quantitative analysis of porous silicon nanoparticles functionalization by 1H NMR. *ACS Biomaterials Science & Engineering* 8 (10): 4132–4139.

Comfort, N., Cai, K., Bloomquist, T.R. et al. (2021). Nanoparticle tracking analysis for the quantification and size determination of extracellular vesicles. *JoVE (Journal of Visualized Experiments)* 169: e62447.

Connolly, J.R., for EPS400-002(2007). *Introduction to X-Ray Powder Diffraction*. Springer.

Cullity, B.D. (1978). *Elements of X-ray Diffraction*. Addison-Wesley Publishing Co. Inc.

Dragovic, R.A., Gardiner, C., Brooks, A.S. et al. (2011). Sizing and phenotyping of cellular vesicles using nanoparticle tracking analysis. *Nanomedicine: Nanotechnology, Biology and Medicine* 7 (6): 780–788.

Ebnesajjad, S. (2011). Surface and material characterization techniques. In: *Handbook of Adhesives and Surface Preparation* (ed. S. Ebnesajjad), 31–48. William Andrew Publishing.

Filipe, V., Hawe, A., and Jiskoot, W. (2010). Critical evaluation of nanoparticle tracking analysis (NTA) by NanoSight for the measurement of nanoparticles and protein aggregates. *Pharmaceutical Research* 27: 796–810.

Gaabour, L.H. (2019). Influence of silica nanoparticles incorporated with chitosan/polyacrylamide polymer nanocomposites. *Journal of Materials Research and Technology* 8 (2): 2157–2163.

Ghorbanpour, M., Yousofi, M., and Lotfiman, S. (2017). Photocatalytic decolorization of methyl orange by silica-supported TiO_2 composites. *Journal Of Ultrafine Grained And Nanostructured Materials* 50 (1): 43–50.

Gkika, D.A., Vordos, N., Mitropoulos, A.C. et al. (2023). Nanomaterials and their properties: thermal analysis, physical, mechanical and chemical properties. In: *Advanced and Innovative Approaches of Environmental Biotechnology in Industrial Wastewater Treatment* (ed. M.P. Shah), 301–331. Singapore: Springer Nature Singapore.

Greish, K., Thiagarajan, G., Herd, H. et al. (2012). Size and surface charge significantly influence the toxicity of silica and dendritic nanoparticles. *Nanotoxicology* 6 (7): 713–723.

Gulin, A., Shakhov, A., Vasin, A. et al. (2019). ToF-SIMS depth profiling of nanoparticles: chemical structure of core-shell quantum dots. *Applied Surface Science* 481: 144–150.

Henrich, V.E. and Cox, P.A. (1994). *The Surface Chemistry of Metal Oxides*. Cambridge: Cambridge University Press.

Hole, P., Sillence, K., Hannell, C. et al. (2013). Interlaboratory comparison of size measurements on nanoparticles using nanoparticle tracking analysis (NTA). *Journal of Nanoparticle Research* 15: 1–12.

Huseynov, E., Garibov, A., and Mehdiyeva, R. (2016). TEM and SEM study of nano SiO_2 particles exposed to influence of neutron flux. *Journal of Materials Research and Technology* 5 (3): 213–218.

Jackman, L.M. and Sternhell, S. (2013). *Application of Nuclear Magnetic Resonance Spectroscopy in Organic Chemistry: International Series in Organic Chemistry*, vol. 10. Elsevier.

Kanemitsu, Y. and Okamoto, S. (1997). Visible luminescence from silicon quantum dots and wells. *Materials Science and Engineering* B48: 108–115.

Kargina, Y.V., Perepukhov, A.M., Kharin, A.Y. et al. (2019). Silicon nanoparticles prepared by plasma-assisted ablative synthesis: physical properties and potential biomedical applications. *Physica Status Solidi (a)* 216 (14): 1800897.

Kim, T. and Lee, J. (2023). Silicon nanoparticles: fabrication, characterization, application and perspectives. *Micro and Nano Systems Letters* 11 (1): 18.

Kolahalam, L.A., Viswanath, I.K., Diwakar, B.S. et al. (2019). Review on nanomaterials: synthesis and applications. *Materials Today Proceedings* 18: 2182–2190.

Korin, E., Froumin, N., and Cohen, S. (2017). Surface analysis of nanocomplexes by X-ray photoelectron spectroscopy (XPS). *ACS Biomaterials Science & Engineering* 3 (6): 882–889.

Kravets, V.G., Meier, C., Konjhodzic, D. et al. (2005). Infrared properties of silicon nanoparticles. *Journal of Applied Physics* 97 (8).

Kravitz, K., Kamyshny, A., Gedanken, A., and Magdassi, S. (2010). Solid state synthesis of water-dispersible silicon nanoparticles from silica nanoparticles. *Journal of Solid State Chemistry* 183 (6): 1442–1447.

Kustov, L.M., Mashkin, P.V., Zakharov, V.N. et al. (2018). Silicon nanoparticles: characterization and toxicity studies. *Environmental Science: Nano* 5 (12): 2945–2951.

Laeter, D. and John, R. (2001). *Mass Spectrometry (Including SIMS, ICP-MS, Accelerator MS, TIMS) Applications of Inorganic Mass Spectrometry*, 474. New York: Wiley.

Larsson, M., Elfving, A., Holtz, P.O. et al. (2003). Photoluminescence study of Si/Ge quantum dots. *Surface Science* 532: 832–836.

Li, Z., Hadioui, M., and Wilkinson, K.J. (2023). Extraction of silicon-containing nanoparticles from an agricultural soil for analysis by single particle sector field and time-of-flight inductively coupled plasma mass spectrometry. *Nanomaterials* 13 (14): 2049.

Manjakkal, L., Sakthivel, B., Gopalakrishnan, N., and Dahiya, R. (2018). Printed flexible electrochemical pH sensors based on CuO nanorods. *Sensors and Actuators, B: Chemical* 263: 50–58.

Mbakaan, C., Ahemen, I., Amah, A.N. et al. (2018). White-light-emitting Dy^{3+}-doped amorphous SiO_2 nanophosphors derived from rice husk. *Applied Physics A* 124 (11): 741.

Moreno, L.A. and Cox, K.L. (2010). Quantification of dsDNA using the Hitachi F-7000 Fluorescence Spectrophotometer and PicoGreen dye. *Journal of Visualized Experiments* 45: e2465.

Morozova, S., Alikina, M., Vinogradov, A., and Pagliaro, M. (2020). Silicon quantum dots: synthesis, encapsulation, and application in light-emitting diodes. *Frontiers in Chemistry* 8: 191.

Murty, B.S., Shankar, P., Raj, B. et al. (2013). Unique properties of nanomaterials. In: *Textbook of Nanoscience and Nanotechnology* (ed. B. Raj), 29–65.

Nishi, Y. (1971). Study of silicon-silicon dioxide structure by electron spin resonance I. *Japanese Journal of Applied Physics* 10 (1): 52.

Nordling, C., Sokolowski, E., and Siegbahn, K. (1957). Precision method for obtaining absolute values of atomic binding energies. *Physical Review* 105 (5): 1676.

Nordling, C., Sokolowski, E., and Siegbahn, K. (1958). Evidence of chemical shifts of the inner electronic levels in a metal relative to its oxides (Cu, Cu_2O, CuO). *Arkiv for Fysik* 13 (5): 483–506.

Patterson, A.L. (1939). The Scherrer formula for X-ray particle size determination. *Physics Review* 56 (10): 978–982.

Perego, M., Ferrari, S., Fanciulli, M. et al. (2004). Characterization of silicon nanocrystals embedded in thin oxide layers by TOF-SIMS. *Applied Surface Science* 231: 813–816.

Petreanu, I., Niculescu, V.C., Enache, S. et al. (2023). Structural characterization of silica and amino-silica nanoparticles by Fourier transform infrared (FTIR) and Raman spectroscopy. *Analytical Letters* 56 (2): 390–403.

Prabha, S., Durgalakshmi, D., Rajendran, S., and Lichtfouse, E. (2021). Plant-derived silica nanoparticles and composites for biosensors, bioimaging, drug delivery and supercapacitors: a review. *Environmental Chemistry Letters* 19 (2): 1667–1691.

Presentato, A., Armetta, F., Spinella, A. et al. (2020). Formulation of mesoporous silica nanoparticles for controlled release of antimicrobials for stone preventive conservation. *Frontiers in Chemistry* 8: 699.

Rafigh, S.M. and Heydarinasab, A. (2017). Mesoporous chitosan–SiO_2 nanoparticles: synthesis, characterization, and CO_2 adsorption capacity. *ACS Sustainable Chemistry & Engineering* 5 (11): 10379–10386.

Rim, Y.S., Bae, S.-H., Chen, H. et al. (2015). Printable ultrathin metal oxide semiconductor-based conformal biosensors. *ACS Nano* 9 (12): 12174–12181.

Rogers, K.L., Cruz-Hernandez, A., and Brown, J.M. (2022). A quantitative method for determining uptake of silica nanoparticles in macrophages by single particle inductively coupled plasma–mass spectrometry. *Current Protocols* 2 (3): e396.

Saleh, T.A. (2020). Nanomaterials: classification, properties, and environmental toxicities. *Environmental Technology & Innovation* 20: 101067.

Sato, S. and Swihart, M.T. (2006). Propionic-acid-terminated silicon nanoparticles: synthesis and optical characterization. *Chemistry of Materials* 18 (17): 4083–4088.

Scherrer, P. (1918). Bestimmung der Größe und der inneren Struktur von Kolloidteilchen mittels Röntgenstrahlen. *Nachrichten von der Gesellschaft der Wissenschaften zu Göttingen, Mathematisch-Physikalische Klasse* 1918: 98–100.

Sharififard, H. and Rezvanpanah, E. (2021). Ultrasonic-assisted synthesis of SiO_2 nanoparticles and SiO_2/chitosan/Fe nanocomposite and their application for vanadium adsorption from aqueous solution. *Environmental Science and Pollution Research* 28 (9): 11586–11597.

Siegbahn, K. (1967). ESCA: atomic, molecular and solid state structure studies by means of electron spectroscopy. *Nova Acta Regiae Societatis Scientiarum Upsaliensis* 20: 1.

Song, Y. and Chen, S. (2014). Janus nanoparticles: preparation, characterization, and applications. *Chemistry–An Asian Journal* 9 (2): 418–430.

Stevie, F.A. and Donley, C.L. (2020). Introduction to x-ray photoelectron spectroscopy. *Journal of Vacuum Science & Technology A* 38 (6): 063204.

Valenta, J., Juhasz, R., and Linnros, J. (2002). Photoluminescence from single silicon quantum dots at room temperature. *Journal of Luminescence* 98 (1-4): 15–22.

Van der Heide, P. (2011). *X-ray Photoelectron Spectroscopy: An Introduction to Principles and Practices*. Wiley.

Vorokhta, M., Khalakhan, I., Vondráček, M. et al. (2018). Investigation of gas sensing mechanism of SnO_2 based chemiresistor using near ambient pressure XPS. *Surface Science* 677: 284–290. https://doi.org/10.1016/j.susc.2018.08.003.

Watts, J.F. and Wolstenholme, J. (2019). *An Introduction to Surface Analysis by XPS and AES*. Wiley.

Wen, X., Dagan, S., and Wysocki, V.H. (2007). Small-molecule analysis with silicon-nanoparticle-assisted laser desorption/ionization mass spectrometry. *Analytical Chemistry* 79 (2): 434–444.

Williamson, G.K. and Hall, W.H. (1953). X-ray line broadening from filed aluminium and wolfram. *Acta Metallurgica* 1: 22–31.

Wilson, A.J.C. (1949). *X-ray Optics*. London: Menthuen.

Zheng, Y. and Chen, M. (2023). Low energy ion scattering (LEIS) spectroscopy. In: *Springer Handbook of Advanced Catalyst Characterization* (ed. I.E. Wachs and M.A. Bañares), 461–484. Cham: Springer Handbooks. Springer. https://doi.org/10.1007/978-3-031-07125-6_22.

Zou, J., Baldwin, R.K., Pettigrew, K.A., and Kauzlarich, S.M. (2004). Solution synthesis of ultrastable luminescent siloxane-coated silicon nanoparticles. *Nano Letters* 4 (7): 1181–1186.

8

Technological Advancements for Characterization of Si/SiO_2 Nanoparticles

Prasann Kumar[1] *and Joginder Singh*[2]

[1] *Department of Agronomy, School of Agriculture, Lovely Professional University, Phagwara, Punjab, India*
[2] *Department of Botany, Nagaland University, Lumami, Nagaland, India*

8.1 Introduction

Nanotechnology is frequently regarded as the forthcoming industrial revolution involving manipulating and regulating matter at the nanoscale, which typically spans from 1 to 100 nm. The field has experienced rapid and significant growth and innovation in recent decades, driven by materials science, chemistry, physics and engineering advancements (Afshari Aghajari et al. 2023; Ermina et al. 2024; Hassan et al. 2023). The capability to manipulate materials at the nanoscale has provided unparalleled prospects in various fields, leading to transformative advancements in electronics, medicine, energy and manufacturing. The inception of nanotechnology can be attributed to the influential lecture delivered by the physicist Richard Feynman in 1959, wherein he conceptualized the potential for manipulating individual atoms and molecules to fabricate diminutive machines. Nevertheless, the emergence of nanotechnology as a separate scientific field in the 1980s and 1990s can be attributed to significant advancements in microscopy, nanofabrication methods and theoretical comprehension. Subsequently, nanotechnology has undergone significant advancements, transforming into a diverse domain encompassing various branches such as nanomaterials, nanoelectronics, nanomedicine and nanobiotechnology. The distinctive physical, chemical and biological properties displayed by materials at the nanoscale are key features of nanotechnology. Nanomaterials exhibit quantum effects, surface phenomena and size-dependent properties, resulting in unique functionalities and improved performance compared to larger materials. Nanoparticles can display modified optical, magnetic, catalytic and mechanical characteristics due to their diminutive

Silicon Nanoparticles: Synthesis, Uptake, and Applications, First Edition.
Edited by Archana Singh, Aryadeep Roychoudhury, and Indrakant K. Singh.

dimensions and elevated surface-to-volume ratio. Nanotechnology has a wide range of applications that cover almost every industry and sector of the economy. Nanoscale materials play a pivotal role in electronics and computing by facilitating the advancement of devices that exhibit enhanced speed, reduced size and improved efficiency. Notable examples include nanoscale transistors, memory storage devices and quantum computers (Khajavi et al. 2021; Mao et al. 2023; Mohammed et al. 2023; Shirke et al. 2024; Zhang and Atchimarungsri 2023). Nanotechnology presents unparallel medical and healthcare prospects, including drug delivery, diagnostics, imaging and regenerative medicine. It can transform personalized medicine and address various diseases, including cancer and neurodegenerative disorders. Furthermore, nanotechnology plays a pivotal role in tackling urgent global issues such as the generation of clean energy, remediation of the environment and the purification of water. In sustainable energy, nanomaterials, including quantum dots, nanowires and nanocomposites exhibit considerable potential in enhancing the efficiency of solar cells, energy storage devices and catalytic converters. Furthermore, using nanotechnology-based sensors and devices enables the continuous monitoring of environmental pollutants, contaminants and pathogens in real time, thereby significantly promoting environmental sustainability and enhancing public health (Roychoudhury and Bhowmik 2023; Singh and Roychoudhury 2022). Silicon (Si) and silicon dioxide (SiO_2) nanoparticles are highly significant nanomaterials extensively used in different industries. Silicon, the second most prevalent element in the Earth's crust, is the basis for contemporary electronics and semiconductor technology. The electronic, optical and mechanical properties of silicon nanoparticles, which typically have sizes ranging from a few to several hundred nanometers, render them essential for a wide range of applications. Si/SiO_2 nanoparticles are utilized as fundamental components in integrated circuits, optoelectronic devices and photovoltaic cells within electronics and photonics. By accurately manipulating silicon nanoparticles at the nanoscale, it is possible to observe quantum confinement phenomena. This phenomenon allows for advancing transistors, light-emitting diodes and photodetectors with improved performance and efficiency. In addition, Si/SiO_2 nanocomposites are utilized in various fields, such as advanced lithography, nanoelectromechanical systems and quantum computing, thereby fostering advancements within the semiconductor industry. In addition to their applications in electronics, Si/SiO_2 nanoparticles have significant implications in biomedical and healthcare fields. Silicon-based nanomaterials possess desirable biocompatibility, stability and adjustable surface chemistry, making them highly suitable for drug delivery systems, biosensors and medical imaging agents. Si/SiO_2 nanoparticles that have been modified can be designed to selectively target particular cells, tissues or biomolecules, facilitating accurate identification and management of conditions like cancer, diabetes and cardiovascular disorders. Moreover, silicon nanoparticles exhibit potential in tissue engineering, regenerative medicine and neural interfacing, facilitating progress in personalized healthcare and therapeutics (Batista et al. 2023; Dhaffouli et al. 2024; Liu et al. 2024; Madani et al. 2022; Wu et al. 2023b). Moreover, utilizing Si/SiO_2 nanoparticles is widespread in the energy and environmental sectors, significantly advancing sustainable technologies and solutions. Silicon-based nanomaterials have demonstrated high efficacy as catalysts, photocatalysts and electrode materials in various energy conversion and storage applications, such as fuel cells, batteries and supercapacitors. The adsorption and photocatalytic properties of Si/SiO_2 nanocomposites are noteworthy, rendering them highly efficient in water purification, air filtration and environmental remediation. Researchers and engineers are tackling significant challenges concerning clean energy production, climate change mitigation and resource sustainability by utilizing the distinctive characteristics of silicon nanoparticles. The exponential growth of nanotechnology and the escalating intricacy of nanomaterials require the implementation of sophisticated methodologies for their identification and characterization. Conventional techniques for characterizing nanomaterials at the nanoscale, such as microscopy, spectroscopy and chromatography, frequently fail to understand their structural, chemical and functional attributes comprehensively. Consequently, there is an increasing need for novel analytical tools and methodologies to investigate and clarify the complex characteristics of nanomaterials with exceptional accuracy and detail. One of the main difficulties encountered in characterizing Si/SiO_2 nanoparticles is their diminutive dimensions and diverse composition, necessitating methodologies that can effectively investigate individual nanoparticles and their aggregates. Transmission electron microscopy (TEM) and scanning electron microscopy (SEM) are advanced imaging techniques that provide exceptional resolution and spatial resolution. These techniques allow scientists to observe nanoparticle morphology, size distribution and crystalline structure with precision at the atomic level. In addition, atomic force microscopy (AFM) enables the visualization of nanoparticle surfaces in three dimensions, thereby facilitating the examination of surface topography and nanoscale characteristics. Spectroscopic methods are essential for characterizing the composition, chemical bonding and surface chemistry of Si/SiO_2 nanoparticles alongside imaging techniques. X-ray photoelectron spectroscopy (XPS) is a technique that offers significant insights into the elemental composition and chemical states of nanoparticles. It enables researchers to effectively detect and characterize dopants, impurities and functional groups on the surface of nanoparticles (Cani et al. 2021; Fatahi et al. 2024; Jafari et al. 2022; Navarro et al. 2022; Rahimi et al. 2024; Shaba et al. 2023). Fourier-transform infrared spectroscopy (FTIR) is a complementary technique to XPS that examines the vibrational modes of chemical bonds. It provides valuable information about surface functionalization, molecular interactions and contaminants. In addition, integrating energy-dispersive X-ray spectroscopy (EDS) and TEM facilitates the quantitative examination of nanoparticle composition, enabling researchers to precisely ascertain the elemental distribution and stoichiometry of Si/SiO_2 nanoparticles. The assessment of nanoparticle colloidal stability and surface charge and the examination of critical parameters that affect their behaviour in solution and interactions with biological systems heavily rely on utilizing dynamic light scattering (DLS) and zeta potential measurements. Developing and applying advanced techniques for their identification and characterization is crucial to advance fundamental understanding, optimize material properties and ensure the safe and effective integration of Si/SiO_2 nanoparticles into various applications. Researchers can harness the complete capabilities of Si/SiO_2 nanoparticles and expedite advancements in multiple fields such as electronics, medicine, energy and environmental science by utilizing a comprehensive approach that includes imaging, spectroscopy and analytical techniques (Table 8.1 and Figure 8.1).

8.2 Techniques for Identifying Si/SiO_2 Nanoparticles

A) **Transmission electron microscopy**: TEM is an essential tool in material science. The exceptional resolution capabilities of this technology offer unparallel insights into the realm of nanotechnology. At its essence, TEM functions based on the fundamental principles of electron optics and wave-particle duality, facilitating the observation of structures at the atomic and subatomic levels. TEM functions on a carefully designed configuration of intricate components that are carefully coordinated to manipulate electrons to generate high-resolution images. The functionality of TEM relies heavily on the electron source, which is commonly a heated tungsten filament or a field emission gun. This source emits a concentrated beam of electrons. The beam above is subsequently accelerated using a sequence of electromagnetic lenses, which converge and concentrate the electrons onto the examined substrate. The ultra-thin specimen is carefully positioned on a specialized holder within the transmission electron microscope chamber, facilitating the passage of electrons through it with minimal interference. Upon interaction between the electron beam and the specimen, a range of imaging modes are activated, each providing distinct perspectives on the structure of the sample (Bahramnia et al. 2021; Ece 2021; Fan et al. 2024; Huang et al. 2023; Namakka et al. 2023; Nguyen et al. 2021; Smitshuysen et al. 2024). In its most traditional form, bright-field imaging utilizes the differential scattering of electrons within the specimen to create an image emphasizing differences in thickness and density. In contrast, dark-field imaging specifically captures electrons scattered at significant angles, highlighting defects and strain fields in the

Table 8.1 Different techniques for identifying and characterizing Si/SiO_2 nanoparticles.

No.	Technology	Description
1	Transmission electron microscopy (TEM)	Provides high-resolution images of nanoparticle size, shape and morphology.
2	Scanning electron microscopy (SEM)	Offers detailed imaging of nanoparticle surfaces and morphology.
3	Atomic force microscopy (AFM)	Enables nanoscale imaging and topographic analysis of nanoparticle surfaces.
4	Fourier-transform infrared spectroscopy (FTIR)	Identifies surface functional groups and chemical composition.
5	Raman spectroscopy	Provides information about molecular vibrations and chemical bonding.
6	X-ray photoelectron spectroscopy (XPS)	Determines elemental composition and chemical states on nanoparticle surfaces.
7	Energy-dispersive X-ray spectroscopy (EDS)	Analyses elemental composition and mapping of nanoparticles.
8	X-ray diffraction (XRD)	Determines crystal structure, phase composition and lattice parameters.
9	Dynamic light scattering (DLS)	Measures particle size distribution and hydrodynamic diameter in colloidal systems.
10	Zeta Potential Measurements	Determines surface charge and stability of colloidal dispersions.
11	Small-angle X-ray scattering (SAXS)	Provides information about nanoparticle size, shape and internal structure.
12	Laser diffraction	Measures particle size distribution in suspensions or powders.
13	Nuclear magnetic resonance (NMR) spectroscopy	Provides insights into nanoparticle structure and surface functionalization.
14	UV-Vis spectroscopy	Analyses absorbance and optical properties of Si/SiO_2 nanoparticles.
15	Inductively coupled plasma mass spectrometry (ICP-MS)	Quantifies elemental concentrations in nanoparticles.
16	Quartz crystal microbalance (QCM)	Measures mass changes on surfaces, helpful in studying nanoparticle adsorption.
17	Gas chromatography-mass spectrometry (GC-MS)	Identifies organic compounds and surface contaminants on nanoparticles.

(Continued)

Table 8.1 (Continued)

No.	Technology	Description
18	Differential scanning calorimetry (DSC)	Analyses thermal behaviour and phase transitions in Si/SiO$_2$ nanoparticles.
19	Thermo-gravimetric analysis (TGA)	Determines the composition and thermal stability of nanoparticles.
20	Nuclear reaction analysis (NRA)	It provides quantitative elemental analysis and is helpful for Si/SiO$_2$ nanoparticle characterization.
21	Electron energy loss spectroscopy (EELS)	Measures the energy loss of electrons to provide chemical composition information.
22	Time-of-flight secondary ion mass spectrometry (ToF-SIMS)	Provides surface chemical composition and depth profiling.
23	Surface plasmon resonance (SPR)	Measures changes in refractive index due to nanoparticle binding.
24	Surface enhanced Raman spectroscopy (SERS)	Enhances Raman signals for sensitive detection of Si/SiO$_2$ nanoparticles.
25	Electrochemical impedance spectroscopy (EIS)	Studies electrical properties and surface interactions of nanoparticles.
26	Nanoparticle tracking analysis (NTA)	Measures nanoparticle size, concentration and behaviour in suspension.
27	Circular dichroism (CD) Spectroscopy	Provides information about nanoparticle chirality and surface interactions.
28	Surface-enhanced infrared absorption (SEIRA)	Enhances infrared absorption signals for surface analysis of nanoparticles.
29	Surface plasmon enhanced fluorescence (SPEF)	Enhances fluorescence signals for sensitive detection of Si/SiO$_2$ nanoparticles.
30	Dual polarization interferometry (DPI)	Measures changes in refractive index and molecular binding kinetics.
31	Super-Resolution Microscopy	Provides nanoscale imaging resolution for detailed characterization of nanoparticles.

Source: Jia et al. (2022), Marimuthu et al. (2023), Miralaei et al. (2023), Niu et al. (2021).

material. High-resolution transmission electron microscopy expands the resolution limits, facilitating the direct observation of atomic configurations within crystalline substances and offering invaluable insights into lattice structures and imperfections. In addition to imaging, TEM also enables the observation of electron diffraction, a phenomenon characterized by the

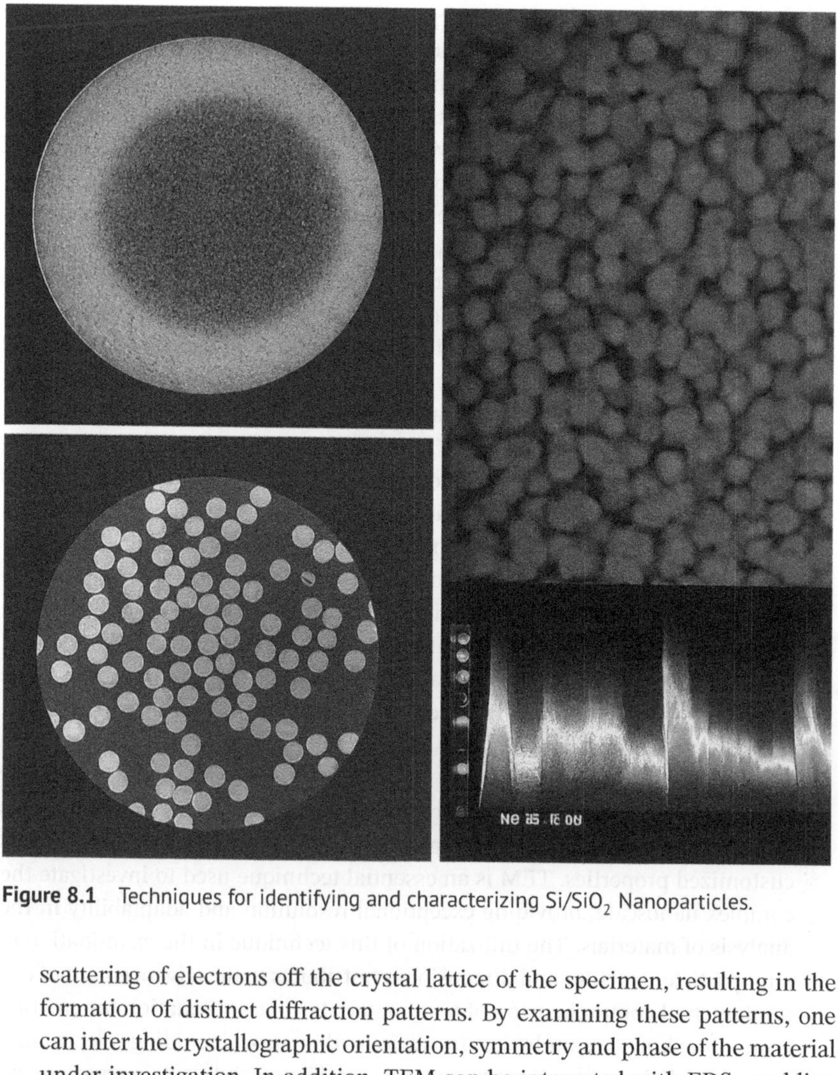

Figure 8.1 Techniques for identifying and characterizing Si/SiO_2 Nanoparticles.

scattering of electrons off the crystal lattice of the specimen, resulting in the formation of distinct diffraction patterns. By examining these patterns, one can infer the crystallographic orientation, symmetry and phase of the material under investigation. In addition, TEM can be integrated with EDS, enabling the analysis of elements and the mapping of the composition of the sample at a resolution of nanometers. TEM is a crucial technique used in nanoparticle research to analyse the structure and size of nanoparticles. These nanoparticles are often complex to observe using traditional optical microscopy because they are tiny. The precise characterization of nanoparticles, which typically have sizes ranging from 1 to 100 nm, is of utmost importance for various applications, including catalysis, drug delivery, electronics and environmental remediation. These particles possess distinct physicochemical properties determined by their size, shape and surface characteristics. Sample preparation is of utmost importance in the field of TEM-based nanoparticle characterization. To achieve a consistent distribution and minimize aggregation, nanoparticles are frequently dispersed onto ultra-thin carbon films or supported on specialized TEM grids. The meticulous management of sample thickness is of utmost importance, as excessive thickness can result in electron scattering and the deterioration of images. After being prepared, nanoparticles undergo TEM analysis, during which their morphology, size and structure are carefully examined (Cheng et al. 2024; Dong et al. 2023; Ghattavi and Homaei 2024; Lasemi et al. 2020; Li et al. 2022; Ren et al. 2020; Steffi et al. 2022; Wang et al. 2023). Bright-field imaging enables the visualization of individual nanoparticles, offering valuable information about their morphology and spatial arrangement. High-resolution TEM reveals the intricate atomic-level characteristics of nanoparticle structure, providing clear insights into phenomena such as crystallinity, lattice defects and surface morphology. In addition to visual representation, TEM enables the quantitative analysis of nanoparticle size distribution using image processing techniques and software tools. The precise quantification of particle size, diameter, aspect ratio and dispersion allows researchers to customize nanoparticle synthesis and fabrication methods to attain specific properties. TEM uses in-situ TEM techniques and static imaging to examine nanoparticle behaviour in different environmental conditions, including temperature, pressure and chemical exposure. The continuous real-time observation of nanoparticle growth, aggregation and phase transformations significantly contributes to understanding their synthesis mechanisms and stability. This knowledge is crucial in developing sophisticated nanomaterials that possess customized properties. TEM is an essential technique used to investigate the complex nanoscale, providing exceptional resolution and adaptability in the analysis of materials. The utilization of this technique in the examination of nanoparticle structure and dimensions contributes to the advancement of our fundamental comprehension of nanomaterials. It is a foundation for creating groundbreaking technologies in various domains, including electronics, energy, medicine and environmental science. Nanotechnology is experiencing significant growth, and TEM continues to play a prominent role in advancing scientific knowledge and pushing the limits of scientific investigation.

B) **Scanning electron microscopy**: SEM is a fundamental technique in microscopy that provides exceptional imaging and characterization abilities, especially in nanotechnology. Fundamentally, SEM functions based on the fundamental principle of systematically moving a concentrated electron beam across the surface of a given sample while carefully examining the interactions between the electrons and the specimen to produce images of exceptional resolution. Comprehending the fundamental principles and functioning of SEM is essential for fully utilizing its capabilities to capture the morphology of nanoparticles accurately. SEM commences by generating an electron beam using an electron source, commonly an electron filament or a field emission gun. Subsequently, the beam accelerates via a sequence of electromagnetic lenses, effectively concentrating and regulating its path towards the examined specimen. Once the electron beam reaches the specimen, it engages with the surface atoms, initiating a series of events that encompass electron scattering, emission of secondary electrons (SEs) and the occurrence of backscattered electrons (Alsaad et al. 2020; Aryanti et al. 2024; Dadashi et al. 2023; Feng et al. 2023; Ghanaatian et al. 2022; Lin et al. 2021; Majeed et al. 2024). The emitted SEs and backscattered electrons are detected by specialized detectors located within the SEM Chamber. Surface emission (SE) detectors can capture low-energy electrons emitted from the surface, offering valuable topographic insights into the morphology of the sample. In the interim, backscattered electrons (BSE) detectors can detect high-energy electrons that experience minimal interaction with the specimen. This enables the acquisition of compositional and contrast information by analysing variations in atomic numbers. The scanning mechanism used in SEM is a meticulous procedure that entails the systematic displacement of the electron beam across the surface of the sample in a raster configuration. During the scanning process of the beam, the signals emitted by the SE and BSE detectors are gathered and analysed to produce images that exhibit remarkable levels of resolution and contrast. In addition, SEM enables the collection of elemental composition data using EDS. EDS is a method that involves the analysis of characteristic X-rays emitted by the sample when electron excitation occurs, the purpose of which is to identify and map the distribution of elements. Within nanoparticle morphology imaging, SEM emerges as an essential instrument, providing distinct functionalities that enable researchers to thoroughly examine the intricate aspects of nanoparticle structure and morphology with unparallel precision. Nanoparticles, characterized by their size range of 1–100 nm, frequently display unique morphologies influenced by their synthesis techniques, crystallinity and surface characteristics. Comprehending these morphological characteristics holds significance in understanding the behaviour of nanoparticles, enhancing fabrication procedures and customizing their properties for particular applications. The role of SEM in accurately imaging the morphology of the nanoparticles encompasses various crucial aspects. To begin with, SEM offers a non-invasive and high-resolution imaging technique, enabling researchers to observe individual nanoparticles and their clusters with remarkable precision. SE imaging allows researchers to acquire topographic data regarding size, shape and surface texture, providing valuable insights into their structural properties. Moreover, using BSE imaging in SEM allows for the comprehensive analysis of nanoparticle composition and density fluctuations, thereby facilitating the distinction of various nanoparticle types and phases in a given sample. BSE imaging utilizes contrast mechanisms to detect heavy elements and compositional gradients in nanoparticles, providing insights into their chemical composition and distribution. Moreover, the combination of SEM and EDS provides a comprehensive method for characterizing nanoparticles, allowing for the simultaneous imaging and elemental analysis of individual nanoparticles (Dong et al. 2023; Fan et al. 2024; Feng et al. 2023; Ghattavi and Homaei 2024; Park et al. 2024; Steffi et al. 2022; Wu et al. 2023a; Zhang et al. 2023). Through the correlation of morphological information derived from SEM images with elemental mapping data obtained via EDS, researchers can understand the spatial arrangement of elements within nanoparticle assemblies. This correlation offers valuable insights into the chemical heterogeneity and surface functionalization of these assemblies. SEM is an indispensable instrument for accurately visualizing the morphology of nanoparticles, providing remarkable resolution and adaptability in the characterization of nanomaterials. SEM plays a crucial role in advancing our comprehension of nanoscience and facilitating the advancement of innovative nanotechnologies in various fields by providing the ability to visualize and analyse the structural and compositional characteristics of nanoparticles with exceptional precision. Nanotechnology is experiencing significant growth, and SEM continues to play a leading role in driving advancements and pushing the limits of scientific investigation.

C) **Atomic force microscopy**: The utilization of AFM has significantly transformed the nanotechnology domain through its ability to generate high-resolution images of surfaces at the atomic level. The fundamental concept underlying AFM is centred on the dynamic interplay between a sharp tip affixed to a pliable cantilever and the surface of a specimen. The measurement of interaction forces between the tip and the sample is conducted as the tip scans the surface, generating intricate topographic images. This section will examine the fundamental concept and functioning of AFM and its ability to capture and represent surface topography in three dimensions. AFM quantifies the forces exerted between a pointed tip and the sample surface. The essential elements of an atomic force microscope configuration consist of a sharp tip affixed to the unrestricted extremity of a flexible cantilever (Batista et al. 2023; Dhaffouli et al. 2024; Fatahi et al. 2024; Jafari et al. 2022; Liu et al. 2024; Madani et al. 2022; Mao et al. 2023; Shaba et al. 2023; Taha et al. 2024; Wu et al. 2023b). The cantilever undergoes bending due to the forces exerted interacting with the sample surface. The laser beam is focused onto the posterior surface of the cantilever, and a photodetector detects the reflected beam. The displacement of the laser spot on the photodetector undergoes alterations in response to the bending of the cantilever, thereby enabling accurate deflection measurement. A piezoelectric scanner mounts the sample, enabling precise movement in the x, y, and z directions. The scanner facilitates surface scanning by displacing the sample at the tip. A feedback mechanism is used to sustain a consistent force between the tip and the surface of the sample. As the scanning tip traverses the surface, the feedback system modulates the height of the scanner to maintain a consistent measured force, thereby generating a topographic representation of the surface. AFM commences by calibrating the cantilever and tip. The laser spot is precisely positioned on the photodetector, and the sensitivity of the photodetector is fine-tuned to measure the deflection of the cantilever accurately. Subsequently, the tip is carefully brought into contact with the surface of the sample, and the feedback loop is activated to sustain a consistent force between the tip and the surface. The cantilever deflection is measured as the scanner moves the sample about the tip, generating a topographic image of the surface through the analysis of height variations. One of the critical benefits of AFM is its ability to accurately depict surface topography in three dimensions, exhibiting resolution at the nanometer scale. The resolution of features smaller than the wavelength of light poses a challenge for traditional imaging techniques, such as optical microscopy, due to the inherent limitations imposed by the diffraction of light. On the other hand, AFM can capture images of surfaces with atomic-level precision, offering intricate details about surface morphology and structure. AFM images are commonly presented as two-dimensional height maps, wherein variations in brightness or colour correspond to differences in height. Moreover, AFM can produce three-dimensional representations of surfaces by quantifying the variations in height at every pixel within the image. AFM can generate a three-dimensional surface depiction by scanning the sample in the x, y and z directions. This allows for observing intricate features such as surface roughness, grain boundaries and individual atoms. Three-dimensional AFM images significantly contribute to understanding the structure and characteristics of materials at the nanoscale (Alsaad et al. 2020; Bahramnia et al. 2021; Cheng et al. 2024; Dong et al. 2023; Fan et al. 2024; Ghattavi and Homaei 2024; Huang et al. 2023; Lasemi et al. 2020; Ren et al. 2020; Smitshuysen et al. 2024; Steffi et al. 2022). For instance, AFM has been used to investigate surface roughness in thin films, the morphology of nanoparticles and the topography of biological samples, including DNA and proteins. AFM allows researchers to gain a deeper understanding of the structure-function relationships of materials and create innovative technologies for various applications by visually representing surface topography in three dimensions. The imaging technique known as AFM is a highly effective method that relies on measuring forces exerted between a sharp tip and the sample surface. AFM can observe surface topography in three dimensions with nanometer resolution, offering comprehensive insights into surface morphology and structure. AFM has become an essential tool in nanotechnology research and development because it can image materials at the atomic scale. Its applications span various fields, such as materials science, biology, electronics and medicine.

D) **X-ray photoelectron spectroscopy**: X-ray Electron Spectroscopy for Chemical Analysis, also known as XPS, is a highly accurate method for analysing surfaces extensively used in materials science, chemistry and other domains. This section thoroughly examines the principles and functioning of XPS, emphasizing its exceptional significance in determining elemental composition and chemical states. XPS functions based on the fundamental principle of the photoelectric effect. X-rays are used to expel electrons from the inner structures of atoms present in a given sample. The photoelectrons, the ejected electrons, possess energy specific to the element they come from. A monochromatic X-ray source emits X-rays with a constant energy. The sample is enclosed in a vacuum chamber to mitigate the potential influence of air molecules. An electron analyser, such as a hemispherical or cylindrical mirror analyser, gathers and examines photoelectrons emitted from the sample surface. The energy of the photoelectrons is quantified through a detector, which facilitates the conversion of electrons into electrical signals. The data acquisition system is responsible for recording and processing the energy spectrum of the photoelectrons, resulting in the generation of an XPS spectrum. The XPS technique adheres to a systematic procedure, commencing with the exposure of the sample surface to X-ray radiation. X-rays can penetrate the surface of a sample and engage with the atoms present, resulting in the expulsion of photoelectrons from the inner atomic shells. The kinetic energy of photoelectrons is determined by the energy of the incident X-rays and the binding energy of the electrons in the sample, guaranteeing a consistent result (Batista et al. 2023; Cani et al. 2021; Dhaffouli et al. 2024; Fatahi et al. 2024; Jafari et al. 2022; Jia et al. 2022; Marimuthu et al. 2023; Navarro et al. 2022; Rahimi et al. 2024; Wu et al. 2023b). The electron analyser collects the photoelectrons, where they are separated according to their kinetic energy. The energy spectrum of the photoelectrons is documented, wherein the observed peaks correspond to the binding energies of electrons originating from distinct atomic orbitals. The correlation between the intensity of the peaks and the concentration of the respective elements in the sample is directly proportional. Analysing the XPS spectrum allows determining the elemental composition and chemical states of the sample surface. Identifying elements and their chemical environment in the XPS spectrum is facilitated by analysing the position and intensity of the peaks. The binding energy of photoelectrons is intrinsic to the element from which they originate. The elemental composition of the sample surface can be ascertained by comparing the binding energies of the peaks in the XPS spectrum to their respective reference values. The sensitivity of XPS to all elements, except hydrogen and helium, renders it highly valuable for analysing a diverse array of materials. In addition to determining the elemental composition, XPS offers insights into the chemical states of the elements in the sample. Differences in chemical bonding or electronic structure can cause chemical shifts in the binding energies of the peaks. The shape and position of peaks in the XPS spectrum can provide insights into the oxidation state, coordination environment and chemical bonding of the elements. This facilitates examination of surface reactions, the detection of contaminants and the observation of chemical transformations in materials. The XPS technique is highly effective for analysing surfaces, which relies on the photoelectric effect. XPS is a technique that offers significant insights into the elemental composition and chemical states of materials through the measurement of photoelectrons emitted from the surface of the sample. XPS has emerged as an essential instrument in materials science, chemistry and other related fields due to its capacity to analyse diverse samples and furnish comprehensive chemical data.

E) **Fourier-transform infrared spectroscopy**: FTIR is a pivotal technique within the field of spectroscopy, providing exceptional capabilities in detecting molecular vibrations and determining the chemical composition of substances. This extensive analysis will explore the fundamental concept and functioning of FTIR, along with its pivotal function in identifying functional groups present on the surfaces of nanoparticles. FTIR is a scientific technique that quantifies how much molecules absorb infrared radiation. An infrared source emits a wide range of wavelengths of infrared radiation. The sample is positioned within a compartment that is exposed to infrared radiation. An interferometer is a device that utilizes modulation techniques to convert infrared radiation into an interferogram. The interferogram is detected by a detector, commonly a mercury cadmium telluride or deuterated triglycine sulphate detector. The interferogram undergoes processing through Fourier-transform algorithms to produce an infrared spectrum. FTIR commences by emitting infrared radiation from the source and directing it towards the sample (Dhaffouli et al. 2024; Fatahi et al. 2024; Jafari et al. 2022; Liu et al. 2024; Mohammed et al. 2023; Navarro et al. 2022; Shaba et al. 2023; Taha et al. 2024). When the sample is exposed to infrared radiation, the molecules selectively absorb particular wavelength of infrared radiation that align with the vibrational modes of the chemical bonds within the molecules. The absorption of infrared radiation induces vibrational transitions between different energy levels. The determination of these energy levels is contingent upon the stiffness of the chemical bonds and the masses of the atoms participating in the bond. Chemical bonds of various types, including C—H, C=O and O—H, demonstrate distinct vibrational frequencies within the infrared spectrum. The interferometer modulates the infrared radiation, generating an interferogram, encompassing data about its amplitude and phase. The interferogram is subsequently subjected to Fourier-transform algorithms to produce an infrared spectrum. This spectrum represents the absorption of infrared radiation of the sample, with the wavelength being the determining factor. The spectrum in the infrared range exhibits peaks corresponding to the absorption of infrared radiation by distinct vibrational modes of the chemical bonds within the sample. The chemical composition of the sample can be determined, and functional groups can be identified by comparing the infrared spectrum of the sample to reference spectra of known compounds. FTIR spectroscopy is of paramount importance in identifying functional groups on the surfaces of nanoparticles, thereby offering valuable insights into the surface chemistry and stability of the said nanoparticles. Nanoparticles with sizes between 1 and 100 nm frequently display distinctive surface characteristics due to their elevated surface-to-volume ratio. Modifying nanoparticle surfaces by adding organic molecules or ligands is a widely used approach to customize their characteristics for particular purposes, including catalysis, drug delivery and sensing. FTIR spectroscopy facilitates the characterization of nanoparticle surface chemistry and identifies functional groups on their surfaces. Researchers can ascertain the chemical bonds on nanoparticle surfaces by comparing the infrared spectra of functionalized nanoparticles with the reference spectra of known functional groups (Cheng et al. 2024; Dadashi et al. 2023; Dong et al. 2023; Fan et al. 2024; Ghattavi and Homaei 2024; Lasemi et al. 2020; Li et al. 2022; Lin et al. 2021; Majeed et al. 2024; Ren et al. 2020; Steffi et al. 2022; Wang et al. 2023). Alkyl or aromatic groups can be inferred from peaks associated with C—H stretching vibrations. In contrast, the presence of carbonyl groups can be inferred from peaks associated with C=O stretching vibrations. FTIR spectroscopy also enables researchers to investigate the interactions between nanoparticles and their surrounding molecules, including solvents and biomolecules. The alterations observed in the infrared spectra of nanoparticles when they come into contact with other molecules offer significant insights into the surface adsorption or chemical reactions taking place on the surfaces of the nanoparticles. In addition, integrating FTIR spectroscopy with other analytical methodologies, such as XPS or TEM can yield supplementary insights into chemical composition and structural characteristics. FTIR is a highly effective methodology utilized for the identification of functional groups present on the surfaces of nanoparticles. This technique offers valuable insights into the surface chemistry and stability of nanoparticles. The characterization of surface functionalization, study of surface interactions and customization of nanoparticle properties for diverse applications in nanotechnology, materials science and biotechnology can be achieved by analysing the infrared spectra of nanoparticles.

F) **Energy-dispersive X-ray spectroscopy**: EDS is a pivotal analytical methodology in materials science, providing exceptional capabilities in analysing and mapping elements. This comprehensive analysis will explore the fundamental principles and functioning of EDS and its crucial significance when used with TEM to examine elemental composition quantitatively. The operational principle of EDS involves the detection of distinctive X-rays emitted by a sample upon exposure to high-energy electrons. TEM produces an electron beam using a thermionic or field emission electron gun. The sample undergoes irradiation with an electron beam, which excites atoms within the sample. The X-ray detector is strategically placed close to the sample to identify the distinctive X-rays it emits. The signal processing system involves the X-ray detector converting incoming X-rays into electrical signals. These signals are then processed and analysed to generate an elemental spectrum. The EDS procedure commences by subjecting the sample to irradiation using high-energy electrons emitted by the electron source. During the interaction between electrons and atoms in the sample, inner-shell electrons can be displaced from their orbitals, forming vacancies in the electron shells. Afterwards, electrons in the outer shell may move to lower energy levels to occupy these empty spaces, resulting in the emission of X-rays (Batista et al. 2023; Dhaffouli et al. 2024; Fatahi et al. 2024; Jafari et al. 2022; Liu et al. 2024; Madani et al. 2022; Navarro et al. 2022; Rahimi et al. 2024; Shaba et al. 2023; Taha et al. 2024; Wu et al. 2023b; Zhang and Atchimarungsri 2023). The emitted X-rays exhibit distinct characteristics specific to the elements present in the sample. Each component within the sample possesses a different set of characteristic X-ray energies. The X-ray detector collects the emitted X-rays and measure their energies. The relationships between the energies of X-rays and the disparities in energy levels among the electron orbitals participating in the transitions are directly proportional. Analysing the energy spectrum of the detected X-rays allows for determining the sample's elemental composition. The X-ray peaks in the spectrum exhibit a direct relationship with the concentration of the respective elements present in the sample. In addition, EDS can produce elemental maps of the sample's surface, which offer spatial insights into the distribution of elements within the sample. The generation of elemental maps with nanometer-scale resolution can be achieved by scanning the electron beam over the sample surface and collecting X-ray spectra at each point. The quantitative analysis of elemental composition in nanoscale samples is commonly achieved by combining EDS and TEM. By using this integrated method, it becomes possible to investigate the elemental makeup of individual nanoparticles, thin films and nanomaterials with exceptional precision in terms of spatial resolution. TEM-EDS analysis uses TEM to capture high-resolution images of the sample, enabling researchers to observe the nanoscale structure and morphology. Simultaneously, EDS is utilized to conduct an elemental analysis of the specimen, thereby yielding insights into the composition of the specimen at every individual location. Calibration of the TEM-EDS system is accomplished by using standard reference materials possessing known elemental compositions, enabling the quantitative analysis of elemental composition. The generation of a calibration curve involves the measurement of X-ray intensities of characteristic peaks originating from reference materials to establish a correlation between X-ray intensity and elemental concentration. Quantitative determination of the elemental composition of the sample can be achieved by measuring the X-ray intensities of characteristic peaks in the EDS spectrum and comparing them to the calibration curve once the system has been calibrated. Scientists can ascertain the comparative concentrations of various elements within the sample and compute atomic ratios through this process (Alsaad et al. 2020; Cheng et al. 2024; Dong et al. 2023; Ghattavi and Homaei 2024; Lasemi et al. 2020; Li et al. 2022; Lin et al. 2021; Ren et al. 2020; Steffi et al. 2022; Wang et al. 2023; Wu et al. 2023a). TEM-EDS analysis is precious in examining nanomaterials, as it enables the meticulous manipulation of elemental composition to customize their properties and functionalities. Through the integration of high-resolution imaging and quantitative elemental analysis, TEM-EDS facilitates researchers to acquire a profound understanding of the interplay between the structure and properties of nanomaterials. This, in turn, propels the progress of innovative nanotechnologies. EDS is a highly reliable material science elemental analysis and mapping application method. When combined with TEM, EDS is a powerful tool for quantitatively examining elemental composition in nanoscale specimens. This combination offers significant insights into the interplay between the structure and properties of nanomaterials, making it an essential technique in nanotechnology research and development. Its applications span various fields, including materials science, electronics, biotechnology and environmental science, underscoring its importance in various scientific disciplines.

8.3 Characterization of Si/SiO$_2$ Nanoparticles

High-resolution images and valuable insights into nanoparticle size distribution, aggregationbehaviour and surface morphology can be obtained through advanced imaging techniques such as TEM, SEM and AFM. Spectroscopic techniques, including FTIR, Raman spectroscopy, and XPS, offer comprehensive insights into the chemical composition, surface functional groups and bonding states. These techniques facilitate the determination of distinct chemical species and their spatial arrangement on the surfaces of nanoparticles. Analytical techniques, such as EDS and XRD, enable the determination of elemental composition, analysis of crystal structure and identification of phases. They aid in comprehending the atomic configuration of the material, crystallinity and phase transitions. TEM and SEM facilitate the quantitative analysis of nanoparticle morphology, size distribution and aggregation behaviour (Batista et al. 2023; Dhaffouli et al. 2024; Fatahi et al. 2024; Jafari et al. 2022; Liu et al. 2024; Madani et al. 2022; Navarro et al. 2022; Rahimi et al. 2024; Shaba et al. 2023; Taha et al. 2024; Wu et al. 2023b; Zhang and Atchimarungsri 2023). The utilization of image analysis software facilitates the quantification of particle size, determination of aspect ratio and statistical analysis of populations of nanoparticles. Cryo-TEM enables the examination of nanoparticles in their original hydrated condition, offering valuable information about their structure and performance in liquid surroundings. AFM provides the ability to observe and analyse surface roughness, topography and surface features at the nanoscale, enabling the visualization and characterization of these properties in three dimensions. The technique known as scanning tunnelling microscopy enables the examination of surface morphology and electronic properties at the atomic level, thereby providing valuable information regarding surface defects and atomic configurations on Si/SiO$_2$ nanoparticles. Elemental analysis of Si/SiO$_2$ nanoparticles can be conducted using EDS in conjunction with SEM or TEM, which allows for quantitatively determining elemental composition and distribution within the sample. XPS provides a means of conducting elemental analysis using surface-sensitive techniques, enabling the detection of surface contaminants, determination of oxidation states and analysis of chemical bonding configurations. FTIR spectroscopy examines the vibrational modes of functional groups present on nanoparticle surfaces, including hydroxyl groups (OH), carbonyl groups (C=O), and silanol groups (Si—OH). This analysis offers valuable information about the surface chemistry and reactivity of these groups. In the realm of Si/SiO$_2$ nanoparticles, Raman spectroscopy provides valuable insights into molecular vibrations and chemical bonding, thereby facilitating the identification of surface functional groups and structural defects. XRD is a technique that offers valuable insights into the crystal structure, lattice parameters and phase composition of Si/SiO_2 nanoparticles. It plays a crucial role in identifying crystalline phases, crystallographic orientation and lattice defects. The utilization of selected area electron diffraction in TEM enables the determination of crystal structure and lattice spacing at the nanoscale, thereby offering valuable insights into the crystallographic properties of Si/SiO_2 nanoparticles. TEM and SEM, in conjunction with EDS, enable the examination of nanoparticle aggregation patterns, interparticle interactions and surface chemistry (Cheng et al. 2024; Dadashi et al. 2023; Fan et al. 2024; Feng et al. 2023; Ghattavi and Homaei 2024; Lasemi et al. 2020; Majeed et al. 2024; Ren et al. 2020; Wu et al. 2023a). In-situ TEM is a dynamic imaging technique that enables the real-time observation of nanoparticle aggregation and surface interactions across diverse environmental conditions. SAXS and DLS provide complementary data on nanoparticle size distribution, aggregation kinetics and colloidal stability, offering valuable insights into the behaviour of nanoparticles in solution. To fully understand Si/SiO_2 nanoparticles, it is crucial to combine sophisticated imaging, spectroscopic and analytical methods. This comprehensive approach enhances comprehension of nanoparticle properties, structure–function relationship and surface chemistry, enabling the development of customized nanomaterials for different applications (Figure 8.2).

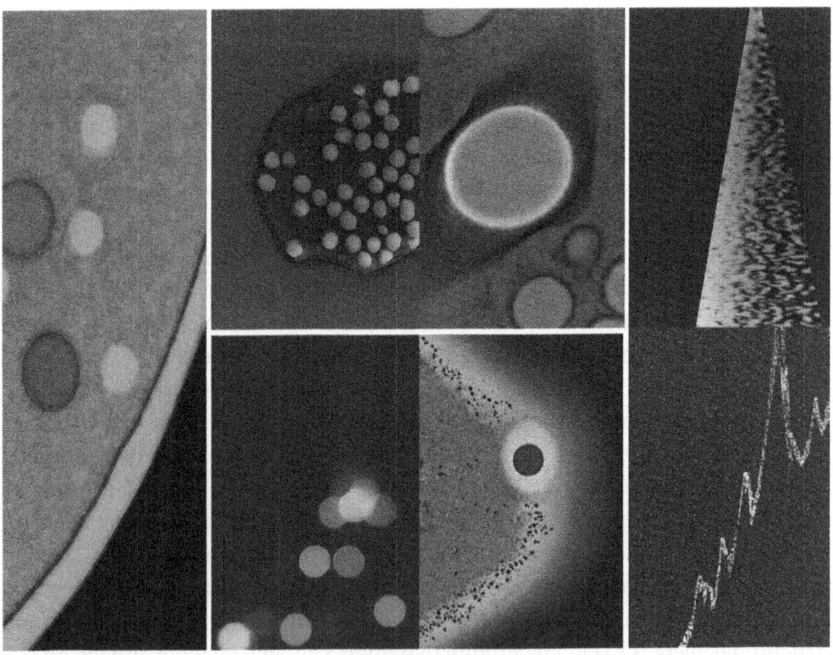

Figure 8.2 Characterization of Si/SiO_2 nanoparticles.

8.4 Importance of Colloidal Stability and Surface Charge

The behaviour, performance and applications of colloidal systems, including nanoparticles, are significantly influenced by fundamental properties such as colloidal stability and surface charge. This extensive analysis delves into the significance of colloidal stability and surface charge, highlighting their functions in diverse domains, ranging from material science to biotechnology. DLS, alternatively called Photon Correlation Spectroscopy or Quasi-Elastic Light Scattering, is a highly effective methodology for analysing particle size distribution within colloidal systems. DLS is predicated upon the fundamental principle of Brownian motion, wherein the stochastic movement of particles within a suspension induces variations in the intensity of scattered light. DLS analyses the intensity fluctuations over time to provide information about particle size distribution and hydrodynamic diameter in the nanometer to micron range. DLS is essential for evaluating colloidal stability as it enables tracking alterations in particle size distribution over time. Particle aggregation or coalescence in colloidal systems frequently undermines stability, resulting in particle size distribution and dynamics alterations. DLS facilitates the continuous monitoring of these alterations, enabling researchers to assess the equilibrium of colloidal dispersions and optimize formulation parameters to improve stability. Furthermore, DLS offers valuable information regarding the polydispersity of colloidal systems, indicating whether the particle size distribution is uniform or heterogeneous. The performance and functionality of colloidal systems in diverse applications, including drug delivery, catalysis and nanoparticle synthesis, can be influenced by polydispersity. Researchers can customize colloidal formulations by analysing particle size distribution using DLS to attain specific properties and enhance performance (Batista et al. 2023; Fatahi et al. 2024; Jafari et al. 2022; Liu et al. 2024; Madani et al. 2022; Mohammed et al. 2023; Navarro et al. 2022; Rahimi et al. 2024; Shaba et al. 2023; Zhang and Atchimarungsri 2023). Zeta potential measurements provide insights into surface charge and electrostatic interactions of colloidal systems. The symbol ι, which represents zeta potential, measures the electric potential at the shear plane surrounding a charged particle within a colloidal dispersion. The phenomenon under consideration pertains to the extent of electrostatic repulsion or attraction among particles, which dictates their stability and conduct within a suspension. Zeta potential is crucial in stabilizing colloids by influencing electrostatic stabilization mechanisms. Colloidal dispersions characterized by high absolute zeta potentials ($>\pm 30\,\text{mV}$) demonstrate the phenomenon of electrostatic solid repulsion among particles, which hinders aggregation and enhances colloidal stability. On the other hand, when the zeta potentials are low (less than $10\,\text{mV}$), particles may gather together because there is not enough electrostatic repulsion. Zeta potential measurements facilitate the evaluation of stability of colloidal dispersions and can be utilized to optimize formulation parameters to improve stability. The control of zeta potential in colloidal systems can be achieved by manipulating pH, ionic strength and surface functionalization, thereby enabling the attainment of desired levels of stability and performance. The zeta potential of nanoparticles is of utmost importance in enhancing their performance across various applications, such as drug delivery, nanomedicine and catalysis. Nanoparticles with high zeta potentials in drug delivery systems can demonstrate extended circulation, heightened cellular uptake and enhanced drug release profiles. Scientists can customize nanoparticle compositions to enhance the effectiveness of drug delivery and improve therapeutic results by manipulating the surface charge through surface functionalization or adjusting the pH. The stability and activity of nanoparticle catalysts in catalysis are influenced by zeta potential, which impacts reaction kinetics, selectivity and product yields. Catalytic performance, stability and recyclability can be significantly improved by nanoparticles possessing optimal zeta potentials, rendering them highly sought-after for application in industrial catalytic processes. In addition, using zeta potential measurements is crucial for comprehending the dynamics of nanoparticles within biological contexts, including blood plasma, cell culture media and physiological fluids (Batista et al. 2023; Dhaffouli et al. 2024; Ermina et al. 2024; Fatahi et al. 2024; Jafari et al. 2022; Mao et al. 2023; Navarro et al. 2022; Rahimi et al. 2024; Shaba et al. 2023; Shirke et al. 2024; Wu et al. 2023b). By analysing their surface charge, researchers can evaluate the biocompatibility of nanoparticles, cellular interactions and potential toxicity. This analysis helps create safe and effective nanomedicines and biomedical devices. The design, synthesis and application of nanoparticles and colloidal systems heavily rely on the crucial factors of colloidal stability and surface charge. The utilization of DLS and Zeta Potential Measurements offers significant contributions to understanding particle size distribution, surface charge and electrostatic interactions. These techniques empower researchers to enhance colloidal formulations for diverse applications. Researchers can harness the complete capabilities of nanoparticles in various fields, such as material science, nanotechnology, biomedicine and environmental remediation, by comprehending and managing colloidal stability and surface charge (Table 8.2).

Table 8.2 Colloidal stability and surface charge.

	Aspect	Description
1	Particle size	Colloidal stability is influenced by particle size, with smaller particles generally more stable.
2	Surface charge	Charged particles repel each other, contributing to colloidal stability.
3	Zeta potential	Measure of surface charge, indicating the potential for particle aggregation or dispersion.
4	Electrostatic repulsion	Charges on particles prevent aggregation by repelling each other.
5	Ionic strength	The concentration of ions in the solution affects the stability of colloids.
6	pH	pH influences the surface charge of colloidal particles and their stability.
7	Electrolyte concentration	Higher concentrations of electrolytes can destabilize colloids by screening surface charges.
8	Steric stabilization	Polymers or surfactants are used to create a barrier preventing particle aggregation.
9	Van der Waals forces	Weak attractive forces between particles can cause aggregation.
10	DLVO theory	Theory explaining colloidal stability based on electrostatic and van der Waals interactions.
11	Coagulation	Aggregation of colloidal particles due to destabilizing factors.
12	Flocculation	Aggregation of colloidal particles into larger, settleable clusters.
13	Hydration shell	A layer of water molecules around colloidal particles influencing their stability.
14	pHpzc (point of zero charge)	pH at which the surface charge of a particle is neutral.
15	Adsorption	Attachment of ions or molecules to the surface of colloidal particles, affecting their charge.
16	Solvation	The interaction of solvent molecules with colloidal particles, affecting their stability.
17	Brownian motion	Random motion of colloidal particles due to collisions with solvent molecules.
18	Rheology	Study of the flow and deformation of colloidal suspensions.

(Continued)

Table 8.2 (Continued)

	Aspect	Description
19	Osmotic pressure	Pressure is exerted by solvent molecules moving through a semipermeable membrane.
20	Surface roughness	Surface irregularities influence the stability of colloidal particles.
21	Agglomeration	Formation of clusters of colloidal particles without significant change in structure.
22	Shear rate	The rate of deformation of a fluid influencing colloidal stability.
23	Particle shape	Irregular shapes may lead to lower stability compared to spherical particles.
24	Temperature	Temperature changes can affect colloidal stability through effects on solvation and mobility.
25	Depletion flocculation	Induced aggregation due to the presence of large solutes reducing steric stabilization.
26	pH buffering capacity	The ability of a system to resist changes in pH influences surface charge and stability.
27	Ostwald ripening	A process where larger particles grow at the expense of smaller ones, affecting stability.
28	Particle concentration	Higher concentration can lead to increased chances of particle interactions and instability.
29	Particle charge density	The density of charges on colloidal particles affects their repulsion and stability.
30	Phase separation	Complete separation of colloidal particles from the dispersion medium due to instability.
31	Surfactant concentration	Concentration of surface-active agents affecting surface charge and stability.

Source: Dong et al. (2023), Ece (2021), Fan et al. (2024), Huang et al. (2023), Namakka et al. (2023), Nguyen et al. (2021), Smitshuysen et al. (2024).

8.5 Conclusion

Si/SiO_2 nanoparticles play a crucial role in various technological applications and their characterization is essential for understanding their properties and behaviour. Several techniques are used for identifying and characterizing Si/SiO_2 nanoparticles, including:

- Imaging techniques such as TEM, SEM and AFM provide high-resolution images of nanoparticle size, morphology and surface topography.

- Spectroscopic methods, including FTIR, Raman spectroscopy and XPS, offer insights into chemical composition, surface functional groups and bonding states.
- Analytical techniques such as EDS and XRD facilitate elemental composition determination and crystal structure analysis.

Adopting a comprehensive approach that combines multiple techniques is crucial for achieving optimal functionality and integration of Si/SiO_2 nanoparticles in various applications. By using diverse characterization methods, researchers can thoroughly understand nanoparticle properties, structure–function relationships and surface chemistry. This enables the optimization of nanoparticle formulations for enhanced performance and integration into specific technological systems. The characterization of Si/SiO_2 nanoparticles has significant implications for various technological applications across diverse fields, including electronics, photonics, catalysis, biomedical engineering and environmental science. By understanding the properties and behaviour of Si/SiO_2 nanoparticles, researchers can develop novel materials, devices and technologies with enhanced functionality, efficiency and performance. Applications range from semiconductor devices and photovoltaic cells to drug delivery systems, biosensors and environmental remediation technologies. The field of nanoparticle characterization is continuously evolving, driven by advancements in materials science, nanotechnology and analytical techniques. Future research and development efforts in nanoparticle characterization are expected to focus on:

- Advancing imaging techniques to achieve higher resolution and sensitivity for detailed nanoparticle structure, morphology and dynamics analysis.
- Developing new spectroscopic and analytical methods for probing nanoscale phenomena, including surface chemistry, interparticle interactions and functionalization.
- Integrating computational modelling and simulation techniques to complement experimental characterization and predict nanoparticle behaviour under different conditions.
- Exploring interdisciplinary approaches and collaborative efforts to address complex challenges in nanoparticle characterization and accelerate technological innovation.

In conclusion, the characterization of Si/SiO_2 nanoparticles is essential for unlocking their full potential and enabling their integration into various technological applications. By adopting a comprehensive approach that combines multiple techniques, researchers can gain valuable insights into nanoparticle properties, optimize their functionality and drive innovation across diverse fields. Continued research and developmental efforts in nanoparticle characterization hold promise for advancing materials science, nanotechnology and technological innovation in the years to come.

References

Aghajari, A., Amirhossein, M.S., Khosrowshahi, S.A. et al. (2023). Effect of multi-walled carbon nanotube/porous carbon/SiO_2 on reinforcing of cement mortar matrix: modeling studies of effective parameters. *Journal of Building Engineering* 78: 107758. https://doi.org/10.1016/j.jobe.2023.107758.

Alsaad, A.M., Ahmad, A.A., Al Dairy, A.R. et al. (2020). Spectroscopic characterization of optical and thermal properties of (PMMA-PVA) hybrid thin films doped with SiO_2 nanoparticles. *Results in Physics* 19: 103463. https://doi.org/10.1016/j.rinp.2020.103463.

Aryanti, N., Nafiunisa, A., Adina, A.R., and Kusworo, T.D. (2024). Synthesis, characterization and anti-fouling properties of poly[vinylidene fluoride]-incorporated SiO_2, TiO_2, ZrO_2 nanoparticle-LiCl pore former ultrafiltration membranes. *Case Studies in Chemical and Environmental Engineering* 9: 100664. https://doi.org/10.1016/j.cscee.2024.100664.

Bahramnia, H., Semnani, H.M., Habibolahzadeh, A., and Abdoos, H. (2021). Epoxy/polyurethane hybrid nanocomposite coatings reinforced with MWCNTs and SiO_2 nanoparticles: processing, mechanical properties and wear behavior. *Surface and Coatings Technology* 415: 127121. https://doi.org/10.1016/j.surfcoat.2021.127121.

Batista, G., dos Santos, A., Takimi, S. et al. (2023). Hardened oil well cement paste modified with TiO_2@SiO_2 nanoparticles: physical and chemical properties. *Construction and Building Materials* 367: 130282. https://doi.org/10.1016/j.conbuildmat.2022.130282.

Cani, D., van der Waal, J.C., and Pescarmona, P.P. (2021). Highly-accessible, doped TiO_2 nanoparticles embedded at the surface of SiO_2 as photocatalysts for the degradation of pollutants under visible and UV radiation. *Applied Catalysis A: General* 621: 118179. https://doi.org/10.1016/j.apcata.2021.118179.

Cheng, Y., Tang, X., Li, Z. et al. (2024). Preparation technology of nano-SiO_2 super-hydrophobic hybrid coating with low cost and high abrasion-resistant. *Materials Letters* 355: 135453. https://doi.org/10.1016/j.matlet.2023.135453.

Dadashi, J., Ghasemzadeh, M.A., and Abdollahi-Basir, M.H. (2023). Cu (II) complex stabilized on Fe_3O_4@SiO_2 nanoparticles: an effective and recoverable catalyst for the preparation of chromene derivatives. *Polycyclic Aromatic Compounds*. https://doi.org/10.1080/10406638.2022.2136217.

Dhaffouli, A., Salazar-Carballo, P.A., Carinelli, S. et al. (2024). Improved electrochemical sensor using functionalized silica nanoparticles (SiO_2-APTES) for high selectivity detection of lead ions. *Materials Chemistry and Physics* 129253. https://doi.org/10.1016/j.matchemphys.2024.129253.

Dong, Y., Liu, Z., and Lin, H. (2023). Hydrophobic modification of pyrite with a composite of sodium oleate and SiO_2 nanoparticles to inhibit its oxidation for controlling acid mine drainage. *Journal of Environmental Chemical Engineering* 11 (2): 109571. https://doi.org/10.1016/j.jece.2023.109571.

Ece, M.Ş. (2021). Synthesis and characterization of activated carbon supported magnetic nanoparticles (Fe_3O_4/AC@SiO_2@Sulfanilamide) and its application in removal of toluene and benzene. *Colloids and Surfaces A: Physicochemical and Engineering Aspects* 617: 126231. https://doi.org/10.1016/j.colsurfa.2021.126231.

Ermina, A.A., Solodovchenko, N.S., Levitskii, V.S. et al. (2024). Plasmonic disordered array of hemispherical AgNPs on SiO_2@c-Si: their optical and SERS properties. *Materials Science in Semiconductor Processing* 169: 107861. https://doi.org/10.1016/j.mssp.2023.107861.

Fan, Y., Jiao, J., Zhao, L. et al. (2024). Nd-doped porous CeO_2 abrasives for chemical mechanical polishing of SiO_2 films. *Materials Science in Semiconductor Processing* 175: 108265. https://doi.org/10.1016/j.mssp.2024.108265.

Fatahi, H., Daoust, K., Akbarzadeh, A. et al. (2024). Thermal conductivity improvement of adipic acid by the addition of AlO(OH) and AlO(OH)@SiO_2 core-shell nanoparticles. *International Journal of Thermofluids* 22: 100617. https://doi.org/10.1016/j.ijft.2024.100617.

Feng, Y., Zhang, L., Li, Z. et al. (2023). Ultralight reduced graphene oxide/SiO_2–C nanofibers composite aerogel decorated with Li0.35Zn0.3Fe_2.35O_4 nanoparticles with a three-dimensional bridging network for efficient electromagnetic wave absorption. *Carbon* 215: 118435. https://doi.org/10.1016/j.carbon.2023.118435.

Ghanaatian, A., Elhambakhsh, A., Bakhtyari, A. et al. (2022). Coating SiO_2 nanoparticles with polyvinyl alcohol for interfacial tension alteration in the system CO_2+polyethylen glycol+water. *Surfaces and Interfaces* 32: 102164. https://doi.org/10.1016/j.surfin.2022.102164.

Ghattavi, S. and Homaei, A. (2024). Synthesis and characterization of ZnO-SiO_2 hybrid nanoparticles as an effective inhibitor for marine biofilm and biofouling. *Journal of Molecular Liquids* 396: 123974. https://doi.org/10.1016/j.molliq.2024.123974.

Hassan, S.S., Hubeatir, K.A., and Al-Haddad, R.M.S. (2023). Characterization and antibacterial activity of silica-coated bismuth (Bi@SiO_2) nanoparticles synthesized by pulsed laser ablation in liquid. *Optik* 273: 170453. https://doi.org/10.1016/j.ijleo.2022.170453.

Huang, R., Chen, L., Chaogang, Y. et al. (2023). In-situ synthesis of well-dispersed Cu/Cu_2O nanoparticles supported on petaloid SiO_2 for efficient degradation of high concentration tetracycline hydrochloride. *Journal of Environmental Chemical Engineering* 11 (4): 110326. https://doi.org/10.1016/j.jece.2023.110326.

Jafari, B., Rezaei, E., Abbasi, M. et al. (2022). Application of mullite-zeolite-alumina microfiltration membranes coated by SiO_2 nanoparticles for separation of oil-in-water emulsions. *Journal of the European Ceramic Society* 42 (13): 6005–6014. https://doi.org/10.1016/j.jeurceramsoc.2022.06.060.

Jia, H., Zhang, X., Wang, Q. et al. (2022). The study of novel amphiphilic janus-SiO_2 nanoparticles for enhanced viscoelasticity of wormlike micelles. *Journal of Molecular Liquids* 366: 120212. https://doi.org/10.1016/j.molliq.2022.120212.

Khajavi, P., Keshtkar, A.R., and Moosavian, M.A. (2021). The optimization of U(VI) removal by a novel amidoximated modified calcium alginate gel bead with entrapped functionalized SiO_2 nanoparticles. *Progress in Nuclear Energy* 140: 103887. https://doi.org/10.1016/j.pnucene.2021.103887.

Lasemi, N., Rentenberger, C., Liedl, G., and Eder, D. (2020). The influence of the fluid nature on femtosecond laser ablation properties of a SiO_2/Si target and synthesis of ultrafine-grained Si nanoparticles. *Nanoscale Advances* 2 (9): 3991–4002. https://doi.org/10.1039/d0na00317d.

Li, J., Cheng, L., Song, W. et al. (2022). In-situ sol-gel generation of SiO_2 nanoparticles inside polyamide membrane for enhanced nanofiltration. *Desalination* 540: 115981. https://doi.org/10.1016/j.desal.2022.115981.

Lin, Y.-S., Jia-Yun, T., and Tsai, D.-H. (2021). Steam-promoted methane-CO_2 reforming by NiPdCeOx@SiO_2 nanoparticle clusters for syngas production. *International Journal of Hydrogen Energy* 46 (49): 25103–25113. https://doi.org/10.1016/j.ijhydene.2021.05.053.

Liu, W., Liu, D., Zhang, Z., and Sun, Z. (2024). Synthesis of core-shell nanostructured SiO_2/TiO_2 photocatalysts via atomic layer deposition in a fluidized bed with central tube. *Particuology* 91: 19–28. https://doi.org/10.1016/j.partic.2023.12.015.

Madani, M., Mansourian, M., Almadari, S. et al. (2022). Enhanced photosensitivity of heterostructure SiO_2/Bi_2WO_6/GO composite nanoparticles. *Physica B: Condensed Matter* 645: 414241. https://doi.org/10.1016/j.physb.2022.414241.

Majeed, S.S., Mydin, M.A.O., Bahrami, A. et al. (2024). Development of ultralightweight foamed concrete modified with silicone dioxide (SiO_2) nanoparticles: appraisal of transport, mechanical, thermal, and microstructural properties. *Journal of Materials Research and Technology* https://doi.org/10.1016/j.jmrt.2024.01.282.

Mao, S., Pei, F., Feng, S. et al. (2023). Detection of trace rhodamine B using stable, uniformity, and reusable SERS substrate based on Ag@SiO_2-Au nanoparticles. *Colloids and Surfaces A: Physicochemical and Engineering Aspects* 657: 130595. https://doi.org/10.1016/j.colsurfa.2022.130595.

Marimuthu, M., Ramalingam, B., Subramanian, B. et al. (2023). Synthesis and characterization of Ag-Au loaded SiO_2 nanocomposites and study of its catalytic applicability in mannich reaction. *Inorganic Chemistry Communications* 153: 110808. https://doi.org/10.1016/j.inoche.2023.110808.

Miralaei, N., Mohammadimehr, M., Farazin, A. et al. (2023). Design, fabrication, evaluation, and in vitro study of green biomaterial and antibacterial polymeric biofilms of polyvinyl alcohol/tannic acid/CuO/SiO_2. *Journal of the Mechanical*

Behavior of Biomedical Materials 148: 106219. https://doi.org/10.1016/j.jmbbm.2023.106219.

Mohammed, S.H., Ahmed, O.K., and Wadullah, H.M. (2023). Enhancement of the efficiency of solar collector by SiO_2, TiO_2, and ZnO thin films layers. *Journal of Engineering Research*. https://doi.org/10.1016/j.jer.2023.11.002.

Namakka, M., Rezaur Rahman, M., Said, K.A.M.B. et al. (2023). A review of nanoparticle synthesis methods, classifications, applications, and characterization. *Environmental Nanotechnology, Monitoring & Management* 20: 100900. https://doi.org/10.1016/j.enmm.2023.100900.

Navarro, M.E., Palacios, A., Jiang, Z. et al. (2022). Effect of SiO_2 nanoparticles concentration on the corrosion behaviour of solar salt-based nanofluids for concentrating solar power plants. *Solar Energy Materials and Solar Cells* 247: 111923. https://doi.org/10.1016/j.solmat.2022.111923.

Nguyen, T.-Q., Tung, K.-L., Lin, Y.-L. et al. (2021). Modifying thin-film composite forward osmosis membranes using various SiO_2 nanoparticles for aquaculture wastewater recovery. *Chemosphere* 281: 130796. https://doi.org/10.1016/j.chemosphere.2021.130796.

Niu, S., Dong, X., Zhang, W. et al. (2021). Antibacterial and photocatalytic properties of SiO_2-CuxO Films with high copper loading and crystallized CuxO nanoparticles. *Surfaces and Interfaces* 23: 101021. https://doi.org/10.1016/j.surfin.2021.101021.

Park, J., Kwon, Y.J., Yun, J. et al. (2024). Ultra-thin SiO_2 nanoparticle layered separators by a surface multi-functionalization strategy for Li-metal batteries: highly enhanced Li-dendrite resistance and thermal properties. *Energy Storage Materials* 65: 103135. https://doi.org/10.1016/j.ensm.2023.103135.

Rahimi, M., Zolgharnein, J., and Farahani, S.D. (2024). Modification of magnetic active carbon nanocomposite with SiO_2 and Zincon for efficient removal of Pb(II) and 201Pb(II); multivariate optimization and adsorption characterization. *Inorganic Chemistry Communications* 162: 112127. https://doi.org/10.1016/j.inoche.2024.112127.

Ren, Z., Chen, L., Liu, X. et al. (2020). Preparation, characterization and simulation of Al@SiO_2 nanoparticle composite films with infrared–visible stealth. *Infrared Physics & Technology* 111: 103472. https://doi.org/10.1016/j.infrared.2020.103472.

Roychoudhury, A. and Bhowmik, R. (2023). Nanoremediation of heavy metals in agricultural soil. In: *Agricultural and Environmental Nanotechnology* (ed. F. Fernandez-Luqueno and J.K. Patra), 433–450. Springer Nature Singapore Pte Ltd.

Shaba, E.Y., Tijani, J.O., Jacob, J.O., and Suleiman, M.A.T. (2023). Effect of mixing ratios of SiO_2 nanoparticles synthesized from metakaolin on the physicochemical properties of ZnO/SiO_2 nanocomposites. *Nano-Structures & Nano-Objects* 35: 101003. https://doi.org/10.1016/j.nanoso.2023.101003.

Shirke, Y.M., Yeong Jun, Y., Chung, J.-W. et al. (2024). Hydrophobic hollow fiber composite membranes based on hexadecyl-modified SiO_2 nanoparticles for toluene separation. *Journal of Environmental Chemical Engineering* 12 (1): 111819. https://doi.org/10.1016/j.jece.2023.111819.

Singh, A. and Roychoudhury, A. (2022). Role of nanoparticles in remediation of environmental contaminants. In: *Hazardous and Trace Materials in Soil and Plants* (ed. M. Naeem, T. Aftab, A.A. Ansari, et al.), 327–340. Academic Press (Elsevier).

Smitshuysen, T.E., Lyck, M.L., Zimina, A. et al. (2024). Using $CoCu_2Ga/SiO_2$ to identify stability-issues in ethanol-selective Co-Cu alloyed catalysts in carbon monoxide hydrogenation. *Applied Catalysis A: General* 675: 119636. https://doi.org/10.1016/j.apcata.2024.119636.

Steffi, A.P., Balaji, R., Prakash, N. et al. (2022). Incorporation of SiO_2 functionalized GC_3N_4 sheets with TiO_2 nanoparticles to enhance the anticorrosion performance of metal specimens in aggressive Cl− environment. *Chemosphere* 290: 133332. https://doi.org/10.1016/j.chemosphere.2021.133332.

Taha, M., Khalid, A., Farid, A. et al. (2024). Controlling the shell thickness of SiO_2 on TiO_2 NPs: characterization, linear and nonlinear optical properties. *Ceramics International* 50 (9, Part B): 15397–15406. https://doi.org/10.1016/j.ceramint.2024.02.011.

Wang, D., Ma, Q., Li, X. et al. (2023). Coupling ultrafine SiO_2 nanoparticles with three-dimensional porous Ti_3C_2Tx MXene as anode materials for high-performance lithium-ion batteries. *Diamond and Related Materials* 139: 110B45379. https://doi.org/10.1016/j.diamond.2023.110379.

Wu, C., Li, S., Chen, Y. et al. (2023a). Tribological properties of chemical composite and physical mixture of ZnO and SiO_2 nanoparticles as grease additives. *Applied Surface Science* 612: 155932. https://doi.org/10.1016/j.apsusc.2022.155932.

Wu, R., Abdulhameed, A.S., Yong, S.K. et al. (2023b). Functionalization of chitosan biopolymer with SiO_2 nanoparticles and benzaldehyde via hydrothermal process for acid red 88 dye adsorption: box-behnken design optimization. *International Journal of Biological Macromolecules* 247: 125806. https://doi.org/10.1016/j.ijbiomac.2023.125806.

Zhang, P.-P. and Atchimarungsri, T. (2023). Direct synthesis of LPG from syngas over Cu modified FeMg@SiO_2 nano-level core@shell catalyst. *Journal of Fuel Chemistry and Technology* 51 (5): 656–663. https://doi.org/10.1016/S1872-5813(22)60064-1.

Zhang, X., Gao, X., Dong, Y. et al. (2023). Nanoporous Mg-doped SiO_2 nanoparticles with tunable infrared emissivity toward effective radiative cooling coatings. *Journal of Alloys and Compounds* 940: 168905. https://doi.org/10.1016/j.jallcom.2023.168905.

9

Uptake, Translocation and Consequences of Application of Si/SiO_2 Nanoparticles (SiNPs) in Plants

Arun Kumar Maurya, Sujata Rathi, Yogendra Prasad Saksena, Anupam Awasthi, Bharti Kaushik and Sheetal

Department of Botany, Multanimal Modi College, Modinagar, Ghaziabad, Uttar Pradesh, India

9.1 Introduction

An estimated global population is expected to surpass 9 billion in 2050 (FAO). However, to meet the population's food requirements, global food production should increase by 50–60% (Falcon et al. 2022; Riggs et al. 2018). Considering climate change effects, the food demand and hunger risk of the population ranges change marginally (+30% to +62% and −91% to +30%, respectively) but with no statistical differences overall (van Dijk et al. 2021). In addition to climate change, biotic and abiotic stressors such as drought, salt, diseases and pests impose negative pressure, leading to 51–82% of crop loss worldwide (Oshunsanya et al. 2019). Excess and indiscriminate application of pesticides and fertilizers has led to soil and water quality degradation, ultimately affecting crop yield (Patle et al. 2019). Therefore, to avoid negative impacts, several green solutions have been innovated to ensure agricultural sustainability (Arora et al. 2018). Among them, the use of the newly emerging field, namely nanotechnology, has given a green solution that helps to promote plant growth and confer stress tolerance in plants to ensure crop yield by mitigating negative impacts (Rajput et al. 2021; Swain et al. 2025; Ullah et al. 2023).

Nanotechnology adopts a top-down or bottom-up approach where metal materials or compounds are cut into small particles ranging from 1 to 100 nm in size or nanoparticles (NPs) generated by fusion of desired material in solution. This can be achieved by controlled precipitation (or crystallization) and evaporation. These tiny particles formed by either method are called nanoparticles (NPs). NPs exhibit unique physical and chemical properties at this nano-size stage, which are not observed at the macro or micro-size (Behera 2022; Chan and Kwok 2011;

Silicon Nanoparticles: Synthesis, Uptake, and Applications, First Edition.
Edited by Archana Singh, Aryadeep Roychoudhury, and Indrakant K. Singh.

Mohapatra et al. 2023). These properties belong to their surface, optical, thermal and electrical properties (Rastogi et al. 2019). The NPs show small size and large surface-area-to-weight ratios, which increases solubility and transportation of metal, material compounds or elements in vivo (Eker et al. 2024). (Figure 9.1) Various oxide nanomaterials (SiO_2, Fe_3O_4, CuO, and TiO_2) have been used for their ability to show promotory plant growth and plant resistance against stress (Li, Yiting et al. 2024; Mohammadi et al. 2023). Si/SiO_2 is used widely as silicon nanoparticles (SiNPs) (Ali 2024). Silicon (Si) is the second most abundant element in the earth's crust (J. A. Bhat et al. 2021). In soil, most Si is found in complex forms as mineral silicates, which are unavailable for plant uptake (Thakral et al. 2024). Silicon is present in plant cell walls, making the physical barrier of plants, promoting disease resistance and imparting mechanical strength that gives the ability to lodging tolerance (Ma and Yamaji 2008). (Table 9.1)

Silicon uptake in plants only occurs in the available form of silicic acid or monosilicic acid [$Si(OH)_4$ or H_4SiO_4] (Zargar et al. 2019) (Figure 9.2). SiNPs can be obtained from green sources or synthesized chemically in the laboratory (Karande et al. 2021) and are reported to affect plant growth and development by regulating the physio-biochemical and molecular responses (Hassim et al. 2021; Ng et al. 2021). The beneficial effects can be listed as promoting seed germination (Li, Yiting et al. 2024; Sembada et al. 2023; Wang et al. 2024), increasing plant resistance, degrading pesticide residues, enhancing soil quality and enhancing fungicidal activity (Ochoa et al. 2025; Yousaf et al. 2024; Zafar et al. 2024). The applications of silicon and SiNPs influence several physio-biochemical responses, such as oxidative stress reduction by priming defence reactions under biotic and abiotic stresses (Falouti et al. 2024; Gupta et al. 2024; Ochoa et al. 2025; Roy et al. 2025; Swain et al. 2025; Wang et al. 2022a). This chapter mainly discusses the uptake, translocation and consequences of SiNPs in plants, aggregating the existing literature. It describes current knowledge, research insights, existing gaps and future research directions on SiNPs' mitigation of abiotic and biotic stress on plants.

9.2 Uptake of the Silicon and SiNPs

NPs are promising carrier agents for targeted and/or controlled delivery of the desired components, such as essential nutrients, pesticides and specific chemicals to the plant bodies (Karuvelan et al. 2025). It requires a thorough understanding of the mechanisms adopted by plants to uptake and transport the NPs (Sembada and Lenggoro 2024). Studies showed that Si is easily absorbed by plants at physiological pH only in the form of the monomeric, uncharged molecule of orthosilicic acid or silicic acid (H_4SiO_4 or $Si(OH)_4$) from soil solution, which contains approximately

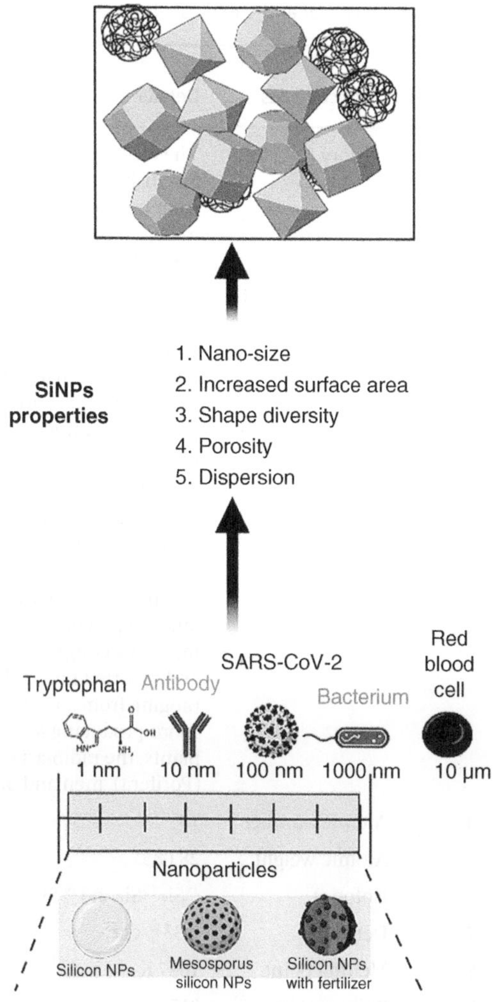

Figure 9.1 Diagrammatic representation of a comparison of NPs with other biomolecules, types of SiNPs and properties. *Source*: Created in https://BioRender.com.

0.1–0.6 mM (Raven 2001) (Figure 9.2). Silica dissolution is dependent upon pH, and $Si(OH)_4$ is the predominant form at pH value ≤ 9.0, and increasing alkalinity in soil solution leads to the formation of monosilicates ($SiO(OH)_3^-$), $SiO_2(OH)_2^{2-}$) and polysilicates (Sjöberg 1996). Si is absorbed and transported to the shoot, where monosilicic acid gets accumulated and polymerized to amorphous silica (opal or phytolith) and deposited in specific silica cells (Liang et al. 2015). Si accumulates in different parts of their body and at different concentrations ranging from 0.1% to 10% of the plant dry weight (Epstein 1994; Ma and Takahashi 2002; Manivannan et al. 2023).

Table 9.1 Physical and chemical characteristics of silicon.

Serial number	Characteristics	Details	References
1.	History	The element was discovered by a Swedish chemist, J.J. Berzelius, in 1823 from potassium fluosilicate, converted to SiO_2 by combustion, and named silicium (Latin *silex*, 'quartz').	
2.	Occurrence and distribution	The second most abundant element in the earth's crust, SiO_2 occurs as oxide and silicates. It occurs mainly in crystalline form as quartz, rarely as tridymite and cristobalite, and as the amorphous mineral opal. A few silicate minerals are asbestos, tremolite, feldspars, clays and micas.	
3.	Si	A grey-in-color Group IV, p-block element, relatively inert but reactive towards halogens and dilute alkalis, making it only sparingly soluble and tends to polymerize. Both organic and inorganic compounds are formed by Si and reported range of organisms ranging from silicate bacteria, algae, spore-producing lower and higher plants, the radiolaria, and some sponges (Porifera), men and other animals.	Carlisle (1984)
4.	Atomic number	14	Coursey and Dragoset (2000); Haynes (2016)
5.	Atomic weight	28.085	
6.	Isotopes	^{28}Si, ^{29}Si, and ^{30}Si	
7.	Density	2.33 g/cm^3	
8.	Melting point	1687 K	
9.	Boiling point	3173 K	
10.	Oxidation state	− 4, +4	
11.	Electronegativity,	1.9	
12.	Electron affinity	133.6 kJ/mol	

Si, an unreactive metalloid, is present in soil and soil solution. Uptake and transport of Si depend upon the soil types and plant species (Mandlik et al. 2020). Plants are classified into active (faster than water uptake), passive (equal to water uptake) and rejective modes (slower/lower than water uptake) based on uptake

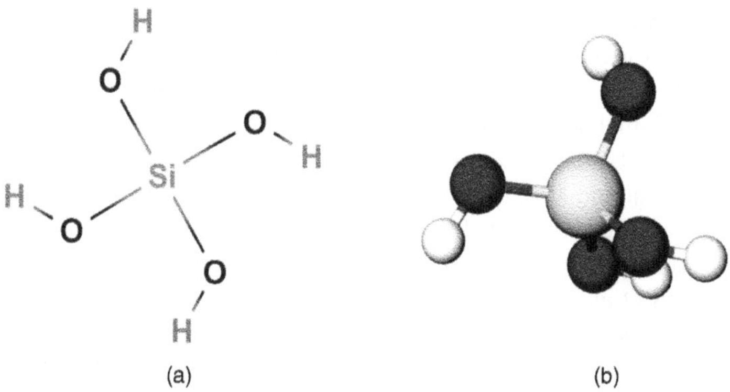

Figure 9.2 Chemical formula (a) and molecular structure (b) of monosilicic acid (H_4SiO_4).

and transport (Raven 2001; Takahashi et al. 1990). The anatomical differences in plant body, such as root anatomy, generate variation in uptake strategy as rice roots possess two Casparian strips at the exodermis and endodermis, whereas, in maize and barley roots, there is one Casparian strip at the endodermis (Kaur and Greger 2019). Besides anatomical differences, Si uptake largely depends upon the transporter proteins (Low silicon transporters, Lsi) present on the plasma membrane of different cells. Interestingly, instead of root hairs, the lateral roots participate in Si uptake in rice (Ma et al. 2001).

Diffusion or facilitated diffusion helps the passive transport of uncharged $SiOH_4$ across the lipid membrane of PM or using a protein aqueous channel, works in an energy-independent manner and is present in all plant species regardless of their ability to accumulate Si or not (Raven 2003). In contrast, active transport requires a metabolically active, energy-dependent component Si transport system that works in various Si-accumulating species such as rice, barley, maize (*Zea mays*), wheat (*Triticum aestivum*), banana (*Musa* sp.) and cucumber (*Cucumis sativus*). Both active and passive transport involved in Si uptake are energy-dependent processes, as metabolic inhibitors and low temperature inhibit Si uptake (Liang et al. 2015).

9.2.1 Higher Plants

The historical work on Si transporters started with the discovery of the first Si transporter, low-silicon rice (Lsi1), an influx transporter identified from rice (*Oryza sativa*), using an induced mutagenesis approach (Ma et al. 2006). These duo transporters are required for sustainable rice productivity, a typical high Si-accumulating monocot species and a staple crop used in various countries (Tamai and Ma 2008). The Lsi1, Si permeable channel, belongs to the Nod26-like major intrinsic protein (NIP) III, which is a subgroup of the aquaporin membrane protein family (Maryam and Gul 2022). Later on, using the same approach, another transporter protein, Lsi2, an efflux transporter, was identified in rice from the same group (Ma et al. 2007). The Lsi2, an anion transport family member, is localized on the proximal side of the same cells opposite to Lsi1 and facilitates Si uptake from the roots that help to translocate to the vascular tissues in rice (Ma et al. 2007; Yamaji and Ma 2011). These two influx-efflux pair transporters are responsible for Si uptake, reported in diverse plant species such as pumpkin (*Cucurbita moschata* Duch.) cultivars (Mitani-Ueno et al. 2011), wheat (*Triticum aestivum*) (Montpetit et al. 2012), barley (*Hordeum vulgare*), maize (*Zea mays*), pumpkin (*Cucurbita moschata*) and the primitive vascular plant, *Equisetum arvense* (horsetail) (Liang et al. 2015). Another important Si influx transporter, named Lsi6, involved in xylem unloading, is identified in the members of the family Poaceae such as rice, barley and maize (Liang et al. 2015). Lsi6 belongs to the NIPIII subgroup of aquaporins involved in the movement of silicic acid across the vascular bundle and expressed in the xylem parenchyma cells of leaf sheath, leaf blades, and root tips (Maryam and Gul 2022; Yamaji et al. 2008). The Lsi1 assists in the uptake of silicic acid from the soil up to the exodermis, effluxed by Lsi2 and transported to the aerial part by Lsi6, where it gets deposited or accumulated in special cells, enabling the complete translocation of Si in the plant body (Ma et al. 2011). (Figure 9.2) Lsi2, Lsi3 and Lsi6 distribute Si into different plant parts involving the vascular transport system. Lsi2 is present in the bundle sheath cell, Lsi2 is in the xylem transfer cell layer and Lsi3 is in the parenchyma tissues (Ma and Yamaji 2015). Different expression patterns are observed in Lsi1 and Lsi2 transporters depending on the plant species. These are categorized as only root-based expression (rice, barley, maize and wheat) or both root and shoot expressions such as OsLsi6, HvLsi6, ZmLsi6, GmNIP2-1, GmNIP2-2, CmLsi1, CSiT1 and CSiT2 homologs (Tamai and Ma 2008).

The uptake of SiNP depends on the particle's mode of application and properties. Three popular ways of SiNPs administration are foliar, soil and seed priming (Naidu et al. 2023). Surrounding environmental conditions, the physiology of the plant species also affects the Si uptake (Wang et al. 2022b). Many NPs can passively show uptake through plant surfaces viz natural openings such as hydathodes, stomata, bark surfaces and stigma (Eichert et al. 2008; Kurepa et al. 2010). However, the uptake of SiNPs via the root is found to be more efficient as compared to the leaf (Suriyaprabha et al. 2014). The foliar-applied NPs followed two distinct pathways through the stomata or cuticular route. The cuticular uptake of SiNPs again follows two sub-pathways depending upon the polar or non-polar nature of SiNPs. The non-polar NPs follow the lipophilic path, which shows uptake via diffusion through the cuticle. However, the polar SiNPs are transported through aquaporins and may uptake through stomata depending on size. The leaf takes the SiNPs and shows further travel through the phloem towards the root (Ma and

Yamaji 2008; Popp et al. 2005). The SiNPs applied through the nutrient solution show absorption and uptake by roots and are found to be transported either passively or actively. The root secretes many compounds, such as amino acids, mucilages, organic acids, phenolics, proteins and secondary metabolites, which also help in the uptake and transport of SiNPs (Laane 2018; Wang et al. 2020).

The expression studies of the *Lsi1* revealed that Lsi1 is involved in Si uptake and controls Si accumulation in rice and is located on the distal side of the exodermis and endodermis cells, but Lsi2 is localized to the proximal side of the same cells (Ma et al. 2006). Similarly, an influx transporter of Si has been identified from the wheat (TaLsi1) (Montpetit et al. 2012), barley (HvLsi1) (Chiba et al. 2009) and maize (ZmLsi1) (Mitani et al. 2009a). An Si influx transporter identified from barley (HvLsi1) shows 81% identity with rice OsLsi1 and facilitates the radial transport of Si through the epidermal and cortical layer of the basal root of barley (Chiba et al. 2009). The expression study revealed the localization of ZmLsi1, which is present on the distal side of the root epidermal and hypodermal cells' plasma membrane in the seminal and crown roots and acts as an influx transporter, transporting Si from the external medium. However, ZmLsi6 was found in the xylem parenchyma cells adjacent to the vessels that help Si in xylem uploading (Figure 9.3). The OsLsi2 transporter is homologous to the efflux transporter of barley (HvLsi2) and maize (ZmLsi2), showing more than 80% similarity with Lsi2. These transporters are endodermis localized and don't show polar localization. ZmLsi6 is a homologue of OsLsi6 of rice (Mitani et al. 2009a). Maize also has an orthologous gene ZmLsi1 in the crown and seminal roots at the epidermis, and sub-epidermal cells control symplastic movements of Si. The ZmLsi2 is expressed in seminal and lateral roots, and the endodermis uploads Si into the stele. A third Si transporter (Lsi6), homologous to Lsi1, is localized in the xylem parenchyma associated with the leaf sheath and leaf blades, allowing Si to move towards the aerial part of the plant body (Markovich et al. 2019). HvLsi1, a Si influx transporter in barley, is homologous to OsLsi1 with 81% identity and belongs to a Nod26-like major intrinsic protein subfamily of aquaporins (Chiba et al. 2009). *PeLsi1-1* and *PeLsi1-2* are also reported from moso bamboo (*Phyllostachys edulis*) that shows Si uptake and accumulation ability (Ge et al. 2024).

There are dicot plant families of angiosperm plants where Si transporters (Lsi1 and Lsi2) are identified, such as Cucurbitaceae, Urticaceae and Asteraceae (Mitani et al. 2011), and species such as pumpkin (*Cucurbita moschata*) (Mitani-Ueno et al. 2011), cucumber (*Cucumis sativus*) (Sun et al. 2017) and soybean (*Glycine max*) (Deshmukh et al. 2013). The CmLsi1 was the first Si influx gene identified in the dicot plant, pumpkin (*Cucurbita moschata*). Si accumulators actively uptake Si against a concentration gradient (Liang et al. 2005; Mitani et al. 2011). The two efflux transporters, CmLsi2-1 and CmLsi2-2, were identified from two pumpkins (*Cucurbita moschata* Duch.) (Mitani-Ueno et al. 2011). CsLsi2 is an

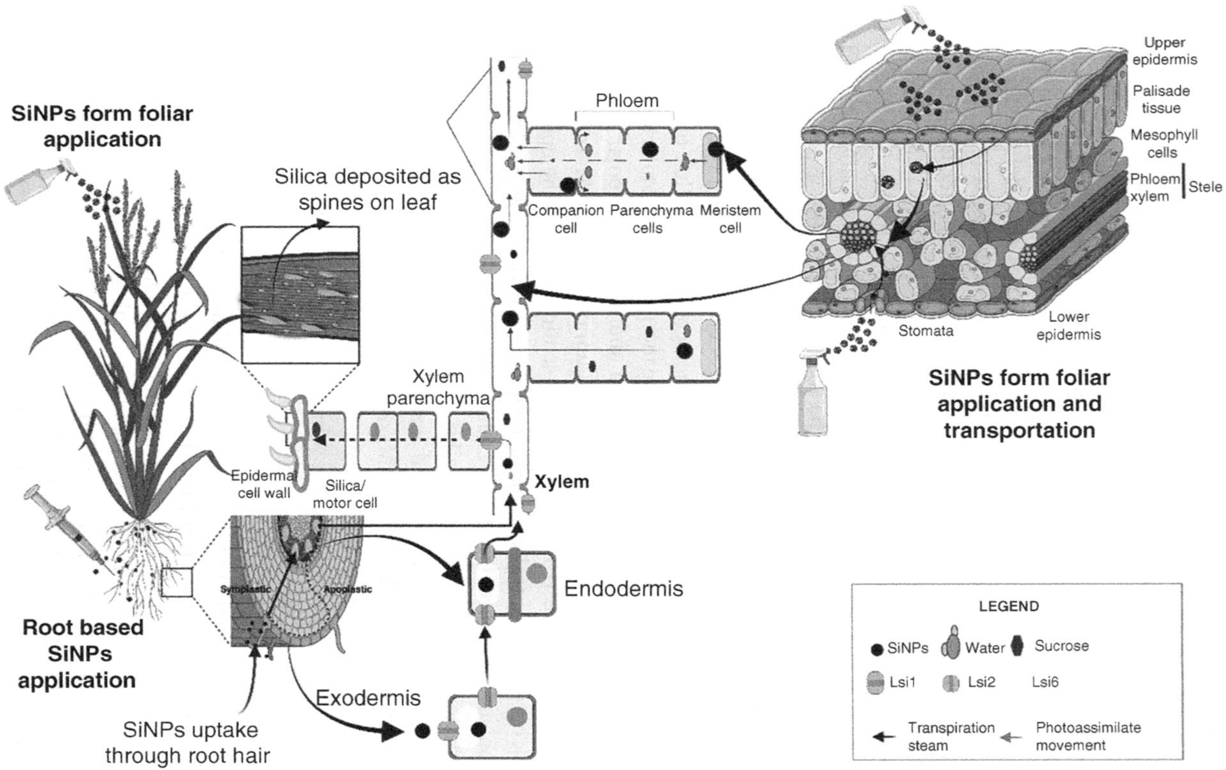

Figure 9.3 Diagrammatic representation of SiNPs uptake and translocation pathways assisted by Si transporters (Lsi1, Lsi2 and Lsi6). *Source*: Created in https://BioRender.com.

energy-dependent efflux transporter of cucumber, localized at the root PM, and shares 55.70% and 90.63% homology with the Lsi1s rice (monocot) and pumpkin (Sun et al. 2017). In coordination with CsLsi1, it participates in the uptake of Si (Sun et al. 2018). In contrast, Brassicaceae and Solanaceae members, such as Arabidopsis (*Arabidopsis thaliana*), canola (*Brassica rapa*), tomato (*Solanum lycopersicum*) and potato (*Solanum tuberosum*) and tobacco (*Nicotiana sylvestris* Speg. & Comes), are known to be low accumulators (Coskun et al. 2019; Zhang et al. 2024) (Table 9.2).

Low Si accumulators show a lack of functional Si efflux transporter (Lsi2) required for active Si uptake as reported in tomato, although Lsi1 is functional (Sun et al. 2020). Soya bean plants also possess Si transporters (GmNIP2-1 and GmNIP2-2 influx transporters) involved in Si uptake (Deshmukh et al. 2013). The putative transporter, StLsi1was identified in potatoes, which helps abiotic stress tolerance and is localized in the roots and leaves (Vulavala et al. 2016).

9.2.2 Cryptogams

Among cryptogams, a pteridophyte known as horsetail (*Equisetum arvense*) shows a great amount of Si accumulation and phylogenetic analysis revealed that the plant possesses a new multigene family of NIPs, differing from identified flowering plant Si transporters (Grégoire et al. 2012). The protein from these genes consists of a STAR pore (EaNIP3;1, EaNIP3;3 and EaNIP3;4) in contrast to previously identified GSGR pores reported in other known Si transporters. Among these proteins, EaNIP3;1 and EaNIP3;4 were found to be the most efficient transporters, exhibiting even higher activity than rice transporters. Two putative Si efflux transporter genes (*EaLsi2-1* and *EaLsi2-2*) have also been identified in horsetail, showing less than 50% homology to rice gene transporters (Trembath-Reichert et al. 2015).

9.2.3 Algae

Silica is also reported in groups of aquatic plants called algae (all brown algae and green Hydrodictyaceae members such as *Pediastrum* and *Tetraëdron*) (Parker 1969). Diatoms are among the most diverse phytoplanktonic groups and productive organisms on earth, as they contribute approximately 20% of global primary production as well as approximately 240 Tmol of biogenic silica precipitation annually (Nelson et al. 1995) and are responsible for a unique ecological and biogeochemical roles in the ocean (Durkin et al. 2016). Si is essential for diatom growth, as it builds the siliceous frustule surrounding the cell wall (Lewin 1955). The role of Si in diatoms is required in DNA and chlorophyll synthesis. To know the essentiality of Si in diatoms (*Navicula pelliculosa*), Si-starvation and supplementation-based studies on the biochemistry and fine structure of silica

Table 9.2 Transporter proteins for Si/SiNPs uptake.

Serial number	Name of transporter	Plant	Localization	Nature & function	References
1.	Lsi1	Rice (*Oryza sativa* L.)	Root exodermis and endodermis (distal sides)	Si influx	Ma et al. (2006); Ma and Yamaji (2008); Ma et al. (2007); Yamaji et al. (2008)
2.	Lsi2		Root exodermis and endodermis (proximal side)	Si efflux from the root cells to the apoplast	
3.	Lsi6		parenchyma cells associated with xylem in leaf sheath and blade	Si transport out of the xylem (xylem unloading)	
4.	Lsi1	Maize (*Zea mays* L.)	The seminal and crown root epidermal and hypodermal cells (distal side) PM lateral roots cortex cells		Mitani et al. (2009b)
5.	Lis2		Root endodermis PM without polar localization	Si influx	
6.	Lsi6		The parenchyma cells adjacent to the vessels in leaf sheaths and blades	Si transporter for xylem unloading	
7.	Lsi1		Root epidermal cells PM (distal side)	Si efflux	
8.	Lsi2		The parenchyma cell adjacent to the transfer cells with opposite polarity	Si influx	
9.	Lsi6	Barley (*Hordeum vulgare* L.)	Root tips epidermis and cortex cells and leaf blades and sheaths as well as at parenchyma cells of vascular bundles	Si efflux Si uptake in the root tips, xylem unloading of Si in leaf blade and sheath, and intervascular transfer of Si in the nodes.	Chiba et al. (2009); Yamaji et al. (2012)

10.	Lis1	Soya bean (*Glycine max*)	PM	Si influx	Deshmukh et al. (2013)
11.	Lis6				
12.	Lsi1	Cucumber (*Cucumis sativus*)	Root endodermis and the cortical cells in the root tips as well as in the root hairs near the tips (distal side).	Si influx	Sun et al. (2017)
13.	Lsi2		Endodermis	Si efflux	Sun et al. (2018)
14.	Lis1	Squash or pumpkin (*Cucurbita moschata*)	The PM of all root cells, without polar localization	Si influx and efflux	Mitani et al. (2011)
15.	Lis2				
16.	TaLsi1,	Wheat (*Triticum aestivum* L.)	PM	Si influx	Montpetit et al. (2012)
17.	Lsi1	Potato (*Solanum tuberosum*)	The roots and leaves	Si influx	Vulavala et al. (2016)
18.	Lsi2-1	Horsetail (*Equisetum arvense*)	Putative Si efflux	Not determined	Trembath-Reichert et al. (2015)
19.	Lsi2-2		Putative Si efflux	Not determined	
20.	Horsetail Major intrinsic protein (MIP) family	Horsetail (*Equisetum arvense*)	Not yet determined	Not determined	Grégoire et al. (2012)

(Continued)

Table 9.2 (Continued)

Serial number	Name of transporter	Plant	Localization	Nature & function	References
21.	The silicic acid transporters (SITs). CfSIT1-5	The marine pennate diatom, *Cylindrotheca fusiformis*	First identified in *C. fusiformis*, contain 10 transmembrane segments (TMS) and a coiled-coil motif at the C-terminus of SITs, 74% identical at a. a. level, highest degree of a. a. conservation; a sodium silicic acid symporter, electrogenic with a 1:1 ratio of $Si(OH)_4$:Na^+ and Zinc also play a role in uptake.	Si influx/efflux	Hildebrand et al. (1998); Thamatrakoln et al. (2006)
22.	TpSIT1-3	The centric diatom, *Thalassiosira pseudonana*	α-helical membrane proteins, a homotetramer, sodium gradient dependent, transport silicic acid:Na^+ stoichiometry of 2:1	Si influx and probably efflux	Curnow et al. (2012)
23.	Silicon Transporter (SIT)-type protein	The siliceous loricate choanoflagellates *Stephanoeca diplocostata* and *Diaphanoeca grandis*.	transmembrane Si transport		Marron et al. (2013)

shell formation were carried out which showed changes in total cellular organic carbon, nucleic acids, proteins, carbohydrates, lipids and chlorophylls during experimentation (Coombs et al. 1967). Further laboratory experiments showed that Si essentiality in diatoms (*Navicula pelliculosa*) increases the biological availability of phosphorus in the presence of aluminium toxicity as it reduces the availability of phosphorus. This was counter-verified by Si-stimulated growth in the presence of aluminium in the non-silicon-dependent green alga, *Chlorella vulgaris*, where an increased growth rate was observed on Si addition, which shows a direct correlation with the increase in the bioavailable phosphorus fraction (Exley et al. 1993).

9.3 Translocation of SiNPs in Plants

Although Si is not classified as an essential element, it is naturally present in plants, especially the grass family members, and has long been recognized for its critical role in plant physiology (Pavlovic et al. 2021). Si helps to impart strength to the cell walls, thereby providing enhanced resistance to biotic and abiotic stresses and promoting plant growth and development (Hassan et al. 2024).

It is well known that most plants accumulate variable amounts of Si, ranging from 0.1% to 10% (Epstein 1999). Based on their Si content, plant species are broadly classified into three categories, namely Si accumulators, Si excluders and intermediate Si accumulators. Silicon accumulators are the group of plants that contain more than 10,000 mg/kg or 1.0% foliar Si in the dry mass of tissue and a Si/Ca^{++} ratio >1, for example, Poaceae members such as rice (*Oryza sativa* L.), wheat (*Triticum aestivum* L.), barley (*Hordeum vulgare* L.), oats (*Avena sativa* L.) and sugarcane (*Saccharum officinarum* L.)]; Cucurbitaceae members such as cucumber (*Cucumis sativa* L.), pumpkin (*Cucurbita pepo* L.), watermelon (*Citrullus lanatus* (Thunb.) Matsum. & Nakai)]; Asteraceae member such as sunflower (*Helianthus annuus* L.) and zinnia (*Zinnia elegans* Jacq.) and Cannabaceae member cannabis (*Cannabis sativa* L.). Si excluders or non-accumulating species contain less than 10,000 mg/kg or 0.5% and a Si/Ca^{++} ratio <0.5, for example, tomato (*Solanum lycopersicum*). However, the plants that do not meet these two criteria are considered intermediate Si accumulator plants, for example, stinging nettle (*Urtica dioica*), which contains between 0.5 and 1.0% Si (Chanchal Malhotra et al. 2016; Epstein 1999; Frantz et al. 2011). These variations in Si accumulation are primarily attributed to genetic differences governing Si uptake mechanisms facilitated by specific transport proteins known as Lsi1, Lsi2 and Lsi6. These are localized in root cells and enable the movement of silicic acid, the primary bioavailable form of Si, which varies between 0.1 and 0.6 mM (Takahashi et al. 1990; Yan et al. 2018). The efficiency of Si transport and accumulation thus depends on the presence and activity of these transporters.

In recent years, advancements in nanotechnology have introduced novel methods for Si delivery using Si and silicon dioxide (SiO_2) NPs. The uptake, translocation and storage of NPs within plants are influenced by multiple factors such as plant species, growth stage and environmental conditions, as well as the NP's intrinsic properties, including size, shape, chemical composition and stability in solution (Liu et al. 2015). However, understanding how Si/SiO_2 NPs translocate within plants is critical for their effective use in agriculture. Once SiNPs penetrate the plant's outer protective layers, they can move into the plant through two primary transport pathways: the apoplastic and symplastic pathways. The former path involves the movement of NPs through the cell walls and extracellular spaces (ECS), bypassing the cytoplasm of individual cells. In contrast, the symplastic pathway facilitates direct transfer between the cytoplasm of adjacent cells through specialized intercellular channels known as plasmodesmata. These two distinct pathways enable NPs to travel efficiently within the plant's tissues. In the apoplastic pathway, NPs can traverse the root cortex, eventually reaching the central cylinder (stele) and entering the vascular system, where they are transported upwards through the xylem towards the plant's aerial parts, including the stems, leaves and apical sections (Sun et al. 2014). However, this movement is often interrupted by the Casparian strip, a highly specialized structure, lignin-rich, making a hydrophobic barrier in the endodermis of plant roots that restricts the radial flow of substances, including NPs, through the apoplastic pathway (Sun et al. 2014). As a result, NPs that fail to bypass this barrier tend to accumulate at the Casparian strip. For NPs that successfully navigate past the Casparian strip, the xylem serves as a conduit, enabling their upward transport to the shoots and other plant parts (Wang et al. 2013). The Casparian strip poses constraints, which are overcome by the transition from apoplastic to symplastic pathway adoption that allows NPs or their associated molecules to circumvent the restrictive movement by allowing movement through the plasmodesmata into the symplast, which makes it easy to sail their journey within the plant. Interestingly, different types of NPs have demonstrated the ability to overcome the barrier posed by the Casparian strip through similar adaptive mechanisms. These findings underscore the potential of NPs as efficient transport vehicles within plants, offering promising applications in agriculture, such as targeted nutrient delivery and enhanced stress resilience (Schwab et al. 2016).

For NPs to utilize the symplastic pathway for transport within plants, they must first penetrate individual plant cells at specific entry points. Unlike in animal cells, this process is more challenging in plant cells due to the cell wall, a rigid and protective structure that serves as a primary barrier to foreign materials. Despite this limitation, NPs with diameters ranging from 3 to 50 nm, including carbon nanotubes and other nanoscale structures, have been shown to successfully traverse the cell wall (Liu et al. 2019). These small dimensions enable such nanoparticles to overcome the physical constraints of the cell wall and enter the intracellular environment effectively. Once inside the cell, the movement of Si and SiO_2 NPs between adjacent cells is facilitated by plasmodesmata, specialized cytoplasmic bridges that connect neighbouring plant cells. These structures maintain cytoplasmic continuity and enable efficient symplastic transport of NPs. Comparative studies indicate that nanoscale silicon is significantly more effective than bulk silicon in uptake and translocation, particularly in young seedlings. This heightened efficiency positions Si/SiO_2 nanoparticles as highly favourable materials for agricultural applications, such as enhancing plant growth and stress resilience (Tripathi et al. 2017; Tripathi et al. 2021).

Nanoparticles can also access the symplast through alternative mechanisms such as endocytosis, where the cell membrane engulfs the particles; carrier protein-mediated transport, which involves specific transport proteins; and pore formation, where transient openings in the membrane allow nanoparticle entry. Once inside the cytoplasm, nanoparticles are transported through plasmodesmata, ensuring their distribution across interconnected cells (Ali et al. 2021; Tripathi et al. 2017).

To elucidate the behaviour and transport pathways of SiNPs, advanced imaging and analytical techniques such as confocal laser scanning microscopy (CLSM), transmission electron microscopy (TEM) and proton-induced X-ray emission (micro-PIXE) elemental analysis were employed (Sun et al. 2014). These methods enabled precise tracking and localization of mesoporous SiNPs (MSiNPs) within plant tissues, revealing their movement through symplastic and apoplastic pathways. Their study demonstrated that MSiNPs entered the roots, navigated through the xylem and reached aerial parts of the plant, such as stems and leaves. Apart from internal structural impediments and support, the critical role of environmental factors, such as pH and surfactant concentration, also influences NPs uptake efficiency. Proper optimization of these parameters was shown to enhance the permeability of the cell wall and promote effective nanoparticle transport. In addition to underground transport through the xylem, SiNPs can also translocate within plants via aerial pathways. The uptake and movement of NPs in the aerial parts of plants, such as leaves, are notably more dynamic than their behaviour in roots. SiNPs can penetrate leaves primarily through the cuticular or stomatal pathways. The cuticle is a waxy, hydrophobic outer layer covering most plant surfaces, providing a protective barrier, preventing water loss and regulating solute exchange. This unique structure positions the cuticle as the plant's first defence against external particles, including nanoparticles (Buick et al. 1993; Buick et al. 1993). The cuticle's selective permeability allows only nanoparticles smaller than approximately 5 nm to pass through its structure effectively (Ali et al. 2021). Thus, the cuticle serves as a critical barrier and influences the efficiency of nanoparticle entry into the plant.

In addition to the cuticular route, Si/SiO_2 nanoparticles can enter leaves through the stomatal pathway. Stomata are microscopic pores on the leaf surface, primarily responsible for gas exchange and water vapour regulation. When Si NPs enter stomata, they pass through the epidermis and reach the leaf mesophyll, a spongy tissue containing chloroplast-rich cells that facilitate photosynthesis. Once within the mesophyll, the nanoparticles integrate into the vascular system, enabling further distribution through the plant's conductive tissues (Bhat et al. 2021). After entering the vascular tissues, SiNPs can be transported via the phloem, which carries sugars and other nutrients produced during photosynthesis. The phloem's translocation pathways facilitate the distribution of NPs to various plant parts, including roots, stems, fruits and young leaves, depending on the plant's physiological needs and growth stage (Raliya et al. 2016; Tripathi et al. 2017). This movement highlights the systemic nature of NP transport, where aerial entry routes contribute to the efficient distribution of NPs throughout the plant. Understanding these aerial pathways is crucial for developing targeted applications of SiNPs in agriculture, such as foliar sprays for nutrient delivery or stress alleviation. The dual entry routes, cuticular and stomatal, offer insights into the potential of NPs to bypass traditional barriers and optimize plant health and productivity.

9.4 Consequences of SiNP Application in Plants

SiNPs have been reported as antioxidants, signalling molecule regulators, anatomical modulators and metabolic ameliorators against diverse biotic and abiotic stresses (Ma 2004 (Du et al. 2022; Mukarram et al. 2022)). However, the benefits of Si are restricted to stressed conditions and reported in various fundamental processes of plant systems. The positive impact of Si on various plants is well demonstrated by several investigators (Fauteux et al. 2005; Khandekar and Leisner 2011). (Figure 9.4)

The physio-biochemical and molecular responses can be seen as follows:

9.4.1 Morphological and Anatomical Changes

Si accumulation was noticed in the aerial parts of the plant, such as the leaf, stem, vascular tissues and cuticle. The deposition of Si on the aerial surface acts as a mechanical barrier to protect the plant from several biotic and abiotic stresses. Si is deposited on the foliar surface upon the foliar spray of Si suspension on leaves in wheat (Merwad et al. 2018), providing physical rigidity to plants. Si was shown to enhance the expression of various genes by several folds, affecting the root's

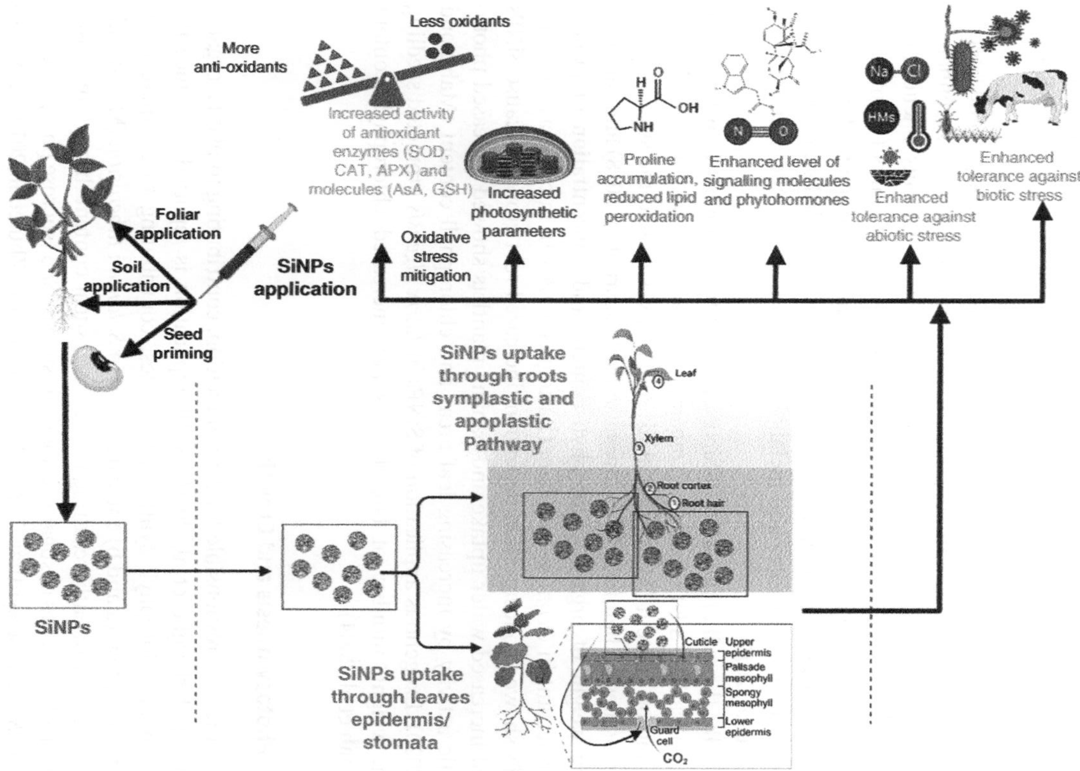

Figure 9.4 SiNPs application modes, uptake, translocation and effects on different physio-biochemical parameters in plants. *Source*: Created in https://BioRender.com.

anatomy in rice. Si is considered an essential element in the Equisetaceae family as well as algae (Diatoms), as plants belonging to this family cannot survive in Si if they lack a nutrient solution (Epstein 1994; Lewin 1955; Luyckx et al. 2017). In *Arabidopsis*, TEM image analysis confirmed the deposition of Si NPs around the stomatal apparatus. This observation further confirmed that SiNPs entered the plants through the stomata and the adjoining epidermal cells (El-Shetehy et al. 2021).

In rice, SiNP application-led accumulation kept the leaves upright, and stretching helped to obtain maximum solar radiation (Hatami et al. 2021). This enhanced the production of various metabolites, such as proteins, pigments, phenols, etc., that play a crucial role in the growth and functioning of plants (Verma et al. 2021). SiNPs provide the structural colour to plants by forming a binary film at the outer surface of the epidermis (Strout et al. 2013). SiNPs can potentially improve anatomical features and metabolism in cucumbers under salinity stress (Avestan et al. 2019).

Generally, the soil application of SiNPs is found to be more effective than the foliar application. SiNPs-treated *Cymbopogon citratus* showed increased plant height and oil content (Mukarram et al. 2021). SiNPs caused the strengthening of plant tissues by lignification in *Avena sativa* (Asgari et al. 2018) and the fresh and dry weight of organs, chlorophyll content and concentration of proline osmoprotectant in *Ocimum basilicum* (Rastogi et al. 2019). In bananas, applying SiO_2NPs showed concentration-dependent increased root growth because SiNPs induced increased water uptake and banana plantlets showed increased photosynthetic activity by increasing leaf areas and chlorophyll content (Mahmoud et al. 2020). Exogenous application of SiNPs in *Larix olgensis* promoted seedling growth, weight, number of lateral roots and increased chlorophyll content (Bao-Shan et al. 2004).

9.4.2 Photosynthesis and Growth

Photosynthesis is responsible for carbon fixation, contributing to plants' organic growth (Vanlerberghe et al. 2020). SiNP application is reported to induce the expression of chlorophyll synthesizing genes, contributing to an increase in chlorophyll content (a and b) and carotenoid in *Solanum tuberosum* (Mushinskiy et al. 2018), consequently resulting in increased sugar content through increased expression of key genes of sugar metabolisms such as *SPS* (sucrose phosphate synthase), *SUS* (sucrose synthase) and *alpha-glucosidase* (Zhu et al. 2016). Si is also reported to increase rice's photosynthetic activity by enhancing the circulation of primary metabolites such as amino acids (Detmann et al. 2012). Exogenous application of SiNPs is reported to increase seed germination percentage and photosynthetic efficiency in *Cucurbita pepo* L. (Siddiqui et al. 2014) and *Zea mays*

(Yuvakkumar et al. 2011). Most reports favour the positive or no impact of SiNPs on plant growth and development, eliminating the negative aspects reported in a few investigations.

9.4.3 Phytohormone Synthesis and Regulation

Exogenous SiNPs application is reported to affect the level of phytohormones in various plant species. A decrease in the cytokinin level is reported by SiNPs application that was mediated by lowering the expression of the IPT (isopentenyltransferase) gene in wheat leaves. On the contrary, SiNPs increased the cytokinin level in the stem to induce internode elongation (Li et al. 2023). Recent investigations revealed the role of Si in reducing the ageing process of leaves by stimulating the cytokinin pathway (Markovich et al. 2017).

9.4.4 Gene Expression and Defence Response

Silicon is reported to influence the gene expression in plants in stressed and unstressed conditions. In the case of wheat, Si contributes to the regulation of nearly 47 genes in unstressed conditions (F Chain et al. 2009). However, the number of genes responding to Si is 221 that take part in primary and secondary metabolic pathways (Brunings et al. 2009). These genes were found to be involved in cell wall formation, carbon and nitrogen metabolism and phytohormone metabolism. Moreover, SiNPs act by altering the route and components of signal transduction (Mukarram et al. 2022) and enhance the expression of salicylic acid-responsive genes and defence-related genes, boosting antioxidant enzymatic activity to support plant growth and development.

9.4.5 Mitigation of Oxidative Stress

All abiotic and biotic stressors induce a negative influence on the cell metabolism that leads to the generation of reactive oxygen species (ROS) and reactive nitrogen species (RNS) in the cells, cumulatively called oxidative stress (Maurya 2020). The impacts of oxidative stress are seen as oxidation of DNA fragments, RNA degradation, oxygenation, modification and destruction of lipids and proteins that ultimately cause membrane damage and leakages and result in the decline in the growth, development and productivity of crop plants (Maurya 2020). Si/SiNPs influence normal plant functioning by making them endure the general oxidative stress (Apel and Hirt 2004) induced by UV-B stress (Tripathi et al. 2017), salinity (Sepasi et al. 2024), water deficit or drought (Mahmoud et al. 2020), temperature stress (Liang et al. 2024), hypoxia (Iqbal et al. 2021), heavy metals (Ahmed et al. 2023; Hussain et al. 2019a) or biotic stress (Wang et al. 2022a; Lei Wang et al. 2022b) by protecting the plant from cell membrane injury caused by ROS, thus lessening the impact of oxidative stress. SiNPs are also known to increase the proline level, an antioxidant molecule which relieves oxidative stress and maintains osmotic equilibrium (Matysik et al. 2002).

9.4.6 Abiotic Stress

Plants respond to several abiotic stresses, such as drought, salinity and heavy metal toxicity, which negatively affect plant growth depending on the severity of the stressor. SiNPs play a crucial role in mitigating the harmful impacts of these abiotic stresses (Roy et al. 2025) and sustaining agricultural production (Ochoa et al. 2025).

9.4.6.1 UV-Stress

Ultraviolet radiation is a non-ionizing radiation making a part of the electromagnetic spectrum and has been classified into three categories based on wavelength: UV-A (320–400 nm), UV-B (280–320 nm) and UV-C (200–280 nm) (Shama 1999). Similar to other organisms, plants sense and respond to UV radiation, especially UV-B and UV-C and are known to damage various plant physiological processes (Abdipour et al. 2019; Ma et al. 2021; Stapleton 1992; Xu et al. 2019). UV-B radiation is reported to cause adverse effects in plants seen as declined photosynthetic performance, altered vital leaf structures and enhanced oxidative stress, increased lipid peroxidation (LPO), electrolyte leakage (EL) and inhibition of superoxide dismutase (SOD) and ascorbate peroxidase (APX) induction of catalase (CAT) and guaiacol peroxidase (POD), and all non-enzymatic antioxidants in wheat (*Triticum aestivum*) seedlings' growth (Tripathi et al. 2017). Although gasotransmitter nitric oxide (NO) content was increased, the maximum response was observed under SiNPs together with UV-B treatment. However, SiNPs appear to be more effective than Si against UV-B stress (Tripathi et al. 2017). Si is reported to improve UV tolerance through Si deposition on the leaf epidermis (Goto et al. 2003) or by reducing the negative consequences of membrane damage induced by UV-B radiation (Shen et al. 2010), making Si a natural protective guard at the morpho-anatomical and physio-biochemical level.

9.4.6.2 Drought Stress

Drought is now becoming a serious global problem in light of the current global warming scenario, which is posing a threat to sustainable agriculture. To overcome such a threat, SiNPs tried to augment drought stress in barley (Alabdallah et al. 2025; Ghorbanpour et al. 2020), maize (Sezgin Muslu et al. 2025), *Fragaria ananassa* (Zahedi et al. 2020), pea (*Pisum sativum* L.) (Sutulienė et al. 2022), lupin (Sun et al. 2016) and *Arabidopsis thaliana* (Tran et al. 2023).

SiNPs proved helpful in recovering from drought stress by changing the physiological responses (MDA, H_2O_2 and antioxidant enzymes) in barley (Ghorbanpour et al. 2020) and *Fragaria ananassa* (Zahedi et al. 2020) and pea. Apart from that pea also showed that SiO_2 NPs (50 ppm) increased relative water content, specific leaf area and decreased root/shoot ratio when sprayed foliar or watered along with enhanced activity of antioxidant enzymes (CAT, SOD, APX, GR), the increased total phenolic compounds and non-enzymatic antioxidant activity (DPPH, ABTS, FRAP) in peas (Sutulienė et al. 2022). Cereal crops, a key component of the food basket for the daily human diet, as well as cattle feed supporting meat production, are affected by drought, becoming a cause of food insecurity. The effects of SiNPs on drought tolerance are well-reviewed (Sulaiman et al. 2023). SiNPs application showed potential for agricultural application in regions of low water availability or recurrent drought growing cereal crops such as wheat (Jasem and Khalil 2022; Rai-Kalal et al. 2021; Raza et al. 2023; Sun et al. 2016), maize (*Zea mays* L.) (Sharf-Eldin et al. 2023), rice (Raja et al. 2021), barley (Ghorbanpour et al. 2020) and millet seedlings (Mokari and Abedinpour 2020).

9.4.6.3 Heavy Metal Stress

Heavy metals (HMs) have been categorized into essential (Co, Cu, Fe, Mn, Mo and Ni) and non-essential elements (As, Ag, Cd, Cr, Hg and Pd). The consequences of HMs are characterised by toxicity and adverse effects on plant growth and development, including inhibition of photosynthesis, chlorosis, reduced biomass, imbalanced cellular nutrient levels and water balance, ultimately leading to senescence and plant death (Bhat et al. 2019). Si is reported to alleviate various mineral or heavy metal stresses, all of which are highly detrimental to crop production (Pontigo et al. 2015). In the case of wheat, seedlings grown in combination with Cr and SiNPs showed better growth and biomass accumulation than seedlings grown in Cr alone, reduced the accumulation of Cr in wheat grain, thereby helping to promote plant growth and development (Manzoor et al. 2022). Similarly, foliar spray of SiNPs in different concentrations in rape seed (*Brassica napus* L.) found to mitigate the Cr toxicity by reduced Cr accumulation and generation of ROS to prevent plants from oxidative damage, effectively increased photosynthetic activity, electron transport rate (ETR), non-photochemical quenching (NPQ), and quantum yields (QY) of Photosystem II (Huang et al. 2024). SiNPs also enhanced plant tolerance to HMs stress by changing the free radicle species, antioxidative enzymes and cellular components (Emamverdian et al. 2020; Tripathi et al. 2017). SiNPs reduce the effect of Pb toxicity in different tissues in *Brassica chinensis* L (Tian et al. 2020).

9.4.6.4 Salinity Stress

Salinity is an abiotic stressor which is increasing with time in response to climate change as well as anthropogenic activities in both fertile terrestrial crop land and aquatic system (Corwin 2021; Harley et al. 2012) influencing negatively on the survival, biomass production and crop yield (Parihar et al. 2015).

Nanosized, monodisperse Si or SiNPs foliar application showed beneficial effects on crop plants, including increased tolerance against salinity induced by NaCl on 'Valencia' sweet orange (*Citrus x sinensis* L. Osbeck) plants budded onto Kuharske (*Citrus sinensis* L. Osbeck x *Poncirus trifoliata* L. Raf.) through an incremental enhancement plant growth with SiNPs applied (400 mM), improved root growth, chlorophyll content and alleviated the osmotic effect, expression of Na^+ cotransporter (*CsSOS1, CsSOS2, CsSOS3* and *CsNHX1*) and aquaporin (*CsPIP1;1, CsPIP2;3*, and *CsTIP4;1*) transcripts in the root tissues. It led to suggest the use of SiNPs in an integrated management program to improve overall citrus plant health against stressful saline conditions (Parihar et al. 2015). The SiNP treatment negated salt stress's impact on soil water availability and promoted vegetative growth and yield as SiNPs were reported to upregulate the expression of genes *AREB*, *TAS14*, *NCED3* genes while downregulating the expression *RBOH1*, *APX2*, *MAPK2*, *ERF5*, *MAPK3* and *DDF2* (Almutairi 2016). On the other hand, SiNPs application against salinity stress showed that SiNPs (200 mg/l) negatively affected Na^+ accumulation and improved growth by regulating photosynthesis and chlorophyll content, mitigating salt-induced oxidative stress by increasing the activity of antioxidant enzymes expression and reducing the H_2O_2 levels and increases in the K^+ and Si content in tomato seedlings (Wang et al. 2025).

9.4.7 Mineral Deficiency

Si or SiNPs are reported to act as 'quasi-essential' element, alleviating mineral deficiency, namely nitrogen (N), potassium (K), phosphorus (P), sulphur (S), magnesium and several micronutrients, making detrimental to crop production (Ali et al. 2020; Pontigo et al. 2015) and promoting plant growth and development by improving nutrient availability and ionic uptake in rhizospheric soil (Sharma et al. 2023). In the case of *Agropyron elongatum*, SiO_2-NPs exerted a positive impact by improving seed and seedling growth (Azimi et al. 2014). Si-based fertilizers have been proven more beneficial for crops due to the controlled release of N and phosphorus (Nair et al. 2010). Si deficiency causes nutritional imbalance and is responsible for the poor growth of plants, and several investigations have favoured the role of SiNPs as fertilizer especially in tropical soils (Ali et al. 2020; Meena et al. 2014). SiNPs act as carriers of N, P and K, thus ensuring the slow release of these macronutrients in soil that help the plants prevent the plant from salt stress (Ali 2024). The effects of SiNPs with a high surface-to-volume ratio are more pronounced in plants as it increases their activity in the plant systems (Adrees et al. 2015).

9.4.8 Biotic Stress

Silicon transporters (Lsi1, Lsi2) play a key role in the accumulation of Si, acting in both the export and import of Si (Frew et al. 2018; Ma et al. 2008). SiNPs also improve plant growth by preventing biotic stress from several fungal, bacterial and nematode infections. In an unstressed *Arabidopsis thaliana* plant, Si is found to change the expression of two genes (Fauteux et al. 2005), nearly 47 genes in wheat (F. Chain et al. 2009), and its application provides defence and promotes growth under biotic stress and unstressed conditions in sugarcane (Frew et al. 2017). Si was reported to improve defence against biotic stress from plant pathogens such as viruses, bacteria, fungi, parasitic plants or animals (vertebrates and arthropod herbivores) (Luyckx et al. 2017). Si diverts the signalling system in a way that induces the expression of defence-related genes and cell wall makeover, and synthesis of phytohormones and antimicrobial compounds (Wang et al. 2017). In *Z. mays*, SiNPs at 20–40 nm in size remarkably modulated various physiological and anatomical behaviours (Suriyaprabha et al. 2012). SiNPs application has shown improved defence of *Solanum lycopersicum* against the pathogenic bacteria *Ralstonia solanecearum* by enhancing the level of 26 proteins, modulating the expression of various defence-related genes (Chen et al. 2015; Ghareeb et al. 2011). Similarly, SiNPs help fight the pathogen *Magnaporthe oryzae* in rice (Brunings et al. 2009). Seed priming with SiNPs increased root and shoot length, improved biomass, vigour indices and the microbial consortium of the soil, thereby increasing the capability of the plants to resist several biotic stresses such as disease management of a complex of *Meloidogyne incognita*, *Pectobacteriam betavasculorum* and *Rhizoctonia solani* in the case of beetroot (*Beta vulgaris* L.) (Khan and Siddiqui 2020). These findings further supported the view that SiNPs decrease plant invasion of various pathogens by upregulating the expression of defence-related genes and by modulating the phytohormone crosstalk involving salicylic acid, jasmonic acid and ethylene, which confers resistance in grapevine against *Plasmophora viticola* (Rashad et al. 2021). Si accumulation produces secondary metabolites that further confer resistance against pathogens (Datnoff et al. 1997) (Table 9.3).

9.4.9 SiNPs Work Synergistically in Association with Other Molecules

Several examples exist where SiNP, along with other materials, has shown positive or synergistic responses against stress or physio-biochemical molecular responses in plant systems such as with other chemicals iron oxide (Fe_3O_4) (Irshad et al. 2024), selenium (Zahedi et al. 2020), lignin-based silicon/magnesia biochar (ALB/Si-Mg) (Li, Xianzhen et al. 2024) and Zn loaded MSiNPs (Ghosh et al. 2024); phytohormones brassinolids (Zhou et al. 2024), salicylic acid (Muhammad

Table 9.3 Si/SiNPs mode of application, concentration and consequences on the growth and development of various species.

Serial number	Plant species	Mode of Si/SiNPs treatment and effective conc.	Type of stress	Consequences/responses of Si/SiNPs application	References
1.	Pepper (*Capsicum annuum* L.),	Nutrient solution (300 mg/l)	Salinity	Improved germination rate, germination index and radicle length; decreased antioxidant enzymes (POD, CAT, SOD), MDA, protein, proline, H_2O_2 and superoxide (O_2^-); increased leaves CTKs, IAA and ABA levels	Li, Yiting et al. (2024)
2.	Marigold (*Calendula officinalis* L.)	Seed priming (SiNPs:1–2 mg/l and Silicates 100–500 mg/l)	Drought	Improved seed germination and seedling growth	Saeedeh et al. (2021)
3.	Basil (*Ocimum basilicum*)	Nutrient solution (10 mg/kg)	Salinity	Alleviated salt stress by enhanced total chlorophyll concentration, proline, POD, CAT, gamma-cadinene synthase (CDS), caffeic acid O-methyltransferase (COMT), the eugenol synthase (EGS1) and fenchol synthase (FES), the geraniol synthase (GES) gene expression; stimulated the R-linalool synthase (LIS), the terpinolene synthase (TES) gene and the phenylalanine ammonia-lyase (PAL) gene	Sepasi et al. (2024)
4.	Pennyroyal (*Mentha pulegium* L.)	Nutrient solution (25, 50, and 100 ppm)	Drought	Improved PEG-induced drought tolerance by accumulating osmolytes	El-Naggar and Osman (2024)
5.	Siberian wild rye (*Elymus sibiricus* L.)	Seed priming (50,100,150, 200, 250 mg/l)	Drought	Increased chlorophyll, osmolyte accumulation and total proteins, phytohormones, SOD, CAT, POD, and APX activities led to decreased oxidative damage, decreased Na^+ and increased K^+ concentrations in response	Khan et al. (2024)

6.	Chinese liquorice (*Glycyrrhiza uralensis*)	Nutrient solution (1.0 mM)	Drought and Salinity	Improved germination rate, germination index, seedling vitality index and biomass, increased radicle length; increased CAT, APX GSH content under salinity	Zhang et al. (2017)
7.	Citrus	Foliar as well as root application (250 mg/l)	Hypoxia	increased photosynthesis, leaf greenness, antioxidant enzymes, and carbohydrate metabolic activities, proline and glycine betaine accumulation under hypoxia	Iqbal et al. (2024)
8.	Rice (*Oryza sativa* L.)	Nutrient application (300, 600 and 900 mg/l)	Drought	Improved plant growth, root vigour and enzymatic antioxidant levels	Ahmad and Hassim (2024); Elshayb et al. (2021); Tripathi et al. (2021)
		Nutrient solution (2.5 mM)	Heavy metal (Cu and As)	Relieve Cu^{2+} stress by decreasing Cu^{2+} accumulation in shoots, MDA and H_2O_2 contents; increased root length and shoot height, non-enzymatic compounds AsA and GSH and enzymes (SOD, POD, CAT, APX) activities	
				Increased plant growth under drought conditions by modulating the root histological traits and enhanced protein, proline, SOD, POD and CAT content	
		Nutrient medium (10 μM)		Favor the formation of adventitious roots under arsenate stress (AsV);AsV + Si regulated NO metabolism genes expression, cell cycle progression, auxin biosynthesis and transport; triggered the and ascorbate recycling	
9.	Muscadine Grape (*Muscadinia rotundifolia* Michx.)	Nutrient solution (250 ppm)	Hypoxia	Improved antioxidant enzymes (SOD, POD, CAT activities, non-enzymatic antioxidants (AsA, GSH), Proline, glycine betaine accumulation; increased N, P, K, and Zn contents and limited Mn and Fe concentration	Iqbal et al. (2020)

(Continued)

Table 9.3 (Continued)

Serial number	Plant species	Mode of Si/SiNPs treatment and effective conc.	Type of stress	Consequences/responses of Si/SiNPs application	References
10.	Wheat (*Triticum aestivum*)	Seed priming (10 mM)	UV-B radiation	Protected seedlings against UV-B by regulating oxidative stress through NO-mediated enhanced antioxidants	Tripathi et al. (2017)
		Seed priming (300, 600, 900, 1200 mg/l)	Heavy metal (Cd)	Increased biomass and yield; reduced Cd uptake	Hussain et al. (2019b)
11.	Fenugreek plants (*Trigonella foenum graceum* L.)	Nutrient solution 0.5, 1, 1.5, 2 and 2.5 mM	Si effects on plants	Increased SOD, POD, CAT level, enhanced xylem cell walls lignification and PAL activity, expression of silicon transporter genes, defensive (Tfgd1) and PEPCK	Nazaralian et al. (2017)
12.	Oat (*Avena sativa* L.)	Nutrient solution (SiNPs and sodium silicate 5 and 10 mM)	Si effect on plant	Enhanced plant growth and silicon transporter (Lsi1) gene expression	Asgari et al. (2018)
13.	Pea (*Pisum sativum* (L.)	Nutrient solution (10 μM)	Heavy metal (Cr)	Protects seedlings against Cr (VI) phytotoxicity by reducing oxidative stress, up-regulating antioxidant defense system and nutrient elements	Tripathi et al. (2015)
14.	Ginger (*Z. officinale Ros.* cv. fengtou)	Nutrient solution (50, 100 and 150 mg/l)	Pathogen (*F. solani*)	enhanced SOD, APX, PPO, CAT and phenylpropanoid pathway-related genes	Peng et al. (2022); Zhou et al. (2022)
15.	Sugar beet (*Beta vulgaris*)	Foliar spraying (12.5 mg/l)	Salinity	Decreased H_2O_2, lipid peroxidation, Na^+ ions and increased SOD, CAT, POX activities, K^+ ions, root and sugar yield	Alharbi et al. (2022)

16.	Barley (*Hordeum vulgare* L)	Soil solution (500 mg/l)	Nutritional defi-ciency stress	Attenuate nutritional stresses, mitigate phosphate deficiency with auxin and NO, and enhance growth, photosynthesis, and redox balance	Kandhol et al. (2024)
		Foliar application (12.5 mg/l)	Salinity	PGPR + SiNPs alleviated salinity stress by enhanced relative chlorophyll content, RWC, stomatal conductance, and K^+/Na^+ ratio), SOD, CAT, POX, and better nutrient uptake (N, P, and K) whereas decreased Na^+, H_2O_2, MDA, EL, and proline contents	Alharbi et al. (2022)
17.	Maize (*Zea mays*)	Soil solution (100% v/v)	Heavy metal	Alleviated as stress through sequestration with PCs, improved root hydraulic conductance, antioxidant activities	David et al. (2024)
		Soil application (4.0 mg/kg)	Al stress	Membrane stability and mitigated Al phytotoxicity in acidic soils	de Sousa et al. (2019)
18.	Tomato (*Lycopersicon esculentum*) on branched broomrape (*Orobanche ramosa*), a root holoparasitic weed	Seed priming (20 mg/l; w/v)	Parasitic plant	Decreased ROS production and improved both enzymatic and non-enzymatic detoxification systems	Madany et al. (2020)
		Nutrient solution (1000 mg/l)	Heavy metal (As)	Reduced the translocation of As^V, H_2O_2, enhanced Fe and Zn accumulation when exposed to As^{III}, increased chlorophyll contents, enhanced phenols, flavonoids and glutathione and APX, CAT, SOD in roots than in leaves	González-Moscoso et al. (2022)

(Continued)

Table 9.3 (Continued)

Serial number	Plant species	Mode of Si/SiNPs treatment and effective conc.	Type of stress	Consequences/responses of Si/SiNPs application	References
19.	White clover (*Trifolium repens*)	Nutrient soil solution (1.0 mM)	Al stress and mineral deficiency	Increased both shoot and root length, decreased EL and MDA content; gradual Si deposition in root tissues prevents the absorption of biologically active Al, reducing the risk of mineral nutrient deficiencies, promoting organic acids exudation, and compartmentalizing Al in the outer layer of root tissues	Yang et al. (2024)
20.	Banana cv. 'Grand Nain'	Foliar application (200, 400 and 600 mg/l)	Salinity	Improved chlorophyll content, induce K^+ uptake, modulate Na^+ levels and decreases cell wall damage	Mahmoud et al. (2022)
21.	Rapeseed (*Brassica napus* L.)	Nutrient solution (250 mg/kg)	Heavy metals Cd and Cr	Improved plant growth, photosynthesis, and antioxidative enzyme activity	Ahmed et al. (2023)
		Foliar application (50, 100 and 150 µM)		Upregulated defence-related gene expression increased Si content and reduced Cr uptake and translocation in roots and leaves	Huang et al. (2024)
22.	*Arabidopsis thaliana*	Foliar application (SiO_{2}-NPs (25, 100, 400 and 1,600 mg); $Si(OH)_4$ (5, 20, 80, 100, 320, 640 and 2,560 mg)	Pathogen	The induction of systemic acquired resistance (SAR) against o *Pseudomonas syringae*	El-Shetehy et al. (2021)

23.	Black goji berry (*Lycium ruthenicum*)	Nutrient solution (1, 4, 8 mmol/l)	Salinity	Increased chlorophyll content, proline and soluble sugar content but reduced membrane permeability and MDA	Shen et al. (2012)
24.	Squash (*Cucurbita pepo* L.)	Nutrient solution (1.5, 3.0, 4.5, 6.0 and 7.5 mM)	Salinity	Increased seed germination and growth characteristics reduced oxidative damage by enhanced expression of antioxidant enzymes CAT, SOD, POD, APX and GR	Siddiqui et al. (2014)
25.	Freesia (*Freesia refracta*)	foliar application (10 mM)	Heavy metal (Cr)	Increased defence mechanism and tolerance, vegetative and reproductive traits, biomass, photosynthetic pigments and gaseous exchange attributes, proline, total protein; enhanced POD, CAT and SOD activity; reduced MDA, H_2O_2 contents, and Cr levels in roots, stems, leaves and flowers	Ahsan et al. (2024)

et al. 2025), kinetin and auxin (Cokuraj et al. 2025; Manokari et al. 2025); and secondary metabolite (Sezgin Muslu et al. 2025).

Spinach seeds primed with a composite of iron oxide (Fe_3O_4) and SiNPs (100 mg/l) helped to mitigate reduced uptake of HMs (Cd and Cr), increased root and shoot dry weight, enhanced gas exchange, protected photosynthetic machinery, induced antioxidant enzymes (CAT, SOD, APX) and antioxidant molecule phenolics, ascorbic acid, total sugar, flavonoids. It also helped mitigate anthocyanin-based oxidative stress and HM-based phytotoxicity in spinach (*Spinacia oleracea* L.) plants from the rooting media (Irshad et al. 2024). Lignin-based silicon/magnesia biochar (LB/Si-Mg) is reported to remove heavy metals such as Cd^{2+}, Pb^{2+}, Cu^{2+} and Zn^{2+} from water and soil environments because of its increased specific surface area. Thus, enhanced ability of the LB/Si-Mg, used for reduced bioavailability of HMs in the soil and Pakchoi, promotes growth and development (Li, Xianzhen et al. 2024). Applying mesoporous silica nanoparticles (MSiNPs) as an innovative micronutrient delivery system in wheat seedlings was tested, which showed Zn-loaded MSiNPs (Zn-MSiNPs) (1 g/l and 5 g/l) have a spherical shape and negatively charged mesoporous surface, making effective responses against augmenting salinity tolerance (200 mM NaCl) by sustained release of Zn^{+2} from Zn-MSiNPs. It brought tolerance by increased chlorophyll, total protein and sugar level with a substantial decline in MDA, electrolyte leakage and ROS accumulation and increased the K^+/Na^+ ratio in shoots (Ghosh et al. 2024). Application of SiNPs with selenium (Se) for imparting drought-stress tolerance in *Fragaria ananassa* showed increased chlorophyll contents MDA, H_2O_2, and antioxidative enzyme activities (Zahedi et al. 2020).

SiNPs and salicylic acid (SA) applied together enhanced plant performance and alleviated Cd stress synergistically in rice grown in a hydroponic condition through minimizing the Cd-induced toxicity by reducing Cd-bioaccumulation by downregulating the Cd-related transporter genes (*OsNramp1*, *OsNramp5*, *OsHMA2* and *OsHMA3*), protecting photosynthetic efficiency, seedling biomass and restoring cellular structures (leaf ultrastructure and surface morphology). The enhanced tolerance was associated with activating antioxidant enzymes, phytohormones biosynthesis and reducing oxidative damages by decreasing the production of ROS, MDA contents and membrane stability (Muhammad et al. 2025). The amalgamation of SiNPs and plant hormones like brassinolide (BR) has shown tolerance to water stress in sugar beets by enhancing the antioxidant defence, photosynthetic capacity, dry matter accumulation, nutrient uptake, the TY, NR activity and water use efficiency and yield (Zhou et al. 2024). The impact of SiNPs on the in vitro proliferation efficiency and micro-structural developments of the *Vitex agnus-castus* L. tree was found synergistic in association with phytohormones such as kinetin (Kn) and indole-3-acetic acid (IAA). The supplementation showed the promotion of shoot proliferation and morphological changes such as shoot number, leaf area and biomass (fresh weight). Leaf and stem growth revealed the formation of tissue systems and increased lignification in sclerenchyma that enhanced the survival rate of plantlets of *V. agnus-castus* during the hardening phase (Cokuraj et al. 2025). Similar potential of SiNPs in improving the regeneration potential for clonal propagation of *Cadaba trifoliata* with *meta*-Topolin (*m*T) and IAA percentage of the shoots and well-developed tissue systems in the shoot (stomatal guard and epidermal cells), thick cuticle and epidermis, and increased tissue density in mesophyll and vascular tissue and lignification (Manokari et al. 2025). Apart from phytohormones, secondary metabolites (Rutin) applied exogenously along with Si showed positive effects on the photosynthetic capacity (enhanced gas exchange, chlorophyll contents, PSII activity, RUBISCO activity, etc.) of maize seedlings grown under osmotic stress (induced by 10% and 15% (w/v) polyethylene glycol) for 48 h and decreased the chlorophyllase gene expression (Sezgin Muslu et al. 2025).

9.5 Conclusion and Future Prospective

Silicon has long been associated with plant systems, but its benefits and utility increased with time, especially after the development of NPs. Since then, it has been applied in agriculture singly or combined with other bioactive materials as an augmenting chemical but eco-friendly substance to enhance stress tolerance or ensure crop production and yield. Apart from agricultural applications, SiNP's chemical nature and properties vis-à-vis their uptake and translocation mechanisms through passive or active transporters have grown, and scientists across the globe have identified various transporters in plants or associated mechanisms. The search is still ongoing to enlighten the domain with more scientific information. SiNP research will also help better utilize fertilizers and pesticides to ensure a healthy soil system through nutrient management. Global warming generates diverse abiotic and biotic stresses, such as UV-B, salinity, drought or water deficit, high or low temperature, hypoxia or pathogens on crop plants. However, SiNPs' promising results, as they can be applied at seed priming, foliar or root zone application, in overcoming or minimizing stresses, provide an avenue for using SiNPs in crop-specific, region-climate-specific recommended doses.

References

Abdipour, M., Hosseinifarahi, M., and Naseri, N. (2019). Combination method of UV-B and UV-C prevents post-harvest decay and improves organoleptic quality of peach fruit. *Scientia Horticulturae* 256: 108564.

Adrees, M., Ali, S., Rizwan, M. et al. (2015). Mechanisms of silicon-mediated alleviation of heavy metal toxicity in plants: a review. *Ecotoxicology and Environmental Safety* 119: 186–197.

Ahmad, A. and Hassim, M.F.N. (2024). Effects of silica nanoparticles on morpho-histological and antioxidant activities of rice seedlings under drought stress. *South African Journal of Botany* 168: 497–508.

Ahmed, T., Masood, H.A., Noman, M. et al. (2023). Biogenic silicon nanoparticles mitigate cadmium (Cd) toxicity in rapeseed (*Brassica napus* L.) by modulating the cellular oxidative stress metabolism and reducing Cd translocation. *Journal of Hazardous Materials* 459: 132070.

Ahsan, M., Radicetti, E., Jamal, A. et al. (2024). Silicon nanoparticles and indole butyric acid positively regulate the growth performance of *Freesia refracta* by ameliorating oxidative stress under chromium toxicity. *Frontiers in Plant Science* 15: https://doi.org/10.3389/fpls.2024.1437276.

Alabdallah, N.M., Saleem, K., Al-Shammari, A.S. et al. (2025). Unfolding the role of silicon dioxide nanoparticles (SiO_2NPs) in inducing drought stress tolerance in *Hordeum vulgare* through modulation of root metabolic, nutritional, and hormonal profiles. *South African Journal of Botany* 177: 397–410.

Alharbi, K., Hafez, E., Omara, A.E. et al. (2022). Plant growth promoting rhizobacteria and silica nanoparticles stimulate sugar beet resilience to irrigation with saline water in salt-affected soils. *Plants (Basel)* 11 (22): https://doi.org/10.3390/plants11223117.

Ali, S. (2024). Exploring silicon nanoparticle uptake, transport, and biological functions in plants. In: *Silicon Advances for Sustainable Agriculture and Human Health: Increased Nutrition and Disease Prevention* (ed. R. de Mello Prado, H. Etesami, and A.K. Srivastava), 237–263. Springer Nature Switzerland. https://doi.org/10.1007/978-3-031-69876-7_12.

Ali, N., Réthoré, E., Yvin, J.C., and Hosseini, S.A. (2020). The regulatory role of silicon in mitigating plant nutritional stresses. *Plants (Basel)* 9 (12): https://doi.org/10.3390/plants9121779.

Ali, S., Mehmood, A., and Khan, N. (2021). Uptake, translocation, and consequences of nanomaterials on plant growth and stress adaptation. *Journal of Nanomaterials* 2021 (1): 6677616.

Almutairi, Z.M. (2016). Effect of nano-silicon application on the expression of salt tolerance genes in germinating tomato ('*Solanum lycopersicum*'L.) seedlings under salt stress. *Plant Omics* 9 (1): 106–114.

Apel, K. and Hirt, H. (2004). Reactive oxygen species: metabolism, oxidative stress, and signal transduction. *Annual Review of Plant Biology* 55 (1): 373–399.

Arora, N.K., Fatima, T., Mishra, I. et al. (2018). Environmental sustainability: challenges and viable solutions. *Environmental Sustainability* 1: 309–340.

Asgari, F., Majd, A., Jonoubi, P., and Najafi, F. (2018). Effects of silicon nanoparticles on molecular, chemical, structural and ultrastructural characteristics of oat (*Avena sativa* L.). *Plant Physiology and Biochemistry* 127: 152–160. https://doi.org/10.1016/j.plaphy.2018.03.021.

Avestan, S., Ghasemnezhad, M., Esfahani, M., and Byrt, C.S. (2019). Application of nano-silicon dioxide improves salt stress tolerance in strawberry plants. *Agronomy* 9 (5): 246.

Azimi, R., Borzelabad, M.J., Feizi, H., and Azimi, A. (2014). Interaction of SiO nanoparticles with seed prechilling on germination and early seedling growth of tall wheatgrass (*Agropyron elongatum* L.). *Polish Journal of Chemical Technology* 16 (3): 25–29.

Bao-Shan, L., shao-qi, D., Chun-Hui, L. et al. (2004). Effect of TMS (nanostructured silicon dioxide) on growth of Changbai larch seedlings. *Journal of Forestry Research* 15: 138–140.

Behera, A. (2022). Nanomaterials. In: *Advanced Materials: An Introduction to Modern Materials Science* (ed. A. Behera), 77–125. Springer International Publishing. https://doi.org/10.1007/978-3-030-80359-9_3.

Bhat, J.A., Shivaraj, S.M., Singh, P. et al. (2019). Role of silicon in mitigation of heavy metal stresses in crop plants. *Plants* 8 (3): 71. https://www.mdpi.com/2223-7747/8/3/71.

Bhat, J.A., Rajora, N., Raturi, G. et al. (2021). Silicon nanoparticles (SiNPs) in sustainable agriculture: major emphasis on the practicality, efficacy and concerns. *Nanoscale Advances* 3 (14): 4019–4028. https://doi.org/10.1039/d1na00233c.

Brunings, A., Datnoff, L., Ma, J. et al. (2009). Differential gene expression of rice in response to silicon and rice blast fungus *Magnaporthe oryzae*. *Annals of Applied Biology* 155 (2): 161–170.

Buick, R.D., Buchan, G.D., and Field, R.J. (1993). The role of surface tension of spreading droplets in absorption of a herbicide formulation via leaf stomata. *Pesticide Science* 38 (2–3): 227–235.

Carlisle, E.M. (1984). Silicon. In: *Biochemistry of the Essential Ultratrace Elements* (ed. E. Frieden), 257–291. Springer US. https://doi.org/10.1007/978-1-4684-4775-0_11.

Chain, F., Côté-Beaulieu, C., Belzile, F. et al. (2009). A comprehensive transcriptomic analysis of the effect of silicon on wheat plants under control and pathogen stress conditions. *Molecular Plant-Microbe Interactions* 22 (11): 1323–1330. https://doi.org/10.1094/mpmi-22-11-1323.

Chan, H.-K. and Kwok, P.C.L. (2011). Production methods for nanodrug particles using the bottom-up approach. *Advanced Drug Delivery Reviews* 63 (6): 406–416.

Chanchal Malhotra, C., Kapoor, R., and Ganjewala, D. (2016). Alleviation of abiotic and biotic stresses in plants by silicon supplementation. *Scientia* 13 (2): 59–73.

Chen, Y., Liu, M., Wang, L. et al. (2015). Proteomic characterization of silicon-mediated resistance against *Ralstonia solanacearum* in tomato. *Plant and Soil* 387: 425–440.

Chiba, Y., Mitani, N., Yamaji, N., and Ma, J.F. (2009). HvLsi1 is a silicon influx transporter in barley. *The Plant Journal* 57 (5): 810–818. https://doi.org/10.1111/j.1365-313X.2008.03728.x.

Cokuraj, M., Manokari, M., Faisal, M. et al. (2025). Synergistic effects of growth regulators and silicon nanoparticles on enhancing morpho-structural stability in *Vitex agnus-castus* L. plantlets. *South African Journal of Botany* 177: 500–509.

Coombs, J., Darley, W.M., Holm-Hansen, O., and Volcani, B.E. (1967). Studies on the biochemistry and fine structure of silica shell formation in diatoms. Chemical composition of *Navicula pelliculosa* during silicon-starvation synchrony 1. *Plant Physiology* 42 (11): 1601–1606. https://doi.org/10.1104/pp.42.11.1601.

Corwin, D.L. (2021). Climate change impacts on soil salinity in agricultural areas. *European Journal of Soil Science* 72 (2): 842–862. https://doi.org/10.1111/ejss.13010.

Coskun, D., Deshmukh, R., Sonah, H. et al. (2019). The controversies of silicon's role in plant biology. *New Phytologist* 221 (1): 67–85.

Coursey, J. and Dragoset, R. (2000). Atomic weights and isotopic compositions (Version 2.0). Electronic Publication. https://www.nist.gov/publications/atomic-weights-and-isotopic-compositions-version-20 (accessed June 30, 2025).

Curnow, P., Senior, L., Knight, M.J. et al. (2012). Expression, purification, and reconstitution of a diatom silicon transporter. *Biochemistry* 51 (18): 3776–3785. https://doi.org/10.1021/bi3000484.

Datnoff, L., Deren, C., and Snyder, G. (1997). Silicon fertilization for disease management of rice in Florida. *Crop Protection* 16 (6): 525–531.

David, O.A., Labulo, A.H., Hassan, I. et al. (2024). Complexation and immobilization of arsenic in maize using green synthesized silicon nanoparticles (SiNPs). *Scientific Reports* 14 (1): 6176. https://doi.org/10.1038/s41598-024-56924-3.

Deshmukh, R.K., Vivancos, J., Guérin, V. et al. (2013). Identification and functional characterization of silicon transporters in soybean using comparative genomics of major intrinsic proteins in Arabidopsis and rice. *Plant Molecular Biology* 83 (4–5): 303–315. https://doi.org/10.1007/s11103-013-0087-3.

Detmann, K.C., Araújo, W.L., Martins, S.C. et al. (2012). Silicon nutrition increases grain yield, which, in turn, exerts a feed-forward stimulation of photosynthetic rates via enhanced mesophyll conductance and alters primary metabolism in rice. *New Phytologist* 196 (3): 752–762.

van Dijk, M., Morley, T., Rau, M.L., and Saghai, Y. (2021). A meta-analysis of projected global food demand and population at risk of hunger for the period 2010–2050. *Nature Food* 2 (7): 494–501. https://doi.org/10.1038/s43016-021-00322-9.

Du, J., Liu, B., Zhao, T. et al. (2022). Silica nanoparticles protect rice against biotic and abiotic stresses. *Journal of Nanobiotechnology* 20 (1): 197. https://doi.org/10.1186/s12951-022-01420-x.

Durkin, C.A., Koester, J.A., Bender, S.J., and Armbrust, E.V. (2016). The evolution of silicon transporters in diatoms. *Journal of Phycology* 52 (5): 716–731.

Eichert, T., Kurtz, A., Steiner, U., and Goldbach, H.E. (2008). Size exclusion limits and lateral heterogeneity of the stomatal foliar uptake pathway for aqueous solutes and water-suspended nanoparticles. *Physiologia Plantarum* 134 (1): 151–160. https://doi.org/10.1111/j.1399-3054.2008.01135.x.

Eker, F., Duman, H., Akdaşçi, E. et al. (2024). A comprehensive review of nanoparticles: from classification to application and toxicity. *Molecules* 29 (15): 3482.

El-Naggar, H.M. and Osman, A.R. (2024). Enhancing growth and bioactive metabolites characteristics in *Mentha pulegium* L. via silicon nanoparticles during in vitro drought stress. *BMC Plant Biology* 24 (1): 657. https://doi.org/10.1186/s12870-024-05313-z.

Elshayb, O.M., Nada, A.M., Ibrahim, H.M. et al. (2021). Application of silica nanoparticles for improving growth, yield, and enzymatic antioxidant for the hybrid rice ehr1 growing under water regime conditions. *Materials* 14 (5): 1150.

El-Shetehy, M., Moradi, A., Maceroni, M. et al. (2021). Silica nanoparticles enhance disease resistance in Arabidopsis plants. *Nature Nanotechnology* 16 (3): 344–353. https://doi.org/10.1038/s41565-020-00812-0.

Emamverdian, A., Ding, Y., Mokhberdoran, F. et al. (2020). Silicon dioxide nanoparticles improve plant growth by enhancing antioxidant enzyme capacity in bamboo (*Pleioblastus pygmaeus*) under lead toxicity. *Trees* 34: 469–481.

Epstein, E. (1994). The anomaly of silicon in plant biology. *Proceedings of the National Academy of Sciences of the United States of America* 91 (1): 11–17. https://doi.org/10.1073/pnas.91.1.11.

Epstein, E. (1999). Silicon. *Annual Review of Plant Biology* 50 (1): 641–664.

Exley, C., Tollervey, A., Gray, G. et al. (1993). Silicon, aluminium and the biological availability of phosphorus in algae. *Proceedings of the Royal Society of London. Series B: Biological Sciences* 253 (1336): 93–99. https://doi.org/10.1098/rspb.1993.0086.

Falcon, W.P., Naylor, R.L., and Shankar, N.D. (2022). Rethinking global food demand for 2050. *Population and Development Review* 48 (4): 921–957.

Falouti, M., Ellouzi, H., Amraoui, S. et al. (2024). Silicon (Si) seed priming improves growth and photosynthetic efficiency of barley plants under salt stress in agricultural field conditions. *Journal of Plant Growth Regulation* 44: 1–20.

Fauteux, F., Rémus-Borel, W., Menzies, J.G., and Bélanger, R.R. (2005). Silicon and plant disease resistance against pathogenic fungi. *FEMS Microbiology Letters* 249 (1): 1–6. https://doi.org/10.1016/j.femsle.2005.06.034.

Frantz, J.M., Khandekar, S., and Leisner, S. (2011). Silicon differentially influences copper toxicity response in silicon-accumulator and non-accumulator species. *Journal of the American Society for Horticultural Science* 136 (5): 329–338.

Frew, A., Allsopp, P.G., Gherlenda, A.N., and Johnson, S.N. (2017). Increased root herbivory under elevated atmospheric carbon dioxide concentrations is reversed by silicon-based plant defences. *Journal of Applied Ecology* 54 (5): 1310–1319.

Frew, A., Weston, L.A., Reynolds, O.L., and Gurr, G.M. (2018). The role of silicon in plant biology: a paradigm shift in research approach. *Annals of Botany* 121 (7): 1265–1273. https://doi.org/10.1093/aob/mcy009.

Ge, B., Liu, Q., Li, B. et al. (2024). Characterization and function of promoters of silicon transporter genes PeLsi1-1 and PeLsi1-2 from moso bamboo (*Phyllostachys edulis*). *Plant Cell Reports* 43 (10): 233.

Ghareeb, H., Bozsó, Z., Ott, P.G. et al. (2011). Transcriptome of silicon-induced resistance against *Ralstonia solanacearum* in the silicon non-accumulator tomato implicates priming effect. *Physiological and Molecular Plant Pathology* 75 (3): 83–89.

Ghorbanpour, M., Mohammadi, H., and Kariman, K. (2020). Nanosilicon-based recovery of barley (*Hordeum vulgare*) plants subjected to drought stress. *Environmental Science: Nano* 7 (2): 443–461. https://doi.org/10.1039/C9EN00973F.

Ghosh, D., Das, T., Paul, P. et al. (2024). Zinc-loaded mesoporous silica nanoparticles mitigate salinity stress in wheat seedlings through silica-zinc uptake, osmotic balance, and ROS detoxification. *Plant Physiology and Biochemistry* 211: 108693. https://doi.org/10.1016/j.plaphy.2024.108693.

González-Moscoso, M., Martínez-Villegas, N., Cadenas-Pliego, G., and Juárez-Maldonado, A. (2022). Effect of silicon nanoparticles on tomato plants exposed to two forms of inorganic arsenic. *Agronomy* 12 (10): 2366. https://www.mdpi.com/2073-4395/12/10/2366.

Goto, M., Ehara, H., Karita, S. et al. (2003). Protective effect of silicon on phenolic biosynthesis and ultraviolet spectral stress in rice crop. *Plant Science* 164 (3): 349–356.

Grégoire, C., Rémus-Borel, W., Vivancos, J. et al. (2012). Discovery of a multigene family of aquaporin silicon transporters in the primitive plant *Equisetum arvense*. *The Plant Journal* 72 (2): 320–330. https://doi.org/10.1111/j.1365-313X.2012.05082.x.

Gupta, N., Rai, S.K., Kumar, R. et al. (2024). Seed priming with engineered nanomaterials for mitigating abiotic stress in plants. In: *Nanotechnology for Abiotic Stress Tolerance and Management in Crop Plants* (ed. R.N. Pudake, R.M. Tripathi, and S.S. Gill), 229–247. Elsevier.

Harley, C.D., Anderson, K.M., Demes, K.W. et al. (2012). Effects of climate change on global seaweed communities. *Journal of Phycology* 48 (5): 1064–1078.

Hassan, K.M., Ajaj, R., Abdelhamid, A.N. et al. (2024). Silicon: a powerful aid for medicinal and aromatic plants against abiotic and biotic stresses for sustainable agriculture. *Horticulturae* 10 (8): 806. https://www.mdpi.com/2311-7524/10/8/806.

Hassim, M.F.N., Chuen, N.L., Ghazali, M.S.M., and Ahmad, A. (2021). Nanotechnology: recent progress in agriculture and effects on physiology of plants. *Journal of Tropical Plant Physiology* 13 (2): 14–14.

Hatami, M., Khanizadeh, P., Bovand, F., and Aghaee, A. (2021). Silicon nanoparticle-mediated seed priming and *Pseudomonas* spp. inoculation augment growth, physiology and antioxidant metabolic status in *Melissa officinalis* L. plants. *Industrial Crops and Products* 162: 113238.

Haynes, W.M. (2016). *CRC Handbook of Chemistry and Physics*. CRC Press.

Hildebrand, M., Dahlin, K., and Volcani, B.E. (1998). Characterization of a silicon transporter gene family in *Cylindrotheca fusiformis*: sequences, expression analysis, and identification of homologs in other diatoms. *Molecular and General Genetics MGG* 260 (5): 480–486. https://doi.org/10.1007/s004380050920.

Huang, Q., Ayyaz, A., Farooq, M.A. et al. (2024). Silicon dioxide nanoparticles enhance plant growth, photosynthetic performance, and antioxidants defence machinery through suppressing chromium uptake in *Brassica napus* L. *Environmental Pollution* 342: 123013. https://doi.org/10.1016/j.envpol.2023.123013.

Hussain, A., Rizwan, M., Ali, Q., and Ali, S. (2019). Seed priming with silicon nanoparticles improved the biomass and yield while reduced the oxidative stress and cadmium concentration in wheat grains. *Environmental Science and Pollution Research* 26: 7579–7588. https://doi.org/10.1007/s11356-019-04210-5.

Iqbal, Z., Sarkhosh, A., Balal, R.M. et al. (2020). Silicon alleviate hypoxia stress by improving enzymatic and non-enzymatic antioxidants and regulating nutrient uptake in muscadine grape (*Muscadinia rotundifolia* Michx.). *Frontiers in Plant Science* 11: 618873. https://doi.org/10.3389/fpls.2020.618873.

Iqbal, Z., Sarkhosh, A., Balal, R.M. et al. (2021). Silicon nanoparticles mitigate hypoxia-induced oxidative damage by improving antioxidants activities and concentration of osmolytes in southern highbush blueberry plants. *Agronomy* 11 (11): 2143. https://www.mdpi.com/2073-4395/11/11/2143.

Iqbal, S., Hussain, M., Sadiq, S. et al. (2024). Silicon nanoparticles confer hypoxia tolerance in citrus rootstocks by modulating antioxidant activities and carbohydrate metabolism. *Heliyon* 10 (1): e22960. https://doi.org/10.1016/j.heliyon.2023.e22960.

Irshad, M.K., Ansari, J.R., Noman, A. et al. (2024). Seed priming with Fe_3O_4-SiO_2 nanocomposites simultaneously mitigate Cd and Cr stress in spinach (*Spinacia oleracea* L.): a way forward for sustainable environmental management. *Ecotoxicology and Environmental Safety* 286: 117195. https://doi.org/10.1016/j.ecoenv.2024.117195.

Jasem, D.H. and Khalil, M.I. (2022). Effect of ZnO, Cu, and SiO_2 nanoparticles on drought resistance in two wheat varieties: kalar1 and kalar2 during seedling stage. *Passer Journal of Basic and Applied Sciences* 4 (2): 205–210. https://doi.org/10.24271/psr.2022.361419.1160.

Kandhol, N., Rai, P., Mishra, V. et al. (2024). Silicon regulates phosphate deficiency through involvement of auxin and nitric oxide in barley roots. *Planta* 259 (6): 144. https://doi.org/10.1007/s00425-024-04364-8.

Karande, S.D., Jadhav, S.A., Garud, H.B. et al. (2021). Green and sustainable synthesis of silica nanoparticles. *Nanotechnology For Environmental Engineering* 6 (2): 29. https://doi.org/10.1007/s41204-021-00124-1.

Karuvelan, M., Raj, S., Gururajan, G. et al. (2025). Perspectives of nanomaterials for crop plants improvement and practices. In: *Agricultural Crop Improvement: Plant and Soil Relationships* (ed. A. Husen), 1–22. CRC Press.

Kaur, H. and Greger, M. (2019). A review on Si uptake and transport system. *Plants (Basel)* 8 (4): https://doi.org/10.3390/plants8040081.

Khan, M.R. and Siddiqui, Z.A. (2020). Use of silicon dioxide nanoparticles for the management of *Meloidogyne incognita, Pectobacterium betavasculorum* and *Rhizoctonia solani* disease complex of beetroot (*Beta vulgaris* L.). *Scientia Horticulturae* 265: 109211.

Khan, I., Awan, S.A., Rizwan, M. et al. (2024). Silicon-induced modulation of photosynthetic pigments, osmolytes, and phytohormonal regulation boosted the drought tolerance in *Elymus sibiricus* L. *Journal of Plant Growth Regulation* 43 (4): 998–1011. https://doi.org/10.1007/s00344-023-11155-0.

Khandekar, S. and Leisner, S. (2011). Soluble silicon modulates expression of *Arabidopsis thaliana* genes involved in copper stress. *Journal of Plant Physiology* 168 (7): 699–705.

Kurepa, J., Paunesku, T., Vogt, S. et al. (2010). Uptake and distribution of ultrasmall anatase TiO_2 Alizarin red S nanoconjugates in *Arabidopsis thaliana*. *Nano Letters* 10 (7): 2296–2302. https://doi.org/10.1021/nl903518f.

Laane, H.-M. (2018). The effects of foliar sprays with different silicon compounds. *Plants* 7 (2): 45.

Lewin, J.C. (1955). Silicon metabolism in diatoms. II. Sources of silicon for growth of *Navicula Pelliculosa*. 123. *Plant Physiology* 30 (2): 129–134. https://doi.org/10.1104/pp.30.2.129.

Li, Y., Xi, K., Liu, X. et al. (2023). Silica nanoparticles promote wheat growth by mediating hormones and sugar metabolism. *Journal of Nanobiotechnology* 21 (1): 2. https://doi.org/10.1186/s12951-022-01753-7.

Li, Y., Zan, T., Li, K. et al. (2024). Silica nanoparticles promote the germination of salt-stressed pepper seeds and improve growth and yield of field pepper. *Scientia Horticulturae* 337: 113570.

Li, X., Zhang, Y., Huang, W. et al. (2024). Silica–magnesium coupling in lignin-based biochar: a promising remediation for composite heavy metal pollution in environment. *Journal of Environmental Management* 363: 121392. https://doi.org/10.1016/j.jenvman.2024.121392.

Liang, Y., Si, J., and Römheld, V. (2005). Silicon uptake and transport is an active process in *Cucumis sativus*. *New Phytologist* 167 (3): 797–804.

Liang, Y., Nikolic, M., Bélanger, R. et al. (2015). Silicon uptake and transport in plants: physiological and molecular aspects. In: *Silicon in Agriculture: From Theory to Practice*, 69–82. Netherlands: Springer. https://doi.org/10.1007/978-94-017-9978-2_4.

Liang, Y., Liu, H., Zhang, Y. et al. (2024). Exogenous application of silica nanoparticles mitigates combined salt and low-temperature stress in cotton seedlings by improving the K^+/Na^+ ratio and antioxidant defense. *Plant Stress* 14: 100597.

Liu, J., Cai, H., Mei, C., and Wang, M. (2015). Effects of nano-silicon and common silicon on lead uptake and translocation in two rice cultivars. *Frontiers of Environmental Science & Engineering* 9: 905–911.

Liu, B., Soundararajan, P., and Manivannan, A. (2019). Mechanisms of silicon-mediated amelioration of salt stress in plants. *Plants* 8 (9): 307.

Luyckx, M., Hausman, J.-F., Lutts, S., and Guerriero, G. (2017). Silicon and plants: current knowledge and technological perspectives. *Frontiers in Plant Science* 8: https://doi.org/10.3389/fpls.2017.00411.

Ma, J.F. (2004). Role of silicon in enhancing the resistance of plants to biotic and abiotic stresses. *Soil Science and Plant Nutrition* 50 (1): 11–18. https://doi.org/10.1080/00380768.2004.10408447.

Ma, J.F. and Takahashi, E. (2002). *Soil, Fertilizer, and Plant Silicon Research in Japan*. Elsevier.

Ma, J.F. and Yamaji, N. (2008). Functions and transport of silicon in plants. *Cellular and Molecular Life Sciences* 65 (19): 3049–3057. https://doi.org/10.1007/s00018-008-7580-x.

Ma, J.F. and Yamaji, N. (2015). A cooperative system of silicon transport in plants. *Trends in Plant Science* 20 (7): 435–442. https://doi.org/10.1016/j.tplants.2015.04.007.

Ma, J.F., Goto, S., Tamai, K., and Ichii, M. (2001). Role of root hairs and lateral roots in silicon uptake by rice. *Plant Physiology* 127 (4): 1773–1780. https://doi.org/10.1104/pp.010271.

Ma, J.F., Tamai, K., Yamaji, N. et al. (2006). A silicon transporter in rice. *Nature* 440 (7084): 688–691. https://doi.org/10.1038/nature04590.

Ma, J.F., Yamaji, N., Mitani, N. et al. (2007). An efflux transporter of silicon in rice. *Nature* 448 (7150): 209–212. https://doi.org/10.1038/nature05964.

Ma, J.F., Yamaji, N., Mitani, N. et al. (2008). Transporters of arsenite in rice and their role in arsenic accumulation in rice grain. *Proceedings of the National Academy of*

Sciences of the United States of America 105 (29): 9931–9935. https://doi.org/10.1073/pnas.0802361105.

Ma, J.F., Yamaji, N., and Mitani-Ueno, N. (2011). Transport of silicon from roots to panicles in plants. *Proceedings of the Japan Academy, Series B* 87 (7): 377–385. https://doi.org/10.2183/pjab.87.377.

Ma, L., Wang, Q., Li, L. et al. (2021). UV-C irradiation delays the physiological changes of bell pepper fruit during storage. *Postharvest Biology and Technology* 180: 111506.

Madany, M.M.Y., Saleh, A.M., Habeeb, T.H. et al. (2020). Silicon dioxide nanoparticles alleviate the threats of broomrape infection in tomato by inducing cell wall fortification and modulating ROS homeostasis. *Environmental Science: Nano* 7 (5): 1415–1430. https://doi.org/10.1039/C9EN01255A.

Mahmoud, L.M., Dutt, M., Shalan, A.M. et al. (2020). Silicon nanoparticles mitigate oxidative stress of in vitro-derived banana (*Musa acuminata* 'Grand Nain') under simulated water deficit or salinity stress. *South African Journal of Botany* 132: 155–163.

Mahmoud, L.M., Shalan, A.M., El-Boray, M.S. et al. (2022). Application of silicon nanoparticles enhances oxidative stress tolerance in salt stressed 'Valencia' sweet orange plants. *Scientia Horticulturae* 295: 110856.

Mandlik, R., Thakral, V., Raturi, G. et al. (2020). Significance of silicon uptake, transport, and deposition in plants. *Journal of Experimental Botany* 71 (21): 6703–6718. https://doi.org/10.1093/jxb/eraa301.

Manivannan, A., Soundararajan, P., and Jeong, B.R. (2023). Editorial: silicon: a "quasi-essential" element's role in plant physiology and development. *Frontiers in Plant Science* 14: 1157185. https://doi.org/10.3389/fpls.2023.1157185.

Manokari, M., Faisal, M., Alatar, A.A., and Shekhawat, M.S. (2025). Unveiling the stimulatory effect of silicon nanoparticles on the direct regeneration and micro-structural developments in Cadaba trifoliata (Roxb.) Wight & Arn. *Environmental Science: Nano* 12 (1): https://doi.org/10.1039/D4EN00640B.

Manzoor, N., Ali, L., Ahmed, T. et al. (2022). Silicon oxide nanoparticles alleviate chromium toxicity in wheat (*Triticum aestivum* L.). *Environmental Pollution* 315: 120391.

Markovich, O., Steiner, E., Kouřil, Š. et al. (2017). Silicon promotes cytokinin biosynthesis and delays senescence in Arabidopsis and Sorghum. *Plant, Cell & Environment* 40 (7): 1189–1196.

Markovich, O., Kumar, S., Cohen, D. et al. (2019). Silicification in leaves of sorghum mutant with low silicon accumulation. *Silicon* 11 (5): 2385–2391. https://doi.org/10.1007/s12633-015-9348-x.

Marron, A.O., Alston, M.J., Heavens, D. et al. (2013). A family of diatom-like silicon transporters in the siliceous loricate choanoflagellates. *Proceedings of the Royal*

Society B: Biological Sciences 280 (1756): 20122543. https://doi.org/10.1098/rspb.2012.2543.

Maryam, S., & Gul, A. (2022). The mechanism of silicon transport in plants. In *Plant Metal and Metalloid Transporters* Kumar, K., Srivastava, S. (eds) (pp. 245-273). Springer Singapore. https://doi.org/10.1007/978-981-19-6103-8_12.

Matysik, J., Alia, B., and B., & Mohanty, P. (2002). Molecular mechanisms of quenching of reactive oxygen species by proline under stress in plants. *Current Science* 82 (5): 525–532.

Maurya, A.K. (2020). Oxidative stress in crop plants. In: *Agronomic Crops: Volume 3: Stress Responses and Tolerance* (ed. M. Hasanuzzaman), 349–380. Springer Singapore. https://doi.org/10.1007/978-981-15-0025-1_18.

Meena, V.D., Dotaniya, M.L., Coumar, V. et al. (2014). A case for silicon fertilization to improve crop yields in tropical soils. *Proceedings of the National Academy of Sciences, India Section B: Biological Sciences* 84 (3): 505–518. https://doi.org/10.1007/s40011-013-0270-y.

Merwad, A.-R.M., Desoky, E.-S.M., and Rady, M.M. (2018). Response of water deficit-stressed Vigna unguiculata performances to silicon, proline or methionine foliar application. *Scientia Horticulturae* 228: 132–144.

Mitani, N., Chiba, Y., Yamaji, N., and Ma, J.F. (2009a). Identification and characterization of maize and barley Lsi2-Like silicon efflux transporters reveals a distinct silicon uptake system from that in rice. *The Plant Cell* 21 (7): 2133–2142. https://doi.org/10.1105/tpc.109.067884.

Mitani, N., Yamaji, N., and Ma, J.F. (2009b). Identification of maize silicon influx transporters. *Plant and Cell Physiology* 50 (1): 5–12. https://doi.org/10.1093/pcp/pcn110.

Mitani, N., Yamaji, N., Ago, Y. et al. (2011). Isolation and functional characterization of an influx silicon transporter in two pumpkin cultivars contrasting in silicon accumulation. *The Plant Journal* 66 (2): 231–240. https://doi.org/10.1111/j.1365-313X.2011.04483.x.

Mitani-Ueno, N., Yamaji, N., and Ma, J.F. (2011). Silicon efflux transporters isolated from two pumpkin cultivars contrasting in Si uptake. *Plant Signaling & Behavior* 6 (7): 991–994. https://doi.org/10.4161/psb.6.7.15462.

Mohammadi, H., Mousavi, Z., Hazrati, S. et al. (2023). Foliar applied titanium dioxide nanoparticles (TiO_2 NPs) modulate growth, physiological and phytochemical traits in *Melissa officinalis* L. under various light intensities. *Industrial Crops and Products* 197: 116642.

Mohapatra, S., Sahu, B.K., and Dash, D.K. (2023). A brief review on basic fundamentals of nanoparticle (NPs). *Nano and Medical Materials* 3 (2): 31–31.

Mokari, M. and Abedinpour, M. (2020). Investigating the effect of spraying silica nanoparticles in increasing the drought resistance of millet seedlings in Kashmar weather conditions. *Iranian Journal of Soil and Water Research* 51 (9): 2307–2322.

Montpetit, J., Vivancos, J., Mitani-Ueno, N. et al. (2012). Cloning, functional characterization and heterologous expression of TaLsi1, a wheat silicon transporter gene. *Plant Molecular Biology* 79 (1–2): 35–46. https://doi.org/10.1007/s11103-012-9892-3.

Muhammad, S., Ulhassan, Z., Munir, R. et al. (2025). Nanosilica and salicylic acid synergistically regulate cadmium toxicity in rice. *Environmental Pollution* 364: 125331. https://doi.org/10.1016/j.envpol.2024.125331.

Mukarram, M., Khan, M.M.A., and Corpas, F.J. (2021). Silicon nanoparticles elicit an increase in lemongrass (Cymbopogon flexuosus (Steud.) Wats) agronomic parameters with a higher essential oil yield. *Journal of Hazardous Materials* 412: 125254.

Mukarram, M., Petrik, P., Mushtaq, Z. et al. (2022). Silicon nanoparticles in higher plants: uptake, action, stress tolerance, and crosstalk with phytohormones, antioxidants, and other signalling molecules. *Environmental Pollution* 310: 119855.

Mushinskiy, A.A., Aminova, E.V., and Korotkova, A.M. (2018). Evaluation of tolerance of tubers *Solanum tuberosum* to silica nanoparticles. *Environmental Science and Pollution Research* 25 (34): 34559–34569.

Naidu, S., Pandey, J., Mishra, L.C. et al. (2023). Silicon nanoparticles: synthesis, uptake and their role in mitigation of biotic stress. *Ecotoxicology and Environmental Safety* 255: 114783.

Nair, R., Varghese, S.H., Nair, B.G. et al. (2010). Nanoparticulate material delivery to plants. *Plant Science* 179 (3): 154–163.

Nazaralian, S., Majd, A., Irian, S. et al. (2017). Comparison of silicon nanoparticles and silicate treatments in fenugreek. *Plant Physiology and Biochemistry* 115: 25–33. https://doi.org/10.1016/j.plaphy.2017.03.009.

Nelson, D.M., Tréguer, P., Brzezinski, M.A. et al. (1995). Production and dissolution of biogenic silica in the ocean: revised global estimates, comparison with regional data and relationship to biogenic sedimentation. *Global Biogeochemical Cycles* 9 (3): 359–372. https://doi.org/10.1029/95GB01070.

Ng, L.C., Hassim, M.F.N., Ghazali, M.S.M., and Ahmad, A. (2021). Nanotechnology: recent progress in agriculture and effects on physiology of plants. *Journal of Tropical Plant Physiology (JTPP) (Malaysia)* 13 (2): 23–36.

Ochoa, L., Shrivastava, M., Srivastava, S. et al. (2025). Nanomaterials for managing abiotic and biotic stress in the soil–plant system for sustainable agriculture. *Environmental Science: Nano* 12 (11): 10.1039/D4EN00789A.

Oshunsanya, S. O., Nwosu, N. J., & Li, Y. (2019). Abiotic stress in agricultural crops under climatic conditions. In: Jhariya, M., Banerjee, A., Meena, R., Yadav, D. (eds) *Sustainable Agriculture, Forest and Environmental Management*, Springer, Singapore. https://doi.org/10.1007/978-981-13-6830-1_3.

Parihar, P., Singh, S., Singh, R. et al. (2015). Effect of salinity stress on plants and its tolerance strategies: a review. *Environmental Science and Pollution Research* 22 (6): 4056–4075. https://doi.org/10.1007/s11356-014-3739-1.

Parker, B.C. (1969). Occurrence of silica in brown and green algae. *Canadian Journal of Botany* 47 (4): 537–540. https://doi.org/10.1139/b69-073.

Patle, P., Kadu, P., Gabhane, A. et al. (2019). Consequences provoked due to excess application of agrochemical on soil health deterioration–a review for Sustainable Agriculture. *Journal of Pharmacognosy and Phytochemistry* 8 (2S): 63–66.

Pavlovic, J., Kostic, L., Bosnic, P. et al. (2021). Interactions of silicon with essential and beneficial elements in plants [Review]. *Frontiers in Plant Science* 12: https://doi.org/10.3389/fpls.2021.697592.

Peng, H., Hu, H., Xi, K. et al. (2022). Silicon nanoparticles enhance ginger rhizomes tolerance to postharvest deterioration and resistance to *Fusarium solani. Frontiers in Plant Science* 13: 816143. https://doi.org/10.3389/fpls.2022.816143.

Pontigo, S., Ribera, A., Gianfreda, L. et al. (2015). Silicon in vascular plants: uptake, transport and its influence on mineral stress under acidic conditions. *Planta* 242 (1): 23–37. https://doi.org/10.1007/s00425-015-2333-1.

Popp, C., Burghardt, M., Friedmann, A., and Riederer, M. (2005). Characterization of hydrophilic and lipophilic pathways of *Hedera helix* L. cuticular membranes: permeation of water and uncharged organic compounds. *Journal of Experimental Botany* 56 (421): 2797–2806. https://doi.org/10.1093/jxb/eri272.

Rai-Kalal, P., Tomar, R.S., and Jajoo, A. (2021). Seed nanopriming by silicon oxide improves drought stress alleviation potential in wheat plants. *Functional Plant Biology* 48 (9): 905–915. https://doi.org/10.1071/FP21079.

Raja, R.K., Surendar, K.K., Ravichandran, V. et al. (2021). Influence of nanosilica on physio-biochemical and antioxidative enzymes in rice under drought. *The Pharma Innovation Journal* 10 (11): 1318–1323.

Rajput, V.D., Minkina, T., Kumari, A. et al. (2021). Coping with the challenges of abiotic stress in plants: new dimensions in the field application of nanoparticles. *Plants* 10 (6): 1221.

Raliya, R., Franke, C., Chavalmane, S. et al. (2016). Quantitative understanding of nanoparticle uptake in watermelon plants. *Frontiers in Plant Science* 7: 1288.

Rashad, Y.M., El-Sharkawy, H.H., Belal, B.E. et al. (2021). Silica nanoparticles as a probable anti-oomycete compound against downy mildew, and yield and quality enhancer in grapevines: field evaluation, molecular, physiological, ultrastructural, and toxicity investigations. *Frontiers in Plant Science* 12: 763365.

Rastogi, A., Tripathi, D.K., Yadav, S. et al. (2019). Application of silicon nanoparticles in agriculture. *3 Biotech* 9 (3): 90. https://doi.org/10.1007/s13205-019-1626-7.

Raven, J. (2001). Chaper 3: Silicon transport at the cell and tissue level. In: *Studies in Plant Science*, vol. 8 (ed. L.E. Datnoff, G.H. Snyder, and G.H. Korndörfer), 41–55. Elsevier.

Raven, J.A. (2003). Cycling silicon: the role of accumulation in plants. *New Phytologist* 158 (3): 419–421.

Raza, M.A.S., Zulfiqar, B., Iqbal, R. et al. (2023). Morpho-physiological and biochemical response of wheat to various treatments of silicon nano-particles under drought stress conditions. *Scientific Reports* 13 (1): 2700. https://doi.org/10.1038/s41598-023-29784-6.

Riggs, P. K., Fields, M. J., & Cross, H. R. (2018). Food and nutrient security for a growing population. *Animal Frontiers* 8(3): 3–4. https://doi.org/10.1093/af/vfy014.

Roy, T.K., Islam, M.S., Mahiddin, N.A. et al. (2025). Application of nanoparticles (NPs) to ameliorate abiotic stress in economically important crop species: a potential review. *Journal of Crop Health* 77 (1): 1–20.

Saeedeh, R., Mehrnaz, H., and Mansour, G. (2021). Silicon-nanoparticle mediated changes in seed germination and vigor index of marigold (*Calendula officinalis* L.) compared to silicate under PEG-induced drought stress. *Gesunde Pflanzen* 73 (4): 575–589.

Schwab, F., Zhai, G., Kern, M. et al. (2016). Barriers, pathways and processes for uptake, translocation and accumulation of nanomaterials in plants–critical review. *Nanotoxicology* 10 (3): 257–278.

Sembada, A.A. and Lenggoro, I.W. (2024). Transport of nanoparticles into plants and their detection methods. *Nanomaterials* 14 (2): 131.

Sembada, A.A., Maki, S., Faizal, A. et al. (2023). The role of silica nanoparticles in promoting the germination of tomato (*Solanum lycopersicum*) seeds. *Nanomaterials* 13 (14): 2110.

Sepasi, M., Iranbakhsh, A., Saadatmand, S. et al. (2024). Silicon nanoparticles (SiNPs) stimulated secondary metabolism and mitigated toxicity of salinity stress in basil (*Ocimum basilicum*) by modulating gene expression: a sustainable approach for crop protection. *Environmental Science and Pollution Research International* 31 (11): 16485–16496. https://doi.org/10.1007/s11356-024-32260-x.

Sezgin Muslu, A., Altuntaş, C., Altansambar, N. et al. (2025). Combined application of rutin and silicon sustains maize seedlings osmotic stress tolerance by improving photosynthetic capacity and chlorophyll metabolism. *Acta Botanica Croatica* 84 (1): 70–80. https://doi.org/10.37427/botcro-2025-015.

Shama, G. (1999). ULTRA-VIOLET LIGHT. In: *Encyclopedia of Food Microbiology* (ed. R.K. Robinson), 2208–2214. Elsevier. https://doi.org/10.1006/rwfm.1999.1615.

Sharf-Eldin, A.A., Alwutayd, K.M., El-Yazied, A.A. et al. (2023). Response of maize seedlings to silicon dioxide nanoparticles (SiO_2NPs) under drought stress. *Plants* 12 (14): 2592. https://www.mdpi.com/2223-7747/12/14/2592.

Sharma, B., Kumawat, K.C., Tiwari, S. et al. (2023). Silicon and plant nutrition—dynamics, mechanisms of transport and role of silicon solubilizer microbiomes in sustainable agriculture: a review. *Pedosphere* 33 (4): 534–555.

Shen, X., Zhou, Y., Duan, L. et al. (2010). Silicon effects on photosynthesis and antioxidant parameters of soybean seedlings under drought and ultraviolet-B radiation. *Journal of Plant Physiology* 167 (15): 1248–1252.

Shen, H., Mi, Y.-W., and Wang, L.-Q. (2012). Effects of exogenous silicon on physiological characteristics of *Lycium ruthenicum* seedling under salt stress. *Acta Agrestia Sinica* 20 (3): 553.

Siddiqui, M.H., Al-Whaibi, M.H., Faisal, M., and Al Sahli, A.A. (2014). Nano-silicon dioxide mitigates the adverse effects of salt stress on *Cucurbita pepo* L. *Environmental Toxicology and Chemistry* 33 (11): 2429–2437.

Sjöberg, S. (1996). Silica in aqueous environments. *Journal of Non-Crystalline Solids* 196: 51–57. https://doi.org/10.1016/0022-3093(95)00562-5.

de Sousa, A., Saleh, A.M., Habeeb, T.H. et al. (2019). Silicon dioxide nanoparticles ameliorate the phytotoxic hazards of aluminum in maize grown on acidic soil. *The Science of the Total Environment* 693: 133636. https://doi.org/10.1016/j.scitotenv.2019.133636.

Stapleton, A.E. (1992). Ultraviolet radiation and plants: burning questions. *Plant Cell* 4 (11): 1353–1358. https://doi.org/10.1105/tpc.4.11.1353.

Strout, G., Russell, S.D., Pulsifer, D.P. et al. (2013). Silica nanoparticles aid in structural leaf coloration in the Malaysian tropical rainforest understorey herb *Mapania caudata*. *Annals of Botany* 112 (6): 1141–1148.

Sulaiman, El-Beltagi, H.S., Shehata, W.F. et al. (2023). Role of silica nanoparticles in enhancing drought tolerance of cereal crops. *Notulae Botanicae Horti Agrobotanici Cluj-Napoca* 51 (4): 13480. https://doi.org/10.15835/nbha51413480.

Sun, D., Hussain, H.I., Yi, Z. et al. (2014). Uptake and cellular distribution, in four plant species, of fluorescently labeled mesoporous silica nanoparticles. *Plant Cell Reports* 33 (8): 1389–1402. https://doi.org/10.1007/s00299-014-1624-5.

Sun, D., Hussain, H.I., Yi, Z. et al. (2016). Mesoporous silica nanoparticles enhance seedling growth and photosynthesis in wheat and lupin. *Chemosphere* 152: 81–91. https://doi.org/10.1016/j.chemosphere.2016.02.096.

Sun, H., Guo, J., Duan, Y. et al. (2017). Isolation and functional characterization of CsLsi1, a silicon transporter gene in *Cucumis sativus*. *Physiologia Plantarum* 159 (2): 201–214. https://doi.org/10.1111/ppl.12515.

Sun, H., Duan, Y., Qi, X. et al. (2018). Isolation and functional characterization of CsLsi2, a cucumber silicon efflux transporter gene. *Annals of Botany* 122 (4): 641–648.

Sun, H., Duan, Y., Mitani-Ueno, N. et al. (2020). Tomato roots have a functional silicon influx transporter but not a functional silicon efflux transporter. *Plant, Cell & Environment* 43 (3): 732–744.

Suriyaprabha, R., Karunakaran, G., Yuvakkumar, R. et al. (2012). Growth and physiological responses of maize (*Zea mays* L.) to porous silica nanoparticles in soil. *Journal of Nanoparticle Research* 14 (12): 1294. https://doi.org/10.1007/s11051-012-1294-6.

Suriyaprabha, R., Karunakaran, G., Kavitha, K. et al. (2014). Application of silica nanoparticles in maize to enhance fungal resistance. *IET Nanobiotechnology* 8 (3): 133–137.

Sutulienė, R., Ragelienė, L., Samuolienė, G. et al. (2022). The response of antioxidant system of drought-stressed green pea (*Pisum sativum* L.) affected by watering and foliar spray with silica nanoparticles. *Horticulturae* 8 (1): 35. https://www.mdpi.com/2311-7524/8/1/35.

Swain, R., Behera, M., Sahoo, S., and Rout, G.R. (2025). Nanoscience in plant stress mitigation: a comprehensive review. *BioNanoScience* 15 (1): 1–28.

Takahashi, E., Ma, J., & Miyake, Y. (1990). The possibility of silicon as an essential element for higher plants 99-102.

Tamai, K. and Ma, J.F. (2008). Reexamination of silicon effects on rice growth and production under field conditions using a low silicon mutant. *Plant and Soil* 307 (1): 21–27. https://doi.org/10.1007/s11104-008-9571-y.

Thakral, V., Raturi, G., Sudhakaran, S. et al. (2024). Silicon, a quasi-essential element: availability in soil, fertilizer regime, optimum dosage, and uptake in plants. *Plant Physiology and Biochemistry* 208: 108459. https://doi.org/10.1016/j.plaphy.2024.108459.

Thamatrakoln, K., Alverson, A.J., and Hildebrand, M. (2006). Comparative sequence analysis of diatom silicon transporters: toward a mechanistic model of silicon transport. *Journal of Phycology* 42 (4): 822–834. https://doi.org/10.1111/j.1529-8817.2006.00233.x.

Tian, L., Shen, J., Sun, G. et al. (2020). Foliar application of SiO$_2$ nanoparticles alters soil metabolite profiles and microbial community composition in the pakchoi (*Brassica chinensis* L.) rhizosphere grown in contaminated mine soil. *Environmental Science & Technology* 54 (20): 13137–13146.

Tran, T.L.C., Guirguis, A., Jeyachandran, T. et al. (2023). Mesoporous silica nanoparticle-induced drought tolerance in *Arabidopsis thaliana* grown under *in vitro* conditions. *Functional Plant Biology* 50 (11): 889–900. https://doi.org/10.1071/FP22274.

Trembath-Reichert, E., Wilson, J.P., McGlynn, S.E., and Fischer, W.W. (2015). Four hundred million years of silica biomineralization in land plants. *Proceedings of the National Academy of Sciences* 112 (17): 5449–5454. https://doi.org/10.1073/pnas.1500289112.

Tripathi, D.K., Singh, V.P., Prasad, S.M. et al. (2015). Silicon nanoparticles (SiNp) alleviate chromium (VI) phytotoxicity in *Pisum sativum* (L.) seedlings. *Plant Physiology and Biochemistry* 96: 189–198. https://doi.org/10.1016/j.plaphy.2015.07.026.

Tripathi, D.K., Singh, S., Singh, V.P. et al. (2017). Silicon nanoparticles more effectively alleviated UV-B stress than silicon in wheat (*Triticum aestivum*) seedlings. *Plant Physiology and Biochemistry* 110: 70–81. https://doi.org/10.1016/j.plaphy.2016.06.026.

Tripathi, D.K., Rai, P., Guerriero, G. et al. (2021). Silicon induces adventitious root formation in rice under arsenate stress with involvement of nitric oxide and indole-3-acetic acid. *Journal of Experimental Botany* 72 (12): 4457–4471. https://doi.org/10.1093/jxb/eraa488.

Ullah, I., Toor, M.D., Basit, A. et al. (2023). Nanotechnology: an integrated approach towards agriculture production and environmental stress tolerance in plants. *Water, Air, & Soil Pollution* 234 (11): 666.

Vanlerberghe, G.C., Dahal, K., Alber, N.A., and Chadee, A. (2020). Photosynthesis, respiration and growth: a carbon and energy balancing act for alternative oxidase. *Mitochondrion* 52: 197–211.

Verma, K.K., Song, X.-P., Verma, C.L. et al. (2021). Functional relationship between photosynthetic leaf gas exchange in response to silicon application and water stress mitigation in sugarcane. *Biological Research* 54 (1): 15.

Vulavala, V.K., Elbaum, R., Yermiyahu, U. et al. (2016). Silicon fertilization of potato: expression of putative transporters and tuber skin quality. *Planta* 243: 217–229.

Wang, W.-N., Tarafdar, J.C., and Biswas, P. (2013). Nanoparticle synthesis and delivery by an aerosol route for watermelon plant foliar uptake. *Journal of Nanoparticle Research* 15: 1–13.

Wang, M., Gao, L., Dong, S. et al. (2017). Role of silicon on plant-pathogen interactions. *Frontiers in Plant Science* 8: 701. https://doi.org/10.3389/fpls.2017.00701.

Wang, Z., Yue, L., Dhankher, O.P., and Xing, B. (2020). Nano-enabled improvements of growth and nutritional quality in food plants driven by rhizosphere processes. *Environment International* 142: 105831–105831. https://doi.org/10.1016/j.envint.2020.105831.

Wang, L., Ning, C., Pan, T., and Cai, K. (2022a). Role of silica nanoparticles in abiotic and biotic stress tolerance in plants: a review. *International Journal of Molecular Sciences* 23 (4): https://doi.org/10.3390/ijms23041947.

Wang, L., Pan, T., Gao, X. et al. (2022b). Silica nanoparticles activate defense responses by reducing reactive oxygen species under *Ralstonia solanacearum* infection in tomato plants. *NanoImpact* 28: 100418.

Wang, T., Long, H., Mao, S. et al. (2024). Silicon nanoparticles improve tomato seed germination more effectively than conventional silicon under salt stress via regulating antioxidant system and hormone metabolism. *Horticulturae* 10 (8): 785.

Wang, S., Shen, X., Guan, X. et al. (2025). Nano-silicon enhances tomato growth and antioxidant defense under salt stress. *Environmental Science: Nano* 12 (1): 315–324.

Xu, Y., Charles, M.T., Luo, Z. et al. (2019). Preharvest UV-C treatment affected postharvest senescence and phytochemicals alternation of strawberry fruit with the possible involvement of abscisic acid regulation. *Food Chemistry* 299: 125138.

Yamaji, N. and Ma, J.F. (2011). Further characterization of a rice silicon efflux transporter, Lsi2. *Soil Science and Plant Nutrition* 57 (2): 259–264.

Yamaji, N., Mitatni, N., and Ma, J.F. (2008). A transporter regulating silicon distribution in rice shoots. *The Plant Cell* 20 (5): 1381–1389.

Yamaji, N., Chiba, Y., Mitani-Ueno, N., and Feng Ma, J. (2012). Functional characterization of a silicon transporter gene implicated in silicon distribution in barley. *Plant Physiology* 160 (3): 1491–1497. https://doi.org/10.1104/pp.112.204578.

Yan, G.-C., Nikolic, M., Ye, M.-J. et al. (2018). Silicon acquisition and accumulation in plant and its significance for agriculture. *Journal of Integrative Agriculture* 17 (10): 2138–2150.

Yang, W., Feng, H., Zhou, J. et al. (2024). Exogenous silicon induces aluminum tolerance in white clover (*Trifolium repens*) by reducing aluminum uptake and enhancing organic acid secretion. *PeerJ* 12: e17472. https://doi.org/10.7717/peerj.17472.

Yousaf, S., Sidrah, A., Asrar, R. et al. (2024). Chapter 17: Nanostructured silica for enhanced fungicidal activity in agriculture. In: *Nanobiotechnology for Plant Protection, NanofungicidesNanofungicides* (ed. K.A. Abd-Elsalam), 349–373. Elsevier.

Yuvakkumar, R., Elango, V., Rajendran, V. et al. (2011). Influence of nanosilica powder on the growth of maize crop (*Zea mays* L.). *International Journal of Green Nanotechnology* 3 (3): 180–190.

Zafar, S., Bilal, M., Ali, M.F. et al. (2024). Nano-biofertilizer an eco-friendly and sustainable approach for the improvement of crops under abiotic stresses. *Environmental and Sustainability Indicators* 24: 100470.

Zahedi, S.M., Moharrami, F., Sarikhani, S., and Padervand, M. (2020). Selenium and silica nanostructure-based recovery of strawberry plants subjected to drought stress. *Scientific Reports* 10 (1): 17672.

Zargar, S.M., Mahajan, R., Bhat, J.A. et al. (2019). Role of silicon in plant stress tolerance: opportunities to achieve a sustainable cropping system. *3 Biotech* 9 (3): 73. https://doi.org/10.1007/s13205-019-1613-z.

Zhang, W., Xie, Z., Wang, L. et al. (2017). Silicon alleviates salt and drought stress of *Glycyrrhiza uralensis* seedling by altering antioxidant metabolism and osmotic adjustment. *Journal of Plant Research* 130 (3): 611–624. https://doi.org/10.1007/s10265-017-0927-3.

Zhang, F., Liu, Y., Liang, Y. et al. (2024). Improving the yield and quality of tomato by using organic fertilizer and silicon compared to reducing chemical nitrogen fertilization. *Agronomy* 14 (5): 966. https://www.mdpi.com/2073-4395/14/5/966.

Zhou, J., Liu, X., Sun, C. et al. (2022). Silica nanoparticles enhance the disease resistance of ginger to rhizome rot during postharvest storage. *Nanomaterials (Basel)* 12 (9): https://doi.org/10.3390/nano12091418.

Zhou, H., Wang, L., Su, J. et al. (2024). Combined application of silica nanoparticles and brassinolide promoted the growth of sugar beets under deficit irrigation. *Plant Physiology and Biochemistry* 216: 109165. https://doi.org/10.1016/j.plaphy.2024.109165.

Zhu, Y., Guo Jia, G.J., Feng Ru, F.R. et al. (2016). The regulatory role of silicon on carbohydrate metabolism in *Cucumis sativus* L. under salt stress. *Plant and Soil* 406 (1-2): 231–249. https://doi.org/10.1007/s11104-016-2877-2.

Section C

Applications of Si/SiO_2 NPs in Plants

10

Role of Si/SiO_2 NPs in Mitigation of Biotic Stress Caused Due to Herbivory

Shrishti Naidu[1], Indrakant K. Singh[2,3] and Archana Singh[1]

[1]*Department of Plant Molecular Biology, University of Delhi South Campus, New Delhi, Delhi, India*
[2]*Department of Zoology, Deshbandhu College, University of Delhi, New Delhi, Delhi, India*
[3]*Dr. B. R. Ambedkar Centre for Biomedical Research, & Delhi School of Public Health, Institute of Eminence, University of Delhi, Delhi, India*

10.1 Introduction

Plants are continually under the threat of biotic stressors like insect pests, which diminish agricultural production as well as quality (Oerke 2006). Plants have acquired both direct and indirect means of pest resistance throughout millions of years of evolution (Kessler and Baldwin 2002; Paré and Tumlinson 1997). Direct defence includes structural features such as thicker cuticle layers, oils, waxes, lignin, trichomes and thorns, which provide an external barrier that affects pest insect habits and performance. Another strategy for protection against herbivory is to release 'herbivore-induced plant volatiles' (HIPVs) and to increase the expression of defence-associated protein molecules and genes (Howe and Jander 2008; Oerke 2006; Singh et al. 2020b). Phytohormones like jasmonic acid (JA), salicylic acid (SA) and ethylene (ET) controls signalling networks that increase the resistance of plants toward attacks from insects (Aljbory and Chen 2018; Dicke et al. 2003; Howe and Jander 2008; Mithöfer and Boland 2012; Verhage et al. 2010; Waterman et al. 2019). This proves that the plants entail a multifaceted defensive mechanism against insect assault that is activated when plant immune-system receptors recognize insect pest effectors called 'herbivore-associated molecular patterns' (HAMPs). Despite such advanced defence mechanisms, insect pests frequently bypass the plant defence systems. In agricultural plants, this has a significant influence on production, yield and food supply (Howe and Jander 2008; Oerke 2006; Verhage et al. 2010). A significant amount of published research has indicated that Si improves the direct as well as indirect means of plant defence mechanisms (Singh et al. 2020a). As a result, using Si to tap into plant resistance

Silicon Nanoparticles: Synthesis, Uptake, and Applications, First Edition.
Edited by Archana Singh, Aryadeep Roychoudhury, and Indrakant K. Singh.

and improve induced defences might be a less ecologically hazardous yet effective substitute to artificial insecticides, as well as an efficient technique in holistic pest control.

Si is a common metalloid ranked as the second most prevalent inorganic element on the Earth (Epstein 1994). Silicon nanoparticles (SiNPs) are nanoscale Si particles whose size ranges from 1 to 100 nm. SiNPs are classified into four categories based on their structures: spherical, hollow, shaped and porous (Mathur and Roy 2020). SiNPs have several benefits over mass Si sources, such as a higher surface by volume ratio, unique charge or ionic characteristics, porous nature, nano size, increased biological availability and nutritional effects (Jeelani et al. 2020). Their synthesis process, which may be physical, chemical or biological (green synthesis) based on the driving force, often determines the features and architectures of SiNPs (Naidu et al. 2023). Because of their unique charge and size features, SiNPs can quickly pass through plant barriers like the cell wall, plasma membrane and epidermis when they are exposed via foliar and root application strategy. They have unique features and advantages, making them flexible in nano-enabled crop farming as a plant development booster, a nanocarrier and a conditioner for soil (Ji et al. 2018; Mahawar et al. 2023; Rastogi et al. 2019). Numerous research laboratories and field experiments have confirmed that SiNPs also act as phyto-stimulants, can improve the resilience of plants against multiple abiotic or biotic stressors (Bansal et al. 2022; Verma et al. 2022; Wang et al. 2022), vehicles for administering chemicals (e.g., pesticides, fertilizers and hormones) and biologically active compounds (e.g., RNA, DNA and proteins) in crop and agricultural biotechnology (Yan et al. 2024).

Herbivory results in accelerated Si accumulation in monocots or grasses, demonstrating that plants utilize silicification as an elicited defensive mechanism (Massey et al. 2007; McNaughton and Tarrants 1983). Plants absorb soil-available silicon and store it between or within cells, enhancing the stiffness and roughness of plant tissues. This enhances plants' physical barriers, resulting in increased damage to insects oral or feeding parts (Massey and Hartley 2009). Furthermore, Si-accumulating plants have a greater percentage of defensive genes and proteins (Cai et al. 2008; Rehman et al. 2015). Furthermore, these plants likewise have an amplified terpenoid synthesis and enhanced phenylpropanoid pathway that leads to the generation of secondary metabolic compounds (Fawe et al. 1998; Rémus-Borel et al. 2005; Rodrigues et al. 2004). Greater levels of Si buildup is also known to be associated with increased amounts of defence-related enzymes (Alhousari and Greger 2018; Chérif et al. 1994). Many studies have shown that providing plants with increased amounts of Si improves their resistance to herbivores (Singh et al. 2020a). This chapter highlights Si's influence on the defence of plants against insect pests/herbivores, the basic mechanism by which it occurs and the possible advantages of using Si as a fertilizer for crop defence.

10.2 Impact of Herbivory on Crop Production

If other stress variables (such as drought, climate and daylight) have no impact on plant production, plant abundance should drop when nutrient cycling is delayed and vice versa. Herbivory has been known to influence nutrient cycle a long ago (Hutchinson and Deevey Jr. 1949). Herbivores influence nutrient cycling by depositing waste products, modifying the nutrient or decomposition rates of plant litter and locking up nutrients in their tissues or body parts (Belovsky and Slade 2000). However, because consumption decreases plant abundance, herbivores may only increase plant abundance if nutrient cycling improves more than consumption reduces. Herbivory has the following consequences for the nutrient cycle: The fast cycle (McNaughton et al. 1988) refers to the quick decomposition of dung and deceased herbivores, which delivers nutrients for plant absorption. The slow cycle (McNaughton et al. 1988) refers to the slow decomposition of plant litter, which gradually delivers nutrients for absorption. Herbivory impacts the slow cycle by altering the amount of plant litter, particularly if herbivores consume plants with varying rates of litter decomposition (Pastor et al. 1988; Pastor and Naiman 1992). Selective feeding on plants that create slower decomposing litter lowers their relative abundance, accelerating the slow cycle, whereas selective feeding on plants that generate fast decomposing litter decelerates the slow cycle (Belovsky and Slade 2000). Changes in the quantity of nutrients released by fast vs. slow cycles may alter the availability of nutrients to plants, influencing plant production and species composition. If selective feeding on slow-decomposing plants outweighs the negative impacts of eating on plants, it may speed nutrient cycling and boost plant output or production (Pastor and Naiman 1992). However, selective feeding on fast-decomposing plants may slow nutrient cycling and reduce plant output (Pastor and Naiman 1992).

Nutrient cycle variations are one of the variables that influence plant species composition (Belovsky and Slade 2000). For example, if slow decaying plants compete better when nutrients are scarce and fast decomposing plants compete better when resources are abundant, herbivore-induced modifications to nutrient cycling may have an impact on plant competition and, hence, communities of vegetation (Pastor and Naiman 1992). Additionally, changes in nutrient cycling, which result in alterations in plant species composition, may further amplify changes in nutrient cycling, plant composition and plant production, generating a favourable positive feedback loop (Belovsky and Slade 2000). Herbivory has been shown to speed nutrient cycling and improve the production of plants (DeAngelis 1992). Insect herbivores, due to their abundance and short life, may have a significant impact on the terrestrial nutrient cycle, perhaps generally accelerating it (Lovett and Ruesink 1995; Seastedt and Crossley Jr. 1984). Herbivory can boost plant abundance as long as they are not consumed in excess

(Belovsky and Slade 2000). If consumed in very less amount, herbivory can have insufficient impact on nutrient cycling to boost plant productivity. If too much of plants are taken in, herbivory will reduce the survival and growth of plants more than increased growth from nutrient cycling. Therefore, ecologically there are two perspectives on the significance of herbivores in the functioning of ecosystems. First, in 'population or community ecology', herbivory minimizes plant abundance. Second, from the standpoint of nutrient cycling in the 'ecosystem ecology', it can impede cycling, reducing plant abundance, or it can speed up cycling, potentially increasing plant abundance (Belovsky and Slade 2000).

10.3 Application of Si/SiO_2 NPs Causes Enhanced Plant Defence Against Herbivory

Insects damage affects 1/5th of worldwide agricultural crop production by eating leaves, extracting plant fluid, as well as by exerting multiple stresses (Aetiba 2015; Sallam 2008). Insect pests also have evolved tolerance to plants poisonous secretions (Damalas 2009; Dos Santos et al. 2015; Korth et al. 2006). The physical health of the plant serving as a host has a significant impact on herbivorous insects host choices (Frew et al. 2018; Huberty and Denno 2004). Inorganic elements like Si can be used to minimize agricultural vulnerability (Meyer and Keeping 2005) and improve plants resistance to insects pests (Debona et al. 2017; Ma and Takahashi 2002). Plants absorb silicon as the compound silicic acid ($Si(OH)_4$) and when pests damage plant tissue, the deposition of Si increases. *Magnaporthe grisea* – affected leaves have a high concentration of Si, which lowers lesion development (Nakata et al. 2008; Sakr 2016) and enhances biotic stress resistance in various plant parts in wheat, rice, maize and sugarcane (Bakhat et al. 2018; Debona et al. 2017). Silicon helps plants withstand insect attack by forming phytoliths in their leaves, which enhance tissue toughness, impair insects mouth parts and diminish foliage digestion (Singh et al. 2020a). It may also have an impact on the insect ability to digest by limiting the carbohydrate and nitrogen supply (Dos Santos et al. 2015; Hunt et al. 2008). The accumulation of Si also induces metabolic modifications that prevent herbivory mediated crop losses (Liang et al. 2015). In general, pesticides containing inactive silica first target insects' midgut cells (Forkpah et al. 2014; Lehane and Billingsley 2012) and making it harder for insects to bite and digest plant tissues (owing to inactive silica) (Coskun et al. 2019; Harrewijn et al. 1990; Massey et al. 2006). Furthermore, Si is also known to play a significant role in phytohormone-mediated plant protection (Singh et al. 2020a). Effects of silicon on most commonly observed herbivory and associated resistance/resilience mechanisms are shown in Table 10.1.

Table 10.1 The effects of using silicon for herbivory and associated resistance mechanisms against herbivory.

S. No	Crop	Scientific name	Insect	Scientific name	Effect	References
1	Bean	*Phaseolus vulgaris* L.	Spider mite	*Tetranychus urticae*	Reduced *T. urticae* egg laying, increased leaflet damage while marginally mitigating *T. urticae* – mediated drops in photosynthesis activities	Islam et al. (2004)
2	Cocoa	*Toxoptera aurantii*	Aphid	*Toxoptera aurantii*	Improved net soluble phenol activity, PSII chlorophyll fluorescence yield (Fv/Fm) and photosynthesis activity. No effect on the structural performance index	Pinto et al. (2012)
3	Collard greens	*Brassica oleracea*	Diamond back moth	*Plutella xylostella*	Nutritional changes caused by silicon and stress in soluble N, glucosinolates, fibre and LWC did not improve insect activity in any feeding guild	Teixeira et al. (2020)
4	Cucumber	*Cucumis sativus* L.	Cotton Bollworm	*Helicoverpa armigera*	The addition of Si increased accumulated Si and the proportion of carbon to nitrogen, while decreasing larval relative intake and growth rate in the in-situ tests	Islam et al. (2020)
5	Grape	*Vitis vinifera* L.	Grapevine moth	*Phalaenoides glycinae*	Si is critical for induced plant defence mechanisms that are activated by herbivore eating or oviposition	Connick (2011)

(Continued)

Table 10.1 (Continued)

S. No	Crop	Scientific name	Insect	Scientific name	Effect	References
6	Maize	*Zea mays* L.	Fall armyworm	*Spodoptera frugiperda*	Si decreased weight (larval, pre-pupal and pupal), survival rate, fecundity and fertility	Nagaratna et al. (2022)
7	Papaya	*Carica papaya* L.	Spotted spider mite	*Tetranychus urticae*	Si promoted the development of plant defence chemicals, which decreased the overall reproduction rates	Christalcatalani et al. (2017)
8	Purple false brome	*Brachypodium distachion*	Cotton bollworm House cricket	*Helicoverpa armigera* *Acheta domesticus*	Changed morphology of specific defensive structures without affecting leaf mechanical properties	Hall et al. (2020)
9	Rice	*Oryza sativa* L.	Asiatic stem borer	*Chilo suppressalis*	Plants supplied with Si had greater Si concentrations than the untreated' plants, which lowered borer weight gain, penetration, stem damage, penetrating time and the larval activity. Plant mortality was positively decreased	Han et al. (2018); Hou and Han (2010)
10	Ryegrass	*Lolium perenne* L.	African Armyworm	*Spodoptera exempta*	Si reduced the ability to digest plant leaves, as well as the functioning of *S. exempta*-ingested food in terms of body mass and nitrogen absorption, resulting in a slower rate of insect development	Massey and Hartley (2009)

11	Soybean	*Glycine max* L.	Budworm	*Helicoverpa punctigera*	Si treatment and herbivory increased the Si percentage in the plant. It also reduced *H. punctigera* relative growth rates	Johnson et al. (2020)
12	Sugarcane	*Saccharum officinarum* L.	Leafhopper	*Pyrilla perpusilla*	The *P. perpusilla* abundance was reduced with Si, whereas parasitism (%) increased. The population was decreased by an increase in *Epiricania melanoleuca* parasitism following Si amendment	Indhumathi et al. (2018)
13	Sugarcane	*Saccharum officinarum* L.	Cane bettle	*Dermolepida albohirtum*	High quantities of silicon in sugarcane roots can impact insect pest feeding, behaviour and performance	Frew et al. (2017)
14	Sunflower	*Helianthus annuus* L.	Bordered patch	*Chlosyne lacinia*	Si application decreased caterpillar weight (first and second week), reduced lignin content, and enhanced element distribution. There was a negative relationship between Si and larvae weight	Assis et al. (2015)
15	Tahiti lime	*Citrus* spp.	Asian citrus psyllid	*Diaphorina citri*	The application of Si for trees and seedlings promoted ACP oviposition, resulting in a 60% loss. Except for the foliar spray, it had little effect on plant macro-micronutrient profiles	Ramírez-Godoy et al. (2018)
16	Tomato	*Solanum lycopersicum* L.	Silver whitefly	*Bemisia tabaci*	Decreased number of immature whiteflies (foliar application more effective than irrigation) with Si treatment	Alyousuf et al. (2022)
17	Wheat	*Triticum aestivum* L.	Pink stem borer	*Sesamia inferens*	Si treatment increased photosynthesis efficiency, productivity and plant biomass	Jeer et al. (2021)

10.4 Cell Signalling During Si-Mediated Induced Plant Defence Against Herbivory

Si accumulation takes place mostly near the cell wall-cell membrane interface, i.e., apoplast, where receptors and effector molecules interact with each other. Plants identify herbivores through the interplay of pattern recognition receptors and HAMPs, which triggers HAMP-mediated immunity. During this process, reactive oxygen species (ROS) and phytohormone signalling pathways are activated, resulting in activated defence responses. Plant-herbivore interactions also incorporates effector-triggered immunity (ETI) (Hogenhout and Bos 2011). ETI is governed by effector molecules present in insects mouth fluids, frass, salivary secretions and oviposition fluids (Hilfiker et al. 2014), and it has been recently discovered that these effector molecules are transported across mesophyll tissue particularly in cytosol and apoplast (Mugford et al. 2016). As a result, it is postulated that Si accumulated in the apoplast might disrupt insect feeding patterns and the release of effectors because effectors are caught in the Si matrix, preventing insects from recognizing the plant as an appropriate host (Coskun et al. 2019). The method through which plants gained this anti-herbivore protection owing to accumulated Si has only been explored in some plants, but the findings indicate that physical, molecular and biochemical mechanisms get activated and react with accumulated Si to boost plant defence.

10.5 Mechanisms Involved in Si-Mediated Plant Defence Against Herbivory

Si improves resistance of plants to insect pests by affecting the direct as well as indirect means of plant resistance. Si accumulation strengthens the plant's physical barriers and also induces molecular and biochemical defence mechanisms by increasing the production of secondary compounds, insecticidal enzymes and defence proteins. When plants are provided with Si, they store it in their numerous body parts (Exley 2015; Guerriero et al. 2016; Ma and Yamaji 2006; McLarnon et al. 2017), which are then eaten by insect pests or herbivores. Greater concentrations of Si in plant tissues lowers plant tissue digestion and palatable qualities, which discourages insect feeding and impairs insect ability by influencing its development and growth (Coskun et al. 2019; Massey and Hartley 2009; Massey et al. 2006; Reynolds et al. 2009). Si also reduces insect herbivores utilization of carbohydrates and amino acids (Dos Santos et al. 2015; Hunt et al. 2008; Massey and Hartley 2009). Furthermore, higher Si levels restrict food availability, causing herbivores to consume more Si-enriched plant parts. Furthermore, it has been found that supplementing maize hybrids with a higher dose of nitrogen fertilizer causes them to deposit lesser amount of Si, thereby making them more sensitive to *Ostrinia nubilalis* (Ramputh et al. 2002).

10.5.1 How Si Stimulates the Direct Mode of Plant Defence Against Phytophagous Insects

One of the first methods of silicon-mediated plant resistance against herbivory is by increasing physical defence by Si deposition (Singh et al. 2020a). Plants absorb and accumulate Si in the form of opaline phytoliths between epidermal walls and beneath stem epidermal cells and leaf cuticle, increasing plant tissue hardness and abrasiveness (Massey et al. 2007). Keeping et al. (2009) used 'energy-dispersive X-ray microanalysis and mapping' to identify Si accumulation in the internodal and root region of Si-treated *Saccharum officinarum* plants. Insects feed on Si-rich plants, which wear down their mouthparts and diminish palatability or ease of digestion, resulting in reduced herbivore development and growth (Singh et al. 2020a). Research on various grass varieties revealed that the form and dimensions of phytoliths and spines impacted the stiffness and palatability of foliage (Hartley et al. 2015). Si buildup in plants also reduced the performance of various herbivores, including true bugs, leaf beetles weevils, bees and stalk borers (Barnes and Giliomee 1992; King et al. 1998; Kvedaras et al. 2009; Roitberg et al. 2005; Schaber et al. 1993). High Si concentration appears to mechanically preserve cells by lowering herbivores capacity to break them: Si-treated grass foliage releases less chlorophyll after crushing, while greater chlorophyll has been found in locusts faeces (*Schistocera gregaria*) that feed on them (Hunt et al. 2008). Si-supplemented plants produce silicified microstructures with varying patterns and distribution throughout the plants tissues (Cooke and Leishman 2011; Hartley et al. 2015; Kumar et al. 2017). Si also improves plant physical or mechanical resistance to insect attacks (Alhousari and Greger 2018; Dorairaj and Ismail 2017; Yang et al. 2017). More recently, the effects of Si-governed leaf mechanistic features were distinguished from Si-governed alterations to leaf surface architecture (Hall et al. 2020). Direct means of Si-based defence is diagrammatically represented in Figure 10.1a.

10.5.2 How Si Stimulates the Indirect Mode of Plant Defence Against Phytophagous Insects

Whenever agricultural crop plants are infested by insect pests, they emit HIPVs which lure predators, pathogens and parasitoids (Mumm and Dicke 2010). This is an induced indirect defence. It has been discovered that Si treatment can strengthen indirect way of inducible protection via boosting and altering HIPVs (Singh et al. 2020a). When attacked by *Helicoverpa armigera*, Si-treated cucumber plants lured greater numbers of predators than untreated plants in both test and control trials (Kvedaras et al. 2010). Similarly, Si-treated *Oryza* sp. infected with *Cnaphalocrocis medinalis* has also lured greater numbers of the parasite *Trathala flavororbitalis* compared to control group (Leroy et al. 2019). Likewise, plants supplied with Si and infested by *Epiphyas postvittana* were able to draw greater number predator's prey, i.e., *Dicranolaius bellulus* (Leroy et al. 2019).

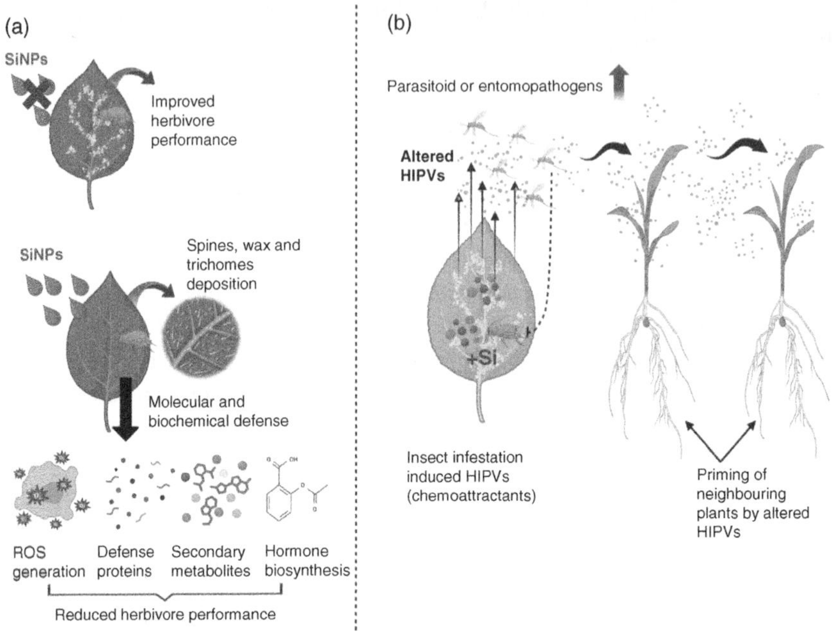

Figure 10.1 Direct and indirect modes of silicon-mediated plant defence against herbivorous insect pests. (a) Direct defence mechanisms: In conditions of limited silicon (Si) uptake, plants show reduced defence capability, resulting in improved herbivore performance. However, with adequate Si supply, plants exhibit enhanced physical barriers through increased deposition of spines, wax and trichomes. Plants also activate molecular and biochemical defences including ROS generation, defence proteins, secondary metabolites and hormone biosynthesis, collectively leading to reduced herbivore performance. (b) Indirect defence mechanisms: Si-enriched plants under insect infestation release altered HIPVs that serve as chemoattractants for natural enemies like parasitoids and entomopathogens. These altered HIPVs also act as warning signals to prime neighbouring plants for potential attacks. *Source*: Created in https://BioRender.com.

Furthermore, HIPVs also serve as a warning signal for damaged areas of plants and surrounding plants, as well as act as insect retardants (Dudareva et al. 2006; Frost et al. 2008; Turlings and Erb 2018). Si increases the formation of indoles, an established defensive priming compound. Si stimulates HIPVs synthesis in both injured and unaffected plant parts, as well as in the nearby plants, via a JA-driven route (Singh et al. 2020a). Indirect means of Si-based defence is diagrammatically represented in Figure 10.1b.

10.5.3 Effect of Si on Molecular and Biochemical Defence Against Herbivory

Although a number of studies have confirmed that Si treatment triggers bio-molecular defences signalling (Fauteux et al. 2006), there has been very less exploration in the terms of Si-mediated changes in the genome expression patterns during herbivory (Reynolds et al. 2016). Si treatment was found to increase the expression or activity of defensive genes and enzymes, including β-1,3-glucanases, polyphenol oxidase (PPO), chitinase, peroxidase (POD) and phenylalanine ammonia-lyase (PAL) (Cai et al. 2008; Correa et al. 2005; Kvedaras et al. 2009; Liang et al. 2003; Reynolds et al. 2009; Singh et al. 2020a). Transcriptome study in Arabidopsis and rice indicated a useful tool for investigating overall transcriptional alterations in the aftermath of Si treatment (Fauteux et al. 2006; Ye et al. 2013). Transcriptome analysis showed that herbivory or insect pest attack resulted in increased expression of OsAOS and OsCOI1 transcript levels, mitogen-activated protein kinases (OsMPK3 and OsMPK6), WRKY transcription factors (OsWRKY53 and OsWRKY70) and defensive genes like peroxidase (POD), PPO and trypsin proteinase inhibitor (TrypPI) in Si-treated plants when compared with plants that were untreated (Singh et al. 2020a). A study in rice revealed that both herbivory and external treatment with methyl JA increased levels of OsAOS and OsCOI1 transcripts compared to control plants, implying that Si-treatment allows a rapid and elevated JA-driven protection against insect attack (Singh et al. 2020a). Si application boosted the levels of Si channels 'OsLsi1, OsLsi2 and OsLsi6', which were then improved by methyl JA application and insect feeding. Omics approaches are now used to analyse defence responses in the primed plants (Balmer et al. 2015); however, this strategy is used very sparingly to investigate the usefulness Si in plant defence mechanisms against insect pest assaults (Singh et al. 2020a).

10.5.4 Effect of Si on Chemical Defence Against Herbivory

Plants have a diverse range of chemical defences involving secondary metabolites like phenolics, terpenoids, flavonoids, alkaloids and defensive proteins (Hartley and Jones 1996). Chemical defences deter insect pests by harming them, which discourages feeding, and reduces their growth and reproduction. Si application promotes the development of leaves rich in secondary metabolic compounds, hence enhancing plant pest resistance (Reynolds et al. 2016). This is accomplished through Si stimulating the activity of genes and proteins associated with the synthesis of secondary metabolic defensive compounds (Cai et al. 2008; Fawe et al. 1998; Guerriero et al. 2018; Liang et al. 2003; Rémus-Borel et al. 2005; Rodrigues et al. 2004). These herbivore deterring proteins and metabolites degrade the nutritional content, efficiency of conversion and uptake of food by insects, reducing their productivity and fertility (Singh et al. 2020a). Si increases both the activity and production of lipoxygenases, peroxidase (POX), 1,3-glucanase proteinase, polyphenol oxidase (PPO) and phenyl ammonia lyase (PAL), all produce secondary metabolites. This promotes lignification and suberinization, as well as the formation of phenols and quinones (Singh et al. 2020a). These metabolites with anti-nutritional, deterring and lethal characteristics improve the mechanical barriers of plants against invaders. Secondary compounds like terpenoids, alkaloids and flavonoids are found in glandular trichomes, where they function to repel encroaching herbivores (Singh et al. 2020a).

10.5.5 Si-Mediated ROS Scavenging During Herbivory

Si-mediated abiotic and biotic stress tolerance is linked to enhanced ROS removal and activity of antioxidants (Van Bockhaven et al. 2013). ROS production is a typical event during insect pest assault, and it plays numerous functions in the plant defensive pathway (Yang et al. 2017). ROS acts as a messenger that stimulates defensive proteins (Van Bockhaven et al. 2013, 2015), activates secondary compound biosynthesis (Thoma et al. 2003) and interacts with plant hormones (SA and JA), thereby reducing herbivore attacks (Glazebrook 2005; León et al. 1995; Ramputh et al. 2002; Torres 2010). However, excess output of ROS can induce oxidative stress; thus, the induction of antioxidative pathway and the upkeep of appropriate ROS levels are required to avoid damage from oxidative stress and to maintain the ROS-governed defence response functional (Domiciano et al. 2015; Yang et al. 2017). Si regulates oxidative homeostasis by increasing the expression of peroxidases, catalases, ascorbate peroxidase and superoxide dismutase. As per Van Bockhaven et al. (2013, 2015), photorespiration and production of ROS play a role in Si's positive effects, which include an increased defensive response to biotic stresses. There is very less research done that illustrate the interaction of Si with antioxidants in response to biotic stresses, indicating that additional research into the workings underlying this phenomenon will be beneficial.

10.6 Conclusion and Future Prospective

SiNPs have emerged as an important component in sustainable agriculture, notably for improving the defence of plants against herbivory via both direct and indirect pathways. Their potential to enhance plant defence while preserving ecological equilibrium makes them appealing alternatives to traditional pesticides. The International Plant Nutrition Institute's designation of silicon as a 'beneficial element', as well as its inclusion as a 'biostimulant' under EU laws, demonstrates its expanding relevance in modern agriculture (Laane 2018). Despite this awareness, the optimum use of silicon/silica nanoparticles (Si/SiO_2 NPs) must address a number of significant issues. A growing body of research shows that Si accumulation varies greatly between plant species and genotypes, depending on parameters such as soil type, weather conditions and treatment methods (Keeping 2017; Tubana et al. 2016). Understanding these differences is critical for creating successful Si-based agricultural methods that can be tailored to certain agricultural systems and specific environmental conditions.

Several key elements must be thoroughly investigated in order to maximize Si/SiO_2 NPs usage in agriculture. The primary goal is to create effective green synthesis procedures using agricultural waste, which might drastically lower production costs while also assisting with agricultural waste management. The intricate connections between plant traits and SiNPs qualities require careful investigation, since variables such as species of plant, stage of development and plant building blocks have a substantial impact on Si absorption and accumulation. Similarly, SiNP features such as size, surface charge and specific modifications impact their absorption and translocation inside plants (Ma et al. 2007; Soininen et al. 2013). The molecular processes behind SiNP-induced broad spectrum resistance deserve further exploration, including the expression and control of Si transport genes such as Lsi1 and Lsi2 (Singh et al. 2020a). Modern genomic, proteomic and metabolomic methods will be required to untangle these complicated relationships (Jang et al. 2018; Reynolds et al. 2016). The genetic base for Si accumulation presents intriguing options for crop enhancement using marker-assisted breeding and genome editing technologies, which might lead to the production of crops with improved Si absorption capacity. This is especially useful for crops that naturally gather low amounts of silicon and yet are vulnerable to herbivores (Guo et al. 2014; McLarnon et al. 2017). Furthermore, standardization of application procedures, assessment of possible phytotoxicity and compatibility with existing pest management systems are critical issues that require study. The function of Si in boosting plant resistance to multiple environmental stresses also shows that it has promise for use in climate-smart agriculture, while further field research is needed to replicate laboratory results under real-world settings (Johnson et al. 2018).

Looking ahead, the emerging area of nanotechnology continues to provide up new opportunities for Si delivery and usage in agriculture. Future research should focus on enhancing delivery systems through surface changes and increased absorption efficiency (Meharg and Meharg 2015). Si/SiO_2 NPs have the potential to be used in sustainable agriculture for uses other than herbivory management, such as plant stress tolerance and productivity augmentation. According to the current research, Si can boost plant tolerance to a variety of environmental challenges, indicating potential implications in climate change adaptation techniques. However, reaping these advantages necessitates close attention to soil chemistry and plant physiology, since the efficacy of Si treatments varies greatly depending on soil conditions and plant type (Haynes 2014). The development of innovative Si-based formulations and application technologies has the potential to alter present agricultural practices, particularly in areas where conventional pest management measures are limited or cause environmental problems. The incorporation of Si/SiO_2 NPs into current pest management tactics necessitates careful evaluation of possible synergistic effects with other control measures, which might result in more sustainable and effective pest management systems.

An interdisciplinary approach that combines molecular biology, materials science and agricultural engineering will be critical to achieving the full potential of Si-based technology. This comprehensive approach, when combined with thorough investigation of environmental effects and economic feasibility, will pave the way for more sustainable and resilient agricultural systems. Future study should concentrate on improving Si delivery methods, understanding the long-term impacts of Si/SiO_2 NPs applications on soil health and ecosystem functioning, and generating crop-specific Si supplementation guidelines. These improvements will be vital in tackling present agricultural difficulties while preserving ecological balance and food availability in the face of changing climates and increased insect pest or herbivory pressure.

References

Aetiba, N. and Peter, J.2015). Field evaluation of levo 2.4 sl (Botanical) for the Management of Insect Pests of Eggplant (Solanum melongena L.) and Okra (Abelmoschus esculentus L. Moench).

Alhousari, F. and Greger, M. (2018). Silicon and mechanisms of plant resistance to insect pests. *Plants* 7: 33.

Aljbory, Z. and Chen, M. (2018). Indirect plant defense against insect herbivores: a review. *Insect Science* 25: 2–23.

Alyousuf, A., Hamid, D., Desher, M.A. et al. (2022). Effect of silicic acid formulation (silicon 0.8%) on two major insect pests of tomato under greenhouse conditions. *Silicon* 14: 3019–3025.

Assis, F.A., Moraes, J.A., Assis, G.A., and Parolin, F.J.T. (2015). Induction of caterpillar resistance in sunflower using silicon and acibenzolar-S-methyl. *Journal of Agricultural Science and Technology* 17: 543–550.

Bakhat, H.F., Bibi, N., Zia, Z. et al. (2018). Silicon mitigates biotic stresses in crop plants: a review. *Crop Protection* 104: 21–34.

Balmer, A., Pastor, V., Gamir, J. et al. (2015). The 'prime-ome': towards a holistic approach to priming. *Trends in Plant Science* 20: 443–452.

Bansal, K., Hooda, V., Verma, N. et al. (2022). Stress alleviation and crop improvement using silicon nanoparticles in agriculture: a review. *Silicon* 14: 10173–10186.

Barnes, B.N. and Giliomee, J.H. (1992). Fruit-feeding behaviour of banded fruit weevil, *Phlyctinus callosus* (Schönherr) (Col., Curculionidae), in apple orchards. *Journal of Applied Entomology* 113: 407–415.

Belovsky, G.E. and Slade, J.B. (2000). Insect herbivory accelerates nutrient cycling and increases plant production. *Proceedings of the National Academy of Sciences of the United States of America* 97: 14412–14417.

Cai, K., Gao, D., Luo, S. et al. (2008). Physiological and cytological mechanisms of silicon-induced resistance in rice against blast disease. *Physiologia Plantarum* 134: 324–333.

Chérif, M., Asselin, A., and Bélanger, R.R. (1994). Defense responses induced by soluble silicon in cucumber roots infected by *Pythium* spp. *Phytopathology* 84: 236–242.

Christalcatalani, G., Vieira, M.R., Camargos, L.S.D.E. et al. (2017). Effects of potassium silicate application on papaya plants in the twospotted spider mite population. *Revista Brasileira de Fruticultura* 39: e-840.

Connick, V.J. (2011). The impact of silicon fertilisation on the chemical ecology of grapevine, *Vitis vinifera*; constitutive and induced chemical defences against arthropod pests and their natural enemies. Master thesis. Charles Sturt University.

Cooke, J. and Leishman, M.R. (2011). Silicon concentration and leaf longevity: is silicon a player in the leaf dry mass spectrum? *Functional Ecology* 25: 1181–1188.

Correa, R.S.B., Moraes, J.C., Auad, A.M., and Carvalho, G.A. (2005). Silicon and acibenzolar-S-methyl as resistance inducers in cucumber, against the whitefly *Bemisia tabaci* (Gennadius) (Hemiptera: Aleyrodidae) biotype B. *Neotropical Entomology* 34: 429–433.

Coskun, D., Deshmukh, R., Sonah, H. et al. (2019). The controversies of silicon's role in plant biology. *The New Phytologist* 221: 67–85.

Damalas, C.A. (2009). Understanding benefits and risks of pesticide use. *Scientific Research and Essays* 4: 945–949.

DeAngelis, D.L. (1992). *Dynamics of Nutrient Cycling and Food Webs*. London, UK: Chapman & Hall.

Debona, D., Rodrigues, F.A., and Datnoff, L.E. (2017). Silicon's role in abiotic and biotic plant stresses. *Annual Review of Phytopathology* 55: 85–107.

Dicke, M., van Poecke, R.M.P., and de Boer, J.G. (2003). Inducible indirect defence of plants: from mechanisms to ecological functions. *Basic and Applied Ecology* 4: 27–42.

Domiciano, G.P., Cacique, I.S., Chagas Freitas, C. et al. (2015). Alterations in gas exchange and oxidative metabolism in rice leaves infected by *Pyricularia oryzae* are attenuated by silicon. *Phytopathology* 105: 738–747.

Dorairaj, D. and Ismail, M.R. (2017). Distribution of silicified microstructures, regulation of cinnamyl alcohol dehydrogenase and lodging resistance in silicon and paclobutrazol mediated *Oryza sativa*. *Frontiers in Physiology* 8: 491.

Dos Santos, M.C., Junqueira, M.R.R., de Sá, V.G.M. et al. (2015). Effect of silicon on the morphology of the midgut and mandible of tomato leafminer *Tuta absoluta* (Lepidoptera: Gelechiidae) larvae. *Invertebrate Survival Journal* 12: 158–165.

Dudareva, N., Negre, F., Nagegowda, D.A., and Orlova, I. (2006). Plant volatiles: recent advances and future perspectives. *Critical Reviews in Plant Sciences* 25: 417–440.

Epstein, E. (1994). The anomaly of silicon in plant biology. *Proceedings of the National Academy of Sciences of the United States of America* 91: 11–17.

Exley, C. (2015). A possible mechanism of biological silicification in plants. *Frontiers in Plant Science* 6: 853.

Fauteux, F., Chain, F., Belzile, F. et al. (2006). The protective role of silicon in the Arabidopsis–powdery mildew pathosystem. *Proceedings of the National Academy of Sciences of the United States of America* 103: 17554–17559.

Fawe, A., Abou-Zaid, M., Menzies, J.G., and Bélanger, R.R. (1998). Silicon-mediated accumulation of flavonoid phytoalexins in cucumber. *Phytopathology* 88: 396–401.

Forkpah, C., Dixon, L.R., Fahrbach, S.E., and Rueppell, O. (2014). Xenobiotic effects on intestinal stem cell proliferation in adult honey bee (*Apis mellifera* L.) workers. *PLoS One* 9: e91180.

Frew, A., Powell, J.R., Hiltpold, I. et al. (2017). Host plant colonisation by arbuscular mycorrhizal fungi stimulates immune function whereas high root silicon concentrations diminish growth in a soil-dwelling herbivore. *Soil Biology and Biochemistry* 112: 117–126.

Frew, A., Weston, L.A., Reynolds, O.L., and Gurr, G.M. (2018). The role of silicon in plant biology: a paradigm shift in research approach. *Annals of Botany* 121: 1265–1273.

Frost, C.J., Mescher, M.C., Carlson, J.E., and De Moraes, C.M. (2008). Plant defense priming against herbivores: getting ready for a different battle. *Plant Physiology* 146: 818–824.

Glazebrook, J. (2005). Contrasting mechanisms of defense against biotrophic and necrotrophic pathogens. *Annual Review of Phytopathology* 43: 205–227.

Guerriero, G., Hausman, J.-F., and Legay, S. (2016). Silicon and the plant extracellular matrix. *Frontiers in Plant Science* 7: 463.

Guerriero, G., Berni, R., Muñoz-Sanchez, J.A. et al. (2018). Production of plant secondary metabolites: examples, tips and suggestions for biotechnologists. *Genes (Basel)* 9: 309.

Guo, L., Gao, Z., and Qian, Q. (2014). Application of resequencing to rice genomics, functional genomics and evolutionary analysis. *Rice* 7 (1): 1–10.

Hall, C.R., Dagg, V., Waterman, J.M., and Johnson, S.N. (2020). Silicon alters leaf surface morphology and suppresses insect herbivory in a model grass species. *Plants* 9: 643.

Han, Y.-Q., Wen, J.-H., Peng, Z.-P. et al. (2018). Effects of silicon amendment on the occurrence of rice insect pests and diseases in a field test. *Journal of Integrative Agriculture* 17: 2172–2181.

Harrewijn, P., Campbell, R.K., and Eikenbary, R.D. (1990). *Aphid-Plant Genotype Interactions*. Amsterdam: Elsevier.

Hartley, S.E. and Jones, C.G. (1996). Plant chemistry and herbivory, or why the world is green. *Plant Ecology* 284–324.

Hartley, S.E., Fitt, R.N., McLarnon, E.L., and Wade, R.N. (2015). Defending the leaf surface: intra-and inter-specific differences in silicon deposition in grasses in response to damage and silicon supply. *Frontiers in Plant Science* 6: 35.

Haynes, R.J. (2014). A contemporary overview of silicon availability in agricultural soils. *Journal of Plant Nutrition and Soil Science* 177: 831–844.

Hilfiker, O., Groux, R., Bruessow, F. et al. (2014). Insect eggs induce a systemic acquired resistance in Arabidopsis. *The Plant Journal* 80: 1085–1094.

Hogenhout, S.A. and Bos, J.I.B. (2011). Effector proteins that modulate plant–insect interactions. *Current Opinion in Plant Biology* 14: 422–428.

Hou, M. and Han, Y. (2010). Silicon-mediated rice plant resistance to the Asiatic rice borer (Lepidoptera: Crambidae): effects of silicon amendment and rice varietal resistance. *Journal of Economic Entomology* 103: 1412–1419.

Howe, G.A. and Jander, G. (2008). Plant immunity to insect herbivores. *Annual Review of Plant Biology* 59: 41–66.

Huberty, A.F. and Denno, R.F. (2004). Plant water stress and its consequences for herbivorous insects: a new synthesis. *Ecology* 85: 1383–1398.

Hunt, J.W., Dean, A.P., Webster, R.E. et al. (2008). A novel mechanism by which silica defends grasses against herbivory. *Annals of Botany* 102: 653–656.

Hutchinson, G.E. and Deevey, E.S. Jr., (1949). Ecological studies on populations. In: *Survey of Biological Progress*, vol. 1, 325–359. Elsevier.

Indhumathi, V.S., Chandramani, P., Jayaraj, J. et al. (2018). Impact of silicon on sugarcane leafhopper, *Pyrilla perpusilla* Walker by enhancing parasitism of *Epiricania melanoleuca* (Fletcher). *Journal of Biological Control* 32: 155–159.

Islam, T., Moore, B.D., and Johnson, S.N. (2020). Novel evidence for systemic induction of silicon defences in cucumber following attack by a global insect herbivore. *Ecological Entomology* 45: 1373–1381.

Islam, T., Moore, B.D., and Johnson, S.N. (2022, 2004). Silicon suppresses a ubiquitous mite herbivore and promotes natural enemy attraction by altering plant volatile blends. *Journal of Pest Science* 95: 423–434.

Jang, S.-W., Kim, Y., Khan, A.L. et al. (2018). Exogenous short-term silicon application regulates macro-nutrients, endogenous phytohormones, and protein expression in *Oryza sativa* L. *BMC Plant Biology* 18: 1–12.

Jeelani, P.G., Mulay, P., Venkat, R., and Ramalingam, C. (2020). Multifaceted application of silica nanoparticles. A review. *Silicon* 12: 1337–1354.

Jeer, M., Yele, Y., Sharma, K.C., and Prakash, N.B. (2021). Exogenous application of different silicon sources and potassium reduces pink stem borer damage and improves photosynthesis, yield and related parameters in wheat. *Silicon* 13: 901–910.

Ji, X., Wang, H., Song, B. et al. (2018). Silicon nanomaterials for biosensing and bioimaging analysis. *Frontiers in Chemistry* 6: 38.

Johnson, S.N., Ryalls, J.M.W., Gherlenda, A.N. et al. (2018). Benefits from below: silicon supplementation maintains legume productivity under predicted climate change scenarios. *Frontiers in Plant Science* 9: 202.

Johnson, S.N., Rowe, R.C., and Hall, C.R. (2020). Silicon is an inducible and effective herbivore defence against *Helicoverpa punctigera* (Lepidoptera: Noctuidae) in soybean. *Bulletin of Entomological Research* 110: 417–422.

Keeping, M.G. (2017). Uptake of silicon by sugarcane from applied sources may not reflect plant-available soil silicon and total silicon content of sources. *Frontiers in Plant Science* 8: 760.

Keeping, M.G., Kvedaras, O.L., and Bruton, A.G. (2009). Epidermal silicon in sugarcane: cultivar differences and role in resistance to sugarcane borer *Eldana saccharina*. *Environmental and Experimental Botany* 66: 54–60.

Kessler, A. and Baldwin, I.T. (2002). Plant responses to insect herbivory: the emerging molecular analysis. *Annual Review of Plant Biology* 53: 299–328.

King, B.H., Crowe, M.L., and Blackmore, M.D. (1998). Effects of leaf age on oviposition and on offspring fitness in the imported willow leaf beetle *Plagiodera versicolora* (Coleoptera: Chrysomelidae). *Journal of Insect Behavior* 11: 23–36.

Korth, K.L., Doege, S.J., Park, S.-H. et al. (2006). *Medicago truncatula* mutants demonstrate the role of plant calcium oxalate crystals as an effective defense against chewing insects. *Plant Physiology* 141: 188–195.

Kumar, S., Milstein, Y., Brami, Y. et al. (2017). Mechanism of silica deposition in sorghum silica cells. *The New Phytologist* 213: 791–798.

Kvedaras, O.L., Byrne, M.J., Coombes, N.E., and Keeping, M.G. (2009). Influence of plant silicon and sugarcane cultivar on mandibular wear in the stalk borer *Eldana saccharina*. *Agricultural and Forest Entomology* 11: 301–306.

Kvedaras, O.L., An, M., Choi, Y.S., and Gurr, G.M. (2010). Silicon enhances natural enemy attraction and biological control through induced plant defences. *Bulletin of Entomological Research* 100: 367–371.

Laane, H.-M. (2018). The effects of foliar sprays with different silicon compounds. *Plants* 7: 45.

Lehane, M. and Billingsley, P. (2012). *Biology of the Insect Midgut*. Springer Science & Business Media.

León, J., Lawton, M.A., and Raskin, I. (1995). Hydrogen peroxide stimulates salicylic acid biosynthesis in tobacco. *Plant Physiology* 108: 1673–1678.

Leroy, N., De Tombeur, F., Walgraffe, Y. et al. (2019). Silicon and plant natural defenses against insect pests: impact on plant volatile organic compounds and cascade effects on multitrophic interactions. *Plants* 8: 444.

Liang, Y., Chen, Q.I.N., Liu, Q. et al. (2003). Exogenous silicon (Si) increases antioxidant enzyme activity and reduces lipid peroxidation in roots of salt-stressed barley (*Hordeum vulgare* L.). *Journal of Plant Physiology* 160: 1157–1164.

Liang, Y., Nikolic, M., Bélanger, R. et al. (2015). Silicon-mediated tolerance to salt stress. In: *Silicon in Agriculture: From Theory to Practice*, 123–142.

Lovett, G.M. and Ruesink, A.E. (1995). Carbon and nitrogen mineralization from decomposing gypsy moth frass. *Oecologia* 104: 133–138.

Ma, J.F. and Takahashi, E. (2002). *Soil, Fertilizer, and Plant Silicon Research in Japan*. Amsterdam: Elsevier.

Ma, J.F. and Yamaji, N. (2006). Silicon uptake and accumulation in higher plants. *Trends in Plant Science* 11: 392–397.

Ma, J.F., Yamaji, N., Tamai, K., and Mitani, N. (2007). Genotypic difference in silicon uptake and expression of silicon transporter genes in rice. *Plant Physiology* 145: 919–924.

Mahawar, L., Ramasamy, K.P., Suhel, M. et al. (2023). Silicon nanoparticles: comprehensive review on biogenic synthesis and applications in agriculture. *Environmental Research* 232: 116292.

Massey, F.P. and Hartley, S.E. (2009). Physical defences wear you down: progressive and irreversible impacts of silica on insect herbivores. *The Journal of Animal Ecology* 78: 281–291.

Massey, F.P., Ennos, A.R., and Hartley, S.E. (2006). Silica in grasses as a defence against insect herbivores: contrasting effects on folivores and a phloem feeder. *The Journal of Animal Ecology* 75: 595–603.

Massey, F.P., Ennos, A.R., and Hartley, S.E. (2007). Grasses and the resource availability hypothesis: the importance of silica-based defences. *Journal of Ecology* 414–424.

Mathur, P. and Roy, S. (2020). Nanosilica facilitates silica uptake, growth and stress tolerance in plants. *Plant Physiology and Biochemistry* 157: 114–127.

McLarnon, E., McQueen-Mason, S., Lenk, I., and Hartley, S.E. (2017). Evidence for active uptake and deposition of Si-based defenses in tall fescue. *Frontiers in Plant Science* 8: 1199.

McNaughton, S.J., Ruess, R.W., and Seagle, S.W. (1988). Large mammals and process dynamics in African ecosystems. *Bioscience* 38: 794–800.

McNaughton, S.J. and Tarrants, J.L. (1983). Grass leaf silicification: natural selection for an inducible defense against herbivores. *Proceedings of the National Academy of Sciences of the United States of America* 80: 790–791.

Meharg, C. and Meharg, A.A. (2015). Silicon, the silver bullet for mitigating biotic and abiotic stress, and improving grain quality, in rice? *Environmental and Experimental Botany* 120: 8–17.

Meyer, J.H. and Keeping, M.G. (2005). Impact of silicon in alleviating biotic stress in sugarcane in South Africa. Sugar Cane International 23, 14–18.

Mithöfer, A. and Boland, W. (2012). Plant defense against herbivores: chemical aspects. *Annual Review of Plant Biology* 63: 431–450.

Mugford, S.T., Barclay, E., Drurey, C. et al. (2016). An immuno-suppressive aphid saliva protein is delivered into the cytosol of plant mesophyll cells during feeding. *Molecular Plant-Microbe Interactions* 29: 854–861.

Mumm, R. and Dicke, M. (2010). Variation in natural plant products and the attraction of bodyguards involved in indirect plant defense. *Canadian Journal of Zoology* 88: 628–667.

Nagaratna, W., Kalleshwaraswamy, C.M., Dhananjaya, B.C. et al. (2022). Effect of silicon and plant growth regulators on the biology and fitness of fall armyworm, Spodoptera frugiperda, a recently invaded pest of maize in India. *Silicon* 14: 783–793.

Naidu, S., Pandey, J., Mishra, L.C. et al. (2023). Silicon nanoparticles: synthesis, uptake and their role in mitigation of biotic stress. *Ecotoxicology and Environmental Safety* 255: 114783.

Nakata, Y., Ueno, M., Kihara, J. et al. (2008). Rice blast disease and susceptibility to pests in a silicon uptake-deficient mutant lsi1 of rice. *Crop Protection* 27: 865–868.

Oerke, E.-C. (2006). Crop losses to pests. *The Journal of Agricultural Science* 144: 31–43.

Paré, P.W. and Tumlinson, J.H. (1997). De novo biosynthesis of volatiles induced by insect herbivory in cotton plants. *Plant Physiology* 114: 1161–1167.

Pastor, J. and Naiman, R.J. (1992). Selective foraging and ecosystem processes in boreal forests. *The American Naturalist* 139: 690–705.

Pastor, J., Naiman, R.J., Dewey, B., and McInnes, P. (1988). Moose, microbes, and the boreal forest. *Bioscience* 38: 770–777.

Pinto, D.G., Aguilar, M.A.G., Souza, C.A.S. et al. (2012). Alterações fisiológicas após aplicação de silício em cacau e sua influência na preferência por pulgões. *Revista Ceres* 59: 360–367.

Ramírez-Godoy, A., del Pilar Vera-Hoyos, M., Jiménez-Beltrán, N., and Restrepo-Díaz, H. (2018). Effect of potassium silicate application on populations of Asian citrus psyllid in tahiti lime. *HortTechnology* 28: 684–691.

Ramputh, A.I., Arnason, J.T., Cass, L., and Simmonds, J.A. (2002). Reduced herbivory of the European corn borer (*Ostrinia nubilalis*) on corn transformed with germin, a wheat oxalate oxidase gene. *Plant Science* 162: 431–440.

Rastogi, A., Tripathi, D.K., Yadav, S. et al. (2019). Application of silicon nanoparticles in agriculture. *3 Biotech* 9: 1–11.

Rehman, M.Z., Rizwan, M., Ghafoor, A. et al. (2015). Effect of inorganic amendments for in situ stabilization of cadmium in contaminated soils and its phyto-availability to wheat and rice under rotation. *Environmental Science and Pollution Research* 22: 16897–16906.

Rémus-Borel, W., Menzies, J.G., and Bélanger, R.R. (2005). Silicon induces antifungal compounds in powdery mildew-infected wheat. *Physiological and Molecular Plant Pathology* 66: 108–115.

Reynolds, O.L., Keeping, M.G., and Meyer, J.H. (2009). Silicon-augmented resistance of plants to herbivorous insects: a review. *The Annals of Applied Biology* 155: 171–186.

Reynolds, O.L., Padula, M.P., Zeng, R., and Gurr, G.M. (2016). Silicon: potential to promote direct and indirect effects on plant defense against arthropod pests in agriculture. *Frontiers in Plant Science* 7: 744.

Rodrigues, F.Á., McNally, D.J., Datnoff, L.E. et al. (2004). Silicon enhances the accumulation of diterpenoid phytoalexins in rice: a potential mechanism for blast resistance. *Phytopathology* 94: 177–183.

Roitberg, B.D., Gillespie, D.R., Quiring, D.M.J. et al. (2005). The cost of being an omnivore: mandible wear from plant feeding in a true bug. *Naturwissenschaften* 92: 431–434.

Sakr, N. (2016). The role of silicon (Si) in increasing plant resistance against fungal diseases. *Hellenic Plant Protection Journal* 9: 1–15.

Sallam, M.N. (2008). *Insect Damage: Damage on Post-Harvest*. AGSI/FAO: INPhO.

Schaber, B.D., Kokko, E.G., Entz, T., and Richards, K.W. (1993). The effect of spring burning of seed alfalfa fields on leaf characteristics and on mandibular wear of female leafcutter bees [*Megachile rotundata* (F.)] (Hymenoptera: Megachilidae). *Canadian Entomologist* 125: 881–886.

Seastedt, T.R. and Crossley, D.A. Jr., (1984). The influence of arthropods on ecosystems. *Bioscience* 34: 157–161.

Singh, A., Kumar, A., Hartley, S., and Singh, I.K. (2020a). Silicon: its ameliorative effect on plant defense against herbivory. *Journal of Experimental Botany* 71 (21): 6730–6743. https://doi.org/10.1093/jxb/eraa300.

Singh, S., Singh, A., Kumar, S. et al. (2020b). Protease inhibitors: recent advancement in its usage as a potential biocontrol agent for insect pest management. *Insect Science* 27: 186–201.

Soininen, E.M., Bråthen, K.A., Jusdado, J.G.H. et al. (2013). More than herbivory: levels of silica-based defences in grasses vary with plant species, genotype and location. *Oikos* 122: 30–41.

Teixeira, N.C., Valim, J.O.S., Oliveira, M.G.A., and Campos, W.G. (2020). Combined effects of soil silicon and drought stress on host plant chemical and ultrastructural quality for leaf-chewing and sap-sucking insects. *Journal of Agronomy and Crop Science* 206: 187–201.

Thoma, I., Loeffler, C., Sinha, A.K. et al. (2003). Cyclopentenone isoprostanes induced by reactive oxygen species trigger defense gene activation and phytoalexin accumulation in plants. *The Plant Journal* 34: 363–375.

Torres, M.A. (2010). ROS in biotic interactions. *Physiologia Plantarum* 138: 414–429.

Tubana, B.S., Babu, T., and Datnoff, L.E. (2016). A review of silicon in soils and plants and its role in US agriculture: history and future perspectives. *Soil Science* 181: 393–411.

Turlings, T.C.J. and Erb, M. (2018). Tritrophic interactions mediated by herbivore-induced plant volatiles: mechanisms, ecological relevance, and application potential. *Annual Review of Entomology* 63: 433–452.

Van Bockhaven, J., De Vleesschauwer, D., and Höfte, M. (2013). Towards establishing broad-spectrum disease resistance in plants: silicon leads the way. *Journal of Experimental Botany* 64: 1281–1293.

Van Bockhaven, J., Spíchal, L., Novák, O. et al. (2015). Silicon induces resistance to the brown spot fungus *Cochliobolus miyabeanus* by preventing the pathogen from hijacking the rice ethylene pathway. *The New Phytologist* 206: 761–773.

Verhage, A., van Wees, S.C.M., and Pieterse, C.M.J. (2010). Plant immunity: it's the hormones talking, but what do they say? *Plant Physiology* 154: 536–540.

Verma, K.K., Zeng, Y., Song, X.-P. et al. (2022). Nanosilicon: an approach for abiotic stress mitigation and sustainable agriculture. *Frontiers in Plant Science* 13: 1025974.

Wang, L., Ning, C., Pan, T., and Cai, K. (2022). Role of silica nanoparticles in abiotic and biotic stress tolerance in plants: a review. *International Journal of Molecular Sciences* 23: 1947.

Waterman, J.M., Cazzonelli, C.I., Hartley, S.E., and Johnson, S.N. (2019). Simulated herbivory: the key to disentangling plant defence responses. *Trends in Ecology & Evolution* 34: 447–458.

Yan, G., Huang, Q., Zhao, S. et al. (2024). Silicon nanoparticles in sustainable agriculture: synthesis, absorption, and plant stress alleviation. *Frontiers in Plant Science* 15: 1393458.

Yang, L., Han, Y., Li, P. et al. (2017). Silicon amendment is involved in the induction of plant defense responses to a phloem feeder. *Scientific Reports* 7: 4232.

Ye, M., Song, Y., Long, J. et al. (2013). Priming of jasmonate-mediated antiherbivore defense responses in rice by silicon. *Proceedings of the National Academy of Sciences of the United State of America* 110: E3631–E3639.

11

Role of Silicon and Silicon-Based Nanoparticles in Mitigation of Heat and Drought Stresses

Adeeb Rahman[1], Venkataraja V. Mokshith[2], Shashi Ranjan[3], Sumit Kumar Kashyap[4], Aman Prakash[5], Ayyagari Ramlal[6,7], Dhandapani Raju[8], Ambika Rajendran[7] and Sreeramanan Subramaniam[6,9,10]

[1]*Plant RNAi Biology Group, International Centre for Genetic Engineering and Biotechnology, New Delhi, Delhi, India*
[2]*Department of Molecular Biology and Biotechnology, S.V Agricultural College, Acharya NG Ranga Agricultural University (ANGRAU), Tirupati, Andhra Pradesh, India*
[3]*Amrita Centre for Nanosciences and Molecular Medicine, Amrita Vishwa Vidyapeetham, Kochi, Kerala, India*
[4]*Department of Molecular and Cellular Engineering, Jacob Institute of Biotechnology and Bioengineering, Sam Higginbottom University of Agriculture, Technology and Sciences, Prayagraj, Uttar Pradesh, India*
[5]*Department of Marine Biology, Leon H. Charney School of Marine Sciences, University of Haifa, Haifa, Israel*
[6]*Plant Biotechnology Laboratory, School of Biological Sciences, Universiti Sains Malaysia (USM), Penang, Malaysia*
[7]*Division of Genetics, ICAR-Indian Agricultural Research Institute (IARI), Pusa Campus, New Delhi, Delhi, India*
[8]*Division of Plant Physiology, ICAR-Indian Agricultural Research Institute, Pusa Campus, New Delhi, Delhi, India*
[9]*Centre for Chemical Biology (CCB), Universiti Sains Malaysia (USM), Penang, Malaysia*
[10]*Department of Biology, Faculty of Science and Technology, Universitas Airlangga, Surabaya, Indonesia*

11.1 Introduction

With the rapid population surge, the demand for food is also increasing. The global population is expected to hit around nine billion by 2050; it requires an increase of 70% in global food production to sustain the growing population (Khandelwal et al. 2022). But various biotic and abiotic challenges, i.e., pests, diseases, drought, salinity and heat, are deteriorating crop production, leading to significant yield losses globally. Nearly 51–82% of yield loss is occurring annually due to the abiotic stressors (Oshunsanya et al. 2019; Wang et al. 2022). To mitigate these losses, mankind has started using fertilizers and pesticides, which are causing irreparable environmental damage. Over time, extensive research has led to the emergence of many eco-friendly technologies to overcome the yield losses due to the stresses. The application of nanotechnology is one of a kind that has

Silicon Nanoparticles: Synthesis, Uptake, and Applications, First Edition.
Edited by Archana Singh, Aryadeep Roychoudhury, and Indrakant K. Singh.

received keen attention in crop production enhancement and stress tolerance (Mukarram et al. 2022). Using nanoparticles (NPs) in farming influenced plant mineral and water uptake, soil quality, productivity and tolerance against many biotic and abiotic stresses (Mukarram et al. 2022). The uses of NPs have greatly increased due to their various properties (chemically stable, small size) and wide applicability. Moreover, green synthesis of NPs is largely preferred over other methods (Dhanda et al. 2024; Parab et al. 2025). Silicon (Si) is the second most abundant element available on earth's crust; while it is considered a non-essential mineral to plants, it has a crucial role in increasing tolerance to various stressors by incorporating itself into the cell wall (Ma and Yamaji 2008; Wang et al. 2022). Silicon nanoparticles (SiNPs) have been identified as a source of silicon for plants to counteract the effects of stresses. Due to their small size and high surface area-to-volume ratio, these can be easily uptaken and distributed among plant parts (Suriyaprabha et al. 2012). More investigations have revealed how SiNPs affect plants, including their metabolism and intracellular communication. Their foliar application also influences gene and protein expression, leading to enhanced stress tolerance, growth and productivity (Mosaedi et al. 2024). Recently, nanomaterials have found extensive applications in agriculture, including as growth stimulators, plant protection agents and nano-fertilizers (Parab et al. 2025). Among these, SiNPs have attracted considerable interest due to their unique properties, such as high pore volume, low toxicity, excellent stability, large surface area, morphology and easy surface modification (Laane 2018; Younis et al. 2020; Zhao et al. 2023). SiNPs are found to be more effective than bulk Si in Si-accumulating and non-accumulating plants (Bhat et al. 2021). These proved to be a better alternative approach to agrochemicals as they do not interfere with fundamental processes (Okeke et al. 2023). NPs can enter using any of the routes such as root hairs, stomata and cracks in the surface of the leaf (Wang et al. 2023). When they are inside the tissues, they start moving from cell to cell with the help of Si transporters like (low-silicon) Lsi1, Lsi2, Lsi3 and Lsi6 to the vascular bundles. Nano-SiO_2 is amorphous and its size ranges from 5 to 100 nm (Garza-Alonso et al. 2024). Si possesses many beneficial effects on plant physiology, especially the water retention ability. Reports have mentioned that the deposition of Si on the cell walls reduces transpiration rate and stomatal opening (Jinger et al. 2020), increases photosynthetic pigments (Parveen et al. 2019), regulates osmotic adjustments and prevents oxidative burst (Ma et al. 2016). Si has hydrophilic properties, it can combine with water and form SiO_2-nH_2O, which restores water from dehydration (Souri et al. 2021). Other than this, Si can also modify the carbon stoichiometry which in turn increases the nutritional use efficiency by converting carbon into dry mass (Rocha et al. 2022). Studies reported the increase in the root:shoot ratio in Si-treated plants that suggests Si can modulate the root configuration, hydraulic conductivity and water absorption (Wang et al. 2015). The other crucial aspect of plant biology is the phytohormones, which fine-tune the plant growth and development. It has been observed that Si can alter phytohormones by up-regulating or down-regulating the gene expression of transcription factors (TFs) involved in phytohormone biosynthesis. In this way, Si can reverse the imbalances created by environmental stresses. The chapter discusses the impact, applications and interactive behaviour of numerous biological processes with respect to SiNPs under the influence of drought and heat stresses.

11.2 Impact of the Role of Si/SiO_2-NPs in Mitigating Drought and Heat Stresses

Silicon is known for its ability to enhance crop productivity by improving nutrient uptake and exhibiting tolerance towards both kinds of stresses (biotic and abiotic) (Mahawar et al. 2023; Sarkar et al. 2022; Sun et al. 2022). The growth and productivity of plants are deteriorated by abiotic stressors, viz., drought, heat, cold, salinity and UV-B radiation as they influence the physiological, molecular and biochemical processes involved in the different stages of their growth (Ranjan et al. 2021). SiNPs might be involved in creating a double layer on the epidermal cell wall and alter the antioxidant defense process making plants more tolerant to drought, high temperature and humidity. Here, an overview of the roles of Si/SiO_2-NPs in mitigating drought and heat stresses has been provided (Figure 11.1).

11.2.1 Drought Stress

Water is vital for the survival of plants as it is required for maintaining cell turgor pressure, transport of nutrients and as a key component in numerous processes. During drought conditions, plants weaken, resulting in restricted growth and high mortality (Raju et al. 2024; Siddiqui et al. 2020). Drought is one of the major threats to food production, affecting approximately 82% of the total agriculture loss (FAO 2021) and severely hindering the growth and yield of field crops (Bodner et al. 2015; Raju et al. 2024). Several studies have highlighted the importance of Si in plant metabolism such as photosynthesis and alleviation of photoinhibition especially during drought stress (Ahmad et al. 2021; Alamri et al. 2020; Malik et al. 2021). Si application also improves the root hydraulic conductance by modifying the root growth, increasing the root:shoot ratio along enhancing aquaporin activity and osmotic driving force. Wang et al. (2015) observed a significant increase in the root:shoot ratio of plants supplied with silicon; this change was due to the modulation of root architecture by Si, which enhanced the water uptake and transport in Si-treated plants. Si plays a defensive role in drought stress in sugarcane (Verma et al. 2021). Tolerance to

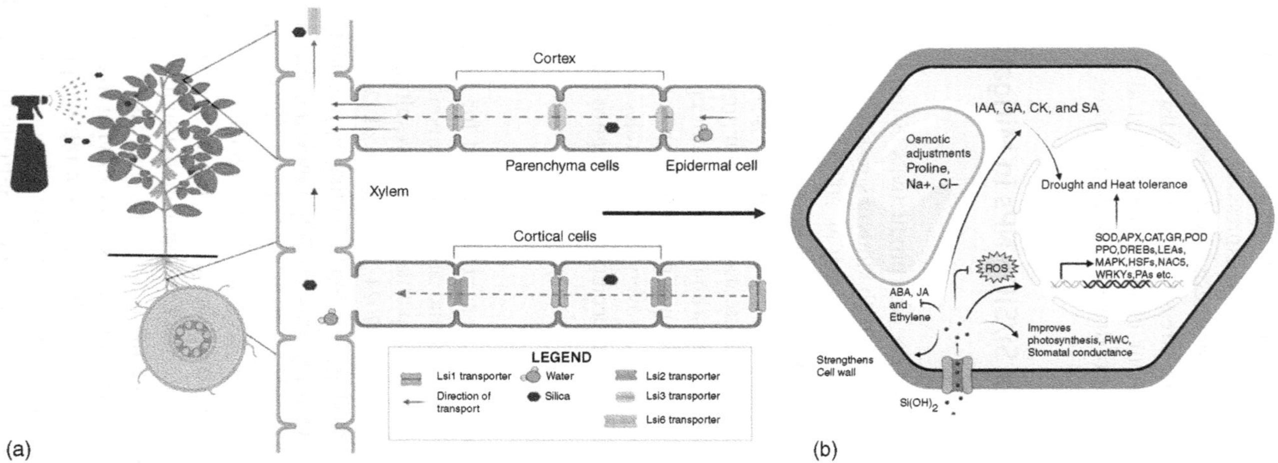

Figure 11.1 (a) Translocation of SiNPs to different parts of the plant: SiNPs uptake by roots and transported to the cortical cells and xylem with the help of Lsi1 (influx) and Lsi2 (efflux) transporters. In leaf mesophyll, Si is transported with the help of Lsi3 and Lsi6 transporters. (b) Modulation of stress-responsive genes and phytohormones, resulting in heat and drought tolerance in plants: Si enters in the form of silicic acid and regulates gene expression of antioxidant enzymes, transcription factors and the biosynthesis of phytohormones. Si has positive effects on IAA, GA, CK and SA while negative effects on ABA, JA and ethylene. Si inhibits ROS generation by increasing its scavengers and also strengthens cell walls and overall plant architecture.

drought stress was reported upon SiNP application in a number of crops such as wheat, barley, strawberries and cherries (Ashkavand et al. 2018; Behboudi et al. 2018; Hellala et al. 2020; Zahedi et al. 2020). The osmotic potential of strawberries and barley was reduced under drought conditions and the ability of tissues to absorb and retain water was improved by the application of SiNPs (Luyckx et al. 2017). SiNPs increase the photosynthetic pigment levels in leaves under drought stress and key metabolites like carbohydrates and proline in strawberry (Zahedi et al. 2020). Drought stress increases reactive oxygen species (ROS) levels, causing over-accumulation of malondialdehyde (MDA), leading plants to suffer oxidative damage (Parveen et al. 2019). Under drought stress, exogenous application of Si has improved seed germination and biochemical processes also protect the seedling from oxidative stress by enhancing antioxidant defense. SiNPs application under drought stress increased the levels of enzymatic antioxidants and decreased the MDA content of sugar beet (Namjoyan et al. 2020). After SiNPs treatment, the photosynthetic efficiency and stomatal conductance of hawthorn plants were found to be enhanced under water-deficit conditions (Ashkavand et al. 2015). Si regulates the expression of genes linked with flavonoid biosynthesis and antioxidant enzymes, thereby enhancing tolerance to drought (Bhardwaj and Kapoor 2021). Si pretreatment up-regulated the expressions of *OsCMO* and *OsRAB16b* that code for choline monooxygenase and dehydrin in drought-stressed rice when compared to the control plants (Khattab et al. 2014). The application of Si directly onto the leaves of drought-affected plants has enhanced the activity of antioxidants, increased the levels of osmolytes and promoted growth, flowering and overall yield in squash (Salim et al. 2021). The studies conducted so far have indicated that Si has a key role in mitigating drought stress.

11.2.2 Heat Stress

Due to global climate change, temperature rise became a serious problem for crop production leading to nearly 15–35% yield loss worldwide (Saha et al. 2021). Heat stress impacts every process and part of a plant's development from seed germination to flower formation and mainly disturbs the photosynthetic activity (I Hussain et al. 2019; Janni et al. 2020; Raju et al. 2024). Exposure of plants to heat stress causes damage to cell membrane and protein structure within the membrane, leading to an increase in the production of ROS that causes oxidative stress (Li et al. 2021). As high temperatures lead to the onset of drought conditions, Si treatment can be considered as an alternative method to surpass the loss caused by heat and drought stress. Major antioxidant enzymes such as glutathione ascorbate peroxidase (APX), S-transferase (GST), catalase (CAT), superoxide dismutase (SOD) and glutathione peroxidase (GPX) are inactivated due to heat stress, which can be induced by application of silicon (Dhakate et al. 2022). Until recent times, only a few studies on silicon-mediated heat tolerance in plants were reported compared to research done on drought and salinity tolerance. Heat stress also retards the production of heat-shock proteins (HSPs), which are essential in the protection of cellular organelles from dehydration caused by heat stress (Sharma et al. 2020). SiNP application has mitigated the abnormalities in chloroplasts and the nucleus of wheat seedlings caused by heat stress (Younis et al. 2020). Si is also involved in the stimulation of heat shock proteins and phytohormone production, and enhancement of the antioxidant system in tomatoes (Khan et al. 2020b). Si-treated barley plants under heat stress exhibited an increase in the chloroplast, carotenoids, leaf area, and root- and shoot-biomass (I Hussain et al. 2019). Silicon is also involved in the regulation of the hormonal balance of plants under heat stress (Khan et al. 2020a, 2020b).

11.3 Application of Si/SiO_2-NPs Causing Tolerance

11.3.1 Mode of Application of Si/SiO_2-NPs

Silicon is available to plants in the form of silicic and mono-silicic acid (H_4SiO_4) and the concentrations range from 0.1 to 0.6 mM (Imtiaz et al. 2016; Mahawar et al. 2023). For plants, SiNPs can be externally applied through foliar spraying, direct soil drenching, seed priming or through hydroponics. Foliar application is a method where SiNPs in liquid form is sprayed on the leaf surfaces. These NPs are absorbed by epidermal cells of leaves or through stomata forming a binary film over the epidermal cell wall creating a distinct colour for the plants (Laane 2018; Roychoudhury 2020; Strout et al. 2013). The effectiveness of soil application depends upon the conditions of soil particles, viz., texture, pH, salinity and strength of NP to get released into the soil. The application of SiNPs to soil was found to be more effective than foliar spray, as root cells have a higher capability for absorption than leaf cells (Suriyaprabha et al. 2014). In hydroponics, the powdered NP is added to the nutrient solution at the required concentration, which is absorbed by the roots immersed in the nutrient solution. Further research should be carried out on the application of SiNPs through soil and hydroponics to assess their effectiveness in enhancing drought and heat stress tolerance. Table 11.1 represents recent studies on SiNPs inducing plant drought and heat stress tolerance.

11.3.1.1 Foliar Application

Foliar application of SiO_2 NPs in grapevine seedlings helped to mitigate drought stress through various mechanisms, including an improved antioxidant defence system, and increased levels of phenolic compounds, proline and proteins

Table 11.1 Summary of some recent reports on SiNP-mediated tolerance against drought and heat stresses in crops.

Stress condition	Species	Mode of application	Mode of regulation	References
Heat stress				
44 °C 6 h	Date palm (*Phoenix dactylifera* L.)	Soil application	Regulated excessive ROS accumulation	Khan et al. (2020a)
43 °C 6 h	Tomato (*Solanum lycopersicum* L.)	Foliar spray	Enhanced activity of antioxidant enzyme encoding genes	Khan et al. (2020b)
33 °C and 41 °C, 48 h	Strawberry (*Fragaria* × *ananassa* (Duchesne ex Weston) Duchesne ex Rozier)	Foliar spray	Improved rate of photosynthesis	Muneer et al. (2017) and Xiao et al. (2022)
35 °C–40 °C	Snake cucumber (*Cucumis sativus* L.)	Foliar spray and seed priming	Improved nutrient uptake, antioxidant activities and chlorophyll content	Hassan Abd-Elrahman et al. (2022)
37 ± 2 °C	Wheat (*Triticum aestivum* L.)	Foliar spray	Improved chlorophyll and carotenoid contents	Mustafa et al. (2021)
Drought stress				
—	Maize (*Zea mays* L.)	Foliar spray	Improved the histological structure of roots	Reyes et al. (2024)
—	Cucumber (*C. sativus*)	Drip irrigation	Improved nitrogen use efficiency, and uptake of sodium and potassium	Alsaeedi et al. (2018)
—	Wheat (*T. aestivum*)	Soil application and Seed priming	Chlorophyll contents increased; uptake of nutrients enhanced	Khan et al. (2020b) and Raza et al. (2023)

(Continued)

Table 11.1 (Continued)

Stress condition	Species	Mode of application	Mode of regulation	References
—	Banana (*Musa acuminata* C.)	Foliar spray and irrigation	Enhanced shoot growth, decreased electrolyte leakage, MDA content and chlorophyll content	Mahmoud et al. (2020)
—	Cumin (*Cuminum cyminum* L.)	Foliar spray	Decreased antioxidant activity of the enzymes and increase in proline levels	Salajegheh et al. (2020)
—	Barley (*Hordeum vulgare* L.)	Soil application	Modifications in the plant chlorophyll content, osmolyte and metabolite Profile	Ghorbanpour et al. (2020)
—	Arabidopsis (*Arabidopsis thaliana* (L.) Heynh)	Soil application	Enhance seed germination and plant biomass	Tran et al. (2023)
—	Sugar beet (*Beta vulgaris* L.)	Foliar spray	Increased leaf area index, photosynthetic rate and glycine betaine	Namjoyan et al. (2020)

(Daler et al. 2024). Application of nano-silica spray at 0.15% concentration on hybrid maize grown under drought stress showed improvement in the plant height and biomass, by enhancing water uptake capacity and translocation through xylem as applied SiNPs modified the expression of proteins and water channels and helped to maintain the turgor pressure (Boora et al. 2023) (Figure 11.1).

11.3.1.2 Seed Priming

Seed priming involves treating seeds with natural or synthetic materials to induce a physiological state before germination (Dahiya et al. 2024). Seed priming with SiNPs has improved the growth and productivity of crops such as wheat (A Hussain et al. 2019). Many studies suggested that applying NPs by seed priming is more beneficial than soil and foliar application (Abou-Zeid and Ismail 2018; Asgari et al. 2018; Rizwan et al. 2019). Seed priming helps maintain a hydrated environment to seed till germination (Bhat et al. 2021; Jisha et al. 2013). Seed germination and seedling growth of wheat, pea and mustard seeds were improved upon seed priming with SiNPs (Chourasiya et al. 2021). SiNPs applied through seed priming repaired the ultrastructural distortions of chloroplasts and the nucleus in heat-stressed wheat seedlings (Younis et al. 2020).

11.4 Effect of SiNPs Foliar Treatment on Physiological Parameters

Drought stress is one of the major atmospheric stresses that directly influence plant growth, which affects global food security and agricultural productivity. Due to less rainfall and lack of water supplies, the condition of drought has been increasing. It has very negative effects on agriculture, the economy and the environment due to poor water quality and soil erosion (Shreshtha et al. 2025). Environmental conditions are crucial in shaping a plant's physiological and molecular processes. These factors may induce stress when they persist at certain intensities or durations. Identified as stress are any adverse environmental conditions that negatively impact a plant's growth, yield, quality or profitability. Factors like drought, frost, extreme temperatures, high humidity, soil acidity or salinity and pesticide pollution are common stressors (Nazim et al. 2024). Drought and salinity stress are the abiotic factors responsible for the global reduction in agriculture (Santisree et al. 2020). Plants affected by drought stress experience some serious effects on their growth and development. To counter these challenges, plants go through some changes, such as stomata closure in day time, reduced turgor pressure, disruption in water potential gradient, reduced cell development, and leaf rolling, and some changes in membrane integrity; these are essential for maintaining life processes (Ranjan et al. 2023).

Drought stress affects the plants' growth and development, which thereby alters biochemical, physiological, and morphological activities (Ramzan et al. 2023). Drought stress is responsible for oxidative stress, which affects the absorption and transportation of several minerals and nutrients in plants (Ahanger and Ahmad 2019; Ranjan et al. 2023).

Until the twentieth century, Si was not considered an important element in plant biology. Even though silica is the major inorganic constituent of plants as well as one of the most abundant elements in the earth's crust, it was ignored. In the early days plant physiologists failed to identify the importance of Si on plant growth and development for these reasons: (1) it is a biologically inert element, and (2) there were no visible signs of its toxicity or deficiency (Coskun et al. 2019). Si is not known as an essential element for plants; instead, it is well recognized for its benefits to plant growth and development, especially under stress conditions (Coskun et al. 2016). Si is the second most abundant mineral in soil, which is up to 50–70% of the soil mass in the form of silicon dioxide. It plays an important ecological role and strongly influences plant processes and interactions with the environment and other organisms. The accumulation of Si varies widely among plant species, which ranges from 0.1% to 10% of dry weight (Wang et al. 2021).

Si which is a non-cellular component accumulates in vascular tissues and the epidermis which provides stiffness, strength and resistance to various stresses (Ahsan et al. 2023). The use of Si plays an important role in improving plant growth, which positively impacts the improvement in leaf chlorophyll content, water relations, water potential and exchange of gases during physiological activity (Ali et al. 2021; Hussain et al. 2024). The positive effects of Si on drought stress have been observed in both monocots (rice, wheat, maize and sorghum) and dicots (tomato, cucumber, sunflower, soybean, cotton, mango and canola). Si neutralizes drought stress in plant species that are poor Si accumulators, such as tomatoes and canola (Wang et al. 2021). Si protects the cell membranes from the adverse effects of drought stress. The ROS can interact with the membrane lipids, proteins, photosynthetic pigments and nucleic acids. Since the concentration of these species is high in plant cells, it should be controlled and eliminated. It has been found that silicon treatment increases the activity of both enzymatic antioxidants (CAT, SOD and peroxidases) and non-enzymatic antioxidants (glutathione, ascorbic acid, carotenoids and tocopherols) under drought stress. Such a type of enhancement helps in scavenging and controlling harmful oxygen species (Nazim et al. 2024; Sutulienė et al. 2021).

In several studies, it has been found that the addition of Si into plants enhances their tolerance against drought stress by improving the water status, antioxidant defence, osmotic adjustment, photosynthetic activity, crop productivity and nutrient absorption balance (Bhardwaj and Kapoor 2021). In maintaining appropriate soil moisture levels for nutrient uptake, silicon foliar plays an important role. This feature of Si prevents nutrient loss through various physiological, biochemical and molecular pathways underlying silicon-mediated drought alleviation in agricultural plants (Nazim et al. 2024). Applying Si-containing solutions such as silicate-sorbitol, monosilicic acid-PEG400 and nanosilica concentrations to leaves is the most effective approach to enhance intermediate Si-accumulative and non-accumulative plants (de Souza Junior et al. 2021).

It has been observed that when drought-stressed rice was supplemented with Si, the expression of TFs such as *OsRDCP1*, *OsRAB16*, *OsCMO*, *OsNAC5* and *OsDREB2A* were elevated (Malik et al. 2021). The application of Si has been reported to up-regulate the expression of four genes in the ASH–GSH cycle, such as *TaDHAR*, *TaGR*, *TaGS* and *TaMDHAR*, in drought-stressed plants. Such enhancement in expressions tends to reduce MDA and H_2O_2 concentrations in drought-stressed plants. Si applications have also raised the transcript level of genes taking part in flavonoid biosynthesis like *TaCHS*, *TaPAL*, *TaDFR*, *TaF3H*, *TaCI*, and *TaANS* at the phase of drought stress through which plants were protected from drought. In addition, Si enhances the expression level of antioxidant enzyme coding genes, such as TaSOD and *TaCAT*, in plants which helps reduce the effect of drought stress and therefore inhibits H_2O_2 accumulation (Ma et al. 2016). Si is capable of enhancing plant growth, biomass production, photosynthesis and nutrient uptake under drought stress. It also increases antioxidant activity. These could be attributed to different biochemical, physiological and molecular mechanisms (Malik et al. 2021).

11.5 Effects of SiNPs Treatment on Gene and Protein Expression

11.5.1 Heat Stress-Responsive Genes Regulated by SiNPs

SiNPs have been used in the modulation of heat stress-responsive gene expression in improving the tolerance of crops to high temperatures. They have emerged as a promising strategy to regulate plant responses to heat stress and other environmental challenges. These have been identified as a promising technique for improving plant responses to heat stress and other environmental problems. These nanoparticles also affect the heat resilience of plants by modifying the expression of heat stress-responsive genes (heat shock factors and proteins; HSFs/Ps), which play essential roles in the plant adaptation mechanisms (Saha et al. 2021; Younis et al. 2020). Furthermore, research suggests that SiNPs increase the expression of stress-related genes such as *ABC1* and *Wdhn13*, which improves plant tolerance against stresses (Ali et al. 2024; Boora et al. 2023; Nazaralian et al. 2017). SiNPs have also been shown to improve the function of antioxidant enzymes such as SOD and APX, which play an important role in reducing oxidative damage under stress circumstances (Zhao et al. 2023). Furthermore, SiNPs can alter the transcription of stress-responsive genes, hence increasing plant resistance to abiotic stresses (Ahmed et al. 2023; Ali et al. 2024). They also fulfil a role in far more than only heat stress; they have been shown to affect genetic pathways related to chlorophyll pigment synthesis and gas exchange characteristics, increasing plant physiological health under stress (Jalil et al. 2023). Their functional influence is evident across several crop systems, demonstrating their broad utility in improving sustainable agriculture methods (Boora et al. 2023; Younis et al. 2020; Weisany et al. 2024).

SiNPs affect the expression and function of heat shock proteins, which are essential molecular chaperones that protect plants from stress. Some essential HSPs, such as HSP70, whose expression is stimulated by SiNPs, aid in protein folding and cellular homeostasis during heat and other abiotic stresses (Ali et al. 2024; Saha et al. 2021; Stanca et al. 2023). SiNPs have also been related to increased heat shock protein synthesis as a general stress tolerance mechanism, as well as plant protection against temperature fluctuations and heavy metal toxicity. This demonstrates that SiNPs have multifunctional benefits in agricultural applications where protein stability and cellular damage prevention are crucial for plant viability (Sidhu et al. 2024; Sonawane et al. 2022). The application of SiNPs improves plant heat tolerance by inducing biochemical changes and increasing the production of HSPs, as well as other stress-response proteins such as Osmoprotectants and antioxidants (Arif et al. 2021; Jalil et al. 2023; Kandhol et al. 2022). For example, SiNPs have been shown to increase HSP activity in barley and *Cannabis sativa*, which is associated with better water management and stress tolerance (Guerriero et al. 2021; Zhao et al. 2023). As a result of HSP regulation, SiNPs are emerging as a highly promising technique for increasing plant tolerance to environmental stressors while also providing a sustainable strategy.

11.5.2 Drought Stress-Responsive Genes Regulated by SiNPs

SiNPs have been demonstrated to up-regulate critical drought-responsive genes in wheat, including *ABC1* and *WDHN13*. These genes are required for stress signal transmission and cellular defence systems. Wheat plants treated with SiNPs showed a significant increase in growth and yield during drought circumstances, revealing silicon regulatory potential on stress-related genetic pathways (Boora et al. 2023; Sulaiman et al. 2023). Likewise, silicon dioxide NPs (SiO_2NPs) raised the drought tolerance of maize by activating genes related to osmotic control and water retention. Because of this gene control, maize was able to sustain vital physiological functions even in situations like that of scarce water supplies (Sharf-Eldin et al. 2023). Furthermore, studies on strawberries exposed to SiNPs showed that Si-induced genetic regulation resulted in improved drought resistance. The negative effects of drought on plant shape and physiological health have been reduced by the upregulation of genes involved in stress signalling pathways (Boora et al. 2023; Zahedi et al. 2023). Similarly, in rice, SiNPs were found to trigger genetic mechanisms that modulate the expression of antioxidant-responsive and transporter genes, further contributing to plant resilience under drought stress (Jalil et al. 2023). The beneficial effects of SiNPs extend to the regulation of water balance and enhancement of nutrient absorption, which are crucial under drought conditions. The modulation of drought-responsive gene expression by SiNPs strengthens the plant's defence mechanisms against water deficit, improving survival and productivity (Sulaiman et al. 2023). A combined study with plant growth-promoting rhizobacteria and SiNPs revealed a synergistic effect on the upregulation of stress-responsive genes in wheat, demonstrating enhanced root architecture, osmoregulation and stress hormone modulation under drought (Akhtar et al. 2021; Jalil et al. 2023).

11.5.3 Proteins Involved in Antioxidant Defence

The antioxidant defence mechanisms of plants are significantly modulated by SiNPs in terms of protein expression, especially in stressed conditions. SiNPs reduce oxidative stress by scavenging ROS and boosting the activity of key antioxidant enzymes (Al-Mokadem et al. 2023; Yan et al. 2023). They also have an impact on proline and other non-enzymatic antioxidant synthesis, which is crucial for preserving cellular homeostasis under stress (Faisal et al. 2024; Saha et al. 2021). SiNPs may also trigger the activation of active oxygen protein-repairing genes and up-regulate polyphenol oxidase and other enzymatic antioxidants, enhancing in this way its capacity to neutralize oxidative damage (Huang et al. 2024). They also enhance the antioxidant response machinery, such as that of the glyoxalase system, through modulation of gene expression related to stress metabolism (Kiany et al. 2022). SiNPs have been shown to significantly boost crop defence responses to stressors such as drought and heavy metals. This reduces lipid and protein peroxidation while enhancing enzymatic activity (Abdelrhim et al. 2021; Al-Mokadem et al. 2023). Furthermore, employing them reduces cellular toxicity and promotes protein stability when exposed to arsenic or cadmium (Ahmed et al. 2023). The integration of SiNPs into the agricultural sector reveals their ability to enhance antioxidant defence systems, making them beneficial for improving plant resilience to environmental stressors.

11.6 Mechanism of Si-Mediated Tolerance

Si is found to manipulate plant machinery via modulation of different components of phytohormones and the redox system.

11.6.1 Modulation of ROS Homeostasis Under Si Treatment

ROS is the most common and immediate effect observed when a plant is subjected to any kind of stress. ROS negatively affects vital processes such as cellular metabolism, molecular processes, growth and development in plants (Gao et al. 2023). During normal conditions, there is a balance in the production and elimination of ROS, which gets disrupted under stress (Mansoor et al. 2022). To counter ROS, plants have evolved antioxidant defence, which can be enzymatic and non-enzymatic (Stoia and Oancea 2022). However, these inherent defences are sometimes not sufficient to combat oxidative bursts and hamper ROS homeostasis. NPs are widely used to improve plant functioning in their environment. Among those, the SiNPs are found to be quite effective in bringing down the ROS levels to normal. Application of SiNPs provides positive outcomes in terms of plant growth, fresh and dry weight of leaf and root, relative water content and overall productivity in *Z. mays*, *Glycine max* (L.) Merr. and *Oryza sativa* L. (Felisberto et al. 2021; Hao et al. 2023). Biochemical assays showed that the Si-treated plants have lower MDA, electrolytic leakage and H_2O_2 content. To activate plant antioxidant machinery, the SiNPs influence the expression of enzymatic genes such as *DREB*, *APX*, *SOD*, *GST*, *POD*, and *CAT* and non-enzymatic genes like *GSH1*, *PCS* and *ABC1* (Kiany et al. 2022; Moradi et al. 2022). Most of the time these genes are up-regulated SiNPs treated plants, but in a few cases, the reverse trend is also observed in which SiNPs at 100 and 250 mg/l reduced the levels of *BnSOD*, *BnCAT*, *BnPOD* and *BnAPX* in *Brassica napus* (Ahmed et al. 2023). Studies observed the outcomes in different SiNP conditions, for instance, SiNPs at 300, 600 and 900 mg/l overcome H_2O_2 and MDA levels and increase the SOD and peroxidase in wheat (Raza et al. 2023). In *Z. mays*, 100, 200, 300 and 400 mg/l conc. of SiO_2NPs found to increase the concentration of carotenoids, proteins, amino acids and proline and also the enzymatic components (peroxidase, CAT, SOD and APX) (Ahmed et al. 2023). Khan et al. (2020b) found that the Si treatment promotes the expression of antioxidant genes while reducing the levels of MDA and superoxide radicals in *S. lycopersicum* under heat stress.

11.6.2 Modulation of Phytohormones Under Si Treatment

Plant hormones play crucial roles in regulating stresses. Their concentration changes and behaves differently depending upon the type of stress. Stress condition disturbs the hormonal balance which can be taken back to normal with the help of silicon. It has been noted that Si can up-regulate or down-regulate the same hormone as per the stress condition. Mainly the regulation of ABA, SA and JA are more emphasized as per their roles in heat and drought conditions (Khan et al. 2020b).

11.6.2.1 Auxin

Auxin is mainly produced in the apical meristem and is responsible for normal cell growth and elongation (Isham et al. 2024). Si adjusts the endogenous levels of auxin which can allow the plant to withstand under different abiotic conditions. More IAA accumulation promotes adventitious roots, which result in better root biomass (Tripathi et al. 2021). In drought, IAA levels are found to decrease which are restored by the application of silicon in *Elymus sibiricus* (Khan et al. 2024). In peach seedlings, silicon alleviates the water deficiency symptoms by increasing the concentration of IAA in roots, thereby promoting the development of the root system (Gao et al. 2022). Combined or isolated heat and drought conditions result in decreased levels of IAA (Naz et al. 2022). One study found that Si treatment up-regulates the DEGs responsible for auxin production in drought stress (Biju et al. 2023). Thus, it can be said that the Si application maintains the normal levels of IAA, which were disrupted due to the stressful environment.

11.6.2.2 Gibberellin

Gibberellin (GA) has diverse roles in plant growth and development. It regulates seed germination, stem elongation, leaf expansion, early flowering, along with other fundamental physiological processes (Babar et al. 2024). It was noted that different GAs behave differently upon Si treatment. For instance, in drought conditions, GA1 levels go up which further reduces if the Si is available, while the GA4 levels reduce during water deficiency but reverse when Si is applied. The same type of observation was reported in mango (Helaly et al. 2017), *E. sibiricus* (Khan et al. 2024) and peach seedlings (Gao et al. 2022). These levels of GAs reverse in flooding or submerged conditions where stress could increase GAs, while Si brings their level to normal (Pan et al. 2022). Khan et al. (2020a) found the interactive effects of Si and GAs on *P. dactylifera* under heat stress. This combined application of Si and GA3 remarkably mitigates the heat stress by boosting antioxidant levels, allowing leaves to retain better relative water content (RWC), development of double-layer cuticles, better photosynthesis, stomatal conductance, down-regulating ABA-responsive genes (*PYL4*, *PYL8* and *PYR1*), and the high levels of SA which also triggers the accumulation of heat shock factors *Hsf3*, *HsfA5* and *Hsf30* and activates proline biosynthesis. So, Si and GA3 regulate this antagonistic cross-talk between ABA and SA, thus combating the effects of heat.

11.6.2.3 Cytokinin

Cytokinin (CK) is largely responsible for cell division and nutrient transport (Nagar et al. 2024). In drought, the levels of CK decrease just like in the case of IAA and GA, but the application of Si brings it to the normal level in mango, *E. sibiricus* and peach seedlings (Gao et al. 2022; Helaly et al. 2017; Khan et al. 2024a,b).

Reports also mentioned that Si may positively regulate the expression of *IPT7* and *ARR5*, which are involved in CK biosynthesis (Markovich et al. 2017). Kandhol et al. (2024) suggested there is a crosstalk between Si, cytokinin and IAA, which is responsible for changes in physiological, biochemical and molecular aspects in wheat under chromium toxicity. Hence, the levels of IAA and CK are mainly up-regulated when plants are subjected to Si treatment which in turn provides tolerance.

11.6.2.4 Ethylene

Ethylene (ET) is the only gaseous hormone that is responsible for climacteric rise in fruits and leaf abscission (Mahto et al. 2024). Ethylene levels also come down during drought stress, but silicon mitigates this reduction by up-regulating the expression of genes involved in ET biosynthesis (Fauteux et al. 2006). Later it was found that Si can also down-regulate ET production to delay senescence and increase chlorophyll production (Biju et al. 2017). In excess water conditions, the levels of ET increase, which further balance out with the help of Si. Si treatment can also decrease the production of ethylene in lentil leaves possibly due to the down-regulation of genes involved in ET biosynthesis (Biju et al. 2023). Thus, Si negatively regulates ET production and hence delayed senescence by preventing chlorophyll degradation (Pan et al. 2022).

11.6.2.5 Abscisic Acid

Abscisic acid (ABA) is known as a growth inhibitor due to its defending nature in plants under stressful conditions (Bagal et al. 2024). ABA levels rise in almost every type of stress such as drought, heat, cold, salt, wounding, etc. (Verma et al. 2016). When Si is applied, the ABA levels decrease and come down to normal, and it is reported in mango (Helaly et al. 2017), *E. sibiricus* (Khan et al. 2023), peach (Gao et al. 2022) and barley (Naz et al. 2022). SiNPs at the dose of 200 mg/l led to the overexpression of *PYR*, *PYL*, *PP2C* and *SnRK2* genes in *T. aestivum* (Li et al. 2023). In the case of heat stress, the Si treatment showed reduced ABA levels in *P. dactylifera* and *S. lycopersicum* which might help provide tolerance to heat (Khan et al. 2020a, 2020b). Therefore, there are high ABA levels in plants under heat stress, which later mitigates by Si.

11.6.2.6 Jasmonic Acid

Jasmonic acid (JA) is involved in plant defence mainly herbivory and response to stress (Ramlal et al. 2024). Its JA-Ile conjugate is found to play the main part in signalling. Generally, the production of JA is higher in drought, while in the presence of Si, it attenuates. This is shown by Biju et al., in the analysis of JA-responsive DEGs (Biju et al. 2023). The same outcome is reported in the case of soybeans also where water deficiency leads to increased JA levels (Hamayun et al. 2010). Thus, decreased JA levels can be correlated with low-stress conditions (Mir et al. 2022). A reverse trend is observed in submerged conditions where the levels of JA get increased when Si is applied so that the elongation of internodes can be inhibited to prevent lodging and reduce carbohydrate consumption (Pan et al. 2022). Si treatment also down-regulates JA under wounding stress by down-regulating the transcripts such as *lipoxygenase*, *allene oxide synthase 1*, *allene oxide synthase 2*, *12-oxophytodienoate reductase 3* and *allene oxide cyclase* (Kim et al. 2014a).

11.6.2.7 Salicylic Acid

Salicylic acid (SA) belongs to the diverse group of phenolic compounds particularly associated with plant–pathogen interactions but also has positive effects on plants under abiotic stresses (Rivas-San Vicente and Plasencia 2011; Shaw et al. 2024). In drought, the increased levels of SA are mitigated by Si, but the reverse is observed in salinity stress (Hamayun et al. 2010; Kim et al. 2014b). Si treatment on *P. dactylifera* under heat stress restores SA levels to a greater extent while the untreated have lower SA (Khan et al. 2020a). This concludes that mainly the Si positively regulates SA levels which might be useful in the activation of heat shock machinery and results in heat tolerance.

11.6.2.8 Brassinosteroids

Brassinosteroids (BR) are also crucial for plants as they are involved in cell elongation (Meena et al. 2024; Nolan et al. 2020). Very few studies are available on the impact of Si on BR regulation. One study conducted on *Elymus sibiricus* showed that BR levels get reduced under drought stress. When different concentrations of Si are provided to the seeds, the difference between treated and untreated plants is found to be lower (Khan et al. 2024). One report suggests BR and Si work synergistically and result in improved plant growth with better tolerance against heavy metal stress (Sadaf et al. 2024).

11.7 Conclusion and Future Prospectives

Silicon applications in agriculture are quite promising in combating abiotic and biotic stresses. Si plays multifaceted roles in a variety of processes which include mechanical support to the cell wall, preventing oxidative bursts by activation of antioxidants, enhancing water and nutrient uptake, better photosynthesis and stomatal conductance. Si changes the altered phytohormone levels to a normal range. For instance, Si is found to up-regulate auxin, GA, cytokinin and salicylic acid while down-regulating abscisic acid, jasmonates and ethylene. It is also noted that Si behaves differently in different stress conditions which makes it a most versatile element in plants. However, there are research gaps which can connect the dots of understanding the detailed mechanism employed by Si. Hence, Si positively regulates overall plant growth and development which gets disrupted under a stressful environment. However, there is also considerable potential for further research to explore the molecular and genetic mechanisms involved. An in-depth analysis of the metabolomic, transcriptomic and proteomic levels will help to unravel the underlying mechanisms of SiNPs involved in the improvement of the antioxidant system in plants during drought conditions. Further study is required to delineate the role of silicon at the molecular level including genes involved in the regulation of silicon absorption, transport and hormonal balance, which could lead to a new pathway for exploring heat stress tolerance mechanisms.

References

Abdelrhim, A.S., Mazrou, Y.S., Nehela, Y. et al. (2021). Silicon dioxide nanoparticles induce innate immune responses and activate antioxidant machinery in wheat against *Rhizoctonia solani*. *Plants* 10 (12): 2758.

Abou-Zeid, H. and Ismail, G. (2018). The role of priming with biosynthesized silver nanoparticles in the response of *Triticum aestivum* L to salt stress. *Egyptian Journal of Botany* 58 (1): 73–85.

Ahanger, M.A. and Ahmad, P. (2019). Chapter 17 – Role of mineral nutrients in abiotic stress tolerance: revisiting the associated signaling mechanisms. In: *Plant Signaling Molecules* (ed. M.I.R. Khan, P.S. Reddy, A. Ferrante, and N.A.B.T.-P.S.M. Khan), 269–285. Woodhead Publishing. https://doi.org/10.1016/B978-0-12-816451-8.00016-2.

Ahmad, A., Aslam, Z., Naz, M. et al. (2021). Exogenous salicylic acid-induced drought stress tolerance in wheat (*Triticum aestivum* L.) grown under hydroponic culture. *PLoS One* 16 (12): e0260556.

Ahmed, T., Masood, H.A., Noman, M. et al. (2023). Biogenic silicon nanoparticles mitigate cadmium (Cd) toxicity in rapeseed (*Brassica napus* L.) by modulating the cellular oxidative stress metabolism and reducing Cd translocation. *Journal of Hazardous Materials* 459: 132070.

Ahsan, M., Valipour, M., Nawaz, F. et al. (2023). Evaluation of silicon supplementation for drought stress under water-deficit conditions: an application of sustainable agriculture. *Agronomy* 13 (2). https://doi.org/10.3390/agronomy13020599.

Akhtar, N., Ilyas, N., Hayat, R. et al. (2021). Synergistic effects of plant growth promoting rhizobacteria and silicon dioxide nano-particles for amelioration of drought stress in wheat. *Plant Physiology and Biochemistry* 166: 160–176.

Alamri, S., Hu, Y., Mukherjee, S. et al. (2020). Silicon-induced postponement of leaf senescence is accompanied by modulation of antioxidative defense and ion homeostasis in mustard (*Brassica juncea*) seedlings exposed to salinity and drought stress. *Plant Physiology and Biochemistry* 157: 47–59.

Ali, M., Afzal, S., Parveen, A. et al. (2021). Silicon mediated improvement in the growth and ion homeostasis by decreasing Na^+ uptake in maize (*Zea mays* L.) cultivars exposed to salinity stress. *Plant Physiology and Biochemistry* 158: 208–218. https://doi.org/10.1016/j.plaphy.2020.10.040.

Ali, L., Manzoor, N., Masood, H.A., and Abbas, A. (2024). Nanotechnology-enabled approaches to mitigating abiotic stresses in agricultural crops. In: *Molecular Dynamics of Plant Stress and its Management* (ed. M. Shahid and R. Gaur), 621–650. Singapore: Springer Nature Singapore.

Al-Mokadem, A.Z., Sheta, M.H., Mancy, A.G. et al. (2023). Synergistic effects of kaolin and silicon nanoparticles for ameliorating deficit irrigation stress in Maize plants by upregulating antioxidant defense systems. *Plants* 12 (11): 2221.

Alsaeedi, A., El-Ramady, H., Alshaal, T. et al. (2018). Exogenous nanosilica improves germination and growth of cucumber by maintaining K^+/Na^+ ratio under elevated Na^+ stress. *Plant Physiology and Biochemistry* 125: 164–171.

Arif, Y., Singh, P., Bajguz, A. et al. (2021). Silicon mediated abiotic stress tolerance in plants using physio-biochemical, omic approach and cross-talk with phytohormones. *Plant Physiology and Biochemistry* 166: 278–289.

Asgari, F., Majd, A., Jonoubi, P., and Najafi, F. (2018). Effects of silicon nanoparticles on molecular, chemical, structural and ultrastructural characteristics of oat (*Avena sativa* L.). *Plant Physiology and Biochemistry* 127: 152–160.

Ashkavand, P., Tabari, M., Zarafshar, M. et al. (2015). *Effect of SiO_2 nanoparticles on drought resistance in hawthorn seedlings*. Sekocin Stary, Poland: Instytut Badawczy Leśnictwa (Forest Research Institute).

Ashkavand, P., Zarafshar, M., Tabari, M. et al. (2018). Application of SiO_2 nanoparticles as pretreatment alleviates the impact of drought on the physiological performance of *Prunus mahaleb* (Rosaceae). *Boletin de la Sociedad Argentina de Botanica* 53 (2): 1–10.

Babar, R., Mane, P., and Chavan, S.B. (2024). Roles of gibberellic acid in mitigating abiotic stresses. In: *Phytohormones in Abiotic Stress* (ed. D. Raju, A. Rajendran, A. Ramlal, and V.P. Singh), 54–63. Boca Raton, FL: CRC Press, Taylor and Francis.

Bagal, D., Vishnoi, R., Rahman, A. et al. (2024). Role of abscisic acid in abiotic stresses. In: *Phytohormones in Abiotic Stress* (ed. D. Raju, A. Rajendran, A. Ramlal, and V.P. Singh), 73–78. Boca Raton, FL: CRC Press, Taylor and Francis.

Behboudi, F., Tahmasebi Sarvestani, Z., Kassaee, M.Z. et al. (2018). Improving growth and yield of wheat under drought stress via application of SiO_2 nanoparticles. *Journal of Agricultural Science and Technology* 20 (7): 1479–1492.

Bhardwaj, S. and Kapoor, D. (2021). Fascinating regulatory mechanism of silicon for alleviating drought stress in plants. *Plant Physiology and Biochemistry* 166: 1044–1053. https://doi.org/10.1016/j.plaphy.2021.07.005.

Bhat, J.A., Rajora, N., Raturi, G. et al. (2021). Silicon nanoparticles (SiNPs) in sustainable agriculture: major emphasis on the practicality, efficacy and concerns. *Nanoscale Advances* 3 (14): 4019–4028. https://doi.org/10.1039/d1na00233c.

Biju, S., Fuentes, S., and Gupta, D. (2017). Silicon improves seed germination and alleviates drought stress in lentil crops by regulating osmolytes, hydrolytic enzymes and antioxidant defense systems. *Plant Physiology and Biochemistry* 119: 250–264.

Biju, S., Fuentes, S., and Gupta, D. (2023). Novel insights into the mechanism(s) of silicon-induced drought stress tolerance in lentil plants revealed by RNA sequencing analysis. *BMC Plant Biology* 23 (1): 498.

Bodner, G., Nakhforoosh, A., and Kaul, H.P. (2015). Management of crop water under drought: a review. *Agronomy for Sustainable Development* 35: 401–442.

Boora, R., Rani, N., Kumari, S. et al. (2023). Efficacious role of silica nanoparticles in improving growth and yield of wheat under drought stress through stress-gene upregulation. *Plant Nano Biology* 6: 100051. https://doi.org/10.1016/j.plana.2023.100051.

Chourasiya, V.K., Nehra, A., Shukla, P.S. et al. (2021). Impact of mesoporous nano-silica (SiO_2) on seed germination and seedling growth of wheat, pea and mustard seed. *Journal of Nanoscience and Nanotechnology* 21 (6): 3566–3572.

Coskun, D., Britto, D.T., Huynh, W.Q., and Kronzucker, H.J. (2016). The role of silicon in higher plants under salinity and drought stress. *Frontiers in Plant Science* 7: 1–7. https://doi.org/10.3389/fpls.2016.01072.

Coskun, D., Deshmukh, R., Sonah, H. et al. (2019). The controversies of silicon's role in plant biology. *New Phytologist* 221 (1): 67–85. https://doi.org/10.1111/nph.15343.

Dahiya, V., Suman, S., Nautiyal, A., and Baweja, B. (2024). Recent discoveries and prospects of phytohormones. In: *Phytohormones in Abiotic Stress* (ed. D. Raju, A. Rajendran, A. Ramlal, and V.P. Singh), 26–49. Boca Raton, FL: CRC Press, Taylor and Francis.

Daler, S., Kaya, O., Canturk, S. et al. (2024). Silicon nanoparticles (SiO_2 NPs) boost drought tolerance in grapevines by enhancing some morphological, physiological, and biochemical traits. *Plant Molecular Biology Reporter* 1–19.

de Souza Junior, J. P., de Mello Prado, R., Soares, M. B., da Silva, J. L. F., de Farias Guedes, V. H., dos Santos Sarah, M. M., & Cazetta, J. O. (2021). Effect of different foliar silicon sources on cotton plants. *Journal of Soil Science and Plant Nutrition*, 21(1), 95–103. https://doi.org/10.1007/s42729-020-00345-4.

Dhakate, P., Kandhol, N., Raturi, G. et al. (2022). Silicon nanoforms in crop improvement and stress management. *Chemosphere* 305: 135165. https://doi.org/10.1016/j.chemosphere.2022.135165.

Dhanda, G., Manjhani, A., Miglani, N. et al. (2024). Antimicrobial gum: a bio-based nanocomposite. In: *Antimicrobial Materials and Coatings*, Woodhead Publishing in Materials (ed. A.K. Singh, M. Dhayal, and C.H. Hussain), 4–24. Elsevier. https://doi.org/10.1016/B978-0-323-95460-0.00001-0.

Faisal, M., Faizan, M., Soysal, S., and Alatar, A.A. (2024). Synergistic application of melatonin and silicon oxide nanoparticles modulates reactive oxygen species generation and the antioxidant defense system: a strategy for cadmium tolerance in rice. *Frontiers in Plant Science* 15: 1484600.

FAO. 2021. Agriculture on the proving grounds: damage and loss. https://www.fao.org/interactive/disasters-in-agriculture/en/ (accessed 15 December 2024).

Fauteux, F., Chain, F., Belzile, F. et al. (2006). The protective role of silicon in the Arabidopsis–powdery mildew pathosystem. *Proceedings of the National Academy of Sciences of the United States of America* 103 (46): 17554–17559.

Felisberto, G., de Mello Prado, R., de Oliveira, R.L.L., and de Carvalho Felisberto, P.A. (2021). Are nanosilica, potassium silicate and new soluble sources of silicon effective for silicon foliar application to soybean and rice plants? *Silicon* 13: 3217–3228.

Gao, H., Yu, W., Yang, X. et al. (2022). Silicon enhances the drought resistance of peach seedlings by regulating hormone, amino acid, and sugar metabolism. *BMC Plant Biology* 22 (1): 422.

Gao, M., Wang, Z., Jia, Z. et al. (2023). Brassinosteroids alleviate nanoplastic toxicity in edible plants by activating antioxidant defense systems and suppressing nanoplastic uptake. *Environment International* 174: 107901.

Garza-Alonso, C.A., González-García, Y., de Jesús Carballo-Méndez, F., and Juárez-Maldonado, A. (2024). Mechanisms of action of nanosilica to reduce oxidative stress in plants. In: *Silicon Advances for Sustainable Agriculture and Human Health: Increased Nutrition and Disease Prevention* (ed. R. de Mello Prado, H. Etesami, and A.K. Srivastava), 325–348. Cham: Springer Nature Switzerland.

Ghorbanpour, M., Mohammadi, H., and Kariman, K. (2020). Nanosilicon-based recovery of barley (*Hordeum vulgare*) plants subjected to drought stress. *Environmental Science: Nano* 7 (2): 443–461.

Guerriero, G., Sutera, F.M., Torabi-Pour, N. et al. (2021). Phyto-courier, a silicon particle-based nano-biostimulant: evidence from *Cannabis sativa* exposed to salinity. *ACS Nano* 15 (2): 3061–3069.

Hamayun, M., Sohn, E.Y., Khan, S.A. et al. (2010). Silicon alleviates the adverse effects of salinity and drought stress on growth and endogenous plant growth hormones of soybean (*Glycine max* L.). *Pakistan Journal of Botany* 42 (3): 1713–1722.

Hao, Y., Yu, Y., Sun, G. et al. (2023). Effects of multi-walled carbon nanotubes and nano-silica on root development, leaf photosynthesis, active oxygen and nitrogen metabolism in maize. *Plants* 12 (8): 1604.

Hassan Abd-Elrahman, S., Abd-Elhamid Hashem, F., and Awny Farag, A. (2022). Mitigation of heat stress effects on snake cucumber plants by silicon additions and rice straw soil cover. *Communications in Soil Science and Plant Analysis* 53 (6): 759–772. https://doi.org/10.1080/00103624.2022.2028814.

Helaly, M.N., El-Hoseiny, H., El-Sheery, N.I. et al. (2017). Regulation and physiological role of silicon in alleviating drought stress of mango. *Plant Physiology and Biochemistry* 118: 31–44.

Hellala, F., Amerb, A.K., El-Sayeda, S., and El-Azab, K. (2020). Mitigation the negative effect of water stress on barley by nano silica application. *Plant Archives* 20: 3224–3231.

Huang, Q., Ayyaz, A., Farooq, M.A. et al. (2024). Silicon dioxide nanoparticles enhance plant growth, photosynthetic performance, and antioxidants defence machinery through suppressing chromium uptake in *Brassica napus* L. *Environmental Pollution* 342: 123013.

Hussain, I., Parveen, A., Rasheed, R. et al. (2019). Exogenous silicon modulates growth, physio-chemicals and antioxidants in barley (*Hordeum vulgare* L.) exposed to different temperature regimes. *Silicon* 11: 2753–2762.

Hussain, A., Rizwan, M., Ali, Q., and Ali, S. (2019). Seed priming with silicon nanoparticles improved the biomass and yield while reduced the oxidative stress and cadmium concentration in wheat grains. *Environmental Science and Pollution Research* 26: 7579–7588.

Hussain, I., Ayub, A., Nayab, A. et al. (2024). Exogenous application of silicon and zinc attenuates drought tolerance in *Eruca sativa* L. through increasing chlorophyll pigments, osmoprotectants, and modulating defense mechanisms. *Journal of Plant Growth Regulation* 43 (9): 3221–3237. https://doi.org/10.1007/s00344-023-11116-7.

Imtiaz, M., Rizwan, M.S., Mushtaq, M.A. et al. (2016). Silicon occurrence, uptake, transport and mechanisms of heavy metals, minerals and salinity enhanced tolerance in plants with future prospects: a review. *Journal of Environmental Management* 183: 521–529.

Islam, F., Shahid, D., Soni, R., & Singh, N. (2024). Auxin in abiotic stress. In: Raju, D., Rajendran, A., Ramlal, A., & Singh, V.P. *Phytohormones in Abiotic Stress*, pp. 54–63, CRC Press, Taylor and Francis, Boca Raton, FL.

Jalil, S., Nazir, M.M., Al-Huqail, A.A. et al. (2023). Silicon nanoparticles alleviate cadmium toxicity in rice (*Oryza sativa* L.) by modulating the nutritional profile and triggering stress-responsive genetic mechanisms. *Ecotoxicology and Environmental Safety* 268: 115699.

Janni, M., Gullì, M., Maestri, E. et al. (2020). Molecular and genetic bases of heat stress responses in crop plants and breeding for increased resilience and productivity. *Journal of Experimental Botany* 71 (13): 3780–3802.

Jinger, D., Dhar, S., Dass, A. et al. (2020). Crop productivity, grain quality, water use efficiency, and soil enzyme activity as influenced by silicon and phosphorus application in aerobic rice (*Oryza sativa*). *Communications in Soil Science and Plant Analysis* 51 (16): 2147–2162.

Jisha, K.C., Vijayakumari, K., and Puthur, J.T. (2013). Seed priming for abiotic stress tolerance: an overview. *Acta Physiologiae Plantarum* 35: 1381–1396.

Kandhol, N., Jain, M., and Tripathi, D.K. (2022). Nanoparticles as potential hallmarks of drought stress tolerance in plants. *Physiologia Plantarum* 174 (2): e13665.

Kandhol, N., Srivastava, A., Rai, P. et al. (2024). Cytokinin and indole-3-acetic acid crosstalk is indispensable for silicon mediated chromium stress tolerance in roots of wheat seedlings. *Journal of Hazardous Materials* 468: 133134.

Khan, A., Bilal, S., Khan, A.L. et al. (2020a). Silicon and gibberellins: synergistic function in harnessing ABA signaling and heat stress tolerance in date palm (*Phoenix dactylifera* L.). *Plants* 9 (5): 620. https://doi.org/10.3390/plants9050620.

Khan, A., Khan, A.L., Imran, M. et al. (2020b). Silicon-induced thermotolerance in *Solanum lycopersicum* L. via activation of antioxidant system, heat shock proteins, and endogenous phytohormones. *BMC Plant Biology* 20: 1–18.

Khan, I., Awan, S.A., Rizwan, M. et al. (2024a). Silicon nanoparticles improved the osmolyte production, antioxidant defense system, and phytohormone regulation in *Elymus sibiricus* (L.) under drought and salt stress. *Environmental Science and Pollution Research* 31 (6): 8985–8999.

Khan, I., Awan, S.A., Rizwan, M. et al. (2024b). Silicon-induced modulation of photosynthetic pigments, osmolytes, and phytohormonal regulation boosted the drought tolerance in Elymus sibiricus L. *Journal of Plant Growth Regulation* 43 (4): 998–1011.

Khandelwal, S., Parwez, S., and Mehra, M. (2022). Climate change: effects on health and nutrition. *Reference Module in Food Science*. https://doi.org/10.1016/B978-0-12-821848-8.00131-1.

Khattab, H.I., Emam, M.A., Emam, M.M. et al. (2014). Effect of selenium and silicon on transcription factors NAC5 and DREB2A involved in drought-responsive gene expression in rice. *Biologia Plantarum* 58: 265–273.

Kiany, T., Pishkar, L., Sartipnia, N. et al. (2022). Effects of silicon and titanium dioxide nanoparticles on arsenic accumulation, phytochelatin metabolism, and antioxidant system by rice under arsenic toxicity. *Environmental Science and Pollution Research* 29 (23): 34725–34737.

Kim, Y.H., Khan, A.L., Waqas, M. et al. (2014a). Regulation of jasmonic acid biosynthesis by silicon application during physical injury to *Oryza sativa* L. *Journal of Plant Research* 127: 525–532.

Kim, Y.H., Khan, A.L., Waqas, M. et al. (2014b). Silicon application to the rice root zone influenced the phytohormonal and antioxidant responses under salinity stress. *Journal of Plant Growth Regulation* 33: 137–149.

Laane, H.M. (2018). The effects of foliar sprays with different silicon compounds. *Plants* 7 (2): 45.

Li, N., Euring, D., Cha, J.Y. et al. (2021). Plant hormone-mediated regulation of heat tolerance in response to global climate change. *Frontiers in Plant Science* 11: 627969.

Li, Y., Xi, K., Liu, X. et al. (2023). Silica nanoparticles promote wheat growth by mediating hormones and sugar metabolism. *Journal of Nanobiotechnology* 21 (1): 2.

Luyckx, M., Hausman, J.F., Lutts, S., and Guerriero, G. (2017). Silicon and plants: current knowledge and technological perspectives. *Frontiers in Plant Science* 8: 411.

Ma, J.F. and Yamaji, N. (2008). Functions and transport of silicon in plants. *Cellular and Molecular Life Sciences* 65: 3049–3057.

Ma, D., Sun, D., Wang, C. et al. (2016). Silicon application alleviates drought stress in wheat through transcriptional regulation of multiple antioxidant defense pathways. *Journal of Plant Growth Regulation* 35 (1): 1–10. https://doi.org/10.1007/s00344-015-9500-2.

Mahawar, L., Ramasamy, K.P., Suhel, M. et al. (2023). Silicon nanoparticles: comprehensive review on biogenic synthesis and applications in agriculture. *Environmental Research* 232: 116292.

Mahmoud, L.M., Dutt, M., Shalan, A.M. et al. (2020). Silicon nanoparticles mitigate oxidative stress of in vitro-derived banana (*Musa acuminata* 'Grand Nain') under simulated water deficit or salinity stress. *South African Journal of Botany* 132: 155–163.

Mahto, R.K., Tripathi, S., Singh, D.P. et al. (2024). Ethylene in abiotic stress. In: *Phytohormones in Abiotic Stress* (ed. D. Raju, A. Rajendran, A. Ramlal, and V.P. Singh), 73–78. Boca Raton, FL: CRC Press, Taylor and Francis.

Malik, M.A., Wani, A.H., Mir, S.H. et al. (2021). Elucidating the role of silicon in drought stress tolerance in plants. *Plant Physiology and Biochemistry* 165: 187–195. https://doi.org/10.1016/j.plaphy.2021.04.021.

Mansoor, S., Ali Wani, O., Lone, J.K. et al. (2022). Reactive oxygen species in plants: from source to sink. *Antioxidants* 11 (2): 225.

Markovich, O., Steiner, E., Kouřil, Š. et al. (2017). Silicon promotes cytokinin biosynthesis and delays senescence in *Arabidopsis* and Sorghum. *Plant, Cell & Environment* 40 (7): 1189–1196.

Meena, S., Yadav, S., Taria, S. et al. (2024). Role of brassinosteriods in abiotic stresses. In: *Phytohormones in Abiotic Stress* (ed. D. Raju, A. Rajendran, A. Ramlal, and V.P. Singh), 89–99. Boca Raton, FL: CRC Press, Taylor and Francis.

Mir, R.A., Bhat, B.A., Yousuf, H. et al. (2022). Multidimensional role of silicon to activate resilient plant growth and to mitigate abiotic stress. *Frontiers in Plant Science* 13: 819658.

Moradi, P., Vafaee, Y., Mozafari, A.A., and Tahir, N.A.R. (2022). Silicon nanoparticles and methyl jasmonate improve physiological response and increase expression of stress-related genes in strawberry cv. Paros under salinity stress. *Silicon* 14 (16): 10559–10569.

Mosaedi, H., Mozafari, H., Sani, B. et al. (2024). Foliar-applied silicon and zinc nanoparticles improve plant growth, biochemical attributes, and essential oil profile of fennel (*Foeniculum vulgare*) under different irrigation regimes. *Functional Plant Biology* 51 (10).

Mukarram, M., Petrik, P., Mushtaq, Z. et al. (2022). Silicon nanoparticles in higher plants: uptake, action, stress tolerance, and crosstalk with phytohormones, antioxidants, and other signalling molecules. *Environmental Pollution* 310: 119855.

Muneer, S., Park, Y.G., Kim, S., and Jeong, B.R. (2017). Foliar or subirrigation silicon supply mitigates high temperature stress in strawberry by maintaining photosynthetic and stress-responsive proteins. *Journal of Plant Growth Regulation* 36: 836–845.

Mustafa, T., Sattar, A., Sher, A. et al. (2021). Exogenous application of silicon improves the performance of wheat under terminal heat stress by triggering physio-biochemical mechanisms. *Scientific Reports* 11 (1): 23170.

Nagar, S., Meena, S., Karwa, S. et al. (2024). Regulatory role of cytokinin in abiotic stress tolerance of plants. In: *Phytohormones in Abiotic Stress* (ed. D. Raju, A. Rajendran, A. Ramlal, and V.P. Singh), 64–72. Boca Raton, FL: CRC Press, Taylor and Francis.

Namjoyan, S., Sorooshzadeh, A., Rajabi, A., and Aghaalikhani, M. (2020). Nano-silicon protects sugar beet plants against water deficit stress by improving the antioxidant systems and compatible solutes. *Acta Physiologiae Plantarum* 42: 1–16.

Naz, R., Gul, F., Zahoor, S. et al. (2022). Interactive effects of hydrogen sulphide and silicon enhance drought and heat tolerance by modulating hormones, antioxidant defence enzymes and redox status in barley (*Hordeum vulgare* L.). *Plant Biology* 24 (4): 684–696.

Nazaralian, S., Majd, A., Irian, S. et al. (2017). Comparison of silicon nanoparticles and silicate treatments in fenugreek. *Plant Physiology and Biochemistry* 115: 25–33.

Nazim, M., Li, X., Anjum, S. et al. (2024). Silicon nanoparticles: a novel approach in plant physiology to combat drought stress in an arid environment. *Biocatalysis and Agricultural Biotechnology* 58: 103190. https://doi.org/10.1016/j.bcab.2024.103190.

Nolan, T.M., Vukašinović, N., Liu, D. et al. (2020). Brassinosteroids: multidimensional regulators of plant growth, development, and stress responses. *The Plant Cell* 32 (2): 295–318.

Okeke, E.S., Nweze, E.J., Ezike, T.C. et al. (2023). Silicon-based nanoparticles for mitigating the effect of potentially toxic elements and plant stress in agroecosystems: a sustainable pathway towards food security. *Science of the Total Environment* 898: 165446.

Oshunsanya, S.O., Nwosu, N.J., and Li, Y. (2019). Abiotic stress in agricultural crops under climatic conditions. In: *Sustainable Agriculture, Forest and Environmental Management* (ed. M.K. Jhariya, A. Banerjee, R.S. Meena, and D.K. Yadav), 71–100. Singapore: Springer Singapore.

Pan, T., Wang, L., Peng, Z. et al. (2022). Silicon enhances the submergence tolerance of rice by regulating quiescence strategy and alleviating oxidative damage. *Plant Physiology and Biochemistry* 182: 124–132.

Parab, A.R., Ramlal, A., Subramaniam, S., and Gopinath, S.C.B. (2025). Forging for future of nanotechnology: embracing greener practices for resilient and a sustainable tomorrow. *Frontiers in Nanotechnology*. https://doi.org/10.3389/fnano.2024.1506665.

Parveen, A., Liu, W., Hussain, S. et al. (2019). Silicon priming regulates morpho-physiological growth and oxidative metabolism in maize under drought stress. *Plants* 8 (10): 431.

Raju, D., Rajendran, R.A., Ramlal, A., and Singh, V.P. (2024). *Phytohormones in Abiotic Stress*. Boca Raton, FL: CRC Press, Taylor and Francis.

Ramlal, A., Jain, R., Miglani, N. et al. (2024). Jasmonic acid. In: *Phytohormones in Abiotic Stress* (ed. D. Raju, A. Rajendran, A. Ramlal, and V.P. Singh), 111–118. Boca Raton, FL: CRC Press, Taylor and Francis.

Ramzan, T., Shahbaz, M., Maqsood, M.F. et al. (2023). Phenylalanine supply alleviates the drought stress in mustard (*Brassica campestris*) by modulating plant growth, photosynthesis, and antioxidant defense system. *Plant Physiology and Biochemistry* 201: 107828. https://doi.org/10.1016/j.plaphy.2023.107828.

Ranjan, A., Sinha, R., Bala, M. et al. (2021). Silicon-mediated abiotic and biotic stress mitigation in plants: underlying mechanisms and potential for stress resilient agriculture. *Plant Physiology and Biochemistry* 163: 15–25.

Ranjan, S., Prakash, A., Singh, R.B. et al. (2023). Effects of drought stress on agricultural plants, and molecular strategies for drought tolerant crop development. In: *[AR13] Environmental Science and Engineering*, vol. Part F1646 (ed. T. Aftab), 267–287. Nature Switzerland. https://doi.org/10.1007/978-3-031-43729-8_10: Springer.

Raza, M.A.S., Zulfiqar, B., Iqbal, R. et al. (2023). Morpho-physiological and biochemical response of wheat to various treatments of silicon nano-particles under drought stress conditions. *Scientific Reports* 13 (1): 2700.

Reyes, I.V., Cerda, L.G., Ortega-Ortega, Y., and Sarmiento-Lopéz, L. (2024). Biochemical responses induced by SiO_2-ZnO nanoparticles in *Zea mays* plants under water deficit conditions. In: *The 4th International Electronic Conference on Agronomy organised on 2 Dec 2024* (ed. M. Cardarelli). Publisher Agronomy, MDPI.

Rivas-San Vicente, M. and Plasencia, J. (2011). Salicylic acid beyond defence: its role in plant growth and development. *Journal of Experimental Botany* 62 (10): 3321–3338.

Rizwan, M., Ali, S., ur Rehman, M.Z. et al. (2019). Effect of foliar applications of silicon and titanium dioxide nanoparticles on growth, oxidative stress, and cadmium accumulation by rice (*Oryza sativa*). *Acta Physiologiae Plantarum* 41: 1–12.

Rocha, J.R., de Mello Prado, R., and de Cassia Piccolo, M. (2022). Mitigation of water deficit in two cultivars of *Panicum maximum* by the application of silicon. *Water, Air, & Soil Pollution* 233 (2): 63.

Roychoudhury, A. (2020). Silicon-nanoparticles in crop improvement and agriculture. *International Journal on Recent Advancement in Biotechnology & Nanotechnology* 3 (1): 2582–1571.

Sadaf, A., Balal, R.M., Jaffar, M.T. et al. (2024). Influence of brassinosteroid and silicon on growth, antioxidant enzymes, and metal uptake of leafy vegetables under wastewater irrigation. *International Journal of Phytoremediation* 26 (6): 936–946.

Saha, G., Mostofa, M.G., Rahman, M.M., and Tran, L.S.P. (2021). Silicon-mediated heat tolerance in higher plants: a mechanistic outlook. *Plant Physiology and Biochemistry* 166: 341–347.

Salajegheh, M., Yavarzadeh, M., Payandeh, A., and Akbarian, M.M. (2020). Effects of titanium and silicon nanoparticles on antioxidant enzymes activity and some biochemical properties of *Cuminum cyminum* L. under drought stress. *Banat's Journal of Biotechnology* 11 (21): 19–25.

Salim, B.B.M., Abou El-Yazied, A., Salama, Y.A.M. et al. (2021). Impact of silicon foliar application in enhancing antioxidants, growth, flowering and yield of squash plants under deficit irrigation condition. *Annals of Agricultural Sciences* 66 (2): 176–183.

Santisree, P., Jalli, L.C.L., Bhatnagar-Mathur, P., and Sharma, K.K. (2020). Emerging roles of salicylic acid and jasmonates in plant abiotic stress responses. In: *Protective [AR14] Chemical Agents in the Amelioration of Plant Abiotic Stress*, 342–373. https://doi.org/10.1002/9781119552154.ch17.

Sarkar, M.M., Mathur, P., and Roy, S. (2022). Silicon and nano-silicon: new frontiers of biostimulants for plant growth and stress amelioration. In: *Silicon and Nano-silicon in Environmental Stress Management and Crop Quality Improvement* (ed. H. Etesami, A.H. Al Saeedi, H. El-Ramady, et al.), 17–36. Academic Press.

Sharf-Eldin, A.A., Alwutayd, K.M., El-Yazied, A.A. et al. (2023). Response of maize seedlings to silicon dioxide nanoparticles (SiO_2NPs) under drought stress. *Plants* 12 (14): 2592.

Sharma, L., Priya, M., Kaushal, N. et al. (2020). Plant growth-regulating molecules as thermoprotectants: functional relevance and prospects for improving heat tolerance in food crops. *Journal of Experimental Botany* 71 (2): 569–594. https://doi.org/10.1093/jxb/erz333.

Shaw, P., Pakhre, R., Devi, R.S., and Tripathi, S. (2024). Role of salicylic acid in abiotic stresses. In: *Phytohormones in Abiotic Stress* (ed. D. Raju, A. Rajendran, A. Ramlal, and V.P. Singh), 162–169. Boca Raton, FL: CRC Press, Taylor and Francis.

Shreshtha, K., Prakash, A., Pandey, P.K. et al. (2025). Isolation and characterization of plant growth promoting rhizobacteria from cacti root under drought condition.

Current Research in Microbial Sciences 8: 100319. https://doi.org/10.1016/j.crmicr.2024.100319.

Siddiqui, H., Ahmed, K.B.M., Sami, F., and Hayat, S. (2020). Silicon nanoparticles and plants: current knowledge and future perspectives. *Sustainable Agriculture Reviews* 41. Nanotechnology for Plant Growth and Development: 129–142. https://doi.org/10.1007/978-3-030-33996-8_7.

Sidhu, A.K., Sharma, M., Bhickchand Agrawal, S. et al. (2024). Nanomaterial strategies for enhancing plant resilience in the face of temperature stress. *CABI Agriculture and Bioscience* 5 (1): 60.

Sonawane, H., Shelke, D., Chambhare, M. et al. (2022). Fungi-derived agriculturally important nanoparticles and their application in crop stress management – prospects and environmental risks. *Environmental Research* 212: 113543.

Souri, Z., Khanna, K., Karimi, N., and Ahmad, P. (2021). Silicon and plants: current knowledge and future prospects. *Journal of Plant Growth Regulation* 40: 906–925.

Stanca, L., Geicu, O.I., Serban, A.I., and Dinischiotu, A. (2023). Interplay of oxidative stress, inflammation, and autophagy in RAW 264.7 murine macrophage cell line challenged with Si/SiO_2 quantum dots. *Materials* 16 (14): 5083.

Stoia, M. and Oancea, S. (2022). Low-molecular-weight synthetic antioxidants: classification, pharmacological profile, effectiveness and trends. *Antioxidants* 11 (4): 638.

Strout, G., Russell, S.D., Pulsifer, D.P. et al. (2013). Silica nanoparticles aid in structural leaf coloration in the Malaysian tropical rainforest understorey herb *Mapania caudata*. *Annals of Botany* 112 (6): 1141–1148.

Sulaiman, El-Beltagi, H.S., Shehata, W.F. et al. (2023). Role of silica nanoparticles in enhancing drought tolerance of cereal crops. *Notulae Botanicae Horti Agrobotanici Cluj-Napoca* 51 (4): 13480–13480.

Sun, S., Yang, Z., Song, Z. et al. (2022). Silicon enhances plant resistance to Fusarium wilt by promoting antioxidant potential and photosynthetic capacity in cucumber (*Cucumis sativus* L.). *Frontiers in Plant Science* 13: 1011859.

Suriyaprabha, R., Karunakaran, G., Yuvakkumar, R. et al. (2012). Growth and physiological responses of maize (*Zea mays* L.) to porous silica nanoparticles in soil. *Journal of Nanoparticle Research* 14: 1–14.

Suriyaprabha, R., Karunakaran, G., Yuvakkumar, R. et al. (2014). Foliar application of silica nanoparticles on the phytochemical responses of maize (*Zea mays* L.) and its toxicological behavior. *Synthesis and Reactivity in Inorganic, Metal-Organic, and Nano-Metal Chemistry* 44 (8): 1128–1131.

Sutulienė, R., Ragelienė, L., Samuolienė, G. et al. (2021). The response of antioxidant system of drought-stressed green pea (*Pisum sativum* L.) affected by watering and foliar spray with silica nanoparticles. *Horticulturae* 8 (1): 35.

Tran, T.L.C., Guirguis, A., Jeyachandran, T. et al. (2023). Mesoporous silica nanoparticle-induced drought tolerance in *Arabidopsis thaliana* grown under in vitro conditions. *Functional Plant Biology* 50 (11): 889–900.

Tripathi, D.K., Rai, P., Guerriero, G. et al. (2021). Silicon induces adventitious root formation in rice under arsenate stress with involvement of nitric oxide and indole-3-acetic acid. *Journal of Experimental Botany* 72 (12): 4457–4471.

Verma, V., Ravindran, P., and Kumar, P.P. (2016). Plant hormone-mediated regulation of stress responses. *BMC Plant Biology* 16: 1–10.

Verma, K.K., Song, X.P., Zeng, Y. et al. (2021). Foliar application of silicon boosts growth, photosynthetic leaf gas exchange, antioxidative response and resistance to limited water irrigation in sugarcane (*Saccharum officinarum* L.). *Plant Physiology and Biochemistry* 166: 582–592.

Wang, S., Liu, P., Chen, D. et al. (2015). Silicon enhanced salt tolerance by improving the root water uptake and decreasing the ion toxicity in cucumber. *Frontiers in Plant Science* 6: 759.

Wang, M., Wang, R., Mur, L.A.J. et al. (2021). Functions of silicon in plant drought stress responses. *Horticulture Research* 8 (1): 1–13. https://doi.org/10.1038/s41438-021-00681-1.

Wang, L., Ning, C., Pan, T., and Cai, K. (2022). Role of silica nanoparticles in abiotic and biotic stress tolerance in plants: a review. *International Journal of Molecular Sciences* 23 (4): 1947.

Wang, X., Xie, H., Wang, P., and Yin, H. (2023). Nanoparticles in plants: uptake, transport and physiological activity in leaf and root. *Materials* 16 (8): 3097.

Weisany, W., Razmi, J., Eshaghabadi, A.H., and Pashang, D. (2024). Silicon nanoparticles (SiNPs): a novel and sustainable strategy for mitigating environmental stresses in plants. *Journal of Soil Science and Plant Nutrition* 24: 1–25.

Xiao, J., Li, Y., and Jeong, B.R. (2022). Foliar silicon spray to strawberry plants during summer cutting propagation enhances resistance of transplants to high temperature stresses. *Frontiers in Sustainable Food Systems* 6: 938128. https://doi.org/10.3389/fsufs.

Yan, G., Jin, H., Yin, C. et al. (2023). Comparative effects of silicon and silicon nanoparticles on the antioxidant system and cadmium uptake in tomato under cadmium stress. *Science of the Total Environment* 904: 166819.

Younis, A.A., Khattab, H., and Emam, M.M. (2020). Impacts of silicon and silicon nanoparticles on leaf ultrastructure and *TaPIP1* and *TaNIP2* gene expressions in heat stressed wheat seedlings. *Biologia Plantarum* 64 (1): https://doi.org/10.32615/bp.2020.030.

Zahedi, S.M., Moharrami, F., Sarikhani, S., and Padervand, M. (2020). Selenium and silica nanostructure-based recovery of strawberry plants subjected to drought stress. *Scientific Reports* 10 (1): 17672.

Zahedi, S.M., Hosseini, M.S., Hoveizeh, N.F. et al. (2023). Comparative morphological, physiological and molecular analyses of drought-stressed strawberry plants affected by SiO_2 and SiO_2-NPs foliar spray. *Scientia Horticulturae* 309: 111686.

Zhao, W., Wu, Z., Amde, M. et al. (2023). Nanoenabled enhancement of plant tolerance to heat and drought stress on molecular response. *Journal of Agricultural and Food Chemistry* 71 (51): 20405–20418.

12

Role of Si/SiO_2 NPs in Mitigation of Heavy Metal Stress

Naveen Kumar[1,2], *Prabhavathi Venkat Raman*[2] *and Archana Singh*[1]

[1] *Department of Plant Molecular Biology, University of Delhi South Campus, Dhaula Kuan, New Delhi, Delhi, India*
[2] *Department of Botany, Shivaji College, University of Delhi, New Delhi, Delhi, India*

12.1 Introduction

Heavy metal pollution is caused by waste from industries, agricultural runoff, e-waste and anthropogenic processes (Adnan et al. 2024; Okeke et al. 2023; Proshad et al. 2024; Yao et al. 2024). Heavy metals are poisonous to both human and plants but some of the heavy metals are needed for the ordinary development of plants and animals (Kiran et al. 2021). Nanomaterials' exceptional sorption capabilities, high surface areas and abundance of active sites have led to their impressive use in heavy metal remediation (Fan et al. 2017). The nanoparticle size varies from 1 to 100 nm and is categorized as semiconductor, metallic, lipid-based, ceramic, carbon-based and polymeric nanoparticles according to their physico-chemical characteristics, size and shape (Naidu et al. 2023). The second greatest abundant substance on earth is silicon; mono-silicic acid is the form which is absorbed by the plants (Siddiqui et al. 2020). Silicon nanoparticles (SiNPs) show minimum toxic/non-toxic effects on plants and maintain soil fertility compared to other nanoparticles (Mahawar et al. 2023). SiNPs are categorized based on their structure into two categories: nonporous and porous. The method of synthesis affects their pore size, shape, size and rate of dissolution, the effectiveness of the SiNPs is affected by the rate of dissolution and cellular uptake is impacted by the shape and size of the SiNPs (Wang et al. 2022).

Direct or indirect interactions between SiO_2NPs and plants can lead to enhanced photosynthetic efficiency, stronger and more efficient shoot systems, greater enzymatic activity and better crop quality (Ashraf et al. 2024). Photosynthetic activity is improved by SiNPs by enhancing enzyme activity, pigment content, photosystem apparatus and photochemical efficiency in plants (Naaz et al. 2023).

Silicon Nanoparticles: Synthesis, Uptake, and Applications, First Edition.
Edited by Archana Singh, Aryadeep Roychoudhury, and Indrakant K. Singh.

The properties and structure of SiNPs depend on the synthesizing process, broadly categorized as physical, biological and chemical (Yan et al. 2024). Silicon and silica nanoparticles (Si/SiO_2 NPs) have quasi-essential effects on plants and environmental stress mitigation is the key feature for special attention (Naidu et al. 2023). Due to its extensive potential for heavy metal mitigation, heavy metal toxicity can be reduced by SiNPs in the plants. SiNPs excite antioxidant enzymatic activity and control processes, like the endogenous level of phytohormones, to trigger the defence system (Asgher et al. 2024). The content of Pb, Al, Cd and Cr in plants is significantly reduced by SiNPs (Yuvaraj et al. 2023). Si dramatically dropped the amount of Cd that collected in wheat roots and leaves (Alzahrani et al. 2018). According to a previous study, Cd stress in *Zea mays* L. decreased by SiNPs, and Cd concentration in seeds decreased by 62.2%. SiNPs significantly enhance the antioxidant enzymes activity, increase the protein and improve photosynthesis (Rahman et al. 2023). In *Arundinaria pygaea*, the growth of plants is improved by silicon oxide nanoparticles (SiO_2NPs) during heavy metal stress of Mn, Cu and Cd, and the concentration of chlorophyll was improved by the treatment of SiONPs during the Cu/Mn stress, but during the Cd stress, the content of chlorophyll does not enhance after SiO_2NPs treatment. By enhancing heavy metal uptake and accumulation on the root surface, silicon dioxide nanoparticles may reduce the build-up of heavy metals in the plant's aerial organs (Emamverdian et al. 2020).

Heavy metals alter plant physiological functions by increasing the production of reactive oxygen species (ROS), which harms plants by directly damaging lipids and nucleic acids (Mansoor et al. 2023). Defence systems against ROS involve enzymes such as superoxide dismutase, peroxidase (POD), glutathione reductase (GR), catalase (CAT) and some other compounds like ascorbic acid, proline, tocopherols, glutathione and carotenoids, but external help is needed because this defence system is not sufficient for oxidative stress tolerance (El-Saadony et al. 2021). By boosting the antioxidant enzyme activity, SiNPs lower H_2O_2 concentrations, thus reducing ROS and enhancing plant stress tolerance (Altaf et al. 2024). In *Oryza glumaepatula* during Cd stress, H_2O_2 concentration is increased, but after the treatment of SiNPs (20 nm average size range), the production of H_2O_2 decreases by 73% and the activity of antioxidant enzymes such as POD, GR, ascorbate peroxidase (APX), CAT and superoxide dismutase (SOD) improves significantly after the treatment of SiNPs (Ashraf et al. 2024). The high degree of lipid peroxidation, increased antioxidant enzyme activities including POD, SOD and CAT, and decreased GSH content are signs of oxidative stress caused by Cd, which also lowers the concentrations of Fe, Zn, Mg, chlorophyll a and GSH. The application of SiNPs topically reduces the build-up and movement of Cd from root to shoot and increases the activity of GSH, other antioxidant enzymes and minerals including Mg, Fe and Zn (Wang et al. 2015).

12.2 Impact of Si/SiO_2 NPs in Mitigation of Heavy Metal Stress on Crop Production

Unpredictable climate change and soil contamination from various contaminants, such as pesticides, fertilizers and heavy metals, have a significant impact on agriculture, which lowers overall agricultural yields (Okeke et al. 2023). Heavy metals alter plant metabolism by increasing the generation of reactive oxygen species (ROS), which disrupt various metabolic processes (Berni et al. 2019). Plants can withstand heavy metal stress by a variety of physiological and metabolic processes, some of which include complex molecular and biochemical processes (Ghori et al. 2019). During absorption from the soil, silicon (Si) immobilizes heavy metals, reducing their bioavailability. Within plants, heavy metals are compartmentalized or distributed into metabolically inactive tissues, minimizing their toxic effects. Under stress conditions, Si enhances antioxidant enzyme activity, modifies plant tissues, alters gene expression and facilitates chelation and coprecipitation with heavy metals. Additionally, nanoparticles (NPs) improve nutrient availability, and Si further enhances nutrient supply while aiding in heavy metal remediation. By acting as a protective barrier, Si helps maintain cell membrane integrity and increases the photosynthetic rate (Khan et al. 2020; Mousavi 2022). Due to their large surface area, SiNPs provide numerous active sites for heavy metal absorption, enhancing their remediation efficiency (Table 12.1). When heavy metals are absorbed by SiO_2, their mobility and leaching can be stopped and their content in readily extractable fractions can be reduced (Samani et al. 2024). SiNPs readily enter cell walls, exit through pores and are extensively absorbed by plants (Bhat et al. 2021). Through the regulation of endogenous hormones and the enhancement of antioxidant enzyme activity, SiNPs activate defence systems (Asgher et al. 2024). The build-up of heavy metals such as Pb and As in agricultural soil is a serious risk to crop yield because it inhibits vital biochemical and metabolic processes, which in turn stops or slows the growth and development of young seedlings. By increasing the chlorophyll pigments, SiNPs maintain the photosynthetic activity still under the As stress, and enhanced chlorophyll content enhances the plant growth, yield and biological productivity, and carbon assimilation (Ahmad et al. 2024). The intake of vital elements including phosphorus (P), potassium (K) and nitrogen (N), which are necessary for crop growth and output, is reduced under cadmium stress. In addition to increasing the intake and concentration of the micronutrients N, K and P, SiO_2NPs improve the expression of the gene for chlorophyll synthesis (CHLG, POR), which raises the amount of chlorophyll in plants.

During cadmium (Cd) stress, the RBOHD gene is less expressed. SiO_2NPs have been found to increase RBOHD expression in melons, contributing to improved

Table 12.1 List of plants in which different sizes of Si/SiO_2 NPs remediate heavy metal stress.

Plant	Heavy metal(s)	Size of Si/SiO_2 NPs	Mode of application	References
Zea mays	As	150–200 nm	Hydroponic medium	Tripathi et al. (2016)
Phaseolus vulgaris	Cd	40–100 nm	Seed priming	Koleva et al. (2022)
Oryza sativa	Cd, Pb	18.4±3 nm	Foliar	Hussain et al. (2020)
Pisum sativum	Cr	10 μm	Hydroponic	Tripathi et al. (2015)
Oryza sativa	Cr	10 μm	Hydroponic	Sharma et al. (2022)
Solanum lycopersicum	As	10–20 nm	Soil	González-Moscoso et al. (2022)
Phaseolus vulgaris	Cd, Ni and Pb	38.78 nm	Foliar	El-Saadony et al. (2021)
Triticum aestivum L.	Cd		Soil	Khan et al. (2020)
Cicer arietinum	Al	160–138 nm		Chandra et al. (2020)
Zea mays	Al	5–15 nm	Soil	de Sousa et al. (2019)
Triticum aestivum	Cd	20–35 nm	Foliar	Thind et al. (2021)
Arundinaria pygmaea	Cu, Mn and Cd	20 nm	Soil	Emamverdian et al. (2020)
Oryza sativa L. japonica	Cd	19 nm, 48 nm and 202 nm	Suspension culture	Cui et al. (2017)
Oryza sativa L.	Cd	60~100 nm	Foliar	Chen et al. (2018)
Coriandrum sativum L.		20–35 nm	Soil	Fatemi et al. (2021)

stress tolerance by promoting root growth and enhancing photosynthetic pigment levels by preventing the generation of superoxide radicals (Malik et al. 2025). Cr affects various molecular, physiological and biochemical mechanisms like water balance, ROS production, seed germination, the content of minerals, gene expression and DNA damage (Joshi et al. 2019; Kapoor et al. 2022; Sharma et al. 2020; Srivastava et al. 2021b). SiNPs control macro- and micronutrients, including Mg, K, P and Ca in roots and shoots, and enhance photosynthetic pigments in

Pisum sativum under Cr stress (Tripathi et al. 2015). In a different investigation, when exposed to SiO_2NPs during Cr stress, *Triticum aestivum* L. demonstrated increased photosynthetic activity, antioxidant enzyme content and decreased ROS (Manzoor et al. 2022). The *Tagetes erecta* L. plant is subjected to Cr stress in another study, which results in a decrease in photosynthetic pigment, fresh and dry plant weight and length of roots and shoots with an increase in malondialdehyde (MDA), H_2O_2 and proline levels. SiNPs increase the absorption of Ca^{2+} and K^+, which helps the defence mechanisms against heavy metals in plants, and under Cd stress, SiNPs increase the chlorophyll content by 71%, the level of MDA decreases by 24%, H_2O_2 level decreases by 37% and the relative membrane permeability is reduced by 45% and the absorption of minerals such as Na^+, K^+ and Ca^{2+} by 51%, 52% and 36%, respectively (Raza et al. 2024). *Chenopodium quinoa* shows increased transcription of PCS1 and RBOHD genes, H_2O_2 burst, decreased lipid peroxidation, increased accumulation of thiol compounds, induced antioxidant enzymes activity and enhanced tolerance index when SiNPs sprayed on leaves during Pb stress (Karimi-Baram et al. 2024).

In *Triticum aestivum* L., Cu stress reduces the shoot and the root length by 29.46% and 27.48%, and chlorophyll a, b and carotenoids decrease by 15.87%, 44.68% and 44%, respectively, and the total chlorophyll content by 23.69%. The length of the roots and shoots grew by 15.61% and 22.04%, respectively, following the SiNPs treatment, and the content of carotenoids, chlorophyll a, b and total chlorophyll was enhanced by 50%, 6.6%, 46.15% and 21.9%, respectively. SiNPs decrease the accumulation of Cu^{2+} in leaves by 26.29% and increase the AsA (ascorbic acid) and GSH (glutathione) by 178% and 43.39% (Riaz et al. 2022). Hg interferes with uptake of micronutrients, binds to nitrogen and/or sulphur-rich ligands, represses plant growth, induces physiological and morphological changes and oxidative stress, and decreases net photosynthesis. Hg also induces geno- and cytotoxicity, producing mutation by interacting with DNA and altering the number and structure of the chromosomes (Azevedo et al. 2018). SiO_2NP treatment in *Glycine max* reduces the build-up of mercury in the roots, shoots and leaves, increases the biomass and chlorophyll content of seedlings and decreases SOD activity (Li et al. 2020).

12.3 Application of Si/SiO_2 NPs Causes Tolerance Against Heavy Metal Stress

The use of nanoparticles in agriculture, an emerging multidisciplinary field, enhances plant growth and improves stress tolerance by mitigating environmental stressors and optimizing nutrient uptake (Wang et al. 2022). SiNPs have positive effects on agriculture under stress because of their size, which allows them to enter the plant system by crossing the cell wall (Li et al. 2024; Wu and Li 2022). Some of the applications of Si/SiO$_2$ NPs are discussed below:

12.3.1 Seed Priming

NPs used in seed priming aids in the germination of seeds under stressful circumstances, and seed priming is an inexpensive and efficient technique for improving seed germination, reducing germination time and boosting uniform seed emergence (Hussain et al. 2019; Johnson and Puthur 2021; Ulhassan et al. 2023). Seed priming is a pre-germination hydration mechanism in which metabolic processes associated with germination are initiated, but radicle emergence is yet to happen. It nullifies the requirement of adding chemicals to the environment because it is directly associated with seeds; moreover, it may replace the agrochemical and fertilizers (Bhatia and Gupta 2022; Srivastava et al. 2021a). In primed seeds, essential components for germination, including dehydrogenase, α-amylase, expansin enzymes, transcription factors and mRNAs, are either readily available or largely synthesized, facilitating a faster and more synchronized germination process. In primed seeds, all the components that are required for germination such as dehydrogenase, α-amylase, expansin, transcription factors and mRNAs are available or largely synthesized (Zhao et al. 2018). Osmoprimed seeds have DNA pre-synthesized prior to germination and that is why cells are in the S-phase and have 4C DNA content. Late embryogenesis is upregulated. Primed seeds' high protein content and upregulation of heat shock protein demonstrate their capacity for stress tolerance (Raj and Raj 2019). The weight of the shoot, root and seedling is greatly increased by seed priming with an appropriate concentration of SiO$_2$NPs; the kind of nanoparticles, their concentration and the application method all affect the response and behaviour of plants. SiO$_2$NP priming can promote pre-germination events like cell elongation, water imbibition, cell division, increased antioxidant enzyme activity, reserve mobilization enzyme activation and an improvement in the seedling emergence parameter (Hasanaklou et al. 2023). The hazardous effects and build-up of heavy metals in plants can be lessened by seed shocking and seed dressing. Additionally, seed shocking with SiNPs increases biomass and yield while lowering the concentration of Cd in wheat (Wu et al. 2024). Grain weight, spike dry weight, shoot length, dry weight of roots and shoots and concentration of chlorophyll a and b all rise in *Triticum aestivum* L. following seed priming, but H_2O_2 and Cd concentrations fall as a result of seed priming with SiNPs (Hussain et al. 2019). In another study in *Brassica napus,* the interaction between seed coat and SiO$_2$NPs is strong which enhances the germination under Cr stress and also enhances the shoot length by 32.24%, germination percentage by 26%, leaf fresh weight by 33.07%, root fresh weight 31.88%, root dry weight 23.73% and root length 35.39%, under Cr stress conditions. BnCAO, BnPOR and BnCHLG (Chlorophyll synthesis genes) are upregulated by SiO_2NPs, BnSAG12 (Senescence) are downregulated and BnPAL, BnMT-1, BnPAO, BnPPO and BnCAD (plant defence gene) are upregulated by SiO_2NPs. SiO_2NPs raise the amount of Cr in the cell wall while lowering the amount in the soluble fraction; seed priming reduces the build-up of Cr and raises the amount of Si in plant parts (Ulhassan et al. 2023).

12.4 Utilizing Si/SiO_2 NPs in Conjunction with Biochar to Reduce Heavy Metal Stress

Biochar is created by pyrolysis of plant-based biomass; it aids in the clean-up of heavy metals from the soil and promotes carbon sequestration (Ghosh and Maiti 2021; Seow et al. 2022). Biochar's strong stability and adsorption capacity improve the chemical and physical characteristics and fertility of the soil, immobilize heavy metals and limit heavy metal absorption by plants (AL-Huqail 2024). Biochar sorption capacity can be improved by adding composites of interest and it could often improve the physicochemical properties (Naeem et al. 2024). Biochar made from rice straw and bamboo is utilized in the decontamination of heavy metals like Zn, Cd, Cu and Pb, and it additionally lowers the bioavailability of heavy metals (Rafi et al. 2022). Nanomaterials are utilized for alleviating a variety of environmental challenges; however, they lose or lessen their effectiveness due to aggregation, bulk structure development, or oxidization. Stabilization is achieved through the use of a variety of support materials, including biochar. Biochar reduces nanomaterial aggregation, and the inclusion of nanomaterial improves the physicochemical properties of biochar, resulting in the formation of a new composite material (Peng et al. 2024). According to a recent study, in *Chenopodium quinoa*, As stress is alleviated by using SiNPs-doped biochar in both saline and normal soil conditions. SiNPs-doped biochar lowers the accumulation of As due to preferential absorption of Si and improves the activity of antioxidant enzymes when compared with undoped biochar (Alsamadany et al. 2022). Another study found that in *Hordeum vulgare* L., SiNPs-loaded biochar reduces As stress and increases biomass, stomatal conductance and leaf relative water content of the plant, as well as yield, when As stress is present. H_2O_2 concentrations decreased, but APX, CAT, SOD and PAT activity rose dramatically. The concentration of As in shoots, roots and grains considerably reduces after the addition of SiNP-loaded biochar (Naeem et al. 2024).

12.4.1 Si/SiO_2 NPs with Other Nanoparticles

Iron nanoparticles (FeNPs) and SiNPs are both used to remove heavy metals from soil; in rice, FeNPs and SiNPs diminish Pb uptake. Because of synergistic effects, utilizing both nanoparticles together offers more potential than using either one alone (Ghouri et al. 2024). Iron oxide nanoparticles (FeONPs) and SiO_2NPs have distinct physicochemical properties and stability, making them potential growth agents. The accumulation and acropetal translocation of As in wheat are reduced by FeONPs and SiO_2NPs, both synergistically raising the percentage of carotenoids and chlorophyll (Manzoor et al. 2022). The development of maize plants is improved by biochar, FeNPs, and SiNPs; when combined, they reduce the heavy metal stress caused due to Cd. All of them absorb more Cd when impregnated, both electrostatically and non-electrostatically. It is mostly dependent on the available adsorption sites, which are influenced by electrostatic or physical forces. ZmHMA2, ZmNramp5 and ZmHMA3 (Cd transport-related gene) are downregulated by biochar, SiNPs, and FeNPs together, and also alleviate stomatal dysfunction, genetic processes, supporting crucial physiological processes induced by Cd stress (Yasin et al. 2024). In *Oryza sativa,* As stress causes lower chlorophyll a and b content, and biomass, and increases the accumulation of H_2O_2, methylglyoxal, malondialdehyde and electron leakage. SiNPs and titanium dioxide nanoparticles (TiO_2NPs) enhance the biomass and chlorophyll content, decrease the H_2O_2, methylglyoxal and malondialdehyde accumulation and reduce electron leakage in As-stressed plants (Kiany et al. 2022). In *Oryza sativa,* selenium nanoparticles (SeNPs) and SiNPs lessen the accumulation of Cd and Pb. The concentration of phytic acid is decreased by the foliar application of a combination of SeNPs and SiNPs, lowering the content of phytic acid improves the absorption of zinc and iron. SiNPs and SeNPs improve biological yield and grain production under Cd and Pb stress conditions (Hussain et al. 2020).

12.4.2 Use of Si/SiO_2 NPs with Other Substances

SiNPs and indole acetic acid (IAA) work together to lessen the Cr stress on *Oryza sativa*. When Cr builds up in rice seedlings, it stunts growth, increases the generation of reactive oxygen species, and damages the macromolecules involved in photosynthesis. The combined effects of SiNPs and IAA reduce the stress effect more efficiently than their separate treatment, accumulation of Cr in seedlings decreases significantly, the content of carotenoids and chlorophyll is enhanced, the level of H_2O_2 decreases significantly, and they enhance the level of CAT and SOD under Cr stress conditions (Sharma et al. 2022). The combined effect of *Trichoderma* metabolites (TM) and SiNPs reduces the toxicity of Cd in tomato plants. SiNPs and TM treatment increase the gaseous exchange, growth and photosynthetic pigment, under Cd stress conditions (Khan et al. 2023).

12.5 Mechanisms of Si-Mediated Tolerance to Heavy Metal Stress

Plants have mainly two general mechanisms tolerance and avoidance when encountering abiotic stresses like heavy metals. Plants acquire mechanisms for avoidance and tolerance to heavy metals, as well as techniques including regulation and adjustment of the permeability of the plasma membrane (Emamverdian et al. 2015). Under stress conditions, phytochelatin is the best detoxifying mechanism for heavy metals. Plants also use metal-binding proteins such as metallothionein (Emamverdian et al. 2018). Excluders, intermediates and accumulators are three classes in which plants are classified: for silicon accumulation, plants need active silicon-uptake mechanisms because silicon uptake is against the concentration gradient (Mukarram et al. 2022). Aquaporin protein family transporter Lsi1 is an influx silicon transporter with NIP (Nod26-like major intrinsic protein) subfamily (Maryam and Gul 2022). The basel zone of the root is the primary location where LSi1 is located, on the plasma membrane, and LSi1 is placed on both the distal sides of the endodermis and exodermis, where the Casparian strip is found, while LSi2 is located on the proximal side of the same cell. LSi2 exhibits efflux transport activity, and both LSi2 and LSi1 work cooperatively to move Si from the root cell to the apoplast (Rao and Susmitha 2017; Coskun et al. 2021; Kaur and Greger 2019; Mitani-Ueno et al. 2011). Most plants absorb silicon in the form of Si(OH)4, while Lsi6 is used to unload in the xylem sap. The intravascular transfer (transport of Si from vascular bundle to panicles) is facilitated by LSi6 (Figure 12.1) (Islam et al. 2020; Maryam and Gul 2022; Mukarram et al. 2023; Rao and Susmitha 2017).

Si is stored in leaf cells to form a double-layer cuticle-silicon-cuticle protective construction consisting of Si and cuticle, and Si alters the physiology and morphology of the plant and enhances its defence system. However, when Si concentration increases, Si is polymerized into silica gel by plants. SiO_2 is the form that is deposited in the plants (Asgher et al. 2024; Balakhnina and Borkowska 2013).

12.6 Influence of Si/SiO_2 NPs on Phytohormone

Si regulates the accumulation, signalling and endogenous production of phytohormones, the level of jasmonic acid (JA), cytokinins (Ck), abscisic acid (ABA), IAA and ethylene (ETH) is enhanced in crops by introducing SiNPs (Khan et al. 2021).

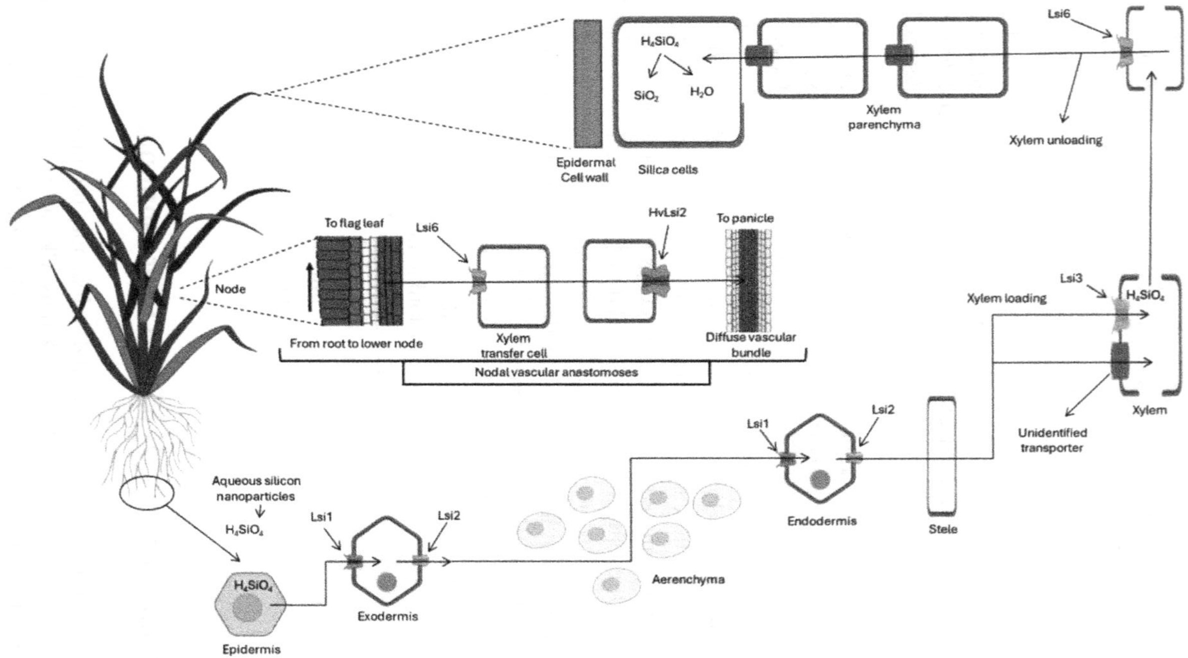

Figure 12.1 Transport of Si nanoparticle in plants involves the participation of many silicon transporters such as LSi1, LSi2, LSi3, and Lsi6. *Source*: Islam et al. 2020; Mukarram et al. 2023.

However, in some circumstances, treatment with SiNPs during abiotic and biotic stress conditions decreases phytohormone production (Asgher et al. 2024). In *Arabidopsis*, Si regulates the signaling of ETH, and Si enhances the biosynthesis of AUR3 and ACS8 genes which up-regulate the response to the toxicity of Cu. These genes have their own pattern of expression. The ACS8 gene is influenced dramatically by Si during high Cu, and the ACS8 gene is responsible for the biosynthesis of ETH (Khandekar and Leisner 2011; Zhao et al. 2009). According to a previous study on *Arabidopsis thaliana*, excessive use of ZnONPs in crops increases ZnO toxicity and the expression of ETR2 (ethylene response 2), ACS (Acetyl-CoA synthase), ERF1 (ethylene response factor 1) and ACO (ACC oxidase). SiO_2NPs increase the expression of ACO but no significant effect is seen on the other three. However, if ZnONPs and SiO_2NPs are used together, the toxic effect of ZnONPs is reduced. ARR7 and ARR15 (cytokinin-regulating genes) have less expression by ZnONPs as compared to the expression by SiO_2NPs. SiO_2 decreases the ETH and increases the CK-related gene expression, and it is nontoxic in nature (Azhar et al. 2021). Silicon and gibberellic acid (GA) have complex interactions where Si enhances the level of GA (Jang et al. 2018). In *Vicia sativa*, SiNPs improve GA levels under Cu stress conditions (Pestovsky and Martínez-Antonio 2017) (Figure 12.1).

12.7 Effect of Si/SiO_2 NPs on Photosynthesis and Reactive Oxygen Species

The fundamental process for acquiring energy in plants is photosynthesis, and heavy metal stress leads to lessening the photosynthetic capacity of the plants. Stomatal conductance is decreased by heavy metals which is the reason for decreasing the photosynthetic capacity (Huihui et al. 2020; Sytar et al. 2018). Build-up of SiNPs in plant leaves keeps them overextended and upright to incarcerate higher amounts of sunlight, thus enhancing the process of photosynthesis and optimizing photosynthesis. Photosystem II activity, net photosynthetic rate, electron transport rate, stomatal conductance and photosynthetic pigment synthesis are involved in the photosynthetic processes like carbonic anhydrase in *Indocalamus barbatus* and lemongrass improved by the SiNPs (Mukarram et al. 2022). Carbonic anhydrase is the enzyme that delivers CO_2 to rubisco. SiNPs improve carbonic anhydrase enzyme activity, as well as carotenoids and photosynthetic pigments (Madany et al. 2020; Sharifi-Rad et al. 2018). Photosynthetic genes induced by SiNPs such as photosystem genes PsbW, PsbP, Psb28, PsbQ, PsbA and PsbA, protein Q7F8E8, P19312 and Q332S1 and electron transport chain PetF and PetE are induced by SiNPs, which support the assembly of the photosystems and thylakoid member and light-harvesting complex (Mukarram et al. 2024).

Si supplementation improves the AsbA, D1 protein of PS II, PsaA and core protein of PS I (Zhang et al. 2019). In *Triticum aestivum* L., the activity of PS II photosystem, including PsbB (encodes PS II), PsbH (regulate PS II stability and assembly), PsbD (for D2 protein) and PS I and PS II, is enhanced by SiNPs (Khattab et al. 2021). Carbon accumulation and carboxylation are intensified by SiNPs by enhancing the stomatal activity in lemongrass plants, and aquaporin activity and hydraulic conductance are the reasons for silicon-induced stomatal activity (Mukarram et al. 2023). A previous study on *Brassica napus* L. found that SiO_2NPs lower the harmful effects of Cr and improve the structure of cells by changing the integrity of the cell membrane, making the thylakoid membrane more stable, and increasing the presence of plastoglobuli and normal starch grains. They also improved stomatal conductance, net photosynthesis, and the rates of transcription and electron transport when Cr stress is present (Huang et al. 2024). According to another study on *Brassica napus* L., SiO_2NPs enhance the total chlorophyll and net photosynthetic rate, carotenoid content and transcription rate and they also maintain the colour of the leaves during the Cd stress (Ahmed et al. 2023). SiNPs enhance photosynthesis by increasing the content of carotenoids, chlorophyll a and b, in *Zea mays* L. during the Pb stress (Maryam et al. 2024). Cd stress in the Ujala wheat variety is alleviated by SiNPs, and it also increases the chlorophyll a, b, and total carotenoids, total chlorophyll, during the Cd stress (Thind et al. 2021).

ROS generation is a normal metabolism of peroxisomes, chloroplasts and mitochondria; nevertheless, environmental stress enhances ROS production (Berni et al. 2019). Chrome and copper are redox-active heavy metals, $OH^{\cdot}$, one of the most reactive ROS they can produce. Lead, cadmium, nickel, zinc, etc., are some of the heavy metals without redox capacity and produce singlet oxygen which is further involved in the production of other ROS superoxides. ROS is essential for plant defence systems and cannot be eliminated completely (Fryzova et al. 2018). Heavy metals alter various physiological processes by inactivation of enzymes at the molecular/cellular level, competing with essential elements, disrupting membrane integrity, increasing the ROS and nitrogen species disrupting the tricarboxylic acid cycle, Calvin cycle and photosynthesis (Rascio and Navari-Izzo 2011). ROS increases the action of enzymes such as SOD, CAT, POD and the anti-oxidant component GSH to remove ROS. In chloroplasts, CAT is lacking to eliminate H_2O_2; hence, predominantly APX and SOD are used to remove $O_2^{\cdot-}$ and H_2O_2 (Figure 12.2) (Huihui et al. 2020).

In *solanum lycopersicum*, Cd stress affects the ROS, SOD, ASA (ascorbate), GR and osmotic protectors like sugars and proline, and some of the non-enzymatic molecules (GSH [glutathione] and ASA [ascorbate]). Plants undergo γ-aminobutyric acid (GABA) production in response to abiotic stress. After treatment of SiNPs, the redox status of GSH and ASA is restored, which protects the cell from

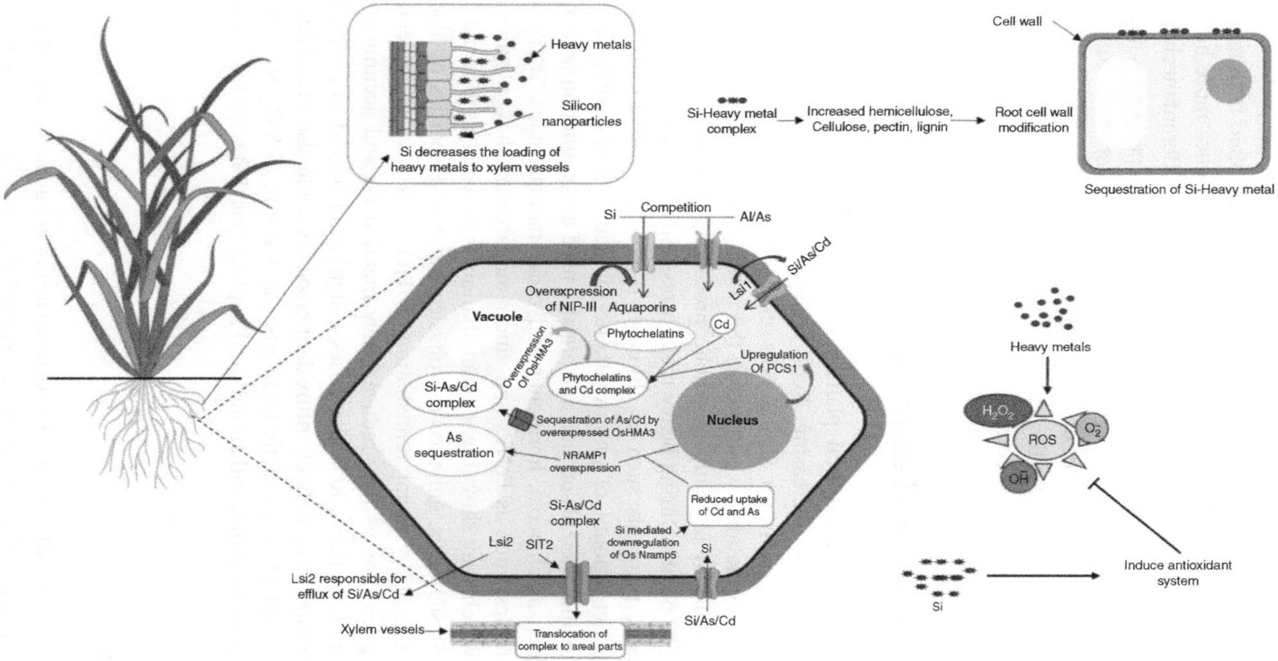

Figure 12.2 The mechanism of Si-mediated heavy metal remediation involves trapping heavy metals in the plant cell's vacuole and forming a compound with the metal to render it biologically unavailable. *Source*: Rachappanavar et al. (2024), Yadav et al. (2023).

oxidative stress. It also boosts the activity of APX, SOD, GR and GST, as well as the content of GABA substantially (Rahmatizadeh et al. 2024). In rice plants, toxicity of Cd is mitigated by applying both TiO_2NPs and SiNPs together; both nanoparticles together increase the root and shoot length and produce more biomass, increase growth and enhance nitrogen uptake, and they use different mechanisms to decrease, detoxify, and mitigate Cd stress. TiO_2NPs have properties to safeguard the membrane from ROS and the level of H_2O_2 declines after the treatment of both SiNPs and TiO_2NPs together during the Cd toxicity. The presence of SiNPs and TiO_2NPs together enhances the antioxidant enzyme activities such as CAT, SOD and POD, during Cd stress conditions (Lai et al. 2023).

12.8 Conclusion and Future Prospective

Heavy metals inhibit agricultural plant output and growth via affecting the photosynthetic process, increasing ROS production, reducing growth and disrupting metabolic pathways. Si/SiO_2 NPs serve to mitigate heavy metal stress and reduce ROS formation by enhancing the activity of antioxidant enzymes. Because ROS is useful in many metabolic processes, entirely removing it is also damaging to plants. During heavy metal stress, SiNPs enhance the content of photosynthetic pigments such as chlorophyll a, b and carotenoids, hence increasing the net amount of photosynthetic pigments. SiNPs compete with heavy metals to limit their absorption and are also used in conjunction with other compounds such as charcoal, hormones and other nanoparticles. SiNPs with other nanoparticles, such as FeNPs and zinc nanoparticles, show a synergetic effect due to which they show higher remediation than when used alone and, with biochar Si/SiO_2 NPs, enhance the physicochemical properties of biochar and increase the heavy metal remediation. Si/SiO_2 NPs have a tremendous potential in agriculture, as they reduce heavy metal stress and accumulate in plants, strengthening them. ROS rise poses a significant hazard to plants, and Si/SiO_2 NPS boosts the activity of antioxidant enzymes while also increasing net photosynthesis and photosynthetic pigments. As a result, Si/SiO_2 NPs are an excellent choice for research and agricultural applications that require enhanced crop output.

References

Adnan, M., Xiao, B., Ali, M.U. et al. (2024). Heavy metals pollution from smelting activities: a threat to soil and groundwater. *Ecotoxicology and Environmental Safety* 274: https://doi.org/10.1016/j.ecoenv.2024.116189.

Ahmad, Z., Younis, R., Ahmad, T. et al. (2024). Modulating physiological and antioxidant responses in wheat cultivars via foliar application of silicon nanoparticles (SiNPs) under arsenic stress conditions. *Silicon* 16 (12): 5199–5211. https://doi.org/10.1007/s12633-024-03078-6.

Ahmed, T., Masood, H.A., Noman, M. et al. (2023). Biogenic silicon nanoparticles mitigate cadmium (Cd) toxicity in rapeseed (Brassica napus L.) by modulating the cellular oxidative stress metabolism and reducing Cd translocation. *Journal of Hazardous Materials* 459: https://doi.org/10.1016/j.jhazmat.2023.132070.

AL-Huqail, A.A. (2024). Combined application of biochar and nanoparticles (silicon and boron) effectively reduced the metal toxicity of sweet basil (Ocimum basilicum L). *Acta Physiologiae Plantarum* 46 (7): https://doi.org/10.1007/s11738-024-03696-y.

Alsamadany, H., Alharby, H.F., Al-Zahrani, H.S. et al. (2022). Silicon-nanoparticles doped biochar is more effective than biochar for mitigation of arsenic and salinity stress in Quinoa: insight to human health risk assessment. *Frontiers in Plant Science* 13: https://doi.org/10.3389/fpls.2022.989504.

Altaf, M.M., Yi, H., Bashir, S. et al. (2024). Mitigating chromium stress in tomato plants using green-silicone nanoparticles: enhancing cellular oxidative stress management and chromium reduction. *Scientia Horticulturae* 338: https://doi.org/10.1016/j.scienta.2024.113635.

Alzahrani, Y., Kuşvuran, A., Alharby, H.F. et al. (2018). The defensive role of silicon in wheat against stress conditions induced by drought, salinity or cadmium. *Ecotoxicology and Environmental Safety* 154: 187–196. https://doi.org/10.1016/j.ecoenv.2018.02.057.

Asgher, M., Rehaman, A., ul Islam, S.N., and Khan, N.A. (2024). Multifaceted roles of silicon nano particles in heavy metals-stressed plants. *Environmental Pollution* 341: https://doi.org/10.1016/j.envpol.2023.122886.

Ashraf, H., Ghouri, F., Liang, J. et al. (2024). Silicon dioxide nanoparticles-based amelioration of Cd toxicity by regulating antioxidant activity and photosynthetic parameters in a line developed from wild rice. *Plants* 13 (12): https://doi.org/10.3390/plants13121715.

Azevedo, R., Rodriguez, E., Mendes, R.J. et al. (2018). Inorganic Hg toxicity in plants: a comparison of different genotoxic parameters. *Plant Physiology and Biochemistry* 125: 247–254. https://doi.org/10.1016/j.plaphy.2018.02.015.

Azhar, B.J., Noor, A., Zulfiqar, A. et al. (2021). Effect of zno, sio2 and composite nanoparticles on arabidopsis thaliana and involvement of ethylene and cytokinin signaling pathways. *Pakistan Journal of Botany* 53 (2): 437–446. https://doi.org/10.30848/PJB2021-2(40).

Balakhnina, T. and Borkowska, A. (2013). Effects of silicon on plant resistance to environmental stresses: review. *International Agrophysics* 27 (2): 225–232. https://doi.org/10.2478/v10247-012-0089-4.

Berni, R., Luyckx, M., Xu, X. et al. (2019). Reactive oxygen species and heavy metal stress in plants: impact on the cell wall and secondary metabolism. *Environmental and Experimental Botany* 161: 98–106. https://doi.org/10.1016/J.ENVEXPBOT.2018.10.017.

Bhat, J. A., Rajora, N., Raturi, G., Sharma, S., Dhiman, P., Sanand, S., Shivaraj, S. M., Sonah, H., & Deshmukh, R. (2021). Silicon nanoparticles (SiNPs) in sustainable agriculture: major emphasis on the practicality, efficacy and concerns. In *Nanoscale Advances* (Vol. 3, Issue 14, pp. 4019–4028). Royal Society of Chemistry. https://doi.org/10.1039/d1na00233c.

Bhatia, P. and Gupta, M. (2022). Micronutrient seed priming: new insights in ameliorating heavy metal stress. *Environmental Science and Pollution Research* 29, Issue 39: 58590–58606. https://doi.org/10.1007/s11356-022-21795-6.

Chandra, J., Chauhan, R., Korram, J. et al. (2020). Silica nanoparticle minimizes aluminium imposed injuries by impeding cytotoxic agents and over expressing protective genes in Cicer arietinum. *Scientia Horticulturae* 260: https://doi.org/10.1016/j.scienta.2019.108885.

Chen, R., Zhang, C., Zhao, Y. et al. (2018). Foliar application with nano-silicon reduced cadmium accumulation in grains by inhibiting cadmium translocation in rice plants. *Environmental Science and Pollution Research* 25 (3): 2361–2368. https://doi.org/10.1007/s11356-017-0681-z.

Coskun, D., Deshmukh, R., Shivaraj, S.M. et al. (2021). Lsi2: a black box in plant silicon transport. *Plant and Soil* 466 (1–2). https://doi.org/10.1007/s11104-021-05061-1.

Cui, J., Liu, T., Li, F. et al. (2017). Silica nanoparticles alleviate cadmium toxicity in rice cells: mechanisms and size effects. *Environmental Pollution* 228: 363–369. https://doi.org/10.1016/j.envpol.2017.05.014.

El-Saadony, M.T., Desoky, E.S.M., Saad, A.M. et al. (2021). Biological silicon nanoparticles improve Phaseolus vulgaris L. yield and minimize its contaminant contents on a heavy metals-contaminated saline soil. *Journal of Environmental Sciences (China)* 106: 1–14. https://doi.org/10.1016/j.jes.2021.01.012.

Emamverdian, A., Ding, Y., Mokhberdoran, F., and Xie, Y. (2015). Heavy metal stress and some mechanisms of plant defense response. *Scientific World Journal* 2015: https://doi.org/10.1155/2015/756120.

Emamverdian, A., Ding, Y., Xie, Y., and Sangari, S. (2018). Silicon mechanisms to ameliorate heavy metal stress in plants. *BioMed Research International* 2018: https://doi.org/10.1155/2018/8492898.

Emamverdian, A., Ding, Y., Mokhberdoran, F. et al. (2020). Determination of heavy metal tolerance threshold in a bamboo species (Arundinaria pygmaea) as treated with silicon dioxide nanoparticles. *Global Ecology and Conservation* 24: https://doi.org/10.1016/j.gecco.2020.e01306.

Fan, H.L., Zhou, S.F., Jiao, W.Z. et al. (2017). Removal of heavy metal ions by magnetic chitosan nanoparticles prepared continuously via high-gravity reactive precipitation method. *Carbohydrate Polymers* 174: 1192–1200. https://doi.org/10.1016/j.carbpol.2017.07.050.

Fatemi, H., Esmaiel Pour, B., and Rizwan, M. (2021). Foliar application of silicon nanoparticles affected the growth, vitamin C, flavonoid, and antioxidant enzyme activities of coriander (Coriandrum sativum L.) plants grown in lead (Pb)-spiked soil. *Environmental Science and Pollution Research* 28: 1417–1425. https://doi.org/10.1007/s11356-020-10549-x/Published.

Fryzova, R., Pohanka, M., Martinkova, P. et al. (2018). Oxidative stress and heavy metals in plants. *Reviews of Environmental Contamination and Toxicology* 245: 129–156.

Ghori, N.H., Ghori, T., Hayat, M.Q. et al. (2019). Heavy metal stress and responses in plants. *International Journal of Environmental Science and Technology* 16 (3): 1807–1828. https://doi.org/10.1007/S13762-019-02215-8.

Ghosh, D. and Maiti, S.K. (2021). Biochar assisted phytoremediation and biomass disposal in heavy metal contaminated mine soils: a review. *International Journal of Phytoremediation* 23 (6): 559–576. https://doi.org/10.1080/15226514.2020.1840510.

Ghouri, F., Sarwar, S., Sun, L. et al. (2024). Silicon and iron nanoparticles protect rice against lead (Pb) stress by improving oxidative tolerance and minimizing Pb uptake. *Scientific Reports* 14 (1): https://doi.org/10.1038/s41598-024-55810-2.

González-Moscoso, M., Juárez-Maldonado, A., Cadenas-Pliego, G. et al. (2022). Silicon nanoparticles decrease arsenic translocation and mitigate phytotoxicity in tomato plants. *Environmental Science and Pollution Research* 29 (23): 34147–34163. https://doi.org/10.1007/s11356-021-17665-2.

Hasanaklou, N.T., Mohagheghi, V., Hasanaklou, H.T. et al. (2023). Seed nano-priming using silica nanoparticles: effects in seed germination and physiological properties of Stevia Rebaudiana Bertoni. *Chemical and Biological Technologies in Agriculture* 10 (1): https://doi.org/10.1186/s40538-023-00445-0.

Huang, Q., Ayyaz, A., Farooq, M.A. et al. (2024). Silicon dioxide nanoparticles enhance plant growth, photosynthetic performance, and antioxidants defence machinery through suppressing chromium uptake in Brassica napus L. *Environmental Pollution* 342: https://doi.org/10.1016/j.envpol.2023.123013.

Huihui, Z., Xin, L., Zisong, X. et al. (2020). Toxic effects of heavy metals Pb and Cd on mulberry (Morus alba L.) seedling leaves: photosynthetic function and reactive oxygen species (ROS) metabolism responses. *Ecotoxicology and Environmental Safety* 195: https://doi.org/10.1016/j.ecoenv.2020.110469.

Hussain, A., Rizwan, M., Ali, Q., and Ali, S. (2019). Seed priming with silicon nanoparticles improved the biomass and yield while reduced the oxidative stress and cadmium concentration in wheat grains. *Environmental Science and Pollution Research* 26 (8): 7579–7588. https://doi.org/10.1007/s11356-019-04210-5.

Hussain, B., Lin, Q., Hamid, Y. et al. (2020). Foliage application of selenium and silicon nanoparticles alleviates Cd and Pb toxicity in rice (Oryza sativa L.). *Science of the Total Environment* 712: https://doi.org/10.1016/j.scitotenv.2020.136497.

Islam, W., Tayyab, M., Khalil, F. et al. (2020). Silicon-mediated plant defense against pathogens and insect pests. *Pesticide Biochemistry and Physiology* 168: https://doi.org/10.1016/j.pestbp.2020.104641.

Jang, S.W., Kim, Y., Khan, A.L. et al. (2018). Exogenous short-term silicon application regulates macro-nutrients, endogenous phytohormones, and protein expression in Oryza sativa L. *BMC Plant Biology* 18 (1): https://doi.org/10.1186/s12870-017-1216-y.

Johnson, R. and Puthur, J.T. (2021). Seed priming as a cost effective technique for developing plants with cross tolerance to salinity stress. *Plant Physiology and Biochemistry* 162: 247–257. https://doi.org/10.1016/j.plaphy.2021.02.034.

Joshi, N., Menon, P., and Joshi, A. (2019). Effect of chromium on germination in some crops of India. *Journal of Agricultural science and Botany* 3 (1): 3. http://www.alliedacademies.org/journal-agricultural-science-botany/.

Kapoor, R.T., Bani Mfarrej, M.F., Alam, P. et al. (2022). Accumulation of chromium in plants and its repercussion in animals and humans. *Environmental Pollution* 301: https://doi.org/10.1016/j.envpol.2022.119044.

Karimi-Baram, A., Amooaghaie, R., Ghorbanpour, M., and Ahadi, A. (2024). RBOH-dependent H_2O_2 burst induced by silicon dioxide nanoparticles establishes systemic acquired acclimation in quinoa under lead toxicity. *Scientia Horticulturae* 335: https://doi.org/10.1016/j.scienta.2024.113366.

Kaur, H. and Greger, M. (2019). A review on si uptake and transport system. *Plants* 8 (4): https://doi.org/10.3390/plants8040081.

Khan, Z.S., Rizwan, M., Hafeez, M. et al. (2020). Effects of silicon nanoparticles on growth and physiology of wheat in cadmium contaminated soil under different soil moisture levels. *Environmental Science and Pollution Research* 27 (5): 4958–4968. https://doi.org/10.1007/s11356-019-06673-y.

Khan, M.I.R., Ashfaque, F., Chhillar, H. et al. (2021). The intricacy of silicon, plant growth regulators and other signaling molecules for abiotic stress tolerance: an entrancing crosstalk between stress alleviators. *Plant Physiology and Biochemistry* 162: 36–47. https://doi.org/10.1016/j.plaphy.2021.02.024.

Khan, R.A.A., Alam, S.S., Najeeb, S. et al. (2023). Mitigating Cd and bacterial wilt stress in tomato plants through trico-synthesized silicon nanoparticles and Trichoderma metabolites. *Environmental Pollution* 333: https://doi.org/10.1016/j.envpol.2023.122041.

Khandekar, S. and Leisner, S. (2011). Soluble silicon modulates expression of Arabidopsis thaliana genes involved in copper stress. *Journal of Plant Physiology* 168 (7): 699–705. https://doi.org/10.1016/j.jplph.2010.09.009.

Khattab, H., Alatawi, A., Abdulmajeed, A. et al. (2021). Roles of Si and SiNPs in improving thermotolerance of wheat photosynthetic machinery via upregulation of PsbH, PsbB and PsbD genes encoding PSII core proteins. *Horticulturae* 7 (2): 1–16. https://doi.org/10.3390/horticulturae7020016.

Kiany, T., Pishkar, L., Sartipnia, N. et al. (2022). Effects of silicon and titanium dioxide nanoparticles on arsenic accumulation, phytochelatin metabolism, and antioxidant system by rice under arsenic toxicity. *Environmental Science and Pollution Research* 29 (23): 34725–34737. https://doi.org/10.1007/s11356-021-17927-z.

Kiran, Bharti, R., and Sharma, R. (2021). Effect of heavy metals: an overview. *Materials Today Proceedings* 51: 880–885. https://doi.org/10.1016/j.matpr.2021.06.278.

Koleva, L., Umar, A., Yasin, N.A. et al. (2022). Iron oxide and silicon nanoparticles modulate mineral nutrient homeostasis and metabolism in cadmium-stressed phaseolus vulgaris. *Frontiers in Plant Science* 13: https://doi.org/10.3389/fpls.2022.806781.

Lai, M., Ghouri, F., Sarwar, S. et al. (2023). Modulation of metal transporters, oxidative stress and cell abnormalities by synergistic application of silicon and titanium oxide nanoparticles: a strategy for cadmium tolerance in rice. *Chemosphere* 345: https://doi.org/10.1016/j.chemosphere.2023.140439.

Li, Y., Li, Y., Zhu, N. et al. (2020). Silica nanoparticles alleviate mercury toxicity: via immobilization and inactivation of Hg(ii) in soybean (Glycine max). *Environmental Science: Nano* 7 (6): 1807–1817. https://doi.org/10.1039/d0en00091d.

Li, Y., Zan, T., Li, K. et al. (2024). Silica nanoparticles promote the germination of salt-stressed pepper seeds and improve growth and yield of field pepper. *Scientia Horticulturae* 337: https://doi.org/10.1016/j.scienta.2024.113570.

Madany, M.M.Y., Saleh, A.M., Habeeb, T.H. et al. (2020). Silicon dioxide nanoparticles alleviate the threats of broomrape infection in tomato by inducing cell wall fortification and modulating ROS homeostasis. *Environmental Science: Nano* 7 (5): 1415–1430. https://doi.org/10.1039/c9en01255a.

Mahawar, L., Ramasamy, K.P., Suhel, M. et al. (2023). Silicon nanoparticles: comprehensive review on biogenic synthesis and applications in agriculture. *Environmental Research* 232: https://doi.org/10.1016/j.envres.2023.116292.

Malik, M.S., Rehman, A., Shah, I.H. et al. (2025). Green synthesized silicon dioxide nanoparticles (SiO_2NPs) ameliorated the cadmium toxicity in melon by regulating antioxidant enzymes activity and stress-related genes expression. *Environmental Pollution* 366: https://doi.org/10.1016/j.envpol.2024.125459.

Mansoor, S., Ali, A., Kour, N. et al. (2023). Heavy metal induced oxidative stress mitigation and ROS scavenging in plants. *Plants* 12 (16): https://doi.org/10.3390/plants12163003.

Manzoor, N., Ali, L., Ahmed, T. et al. (2022). Silicon oxide nanoparticles alleviate chromium toxicity in wheat (Triticum aestivum L.). *Environmental Pollution* 315: https://doi.org/10.1016/j.envpol.2022.120391.

Maryam, S. and Gul, A. (2022). The mechanism of silicon transport in plants. In: *Plant Metal and Metalloid Transporters* (ed. K. Kumar and S. Srivastava), 245–273. Springer Nature. https://doi.org/10.1007/978-981-19-6103-8_12.

Maryam, H., Abbasi, G.H., Waseem, M. et al. (2024). Preparation and characterization of green silicon nanoparticles and their effects on growth and lead (Pb) accumulation in maize (Zea mays L.). *Environmental Pollution* 346: https://doi.org/10.1016/j.envpol.2024.123691.

Mitani-Ueno, N., Yamaji, N., and Ma, J.F. (2011). Silicon efflux transporters isolated from two pumpkin cultivars contrasting in Si uptake. *Plant Signaling & Behavior* 6 (7): 991–994. https://doi.org/10.4161/psb.6.7.15462.

Mousavi, S.M. (2022). Silicon and nano-silicon mediated heavy metal stress tolerance in plants. In: *Silicon and Nano-silicon in Environmental Stress Management and Crop Quality Improvement: Progress and Prospects* (ed. H. Etesami, A.H. Al Saeedi, H. El-Ramady, et al.), 181–191. Elsevier. https://doi.org/10.1016/B978-0-323-91225-9.00012-1.

Mukarram, M., Petrik, P., Mushtaq, Z. et al. (2022). Silicon nanoparticles in higher plants: uptake, action, stress tolerance, and crosstalk with phytohormones, antioxidants, and other signalling molecules. *Environmental Pollution* 310: https://doi.org/10.1016/j.envpol.2022.119855.

Mukarram, M., Petrik, P., Mushtaq, Z. et al. (2022). Silicon nanoparticles in higher plants: Uptake, action, stress tolerance, and crosstalk with phytohormones, antioxidants, and other signalling molecules. *Environmental Pollution* 310: https://doi.org/10.1016/j.envpol.2022.119855.

Mukarram, M., Khan, M.M.A., Kurjak, D. et al. (2023). Silicon nanoparticles (SiNPs) restore photosynthesis and essential oil content by upgrading enzymatic antioxidant metabolism in lemongrass (Cymbopogon flexuosus) under salt stress. *Frontiers in Plant Science* 14: https://doi.org/10.3389/fpls.2023.1116769.

Mukarram, M., Ahmad, B., Choudhary, S. et al. (2024). Silicon nanoparticles vs trace elements toxicity: modus operandi and its omics bases. *Frontiers in Plant Science* 15: https://doi.org/10.3389/fpls.2024.1377964.

Naaz, H., Rawat, K., Saffeullah, P., and Umar, S. (2023). Silica nanoparticles synthesis and applications in agriculture for plant fertilization and protection: a review. *Environmental Chemistry Letters* 21 (1): 539–559. https://doi.org/10.1007/s10311-022-01515-9.

Naeem, M.A., Shabbir, A., Imran, M. et al. (2024). Silicon-nanoparticles loaded biochar for soil arsenic immobilization and alleviation of phytotoxicity in barley: implications for human health risk. *Environmental Science and Pollution Research* 31 (16): 23591–23609. https://doi.org/10.1007/s11356-024-32580-y.

Naidu, S., Pandey, J., Mishra, L.C. et al. (2023). Silicon nanoparticles: synthesis, uptake and their role in mitigation of biotic stress. *Ecotoxicology and Environmental Safety* 255. https://doi.org/10.1016/j.ecoenv.2023.114783.

Okeke, E.S., Nweze, E.J., Ezike, T.C. et al. (2023). Silicon-based nanoparticles for mitigating the effect of potentially toxic elements and plant stress in agroecosystems: a sustainable pathway towards food security. *Science of the Total Environment* 898: https://doi.org/10.1016/j.scitotenv.2023.165446.

Peng, S., Liu, J., Pan, G. et al. (2024). Combining SiO_2 NPs with biochar: a novel composite for enhanced cadmium removal from wastewater and alleviation of soil cadmium stress. *Environmental Geochemistry and Health* 46 (11): 456. https://doi.org/10.1007/s10653-024-02243-5.

Pestovsky, Y.S. and Martínez-Antonio, A. (2017). The use of nanoparticles and nanoformulations in agriculture. *Journal of Nanoscience and Nanotechnology* 17 (12): 8699–8730. https://doi.org/10.1166/jnn.2017.15041.

Proshad, R., Rahim, M.A., Rahman, M. et al. (2024). Utilizing machine learning to evaluate heavy metal pollution in the world's largest mangrove forest. *Science of the Total Environment* 951: https://doi.org/10.1016/j.scitotenv.2024.175746.

Rachappanavar, V., Gupta, S.K., Jayaprakash, G.K., and Abbas, M. (2024). Silicon mediated heavy metal stress amelioration in fruit crops. *Heliyon*.

Rafi, M.N., Imran, M., Nadeem, H.A. et al. (2022). Comparative influence of biochar and doped biochar with Si-NPs on the growth and anti-oxidant potential of Brassica rapa L. under cd toxicity. *Silicon* 14 (17): 11699–11714. https://doi.org/10.1007/s12633-022-01878-2.

Rahman, S., Ahmad, I., and Nafees, M. (2023). Mitigation of heavy metal stress in maize (Zea mays L.) through application of silicon nanoparticles. *Biocatalysis and Agricultural Biotechnology* 50: https://doi.org/10.1016/j.bcab.2023.102757.

Rahmatizadeh, R., Jamei, R., and Arvin, M.J. (2024). Silicon nanoparticles (SiNPs) mediate GABA, SOD and ASA-GSH cycle to improve cd stress tolerance in Solanum lycopersicum. *Scientific Reports* 14 (1): https://doi.org/10.1038/s41598-024-72908-9.

Raj, A.B. and Raj, S.K. (2019). Seed priming: an approach towards agricultural sustainability. *Journal of Applied and Natural Science* 11 (1): 227–234. https://doi.org/10.31018/jans.v11i1.2010.

Rao, G.B. and Susmitha, P. (2017). Silicon uptake, transportation and accumulation in Rice. *~ 290 ~. Journal of Pharmacognosy and Phytochemistry* 6 (6): 290–293.

Rascio, N. and Navari-Izzo, F. (2011). Heavy metal hyperaccumulating plants: how and why do they do it? And what makes them so interesting? *Plant Science* 180 (2): 169–181. https://doi.org/10.1016/j.plantsci.2010.08.016.

Raza, H.Z., Shah, A.A., Noreen, Z. et al. (2024). Unravelling the role silica nanoparticles to ameliorate chromium heavy metal stress in Mexican marigold (Tagetes erecta L.). https://doi.org/10.21203/rs.3.rs-4924549/v1.

Riaz, M., Zhao, S., Kamran, M. et al. (2022). Effect of nano-silicon on the regulation of ascorbate-glutathione contents, antioxidant defense system and growth of copper stressed wheat (Triticum aestivum L.) seedlings. *Frontiers in Plant Science* 13: https://doi.org/10.3389/fpls.2022.986991.

Samani, M., Ahlawat, Y.K., Golchin, A. et al. (2024). Nano silica-mediated stabilization of heavy metals in contaminated soils. *Scientific Reports* 14 (1): https://doi.org/10.1038/s41598-024-69182-0.

Seow, Y.X., Tan, Y.H., Mubarak, N.M. et al. (2022). A review on biochar production from different biomass wastes by recent carbonization technologies and its sustainable applications. *Journal of Environmental Chemical Engineering* 10 (1): https://doi.org/10.1016/j.jece.2021.107017.

Sharifi-Rad, J., Sharifi-Rad, M., and Teixeira Da Silva, J.A. (2018). Morphological, Physiological and Biochemical Respon ses of Crops (Zea mays L., Phaseolus vulgaris L.), Medicinal Plants (Hyssopus officinalis L., Nigella sativa L.), and Weeds (Amaranthus retroflexus L., Taraxacum officinale FH Wigg) Exposed to SiO_2 Nanoparticles.

Sharma, A., Kapoor, D., Wang, J. et al. (2020). Chromium bioaccumulation and its impacts on plants: an overview. *Plants* 9 (1): https://doi.org/10.3390/plants9010100.

Sharma, A., Vishwakarma, K., Singh, N.K. et al. (2022). Synergistic action of silicon nanoparticles and indole acetic acid in alleviation of chromium (CrVI) toxicity in Oryza sativa seedlings. *Journal of Biotechnology* 343: 71–82. https://doi.org/10.1016/j.jbiotec.2021.09.005.

Siddiqui, H., Ahmed, K.B.M., Sami, F., and Hayat, S. (2020). *Silicon Nanoparticles and Plants: Current Knowledge and Future Perspectives*, 129–142. Springer. https://doi.org/10.1007/978-3-030-33996-8_7.

de Sousa, A., Saleh, A.M., Habeeb, T.H. et al. (2019). Silicon dioxide nanoparticles ameliorate the phytotoxic hazards of aluminum in maize grown on acidic soil. *Science of the Total Environment* 693: https://doi.org/10.1016/j.scitotenv.2019.133636.

Srivastava, A.K., Suresh Kumar, J., and Suprasanna, P. (2021a). Seed 'primeomics': plants memorize their germination under stress. *Biological Reviews* 96 (5): 1723–1743. https://doi.org/10.1111/brv.12722.

Srivastava, D., Tiwari, M., Dutta, P. et al. (2021b). Chromium stress in plants: toxicity, tolerance and phytoremediation. *Sustainability (Switzerland)* 13 (9). https://doi.org/10.3390/su13094629.

Sytar, O., Kumari, P., Yadav, S. et al. (2018). Phytohormone priming: regulator for heavy metal stress in plants. *Journal of Plant Growth Regulation* 38 (2): 739–752. https://doi.org/10.1007/S00344-018-9886-8.

Thind, S., Hussain, I., Rasheed, R. et al. (2021). Alleviation of cadmium stress by silicon nanoparticles during different phenological stages of Ujala wheat variety.

Arabian Journal of Geosciences 14 (11): 1028. https://doi.org/10.1007/s12517-021-07384-w/Published.

Tripathi, D.K., Singh, V.P., Prasad, S.M. et al. (2015). Silicon nanoparticles (SiNp) alleviate chromium (VI) phytotoxicity in Pisum sativum (L.) seedlings. *Plant Physiology and Biochemistry* 96: 189–198. https://doi.org/10.1016/j.plaphy.2015.07.026.

Tripathi, D.K., Singh, S., Singh, V.P. et al. (2016). Silicon nanoparticles more efficiently alleviate arsenate toxicity than silicon in maize cultiver and hybrid differing in arsenate tolerance. *Frontiers in Environmental Science* 4 (JUL): https://doi.org/10.3389/fenvs.2016.00046.

Ulhassan, Z., Yang, S., He, D. et al. (2023). Seed priming with nano-silica effectively ameliorates chromium toxicity in Brassica napus. *Journal of Hazardous Materials* 458: https://doi.org/10.1016/j.jhazmat.2023.131906.

Wang, S., Wang, F., and Gao, S. (2015). Foliar application with nano-silicon alleviates Cd toxicity in rice seedlings. *Environmental Science and Pollution Research* 22 (4): 2837–2845. https://doi.org/10.1007/s11356-014-3525-0.

Wang, L., Ning, C., Pan, T., and Cai, K. (2022). Role of silica nanoparticles in abiotic and biotic stress tolerance in plants: a review. *International Journal of Molecular Sciences* 23 (4). https://doi.org/10.3390/ijms23041947.

Wu, H. and Li, Z. (2022). Nano-enabled agriculture: how do nanoparticles cross barriers in plants? *Plant Communications* 3 (6).

Wu, T., Zhou, J., and Zhou, J. (2024). Comparison of soil addition, foliar spraying, seed soaking, and seed dressing of selenium and silicon nanoparticles effects on cadmium reduction in wheat (Triticum turgidum L.). *Chemosphere* 362: https://doi.org/10.1016/j.chemosphere.2024.142681.

Yadav, M., George, N., and Dwibedi, V. (2023). Emergence of toxic trace elements in plant environment: Insights into potential of silica nanoparticles for mitigation of metal toxicity in plants. *Environmental Pollution* 333: 122112.

Yan, G., Huang, Q., Zhao, S. et al. (2024). Silicon nanoparticles in sustainable agriculture: synthesis, absorption, and plant stress alleviation. *Frontiers in Plant Science* 15. https://doi.org/10.3389/fpls.2024.1393458.

Yao, C., Yang, Y., Li, C. et al. (2024). Heavy metal pollution in agricultural soils from surrounding industries with low emissions: assessing contamination levels and sources. *Science of the Total Environment* 917: https://doi.org/10.1016/j.scitotenv.2024.170610.

Yasin, M.U., Hannan, F., Munir, R. et al. (2024). Interactive mode of biochar-based silicon and iron nanoparticles mitigated Cd-toxicity in maize. *Science of the Total Environment* 912: https://doi.org/10.1016/j.scitotenv.2023.169288.

Yuvaraj, M., Sathya Priya, R., Jagathjothi, N. et al. (2023). Silicon nanoparticles (SiNPs): challenges and perspectives for sustainable agriculture. *Physiological and Molecular Plant Pathology* 128: https://doi.org/10.1016/j.pmpp.2023.102161.

Zhang, Y., Liang, Y., Zhao, X. et al. (2019). Silicon compensates phosphorus deficit-induced growth inhibition by improving photosynthetic capacity, antioxidant potential, and nutrient homeostasis in tomato. *Agronomy* 9 (11): https://doi.org/10.3390/agronomy9110733.

Zhao, C.R., Ikka, T., Sawaki, Y. et al. (2009). Comparative transcriptomic characterization of aluminum, sodium chloride, cadmium and copper rhizotoxicities in Arabidopsis thaliana. *BMC Plant Biology* 9: https://doi.org/10.1186/1471-2229-9-32.

Zhao, M., Zhang, H., Yan, H. et al. (2018). Mobilization and role of starch, protein, and fat reserves during seed germination of six wild grassland species. *Frontiers in Plant Science* 9: https://doi.org/10.3389/fpls.2018.00234.

13

Use of Nanomaterials as Nanofertilizers for Crop Betterment

Ankur Singh[1], *Subhransu Panda*[1,*], *Sumali Sucharita*[1,*] *and Aryadeep Roychoudhury*[2]

[1] *KIIT-Technology Business Incubator (KIIT-TBI), KIIT University, Bhubaneswar, Odisha, India*
[2] *Discipline of Life Sciences, School of Sciences, Indira Gandhi National Open University, Maidan Garhi, New Delhi, Delhi, India*

13.1 Introduction

The global population is increasing day by day. As a result, it severely impacts our economy and food necessity. Nanofertilizers have emerged as a promising solution to address such challenges, offering higher efficiency and reduced environmental impacts. Some of the chemical fertilizers such as urea, nitrogen, phosphorous, potassium, monoammonium phosphate and diammonium phosphate are widely used to supply essential nutrients in soil. Conventional fertilizers suffer from low nutrient utilization due to leaching, leading to substantial economic losses and decreased soil fertility. Before being used, soil loses 50–70% of its valuable nutrient content, which includes nitrogen, phosphorus and potassium. Scientists are already working on creating chemical fertilizers with a slow release. Yadav et al. (2023) combined urea and hydroxyapatite to create fertilizers that delayed the release of plant nutrients. In agriculture, nanostructured systems have been utilized for plant nutrition, soil health monitoring and controlled release of pesticides. This technique is a component of precision agriculture, an emerging field that supports sustainable agriculture and lowers energy consumption and waste production. Nanotechnology will help the farmers to use fertilizer, water and other resources more effectively.

Nanoparticles are particles with at least one external dimension and a size range of less than 100 nm (Miernicki et al. 2019) or with a volume-specific surface area >60 m^2/cm^3 (Wohlleben et al. 2017). Counting the number of dimensions

* Both authors have equal contributions to the chapter.

Silicon Nanoparticles: Synthesis, Uptake, and Applications, First Edition.
Edited by Archana Singh, Aryadeep Roychoudhury, and Indrakant K. Singh.

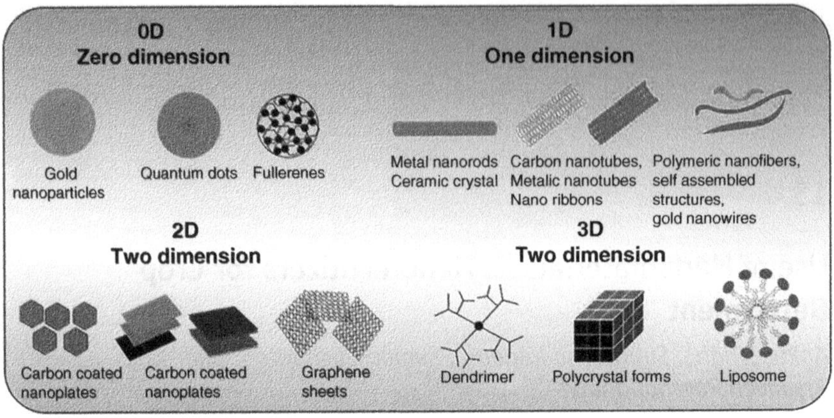

Figure 13.1 Different dimensions of nanoparticles.

outside the nanoscale range is a frequent method of classifying nanomaterials. 0D (dimension) nanomaterials are nanoparticles with all dimensions ≤100 nm, 1D nanomaterials are nanofibers, nanowires, nanotubes and nanorods with one dimension >100 nm, 2D nanomaterials are nanolayers, graphene, nanocoatings and nanofilms with two dimensions >100 nm, 3D nanomaterials are zeolites and other porous materials, powders and dispersions with three dimensions >100 nm (Pokropivny and Skorokhod 2017) (Figure 13.1). Currently, one of the most important applications of nanotechnology in agriculture is nanofertilizers, which, in contrast to conventional fertilizers, gradually feed plants in a controlled manner. Nanofertilizers are more efficient and economical than conventional fertilizers because they use nanoparticles to regulate the delivery of nutrients. Nanomaterials such as carbon nanotubes, graphene and quantum dots are utilized in the manufacturing of nanofertilizers (Yadav et al. 2023). It was found that zinc oxide nanoparticles are efficient against arsenic toxicity in plants like soybean (Ahmed et al. 2020). Activities of ascorbate peroxidase, superoxide dismutase, catalase and glutathione reductase can be increased by using zinc oxide nanoparticles (Ahmed et al. 2020). Three primary nutrients for plants, i.e., nitrogen, phosphorus and potassium can be supplied to the plants by using chemical fertilizers that are synthetic and inorganic materials. They find widespread use in agriculture due to their high nutrient content, easy availability and simple application. According to Nongbet et al. (2022), nano-sized particles of titanium dioxide, ferric oxide and zinc oxide can offer plants vital micronutrients and enhance their growth and productivity. Nanoemulsions, nanogels and polymer-based nanoparticles are the examples of nanoencapsulated nutrients that supply macro and micronutrients to plants. Materials like graphene, carbon nanotubes and fullerenes that come from carbonaceous sources are known as carbon-based nanofertilizers. Nongbet et al. (2022) demonstrated that nanofertilizers might boost crop production by stimulating seed germination, nitrogen metabolism, photosynthesis, protein and carbohydrate synthesis and stress tolerance. The efficacy of nanomaterials as plant growth enhancers is influenced by factors like size, content, concentration, chemical properties and crop type (Nongbet et al. 2022).

Bairwa et al. (2023) examined the response of plants towards chemical nanoparticles like aluminium, cerium, lanthanum and titanium and to metallic nanofertilizers like copper, manganese, zinc and iron. Encapsulating nanofertilizers can reduce nutrient loss, encourage healthy plant development and improve crop quality by boosting nutrient uptake. Yadav et al. (2023) identified three distinct kinds of nanofertilizers by using various formulations: (1) regular fertilizer that is shrunken in size and frequently appears as nanoparticles is known as nanoscale fertilizer, (2) nanoscale additive fertilizer, in which fertilizer is coated on nanoparticles; (3) nanoscale coating fertilizer, in which nutrients are introduced into the nanoscale pores of a host material. Some other categories of nanofertilizers include controlled-release nanofertilizers, nanofertilizers for targeted delivery, plant growth stimulating nanofertilizers, water and nutrient loss controlling fertilizers, inorganic and organic nanofertilizers, hybrid nanofertilizers and nutrient loaded nanofertilizers (Yadav et al. 2023). Similarly, various consistency-based nanofertilizers are surface-coated, synthetic polymer-coated, biological product-coated and nano-carrier-based nanofertilizers (Yadav et al. 2023). Controlled-release fertilizers are structurally granular in shape that helps provide nutrients to the plants over a long time period. By limiting the release of nutrients into the environment, controlled-release fertilizers can increase the environmental sustainability of agriculture. Due to their special qualities, nanomaterials like graphene, carbon nanotubes and quantum dots are perfect as controlled-release nanofertilizers. The effectiveness of nutrient delivery is increased by their tiny size, high surface area to volume ratio and capacity to be coated with different materials to adjust the release rate. Another application of nanotechnology includes monitoring plant growth, modifying the genome quickly and expressing transgenes in intact plant cells (Kwak et al. 2017).

Biotic and abiotic stress majorly affect crop production. Engineered nanomaterials enhance the tolerance level of the crops because of their nanosize and large surface area, which help plant growth and development under biotic and abiotic stresses (Adisa et al. 2019; Kah et al. 2018; Rajput et al. 2021; Rawat et al. 2017; Zulfikar et al. 2019). Abiotic stress includes drought, salinity, alkalinity, temperature, metal toxicity, etc. Adisa et al. (2019) found that photosynthesis is most vulnerable to stress like nutrient deficiency, salinity, drought and heat because major cellular components such as photosystem II, RuBisCO and ATP synthase are the primary targets of stressed conditions. Rajput et al. (2021) showed that the application of silicon dioxide nanoparticles enhanced transpiration rate, water use efficacy, photosynthetic pigments and carbonic anhydrase movement in pumpkin plants. Mingyu et al. (2008) stated that nanoparticles are a convenient tool to reduce the damage caused by abiotic stress. Many applications of metal nanoparticles have been demonstrated in plants. Silica nanoparticles have been shown to increase plant growth and boost up resistance against biotic stress (Wang et al. 2022). Implementation of nanoparticle boosts plant stress adaptive capacity through alteration of the important metabolic cascades and improves the antioxidative defense mechanism and free radical detoxification process (Jalil and Ansari 2019). Improvement and utilization of nanosensors in precision farming enable monitoring of crop growth, soil conditions and disease management. The use of nanotechnology has significantly improved human control over soil and plant health, quality control and safety assurance, contributing to sustainable agriculture (Chen et al. 2016a). Nanocore shell biosensors with protein-ligand and protein nanoparticles can be used for the detection of pathogens associated with seeds. According to Malik et al. (2021), nanopriming enhances seed storage, germination, plant growth, resistance to stress and reduces pesticide and fertilizer usage and crop stress. Some new research indicates that seed nanopriming can activate various genes during germination, particularly those related to plant stress resistance (Mahakham et al. 2017).

13.2 Nanofertilizer and Agriculture

Nanofertilizers play a crucial role in modern agriculture, requiring suitable formulations and delivery mechanisms for optimal plant uptake and usage. Chhipa (2017) described that the use of nano-enabled fertilizers may improve nutrient delivery efficiency in plants, as nanofertilizers enhance productivity by delivering nutrients at a gradually slower rate, reducing fertilizer application and increasing nutrient use efficiency (Kah et al. 2018). Physical and chemical properties of macromolecules remain constant, but nanomaterials exhibit distinct physical, chemical and biological properties compared to bulk materials, bridging the gap between bulk and nanomaterials (Rajput 2015). Nanoparticles are commonly used in plant systems in the form of metalloids, metallic oxides, nonmetals, carbon nanomaterials, quantum dots, liposomes and dendrimers (Elmer and White 2018). Biologically synthesized nanoparticles offer a promising solution for developing nanofertilizers, nanoinsecticides and nanopesticides, also minimizing toxic environmental pollutants by being biodegradable (Ghidan et al. 2017; Shang et al. 2019). *Medicago sativa* was utilized for creating gold nanoparticles, while *Rhynchotechum ellipticum* was used for formulating silver nanoparticles (Hazarika et al. 2014; Panpatte et al. 2016).

With the advent of nanotechnology, more efficient and fewer contaminant nanoformulations are continuously produced for sustainable agriculture

(Fraceto et al. 2016). There have been several physical, biological and chemical methods suggested for the synthesis of nanoparticles. Precipitation method, microemulsion, ultrasound, hydrothermal synthesis, microwave synthesis, inert gas condensation, laser ablation, sputtering, sol–gel, mechanical milling, biosynthesis, etc., are among the ones that have been extensively used (Jamkhade et al. 2019; Rane et al. 2018). Biosynthesis of nanoparticles using plants has proven to be green chemistry that connects plant sciences with nanotechnology. This method also involves the synthesis of nanoparticles at room temperature and neutral pH at low cost without the use of chemicals harmful to the environment (Parveen et al. 2016). The hydroxyl groups in carbohydrates, amino acids, proteins and nucleic acids of plants are reported to stabilize the environmental nanoparticles (Dorjnamjin et al. 2008).

13.3 Types of Nanoparticles Used in Agriculture

13.3.1 Silver Nanoparticles

These materials exhibit various optical, electrical and thermal properties, including high electrical conductivity, antimicrobial and catalytic properties (Zhang et al. 2016). Additionally, they have been used in cosmetic products, cryogenic superconducting materials, composite fibers, the food industry and electronic components due to their extraordinary properties such as chemical stability, good conductivity, catalytic nature and most importantly antibacterial, antiviral, antifungal and anti-inflammatory activities (Ahmed et al. 2016). Silver nanoparticles are also able to enhance the germination of plant seeds (Kale et al. 2021). Studies conducted by Shelar and Chavan (2015) showed that the ideal concentration of silver nanoparticles is essential for encouraging seed germination and plant growth. There has been an improvement in photosynthetic quantum efficiency and chlorophyll concentration in *Brassica juncea* on being treated with silver nanoparticles exposed to antioxidative stress (Sharma et al. 2012).

13.3.2 Zinc Oxide

Zinc oxide nanoparticles have been included in solar cell preparation varistors, gas sensing, chemical absorbent, solar cell preparation, hydrogen catalyst, photocatalytic degradation and also in electrical and optical devices (Hong et al. 2006). In a study conducted by Umavathi et al. (2021), the application of zinc oxide nanoparticles on *Sesamum indicum* significantly enhanced both the germination of seeds and the vegetative growth of plants. Synthesis of zinc oxide nanoparticles with different sizes and shapes has been achieved using various plant species or their tissues like dry ginger rhizome (*Zingiber officinale*), leaf of *Agathosma betulina* and *Aspalathus linearis*, orange and pomegranate fruit peel (Diallo et al. 2015; Thema et al. 2015; Thi et al. 2020; Verbic et al. 2021). Zinc is a regulatory co-factor and structural component of many enzymes and proteins and plays an important role in plant metabolic activity, particularly photosynthesis, phytohormone biosynthesis and antioxidant mechanisms (Umair Hassan et al. 2020). Zinc oxide nanoparticles have been found to play a promising role in maintaining the required concentration of zinc in plants because of their remarkable characteristics (Milani et al. 2015). The fresh and dry weight of *Cicer arietinum* seedlings were increased by the application of zinc oxide nanoparticles (Pavani et al. 2014). A study conducted by Shaik et al. (2020) has shown that fenugreek (*Trigonella foenum-graecum*) plants showed improved plant growth, root development and seed germination. A similar result was found in Indian mustard (*Brassica juncea*) where the application of zinc oxide nanoparticles enhanced seed germination (Mazumder et al. 2020). Applying zinc oxide nanoparticles resulted in an increase in protein content. It also facilitated photosynthesis, which supported the growth and vitality of maize (*Zea mays* L.) plants (Sabir et al. 2020). It was shown that treating seeds with zinc oxide nanoparticles at a concentration of 1000 ppm improved seed germination and seedling health which resulted in early development of seed, as shown by early blooming and elevated leaf chlorophyll concentration (Prasad et al. 2012).

13.3.3 Copper Nanoparticles

Copper-based nanoparticles offer a broad-spectrum antimicrobial property for protecting crops against pests and diseases. Their small size facilitates easy absorption by plants and prevents hazards due to excessive usage (Ingle et al. 2014). This could potentially mitigate the detrimental effects of agricultural activities on the agroecosystem, particularly the soil system. The synthesis of copper-chitosan nanoparticles as shown by Saharan et al. (2015) involved the cross-linking of chitosan and tripolyphosphate ions through ionic gelation. This study investigated the antifungal properties and growth promotion of these compounds against *Alternaria solani* and *Fusarium oxysporum* in tomato plants. Copper-based nanoparticles have some antimicrobial properties. These properties increase their agricultural importance and have also been found to reduce toxicity and stop microbial infection in soil (Ameh and Sayes 2019).

13.3.4 Iron Oxide

The application of iron oxide nanoparticles boosts peanut growth by controlling antioxidant enzymes and phytohormone activity (Rui et al. 2016). Afzal et al. (2021) showed that iron oxide nanoparticles can be produced from *Cassia occidentalis* flower extract. They further showed that the application of extracted iron oxide nanoparticles reduces rice seed dormancy due to higher starch metabolism. Li et al. (2021a) described that low doses of zero-valent iron and iron oxide nanoparticles can improve plant growth under iron-deficient conditions by reducing oxidative stress and regulating phytohormones. Iron is essential for enzyme reactions, photosynthesis, DNA translation, RNA synthesis and auxin activities, thereby enhancing plant development (Sheykhbaglou et al. 2018). Nanoparticles are being explored as an alternative to iron shortage that can enhance crop production and reduce environmental stress (Askary et al. 2017). By interacting with chemicals within plant cells, iron oxide nanoparticles can improve the uptake of nutrients (Rahmatizadeh et al. 2019). Shankramma et al. (2016) demonstrated that the growth of tomato plants was improved by the application of iron oxide nanoparticles. Tawfi et al. (2021) noted that when iron oxide nanoparticles were sprayed on *Moringa* plants, the growth performance of the plants, photosynthetic pigments, indole acetic acid level, proline concentration, free amino acids and total soluble sugars were all greatly improved.

13.3.5 Magnesium Nanoparticles

Essential phenols, amino acids, proteins, vitamins, β-carotene, phytohormones, antioxidant compounds, polyphenols, isothiocyanate, glucosinolates, phenolic acids and flavonoids, found in *Moringa oleifera*, makes it ideal for nanoparticle biosynthesis (Anwar et al. 2007; Foster et al. 2015). Meena et al. (2007) demonstrated that soil magnesium deficiency resulted in a significant decrease in the oil content of groundnut. Peanut seeds showed favourable responses towards magnesium oxide nanoparticles, which promote germination, growth and photosynthetic pigments (Jhansi et al. 2017). Application of magnesium oxide nanoparticles at a dosage of 4 mg/l significantly enhanced mung bean seedling growth, resulting in faster germination, compared to other treatments (Ashok et al. 2016). Similarly, green gram seeds showed maximum germination, seedling growth and vigour index, on being treated with magnesium oxide nanoparticles (Anand et al. 2020). Magnesium oxide nanoparticle application significantly improved the germination rate, growth and photosynthetic pigments of peanut seeds (Jhansi et al. 2017). When compared to other treatments, the effects of applying 4 mg/l of magnesium oxide nanoparticles on the growth of mung bean seedlings were highly positive as compared to that of other treatments (Ashok et al. 2016).

13.3.6 Titanium Dioxide

Jaberadeh et al. showed that the application of titanium dioxide nanoparticles to wheat seedlings improved their growth and production traits, and also the yield of the plant. Furthermore, Mahmoodzadeh et al. (2013) noted an enhanced rate of germination and increased radicle and plumule growth in canola plants that were treated with titanium dioxide nanoparticles.

13.3.7 Calcium Carbonate

When calcium carbonate was sprayed on citrus tankan leaves, it effectively resisted the damage caused by California red scales and oriental fruit flies (Hua et al. 2015). Compared to other treatments, the combined use of calcium carbonate and hydroxyl apatite nanoparticles in soybean plants resulted in higher yield (El-Hady and Hussein 2021). On being applied to groundnut seedlings, calcium carbonate nanoparticles (varying in size from 20 to 80 nm) significantly increased biomass and seedling growth, as compared to the control (Liu and Lal 2015).

13.4 Forms of Nanomaterial in Agriculture

Farmers currently choose chemical fertilizers over other types because they are more cost-effective and efficient. However, excessive and wasteful use of chemical fertilizers can harm the ecosystem, reduce food production and damage the soil (Chen et al. 2020). Urea which is highly water-soluble and subject to loss, provides plants with 80% of the nitrogen fertilizers they require (Zhu et al. 2020). Approximately 30–50% of the total nitrogen loss is caused by leaching, mostly as nitrate or nitrite and around 2–30% as ammonia (Lowe et al. 2008).

Nanomaterials are present in a variety of fertilizers. Nanomaterials improve soil health by improving soil nutrient management, facilitating the creation of a nutrient cycle in agriculture and reducing the depletion of nutrient resources. With specialized polymers or techniques, nanomaterial-mediated nanofertilizers can be used in the agricultural field. These fertilizers are triggered by signals that plant roots release into the surrounding environment (Singh 2017). These days, nanomaterials are used especially as biofertilizers. Nanomaterials are also utilized in organic fertilizers. In place of traditional iron fertilizers, Haydar et al. (2022) analysed the biological impacts of iron oxide nanoparticles and those specimens functionalized with ethylene diamine tetraacetic acid (EDTA) as a nano-micronutrient fertilizer (Haydar et al. 2022).

Through both in-vitro and in-vivo trials, Li et al. (2021a,b) found that long-term pesticide use in tea gardens resulted in excessive reactive oxygen species (ROS) accumulation in plants, which led to oxidative damage and affected the nutritional value of tea. Alimohammadi et al. (2020) studied the impact of urea and nano-nitrogen chelate fertilizers on soil nitrate leaching and sugarcane yield. They used urea and nano-nitrogen chelate having five different concentrations of nitrogen, i.e., 0, 80, 112, 137 and 161 kg N/ha. The average soil nitrate concentrations during sugarcane growth were found to be 10.2 and 12.8 mg/kg for urea and nano-nitrogen chelate treatments, respectively. With urea fertilizer (699.0 mg/l), the nitrate leaching level was high, while with nano-nitrogen chelate fertilizer (183.0 mg/l), it was low. Additionally, this study showed that nano-nitrogen chelates greatly increased the amount of sugar in sugarcane. Nitrate leaching and its effects on the environment and human health make nano-nitrogen chelate potentially useful.

Chen et al. (2016a,b) revealed that calcium sulphate reduced phosphorus loss from farmland. A common soil conditioner that exhibits high complexation to orthophosphate is calcium sulphate. Nano-calcium sulphate offers more surface area, greater solubility and better soil-fertilizer integration than typical crude calcium phosphate, all of which contribute to a decrease in soil phosphorus loss (Chen et al. 2016a,b).

Using the air drying and grinding method, Ahanger et al. (2021) prepared a nano-vermicompost organic fertilizer with particle sizes, ranging from 1 μm to 200 nm. This fertilizer has beneficial effects on growth, oxidative stress parameters, antioxidants, nitrogen metabolism, osmotic accumulation and mineral elements under drought stress in tomato plants. The findings indicated that this fertilizer may help protect tomatoes from drought-induced damage. Thus, it can be said that nanoparticles promote the growth of plants by enhancing nutrient solubility and capacity for dispersion. Additionally, because of their small size, nanoparticles have the unusual ability to easily pass through plant cell wall membrane barriers, changing the rate at which the applied nanoform fertilizers are absorbed. Last but not least, the nanomaterials save fertilizer by covering it to decrease volatilization, stabilizing it in the soil and lowering leaching losses, all of which improve the efficacy of fertilizers and minimize harm to the environment. Fertilizers using nanomaterials and nanotechnology can more efficiently encourage fertilizer uptake and absorption, enhancing crop agronomic parameters (Ahanger et al. 2021). Additionally, it could decrease pollution that is caused by the excessive use of chemical fertilizers in the agricultural field (Figure 13.2).

13.5 Positive Effects of Nanomaterials

It has been observed that the application of nanoparticles can improve the germination process and yield of various crop plants (Younes et al. 2019). Nanoparticles have demonstrated notable effects in inducing tolerance to both biotic and abiotic stress. Hernandez-Hernandez et al. (2018) showed that by raising the concentration of β-carotene, phenols, vitamin C and lycopene in stressed plants, chitosan-polyvinyl alcohol (CsPVA)+ hydrogel copper nanoparticles

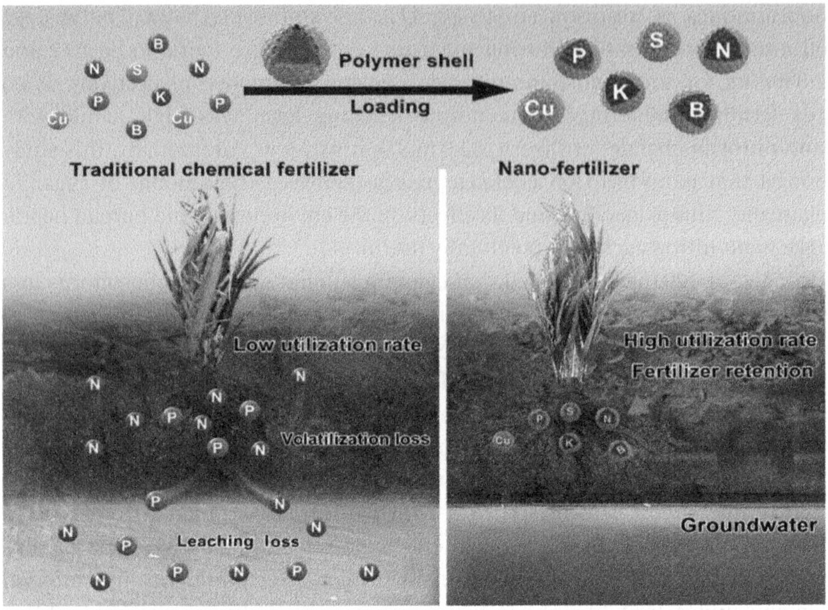

Figure 13.2 Nanomaterials loaded with fertilizers. *Source*: An et al., 2022 / Springer Nature / CC BY 4.0.

enhanced the salt tolerance of tomato plants. Furthermore, (CsPVA)+ hydrogel copper nanoparticles also enhanced the activity of antioxidant enzymes, such as glutathione peroxidase, ascorbate peroxidase, catalase, superoxide dismutase and phenylalanine ammonia-lyase (Hernandez-Hernandez et al. 2018). In a different study, cultivars of bread and durum wheat (*Triticum aestivum*), exposed to drought, showed higher absolute and relative growth rates, unit leaf rates and net assimilation rates when subjected to titanium dioxide nanoparticles (Dawood et al. 2019). It was discovered that the applied nanoparticle increased the biomass and leaf area by increasing the chlorophyll content and the net photosynthetic rate (Dawood et al. 2019). Silicon nanoparticles may improve the nutritional status, photosynthesis, morphology and physiology of crops, hence promoting the growth of metal-stressed plants (Hussain et al. 2019). Zinc oxide nanoparticles decrease cadmium (Cd^{2+}) concentration while increasing antioxidant enzymes, zinc (Zn^{2+}) and the concentration of photosynthetic pigments in wheat grains. In a different research work, the addition of biochar to zinc oxide nanoparticles reduced the Cd^{2+} content in rice and maize within a short growth phase (Rizwan et al. 2019a). According to research by Gil-Díaz et al. (2017), the application of zero-valent iron (Fe^0) reduced the availability of heavy metals in calcareous or acidic soil. Wang et al. (2020) investigated the possibility of using cerium nanoparticles as an alternative to nitrogen treatment in rice to mitigate low (nitrogen shortage) and high (excess nitrogen) stress. They demonstrated how cerium nanoparticles raised nitrogen levels in roots and shoots during times of nitrogen deficit and decreased them during times of acute nitrogen stress (Wang et al. 2020). Additionally, treatment with cerium dioxide nanoparticles improved the absorption of nitrogen by upregulating the activity of enzymes such as glutamine synthetase, glutamine oxoglutarate aminotransferase and glutamate dehydrogenase.

In contrast, in conditions of extreme nitrogen stress, cerium nanoparticles decreased the activities of glutamine synthetase and glutamate dehydrogenase (Wang et al. 2020). Additionally, proline and phytohormone levels and the activity of antioxidant enzyme system were all regulated by cerium nanoparticles during nitrogen stress, which prevented oxidative damage to DNA and cell membranes.

Tripathi et al. (2017) observed that the pre-addition of silicon nanoparticles protected wheat seedlings against UV-B stress through nitric oxide mediated triggering of antioxidant systems which counteracted the damaging effects of ROS on photosynthesis. When alumina (Al_2O_3) nanoparticles measuring 30–60 nm were applied to soybean plants during flooding, the plant root length increased, while mitochondrial proteins involved in glycolysis were reduced (Mustafa and Komatsu 2016). Additionally, different sized and shaped Al_2O_3 efficiently controlled the ascorbate/glutathione pathway (Mustafa et al. 2015a).

According to Mustafa et al. (2015b), treating seedlings with silver nanoparticles (2 ppm and 15 nm) improved their development during flooding stress. Treatment with silver nanoparticles significantly decreased the transcript levels and abundances of glyoxalase II and fermentation-related proteins, indicating that soybean plants treated with silver nanoparticles produce fewer cytotoxic by-products of glycolysis than soybean plants that are flooded (Mustafa et al. 2015b). Furthermore, compared to flooding stressed plants, the genes for *pyruvate decarboxylase 2* and *alcohol dehydrogenase 1* in silver nanoparticle-treated stressed plants responded more slowly, indicating a metabolic shift toward regular cellular functions (Mustafa et al. 2015b). N-heterocyclic coatings on gold nanomaterials showed remarkable efficacy against Gram-positive and multidrug-resistant types of bacteria (Feng et al. 2016). Hao et al. (2019) showed that by raising phytohormone levels, metal nanoparticles and carbon nanomaterials decreased turnip mosaic virus infection. Zinc oxide nanoparticles have been shown by Cai et al. (2019) to have an antiviral protective function against tobacco mosaic virus by modulating immunity and promoting nutritional synergism in tobacco (*Nicotiana benthamiana*) plants. Abbai et al. (2019) reported that the application of silicon nanoparticles to ginseng (*Panax ginseng*) seedlings, infected by *Ilyonectria mors-panacis* (the fungal causal agent of root rot disease), decreased disease-severity index and

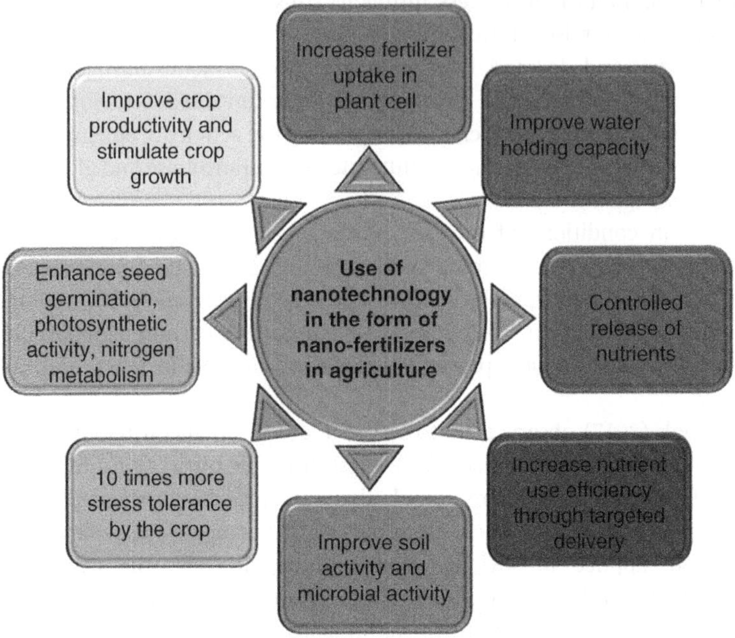

Figure 13.3 Some advantages of nanotechnology in agriculture.

enhanced the tolerance against the pathogen, thereby improving root quality and yield. Younis et al. (2019) developed silica (SiO_2) nanoparticles and chitosan-silica nanocomposites. These two nanoparticles were found to control gray mold disease, caused by *Botrytis cinerea* in grapes (Figure 13.3).

The creation of genetically modified plants is the result of advanced studies in the field of nanoparticles. Due to their higher surface area, configurable pore diameters, three-dimensional open pore structure and well-defined surface characteristics, mesoporous silicon nanoparticles can take a variety of foreign molecules, including DNA, proteins and agrochemicals (Torney et al. 2007). Mesoporous silica nanoparticles can be used to create genetically modified plants, by delivering DNA and small-interfering RNA using surface-coated mesoporous silica nanoparticles and the recombinase enzyme via gold-plated mesoporous silica nanoparticles using particle bombardment technique in maize (Martin et al. 2014). The hormone jasmonate stimulates the production of protective secondary metabolites and a number of plant defense mechanisms (Mishra et al. 2016). In order to induce the secondary metabolite artemisinin in *Artemisia annua* cell suspension cultures, cobalt nanoparticles have been used. Different cobalt nanoparticle concentrations resulted in a decrease in the expression levels of the *SQS* and *DBR2* genes. At high concentrations of cobalt nanoparticles, there may be an increased generation of artemisinin content due to a decrease in *SQS* and *DBR2* gene expression (Ghasemi et al. 2015). By activating RuBisCO, which may cause RuBisCO carboxylation, nano-anatase titanium dioxide improves photosynthetic carbon uptake and increases plant development (Gao et al. 2006).

13.6 Nanotechnology Delivery System

Agrochemicals are often applied to crops using spraying or broadcasting methods. Because of this, very little pesticide is applied to crop target areas which is far less than the minimal effective concentration needed for healthy plant growth. Chemical leaching, photolysis, hydrolysis and microbiological degradation are the causes of the losses (Nair et al. 2010; Yang et al. 2016). The main objective of controlled delivery approaches is to achieve complete biological competency while reducing loss and negative effects by releasing a defined amount of necessary and sufficient quantities of agrochemicals throughout time (Shojaei et al. 2019). Because of their small surface area, ease of attachment and quick mass transfer, nanoparticles provide the benefits of effective pesticide delivery (Ghormade et al. 2011). Because of these factors, agrochemicals contain micronic or submicronic particles that are introduced via a variety of processes, like trapping within the nanomatrix of active substances, ionic surface attachment, absorption and capsulation (Nuruzzaman et al. 2016). Potassium nitrate capsulation by graphene oxide sheets significantly extends the fertilizer release process, and large-scale production appears to be achievable for such a formulation at a reasonable cost (Zhang et al. 2014).

Researchers now have more opportunities due to recent advances in nanobiotechnology to fully replace a species genetic makeup with that of another (Torney et al. 2007). Through the use of genetic engineering, silicon dioxide nanoparticles have been developed to safely transfer DNA fragments and/or sequences to target species, such as maize and tobacco plants (Galbraith 2007). Novel crop varieties resistant to insects have also been created using a nanoparticle-enabled delivery approach. In the case of gene-gun technology, DNA-coated nanoparticles are used as projectiles to blast cells or tissues in order to transmit the required genes to the target plants (Vijayakumar et al. 2010). Recent developments in chitosan nanoparticle-entrapped small interfering RNA delivery vehicles have opened up new agricultural opportunities for the targeted management of insect pests. This is because chitosan exhibits both an effective RNA binding capacity and cell membrane penetration (Zhang et al. 2010). Natural plant apertures such as stomata, hydathodes, stigma and bark texture that have nano- or microscale exclusion sizes can passively absorb nanoparticles at the shooting surface (Eichert et al. 2008). In both their aerial and hypogeal sections, damages and wounds may serve as effective pathways for nanoparticle uptake in plants (Al-Salim et al. 2011). Plant nanoparticle uptake efficiency appears to be influenced by delivery strategies as well. According to recent reports, the use of aerosols encourages a greater rate of internalization of certain nanoparticles in relation to that of nanoparticle drop cast in watermelon (Raliya et al. 2016). Additionally, leaf lamina infiltration techniques may force the penetration of carbon nanotubes in plant tissues (Demirer et al. 2018; Giraldo et al. 2014). Rhizodermis lateral root connections may facilitate simple access to nanomaterials at the root level, particularly close to the root tip, but suberin containing upper portions are impermeable (Chichiriccò and Poma 2015). In general, compared to the aerial portion, the dynamics of nanoparticle uptake seem to be more complex in the soil. The availability of nanoparticles may be influenced by a number of variables, such as the presence of mucilage and exudates, symbiotic organisms and soil organic matter. As an example, exudates and root mucilage that are often expelled into the rhizosphere have two purposes. Either they increase the rate at which nanoparticles are absorbed by encouraging nanoparticle attachment to the root surface, or these gel-like materials can also cause nanoparticles to get trapped and aggregate (Avellan et al. 2017; Milewska-Hendel et al. 2017). It has been shown by Avellan et al. (2017) using X-ray computed nanotomography improved dark-field microscopy and hyperspectral imaging that root border cells and related mucilage have a tendency to capture gold nanoparticles, regardless of particle charge, whereas negatively charged nanoparticles move straight into the root tissue of *Arabidopsis thaliana* without being isolated by the mucilage of the root cap (Avellan et al. 2017). It has been shown that apoplastic transport facilitates the radial movement of nanomaterials, which may carry nanoparticles to the vascular tissues and to the root core cylinder. Likewise, it promotes the flow of these particles upward into the aerial portion (Gonzalez-Melendi et al. 2008; Larue et al. 2014). Water and another solute switch from the apoplastic to the symplastic channel in order to move beyond this natural barrier. Similar capacities to move around the casparian strip block have been identified for a number of nanoparticles (Schwab et al. 2016). Once in the cytoplasm, plasmodesmata membrane-lined cytoplasmic bridges with a flexible diameter of 20–50 nm, allow nanoparticles to migrate from cell to cell inside plant tissues by maintaining membrane and cytoplasmic continuity between cells. Various plant species, including *Arabidopsis*, rice and poplar, have been reported to transport nanoparticles of varying sizes via plasmodesmata (Geisler-Lee et al. 2013; Lin et al. 2009).

Nanomaterials can be used as enhanced delivery systems to increase the uptake and effectiveness of conventional fertilizers (nutrients and phosphates) or applied to the soil as nanofertilizers like iron, manganese, zinc, copper and molybdenum nanoparticles (Liu and Lal 2015). Utilizing hydroxyapatite nanoparticles as phosphorous nanofertilizers increases the seed production and growth rate of soybeans by 33% and 20%, respectively, in comparison to standard phosphorus fertilizers (Liu and Lal 2015).

Multiwalled carbon nanotubes on being added to the soil, doubled the number of flowers and fruits in tomato plants by activating genes and proteins necessary for plant growth and development (Khodakovskaya et al. 2013). In general, pesticides transported by nanomaterials exhibit enhanced stability and solubility, facilitating both targeted and gradual release for pest management (Duhan et al. 2017). Herbicide distribution has made use of organic and polymeric nanoparticles as nanocarriers in the form of nanospheres or nanocapsules (Tanaka et al. 2012). Specifically, polymeric nanoparticles, such as poly(epsilon-caprolactone), show favourable biocompatible characteristics and have been utilized frequently for the encapsulation of the herbicide, atrazine (Tanaka et al. 2012). In a different study, chitosan nanoparticles, loaded with three triazine herbicides, showed low genotoxicity and a decreased environmental impact in *Allium cepa* (Grillo et al. 2015). In another recent research, plasmid and linear DNA were delivered via carbon nanotube scaffolds inserted into external plant tissue. Also, siRNA caused major short-term green fluorescent protein expression in *Nicotiana benthamiana*, *Eruca sativa*, *Triticum aestivum* and *Gossypium hirsutum* leaves as well as in *Eruca sativa* protoplasts. In the immature embryos of *Zea mays*, mesoporous silica nanoparticles have been used as carriers to introduce Cre recombinase. It also carries loxP sites incorporated into chromosomal DNA (Valenstein et al. 2013).

Differently designed nanomaterials such as zinc, calcium phosphate, silica, gold, magnetite, strontium phosphate, magnesium phosphate and manganese phosphate can be used in plants for nanoparticle-mediated DNA transfer, utilizing gene-nanoparticle binding technique (Mahendra et al. 2012). Carbon-based nanomaterials include multi-walled carbon nanotubes, single-walled carbon nanohorns, single-walled carbon nanotubes and starch fullerenes (Burlaka et al. 2015). According to reports, the usage of positively charged nanoparticles causes plant cells to absorb them more quickly than negatively charged ones. This could be because the negatively charged cell wall prefers cations.

13.7 Application of Nanoparticles in Plant Growth and Seed Germination

The size, chemical makeup, functional groups on the surface and kind of coating of nanoparticles determine how well they are absorbed by the plants. The molecular alterations brought about by the interaction and uptake of nanomaterials have an impact on the general physiology of plants (Jin et al. 2017). Furthermore, the use of nanotechnology in the seed priming process is a promising approach that further supports the potential of high-yield value crops (Acharya et al. 2020; Anand et al. 2020). Mesoporous silica nanoparticles have the potential to boost photosynthesis through their interaction with chloroplasts, leading to higher levels of total protein, chlorophyll content and improved seed germination. Mesoporous silica nanoparticles at the greatest concentration (2000 mg/l) did not cause stress in any of the plants, showing a safer use of the material as a smart delivery approach (Sun et al. 2018). Silica is crucial for nutrient production since a lack of it weakens plants and increases their reactivity to both biotic and abiotic stress. Rastogi et al. (2019) have reviewed in great detail the precise effects of silicon dioxide nanoparticles on agriculture. It has been noted that other metal oxide nanoparticles, such as zinc oxide and copper, have an impact on the seed germination of *Vigna mungo* (Raja et al. 2019). The growth-promoting and seed germination characteristics of less-researched platinum nanoparticles protected by polyvinylpyrrolidone were demonstrated by Rahman et al. (2020) in *Pisum sativum*.

13.8 Nanoparticles in Abiotic Stress Response

Nano-enzymes that can scavenge oxidative stress are one way by which nanoparticles can mimic the function of antioxidative enzymes (Sharifi et al. 2020). As a defense mechanism against salt stress, silicon dioxide nanoparticles increased the rate of plant transpiration, water use efficiency, total chlorophyll and carbonic anhydrase activity in *Cucurbita pepo* plants (Siddiqui et al. 2014) (Figure 13.4).

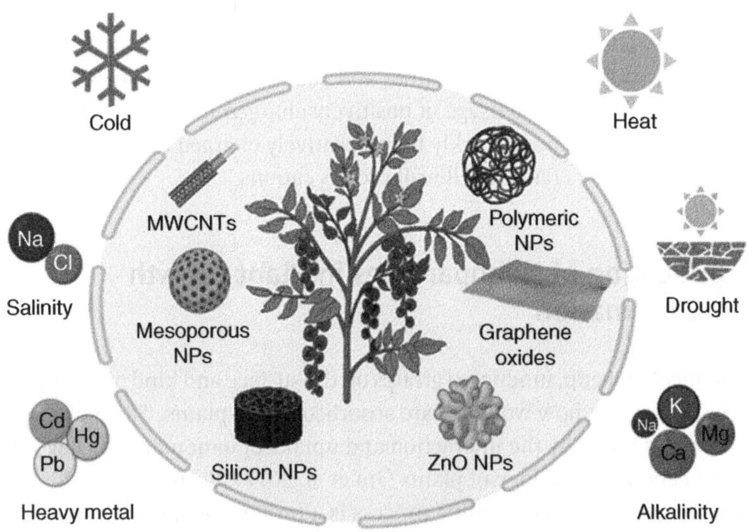

Figure 13.4 Different types of stress can be mitigated by nanoparticles.

13.8.1 Cerium Nanoparticles

The effects of nanocerium on plant health can be both beneficial and detrimental, which is mostly controlled by factors such as the type of plant, growing conditions, coating, concentration of exposure and surface charge (Milenkovic et al. 2019). A class of cerium oxide nanoparticles known as nanocerium is widely used in the biomedical sector as an antioxidant (Liu and Shi 2019). These nanoparticles have the ability to interact at the nano-bio interface, increasing plant resistance to different abiotic stresses by controlling vital pathways (Saxena et al. 2016). Abiotic stress results in an excessive level of ROS, which lowers photosynthetic efficiency and may lead to biomolecule oxidation (Wakeel et al. 2020). The particular redox potential of nano-cerium is based on the simple transition between Ce^{3+} and Ce^{4+} oxidation states (Collin et al. 2014). It is a good choice for reducing the effects of ROS damage since it scavenges ROS. Moreover, it causes alterations in plant shape, a decrease in photosystem II abundance, photochemical efficiency and chlorophyll concentration (Chen et al. 2020; Gao et al. 2020). Wu et al. (2017) described an explanation for the strongest protective impact of anionic cerium oxide nanoparticles against abiotic stress such as excessive heat, light, dark and cold in the *A. thaliana* plant. It was discovered that cerium oxide nanoparticles may reduce oxidative stress in *Brassica napus*. At concentrations of 200 and 1000 mg/kg cerium oxide nanoparticles, altered physiobiochemical characteristics, such as enhanced plant biomass, chlorophyll content (which raises Mg^{2+} uptake) and an effective photosystem were observed in *Brassica napus* plants exposed to NaCl (100 mM) stress (Rossi et al. 2016). Salinity stress has been shown to enhance the cytosolic Na^+/K^+ ratio which is effectively maintained by the nanoparticles (Hauser and Horie 2010). Seeds primed with polyacrylic acid-coated cerium oxide nanoparticles (500 mg/l in water for 24 hours) exhibited significant effects on the root systems, including increased length (56%), weight (41%) and root vitality (114%) in comparison to the seedlings that were germinated under salinity stress (200 mM NaCl). The pathways associated with terpene production, ion binding and Ca^{2+} signalling and the antioxidative enzyme system reduces oxidative stress and increases tolerance to salinity stress (An et al. 2020).

13.8.2 Silicon Nanoparticles

Silicon is regarded as an element that falls somewhere between essential and nonessential for plants (Luyckx et al. 2017). By interacting with plants either directly or indirectly, silicon nanoparticles can alter their morphology and physiology and help them become more resistant to stress. It has positive impacts on plant growth, biomass and physiology, alters tissue differentiation, enhances defense mechanisms and promotes plant adaptation to stressful environment (Babajani et al. 2019). At varying concentrations, silicon nanoparticles showed antistress benefits against drought stress in *Crataegus monogyna*. The responses of the seedlings differed, based on the concentration of nanoparticles applied at different phases of drought stress, i.e., moderate to severe. Silicon dioxide nanoparticles have been demonstrated to enhance water consumption efficiency, stomatal conductance, transpiration rate and reduce chlorophyll degradation during salt stress, which promote tolerance against external stimuli (Haghighi and Pessarakli 2013). When strawberry plants were treated with silicon nanoparticles, their epicuticular wax structure and thickness improved in comparison to the plants under salt stress (Avestan et al. 2019). Another investigation on the effects of silicon nanoparticles to reduce salt stress on sweet pepper plants (*Capsicum annum* L.) showed notable differences when compared to their stressed equivalent (Tantawy et al. 2015). The concentration-dependent impact of silicon dioxide nanoparticles was also seen in potato plants under various salt stress conditions (50 and 100 mg/l). The nanoparticles were able to demonstrate superior stress amelioration at both lower and higher concentrations (50 and 100 mg/l) (Gowayed et al. 2017). The effects of mesoporous silica nanoparticles entrapping abscisic acid were evaluated in *A. thaliana* which clearly demonstrated the extended release of hormones and enhanced seedling capacity to deal with dryness (Sun et al. 2018).

13.8.3 Titanium Dioxide Nanoparticles

Exposure to titanium dioxide nanoparticles (0.01%) induced a rise in biomass and chlorophyll content through the activation of antioxidant enzymes. It produces proline and soluble carbohydrates, reduces malondialdehyde and hydrogen peroxide and regulates the osmotic equilibrium (Abdel Latef et al. 2018). In tobacco seedlings, 11 miRNAs were expressed in response to titanium dioxide nanoparticles (0.1%, 1%, 2.5% and 5%), which were necessary to protect plants against heavy metal stress (Frazier et al. 2014). Gohari et al. (2020) conducted a greenhouse experiment to assess the impact of nano-titanium dioxide (0, 50, 100 and 200 mg/l) on Moldavian balm plants cultivated under aggravated salinity stress (0, 50 and 100 mM NaCl). Under typical circumstances, the plants treated with 100 mg/l of titanium dioxide demonstrated a significant 1.2% increase in the production of essential oils, including geranial, z-citral, geranyl acetate and geraniol. Another study conducted by Karamian et al. (2020) in medicinal plants demonstrated that methyl jasmonate (200 μM), salicylic acid (100 μM) and titanium dioxide nanoparticles (20 ppm) helped to alleviate the effects of drought stress by triggering both enzymatic and nonenzymatic antioxidant defense mechanisms.

13.9 Role of Nanotechnology in Plant Nutrition

The low usage efficiency of chemical fertilizers due to its loss through leaching and volatilization pollutes the environment and raises production costs (FAO 2017).

So, scientists created alternate nanotechnology methods to ensure the sustainable usage of nutrients. Plant micronutrients, including iron, zinc, manganese, copper, nickel and molybdenum, are the focus of most researchers (Al-Mamun et al. 2021; Arya et al. 2022; Dimkpa and Bindraban 2017; Sashidhar et al. 2020; Soni et al. 2023). The type and variety of the plants, as well as a few additional factors including the amount, concentration and chemical characteristics of the nanomaterials will determine how well they work as fertilizers for plant growth (Thakur et al. 2018). Utilizing the nano-porous surfaces of plant components, nanotechnology provides a platform for unique and effective nutrient delivery. Additionally, their characteristics might facilitate a lower fertilizer loss while promoting optimal nutrient uptake by crops (Manjunatha et al. 2016). Nanofertilizers are quickly absorbed by the soil, enhancing their quality and promoting better plant performance. Because the particles of nanofertilizers are smaller than 100 nm, most of these particles can enter the plant from external areas like soil or leaves (Meena et al. 2017). Fertilizers that are coated in nanoparticles will improve the availability of nutrients and their absorption by crop plants (Liscano et al. 2000). The gradual delivery of nutrients to the crop plants facilitated by zeolite-based nanofertilizers improved nutrient availability throughout the development period and inhibited nutrient loss due to denitrification, vaporization, evaporation and fixing in the soil (Qureshi et al. 2018).

13.10 Application of Nanotechnology in Seed Science

The reproductive phase is the most sensitive stage in the life cycle of a plant. A variety of factors, including the fertility of the soil, moisture accessibility, inherited characteristics and environmental conditions, significantly impact seed germination (Manjaiah et al. 2019). Seeds are self-replicating life forms of plants that can persist on their own in adverse circumstances. Producing and collecting seeds can be a challenging task, particularly for crops that are fertilized by the wind. Temperature, humidity and wind velocity all affect the dispersion of pollen to the surroundings. By using bio-nanosensors designed to detect contaminated pollen, field crops can be protected against pollen escape from transgenic crops. In order to distinguish unprofitable and contaminated seeds, a technology was designed called quantum dots as a fluorescence marker combined with immuno-magnetic differentiation for *Escherichia coli* 0157:H7 (Su and Li 2004). Additionally, seeds can be treated with nano-encapsulations encoding a particular bacterial strain known as 'smart seed'. For reforestation purposes, a smart seed that has been programmed to germinate when sufficient moisture is present can be scattered throughout a mountain range. Seeds can be coated with a nanomembrane that recognizes water availability and only permits seeds to germinate at the appropriate time. Nanofertilizers are easily absorbed by seeds that improves the supply of nutrients for developing plant seeds, resulting in healthier shoots and roots. However, if nanoparticles are absorbed higher than optimal level, it may have adverse impacts on the development of plants. A number of researches have demonstrated that nanofertilizers have a major impact on the germination of seeds and the development of seedlings (Meena et al. 2017). By enhancing nutrient availability to the developing plant, nanofertilizers enhance the formation of chlorophyll and photosynthesis, all of which contribute to the higher growth of plants (Salama 2012; Suriyaprabha et al. 2012). Plants grown from seeds treated with nanotitanium dioxide showed increased dry mass, photosynthetic rate and chlorophyll-a production in comparison with the control group (Zheng et al. 2005). Furthermore, nanoparticles can penetrate into the seed layer to enhance water absorption and utilization, which activates the enzymatic system that supports plant formation and development (Salama et al. 2021). Multi-walled carbon nanotubes have a positive effect on the development of seeds of garlic, maize and tomato (Anita and Rao 2014). Higher fresh and dry weight, root and shoot length and other seedling growth parameters may be associated with the positive role of nanoparticles in significantly reducing oxidative damage and ROS production and rapid absorption of oxygen and water (Raskar and Laware 2013; Zimmermann and Zentgraf 2005). The ability of nanomaterials to absorb more water and nutrients, which in turn helps to enhance the strength of root systems with increased enzymatic activity, may partially explain higher plant growth and enriched quality, even though the exact process underlying it is unclear (Dubey and Mailapalli 2016; Shojaei et al. 2019). When compared with bulk zinc sulphate, nano-zinc oxide showed higher root growth and germination percentages in peanut seeds (Prasad et al. 2012). In a similar way, favourable effects of nanoparticles such as silicon dioxide and titanium dioxide on soybean germination have been identified (Mahmoodzadeh et al. 2013). The development and photosynthesis of the peanut plant were enhanced by the exogenous application of nanoferric oxide (Liu et al. 2005). The plant that received the nano-iron treatment had a higher iron content than that of control plants (Nadi et al. 2013).

Nanoparticles can also enhance cellular signaling pathways to promote crop development, seed metabolic processes and seedling vigor (Abbasi Khalaki et al. 2020; Acharya et al. 2020; Mahakham et al. 2017).

Seed nano-priming is a novel technique that uses nanomaterials, especially nanoparticles, for seed priming (Kasote et al. 2019; Mahakham et al. 2017; Shukla et al. 2019). Seed priming is an ancient method applied to agricultural land to enhance the germination of seeds and plant maintenance (Abbasi Khalaki et al. 2020; Bourioug et al. 2020; Carrillo-Reche et al. 2018; Ibrahim 2016; Jisha et al. 2013; Thomas and Puthur 2017). Typically, it uses a water-based technique in which seeds are hydrated and then dried, although it may involve physical techniques like UV priming (Lemmens et al. 2019). Due to the antibacterial properties of many nanoparticles and their ability to carry antimicrobial compounds, seed nanopriming can also be used to protect seeds (Abbasi Khalaki et al. 2020; Pirzada et al. 2020). Moreover, biofortification of seeds can be achieved through nanopriming, which will help to raise food production and quality (Abbasi Khalaki et al. 2020; De La Torre-Roche et al. 2020; Malik et al. 2021). Various nanomaterials have demonstrated a promising role in seed nano-priming, which leads to enhanced morphological and metabolic features in plants as well as increased plant growth. Plant development was enhanced by seed priming with zinc oxide and iron oxide nanoparticles due to improved green leaf photosynthetic rate, chlorophyll levels and plant biomass as compared to the impacts of their bulk counterparts like ferric chloride and ferric oxide (Rizwan et al. 2019b). When maize seeds were primed with copper-containing chitosan nanoparticles, α-amylase and protease activities were stimulated, and reserve mobilization was encouraged, all of which favoured the growth of seedlings (Saharan et al. 2016). Nanoparticles have the ability to directly attack pathogens and modify the metabolic processes of seeds and plants. This can improve the innate immune system that increases the resistance of plants to infections and stress caused by abiotic factors (Panyuta et al. 2016).

13.11 Positive Effects of Nanomaterials in Crop Protection

Based on polymers, nanomaterials for crop protection come in a variety of forms, including micelles, nanogels, nanocapsules and nanosphere. Nanocapsules are made up of a polymeric protective shell and an interior solid or liquid core that contains active ingredients or insecticides. In order to generate polymeric nanospheres, acetone was used to dissolve poly(G-caprolactone) and an active component, which were then injected into an aqueous solution containing a surfactant while being stirred moderately (Ojha et al. 2018). Because of their small size, the nanospheres showed higher absorption (Boehm et al. 2003). Water-insoluble agrochemicals are delivered by micellar systems. Covalent or van der Waals forces can crosslink hydrophilic polymers to create nanogels (Ojha et al. 2018). According to Sultana et al. (2013), there are several methods for obtaining microgel polymers like hyaluronan and chitosan, including spray drying, chemical crosslinking, aqueous homogeneous gelation and water-in-oil heterogeneous emulsion. According to Bhagat et al. (2013), *Bactrocera dorsalis* is a polyphagous pest that contaminates a wide range of fruits and vegetables, including guava, banana, mango and orange. At present, pheromones are used to attract insects due to their species-specificity and lack of toxicity to other species.

The hydrogels that were used to transport these pheromones proved ineffective due to their tendency to alter in response to variations in climate (Bhagat et al. 2013). The most popular tool for controlling pests is an inorganic pesticide; however, as we have seen, using large amounts of these compounds can lead to pollution of the environment and insect resistance. Pesticides can be incorporated into nanoparticles to enhance their solubility and regulate the release of active chemicals (Debnath et al. 2011; Kah et al. 2013; Lauterwasser 2005; Scrinis and Lyons 2007). One effective method of managing pests and diseases while minimizing the amount of residue that harms the environment is the encapsulation of insecticides as well as helpful microbes. These methods can reduce leaching while simultaneously enhancing plant nutrient uptake and reducing eutrophication, which minimizes the amount of nitrogen that is transferred to aquifers (Liu and Lal 2015). By promoting solubility and susceptibility to hydrolysis, enhancing photo-decomposition and/or enabling a more effective and regulated release toward target species, nanomaterials may potentially be used to improve both the structure and the function of pesticides (Grillo et al. 2016; Mishra et al. 2014). According to multiple studies, copper nanoparticles have demonstrated antibacterial properties against a variety of plant-damaging microbial species and fungi. In order to fight *Phytophthora infestans*, Giannousi et al. (2013) used copper nanoparticles with varying dimensions (11–14 nm) and chemical structures (cuprous oxide and copper dioxide) in tomato plants (*Lycopersicon esculentum*) in a greenhouse. Campos et al. (2018) analysed that strong photosensitivity, high volatility and low aqueous solubility are the characteristics of chitosan nanoparticles modified with β-cyclodextrin containing carvacrol and linalool. The insect-fighting activity of the generated nanoparticles was shown against *Tetranychus urticae* and *Helicoverpa armigera*. The plant pathogen *Xanthomonas axonopodis pv. vesicatoria*, which causes bacterial spot disease in tomatoes and peppers, was identified through the use of fluorescent silica nanoparticles and antibody molecules (Yao et al. 2009).

Agricultural losses due to pests and insects are over 13% of total crop production and around 3 billion kg of pesticides are used annually to eradicate pests (Pimentel 2009). The primary drawbacks of chemical pesticides include their severe toxicity to nontarget organisms, including human, fish, amphibian, bird and insect pollinators and their inability to biodegrade (Gill and Garg 2014). The goal of creating nanopesticides is to completely reduce the use of pesticides derived from chemical compounds that might be dangerous (de Oliveira et al. 2014; Madbouly et al. 2017; Ureña Saborio et al. 2017). In order to improve intelligent nanosystems that can keep and disable nutrients and allow their slow release in the soil to increase nutrient efficiency, new nanomaterials have been produced using metal-based, polymerized and inorganic nanoparticles as components (Medina-Pérez et al. 2019). Numerous nanoparticles made of metals, including silver, iron, copper and zinc have two potential applications: first, as nanofertilizers that enhance seed germination and plant growth, and second as fungicides or pesticides that attack certain pathogenic microbes (Le Van et al. 2016). By developing nanostructured materials, the adherence of bacteria to surfaces can be prevented (Rodrigues et al. 2017).

13.12 Nanotechnology and Safety

In the field of nanotechnology, when food and agricultural products enhanced by nanotechnology are introduced to the public and consumed, the risks to health increase. One of the most important variables contributing to the toxicology of manufactured nanomaterials has long been assigned to size at the nanoscale (Hwang et al. 2012). Copper nanoparticles exhibit size-dependent neurotoxicity based on exposure; at smaller diameters (40 nm), lactate dehydrogenase release occurs at a faster rate than at larger diameters (60–80 nm) (Prabhu et al. 2010). According to Gliga et al. (2014), smaller silver nanoparticles (10 nm) dissolve and release a significant number of silver ions in 24 hours in a cell medium than larger silver nanoparticles (40, 50 and 75 nm). The toxicity was caused by the interior silver release rather than variations in cellular absorption. When evaluating nanotoxicology, surface modification is still an additional significant factor. Surface modification with organic or neutral chemicals makes it easy to change the surface characteristics of the manufactured nanomaterials. This kind of modification frequently results in increased consistency and distribution. Surface alteration can generally improve biological compatibility with lowered toxicity of synthetic nanomaterials when designed correctly (He et al. 2019). According to Sotiriou and Pratsinis (2010), Deng et al. (2016) and Xiu et al. (2012), silver nanoparticles may produce silver ions when released into the aqueous environment. These ions are believed to be highly toxic and are likely the source of the antibacterial activity of the silver nanoparticles. Therefore, an in-depth study of synthesized nanomaterials with respect to the nano-bio-eco interactions is required to check its effects on the environment. This is an important stage to evaluate the toxicological potential and dangers associated with manufactured nanomaterials. Hu et al. (2014) showed that arsenic, a common pollutant in wheat and other agricultural plants, might have a significant phytotoxicity when combined with graphite oxide nanoparticles. Using graphite oxide may result in oxidative stress, lower wheat biomass and fewer roots. Arsenic and graphite oxide inhibited the metabolism of carbohydrates, fatty acids and the urea cycle, although they enhanced the metabolism of amino acids and secondary metabolism. In a recent report, graphene oxide was reported to significantly reduce the hydrophilic conductivity of roots and the thickness of the cell wall (Zhou and Hu 2017).

Graphite oxide may also have an impact on the expression of genes encoding plant phosphate transporter gene and arsenic reductase, which could ultimately lead to a higher absorption of arsenic (Hu et al. 2014). According to Zhou and Hu (2017), graphene oxide may alter peroxidases, secondary metabolic processes and phenylalanine metabolism, thereby interfering with the stress-induced wellness mechanisms of rice roots. A different study found that poly (methacrylic acid) nanoparticles containing chitosan, which were used in food packaging sheets, may be harmful to cells (De Lima et al. 2010).

It is predicted that due to food additives, which are mainly composed of titanium dioxide and alum inosilicates, the average individual consumes 1×10^{12} submicron-sized particles per day in a normal diet (Lomer et al. 2002). Common sources of incidental nanoparticles include wood smoke, vehicle exhaust and furnace exhaust (Barregard et al. 2006; Chang et al. 2004; Fang et al. 2005). In outdoor environments close to high-traffic regions, accidental nanoparticle levels can vary from 5000 to 3,000,000 particles/cm^3 (Utell and Frampton 2000). Particulate toxicology, the study of the harmful effects of these particles, is in prevalence for many years. There have been a lot of discussions about the potential negative impacts of nanoparticle interactions with the environment.

13.13 Safety and Regulations for Nanotechnology-Based Agri-products in India

The application of nanotechnology in agricultural products, however, has led to concerns about potential risks to the environment and human health (Khot et al. 2020; Mishra and Singh 2021). Regulations and standards have been formulated by regulatory agencies worldwide to solve such concerns and secure the safe application of nanotechnology to crops and livestock. Following are some important rules and directives concerning agricultural products based on nanotechnology (Kumari et al. 2023).

Regulatory Oversight: The Department of Biotechnology, the Ministry of Environment, Forests and Climate Change, the Food Safety and Standards Authority of India and the Indian Council of Agricultural Research are among the agencies that monitor nanotechnology-based agricultural products in India.

Risk Assessment: Before approving any agricultural product based on nanotechnology, regulatory agencies need a risk evaluation. This includes assessing the effects on human health, the possibility of environmental release and the toxicity of the nanomaterials used.

Labelling: Agricultural products using nanomaterials must be labelled, according to regulatory agencies. Customers can use this information to make more informed choices regarding the items that they buy and use.

International Standards: In order to guarantee the security and excellence of agricultural products based on nanotechnology, international standards have been created. A number of nanotechnology-related standards have been created by the International Organization for Standardization, such as ISO/TS 80004-1, which defines the keywords for nanoparticles.

Research and Development: Regulatory agencies support the development of agricultural products based on nanotechnology in order to guarantee that the final products offer no risks to the environment or public health. In general, makers, customers, regulatory agencies and investigators must work together to ensure the safe application of nanotechnology in agricultural products.

13.14 Negative Effects of Nanomaterials on Plant Health

The negative effects of synthetic nanomaterials on crop health are an important issue, despite their major advantages (Alshatwi et al. 2013; Lowry and Casman 2009). Therefore, it is crucial to decipher the toxicity of nanoparticles, which is usually determined by the size, shape and chemical nature of these particles (Nowack and Bucheli 2007). Certain growth conditions may allow plants to consume both necessary and nonessential elements, which may be potentially dangerous at concentrations over the threshold. The nanoparticles that are used in plants typically have a toxicological factor as well, that partially causes damage to the plants but is more deleterious to the microflora or microbial life that exists in the soil. Plants undergo mutations in DNA and lipid peroxidation as a result of nanoparticles and free radicals, which are normally dangerous (Mwaanga 2018). Exposure to copper hydroxide pesticide spray and copper oxide nanomaterials over short periods of time has been shown to have negative effects on plants (Du et al. 2017; Xiong et al. 2017; Zhao et al. 2017). Rico et al. (2013) reported that rice grains from plants grown in soil treated with cerium oxide nanoparticles exhibited decreased nutritional value in terms of minerals, fatty acids, proteins, carbohydrates and antioxidants. It is possible that altered gene expression is the cause of this decrease in nutritional value. DNA changes and mitotic abnormalities were also seen in *Vicia faba* seedlings treated with silver nanoparticles, *Zea mays* and *Vicia narbonensis* treated with titanium dioxide nanoparticles (Abdel-Azeem and Elsayed 2013; Ruffini et al. 2011). Atha et al. (2012) found that radish and ryegrass seedlings treated with copper oxide nanoparticles showed higher oxidative damage that was also noticeable at low concentrations of the nanoparticle (10 mg/l). When Begum and Fugetsu (2012) applied multi-walled carbon nanotubes to spinach plants, they were seen to elevate ROS and induce cell death processes in the plant tissues. Shen et al. (2010) treated both protoplasts and essential leaves with single-walled carbon nanotubes and noted signs of oxidative stress (ROS accumulation) and a dose-dependent programmed cell death in the *Arabidopsis* and *Oryza sativa* plant species. Giraldo et al. (2014) showed that after single-walled carbon nanotubes penetrated the spinach chloroplast membranes, they increased electron transport and photosynthetic activity. This was most likely due to a stimulating effect on the absorption of light with near-infrared wavelengths. Additionally, it was demonstrated that single-walled carbon nanotubes were sensitive to nitric oxides, indicating that plants that contain nanotubes may be used as nitric oxide detectors. However, Santos et al. (2013) demonstrated that fullerene had a harmful effect on the aquatic plant *Lemna gibba*, resulting in a reduction in both plant growth and photosynthetic activity. When the seedlings were treated with graphene oxide at concentrations of 500–2000 mg/l, the size of the leaves dropped in cabbage, spinach and tomato, and the number of leaves reduced in all of these plants (Begum et al. 2011). Using transmission electron microscopy, Du et al. (2011) found that both titanium dioxide and zinc oxide nanoparticles (predominant sizes 20 nm, 5 nm, 40 nm and 10 nm, respectively) decreased the activity of proteases, catalase and peroxidase enzymes in soil; yet, only the titanium dioxide nanoparticles (20 nm) were able to infiltrate the major root cortical cells and rhizoderm, where they formed vacuoles.

13.15 Conclusion

In spite of its few negative effects, nanotechnology can be widely used as an innovative tool in agriculture. The benefits of using nanomaterials in agriculture, especially for crop protection, are controlled release of agrochemicals that extend their efficacy. Nanotechnology is an invaluable ally in the pursuit of sustainable agricultural goals, including crop yield increases and pollution reduction. One possible spark for a new green revolution is the relatively quick advancement of nanotechnology. Nanoparticles have many applications, including high-throughput genome editing, antimicrobial activities to reduce contamination, improved nutrient levels through nanofertilizers, early pathogen detection through nano-biosensors, fast pathogen detection through engineered nanoparticles and biosensors, and pathogen detection through nanoparticle-based genome editing. Research interest in the environment friendly and highly bioactive production of nanoparticles by biological organisms is growing. In order to have thorough knowledge, more research, both scientific and practical, is necessary, as nanotechnology implementation in agriculture is still in its early phases. Despite the wealth of studies demonstrating the applications of nanoparticles in plants, there are still issues with the uptake and transport of nanoparticles, their interaction with plant surfaces and the causes of some plant species resistance to nanoparticles. In future, it is expected that there will be a major focus to integrate biotechnology and nanotechnology in the development of disease diagnostic tools. This includes the transfer of genetic material and antimicrobial compounds via nanocapsules with the purpose of improving disease resistance. It is noted that more research is required to further our knowledge of the potential uses, methods and applications of nanoparticles in plant pathology, even though the reviews that are already available show a variety of uses for these particles in crop protection and disease control. In brief, current investigations are concentrated on creating nanocomposites with intelligent delivery methods to supply vital nutrients to crops in proportionately balanced amounts. It is crucial to research how beneficial microbes, soil and plant systems are affected by nanofertilizers, as well as what happens to them after they are released into the environment, in order to optimize their influence. For the best outcomes, it is essential to precisely manage agriculture and standardize concentration and dosage for various crops. It is possible to increase crop yield, lower the costs and reduce pollution with nanofertilizers. Nevertheless, because of social considerations, their implementation needs to be carefully considered. Crop productivity may be benefited from more effective usage of nanofertilizers; however, for best outcomes, more investigation and standardization are required. Minimizing pollution risks, cutting fertilizer costs and increasing crop yield are all impacted by the use of different types of nanofertilizers. On the other hand, the application of nanofertilizers in agriculture raises societal problems that must be carefully considered.

In conclusion, while nanomaterials hold promises as innovative tools for enhancing crop productivity and sustainability, their deployment as nanofertilizers requires careful evaluation, regulation and integration into holistic agricultural systems. Collaboration among researchers, policymakers, farmers and other stakeholders is essential to realize the full potential of nanofertilizers while minimizing potential risks and maximizing their benefits for global food security and environmental stewardship.

References

Abbai, R., Kim, Y.J., Mohanan, P. et al. (2019). Silicon confers protective effect against ginseng root rot by regulating sugar efflux into apoplast. *Scientific Reports* 9: 18259.

Abbasi Khalaki, M., Moameri, M., Asgari Lajayer, B., and Astatkie, T. (2020). Influence of nano priming on seed germination and plant growth of forage and medicinal plants. *Plant Growth Regulation* 93: 13–28.

Abdel Latef, A.A.H., Srivastava, A.K., El-sadek, M.S.A. et al. (2018). Titanium dioxide nanoparticles improve growth and enhance tolerance of broad bean plants under saline soil conditions. *Land Degradation & Development* 29: 1065–1073.

Abdel-Azeem, E.A. and Elsayed, B.A. (2013). Phytotoxicity of silver nanoparticles on *Vicia faba* seedlings. *New York Science Journal* 6: 148–156.

Acharya, P., Jayaprakasha, G.K., Crosby, K.M. et al. (2020). Nanoparticle-mediated seed priming improves germination, growth, yield, and quality of watermelons (*Citrullus lanatus*) at multi-locations in texas. *Scientific Reports* 10: 5037.

Adisa, I.O., Pullagurala, V.L.R., Peralta-Videa, J.R. et al. (2019). Recent advances in nano-enabled fertilizers and pesticides: a critical review of mechanisms of action. *Environmental Science: Nano* 6: 2002–2030.

Afzal, S., Sharma, D., and Singh, N.K. (2021). Eco-friendly synthesis of phytochemical-capped iron oxide nanoparticles as nano priming agent for boosting seed germination in rice (*Oryza sativa* L.). *Environmental Science and Pollution Research* 28: 40275–40287.

Ahanger, M.A., Qi, M., Huang, Z. et al. (2021). Improving growth and photosynthetic performance of drought stressed tomato by application of nano organic fertilizer involves upregulation of nitrogen, antioxidant and osmolyte metabolism. *Ecotoxicology and Environmental Safety* 216: 112195.

Ahmad, P., Alyemeni, M.N., Al-Huqail, A.A. et al. (2020). Zinc oxide nanoparticles application alleviates arsenic (As) toxicity in soybean plants by restricting the uptake of as and modulating key biochemical attributes, antioxidant enzymes, ascorbate-glutathione cycle and glyoxalase system. *Plants* 9: 825.

Ahmed, S., Ahmad, M., Swami, B.L., and Ikram, S. (2016). A review on plants extract mediated synthesis of silver nanoparticles for antimicrobial applications: a green expertise. *Journal of Advanced Research* 7: 17–28.

Alimohammadi, M., Panahpour, E., and Naseri, A. (2020). Assessing the effects of urea and nano-nitrogen chelate fertilizers on sugarcane yield and dynamic of nitrate in soil. *Soil Science and Plant Nutrition* 66: 352–359.

Al-Mamun, M.R., Hasan, M.R., Ahommed, M.S. et al. (2021). Nano fertilizers towards sustainable agriculture and environment. *Environmental Technology & Innovation* 23: 101658.

Al-Salim, N., Barraclough, E., Burgess, E. et al. (2011). Quantum dot transport in soil, plants, and insects. *Science of the Total Environment* 409: 3237–3248.

Alshatwi, A.A., Subbarayan, P.V., Ramesh, E. et al. (2013). Aluminium oxide nanoparticles induce mitochondrial-mediated oxidative stress and alter the expression of antioxidant enzymes in human mesenchymal stem cells. *Food Additives & Contaminants: Part A* 30: 1–10.

Ameh, T. and Sayes, C.M. (2019). The potential exposure and hazards of copper nanoparticles: a review. *Environmental Toxicology and Pharmacology* 71: 103220.

An, J., Hu, P., Li, F. et al. (2020). Emerging investigator series: molecular mechanisms of plant salinity stress tolerance improvement by seed priming with cerium oxide nanoparticles. *Environmental Science: Nano* 7: 2214–2228.

An, C., Sun, C., Li, N. et al. (2022). Nanomaterials and nanotechnology for the delivery of agrochemicals: strategies towards sustainable agriculture. *Journal of Nanobiotechnology* 20, Article number: 11.

Anand, K.V., Anugraga, A.R., Kannan, M. et al. (2020). Bio-engineered magnesium oxide nanoparticles as nano-priming agent for enhancing seed germination and seedling vigor of green gram (*Vigna radiata* L.). *Materials Letters* 8: 127792.

Anita, S. and Rao, D. (2014). Enhancement of seed germination and plant growth of wheat, maize, peanut and garlic using multi walled carbon nanotubes. *European Chemical Bulletin* 3: 502–504.

Anwar, F., Latif, S., Ashraf, M., and Gilani, A.H. (2007). *Moringa oleifera*: a food plant with multiple medicinal uses. *Phototherapy Research: An International Journal Devoted to Pharmacological and Toxicological Evaluation of Natural Product Derivatives* 21: 17–25.

Arya, S.S., Rookes, J.E., Cahill, D.M., and Lenka, S.K. (2022). Chitosan nanoparticles and their combination with methyl jasmonate for the elicitation of phenolic and flavonoids in plant cell suspension cultures. *International Journal of Biological Macromolecules* 214: 632–641.

Ashok, C., Rao, K.V., Chakra, C.S., and Rao, K.G. (2016). Magnesium oxide nanoparticles prepared by microwave-irradiation technique and its seed germination application. *Nano Trends – A Journal of Nano Technology & Its Applications* 18: 10–17.

Askary, M., Talebi, S.M., Amini, F., and Bangan, A.D.B. (2017). Effects of iron nanoparticles on *Mentha piperita* L. under salinity stress. *Biologija* 63: 65–75.

Atha, D.H., Wang, H., Ptersen, E.J. et al. (2012). Copper oxide nanoparticle mediated DNA damage in terrestrial plant models. *Environmental Science & Technology* 46: 1819–1827.

Avellan, A., Schwab, F., Masion, A. et al. (2017). Nanoparticle uptake in plants: gold nanomaterial localized in roots of *Arabidopsis thaliana* by X-ray computed nano tomography and hyperspectral imaging. *Environmental Science & Technology* 51: 8682–8691.

Avestan, S., Ghasemnezhad, M., Esfahani, M., and Byrt, C.S. (2019). Application of nano-silicon dioxide improves salt stress tolerance in strawberry plants. *Agronomy* 9: 246.

Babajani, A., Iranbakhsh, A., Ardebili, Z.O., and Eslami, B. (2019). Differential growth, nutrition, physiology, and gene expression in *Melissa officinalis* is mediated by zinc oxide and elemental selenium nanoparticles. *Environmental Science and Pollution Research* 26: 24430–24444.

Bairwa, P., Kumar, N., Devra, V., and Abd-Elsalam, K.A. (2023). Nano-biofertilizers synthesis and applications in agroecosystems. *Agrochemicals* 2: 118–134.

Barregard, L., Sallsten, G., Gustafson, P. et al. (2006). Experimental exposure to wood-smoke particles in healthy humans: effects on markers of inflammation, coagulation, and lipid peroxidation. *Inhalation Toxicology* 18: 845–853.

Begum, P. and Fugetsu, B. (2012). Phytotoxicity of multi-walled carbon nanotubes on red spinach (*Amaranthus tricolor* L) and the role of ascorbic acid as an antioxidant. *Journal of Hazardous Materials* 243: 212–222.

Begum, P., Ikhtiari, R., and Fugetsu, B. (2011). Graphene phytotoxicity in the seedling stage of cabbage, tomato, red spinach, and lettuce. *Carbon* 49: 3907–3919.

Bhagat, D., Samanta, S.K., and Bhattacharya, S. (2013). Efficient management of fruit pests by pheromone nano gels. *Scientific Reports* 3: 1294.

Boehm, A.L., Martinon, I., Zerrouk, R. et al. (2003). Nanoprecipitation technique for the encapsulation of agrochemical active ingredients. *Journal of Microencapsulation* 20: 433–441.

Bourioug, M., Ezzaza, K., Bouabid, R. et al. (2020). Influence of hydro and osmo priming on sunflower seeds to break dormancy and improve crop performance under water stress. *Environmental Science and Pollution Research* 27: 13215–13226.

Burlaka, O.M., Pirko, Y.V., Yemets, A.I., and Blume, Y.B. (2015). Plant genetic transformation using carbon nanotubes for DNA delivery. *Cytology and Genetics* 49: 349–357.

Cai, L., Liu, C., Fan, G. et al. (2019). Preventing viral disease by zinc oxide nanoparticles through directly deactivating tobacco mosaic virus and activating plant immunity in *Nicotiana benthamiana*. *Environmental Science: Nano* 6: 3653–3669.

Campos, E.V.R., Proença, P.L.F., Oliveira, J.L. et al. (2018). Carvacrol and linalool co-loaded in β-cyclodextrin-grafted chitosan nanoparticles as sustainable bio pesticide aiming pest control. *Scientific Reports* 8: 1–15.

Carrillo-Reche, J., Vallejo-Marín, M., and Quilliam, R.S. (2018). Quantifying the potential of 'on-farm' seed priming to increase crop performance in developing countries: a meta-analysis. *Agronomy for Sustainable Development* 38: 64.

Chang, M.C., Chow, J.C., Watson, J.G. et al. (2004). Measurement of ultrafine particle size distributions from coal-, oil-, and gas-fired stationary combustion sources. *Journal of the Air & Waste Management Association* 54: 1494–1505.

Chen, D., Szostak, P., Wei, Z., and Xiao, R. (2016a). Reduction of orthophosphates loss in agricultural soil by nano calcium sulphate. *Science of the Total Environment* 539: 381–387.

Chen, Y.W., Lee, H.V., Juan, J.C., and Phang, S.M. (2016b). Production of new cellulose nanomaterial from red algae marine biomass *Gelidium elegans*. *Carbohydrate Polymers* 151: 1210–1219.

Chen, Y.E., Mao, H.T., Wu, N. et al. (2020). Salicylic acid protects photosystem II by alleviating photo inhibition in *Arabidopsis thaliana* under high light. *International Journal of Molecular Sciences* 21: 1229.

Chhipa, H. (2017). Nano fertilizers and nano pesticides for agriculture. *Environmental Chemistry Letters* 15: 15–22.

Chichiriccò, G. and Poma, A. (2015). Penetration and toxicity of nanomaterials in higher plants. *Nanomaterials* 5: 851–873.

Collin, B., Auffan, M., Johnson, A.C. et al. (2014). Environmental release, fate and eco toxicological effects of manufactured ceria nanomaterials. *Environmental Science: Nano Journal* 1: 533–548.

Dawood, M.F.A., Abeed, A.H.A., and Aldaby, E.E.S. (2019). Titanium dioxide nanoparticles model growth kinetic traits of some wheat cultivars under different water regimes. *Indian Journal of Plant Physiology* 24: 129–140.

De La Torre-Roche, R., Cantu, J., Tamez, C. et al. (2020). Seed bio fortification by engineered nanomaterials: a pathway to alleviate malnutrition? *Journal of Agricultural and Food Chemistry* 68: 12189–12202.

De Lima, R., Feitosa, L., Pereira, A.D.E.S. et al. (2010). Evaluation of the geno toxicity of chitosan nanoparticles for use in food packaging films. *Journal of Food Science* 75: N89–N96.

Debnath, N., Das, S., Seth, D. et al. (2011). Entomotoxic effect of silica nanoparticles against *Sitophilus oryzae* L.). *Journal of Pest Science* 84: 99–105.

Demirer, G.S., Zhang, H., Matos, J. et al. (2018). High aspect ratio nanomaterials enable delivery of functional genetic material without DNA integration in mature plants. *bioRxiv* 10: 1–32.

Deng, H., McShan, D., Zhang, Y. et al. (2016). Mechanistic study of the synergistic antibacterial activity of combined silver nanoparticles and common antibiotics. *Environmental Science & Technology* 50: 8840–8848.

Diallo, A., Ngom, B.D., Park, E., and Maaza, M. (2015). Green synthesis of ZnO nanoparticles by *Aspalathus linearis*: structural & optical properties. *Journal of Alloys and Compounds* 646: 425–430.

Dimkpa, C.O. and Bindraban, P.S. (2017). Nano fertilizers: new products for the industry? *Journal of Agricultural and Food Chemistry* 66: 6462–6473.

Dorjnamjin, D., Ariunaa, M., and Shim, Y.K. (2008). Synthesis of silver nanoparticles using hydroxyl functionalized ionic liquids and their antimicrobial activity. *International Journal of Molecular Sciences* 9: 807–820.

Du, W., Sun, Y., Ji, R. et al. (2011). TiO_2 and ZnO nanoparticles negatively affect wheat growth and soil enzyme activities in agricultural soil. *Journal of Environmental Monitoring* 13: 822–828.

Du, W., Tan, W., Peralta-Videa, J.R. et al. (2017). Interaction of metal oxide nanoparticles with higher terrestrial plants: physiological and biochemical aspects. *Plant Physiology and Biochemistry* 110: 210–225.

Dubey, A. and Mailapalli, D.R. (2016). Nano fertilizers, nano pesticides, nano sensors of pest and nano toxicity in agriculture. *Sustainable Agriculture Reviews* 19: 307–330.

Duhan, J.S., Kumar, R., Kumar, N. et al. (2017). Nanotechnology: the new perspective in precision agriculture. *Biotechnology Reports* 15: 11–23.

Eichert, T., Kurtz, A., Steiner, U., and Goldbach, H.E. (2008). Size exclusion limits and lateral heterogeneity of the stomatal foliar uptake pathway for aqueous solutes and water-suspended nanoparticles. *Plant Physiology* 134: 151–160.

El-Hady, A. and Hussein, H. (2021). Effect of foliar nano fertilizers and irrigation intervals on soybean productivity and quality. *Journal of Plant Production* 12: 1007–1014.

Elmer, W. and White, J.C. (2018). The future of nanotechnology in plant pathology. *Annual Review of Phytopathology* 56: 111–133.

Fang, G.C., Wu, Y.S., Wen, C.C. et al. (2005). Concentrations of nano and related ambient air pollutants at a traffic sampling site. *Toxicology and Industrial Health* 21: 259–271.

FAO (2017). *The Future of Food and Agriculture: Trends and Challenges*. FAO. Bergeret, P. (2017). The Future of Food and Agriculture: Trends and Challenges [Note de lecture]. 3.

Feng, Y., Chen, W., Jia, Y. et al. (2016). N-Heterocyclic molecule capped gold nanoparticles as effective antibiotics against multi drug resistant bacteria. *Nanoscale* 8: 13223–13227.

Forster, N., Ulrichs, C., Schreiner, M. et al. (2015). Ecotype variability in growth and secondary metabolite profile in *Moringa oleifera*: impact of sulphur and water availability. *Journal of Agricultural and Food Chemistry – ACS Publications* 63: 2852–2861.

Fraceto, L.F., Grillo, R., de Medeiros, G.A. et al. (2016). Nanotechnology in agriculture: which innovation potential does it have? *Frontiers in Environmental Science* 4: 20.

Frazier, T.P., Burklew, C.E., and Zhang, B. (2014). Titanium dioxide nanoparticles affect the growth and microRNA expression of tobacco (*Nicotiana tabacum*). *Functional & Integrative Genomics* 14: 75–83.

Galbraith, D.W. (2007). Nano biotechnology-silica breaks through in plants. *Nature Nanotechnology* 2: 272–273.

Gao, F., Hong, F., Liu, C. et al. (2006). Mechanism of nano anatase TiO_2 on promoting photosynthetic carbon reaction of spinach: inducing complex of rubisco-rubisco activase. *Biological Trace Element Research* 111: 239–253.

Gao, G., Tester, M.A., and Julkowska, M.M. (2020). The use of high-throughput phenotyping for assessment of heat stress induced changes in *Arabidopsis*. *The Plant Phenome Journal* 2020: 3723916.

Geisler-Lee, J., Wang, Q., Yao, Y. et al. (2013). Phytotoxicity, accumulation and transport of silver nanoparticles by *Arabidopsis thaliana*. *Nano Toxicology* 7: 323–337.

Ghasemi, B., Hosseini, R., and Dehghan Nayeri, F. (2015). Effects of cobalt nanoparticles on artemisinin production and gene expression in *Artemisia annua*. *Turkish Journal of Botany* 39: 769–777.

Ghidan, A.Y., Al-Antary, T.M., and Awwad, A.M. (2017). Aphidicidal potential of green synthesized magnesium hydroxide nanoparticles using *Olea europaea* leaves extract. *ARPN Journal of Agricultural and Biological Science* 12: 293–301.

Ghormade, V., Deshpande, M.V., and Paknikar, K.M. (2011). Perspectives for nano-biotechnology enabled protection and nutrition of plants. *Biotechnology Advances* 29: 792–803.

Giannousi, K., Avramidis, I., and Dendrinou-Samara, C. (2013). Synthesis, characterization and evaluation of copper based nanoparticles as agrochemicals against *Phytophthora infestans*. *RSC Advances* 3: 21743–21752.

Gil-Díaz, M., Pinilla, P., Alonso, J., and Lobo, M.C. (2017). Viability of a nano remediation process in single or multi-metal(loid) contaminated soils. *Journal of Hazardous Materials* 321: 812–819.

Gill, H.K. and Garg, H. (2014). Pesticides: environmental impacts and management strategies. In: *Pesticides Toxic Aspects* (ed. M.L. Larramendy and S. Soloneski), 187–230. InTech.

Giraldo, J.P., Landry, M.P., Faltermeier, S.M. et al. (2014). Plant nano bionics approach to augment photosynthesis and biochemical sensing. *Nature Materials* 13: 400–408.

Gliga, A.R., Skoglund, S., Wallinder, I.O. et al. (2014). Size-dependent cytotoxicity of silver nanoparticles in human lung cells: the role of cellular uptake, agglomeration and Ag release. *Particle and Fibre Toxicology* 11: 1–17.

Gohari, G., Mohammadi, A., Akbari, A. et al. (2020). Titanium dioxide nanoparticles promote growth and ameliorate salinity stress effects on essential oil profile and biochemical attributes of *Dracocephalum moldavica*. *Scientific Reports* 10: 912.

González-Melendi, P., Fernández-Pacheco, R., Coronado, M.J. et al. (2008). Nanoparticles as smart treatment-delivery systems in plants: assessment of different techniques of microscopy for their visualization in plant tissues. *Annals of Botany* 101: 187–195.

Gowayed, M.H., Al-Zahrani, H.S., and Metwali, E.M. (2017). Improving the salinity tolerance in potato (*Solanum tuberosum*) by exogenous application of silicon dioxide nanoparticles. *International Journal of Agriculture and Biology* 19: 183–194.

Grillo, R., Clemente, Z., Oliveira, J.L. et al. (2015). Chitosan nanoparticles loaded the herbicide paraquat: the influence of the aquatic humic substances on the colloidal stability and toxicity. *Journal of Hazardous Materials* 286: 562–572.

Grillo, R., Abhilash, P.C., and Fraceto, L.F. (2016). Nanotechnology applied to bio-encapsulation of pesticides. *Journal of Nanoscience and Nanotechnology* 16: 1231–1234.

Haghighi, M. and Pessarakli, M. (2013). Influence of silicon and nano-silicon on salinity tolerance of cherry tomatoes (*Solanum lycopersicum* L.) at early growth stage. *Scientia Horticulture* 161: 111–117.

Hao, Y., Xu, B., Ma, C. et al. (2019). Synthesis of novel mesoporous carbon nanoparticles and their phytotoxicity to rice (*Oryza sativa* L.). *Journal of Saudi Chemical Society* 23: 75–82.

Hauser, F. and Horie, T. (2010). A conserved primary salt tolerance mechanism mediated by HKT transporters: a mechanism for sodium exclusion and maintenance of high K^+/Na^+ ratio in leaves during salinity stress. *Plant, Cell & Environment* 33: 552–565.

Haydar, M.S., Ghosh, S., and Mandal, P. (2022). Application of iron oxide nanoparticles as micronutrient fertilizer in mulberry propagation. *Journal of Plant Growth Regulation* 41: 1726–1746.

Hazarika, D., Phukan, A., Saikia, E., and Chetia, B. (2014). Phytochemical screening and synthesis of silver nanoparticles using leaf extract of *Rhynchotechum ellipticum*. *International Journal of Pharmacy and Pharmaceutical Sciences* 6: 672–674.

He, X., Deng, H., Aker, W.G., and Hwang, H.M. (2019). Regulation and safety of nanotechnology in the food and agriculture industry. In: *Food Applications of Nanotechnology* (ed. G. Molina, P. IFM, and A.M. Asiri), 525–536. CRC Press.

Hernandez-Hernandez, H., Gonzalez-Morales, S., Benavides-Mendoza, A. et al. (2018). Effects of chitosan–PVA and copper nanoparticles on the growth and antioxidant capacity of tomato under saline stress. *Molecules* 23: 178.

Hong, R., Pan, T., Qian, J., and Li, H. (2006). Synthesis and surface modification of ZnO nanoparticles. *Chemical Engineering Journal* 119: 71–81.

Hu, X., Kang, J., Lu, K. et al. (2014). Graphene oxide amplifies the phytotoxicity of arsenic in wheat. *Scientific Reports* 4: 6122.

Hua, K.H., Wang, H.C., Chung, R.S., and Hsu, J.C. (2015). Calcium carbonate nanoparticles can enhance plant nutrition and insect pest tolerance. *Journal of Pesticide Science* 40: 208–213.

Hussain, H.A., Men, S., Hussain, S. et al. (2019). Interactive effects of drought and heat stresses on morpho-physiological attributes, yield, nutrient uptake and oxidative status in maize hybrids. *Scientific Reports* 9: 3890.

Hwang, H.M., Ray, P.C., Yu, H., and He, X. (2012). Toxicology of designer/engineered metallic nanoparticles. In: *Sustainable Preparation of Metal Nanoparticles: Methods and Applications* (ed. R. Luque and R. Varma), 190–212. Cambridge: Royal Society of Chemistry.

Ibrahim, E.A. (2016). Seed priming to alleviate salinity stress in germinating seeds. *Journal of Plant Physiology* 192: 38–46.

Ingle, A.P., Duran, N., and Rai, M. (2014). Bioactivity, mechanism of action, and cytotoxicity of copper-based nanoparticles: a review. *Applied Microbiology and Biotechnology* 98: 1001–1009.

Jalil, S.U. and Ansari, M.I. (2019). Nanoparticles and abiotic stress tolerance in plants: synthesis, action, and signalling mechanisms. In: *Plant Signalling Molecules* (ed. M.I.R. Khan, P.S. Reddy, A. Ferrante, and N.A. Khan), 549–561. Elsevier.

Jamkhande, P.G., Ghule, N.W., Bamer, A.H., and Kalaskar, M.G. (2019). Metal nanoparticles synthesis: an overview on methods of preparation, advantages and disadvantages, and applications. *Journal of Drug Delivery Science and Technology* 53: 101174.

Jhansi, K., Jayarambabu, N., Reddy, K.P. et al. (2017). Biosynthesis of magnesium oxide nanoparticles using mushroom extract: effect on peanut (*Arachis hypogaea* L.). seed germination. *3 Biotech* 7: 263.

Jin, Y., Fan, X., Li, X. et al. (2017). Distinct physiological and molecular responses in Arabidopsis thaliana exposed to aluminium oxide nanoparticles and ionic aluminium. *Environmental Pollution* 228: 517–527.

Jisha, K.C., Vijayakumari, K., and Puthur, J.T. (2013). Seed priming for abiotic stress tolerance: an overview. *Acta Physiologiae Plantarum* 35: 1381–1396.

Kah, M., Beulke, S., Tiede, K., and Hofmann, T. (2013). Nano pesticides: state of knowledge, environmental fate, and exposure modelling. *Critical Reviews in Environmental Science and Technology* 43: 1823–1867.

Kah, M., Kookana, R.S., Gogos, A., and Bucheli, T.D. (2018). A critical evaluation of nano pesticides and nano fertilizers against their conventional analogues. *Nature Nanotechnology* 13: 677–684.

Kale, S.K., Parishwad, G.V., and Patil, A.S. (2021). Emerging agriculture applications of silver nanoparticles. *ES Food Agroforestry Journal* 3: 17–22.

Karamian, R., Ghasemlou, F., and Amiri, H. (2020). Physiological evaluation of drought stress tolerance and recovery in *Verbascum sinuatum* plants treated with methyl jasmonate, salicylic acid and titanium dioxide nanoparticles. *Plant Biosystems – An International Journal Dealing with all Aspects of Plant Biology* 154: 277–287.

Kasote, D.M., Lee, J.H.J., Jayaprakasha, G.K., and Patil, B.S. (2019). Seed priming with iron oxide nanoparticles modulate antioxidant potential and defense linked hormones in watermelon seedlings. *ACS Sustainable Chemistry & Engineering* 7: 5142–5151.

Khodakovskaya, M.V., Kim, B.S., Kim, J.N. et al. (2013). Carbon nanotubes as plant growth regulators: effects on tomato growth, reproductive system, and soil microbial community. *Small* 9: 115–123.

Khot, L.R., Sankaran, S., Maja, J.M. et al. (2020). Applications of nanomaterials in agricultural production and crop protection: a review. *Crop Protection* 133: 105221.

Kumari, R., Suman, K., Karmakar, S. et al. (2023). Regulation and safety measures for nanotechnology-based agri-products. *Frontiers in Genome Editing* 5: 1200987.

Kwak, S.Y., Wong, M.H., Lew, T.T.S. et al. (2017). Nano sensor technology applied to living plant systems. *Annual Review of Analytical Chemistry* 10: 113–140.

Larue, C., Castillo-Michel, H., Sobanska, S. et al. (2014). Foliar exposure of the crop *Lactuca sativa* to silver nanoparticles: evidence for internalization and changes in Ag speciation. *Journal of Hazardous Materials* 264: 98–106.

Lauterwasser, C. (2005). Small Sizes that Matter: Opportunities and Risks of Nanotechnologies. Organization for Economic Co-Operation and Development, and Allianz Centre for Technology, Paris, p. 46.

Le Van, N., Ma, C., Shang, J. et al. (2016). Effects of CuO nano particles on insecticidal activity and phytotoxicity in conventional and transgenic cotton. *Chemosphere* 144: 661–670.

Lemmens, E., Deleu, L.J., De Brier, N. et al. (2019). The impact of hydro-priming and osmo priming on seedling characteristics, plant hormone concentrations, activity of selected hydrolytic enzymes, and cell wall and phytate hydrolysis in sprouted wheat (*Triticum aestivum* L.). *ACS Omega* 4: 22089–22100.

Li, D., Zhou, C., Zou, N. et al. (2021a). Nano selenium foliar application enhances biosynthesis of tea leaves in metabolic cycles and associated responsive pathways. *Environmental Pollution* 273: 116503.

Li, M., Zhang, P., Adeel, M. et al. (2021b). Physiological impacts of zero valent iron, Fe_3O_4 and Fe_2O_3 nanoparticles in rice plants and their potential as Fe fertilizers. *Environmental Pollution* 269: 116134.

Lin, S., Reppert, J., Hu, Q. et al. (2009). Uptake, translocation, and transmission of carbon nanomaterials in rice plants. *Small* 5: 1128–1132.

Liscano, J.F., Wilson, C.E., Norman, R.J., and Slaton, N.A. (2000). Zinc availability to rice from seven granular fertilizers. *Research Bulletin Arkansas Agricultural Experiment Station* 963: 1–13.

Liu, R. and Lal, R. (2015). Potentials of engineered nanoparticles as fertilizers for increasing agronomic productions. *Science of The Total Environment* 514: 131–139.

Liu, Y. and Shi, J. (2019). Antioxidative nanomaterials and biomedical applications. *Nano Today* 27: 146–177.

Liu, X.M., Zhang, F.D., Zhang, S.Q. et al. (2005). Effects of nano-ferric oxide on the growth and nutrients absorption of peanut. *Plant Nutrition and Fertilizer Science* 11: 14–18.

Lomer, M.C., Thompson, R.P., and Powell, J.J. (2002). Fine and ultrafine particles of the diet: influence on the mucosal immune response and association with Crohn's disease. *Proceedings of the Nutrition Society* 61: 123–130.

Lowe, B.S., Leer, D.R., Frey, J.W., and Caskey, B.J. (2008). *Occurrence and Distribution of Algal Biomass and Its Relation to Nutrients and Selected Basin Characteristics in Indiana Streams*, 201–205. Sci Invest Rep.

Lowry, G.V. and Casman, E.A. (2009). Nanomaterial transport, transformation, and fate in the environment. *Nanomaterials: Risks and Benefits* 125–137.

Luyckx, M., Hausman, J.F., Lutts, S., and Guerriero, G. (2017). Silicon and plants: current knowledge and technological perspectives. *Frontiers in Plant Science* 8: 411.

Madbouly, A.K., Abdel-Aziz, M.S., and Abdel-Wahhab, M.A. (2017). Biosynthesis of nano silver using *Chaetomium globosum* and its application to control Fusarium wilt of tomato in the greenhouse. *IET Nano Biotechnology* 11: 702–708.

Mahakham, W., Sarmah, A.K., Maensiri, S., and Theerakulpisut, P. (2017). Nano priming technology for enhancing germination and starch metabolism of aged rice seeds using photosynthesized silver nanoparticles. *Scientific Reports* 7: 8263.

Mahendra, R., Shivaji, D., and Mahendra, G. (2012). Strategic nanoparticles mediated gene transfer in plants and animals. *Current Nanoscience* 8: 170–179.

Mahmoodzadeh, H., Nabavi, M., and Kashefi, H. (2013). Effect of nanoscale titanium dioxide particles on the germination and growth of canola (*Brassica napus*). *Journal of Horticultural Science & Ornamental Plants* 3: 25–32.

Malik, A., Mor, V.S., Tokas, J. et al. (2021). Bio stimulant treated seedlings under sustainable agriculture: a global perspective facing climate change. *Agronomy* 11: 14.

Manjaiah, K.M., Mukhopadhyay, R., Paul, R. et al. (2019). Clay minerals and zeolites for environmentally sustainable agriculture. In: *Modified Clay and Zeolite Nanocomposite Materials* (ed. M. Mariano, S. Binoy, and L. Alessio), 309–329. Amsterdam: Elsevier.

Manjunatha, S.B., Biradar, D.P., and Aladakatti, Y.R. (2016). Nanotechnology and its applications in agriculture: a review. *Journal of Farm Sciences* 29: 1–3.

Martin-Ortigosa, S., Peterson, D.J., Valenstein, J.S. et al. (2014). Mesoporous silica nanoparticle-mediated intracellular cre protein delivery for maize genome editing via loxP site excision. *Plant Physiology* 164: 537–547.

Mazumder, J.A., Khan, E., Perwez, M. et al. (2020). Exposure of biosynthesized nanoscale ZnO to *Brassica juncea* crop plant: morphological, biochemical and molecular aspects. *Scientific Reports* 10: 8531.

Medina-Pérez, G., Fernández-Luqueno, F., Campos-Montiel, R.G. et al. (2019). Nanotechnology in crop protection: status and future trends. In: *Nano-biopesticides Today and Future Perspectives* (ed. O. Koul), 17–45.

Meena, S., Malarkodi, M., and Senthilvalavan, P. (2007). Secondary and micronutrients for groundnut – a review. *Agricultural Reviews* 28: 295–300.

Meena, D.S., Gautam, C., Patidar, O.P. et al. (2017). Nano-fertilizers is a new way to increase nutrients use efficiency in crop production. *International Journal of Agriculture Sciences ISSN* 9: 3831–3833.

Miernicki, M., Hofmann, T., Eisenberger, I. et al. (2019). Legal and practical challenges in classifying nanomaterials according to regulatory definitions. *Nature Nanotechnology* 14: 208–216.

Milani, N., Hettiarachchi, G.M., Kirby, J.K. et al. (2015). Fate of zinc oxide nanoparticles coated onto macronutrient fertilizers in an alkaline calcareous soil. *PLoS One* 10: 0126275.

Milenkovic, I., Mitrovic, A., Algarra, M. et al. (2019). Interaction of carbohydrate coated cerium oxide nanoparticles with wheat and pea: stress induction potential and effect on development. *Plants* 8: 478.

Milewska-Hendel, A., Zubko, M., Karcz, J. et al. (2017). Fate of neutral-charged gold nanoparticles in the roots of the *Hordeum vulgare* L. cultivar karat. *Scientific Reports* 7: 3014.

Mingyu, S., Chao, L., Chunxiang, Q. et al. (2008). Nano-anatase relieves the inhibition of electron transport caused by linolenic acid in chloroplasts of spinach. *Biological Trace Element Research* 122: 73–81.

Mishra, A. and Singh, H.B. (2021). Nanoparticles in agro-products: benefits, applications and concerns. In: *Nanomaterials for Agriculture and Forestry Applications* (ed. A. Husen and M. Jawaid), 73–88. Singapore: Springer.

Mishra, V., Mishra, R.K., Dikshit, A., and Pandey, A.C. (2014). Interactions of nanoparticles with plants: an emerging prospective in the agriculture industry. In: *Emerging Technologies and Management of Crop Stress Tolerance: Biological Techniques* (ed. P. Ahmad and S. Rasool), 159–180. Academic Press.

Misra, P., Shukla, P.K., Pramanik, K. et al. (2016). Nanotechnology for crop improvement. In: *Plant Nanotechnology* (ed. C. Kole, D. Kumar, and M. Khodakovskaya), 219–256. Springer.

Mustafa, G. and Komatsu, S. (2016). Insights into the response of soybean mitochondrial proteins to various sizes of aluminium oxide nanoparticles under flooding stress. *Journal of Proteome Research* 15: 4464–4475.

Mustafa, G., Sakata, K., Hossain, Z., and Komatsu, S. (2015a). Proteomic study on the effects of silver nanoparticles on soybean under food stress. *Journal of Proteomics* 122: 100–118.

Mustafa, G., Sakata, K., and Komatsu, S. (2015b). Proteomic analysis of flooded soybean root exposed to aluminium oxide nanoparticles. *Journal of Proteomics* 128: 280–297.

Mwaanga, P. (2018). Risks, uncertainties, and ethics of nanotechnology in agriculture. *New Visions in Plant Science* 22: 143.

Nadi, E., Aynehband, A., and Mojaddam, M. (2013). Effect of nano-iron chelate fertilizer on grain yield, protein percent and chlorophyll content of Faba bean (*Vicia faba* L.). *International Journal of Biosciences* 3: 267–272.

Nair, R., Varghese, S.H., Nair, B.G. et al. (2010). Nanoparticulate material delivery to plants. *Plant Science* 179: 154–163.

Nongbet, A., Mishra, A.K., Mohanta, Y.K. et al. (2022). Nanofertilizers: a smart and sustainable attribute to modern agriculture. *Plants* 11: 2587.

Nowack, B. and Bucheli, T.D. (2007). Occurrence, behaviour and effects of nanoparticles in the environment. *Environmental Pollution* 150: 5–22.

Nuruzzaman, M., Rahman, M.M., Liu, Y.J., and Naidu, R. (2016). Nano encapsulation, nano-guard for pesticides: a new window for safe application. *Journal of Agricultural and Food Chemistry* 64: 1447–1483.

Ojha, S., Singh, D., Sett, A. et al. (2018). Nanotechnology in crop protection. In: *Nanomaterials in Plants, Algae, and Microorganisms* (ed. D.K. Tripathi, P. Ahmad, S. Sharma, et al.), 345–391. Academic Press.

de Oliveira, J.L., Campos, E.V.R., Bakshi, M. et al. (2014). Application of nanotechnology for the encapsulation of botanical insecticides for sustainable agriculture: prospects and promises. *Biotechnology Advances* 32: 1550–1561.

Panpatte, D.G., Jhala, Y.K., Shelat, H.N., and Vyas, R.V. (2016). Nanoparticles: the next generation technology for sustainable agriculture. In: *Microbial Inoculants in Sustainable Agricultural Productivity*, vol. 2, 289–300. Springer.

Panyuta, O., Belava, V., Fomaidi, S. et al. (2016). The effect of pre-sowing seed treatment with metal nanoparticles on the formation of the defensive reaction of wheat seedlings infected with the eyespot causal agent. *Nanoscale Research Letters* 11: 92.

Parveen, K., Banse, V., and Ledwani, L. (2016). Green synthesis of nanoparticles: their advantages and disadvantages. *American Institute of Physics Conference Proceedings, American Institute of Physics Publishing LLC*, Melville, NY, USA 1724: 20048.

Pavani, K., Divya, V., Veena, I. et al. (2014). Influence of bioengineered zinc nanoparticles and zinc metal on *Cicer arietinum* seedlings growth. *Asian Journal of Agriculture and Biology* 2: 216–223.

Pimentel, D. (2009). Integrated pest management: innovation-development process. In: *Pesticides and Pest Control*, vol. 1 (ed. R. Peshin and A.K. Dhawan), 83–87. Springer.

Pirzada, T., de Farias, B.V., Mathew, R. et al. (2020). Recent Advances in biodegradable matrices for active ingredient release in crop protection: towards attaining sustainability in agriculture. *Current Opinion in Colloid & Interface Science* 48: 121–136.

Pokropivny, V. and Skorokhod, V. (2017). Classification of nanostructures by dimensionality and concept of surface forms engineering in nanomaterial science. *Materials Science and Engineering: C* 27: 990–993.

Prabhu, B.M., Ali, S.F., Murdock, R.C. et al. (2010). Copper nanoparticles exert size and concentration dependent toxicity on somatosensory neurons of rat. *Nano toxicology* 4: 150–160.

Prasad, T.N.V.K.V., Sudhakar, P., Sreenivasulu, Y. et al. (2012). Effect of nanoscale zinc oxide particles on the germination, growth and yield of peanut. *Journal of Plant Nutrition* 35: 905–927.

Qureshi, A., Singh, D.K., and Dwivedi, S. (2018). Nano-fertilizers: a novel way for enhancing nutrient use efficiency and crop productivity. *International Journal of Current Microbiology and Applied Sciences* 7: 3325–3335.

Rahman, M.S., Chakraborty, A., Mazumdar, S. et al. (2020). Effects of poly (vinylpyrrolidone) protected platinum nanoparticles on seed germination and growth performance of *Pisum sativum*. *Nano-Structures & Nano-Objects* 21: 100408.

Rahmatizadeh, R., Arvin, S.M.J., Jamei, R. et al. (2019). Response of tomato plants to interaction effects of magnetic (Fe_3O_4) nanoparticles and cadmium stress. *Journal of Plant Interactions* 14: 474–481.

Raja, K., Sowmya, R., Sudhagar, R. et al. (2019). Biogenic zinc oxide and copper nanoparticles to improve seed germination quality in black gram (*Vigna mungo*). *Materials Letters* 235: 164–167.

Rajput, N. (2015). Methods of preparation of nanoparticles – a review. *International Journal of Advances in Engineering and Technology* 7: 1806–1811.

Rajput, V.D., Minkina, T., Feizi, M. et al. (2021). Effects of silicon and silicon-based nanoparticles on rhizosphere microbiome, plant stress and growth. *Biology* 10: 791.

Raliya, R., Franke, C., Chavalmane, S. et al. (2016). Quantitative understanding of nanoparticle uptake in watermelon plants. *Frontiers in Plant Science* 7: 1288.

Rane, A.V., Kanny, K., Abitha, V.K., and Thomas, S. (2018). Methods for synthesis of nanoparticles and fabrication of nanocomposites. In: *Synthesis of Inorganic Nanomaterials* (ed. S.M. Bhagyaraj, O.S. Oluwafemi, N. Kalrikkal, and S. Thomas), 121–139. Sawston: Woodhead Publishing.

Raskar, S. and Laware, S. (2013). Effect of titanium dioxide nano particles on seed germination and germination indices in onion. *Plant Sciences Feed* 3: 103–107.

Rastogi, A., Tripathi, D.K., Yadav, S. et al. (2019). Application of silicon nanoparticles in agriculture. *3 Biotech* 9: 90.

Rawat, M., Nayan, R., Negi, B. et al. (2017). Physio-biochemical basis of iron-sulphide nanoparticle induced growth and seed yield enhancement in *B. juncea*. *Plant Physiology and Biochemistry* 118: 274–284.

Rico, C.M., Morales, M.I., Barrios, A.C. et al. (2013). Effect of cerium oxide nanoparticles on the quality of rice (*Oryza sativa* L.) grains. *Journal of Agricultural and Food Chemistry* 61: 11278–11285.

Rizwan, M., Ali, S., Zia Ur Rehman, M. et al. (2019a). Alleviation of cadmium accumulation in maize (*Zea mays* L.) by foliar spray of zinc oxide nanoparticles and bio char to contaminated soil. *Environmental Pollution* 248: 358–367.

Rizwan, M., Ali, S., Ali, B. et al. (2019b). Zinc and iron oxide nanoparticles improved the plant growth and reduced the oxidative stress and cadmium concentration in wheat. *Chemosphere* 214: 269–277.

Rodrigues, S.M., Demokritou, P., Dokoozlian, N. et al. (2017). Nanotechnology for sustainable food production: promising opportunities and scientific challenges. *Environmental Science: Nano* 4: 767–781.

Rossi, L., Zhang, W., Lombardini, L., and Ma, X. (2016). The impact of cerium oxide nanoparticles on the salt stress responses of *Brassica napus* L. *Environmental Pollution* 219: 28–36.

Ruffini, C.M., Giorgetti, L., Geri, C., and Cremonini, R. (2011). The effects of nano-TiO_2 on seed germination, development and mitosis of root tip cells of *Vicia narbodensis* L. and *Zea mays* L. *Journal of Nanoparticle Research* 13: 2443–2449.

Rui, M., Ma, C., Hao, Y. et al. (2016). Iron oxide nanoparticles as a potential iron fertilizer for peanut (*Arachis hypogaea*). *Frontiers in Plant Science* 7: 815.

Sabir, S., Zahoor, M., Waseem, M. et al. (2020). Biosynthesis of ZnO nanoparticles using bacillus subtilis: characterization and nutritive significance for promoting plant growth in *Zea mays* L. *Dose-Response* 18: 1559325820958911.

Saharan, V., Sharma, G., Yadav, M. et al. (2015). Synthesis and in vitro antifungal efficacy of Cu-chitosan nanoparticles against pathogenic fungi of tomato. *International Journal of Biological Macromolecules* 75: 346–353.

Saharan, V., Kumaraswamy, R.V., Choudhary, R.C. et al. (2016). Cu-chitosan nanoparticle mediated sustainable approach to enhance seedling growth in maize by mobilizing reserved food. *Journal of Agricultural and Food Chemistry* 64: 6148–6155.

Salama, H.M. (2012). Effects of silver nanoparticles in some crop plants, common bean (*Phaseolus vulgaris* L.) and corn (*Zea mays* L.). *International Research Journal of Biotechnology* 3: 190–197.

Salama, D.M., Abd El Aziz, M.E., Rizk, F.A., and Abd Elwahed, M.S.A. (2021). Applications of nanotechnology on vegetable crops. *Chemosphere* 266: 129026.

Santos, S.M.A., Dinis, A.M., Rodrigues, D.M.F. et al. (2013). Studies on the toxicity of an aqueous suspension of C_{60} nanoparticles using a bacterium (*gen. Bacillus*) and an aquatic plant (*Lemna gibba*) as in vitro model system. *Aquatic Toxicology* 142: 347–354.

Sashidhar, P., Kochar, M., Singh, B. et al. (2020). Bio char for delivery of agri-inputs: current status and future perspectives. *Science of the Total Environment* 703: 134892.

Saxena, R., Tomar, R.S., and Kumar, M. (2016). Exploring nanobiotechnology to mitigate abiotic stress in crop plants. *Journal of Pharmaceutical Sciences and Research* 8: 974.

Schwab, F., Zhai, G., Kern, M. et al. (2016). Barriers, pathways and processes for uptake, translocation and accumulation of nanomaterials in plants – critical review. *Nanotoxicology* 10: 257–278.

Scrinis, G. and Lyons, K. (2007). The emerging nano-corporate paradigm: nanotechnology and the transformation of nature, food and agri-food systems. *International Journal of Sociology of Agriculture and Food* 15: 22–44.

Shaik, A.M., David Raju, M., and Reddy, D.R.S. (2020). Green synthesis of zinc oxide nanoparticles using aqueous root extract of *Sphagneticola trilobata* Lin and investigate its role in toxic metal removal, sowing germination and fostering of plant growth. *Inorganic and Nano-Metal Chemistry* 50: 569–579.

Shang, Y., Hasan, M.K., Ahammed, G.J. et al. (2019). Applications of nanotechnology in plant growth and crop protection: a review. *Molecules* 24: 2558.

Shankramma, K., Yallappa, S., Shivanna, M.B., and Manjanna, J. (2016). Fe_2O_3 magnetic nanoparticles to enhance S. lycopersicum (tomato) plant growth and their biomineralization. *Application of Nanoscience* 6: 983–990.

Sharifi, M., Faryabi, K., Talaei, A.J. et al. (2020). Antioxidant properties of gold nanozyme: a review. *Journal of Molecular Liquids* 297: 112004.

Sharma, P., Bhatt, D., Zaidi, M.G.H. et al. (2012). Silver nanoparticle-mediated enhancement in growth and antioxidant status of *Brassica juncea*. *Applied Biochemistry and Biotechnology* 167: 2225–2233.

Shelar, G.B. and Chavan, A.M. (2015). Myco-synthesis of silver nanoparticles from *Trichoderma harzianum* and its impact on germination status of oil seed. *Biolife* 3: 109–113.

Shen, C.X., Zhang, Q.F., Li, J. et al. (2010). Induction of programmed cell death in *Arabidopsis* and rice by single wall carbon nanotubes. *American Journal of Botany* 97: 1602–1609.

Sheykhbaglou, R., Sedghi, M., and Fathi-Achachlouie, B. (2018). The effect of ferrous nano-oxide particles on physiological traits and nutritional compounds of soybean (*Glycine max* L.) seed. *Anais da Academia Brasileira de Ciências* 90: 485–494.

Shojaei, T.R., Salleh, M.A.M., Tabatabaei, M. et al. (2019). Applications of nanotechnology and carbon nanoparticles in agriculture. In: *Synthesis, Technology and Applications of Carbon Nanomaterials* (ed. A.R. Suraya, N.I.R.O. Raja, and Z.H. Mohd), 247–277. Amsterdam: Elsevier.

Shukla, P., Chaurasia, P., Younis, K. et al. (2019). Nanotechnology in sustainable agriculture: studies from seed priming to post-harvest management. *Nanotechnology For Environmental Engineering* 4: 11.

Siddiqui, M.H., Al-Whaibi, M.H., Faisal, M., and Al Sahli, A.A. (2014). Nanosilicon dioxide mitigates the adverse effects of salt stress on *Cucurbita pepo* L. *Environmental Toxicology and Chemistry* 33: 2429–2437.

Singh, M.D. (2017). Nano-fertilizers is a new way to increase nutrients use efficiency in crop production. *International Journal of Agriculture Sciences* 9: 3831–3883.

Soni, A.T., Rookes, J.E., and Arya, S.S. (2023). Chitosan nanoparticles as seed priming agents to alleviate salinity stress in rice (*Oryza sativa* L.) seedlings. *Polysaccharides* 4: 129–141.

Sotiriou, G.A. and Pratsinis, S.E. (2010). Antibacterial activity of nano silver ions and particles. *Environmental Science & Technology* 44: 5649–5654.

Su, X.L. and Li, Y. (2004). Quantum dot bio labelling coupled with immune magnetic separation for detection of *Escherichia coli* 0157:H7. *Analytical Chemistry* 76: 4806–4810.

Sultana, F., Imran-Ul-Haque, M., Arafat, M., and Sharmin, S. (2013). An overview of nano gel drug delivery system. *Journal of Applied Pharmaceutical Science* 3: 95–105.

Sun, D., Hussain, H.I., Yi, Z. et al. (2018). Delivery of abscisic acid to plants using glutathione responsive mesoporous silica nanoparticles. *Journal of Nanoscience and Nanotechnology* 18: 1615–1625.

Suriyaprabha, R., Karunakaran, G., Yuvakkumar, R. et al. (2012). Silica nanoparticles for increased silica availability in maize (*Zea mays*. L) seeds under hydroponic conditions. *Current Nanoscience* 8: 902–908.

Tanaka, Y., Kimura, T., Hikino, K. et al. (2012). Gateway vectors for plant genetic engineering: overview of plant vectors, application for Bimolecular Fluorescence Complementation (BiFC) and multigene construction. In: *Genetic Engineering – Basics, New Applications and Responsibilities* (ed. H. Barrera-Saldaña), 64. Mexico: Universidad Autónoma de Nuevo León.

Tantawy, A.S., Salama, Y.A.M., El-Nemr, M.A., and Abdel-Mawgoud, A.M.R. (2015). Nano silicon application improves salinity tolerance of sweet pepper plants. *International Journal of ChemTech Research* 8: 11–17.

Tawfik, M.M., Mohamed, M.H., Sadak, M.S., and Thalooth, A.T. (2021). Iron oxide nanoparticles effect on growth, physiological traits and nutritional contents of *Moringa oleifera* grown in saline environment. *Bulletin of the National Research Centre* 45: 177.

Thakur, S., Thakur, T., and Kumar, R. (2018). Bio nanotechnology and its role in agriculture and food industry. *Journal of Molecular and Genetic Medicine* 12: 1–5.

Thema, F.T., Manikandan, E., Dhlamini, M.S., and Maaza, M. (2015). Green synthesis of ZnO nanoparticles via *Agathosma betulina* natural extract. *Materials Letters* 161: 124–127.

Thi, T.U.D., Nguyen, T.T., Thi, Y.D. et al. (2020). Green synthesis of zinc oxide nanoparticles using orange fruit peel extract for antibacterial activities. *Royal Society of Chemistry Advances* 10: 23899–23907.

Thomas, T.T.D. and Puthur, J.T. (2017). UV radiation priming: a means of amplifying the inherent potential for abiotic stress tolerance in crop Plants. *Environmental and Experimental Botany* 138: 57–66.

Torney, F., Trewyn, B.G., Lin, V.S.Y., and Wang, K. (2007). Mesoporous silica nanoparticles deliver DNA and chemicals into plants. *Nature Nanotechnology* 2: 295–300.

Tripathi, D.K., Singh, S., Singh, V.P. et al. (2017). Silicon nanoparticles more effectively alleviated UV-B stress than silicon in wheat *(Triticum aestivum)* seedlings. *Plant Physiology and Biochemistry* 110: 70–81.

Umair Hassan, M., Aamer, M., Umer Chattha, M. et al. (2020). The critical role of zinc in plants facing the drought stress. *Agriculture* 10: 396.

Umavathi, S., Mahboob, S., Govindarajan, M. et al. (2021). Green synthesis of ZnO nanoparticles for antimicrobial and vegetative growth applications: a novel approach for advancing efficient high quality health care to human wellbeing. *Saudi Journal of Biological Sciences* 28: 1808–1815.

Ureña-Saborio, H., Madrigal-Carballo, S., Sandoval, J. et al. (2017). Encapsulation of bacterial metabolic infiltrates isolated from different Bacillus strains in chitosan nanoparticles as potential green chemistry-based biocontrol agents against *Radopholus similis*. *Journal of Renewable Materials* 5: 290–299.

Utell, M.J. and Frampton, M.W. (2000). Acute health effects of ambient air pollution: the ultrafine particle hypothesis. *Journal of Aerosol Medicine and Pulmonary Drug Delivery* 13: 355–359.

Valenstein, J.S., Lin, V.S.-Y., Lyznik, L.A. et al. (2013). Mesoporous silica nanoparticle-mediated intracellular cre protein delivery for maize genome editing via loxP site excision. *Plant Physiology* 164: 537–547.

Verbic, A., Šala, M., Jerman, I., and Gorjanc, M. (2021). Novel green in situ synthesis of ZnO nanoparticles on cotton using pomegranate peel extract. *Materials* 14: 4472.

Vijayakumar, P.S., Abhilash, O.U., Khan, B.M., and Prasad, B.L.V. (2010). Nano gold loaded sharp edged carbon bullets as plant gene carriers. *Advanced Functional Materials* 20: 2416–2423.

Wakeel, A., Xu, M., and Gan, Y. (2020). Chromium-induced reactive oxygen species accumulation by altering the enzymatic antioxidant system and associated cytotoxic, genotoxic, ultrastructural, and photosynthetic changes in plants. *International Journal of Molecular Sciences* 21: 728.

Wang, Y., Peng, Z., Mingshu, L. et al. (2020). Alleviation of nitrogen stress in rice *(Oryza sativa)* by ceria nanoparticles. *Environmental Science: Nano* 7: 2930.

Wang, L., Ning, C., Pan, T., and Cai, K. (2022). Role of silica nanoparticles in abiotic and biotic stress tolerance in plants: a review. *International Journal of Molecular Sciences* 23: 1947.

Wohlleben, W., Mielke, J., Bianchin, A. et al. (2017). Reliable nanomaterial classification of powders using the volume-specific surface area method. *Journal of Nanoparticle Research* 19: 61.

Wu, H., Tito, N., and Giraldo, J.P. (2017). Anionic cerium oxide nanoparticles protect plant photosynthesis from abiotic stress by scavenging reactive oxygen species. *American Chemical Society* 11: 11283–11297.

Xiong, T., Dumat, C., Dappe, V. et al. (2017). Copper oxide nanoparticle foliar uptake, phytotoxicity, and consequences for sustainable urban agriculture. *Environmental Science & Technology* 51: 5242–5251.

Xiu, Z.M., Zhang, Q.B., Puppala, H.L. et al. (2012). Negligible particle-specific anti-bacterial activity of silver nanoparticles. *Nano Letters* 12: 4271–4275.

Yadav, A., Yadav, K., and Abd-Elsalam, K.A. (2023). Nanofertilizers: types, delivery and advantages in agricultural sustainability. *Agrochemicals* 2: 296–336.

Yang, H., Xu, M., Koide, R.T. et al. (2016). Effects of ditch-buried straw return on water percolation, nitrogen leaching and crop yields in a rice-wheat rotation system. *Journal of the Science of Food and Agriculture* 96: 1141–1149.

Yao, K.S., Li, S.J., Tzeng, K.C. et al. (2009). Fluorescence silica nano probe as a biomarker for rapid detection of plant pathogens. *Advanced Materials Research* 79: 513–516.

Younes, N.A., Dawood, M.F.A., and Wardany, A.A. (2019). Biosafety assessment of graphene nanosheets on leaf ultrastructure, physiological and yield traits of *Capsicum annuum* L. and *Solanum melongena* L. *Chemosphere* 228: 318–327.

Zhang, X., Zhang, J., and Zhu, Y. (2010). Chitosan/double-stranded RNA nanoparticles-mediated RNA interference to silence chitin synthase genes through larval feeding in the African malaria mosquito (*Anopheles gambiae*). *Insect Molecular Biology* 19: 683–693.

Zhang, M., Gao, B., Chen, J.J. et al. (2014). Slow-release fertilizer encapsulated by graphene oxide films. *Chemical Engineering Journal* 255: 107–113.

Zhang, X.F., Liu, Z.G., Shen, W., and Gurunathan, S. (2016). Silver nanoparticles: synthesis, characterization, properties, applications, and therapeutic approaches. *International Journal of Molecular Sciences* 17: 1534.

Zhao, L., Huang, Y., Adeleye, A.S., and Keller, A.A. (2017). Metabolomics reveals Cu $(OH)_2$ nano pesticide-activated anti-oxidative pathways and decreased beneficial antioxidants in spinach leaves. *Environmental Science & Technology* 51: 10184–10194.

Zheng, L., Hong, F., Lu, S., and Liu, C. (2005). Effect of nano-TiO_2 on strength of naturally aged seeds and growth of spinach. *Biological Trace Element Research* 104: 83–91.

Zhou, Q.X. and Hu, X.G. (2017). Systemic stress and recovery patterns of rice roots in response to graphene oxide nano sheets. *Environmental Science & Technology* 51: 2022–2030.

Zhu, Q., Liu, X., Hao, T. et al. (2020). Cropland acidification increases risk of yield losses and food insecurity in China. *Environmental Pollution* 256: 113145.

Zimmermann, P. and Zentgraf, U. (2005). The correlation between oxidative stress and leaf senescence during plant development. *Cellular & Molecular Biology Letters* 10: 515–534.

Zulfqar, F., Navarro, M., Ashraf, M. et al. (2019). Nano fertilizer use for sustainable agriculture: advantages and limitations. *Plant Science* 28: 110270.

14

Benefits of Si/SiO_2 Nanoparticles in Developing Climate Resilient Crop Plants: Opportunities and Challenges

Sagnik Goswami, Sayan Sadhukhan, Kalpak Majumdar,
Debjyoti Bandhu Banerjee and Anirban Bhar

Department of Botany, Ramakrishna Mission Vivekananda Centenary College, Kolkata, West Bengal, India

14.1 Introduction

Climate is described as the pattern of weather including the parameters, such as temperature, precipitation and storms, over a longer time scale. Climate change is the shifting of this long-term average of weather parameters towards extremes. According to the 2018 US National Climate Assessment, human activity is changing the Earth's climate more quickly than it has ever done before in the history of modern civilization. Ocean acidification, the rise of sea level, loss of biodiversity and biogeochemical cycle modifications are also linked to climate change (Dietz et al. 2020). The concentration of major greenhouse gases (GHGs), e.g., methane (CH_4), nitrous oxide (N_2O) and carbon dioxide (CO_2), has increased by 150%, 20% and 40%, respectively, which ultimately leads to global warming. CO_2 emissions are increasing in developed and developing countries gradually, mainly due to excessive combustion of fossil fuels (Malhi et al. 2021). Climate change also has a severe impact on agricultural production by 2050, it is estimated that production in Africa might drop by 30%. This would disrupt food security and result in a significant loss of biodiversity. According to the World Bank report, India will also be affected most brutally by climate change with major socio-economic damage (https://www.worldbank.org/en/news/feature/2013/06/19/india-climate-change-impacts accessed on 8th April, 2024). Endemic species are also vulnerable to extinction as a result of massive climatic shifts (Muluneh 2021). Anthropogenic climate change is responsible for the extinction and invasiveness of some species and indirectly affects the territory of a species, habitat alteration, natality and mortality by climate-regulated changes (Thomas 2010). Moreover, several

Silicon Nanoparticles: Synthesis, Uptake, and Applications, First Edition.
Edited by Archana Singh, Aryadeep Roychoudhury, and Indrakant K. Singh.

pathogenic infections may undermine plant defensive mechanisms, rendering the plant susceptible to future secondary infection (Abdullah et al. 2017). This may lead to the emergence of novel infectious diseases. As a result of climate change, in Europe, many extremophilic, late-emerging and opportunistic weeds disrupt agriculture in cropping season (Peters et al. 2014). As a consequence of climate change, there will be a major loss in cereal crop productivity. The loss of productivity has been estimated at 20–45% in maize, 5–50% in wheat and 20–30% in rice by the year 2100 (Arora 2019; Wheeler and Von Braun 2013). In this challenging situation, nano-agrochemicals come out as a potential alternative due to their efficacy and low toxicity to nature. Silver (Ag), copper (Cu) and Zinc (Zn) nanoparticles are extensively used for their antimicrobial and antifungal activities (Du et al. 2022). Silicon (Si) is a metalloid with intermediate characteristics of metals and non-metals. Silicon nanoparticle (SiNP) is primarily classified into mesoporous and nanoporous SiNPs. Recently SiNPs have been extensively used in food industry, agricultural field and drug delivery due to their biocompatibility, pore size, large specific surface area, hydrophilic nature, etc. (Jeelani et al. 2020). Mesoporus silica nanoparticles (MSN) are traditionally produced through the sol–gel method and have pivotal importance in biomedical industries, specifically in drug delivery, for their tuneable properties, particle size, etc. (Manzano and Vallet-Regí 2020). Although silicon is not an essential compound for plants, it has many positive attributes in plant growth and development, thus mitigating several environmental stress conditions; it is also used as nanopesticides, nanofertilizers and nanoherbicides. Plants can uptake and dissolve Si as monosilicic acid (Rastogi et al. 2019). SiNPs have immense importance in mitigating biotic and abiotic stress like drought, salinity and metal toxicity in plants (Tripathi et al. 2017). Silicon (Si) promotes disease tolerance by the deposition on the plant cell wall and reducing oxidative stress. Moreover, SiNPs also mitigate stress by up-regulating antioxidant enzymes, phenolic compounds, proline and flavonols (L. Wang et al. 2022). In *Glycine max*, exogenous treatment of Si induces antioxidant activities and K^+ concentration, whereas it down-regulates the production of reactive oxygen species (ROS), lipid peroxidation and Na^+ concentration in response to salt stress. Furthermore, silicon dioxide (SiO_2) also promotes seed germination, photosynthetic rate, electron transport rate and stomatal conductance in different plant species (Rajput et al. 2021). SiNPs can also regulate phytohormones such as auxin (IAA), gibberellic acid (GA), abscisic acid (ABA), ethylene (ET) and jasmonic acid (JA) that alter plant growth and development (Mukarram et al. 2023). A particular influx transporter belonging to the aquaporin family allows plant roots to absorb silicon. Lsi1 transporter and efflux protein Lsi2 carries Si from the xylem to the root. Distribution of Si to other plant components is mediated by Lsi6 (Mukarram et al. 2023). In this regard, SiNPs have immense potential in improving plant fitness against diverse biotic and abiotic stress and fostering towards climate smart agricultural systems.

In this chapter, the beneficial role of Si and SiO_2 NPs in mitigating different biotic and abiotic stress has been extensively discussed with special reference to drought, salinity, heavy metal toxicity and biotic stress in plants. The potentiality and challenges of SiNPs in developing climate-resilient future crops have also been highlighted.

14.2 Climate Change: Facts and Consequences

As previously discussed, one of the main issues the world is facing today is increasing temperature and discrete precipitation, which are due to a significant variation in the average value of meteorological parameters over time. Dynamicity of the climate change affects biotic and abiotic components of the ecosystem. It has effects on the alteration of CO_2 concentration, heat waves, rainfall, pests, pathogens and weed population dynamics. CO_2 along with methane, water vapour and nitrous oxide primarily contribute to the greenhouse effect by absorbing infrared radiation (Chaudhry and Sidhu 2022). Carbon dioxide emissions increased from 22.15 billion metric tons to 36.14 billion metric tons between 1990 and 2014, which subsequently contributed to a higher percentage of GHGs, which ultimately increased atmospheric temperature (Malhi et al. 2021). Atmospheric carbon dioxide concentration has increased from 280 to 410 ppm since the Industrial Revolution. CO_2 absorbs a higher percentage of thermal infrared radiation emitted from the surface of the earth. The absorbance of the thermal infrared radiation is directly proportional to the emitted radiation of the earth's surface to the atmosphere; therefore, higher energy heat flux increases the temperature of the air. Increased temperature affects reproduction, development, survival and dispersal patterns, and population dynamics of agricultural pests (Skendžić et al. 2021). According to estimates, anthropogenic activities will positively accelerate the production of greenhouse gases, causing an average temperature increase of 1.5 °C by 2050 (Arora 2019). According to the sixth assessment report of the Intergovernmental Panel for Climate Change (IPCC), 70% of the world's population lives in Asia, which is more vulnerable to flood and drought disasters due to climate change. Unprecedented natural disasters including drought, floods and disease infestations reduce agricultural productivity. Therefore, several negative consequences of climate change, such as an increase in heat waves and unpredictable rainfall, would eventually have a negative impact on economic growth and development (Habib-ur-Rahman et al. 2022). Additional climatic variables such as wind speed, humidity and sunshine duration also have significant effects on crop yield.

14.2.1 Impact of Climate Change on Agriculture

Climate change imposes different deleterious effects on the quality and quantity of crop plants. The frequency and intensity of different abiotic stresses like extreme temperature, drought, heavy metal toxicity and salinity are also enhanced by climatic variability (Zandalinas et al. 2022). Several studies revealed that increased CO_2 concentration has different positive attributes on the physiology of plants by increasing photosynthetic rate, carbohydrate accumulation and carboxylation efficiency, whereas long-term elevated CO_2 down-regulates stomatal density and conduction. On the other hand, germination frequency, radicle, plumule growth and viability are negatively affected by heat stress; moreover, it also disrupts membrane structure and denatures cellular structures and protein, subsequently leading to oxidative stress and reactive oxygen species (ROS) production (Chaudhry and Sidhu 2022). The climate change has already affected the agricultural damage across the globe. The major economic loss in Cameron was reported as a result of the decline in agricultural production due to temperature and precipitation (Molua 2007). A similar effect of climate change on cotton production has also been reported in the Savannah Region, Northern Togo (Soviadan et al. 2019). The impact of weather is expected to be more pronounced in African and Asian countries with frequent drought and altered precipitation patterns (Malhi et al. 2021). The weather change is projected to cause a decrease in real per capita income from the agriculture and food industries in Ghana (Botchway et al. 2021). The impact of climate change on the economy of Iran has also sought concern recently (Malaekeh et al. 2024). It has been demonstrated that wheat protein, starch and soluble sugar content are decreased due to drought and heat stress (Habib-ur-Rahman et al. 2022). Wheat production is reduced by 6% for a rise of each 1 °C temperature. Drought imposes several detrimental effects on the reproductive phase, floral development, fertilization process and anthesis in cereal crops (Raza et al. 2019). Irrigated rice production is predicted to be lower by around 7% by 2050 as a result of global warming (Kumar et al. 2021). Potato (*Solanum tuberousum* L), one of the most important nongrain food crops, was also adversely affected by drought and heat stress. The number and size of potato tubers are drastically reduced, which also increases reducing sugar content in the tuber. Potato production in India may be reduced by 11% due to the effect of climate change (George et al. 2017). Plants have developed several sophisticated intrinsic measures to mitigate abiotic stress like accumulation of primary and secondary metabolites, for example, major carbohydrates like glucose, fructose and sucrose are increased in soya bean in response to drought stress. Furthermore, proline accumulation is one of the most prominent effects of drought and heat stress, and γ amino butyric acid (GABA) is accumulated during drought and salinity stress *in planta* (Zandalinas et al. 2022).

On the other hand, the mobility, metabolism and metamorphism of insect pests along with the herbivory are also directly linked to environmental temperature. Insect pest population is likely to increase in the temperate zone and decrease in the tropical zone, in response to rising temperature. An increase in average temperature also intensified the severity of pest infestation. Furthermore, invasive alien species (IAS) is another alarming threat to global biodiversity. Climate change also positively affects these species by expanding their geographical distribution and population structure (Skendžić et al. 2021). The population of corn pest, European corn borer (*Ostinia nubilalis*), corn earworm (*Helicoverpa zea*) and corn rootworm (*Diabrotica* spp.) are positively affected by climate change; fungus infection in maize is also enhanced in warm, dry conditions (Vaughan et al. 2018). Pollination is one of the most important factors of flowering plants; insect pollination is also hampered due to climate-changing parameters, e.g., increased CO_2 concentration, temperature rise and nitrogen sedimentation, which change the composition of nectar (Shahzad et al. 2021).

14.3 'Climate Change–Loss of Productivity–Yield Improvement': the Self-Damaging Loop

Climate change and associated factors negatively affect crop yield globally. Besides, the population outburst increases the demand for global food supply. According to WHO, the earth's population is projected to reach from 8.8 billion to 11.6 billion within 2100, increasing the demand for energy-rich food to sustain the global population (Seppelt et al. 2022). Therefore, the scientific community tirelessly emphasized increasing crop yield along with overcoming biotic and abiotic stress responses. The concerns regarding yield improvements and food security had started long back during the 'green revolution' in 1960s. The work of Norman Borlaug and M.S. Swaminathan towards protecting the world population from hunger and increasing agricultural productivity in developing countries is the landmark for modern agriculture. Among several steps, selecting the superior crop varieties for quantity, quality, disease resistance and response to fertilizers are principal. Since then, global agriculture has shifted towards industrial agriculture where crops are cultivated multiple times in large farms using huge amounts of pesticides, fertilizers, hormones and chemicals to improve yield (Pareek et al. 2020). The excessive use of these chemical fertilizers and pesticides not only damages the soil chemistry but also drains out to rivers or streams and pollutes the same. The overuse of chemical fertilizers imbalances the soil nitrogen–phosphorus–potassium (NPK) ratio, and hampers soil pH, porosity and texture (Pahalvi et al. 2021). Along with soil nutrient depletion, excessive fertilizers also impact environmental pollution, leading to agricultural loss (Nadarajan and Sukumaran 2021). Climate change also has impact on soil chemical composition and plant nutrition. The increase in global carbon footprints directly impacts the carbon sequestration pattern in soil, which dramatically affects soil nutritional parameters (Elbasiouny et al. 2022). Climate change also alters soil salinity in agricultural fields (Corwin 2021); which together with fertilizer/pesticide altered soil chemical composition, hampers agricultural production. Additionally, weather variables are important for the disease prediction of different crop plants. Due to the abrupt shift of climatic variables, it is difficult as well as complicated to predict the risk of different diseases (Garrett et al. 2021). The soil houses a plethora of microorganisms that play crucial roles in the soil ecosystem, fertility and fitness of crop plants. These microbes also contribute to the phyllosphere and rhizosphere microbial populations and crop yield (Das et al. 2023). Climate change and anthropogenic activities readily affect soil microflora, which significantly influence soil ecology, plant interactions and stress response (Trivedi et al. 2022). The disruption of these invisible 'enhanced plant holobionts (EPH)' is considered to be the most important regulator of soil ecological functioning (Harman et al. 2021). The climate change-induced loss of these EPHs occupies the principal node in the agricultural production vs loss feedback loop in recent years.

14.4 Emerging Infectious Diseases (EIDs) in Plants due to Climate Change

Infectious diseases occur due to pathogenic microorganisms like bacteria, viruses, parasites or fungi. These diseases have the potential to be spread directly or indirectly from organism to organism. Their rapid and unpredictable spread can cause significant global problems. Emerging infectious diseases (EIDs) have a great impact on both global economies and public health (Morens et al. 2004). Initially, the application of the term EID in plant science was rare. The anthropogenic and weather-induced emerging diseases of both crop and wild plants along with the conservation strategies were observed later (Anderson et al. 2004). Climate change and plant diseases have been long-debated issues since the sudden Oak death in California forests due to warmer seasons. Several studies have emphasized the synchroneity between host cambium activity and pathogen colonization rate in different plants (Pautasso et al. 2012). The changing climatic conditions, e.g., temperature, air, precipitation, soil composition, UV incidence, ozone contamination, carbon footprint and GHGs, not only induce several pathogens, e.g., virus, bacteria, fungus, nematodes and insect pests, but also drive unpredicted epidemics in new non-native areas (Priyadi and Upadhyay 2021). A particular population with a high prevalence of a certain type of pathogen, when comes in contact with a novel potential population to infect, the phenomenon is called pathogen spillover. Climate change exaggerates the pathogen spillover events along with the evolutionary emergence of novel pathogens (Ristaino et al. 2021). 'ProMED-Mail' is an important e-mail-based surveillance system launched in 1994, which identifies unusual health issues related to emerging/re-emerging infectious diseases of humans, animals and plants (Carrion and Madoff 2017). The ProMED-based analysis exhibited that temperature (low and high) and humidity are the two key factors in predicting fungal disease outbreaks in plants globally (Romero et al. 2022). The reports also predicted that climate change will favour the pre-existence of intermediate vectors and hence the prevalence of vector-borne diseases. Additionally, the severity and re-emergence of neglected plant diseases would complicate the situation (El-Sayed and Kamel 2020). Hence, a novel sustainable means of the agricultural system is urgently necessary, which is capable of enhancing productivity and overcome the climate-induced stress. NP-based agro-chemicals may provide an intriguing alternative and support climate smart futuristic agriculture.

14.5 Nanoparticles in Agriculture: A Way Towards Climate-Smart Stress-Tolerant Plants

As discussed above, plants are facing a multitude of stress in response to climate change, and crop plants are more vulnerable to this situation. It has recently been reported that only about 9% of the global agricultural land is conducive to crop production, while the rest is facing different environmental stresses (Minhas et al. 2017). The imminent danger of climate change has accelerated the frequency of these environmental stresses; the cumulative effects of worldwide population rise, urbanization and desertification have posed the challenge of global food security (Fedoroff et al. 2010). Abiotic stresses such as drought, salinity, heavy metal accumulation, mineral deficiencies, flood and heat waves significantly reduce crop productivity. Major abiotic stresses namely drought, cold and salinity are reported to affect 64%, 57% and 6% regions of the earth, respectively (Yadav et al. 2020). Traditional practices of plant breeding have shown limited effectiveness in generating resilience to these abiotic stresses over the years due to the multigenic nature of stress resistance mechanisms (Flowers et al. 2000). Studies on mechanisms of abiotic stress tolerance along with the approaches based on modern genomics have resulted in the discovery of a number of genes, receptors and proteins related to tolerance. Nanotechnology is a novel approach of stress mitigation in plants, and different NPs have been accepted widely due to their potential to alleviate different environmental stressors. The application of nano-agrochemicals as carriers of herbicides, pesticides, fertilizers and even genes is a potential future prospect. Additionally, it can be helpful to target specific intracellular organelles (Siddiqui et al. 2015; Singh and Roychoudhury 2022). However, the major focus has been given to understand the NP toxicity on the flora and fauna, while the mechanistic study of these NPs in inducing plant stress response has largely been obscured (Rajput et al. 2021; Samanta and Roychoudhury 2021). More research in this regard is needed to completely unveil the specific role and mechanisms of NPs in mitigating environmental stresses in plants.

14.5.1 Role of Si/SiO_2 NPs to Mitigate Climate Change-Induced Drought Stress in Plants

Drought is an environmental situation that sets in due to the prolonged low average precipitation, lack of sufficient moisture in the soil, significant rise in temperature and frequent water loss as a result of increased transpiration (Mishra and Cherkauer 2010). The emerging problem of climate change has impacted many regions on earth to become drought-prone. Several studies and observations have demonstrated that with the rapid increase in earth's surface temperature, the intensity of precipitation also subsequently increases, although the total precipitation fails to contribute to the soil hydrological parameter (Mukherjee et al. 2018). The application of NPs has reportedly increased drought tolerance of plants by improving osmolyte, phenolic and chlorophyll content and lowering the accumulation of harmful metabolites, electrolyte leakage, etc. The foliar and rhizosphere application of SiO_2 NPs improves the yield of wheat by 25.35% and 17.81%, respectively, under drought conditions. The leaf greenness and water content index of the same plants increased by 12.54% and 84.04% upon successful application of SiNPs (Behboudi et al. 2018). In marigold (*Calendula officinalis*), priming with SiNPs has been reported to increase the dry weight and vigour index of seeds against polyethylene glycol (PEG)-induced drought stress. It was observed that applying SiNPs at a high concentration to seeds under severe drought stress showed a larger deposition of SiNPs on root cell walls, in comparison to moderate stress (Rahimi et al. 2021). Additionally, studies conducted on maize and strawberry suggest that SiNPs increase the activity of antioxidant enzymes such as superoxide dismutase (SOD), catalase (CAT) and guaiacol peroxidase (G-POX) in response to drought stress as well as promote total phenolic content, membrane stability index, etc. qRT-PCR-based expression analysis indicated that SiNPs down-regulate the expression of genes for osmotic proteins (OSM-34) and aquaporin (AQP), while up-regulating the expression of genes that code for the D2 protein of photosystem II (*PsbD*) (Sharf-Eldin et al. 2023; Zahedi et al. 2020). It has been demonstrated that SiNPs play a role in repairing membrane damage under drought stress and establishing the seedling by promoting phytohormone level (Miao et al. 2010). Some studies have also reported that the co-application of SiNPs with plant growth-promoting rhizobacteria effectively increased the content of some phytohormones like auxin and cytokinin and decreased the ABA (abscisic acid) content, thus contributing to mitigating drought stress by increasing cell division, elongation and reducing stomatal conductance

(Akhtar et al. 2021). Some more interesting studies of implication of SiNPs/SiO_2 NPs on drought tolerance in plants have been documented in Table 14.1.

14.5.2 Climate Change and Salt Stress: The Protecting Role of Si/SiO_2 NPs as Agro-nanochemicals

Contamination of soil and water with high concentrations of water-soluble salts causes salinity stress which has become one of the major issues from the onset of the twenty-first century. It has been reported that salt stress reduces crop productivity by almost 50% globally (Zhao et al. 2020). If the electrical conductivity (EC) of the saturation extract (EC_e) in a plant rhizosphere region measures more than 4 dS/m (approximately 40 mM NaCl) at room temperature and contains 15% exchangeable sodium, it is considered to be saline soil. By 2050, half of the total agricultural land will be salinized due to massive climatic shifts (Munns 2005; Shrivastava and Kumar 2015). Salinity has harmful effects like cell hydration and ion accumulation in plants (Nivas et al. 2011). The impacts of salt stress on a plant have been broadly classified into primary and secondary effects. The ionic and osmotic potential are considered to be the primary effects, while nutritional and hormonal imbalance and oxidative stress are secondary consequences (Yang and Guo 2018). The studies emphasizing the effects of NPs to combat salinity stress are scarce. Elements like zinc, silver, copper, silicon, manganese, cerium and sulphur and chemical compounds like silicon dioxide, iron oxide and titanium dioxide have been reported to be applied on plants as nanomaterials to combat salinity. Both bulk and nano-SiO_2 at an optimum concentration improve growth parameters, e.g., chlorophyll and NPK content of leaf of fenugreek (*Trigonella foenum-graceum*) and capsicum (*Capsicum annum*) under salt stress (Ivani et al. 2018; Tantawy et al. 2015). In tomatoes, phenylalanine lyase activity (PAL), vitamin C content in fruits and glutathione reductase (GSH) were enhanced and thoroughly maintained in nano-silica-treated plants, as compared to the control and K_2SiO_3-treated plants under salt stress (Pinedo-Guerrero et al. 2020). In basil leaf (*Ocimum basilicum*), salinity-induced proline surge is reported to decrease after SiNP treatment (Kalteh et al. 2018). In *in vitro* derived banana (*Musa acuminata*), it was found that the application of SiO_2 NPs improves chlorophyll content, promotes K^+ uptake, reduces Na^+ toxicity by limiting Na^+ absorption, increases malondialdehyde (MDA) content and reduces electrolyte leakage in response to polyethylene glycol (PEG)-mediated water deficit and salinity stress (Mahmoud et al. 2020). Antioxidant enzymes like CAT, SOD, POX, ascorbate peroxidase (APX) and polyphenol oxidase (PPO) play crucial roles in ROS (reactive oxygen species) scavenging. Interestingly, in soya bean (*Glycine max*) and lemongrass (*Cymbopogon flexuosus*), SiNPs down-regulate the accumulation of ROS and associated membrane lipid peroxidation and intracellular Na^+

Table 14.1 Drought stress alleviation by SiNP application.

Plants	Nanoparticle type	Conc. of SiNPs	Effect	Mechanisms of action	References
Triticum aestivum (Wheat)	Nano-Si	0, 300, 600, 900 mg/l	↑ Plant height by 2.5%, 3.2% and 6.9% respectively ↑ Spike length, number of grains per spike, weight of grains ↑ Chlorophyll content ↑ Mean of relative water content (RWC) in leaves (4.6–13.5%) ↑ Transpiration and photosynthesis rate, stomatal conductance	SiNPs help in accumulation of water and nutrient inside the seed. SiNPs improve protein content and nutrient uptake. SiNPs influence on photosynthetic pigments by increasing cytokinin, and thus improving chlorophyll formation and chloroplast ultrastructure.	M. A. S. Raza et al. (2023)
Zea mays (Maize)	SiO_2	0.25 g/l	↑ Fresh weight and overall growth ↑ Photosynthetic pigments like chlorophyll a and chlorophyll b ↓ Concentration of H_2O_2 ↓ Malondialdehyde (MDA) content ↑ Cell membrane stability index (CMSI)	NPs elevate the relative expression of the D2 protein of photosystem II (*PsbD*). NPs increase water and osmotic potential and maintain higher turgor pressure SiO_2 influences genes TaSOD, TaCAT and TaAPX137 that are involved in synthesizing and activating antioxidant enzymes. NPs down-regulate the expression of the osmotic like protein (*OSM-34*) and aquaporin (*AQPs*).	Sharf-Eldin et al. (2023)

Hordeum vulgare (Barley)	Nano-Si	125 and 250 mg/l	↑ In shoot biomass by maximum 27.3% ↑ Total chlorophyll up to 17.1% ↑ Carotenoid content up to 24.1%	SiNPs are effective in altering plant osmolyte and metabolite profile, cellular injury and membrane stability indices, and the activity of antioxidant enzymes.	Ghorbanpour et al. (2020)
Fragaria × *ananassa* (Strawberry)	SiO_2 and Se NPs	50 and 100 mg/l	Improvement in different growth and yield parameters such as fruit size, number of leaves, total phenolic content, anthocyanin and ascorbic acid content in fruit.	SiNPs preserve more photosynthetic pigments and increases relative water content, membrane stability index, and water use efficiency. NPs increase the activity of enzymes like CAT, APX, GPX, and SOD while decrease lipid peroxidation and H_2O_2 content.	Zahedi et al. (2020)

accumulation by stimulating the antioxidant scavenging system. Additionally, in lemongrass, SiNPs are reported to restore photosynthesis and up-regulate the activity of the geraniol dehydrogenase (GeDH) enzyme, responsible for essential oil biosynthesis under salinity stress (Farhangi-Abriz and Torabian 2018; Mukarram et al. 2023). The recent updates on SiNP/SiO_2 NP application in the mitigation of salinity stress in different plants are discussed in Table 14.2.

14.5.3 Role of Si/SiO_2 NPs to Mitigate Abiotic Stresses

The cumulative effects of industrialization and climate change have resulted in the accumulation of heavy metals in agricultural fields and in or near water bodies, exposing plants to heavy metal stress, which poses a serious threat. Some of the heavy metals are needed in trace amounts for the growth and development of plants. These include cadmium, copper, lead, chromium and mercury, which on the other hand act as major environmental pollutants, especially in areas with human activities and interference (Zamora-Ledezma et al. 2021). Several studies conducted in China reported that the total standard rate of soil pollution is 16.1%, among which Cd, As, Hg, Pb and Cr contribute 7.00%, 2.70%, 1.60%, 1.50% and 1.10%, respectively (Source: Ministry of Ecology & Environment, 2019 bulletin) (Zhao et al. 2022). In the last 50 years, globally around 30,000 tons of Cr and 800,000 tons of Pb have been released into the environment (Cheng et al. 2014). It has been well-established that the contamination of heavy metals in soil and water is a worldwide concern regarding crop yield and food sustainability. Heavy metal accumulation induces changes in soil properties such as pH and microbial diversity, causes plants to exhibit altered metabolism, lowers growth and biomass, and affects several physiological and biochemical processes (Nagajyoti et al. 2010). There have been some traditional strategies to curb the accumulation of heavy metals for a long time. The in situ remediation techniques are surface capping, soil flushing, encapsulation, bioremediation, etc., whereas some ex-situ methods include landfilling, vitrification soil washing, etc. (Khan et al. 2021). Most of these methods are time-consuming, expensive, and not beneficial for the environment. Due to the deep-rooted problems such as high toxicity of heavy metals even at low concentrations, slow and incomplete elimination of contaminants using conventional approaches and non-biodegradable nature (Sall et al. 2020), advanced scientific technologies are to be developed and deployed to mitigate such situations. Over the years, nano-remediation has offered a novel and efficient alternative approach against the conventional heavy metal stress-mitigating strategies. Metallic or non-metal-based nanoparticles are cost-effective, environment friendly, easily available and small in size. They also have the potential to penetrate through tiny soil pores, a large surface area beneficial for adsorption, and uses photo-catalytic methods to degrade pollutants

Table 14.2 Salinity stress amelioration by the application of SiNPs.

Plants	NP type	Stress	Concentration of SiNPs	Effect	Possible mechanisms	References
Trigonella foenum-graceum (Fenugreek)	SiO_2	NaCl 0, 50, 100 and 150 mM	50 ppm 100 ppm	↑ Seed germination to 99% from 95.50%. ↑ Dry weight of shoot, root and seedling by 26%, 36% and 62%, respectively ↑ Photosynthesis rate ↓ Mean generation time (lowest for 50 ppm NP treatment)	Water absorption and its transportation in fenugreek seedling is facilitated by NP. NP promotes the Activity of RUBISCO enzyme and modifies the leaf's ultra-structure	Ivani et al. (2018)
Capsicum annum (Capsicum)	Nano-Si (25%)	Saline-well water (pH = 7.8) (3, 6 and 9 wk)	1.0 and 2.0 cm^3/l	↑ Plant height, number of branches, dry and fresh weight of leaves. ↑ Fruit weight, total yield ↑ Total chlorophyll (SPAD) ↑ Percentage in N, P,K ↑ Turgor pressure (42%)	Plant water status under saline conditions is maintained which results in cell expansion and increased leaf area. Uptake of Na^+ is reduced by improving K^+: Na^+	Tantawy et al. (2015)
Musa acuminata 'Grand Nain' (In-vitro made banana)	SiO_2	0, 1, 2, 3% Polyethylene Glycol (PEG-8000) (4 wk)	0, 50, 100, 150 mg/l	↑ Chlorophyll content ↓ Electrolyte leakage (EL) ↓ Malondialdehyde (MDA) ↓ Total phenolic content with ↑ in NP concentration	Protects cell membrane integrity by decreasing the production of free oxygen radicals. MDA is less produced from the lipid peroxidation pathway	Mahmoud et al. (2020)
Solanum lycopersicum (Tomato)	Si	250 mM NaCl	100 mg/l	↑ Net photosynthetic rate, stomatal conductance, transpiration rate ↑ Chlorophyll index ↑ Glutathione (GSH), Proline, Ascorbate (AsA), Catalase (CAT), Superoxide dismutase (SOD)	SiNPs enhances the cytochrome b6/f and ATP synthase. It reduces Na^+ allocation and incorporation.	Alam et al. (2022)

(Continued)

Table 14.2 (Continued)

Plants	NP type	Stress	Concentration of SiNPs	Effect	Possible mechanisms	References
				↑ RUBISCO, Fumarase, Hexokinase activity	Si up-regulates salt stress inducible genes, *AREB, TAS14, NCED3, CRK1* and down regulates the *RBOH1, APX2, MAPK2, ERF5, MAPK3* and *DDF2* genes	
Cucumis sativus (Cucumber)	Si	Na^+ solution 1000, 2000, 3000, 4000, 5000 mg/dl	0, 100, 200, 300 mg/l	↑ Final germination percentage of seeds up to 200 mg/l NP and then ↓ FGP for 300 mg/l ↓ Mean Germination Time (MGT) lowest for 200 mg/l Si-NP ↑ Co-efficient of velocity of germination (CVG) ↑ K^+/Na^+ ratio	SiNP increases K^+ content, decreases Na^+ content, detoxifies ROS through enzyme activations. SiNPs remove free radicals by altering oxidation–reduction reactions and producing superoxide ion radical. SiNPs enhance synthesis of gibberellin and inhibit ABA (abscisic acid) production	Alsaeedi et al. (2018)
Oryza sativa (Rice)	Nano-Si (extracted from rice Straw)	NaCl 0, 100, 200, 400 and 800 mM	150 g/l	↑ Calcium (Ca), potassium (K) content ↓ Na^+ content ↓ H_2O_2 content ↓ Shikimic acid ↑ Free soluble phenol ↑ Jasmonic acid	Jasmonic acid (JA), a stress phytohormone, up-regulates the Si uptake genes *OsLsi1* and *OsLsi2*, activates antioxidant defence system and induces osmolyte production. SiNPs induce phenol production by using shikimic acid	Abdel-Haliem et al. (2017)
Pisum sativum (Pea)	Nano-Si (NSi)	NaCl 100, 150, 200, 250 mM	3 mM	↑ Harvest index (HI) ↑ Crop index (CI) ↓ Na^+/K^+ ratio	SiNPs down-regulate SOD (superoxide dismutase), POD (peroxidase) and Pslec RLK SiNPs up-regulates CAT (catalase)	Ismail et al. (2022)

(A. Hussain et al. 2022). Moreover, the use of SiNPs, due to their tuneable pore size, enhances plant growth, inducing positive physiological and biochemical impacts, and ultimately alleviating heavy metal toxicity (Adrees et al. 2015). Application of SiNPs has been demonstrated to benefit rice (*Oryza sativa*) and wheat (*Triticum aestivum*) against cadmium (Cd) stress by enhancing the biomass of shoots and grains, chlorophyll content of leaf, activities of enzymes such as superoxide dismutase, catalase and ascorbate peroxidase and thus decreasing electrolyte leakage and malondialdehyde content (Ali et al. 2019; Rizwan et al. 2019). Soya bean (*Glycine max*) plants after being treated with SiNPs showed better growth and increased chlorophyll content under Hg stress (Li et al. 2020) (Table 14.3).

14.5.4 Role of Si/SiO_2 to Mitigate Emerging Biotic Stress

Biotic stress is caused by pathogens, e.g., bacteria, fungi, viruses, nematodes, parasites and insects in host organisms (Singla and Krattinger 2016). Biotic agents can be divided into two major groups – biotrophs and necrotrophs. Biotrophs collect nutrients from living cells and necrotrophs accumulate materials from infected dead tissues. In plants, salicylic acid (SA)-dependent pathway offers resistance to biotroph pathogens, and in the case of necrotrophs, the resistance is conferred by jasmonic acid (JA) and ethylene (ET)-dependent pathways (Robert-Seilaniantz et al. 2009). Plants contain a specific genome that encodes stress-resistant genes which regulate the level of metabolites. In response to the biotic stress, the plant increased lignification and chemical or enzymatic defence mechanisms. Chemical compounds like beta-aminobutyric acid (BABA) and benzothiadiazole (BTH), together with plant hormones, play a significant role in mediating biotic stress (Madani et al. 2019). Insects, mites and pests procure food from plant tissue, which directly interrupts their physicochemical properties, causing early flowering, loss in seed set and their photosynthetic activity. Plants release their secondary metabolites to get rid of these agents (Bhar et al. 2021; Mertens et al. 2021). Weeds, which compete for crop plants, release some chemical substances that can be harmful to photosynthesis, respiration and other enzymatic activities of plants (Saudy et al. 2021). In response to weeds, plants also release some allelopathic compounds that can retard weed germination and growth (Bhar et al. 2021). Small RNAs such as siRNA, miRNA and circRNA play a key role in overcoming the infection (Raja and Tennyson 2023). The pathogens produce several symptoms like chlorosis, stunting, lesions, rotting, etc. (Balodi et al. 2017), causing a significant loss of vigour in crop plants and posing a risk in global food security (Sapre et al. 2021). The biotic stresses are responsible for approximately 28.2%, 37.4%, 31.2%, 40.3%, 26.3% and 28.8% yield losses in wheat, rice, maize, potato, soya bean and cotton crops annually, respectively (Oerke 2006).

Table 14.3 Heavy metal stress remediation by the application of SiNPs.

Plants	Nanoparticle type	Specific stressor and their exposure	Concentration of SiNPs	Effect	Mechanism of action	References
Triticum aestivum (Ujala-2016 Wheat cultivar)	Nano Si	Cadmium (Cd) 25 mg/kg Soil	3 mM	↑ Shoot length, shoot fresh and dry weight, leaf area ↑ Total chlorophyll and carotenoid content ↑ Total soluble sugar (TSS), total soluble protein (TSP) and total free amino acid. ↓ Electrolyte leakage by 60% at seedling	Si induces the decrease in Cd uptake and its transport in plant tissues by chelation process. By controlling CAT, SOD, POD activities, SiNPs help in scavenging ROS, mitigate lipid peroxidation and stabilize the cell membrane	Thind et al. (2021)
Pleioblastus pygmaeus (Bamboo)	SiO_2	Lead (Pb) 50, 250, 500, 1000, 1500 µM $PbSO_4$	100 µM and 500 µM	↑ Chlorophyll a content (28%) ↑ Chlorophyll b content (50%) ↑ Carotenoid content (12.7%) ↑ Biomass of root and shoot	SiO_2 activates antioxidant enzymes superoxide dismutase (SOD), catalase (CAT), glutathione reductase (GR) and phenylalanine ammonia-lyase (PAL). Maintains cellular integrity by reducing the contents of hydrogen peroxide and soluble protein (SP), and polyphenol oxidase (PPO) activity	Emamverdian et al. (2020)
Zea mays (Maize)	Nano-Si	Arsenate (AsO_3^{4-}) 25 and 50 µM	10 µM	↑ Fresh mass ↑ Photosynthetic pigments	Si makes complex or helps in co-precipitation of toxic ions. Si promotes compartmentalization and immobilization of ions in plants. Si forms barrier in root endodermis to prevent AsA uptake.	Tripathi et al. (2016)

				↓ Levels of oxidative stress markers (ROS, H_2O_2, MDA) ↑ Ratio of AsA/DHA (Ascorbate to Dehydro-ascorbate) and GSH/GSSG (Glutathione to Glutathione disulphide)	Si promotes ascorbate-glutathione (AsA-GSH) cycle which is a scavenging mechanism of ROS	Tripathi et al. (2016)
Pisum sativum (Pea)	Nano-Si	Chromium (Cr-VI) 100 μM	10 μM	↑ Growth parameters (seed fresh weight, dry weight, shoot length) ↑ Non-photochemical quenching (NPQ) by 37% ↓ H_2O_2, MDA content	Si increases the antioxidant mechanism regulated by SOD, APX, CAT, GR and DHAR. SiNPs prevent the inhibition of plasma membrane H^+-ATPase activity by Cr stress and enhances mineral uptake	Tripathi et al. (2015)
Glycine max (Soya bean)	Nano-SiO_2	Mercury (Hg) 0, 1 and 5 mg/l	0, 100, 500, 1000, 2000 mg/l	↑ Fresh weight ↓ Hg accumulation in root and shoot ↓ Decreased Hg content ↑ Chlorophyll content	SiNPs promote the transport of potassium, magnesium and iron. SiNPs immobilize and inactivate Hg^{2+} by reducing the bioavailability of Hg^{2+} in growth media	Li et al. (2020)

Alleviation of biotic stress includes lots of strategies. With the help of traditional breeding or genetic transformation, stress-resistant plant varieties are developed (Zain et al. 2023). Cultural practices such as crop rotation can also be done. Integrated pest management by chemical pesticides and biological control agents is effective in controlling biotic stress (Hameed et al. 2023). Microorganisms, causing biotic stress, become resistant to antimicrobials because of the extensive use of pesticides and herbicides (Varah et al. 2019). Climate change also causes severe stress that cannot be managed by these techniques. For modern agriculture, many new strategies are evolving such as the use of nanoparticles, nanotubes and quantum dots. These devices are more effective than previous strategies, considering selectivity, accuracy and reactivity (Hameed et al. 2023). Nanoparticles are beneficial for their particular physical and chemical properties. The uptake of these NPs by root and foliar systems is easy (Tortella et al. 2023). Nanoparticles are used as nanofertilizers and nanopesticides, which gives high productivity in crop plants (Shen et al. 2023). Nanomaterial-based sensors are useful in managing biotic stress.

SiNPs can be a better option as it has low toxicity and does not hamper the physiochemical process of plants (Ahamad and Siddiqui 2021). Different types of silica are broken down by hydrolysis to form nontoxic mono-silicic acid, which then gets released into the environment (Croissant et al. 2020; Wang et al. 2022). Pathogens have to penetrate physical barriers such as cell walls and waxy layers for successful infection. SiNPs strengthen physical barriers that inhibit the invasion of pathogens. Moreover, it also reinforces the epidermal coat and cell wall by forming silica papillae and promotes mechanical rigidity by binding with pectin and hemicellulose that prohibit the colonization of the pathogen (Naidu et al. 2023). It was reported that the penetrative appressorial sites for *Pyricularia oryzae* were reduced following silicon treatment in rice. Furthermore, Si induces lignin biosynthesis to prevent the growth of parasitic weed (L. Wang et al. 2022). Lignin can be characterized as an aromatic polymer with guaiacyl (G), *p*-hydroxyphenyl (H) and syringyl (S) monomeric units. Cell wall rigidity is directly proportional to three monomeric unit of lignin. Angiosperms mainly contain G and S units as monomer (Yoon et al. 2015). Si up-regulated the key lignin biosynthesis enzymes namely phenylalanine ammonium lyase (PAL), peroxidase (POD), 4 coumaric ligase (4CL) and cinnamyl alcohol dehydrogenase (CAD) that ultimately escalate stem strength and resist lodging in soya bean (*Glycine max* (L.) Merr.) (S. Hussain et al. 2021). An interesting investigation using SiNPs revealed that they significantly inhibit egg hatching and also induce mortality of root-knot nematode caused by *Meloidogyne incognita* at second-stage juvenile in *Solanum melongena* L. by 87.0–98.50% after 72 hour exposure to 100–200 ppm of SiNPs (El-Ashry et al. 2022). On the other hand, silica-based NPs lower synthesis and accumulation of ROS and lipid peroxidation that subsequently helps to mitigate

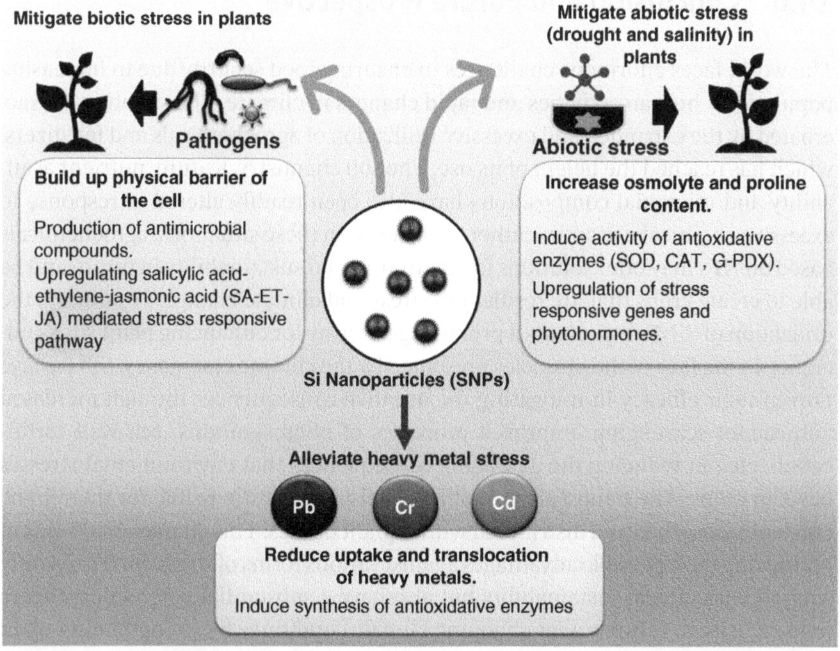

Figure 14.1 Beneficial role of silicon nanoparticles (Si-NPs) in mitigating abiotic, biotic and heavy metal stress in plants.

biotic and abiotic stresses (Tortella et al. 2023). Another study with SiNPs showed that it mitigates root rot disease caused by *Rhizoctonia solani* in *Avena sativa* L. by reducing proline content and increasing membrane stability index, soluble sugar content, antioxidant enzymes and antifungal compounds. Si-inoculated plants exhibit a higher rate of germination, radicle and plumule length, and root length (Ahmad et al. 2023). SiNPs also induce expression of defence-related genes in response to stress conditions, for example, glutathione reductase, β 1, 3 glucanase and lipoxygenase. Moreover, it is also involved in jasmonic acid (JA)-mediated plant defence signalling by regulating the biosynthesis of JA (Naidu et al. 2023). JA along with other derivatives methyl jasmonate (MeJA) and its isoleucine conjugate (JA-lle) plays a significant role in plant stress tolerance. Upon exposure to a pathogen, different molecular patterns, viz., damage-associated molecular pattern (DAMP) and microbe-associated molecular pattern (MAMP) are recognized by cell surface receptors. Interactions of these molecular patterns with cell surface receptors induce *de novo* synthesis of JA-lle that subsequently interacts with different transcription factors and induce protective mechanisms in plants (Y. Wang et al. 2021) (Figure 14.1).

14.6 Conclusion and Future Prospective

The world faces enormous challenges in ensuring food security due to increasing populations, human activities and rapid changes in climate. The situation is exacerbated by the expanded and excessive utilization of agrochemicals and fertilizers, which has reached the height of its use. The soil chemistry, texture, nutrient availability and microbial compositions have also been readily altered in response to excessive chemical use and weather conditions. In these situations, agrochemicals based on NPs may offer solutions for environmental sustainability in future and be able to create crops that are resilient to stress and climate change. To sum up, the utilization of SiNPs represents a promising pathway for enhancing plant stress tolerance in the face of the obstacles presented by the climate emergency. SiNPs have proven their efficacy in mitigating the negative consequences through increased antioxidant scavenging, improved processes of photosynthesis, cell wall fortification, etc., in reducing the damaging consequences that environmental stresses have on crops. The nanoscale size of SiNPs also makes it possible for them to be efficiently absorbed and distributed within plant tissues. This characteristic aids in optimizing their possible advantages against various forms of stress. SiNPs not only support agricultural sustainability but also have a substantial potential to ensure food security in future under changing climate conditions by strengthening plant resilience to abiotic stresses such as heat stress, chilling stress, drought, salinity, heavy metal accumulation and UV radiation. More research is necessary to completely understand the mechanistic behaviour of SiNPs, the best doses for external and internal administration, the long-term effects on plant and soil health, and the possible ecological ramifications of SiNP application (Roychoudhury 2020). However, its integration into farming methods requires more legislation, appropriate regulation and farmer-centric understanding.

Acknowledgement

The authors are extremely thankful to the editors for inviting them to contribute the chapter and giving the opportunities to share the knowledge.

Glossary

Abiotic Stress Detrimental effects of non-living factors such as drought, salinity, heavy metals toxicity are known as abiotic stress.

Biodiversity Biological diversity' or biodiversity is that part of nature which includes the differences in genes among the individuals of a species, the variety and richness of all the plant and animal species at different scales in space, locally, in a region, in the country and the world, and various types of ecosystems, both terrestrial and aquatic, within a defined area.

Biogeochemical Cycle As the earth essentially a closed system with respect to matter, every matter that is used by living organisms passes between the biotic and abiotic components of the biosphere. The cyclic pathways through which chemical elements are transforming from the environment into the organism and from there to environment regularly and continuously known as biogeochemical cycle.

Biotic Stress Biotic stress may be defined as the disease condition arise due to infestation of living organism such as bacteria, viruses and fungi. These living organisms are known as pathogens.

Climate Change Climate change may be defined as the long term change in weather pattern and temperature, primarily caused due to anthropogenic activities such as burning of fossil fuel, greenhouse gas emission and also due to natural activities for example volcanic eruption.

Ecosystem An ecosystem is a basic functional unit of nature comprising the whole community in a given area interacting with one another and with their physical environment. It is an open system that energy and matter transfer in and out.

Emerging Infectious Disease (EID) Emerging Infectious Disease (EID) is referred to the infectious disease recognized recently and increasing its prevalence in new geographical area rapidly.

Endemic Species Endemism means the confinement of a particular species, genus, or groups of plants and animals to a particular area. Taxa occurring only a single restricted geographical area are known as endemics, Endemism normally applied only where there is a considerable restriction in the area of distribution.

Green House Gas (GHG) Greenhouse gases referred to that gaseous content that absorbs and re-emit infrared radiation. Carbon dioxide (CO_2), Water vapour, methane (CH_4) and Nitrous oxide (N_2O) are primary GHG.

Global Warming Increase of annual temperature of the planet is known as global warming. Greenhouse gases absorbs incident ray from sunlight and prevent them to leave the atmosphere.

Habitat Habitat refers to a specific place where a species normally lives. For example, habitat of a tiger is the forest; of a shark is the sea.

Infrared Radiation Radiation from the band of electromagnetic spectrum of longer wavelength ranges from 760 nm to 1 mm.

Invasive Alien Species (IAS) Invasive Alien Species (IAS) are nonnative plant, animals or other organisms that are introduced intentionally or accidentally. IAS impose several economical or environmental deleterious effects on the native ecosystem.

Ocean Acidification Decrease of ocean pH (hydrogen ion concentration increase) due to uptake of CO_2 from atmosphere is known as ocean acidification.

$$CO_2 + H_2O\ H_2CO_3\ \text{(Carbonic acid)} \leftrightarrow H^+\ \text{(Hydrogen ion)} \leftrightarrow HCO_3^-\ \text{(Bicarbonate ions)}$$

Oxidative Stress Imbalance between synthesis and accumulation of reactive oxygen species (ROS) leads to a stress condition that subsequently damage cell or tissue, known as oxidative stress.

Reactive Oxygen Species (ROS) A type of unstable, highly reactive oxygen containing radicals, which are synthesized as a byproduct of cellular metabolism is referred as reactive oxygen species (ROS).

References

Abdel-Haliem, M.E.F., Hegazy, H.S., Hassan, N.S., and Naguib, D.M. (2017). Effect of silica ions and nano silica on rice plants under salinity stress. *Ecological Engineering* 99 (February): 282–289. https://doi.org/10.1016/j.ecoleng.2016.11.060.

Abdullah, A.S., Moffat, C.S., Lopez-Ruiz, F.J. et al. (2017). Host–multi-pathogen warfare: pathogen interactions in co-infected plants. *Frontiers in Plant Science* 8 (October): 1806. https://doi.org/10.3389/fpls.2017.01806.

Adrees, M., Ali, S., Rizwan, M. et al. (2015). Mechanisms of silicon-mediated alleviation of heavy metal toxicity in plants: a review. *Ecotoxicology and Environmental Safety* 119 (September): 186–197. https://doi.org/10.1016/j.ecoenv.2015.05.011.

Ahamad, L. and Siddiqui, Z.A. (2021). Effects of silicon dioxide, zinc oxide and titanium dioxide nanoparticles on *Meloidogyne incognita, Alternaria dauci* and *Rhizoctonia solani* disease complex of carrot. *Experimental Parasitology* 230 (November): 108176. https://doi.org/10.1016/j.exppara.2021.108176.

Ahmad, F., Jabeen, K., Iqbal, S. et al. (2023). Influence of silicon nano-particles on *Avena sativa* L. to alleviate the biotic stress of *Rhizoctonia solani*. *Scientific Reports* 13 (1): 15191. https://doi.org/10.1038/s41598-023-41699-w.

Akhtar, N., Ilyas, N., Mashwani, Z.-u.-R. et al. (2021). Synergistic effects of plant growth promoting rhizobacteria and silicon dioxide nano-particles for amelioration of drought stress in wheat. *Plant Physiology and Biochemistry* 166 (September): 160–176. https://doi.org/10.1016/j.plaphy.2021.05.039.

Alam, P., Arshad, M., Al-Kheraif, A.A. et al. (2022). Silicon nanoparticle-induced regulation of carbohydrate metabolism, photosynthesis, and ROS homeostasis in

Solanum lycopersicum subjected to salinity stress. *ACS Omega* 7 (36): 31834–31844. https://doi.org/10.1021/acsomega.2c02586.

Ali, S., Rizwan, M., Hussain, A. et al. (2019). Silicon nanoparticles enhanced the growth and reduced the cadmium accumulation in grains of wheat (*Triticum aestivum* L.). *Plant Physiology and Biochemistry* 140 (July): 1–8. https://doi.org/10.1016/j.plaphy.2019.04.041.

Alsaeedi, A., El-Ramady, H., Alshaal, T. et al. (2018). Exogenous nanosilica improves germination and growth of cucumber by maintaining K^+/Na^+ ratio under elevated Na+ stress. *Plant Physiology and Biochemistry* 125 (April): 164–171. https://doi.org/10.1016/j.plaphy.2018.02.006.

Anderson, P.K., Cunningham, A.A., Patel, N.G. et al. (2004). Emerging infectious diseases of plants: pathogen pollution, climate change and agrotechnology drivers. *Trends in Ecology & Evolution* 19 (10): 535–544. https://doi.org/10.1016/j.tree.2004.07.021.

Arora, N.K. (2019). Impact of climate change on agriculture production and its sustainable solutions. *Environmental Sustainability* 2 (2): 95–96. https://doi.org/10.1007/s42398-019-00078-w.

Balodi, R., Bisht, S., Ghatak, A., and Rao, K.H. (2017). Plant disease diagnosis: technological advancements and challenges. *Indian Phytopathology* 70 (3): https://doi.org/10.24838/ip.2017.v70.i3.72487.

Behboudi, F., Tahmasebi Sarvestani, Z., Kassaee, M.Z. et al. (2018). Improving growth and yield of wheat under drought stress via application of SiO_2 nanoparticles. *Journal of Agricultural Science and Technology* 20 (7): 1479–1492.

Bhar, A., Chakraborty, A., and Roy, A. (2021). Plant responses to biotic stress: old memories matter. *Plants* 11 (1): 84. https://doi.org/10.3390/plants11010084.

Botchway, Ebo, Fred Kofi Asiedu, Peter Asare-Nuamah, and Michael Insaidoo. 2021. "Factors influencing risk in per capita income from food crop production in Ghana." *Journal of Economic Dynamics & Control* Available at SSRN 3857431. https://ssrn.com/abstract=3857431.

Carrion, M. and Madoff, L.C. (2017). ProMED-Mail: 22 years of digital surveillance of emerging infectious diseases. *International Health* 9 (3): 177–183. https://doi.org/10.1093/inthealth/ihx014.

Chaudhry, S. and Sidhu, G.P.S. (2022). Climate change regulated abiotic stress mechanisms in plants: a comprehensive review. *Plant Cell Reports* 41 (1): 1–31. https://doi.org/10.1007/s00299-021-02759-5.

Cheng, K., Tian, H.Z., Zhao, D. et al. (2014). Atmospheric emission inventory of cadmium from anthropogenic sources. *International Journal of Environmental Science and Technology* 11 (3): 605–616. https://doi.org/10.1007/s13762-013-0206-3.

Corwin, D.L. (2021). Climate change impacts on soil salinity in agricultural areas. *European Journal of Soil Science* 72 (2): 842–862. https://doi.org/10.1111/ejss.13010.

Croissant, J.G., Butler, K.S., Zink, J.I., and Jeffrey Brinker, C. (2020). Synthetic amorphous silica nanoparticles: toxicity, biomedical and environmental implications. *Nature Reviews Materials* 5 (12): 886–909. https://doi.org/10.1038/s41578-020-0230-0.

Das, T., Bhattacharyya, A., and Bhar, A. (2023). Linking phyllosphere and rhizosphere microbiome to the plant–insect interplay: the new dimension of tripartite interaction. *Physiologia* 3 (1): 129–144. https://doi.org/10.3390/physiologia3010011.

Dietz, T., Shwom, R.L., and Whitley, C.T. (2020). Climate change and society. *Annual Review of Sociology* 46 (1): 135–158. https://doi.org/10.1146/annurev-soc-121919-054614.

Du, J., Liu, B., Zhao, T. et al. (2022). Silica nanoparticles protect rice against biotic and abiotic stresses. *Journal of Nanobiotechnology* 20 (1): 197. https://doi.org/10.1186/s12951-022-01420-x.

El-Ashry, R.M., El-Saadony, M.T., El-Sobki, A.E.A. et al. (2022). Biological silicon nanoparticles maximize the efficiency of nematicides against biotic stress induced by *Meloidogyne incognita* in eggplant. *Saudi Journal of Biological Sciences* 29 (2): 920–932. https://doi.org/10.1016/j.sjbs.2021.10.013.

Elbasiouny, H., El-Ramady, H., Elbehiry, F. et al. (2022). Plant nutrition under climate change and soil carbon sequestration. *Sustainability* 14 (2): 914. https://doi.org/10.3390/su14020914.

El-Sayed, A. and Kamel, M. (2020). Climatic changes and their role in emergence and re-emergence of diseases. *Environmental Science and Pollution Research* 27 (18): 22336–22352. https://doi.org/10.1007/s11356-020-08896-w.

Emamverdian, A., Ding, Y., Mokhberdoran, F. et al. (2020). Silicon dioxide nanoparticles improve plant growth by enhancing antioxidant enzyme capacity in bamboo (*Pleioblastus pygmaeus*) under lead toxicity. *Trees* 34 (2): 469–481. https://doi.org/10.1007/s00468-019-01929-z.

Farhangi-Abriz, S. and Torabian, S. (2018). Nano-silicon alters antioxidant activities of soybean seedlings under salt toxicity. *Protoplasma* 255 (3): 953–962. https://doi.org/10.1007/s00709-017-1202-0.

Fedoroff, N.V., Battisti, D.S., Beachy, R.N. et al. (2010). Radically rethinking agriculture for the 21st century. *Science* 327 (5967): 833–834. https://doi.org/10.1126/science.1186834.

Flowers, T.J., Koyama, M.L., Flowers, S.A. et al. (2000). QTL: their place in engineering tolerance of rice to salinity. *Journal of Experimental Botany* 51 (342): 99–106. https://doi.org/10.1093/jexbot/51.342.99.

Garrett, K.A., M. Nita, E.D. De Wolf, P.D. Esker, L. Gomez-Montano, and A.H. Sparks. 2021. "Plant pathogens as indicators of climate change." In *Climate Change*, 499–513. Trevor M. Letcher Elsevier. https://doi.org/10.1016/B978-0-12-821575-3.00024-4.

George, T.S., Taylor, M.A., Dodd, I.C., and White, P.J. (2017). Climate change and consequences for potato production: a review of tolerance to emerging abiotic stress. *Potato Research* 60 (3–4): 239–268. https://doi.org/10.1007/s11540-018-9366-3.

Ghorbanpour, M., Mohammadi, H., and Kariman, K. (2020). Nanosilicon-based recovery of barley (*Hordeum vulgare*) plants subjected to drought stress. *Environmental Science: Nano* 7 (2): 443–461. https://doi.org/10.1039/C9EN00973F.

Habib-ur-Rahman, M., Ahmad, A., Raza, A. et al. (2022). Impact of climate change on agricultural production; issues, challenges, and opportunities in Asia. *Frontiers in Plant Science* 13 (October): 925548. https://doi.org/10.3389/fpls.2022.925548.

Hameed, A., Saif, M.J., Qayyum, M.A. et al. (2023). Nanomaterial-based sensors for real-time monitoring of crop plants growth, development, production, and protection. In: *Engineered Nanomaterials for Sustainable Agricultural Production, Soil Improvement and Stress Management* (ed. A. Husen), 357–385. Elsevier. https://doi.org/10.1016/B978-0-323-91933-3.00007-6.

Harman, G., Khadka, R., Doni, F., and Uphoff, N. (2021). Benefits to plant health and productivity from enhancing plant microbial symbionts. *Frontiers in Plant Science* 11 (April): 610065. https://doi.org/10.3389/fpls.2020.610065.

Hussain, A., Rehman, F., Rafeeq, H. et al. (2022). In-situ, Ex-situ, and nano-remediation strategies to treat polluted soil, water, and air – a review. *Chemosphere* 289 (February): 133252. https://doi.org/10.1016/j.chemosphere.2021.133252.

Hussain, S., Shuxian, L., Mumtaz, M. et al. (2021). Foliar application of silicon improves stem strength under low light stress by regulating lignin biosynthesis genes in soybean (*Glycine max* (L.) Merr.). *Journal of Hazardous Materials* 401 (January): 123256. https://doi.org/10.1016/j.jhazmat.2020.123256.

Ismail, L.M., Soliman, M.I., Abd El-Aziz, M.H., and Abdel-Aziz, H.M. (2022). Impact of silica ions and nano silica on growth and productivity of pea plants under salinity stress. *Plants* 11 (4): 494. https://doi.org/10.3390/plants11040494.

Ivani, R., Nejad, S.H.S., Ghahraman, B. et al. (2018). Role of bulk and Nanosized SiO_2 to overcome salt stress during Fenugreek germination (*Trigonella Foenum-Graceum* L.). *Plant Signaling & Behavior* 13 (7): e1044190. https://doi.org/10.1080/15592324.2015.1044190.

Jeelani, P.G., Mulay, P., Venkat, R., and Ramalingam, C. (2020). Multifaceted application of silica nanoparticles. A review. *Silicon* 12 (6): 1337–1354. https://doi.org/10.1007/s12633-019-00229-y.

Kalteh, M., Alipour, Z.T., Ashraf, S. et al. (2018). Effect of silica nanoparticles on basil (*Ocimum basilicum*) under salinity stress. *Journal of Chemical Health Risks* 4 (3): https://doi.org/10.22034/jchr.2018.544075.

Khan, S., Naushad, M., Lima, E.C. et al. (2021). Global soil pollution by toxic elements: current status and future perspectives on the risk assessment and remediation strategies – a review. *Journal of Hazardous Materials* 417 (September): 126039. https://doi.org/10.1016/j.jhazmat.2021.126039.

Kumar, M., Sandhya, S., Kumar, P. et al. (2021). Abiotic stress tolerance in rice: insight in climate change scenario. In: *Integrative Advances in Rice Research* (ed. M. Huang), 1–19. IntechOpen.

Li, Y., Zhu, N., Xujun Liang, X. et al. (2020). Silica nanoparticles alleviate mercury toxicity *via* immobilization and inactivation of Hg(II) in soybean (*Glycine max*). *Environmental Science: Nano* 7 (6): 1807–1817. https://doi.org/10.1039/D0EN00091D.

Madani, B., Mirshekari, A., and Imahori, Y. (2019). Physiological responses to stress. In: *Postharvest Physiology and Biochemistry of Fruits and Vegetables* (ed. E.M. Yahia), 405–423. Elsevier. https://doi.org/10.1016/B978-0-12-813278-4.00020-8.

Mahmoud, L.M., Dutt, M., Shalan, A.M. et al. (2020). Silicon nanoparticles mitigate oxidative stress of in vitro-derived banana (*Musa acuminata* 'Grand Nain') under simulated water deficit or salinity stress. *South African Journal of Botany* 132 (August): 155–163. https://doi.org/10.1016/j.sajb.2020.04.027.

Malaekeh, S.M., Shiva, L., and Safaie, A. (2024). Investigating the economic impact of climate change on agriculture in iran: spatial spillovers matter. *Agricultural Economics*. https://doi.org/10.1111/agec.12821.

Malhi, G.S., Kaur, M., and Kaushik, P. (2021). Impact of climate change on agriculture and its mitigation strategies: a review. *Sustainability* 13 (3): 1318. https://doi.org/10.3390/su13031318.

Manzano, M. and Vallet-Regí, M. (2020). Mesoporous silica nanoparticles for drug delivery. *Advanced Functional Materials* 30 (2): 1902634. https://doi.org/10.1002/adfm.201902634.

Mertens, D., Boege, K., Kessler, A. et al. (2021). Predictability of biotic stress structures plant defence evolution. *Trends in Ecology & Evolution* 36 (5): 444–456. https://doi.org/10.1016/j.tree.2020.12.009.

Miao, B.-H., Han, X.-G., and Zhang, W.-H. (2010). The ameliorative effect of silicon on soybean seedlings grown in potassium-deficient medium. *Annals of Botany* 105 (6): 967–973. https://doi.org/10.1093/aob/mcq063.

Minhas, P.S., Rane, J., and Pasala, R.K. (2017). Abiotic stresses in agriculture: an overview. In: *Abiotic Stress Management for Resilient Agriculture* (ed. P.S. Minhas, J. Rane, and R.K. Pasala), 3–8. Singapore: Springer Singapore. https://doi.org/10.1007/978-981-10-5744-1_1.

Mishra, V. and Cherkauer, K.A. (2010). Retrospective droughts in the crop growing season: implications to corn and soybean yield in the Midwestern United States. *Agricultural and Forest Meteorology* 150 (7–8): 1030–1045. https://doi.org/10.1016/j.agrformet.2010.04.002.

Molua, Ernest L. 2007. "The economic impact of climate change on agriculture in cameroon." *World Bank Policy Research Working Paper*, no. 4364.

Morens, D.M., Folkers, G.K., and Fauci, A.S. (2004). The challenge of emerging and re-emerging infectious diseases. *Nature* 430 (6996): 242–249. https://doi.org/10.1038/nature02759.

Mukarram, M., Masroor, M., Khan, A. et al. (2023). Silicon nanoparticles (SiNPs) restore photosynthesis and essential oil content by upgrading enzymatic antioxidant metabolism in lemongrass (*Cymbopogon flexuosus*) under salt stress. *Frontiers in Plant Science* 14 (February): 1116769. https://doi.org/10.3389/fpls.2023.1116769.

Mukherjee, S., Mishra, A., and Trenberth, K.E. (2018). Climate change and drought: a perspective on drought indices. *Current Climate Change Reports* 4 (2): 145–163. https://doi.org/10.1007/s40641-018-0098-x.

Muluneh, M.G. (2021). Impact of climate change on biodiversity and food security: a global perspective–a review article. *Agriculture & Food Security* 10 (1): 36. https://doi.org/10.1186/s40066-021-00318-5.

Munns, R. (2005). Genes and salt tolerance: bringing them together. *New Phytologist* 167 (3): 645–663. https://doi.org/10.1111/j.1469-8137.2005.01487.x.

Nadarajan, S. and Sukumaran, S. (2021). Chemistry and toxicology behind chemical fertilizers. In: *Controlled Release Fertilizers for Sustainable Agriculture* (ed. F.B. Lewu, T. Volova, S. Thomas, and K.R. Rakhimol), 195–229. Elsevier. https://doi.org/10.1016/B978-0-12-819555-0.00012-1.

Nagajyoti, P.C., Lee, K.D., and Sreekanth, T.V.M. (2010). Heavy metals, occurrence and toxicity for plants: a review. *Environmental Chemistry Letters* 8: 199–216.

Naidu, S., Pandey, J., Mishra, L.C. et al. (2023). Silicon nanoparticles: synthesis, uptake and their role in mitigation of biotic stress. *Ecotoxicology and Environmental Safety* 255 (April): 114783. https://doi.org/10.1016/j.ecoenv.2023.114783.

Nivas, D., Gaikwad, D.K., and Chavan, P.D. (2011). Physiological responses of two morinda species under saline conditions. *American Journal of Plant Physiology* 6 (3): 157–166. https://doi.org/10.3923/ajpp.2011.157.166.

Oerke, E.-C. (2006). Crop losses to pests. *The Journal of Agricultural Science* 144 (1): 31–43. https://doi.org/10.1017/S0021859605005708.

Pahalvi, H.N., Rafiya, L., Rashid, S. et al. (2021). Chemical fertilizers and their impact on soil health. In: *Microbiota and Biofertilizers*, vol. 2 (ed. G.H. Dar, R.A. Bhat, M.A. Mehmood, and K.R. Hakeem), 1–20. Cham: Springer International Publishing. https://doi.org/10.1007/978-3-030-61010-4_1.

Pareek, A., Dhankher, O.P., and Foyer, C.H. (2020). Mitigating the impact of climate change on plant productivity and ecosystem sustainability. *Journal of Experimental Botany* 71 (2): 451–456. https://doi.org/10.1093/jxb/erz518.

Pautasso, M., Döring, T.F., Garbelotto, M. et al. (2012). Impacts of climate change on plant diseases–opinions and trends. *European Journal of Plant Pathology* 133 (1): 295–313. https://doi.org/10.1007/s10658-012-9936-1.

Peters, K., Breitsameter, L., and Gerowitt, B. (2014). Impact of climate change on weeds in agriculture: a review. *Agronomy for Sustainable Development* 34 (4): 707–721. https://doi.org/10.1007/s13593-014-0245-2.

Pinedo-Guerrero, Z.H., Cadenas-Pliego, G., Ortega-Ortiz, H. et al. (2020). Form of silica improves yield, fruit quality and antioxidant defense system of tomato plants under salt stress. *Agriculture* 10 (9): 367. https://doi.org/10.3390/agriculture10090367.

Priyadi, M. and Upadhyay, P. (2021). Emerging plant diseases under changing climate scenario. In: *Emerging Trends in Plant Pathology* (ed. K.P. Singh, S. Jahagirdar, and B.K. Sarma), 19–31. Singapore: Springer Singapore. https://doi.org/10.1007/978-981-15-6275-4_2.

Rahimi, S., Hatami, M., and Ghorbanpour, M. (2021). Silicon-nanoparticle mediated changes in seed germination and vigor index of marigold (*Calendula officinalis* L.) compared to silicate under PEG-induced drought stress. *Gesunde Pflanzen* 73 (4): 575–589. https://doi.org/10.1007/s10343-021-00579-x.

Raja, I. and Tennyson, J. (2023). Small RNA – regulator of biotic stress and pathogenesis in food crops. In: *Plant Small RNA in Food Crops* (ed. P. Guleria, V. Kumar, and B. Mo), 233–269. Elsevier. https://doi.org/10.1016/B978-0-323-91722-3.00010-5.

Rajput, V.D., Minkina, T., Kumari, A. et al. (2021). Coping with the challenges of abiotic stress in plants: new dimensions in the field application of nanoparticles. *Plants* 10 (6): 1221. https://doi.org/10.3390/plants10061221.

Rastogi, A., Tripathi, D.K., Yadav, S. et al. (2019). Application of silicon nanoparticles in agriculture. *3 Biotech* 9 (3): 90. https://doi.org/10.1007/s13205-019-1626-7.

Raza, A., Razzaq, A., Mehmood, S. et al. (2019). Impact of climate change on crops adaptation and strategies to tackle its outcome: a review. *Plants* 8 (2): 34. https://doi.org/10.3390/plants8020034.

Raza, M.A., Sammar, B.Z., Iqbal, R. et al. (2023). Morpho-physiological and biochemical response of wheat to various treatments of silicon nano-particles under drought stress conditions. *Scientific Reports* 13 (1): 2700. https://doi.org/10.1038/s41598-023-29784-6.

Ristaino, J.B., Anderson, P.K., Bebber, D.P. et al. (2021). The persistent threat of emerging plant disease pandemics to global food security. *Proceedings of the National Academy of Sciences* 118 (23): e2022239118. https://doi.org/10.1073/pnas.2022239118.

Rizwan, M., Ali, S., Rehman, M.Z.U. et al. (2019). Effect of foliar applications of silicon and titanium dioxide nanoparticles on growth, oxidative stress, and cadmium accumulation by rice (*Oryza sativa*). *Acta Physiologiae Plantarum* 41 (3): 35. https://doi.org/10.1007/s11738-019-2828-7.

Robert-Seilaniantz, A., Bari, R., and Jones, J.D.G. (2009). A biotic or abiotic stress? In: *Abiotic Stress Adaptation in Plants* (ed. P. Ashwani, S.K. Sopory, and H.J. Bohnert),

103–122. Dordrecht: Springer Netherlands. https://doi.org/10.1007/978-90-481-3112-9_6.

Romero, F., Cazzato, S., Walder, F. et al. (2022). Humidity and high temperature are important for predicting fungal disease outbreaks worldwide. *New Phytologist* 234 (5): 1553–1556. https://doi.org/10.1111/nph.17340.

Roychoudhury, A. (2020). Silicon-nanoparticles in crop improvement and agriculture. *International Journal on Recent Advancement in Biotechnology & Nanotechnology* 3: 54–65.

Sall, M.L., Diaw, A.K.D., Gningue-Sall, D. et al. (2020). Toxic heavy metals: impact on the environment and human health, and treatment with conducting organic polymers, a review. *Environmental Science and Pollution Research* 27 (24): 29927–29942. https://doi.org/10.1007/s11356-020-09354-3.

Samanta, S. and Roychoudhury, A. (2021). Recent trend in nanoparticle research in regulating arsenic bioaccumulation and mitigating arsenic toxicity in plant species. *Journal of Plant Biochemistry and Biotechnology* 30 (4): 793–812.

Sapre, S., Gontia-Mishra, I., Thakur, V.V. et al. (2021). Molecular techniques used in plant disease diagnosis. In: *Food Security and Plant Disease Management* (ed. A. Kumar and S. Droby), 405–421. Elsevier. https://doi.org/10.1016/B978-0-12-821843-3.00001-5.

Saudy, H.S., El-Bially, M., Ramadan, K.A. et al. (2021). Potentiality of soil mulch and sorghum extract to reduce the biotic stress of weeds with enhancing yield and nutrient uptake of maize crop. *Gesunde Pflanzen* 73 (4): 555–564. https://doi.org/10.1007/s10343-021-00577-z.

Seppelt, R., Klotz, S., Peiter, E., and Volk, M. (2022). Agriculture and food security under a changing climate: an underestimated challenge. *iScience* 25 (12): 105551. https://doi.org/10.1016/j.isci.2022.105551.

Shahzad, A., Ullah, S., Dar, A.A. et al. (2021). Nexus on climate change: agriculture and possible solution to cope future climate change stresses. *Environmental Science and Pollution Research* 28 (12): 14211–14232. https://doi.org/10.1007/s11356-021-12649-8.

Sharf-Eldin, A.A., Alwutayd, K.M., El-Yazied, A.A. et al. (2023). Response of maize seedlings to silicon dioxide nanoparticles (SiO_2NPs) under drought stress. *Plants* 12 (14): 2592. https://doi.org/10.3390/plants12142592.

Shen, M., Liu, S., Jiang, C. et al. (2023). Recent advances in stimuli-response mechanisms of nano-enabled controlled-release fertilizers and pesticides. *Eco-Environment & Health* 2 (3): 161–175. https://doi.org/10.1016/j.eehl.2023.07.005.

Shrivastava, P. and Kumar, R. (2015). Soil salinity: a serious environmental issue and plant growth promoting bacteria as one of the tools for its alleviation. *Saudi Journal of Biological Sciences* 22 (2): 123–131. https://doi.org/10.1016/j.sjbs.2014.12.001.

Siddiqui, M.H., Al-Whaibi, M.H., Firoz, M., and Al-Khaishany, M.Y. (2015). Role of nanoparticles in plants. In: *Nanotechnology and Plant Sciences* (ed. M.H. Siddiqui, M.H. Al-Whaibi, and F. Mohammad), 19–35. Cham: Springer International Publishing. https://doi.org/10.1007/978-3-319-14502-0_2.

Singh, A. and Roychoudhury, A. (2022). Role of nanoparticles in remediation of environmental contaminants. In: *Hazardous and Trace Materials in Soil and Plants* (ed. M. Naeem, T. Aftab, A.A. Ansari, et al.), 327–340. Academic Press (Elsevier).

Singla, J. and Krattinger, S.G. (2016). Biotic stress resistance genes in wheat. In: *Reference Module in Food Science*, 388–392. Elsevier. https://doi.org/10.1016/B978-0-08-100596-5.00229-8.

Skendžić, S., Zovko, M., Živković, I.P. et al. (2021). The impact of climate change on agricultural insect pests. *Insects* 12 (5): 440. https://doi.org/10.3390/insects12050440.

Soviadan, M.K., Koffi-Tessio, E.M., Enete, A.A., and Nweze, N.J. (2019). Impact of climate change on cotton production: case of Savannah region, Northern Togo. *Agricultural Sciences* 10 (07): 927–947. https://doi.org/10.4236/as.2019.107071.

Tantawy, A.S., Salama, Y.A.M., El-Nemr, M.A., and Abdel-Mawgoud, A.M.R. (2015). Nano silicon application improves salinity tolerance of sweet pepper plants. *International Journal of ChemTech Research* 8 (10): 11–17.

Thind, S., Hussain, I., Rasheed, R. et al. (2021). Alleviation of cadmium stress by silicon nanoparticles during different phenological stages of Ujala wheat variety. *Arabian Journal of Geosciences* 14 (11): 1028. https://doi.org/10.1007/s12517-021-07384-w.

Thomas, C.D. (2010). Climate, climate change and range boundaries. *Diversity and Distributions* 16 (3): 488–495. https://doi.org/10.1111/j.1472-4642.2010.00642.x.

Tortella, G., Rubilar, O., Pieretti, J.C. et al. (2023). Nanoparticles as a promising strategy to mitigate biotic stress in agriculture. *Antibiotics* 12 (2): 338. https://doi.org/10.3390/antibiotics12020338.

Tripathi, D.K., Singh, V.P., Prasad, S.M. et al. (2015). Silicon nanoparticles (SiNp) alleviate chromium (VI) phytotoxicity in *Pisum sativum* (L.) seedlings. *Plant Physiology and Biochemistry* 96 (November): 189–198. https://doi.org/10.1016/j.plaphy.2015.07.026.

Tripathi, D.K., Singh, S., Singh, V.P. et al. (2016). Silicon nanoparticles more efficiently alleviate arsenate toxicity than silicon in maize cultiver and hybrid differing in arsenate tolerance. *Frontiers in Environmental Science* 4 (July): https://doi.org/10.3389/fenvs.2016.00046.

Tripathi, D.K., Singh, S., Singh, V.P. et al. (2017). Silicon nanoparticles more effectively alleviated UV-B stress than silicon in wheat (*Triticum aestivum*) seedlings. *Plant Physiology and Biochemistry* 110 (January): 70–81. https://doi.org/10.1016/j.plaphy.2016.06.026.

Trivedi, P., Batista, B.D., Bazany, K.E., and Singh, B.K. (2022). Plant–microbiome interactions under a changing world: responses, consequences and perspectives. *New Phytologist* 234 (6): 1951–1959. https://doi.org/10.1111/nph.18016.

Varah, A., Ahodo, K., Coutts, S.R. et al. (2019). The costs of human-induced evolution in an agricultural system. *Nature Sustainability* 3 (1): 63–71. https://doi.org/10.1038/s41893-019-0450-8.

Vaughan, M.M., Block, A., Christensen, S.A. et al. (2018). The effects of climate change associated abiotic stresses on maize phytochemical defenses. *Phytochemistry Reviews* 17 (1): 37–49. https://doi.org/10.1007/s11101-017-9508-2.

Wang, Y., Mostafa, S., Zeng, W., and Jin, B. (2021). Function and mechanism of jasmonic acid in plant responses to abiotic and biotic stresses. *International Journal of Molecular Sciences* 22 (16): 8568. https://doi.org/10.3390/ijms22168568.

Wang, L., Ning, C., Pan, T., and Cai, K. (2022). Role of silica nanoparticles in abiotic and biotic stress tolerance in plants: a review. *International Journal of Molecular Sciences* 23 (4): 1947. https://doi.org/10.3390/ijms23041947.

Wheeler, T. and Von Braun, J. (2013). Climate change impacts on global food security. *Science* 341 (6145): 508–513. https://doi.org/10.1126/science.1239402.

Yadav, S., Modi, P., Dave, A. et al. (2020). Effect of abiotic stress on crops. In: *Sustainable Crop Production 3* (ed. M. Hasanuzzaman, M.C.M.T. Filho, M. Fujita, and T.A.R. Nogueira), 3–23. London: IntechOpen.

Yang, Y. and Guo, Y. (2018). Elucidating the molecular mechanisms mediating plant salt-stress responses. *New Phytologist* 217 (2): 523–539. https://doi.org/10.1111/nph.14920.

Yoon, J., Choi, H., and An, G. (2015). Roles of lignin biosynthesis and regulatory genes in plant development. *Journal of Integrative Plant Biology* 57 (11): 902–912. https://doi.org/10.1111/jipb.12422.

Zahedi, S.M., Moharrami, F., Sarikhani, S., and Padervand, M. (2020). Selenium and silica nanostructure-based recovery of strawberry plants subjected to drought stress. *Scientific Reports* 10 (1): 17672. https://doi.org/10.1038/s41598-020-74273-9.

Zain, M., Haijiao Ma, M., Nuruzzaman, S.C. et al. (2023). Nanotechnology based precision agriculture for alleviating biotic and abiotic stress in plants. *Plant Stress* 10 (December): 100239. https://doi.org/10.1016/j.stress.2023.100239.

Zamora-Ledezma, C., Negrete-Bolagay, D., Figueroa, F. et al. (2021). Heavy metal water pollution: a fresh look about hazards, novel and conventional remediation methods. *Environmental Technology & Innovation* 22 (May): 101504. https://doi.org/10.1016/j.eti.2021.101504.

Zandalinas, S.I., Balfagón, D., Gómez-Cadenas, A., and Mittler, R. (2022). Plant responses to climate change: metabolic changes under combined abiotic stresses. Edited by Diane Beckles. *Journal of Experimental Botany* 73 (11): 3339–3354. https://doi.org/10.1093/jxb/erac073.

Zhao, H., Yan, W., Lan, X. et al. (2022). Comprehensive assessment of harmful heavy metals in contaminated soil in order to score pollution level. *Scientific Reports* 12 (1): 3552. https://doi.org/10.1038/s41598-022-07602-9.

Zhao, C., Zhang, H., Song, C. et al. (2020). Mechanisms of plant responses and adaptation to soil salinity. *The Innovation* 1 (1): 100017. https://doi.org/10.1016/j.xinn.2020.100017.

Section D

Applications of Si/SiO_2NPs in Medicine and Remediation

15

Si/SiO_2 NPs as a Versatile Tool for the Treatment of Cancer

Aarti Sharma[1], Disha Mittal[1], Archana Singh[2] and Indrakant K. Singh[3]

[1]*Department of Life Sciences, School of Biosciences and Technology, Galgotias University, Greater Noida, Uttar Pradesh, India*
[2]*Department of Plant Molecular Biology, University of Delhi South Campus, New Delhi, Delhi, India*
[3]*Dr. B. R. Ambedkar Centre for Biomedical Research, & Delhi School of Public Health, Institute of Eminence, University of Delhi, Delhi, India*

15.1 Introduction

Since cancer is responsible for around one in four deaths from non-communicable diseases (NCDs) worldwide (22.8%) and one in six deaths (16.8%), it is a substantial social, public health and economic matter in the twenty-first century. Three out of ten premature deaths worldwide are caused by NCDs (30.3% in those aged 30–69), and in 177 out of 183 nations, it is among the top three causes of mortality for this age group. Cancer is linked to significant social and macroeconomic expenses, which differ in degree depending on the kind of cancer, location and gender, furthermore being an important obstacle to extending life expectancy (Bray et al. 2024).

Conventional cancer therapy techniques include radiation therapy, chemotherapy, immunotherapy, targeted therapy, hormone therapy and surgery (Jovčevska and Muyldermans 2020; Park et al. 2018). Though chemotherapy and radiation therapy have cytostasis and cytotoxicity abilities (Zitvogel et al. 2008), these techniques are commonly associated with severe side effects and a significant likelihood of relapse. Neuropathy, bone marrow suppression, gastrointestinal and skin diseases, hair loss and exhaustion are some of the most prevalent adverse effects (Chan and Ismail 2014). The primary cause of the low treatment efficacy is the therapies' low targeting ratio, which can also have serious adverse effects on healthy tissues (O'Brien et al. 2022; Timin et al. 2022). Consequently, there is an urgent need for therapeutic drugs to be targeted to the tumour site specifically. As a result, an enormous amount of research has focused on nanotechnology-based

Silicon Nanoparticles: Synthesis, Uptake, and Applications, First Edition.
Edited by Archana Singh, Aryadeep Roychoudhury, and Indrakant K. Singh.

formulations for potential techniques for addressing the blockages of undirected biodistribution, adverse reactions, and high-dose administrations (Cheng et al. 2022). These materials have grown in relevance in a variety of disciplines because of their unique properties, such as sub-micron size, high surface-to-volume ratio, variation and improved targeting systems. NPs have been exposed to enter deep tissues, snowballing the increased permeability and retention (EPR) impact. Furthermore, the surface characteristics impact bioavailability and half-life by successfully bridging epithelial fenestration (Gavas et al. 2021). As nanomedicine has grown in popularity, various versatile nanoformulations with exceptional biocompatibility and pharmacokinetic characteristics, like liposomes, micelles, nanoemulsion, and NPs, have shown great promise for the delivery of novel anti-cancer drugs (Cheng et al. 2021; Dutta et al. 2020). Their use as therapeutic agents for tumour therapies is greatly facilitated by these NPs, which can efficiently address the poor water solubility and undesirable side effects frequently seen during the delivery of therapeutic agents and prolong their blood circulation time for enhanced tumour accumulation (Ge and Liu 2013; Karimi et al. 2016). NPs provide several benefits in the advancement of nanomedicine, much like a blank canvas including the ability to interact better with biological systems of the same size, improve drug bioavailability, large surface area and functionality, and have the potential to change their physical properties and develop a multifunctional system with various properties (Nichols and Bae 2013).

Presently, NPs utilized in cancer treatments include lipid NPs, polymeric NPs, dendrimers, micelles, liposomes, metallic NPs, viral NPs, protein NPs, ceramic NPs and CNTs (Jabir et al. 2012). Overall, these NPs can be classified into two categories, which are inorganic and organic. In terms of stability, inorganic NPs have several benefits over organic NPs. Mesoporous silica NPs (MSNs) is a prominent inorganic material for the fabrication of anti-cancer drug delivery agents. MSNs have received a lot of interest for cancer applications owing to its advantages, including as facile surface modification, huge surface area, tunability of pore size and shape, significant loading capacity, low toxicity and biodegradability. Over time, MSNs have recognized their own space as transporters for a numerous of anti-cancer agents, including as small molecules, macromolecules, proteins, radionuclides and genes (Dutta Gupta et al. 2024).

15.2 Si/SiO$_2$ NPs-Based Platforms for Cancer Therapy

Over the last few years, there has been a lot of interest in using MSNs to treat cancer. It offers a complete approach to detecting, preventing and treating this complex illness that impacts several biological systems. The objective of every anti-cancer therapy is to eradicate malignant cells while causing little harm to normal cells. Efforts have been undertaken to establish tumour-specific delivery and controlled release of anti-cancer medicines, despite the common issue of early drug release before reaching target locations (Rastegari et al. 2021). According to Vallet-Regí et al. (2022), MSNs' potential for clinical success in the treatment of cancer is reliant upon their (1) stability in physiological environments and bloodstream circulation time, (2) ability to pass through the various physiological barriers and reach the impacted anatomic sites, (3) ability to accumulate at the disease site, (4) safety and (5) optimal pharmacokinetic behaviour. However, the release rate of pharmaceuticals loaded in MSNP is rather quick, which may be ascribed to MSN pore size openness, severely restricting the use of MSNP as an anti-cancer drug delivery carrier. The concept of 'gatekeeper' was also introduced by researchers, and drug molecules were stopped in the MSNP channel by filling the pores with gatekeeper. The creation of an intelligent MSN medication delivery system is currently one of the most active research areas (Zhu et al. 2023).

According to Lombardo et al. (2019), MSNs have shown to be effective nanocarriers for drugs targeting applications in anti-cancer treatment. For example, drugs like camptothecin (CPT) (Lu et al. 2007b) and paxlitaxel (He et al. 2017) exhibit outstanding anti-cancer activity in vitro but have minimal effectiveness in vivo due to their restricted solubility. Bhavsar et al. (2020) developed DOX-MSN-SS-CH-FA, which aggressively targets breast cancer cells in vitro. In this approach, MSN holes were filled with the anti-cancer medication doxorubicin (DOX) and capped by folate gatekeepers and chitosan that respond to pH stimuli and redox, respectively. MSNs produce DOX when they are subjected to acidic redox conditions, which are common in the cancer environment (pH 5.5, 10 mM GSH). In addition, they demonstrated that this MSN-based design is suitable for intravenous distribution.

MSNs produce DOX when exposed to acidic-redox circumstances (pH 5.5, 10 mM GSH) like those seen in the cancer environment. Furthermore, they demonstrated that this MSN-based design is safe for intravenous delivery and has better biocompatibility (Bhavsar et al. 2020).

Additionally, some researchers have investigated the use of MSNPs in the co-delivery of drug combinations, such as gemcitabine (GEM) and Paclitxel (PTX), for the treatment of pancreatic tumours. In orthotopic animal models and xenograft, co-delivery of drugs demonstrated greater efficacy in suppressing the stromal volume and tumour size than monotherapy. Furthermore, MSNPs have been used to administer chemotherapeutic drugs such as cisplatin, irinotecan, 5-fluorouracil and temozolomide for a variety of cancer therapy applications (Barkat et al. 2021) (Tables 15.1–15.3).

Table 15.1 Applications of various silica-based NPs against cancer.

S. No.	Silica particle framework	Drug loaded	Treatment model	References
1.	HMSNPs	Glucose oxidase (GOx), Ce6, CPPO, PFC	Melanoma cell line	Yu et al. (2018)
2.	Mesoporousorgano-silica NP	Dabrafenib, trametinib, phthalocyanine	Human melanoma cell line	Tham et al. (2018)
3.	Magnetic MSNPs	DOX	Human glioblastoma cell line	Wu et al. (2018)
4.	MSNPs	DOX	Breast cancer cell line	Wei et al. (2018)
5.	Diselenide-bridge MSNPs	Cytotoxic ribonuclease A (RNasNe A)	Cervical cancer cell line	Shao et al. (2018)
6.	MSNPs	Axitinib celastrol	Cutaneous squamous cell carcinoma	Choi et al. (2016)
7.	MSNPs	PEG/polydopamine (PDA), AS-1411 aptamer enveloping	Cervical cell line	Duo et al. (2018)
8.	HMSNPs	DOX	Human glioblastoma cell line	Li et al. (2017)
9.	Gd-doped MSNPs	DOX	Mouse breast cancer cell line	Sun et al. (2018)
10.	MSNPs	sorafenib	Mouse hepatocellular carcinoma cell line	Mu et al. (2017)
11.	MSNPs	Paclitaxel	Glioblastoma	Mundžić et al. (2025)
12.	PLG/GEM@MSNPs/MTX@MSNPs	Gemcitabine and mitoxantrone	Human pancreatic cancer cell line	Wu et al. (2025)
13.	Core shell MSNPs	Cabazitaxel	Prostate cancer cell line	Mohanan et al. (2024)
14.	Functionalized MSNPs	Camptothecin	Skin cancer cell line	Qureshi et al. (2024)
15.	Hollow MSNPs	Guanidine carbonate and Curcumin	Human breast cancer cell line	Viswanathan et al. (2022)
16.	MSNPs	Albendazole	Human liver cancer cell	Ghaferi et al. (2022)

15.3 Active Targeting Tumour Tissues Using Si/SiO_2 NPs

Active targeting, an additional approach to the EPR effect that improves the effectiveness of cancer treatment, entails functionalizing the surface of NPs with ligands that have a particular affinity to the receptors that are overexpressed on the cancer cells (Figure 15.1). The first drug delivery nanocarriers that used antibody-modified liposomes for active targeting were published in 1980. Numerous ligands, such as small compounds, antibodies, antibody fragments, peptides and nucleic acids (such as aptamers), have been produced thus far for effective active targeting and have received thorough reviews elsewhere. Three main categories of site-specific active targeting strategies can be distinguished: vascular targeting, subcellular organelle targeting, and tumour cell targeting. Recent developments on active targeting ligands and modified silica-based drugs are discussed in this section.

A significant approach in active targeting is tumour cell targeting, which increases the ligand-receptor interactions that silica NPs have for cancer cells, hence increasing cellular absorption and the effectiveness of drug delivery. Early research into tumour cell targeting with MSNs revealed that folate-modified MSNs were more efficiently internalized by folate receptor-positive cells than others (Slowing et al. 2006). This method was expanded with multifunctional folate-modified NPs for drug delivery and imaging, which demonstrated improved

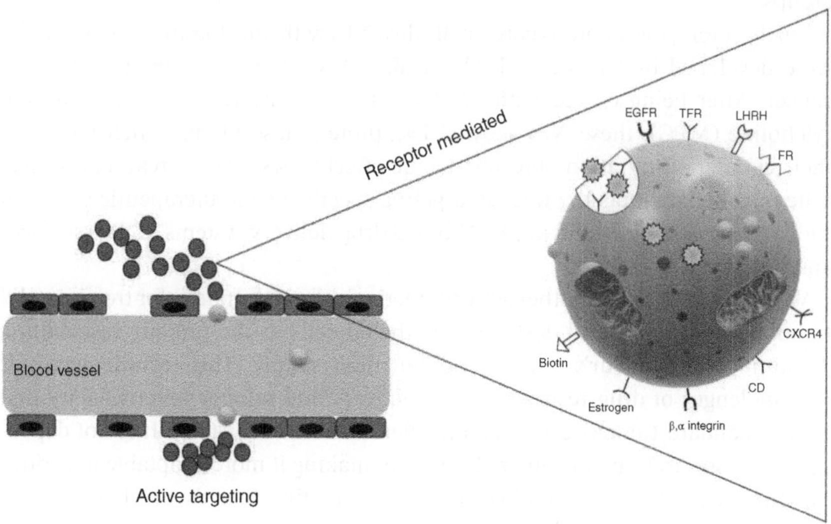

Figure 15.1 Active targeting of tumour tissues using Si/SiO_2 NPs.

cellular uptake and therapeutic efficacy (Liong et al. 2008). Other ligands, including hyaluronic acid (HA) and aptamers, have been investigated for targeting overexpressed receptors, hence enhancing cytotoxicity and cellular uptake (Luo et al. 2011; Yu et al. 2013).

Despite these advancements, the significance of ligand targeting in tumour-specific accumulation is still questioned. Ligands typically improve cellular absorption of NPs that have already reached the tumour site via extravasation. The efficacy of tumour targeting is heavily reliant on the EPR effect, and ligands primarily boost drug concentration and cytotoxicity rather than tumour accumulation. Key parameters impacting targeting efficiency include receptor density on tumour cells, ligand density on NPs, and ligand shielding owing to opsonization (Farokhzad and Langer 2009; Kwon et al. 2012). In vivo research is required to further understand these characteristics and optimize tumour-specific delivery systems.

The current research has investigated a variety of approaches to improve the targeting efficacy of MSNs in cancer therapy. Martínez-Carmona et al. (2018) used lectin-conjugated, pH-responsive MSNs to treat bone cancer. These NPs showed higher internalization in osteosarcoma cells that overexpressed sialic acids, resulting in greater cytotoxicity as compared to free medicines. To enhance active targeting capabilities, Cabañas et al. (2018) conducted another study using synthetic aminopropyl-modified MSNs. The study discovered that the presence of aminopropyl groups, when combined with folic acid, increased cellular absorption in malignant cells, indicating a cooperative impact between the functional groups.

Furthermore, mesoporous core-shell silica NPs with anti-fouling characteristics were developed by Sanchez-Salcedo et al. (2018) for the treatment of ovarian cancer. After being coated with zwitterionic 2-methacryloyloxyethyl phosphorylcholine (MPC), these NPs showed less protein adsorption, which increased their circulation duration and therapeutic effectiveness. These researches illustrate ongoing attempts to increase targeting specificity and therapeutic effects in cancer treatment by modifying MSN-based drug delivery systems' (DDS) surfaces and functionalizing them.

Vascular targeting is another active-targeting potential strategy for treating solid tumours that targets and destroys endothelial cells in the tumour vasculature, disrupting the tumour's oxygen and nutrition supply. This technique avoids the challenges of drug resistance, poor delivery and tumour heterogeneity that plague standard tumour cell targeting (Neri and Bicknell 2005). It is not dependent on extravasation or cellular absorption, making it more adaptable to various tumour forms. It is the binding affinity of targeting ligands to blood vessels, not the EPR effect, that determines the effectiveness of tumour accumulation in vascular targeting.

Though the perception of vascular targeting was first proposed in 1920s, Burrows and Thorpe did not establish its efficacy in vivo until 1990s (Burrows and Thorpe 1993). Since then, to increase the effectiveness of treatment, ligands like VEGF and RGD peptides have been combined with NPs including quantum dots, carbon nanotubes and nanographene oxide.

Chen et al. (2013a) reported the initial in vivo application of MSNs for tumour vasculature targeting. TRC105 antibodies coupled to MSNs targeted CD105 receptors in tumour vasculature, tripling tumour uptake over passive EPR-based targeting. Similarly, RGD-conjugated luminous MSNs effectively target malignancies through integrin $\alpha v\beta 3$. Recently, multifunctional MSNs with in vivo tracking for chemotherapy and phototherapy were developed (Chen et al. 2014a, 2014b).

Nevertheless advancements, the targeting efficiency is still less than 10% of the administered dosage per gram of animal weight (id/g), and there is a notable build-up in significant organs, such as the liver, kidney and lung (Chen et al. 2014a). This emphasizes the need of high-specificity ligands and NPs with 'stealth' qualities in preventing organ capture and improving therapeutic use. Overcoming these obstacles requires an amalgamation of effective targeting ligands and well-designed NPs that consider size, shape, surface charge and functionalization.

15.3.1 Nuclear Targeting

To maximize their effectiveness, nuclear-targeted DDSs are necessary to transport therapeutics – such genes or medications like DOX – directly to the cell nucleus. This method bypasses biobarriers, which decrease therapeutic efficacy after reaching the nucleus (Mizutani et al. 2005). Several nuclear-targeted delivery methods have been devised, utilizing nuclear localization signals (NLS) coupled to carriers such as silver NPs (Austin et al. 2011), quantum dots (Chen and Gerion 2004), magnetic NPs (Xu et al. 2008) and gold NPs (Tkachenko et al. 2003).

By demonstrating that TAT-conjugated MSNs (diameters of $\leq$50 nm) may pass through the nuclear membrane and carry medications straight to the nucleus, Pan et al.'s first MSN-based nuclear-targeted device from 2012 greatly improved therapeutic effects (Pan et al. 2012). This approach was also useful in reducing multi-drug resistance by allowing intra-nuclear drug release (Pan et al. 2013). Ultra-small mesoporous organosilica NPs (MONs, ~30 nm) with huge pores (~6.2 nm) were recently produced by Wu et al. for gene therapy. These MONs, when functionalized with PEI and TAT, demonstrated increased DNA payload, nuclease degradation resistance and excellent nuclear transfection efficiency (Wu et al. 2015).

Recent developments in nuclear-targeted DDSs have centred on increasing the efficacy of therapeutic medicines by assuring direct passage to the nucleus.

MSNs have dominated this study because to their adjustable surface characteristics and biocompatibility. Xu et al.'s (2023) review covers MSNs' multifunctional uses in biological contexts, highlighting the necessity of surface functionalization in improving biocompatibility and targeting capabilities. This covers nuclear targeting techniques for medicines that need direct contact with nuclear DNA. Chen et al. (2024) conducted another investigation in 2023 to investigate receptor ligand-free. The discovery suggests that MSNs can carry therapeutic drugs to particular cellular compartments, such as the nucleus, without depending on standard receptor-mediated routes. Xie et al. (2023) also discusses current breakthroughs in macromolecular medicines delivery using MSNs. The work focuses on improving cellular uptake and endosomal escape to efficiently deliver therapeutic drugs to the nucleus.

To improve EPR-based drug delivery, active targeting was first used as a supplement to passively targeted drugs. This was done to increase the targeting efficiency and improve the retention of NPs at targeted tumours (Susanne et al. 2018). Drugs that are passively targeted and rely on the EPR effect might not be enough to effectively target specific locations. A median of only roughly 0.7% of the administered dose (ID) of NPs reaches the target tumour, according to a meta-analysis of preclinical data. Only 0.1–0.2% of the ID are operative against cancer cells, according to several pre-clinical investigations, which also demonstrate anti-cancer therapy with notable patient benefits (Daniel et al. 2018; Susanne et al. 2018).

Approaches to active targeting are inherently far more complicated than those that are passive.

Physiological barriers, tumour heterogeneity and complicated engineering and design required for these DDSs are some of the difficulties related to these active targeting strategies. The latter may provide considerable challenges, making it more difficult to develop and scale up pharmaceuticals under GMP manufacture, and greatly increase the therapy's overall cost. The capacity to target locations spread across the body, such as metastatic cancers and haematological malignancies where the EPR is ineffective, is a significant benefit of active targeting despite several challenges (Subhan et al. 2021).

To attain therapeutic success, both passive and active targeting approaches must overcome their constraints. Significant advances in pre-clinical tumour models are required for active targeting to improve efficacy against solid tumours, haematological malignancies and metastases. Because passive targeting cannot accurately differentiate between healthy and sick tissues, it has several drawbacks, such as poor tumour accumulation and decreased therapeutic efficacy. Active targeting, despite boosting tumour cell accumulation, may fail to effectively deliver therapeutic drugs due to intracellular hurdles such as drug release inhibition, endosomal escape challenges and drug mechanism activation. Furthermore, conjugated NPs risk losing their stealth capabilities, especially if PEGylation is insufficient, which leads to quick clearance by the spleen, liver and other reticuloendothelial system (RES) organs, resulting in poor tumour drug accumulation. Targeted delivery encounters challenges in successfully interacting with tumour cells that overexpress receptor proteins. For both targeting systems, the development of companion diagnostic imaging technology is crucial for assessing targeted effectiveness. Personalized therapy options, such as patient selection and individualized medications, can improve tumour growth, therapeutic efficacy and results while reducing side effects, unneeded treatments and healthcare costs. Combining active and passive targeting strategies may provide additional benefits for better treatment outcomes.

15.4 Passive Targeting of Tumour Tissues Using Si/SiO$_2$ NPs

Passive targeting depends on the EPR effect (Figure 15.2). EPR effect was first described in 1986 by Hiroshi Maeda and his co-workers at Kumamoto University School of Medicine (Matsumura and Maeda 1986). The EPR effect occurs when tumours grow rapidly, causing the development of new blood vessels (angiogenesis) to give oxygen and nutrients for continued proliferation. These aberrant blood arteries feature a leaky epithelium, discontinuous micro-vasculatures and fenestrations in the range of 200–2000 nm, depending on

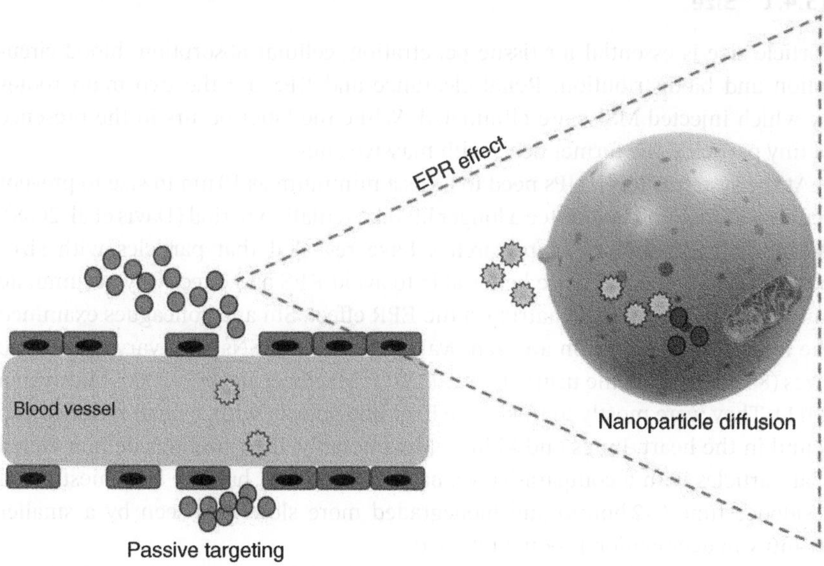

Figure 15.2 Passive targeting of tumour tissues using Si/SiO$_2$ NPs.

the kind of tumour and location (Jain and Stylianopoulos 2010). Fenestrations increase the permeability of arteries around tumours, allowing NPs in the bloodstream to aggregate in the tumour matrix. In contrast, normal tissues have fluid in their blood (Baeza et al. 2015).

The shape, size and surface characteristics of the drug particles affect a few aspects that affect EPR-based drug delivery, including targeting, circulation time and the capacity to get past obstacles. The primary foundation of passive-targeting is a diffusion process. Size is therefore an important consideration in the EPR-dependent delivery method. According to research, NPs ranging in size from 40 to 400 nm are suitable for assuring longer circulation and enhanced accumulation in tumours with impaired renal clearance (Subhan et al. 2021).

In passive targeting, shape and morphology are equally crucial. Rigid, spherical particles between 50 and 200 nm in size often have the strongest propensity for prolonged circulation, preventing absorption by the liver and spleen while yet being big enough to evade renal clearance (Hillaireau and Couvreur 2009; Rejman et al. 2004). The way that the drug particles internalize into the target cell is largely determined by their surface characteristics. PEG surface alteration of polymers can be somewhat successful in preventing opsonization and the RES's subsequent clearance of the polymers. As a result, the size, shape and occasionally surface modification of the NPs may be used to adjust the EPR-based drug delivery.

15.4.1 Size

Particle size is essential for tissue penetration, cellular absorption, blood circulation and biodistribution. Renal clearance and RES are the two main routes by which injected MSNs are eliminated. While the latter occurs in the presence of tiny particles, the former deals with massive ones.

At the systemic level, NPs need to be as a minimum as 10 nm in size to prevent renal clearance and guarantee a longer blood circulation period (Davis et al. 2008). Additionally, a number of researches have revealed that particles with sizes between 100 and 200 nm are better able to avoid RES and effectively accumulate and retain in the tumour matrix via the EPR effect. Shi and colleagues examined the in vivo biodistribution and removal of spherical MSNs with varying particle sizes (80–360 nm) in the urine (He et al. 2011; Maeda et al. 1992, 2000; Maruyama 2011). They were mostly found in the liver and spleen, with a small number also found in the heart, lungs and kidney. Most notably, their findings demonstrated that particles with a comparatively small size of 80 nm had the lengthiest mean residence time (~2 hours) and biodegraded more slowly, as seen by a smaller quantity of degradation product released.

The maximum size of NPs at the tissue level is around 400 nm; if this is exceeded, there may be inadequate diffusion across the tumour interstitium to produce a clinical or therapeutic effect (Yang and Yu 2015). Mou and colleagues investigated the size influence of silica NPs on cellular absorption in Hela cells at the cellular level. They found that the particles' sizes varied in a manner that was approximately $50 > 30 > 110 > 280 > 170$ nm. This finding is in line with the tendency of other NPs that indicate the ideal particle size is 50 nm, indicating that a relatively small particle size is preferred for effective cellular uptake (Lu et al. 2009). Because they are hard to fabricate and tend to clump, NPs smaller than 50 nm have less cellular absorption.

15.4.2 Shape

Recent studies have shown that the shape of NPs has a substantial impact on enhanced permeability and retention (EPR)-based tumour passive targeting. Freund and colleagues established a theoretical model in 2005 that suggests that spherical NPs endocytose more quickly than cylindrical NPs due to shorter 'wrapping times' at the cell membrane. Using dissipative particle dynamics simulations, Yang and Ma highlighted the significance of shape asymmetry and initial particle orientation in interactions with biological membranes.

Christian et al. (2009) found that filomicelles, which are flexible, worm-like NPs, have longer circulation durations than spherical micelles. The longer circulation increases the risk of tumour accumulation due to the EPR effect.

According to Geng et al. (2007), filamentous NPs accumulated more tumours in murine models and remained in the bloodstream for longer, indicating that longer forms could increase targeting efficiency and circulation time. Non-spherical NPs, such as disks and rods, may have distinct flow dynamics and margination characteristics in the bloodstream, which may impact their interaction with the vascular endothelium and consequent tumour growth. According to a 2010 study by Decuzzi et al., additionally, compared to their spherical counterparts, rod-shaped NPs show better rates of internalization in endothelial cells, suggesting that non-spherical forms can improve cellular uptake. Researchers have investigated on the biodistribution of rod-shaped NPs in murine tumour models. When associated to spherical NPs of the similar volume, the results showed a 30% increase in tumour formation for rod-shaped NPs, indicating that shape anisotropy is an important factor in designing NPs for passive targeting techniques.

15.4.3 Surface Characteristic

MSNs often possess a surface-to-volume ratio and a very high surface area. This is significant because NPs' surfaces are where interactions with their surroundings take place. Consequently, MSNs' cell binding, internalization, biodistribution, circulation half-life and extravasation are all significantly impacted by their surface characteristics, such as charges, functional groups and hydrophilicity/hydrophobicity (Nel et al. 2009; Zhang and Feng 2006). Because they are negatively charged under normal settings (pH ~7.4), bare silica NPs with exposed silanol groups on their surface interact extensively with the membrane of red blood cells, ultimately leading to hemolysis. Silica NPs' hemolytic activity reduces their biocompatibility and makes it more difficult to use them for passively targeting tumour tissues.

Organosilica nanoparticles (NPs) hold great potential for drug delivery applications, even though the fundamental cause of hemolysis is still not entirely understood. It has been established that adding organic components to the silica framework can significantly reduce hemolytic activity (Chen et al. 2013a; Teng et al. 2014; Urata et al. 2011).

Silica NPs functionalized with zwitterionic copolymers have shown outstanding protein-repelling capabilities and sustained stability in solutions that are relevant to biochemistry (Jia et al. 2009; Rosen and Gu 2011). According to a recent study, MSNs coated with zwitterionic copolymer demonstrated minimal cytotoxicity and markedly decreased macrophage uptake, underscoring their potential as drug delivery and diagnostic nanocarriers (Zhu et al. 2014).

Lipid coating is an additional strategy to delay blood circulation and inhibit the RES's quick clearance. When compared to unmodified NPs, this alteration can increase the half-life of silica NPs by more than ten times and decrease the amount of time that they are captured by organs relevant to RES (Van Schooneveld et al. 2008). The surface charge of NPs also influences their passive targeting performance (Yang and Yu 2015). Because cell membranes are negatively charged, cationic NPs have higher cellular internalization rates because of electrostatic interactions than neutral or negatively charged NPs (Yuan et al. 2012). However, positively charged NPs are frequently more cytotoxic and immunogenic due to platelet aggregation and enhanced opsonization, resulting in fast clearance by the RES effect (Albanese et al. 2012).

In contrast, neutral NPs have less interactions with plasma proteins, resulting in smaller hydrodynamic diameters than positively and negatively charged NPs. This leads to a prolonged circulatory lifetime and improved interstitial transport within tumours (Verma et al. 2010). Regardless of their original charge, NPs' surface charges tend to become negative as a result of plasma protein coating. The extent of this coating is determined by elements such as surface charge density, surface topology, composition and nanoparticle porosity.

The complexity of these interactions, as well as the use of different tumour models, are likely to contribute to the inconsistent findings in the research. More extensive and systematic research into how surface features influence passive targeting is required to further understand and develop nanoparticle-based therapeutics.

15.4.4 Surface Roughness

It has long been recognized that viral vectors having rough surfaces have a high delivery efficiency. The impact of surface roughness on silica NPs' cellular absorption and drug delivery effectiveness has recently been investigated. To replicate the topography of viruses, Niu et al. (2013) produced rough silica NPs by affixing tiny shell particles to larger core particles. When compared to smooth NPs, these rough ones demonstrated improved cellular absorption, prolonged biomolecule release and enhanced binding, greatly boosting transport efficiency. This idea was expanded to MSNs, which improved therapeutic efficacy and cellular absorption by 37% (Xu et al. 2014). By lowering repulsive forces like hydrophilic and electrostatic interactions and encouraging adhesion and internalization, surface roughness improves particle-membrane contact (Hoek and Agarwal 2006). But further research is necessary to fully comprehend the impact of surface roughness, as well as the effects of hydrophobicity and other variables including spike size, density and composition. To evaluate the behaviour of spiking NPs, in vivo investigations are essential (Table 15.2).

Table 15.2 Properties of NPs that affect the EPR effect.

S. No.	Property	Impact
1	Size	Smaller molecules (3.3–10 kDa; 2–3 nm) permeate deeper into tumours compared to larger molecules (Dreher et al. 2006). NPs >70 nm tend to assemble in tumours if extremely permeable (Cabral et al. 2011).
2	Charge	Surface charge affects nanoparticle clearance and residence time in the body (Salvador-Morales et al. 2009). Cationic, sterically stabilized NPs show increased tumour accumulation (Meng et al. 2011). NPs (positively and negatively charged) bind to extracellular matrix components, reducing diffusion in the interstitium (Ruggiero et al. 2010).
3	Shape	Elongated NPs (carbon nanotubes) with a high aspect ratio (100:1 to 500:1) promotes tumour accumulation (Ruggiero et al. 2010). Nanorods (44 nm) exhibit four times greater extravasation compared to shorter rods (35 nm) (Liechty and Peppas 2012).
4	Surface roughness	Nanoscale surface roughness reduces hydrophilic and electrostatic repulsion, promoting internalization and adherence (Hoek and Agarwal 2006).

15.5 Stimuli-Responsive Therapy Using MSNPs

One innovative method in nanomedicine is the use of mesoporous silica NPs (MSNPs) in stimuli-responsive treatment. MSNPs' intrinsic properties, together with large surface area, variable pore size, biocompatibility and functionalization flexibility, make them ideal carriers for targeted drug delivery (Dutta Gupta et al. 2024). Stimulus-responsive MSNPs are designed to respond to specific internal or external stimuli, resulting in regulated and exact therapeutic effects (Figure 15.3).

15.5.1 Internal Stimuli-Responsive MSNPs

Internal stimuli include pH variations, redox gradients, enzyme activity and hypoxic conditions in the tumour microenvironment (TME).

15.5.2 pH-Responsive

Studies have indicated that MSNPs can be designed to release therapeutic drugs in reaction to the tumour's acidic environment. For example, a study conducted

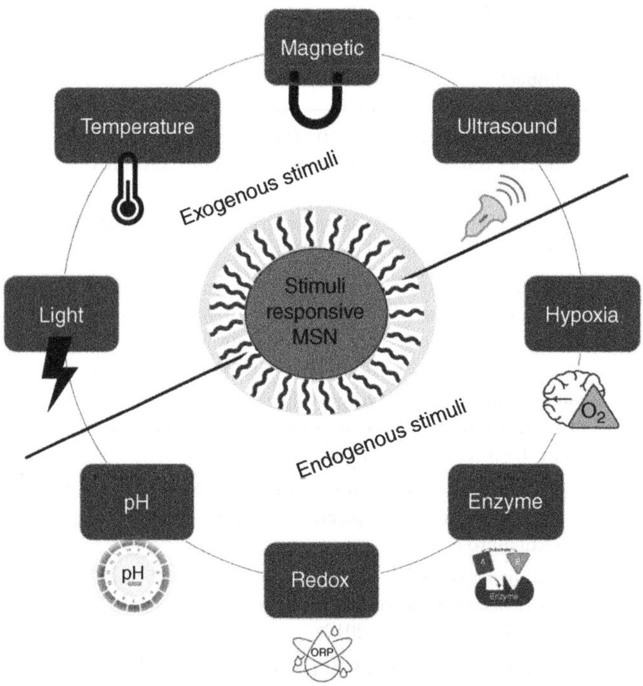

Figure 15.3 Internal and external stimuli-triggered mesoporous silica nanoparticles (MSNs) for controlled and targeted therapeutic effects.

in 2024 demonstrated the prospect of pH-sensitive silica NPs in tailored cancer therapy by developing them for colon-specific drug delivery. Givarian et al. (2024) created gelatin/perfluorohexane (PFH) nanodroplets loaded with berberine chloride that are pH-sensitive and ultrasonic-responsive. The nanodroplets demonstrated pH-responsive activity, releasing 15.4% of the drug passively in a 24-hour period. Drug release was greatly increased by ultrasonic stimulation; 150 kHz ultrasound was more successful than 1 MHz at causing acoustic droplet vaporization while producing the least amount of heat. The ideal formulation demonstrated 89.4% cumulative drug release after 10 minutes of radiation, suggesting the possibility of dual-triggered nanodroplets for regulated drug delivery in cancer treatment.

For cancer treatment, Ling et al. (2023) created a pH-responsive multifunctional nanoplatform that may produce reactive oxygen species (ROS). In acidic TMEs, the therapeutic impact was enhanced by the nanoplatform's pH-sensitive drug release and ROS production. This method highlights the potential of pH-responsive nanoplatforms in cancer treatment by opening a new channel for ROS-involved and TME-responsive therapeutic methods. In another study, Li et al. (2023) provide a thorough analysis of the development and use of pH-responsive nanoplatforms that combine several treatment modalities with magnetic resonance imaging (MRI) diagnostics. By releasing therapeutic drugs in acidic TMEs through pH-sensitive components, these nanoplatforms improve therapeutic effectiveness and imaging contrast. The work highlights how crucial pH-responsive processes are to developing multipurpose nanoplatforms for cancer treatment.

Martínez-Carmona et al. (2018) developed a nanodevice using DOX-loaded MSNs to treat bone cancer. The pH-responsive drug delivery was made possible by the nanoplatform's polyacrylic acid (PAA) capping layer, which was grafted on MSNs using an acid-cleavable acetal linker. Moreover, the NPs were coupled with the plant lectin concanavalin A (ConA) to specifically identify and internalize osteosarcoma cells that overexpress sialic acids. This focused strategy demonstrated the efficacy of pH-responsive and tailored nanoplatforms in cancer treatment by producing noticeably more cytotoxicity against tumour cells in comparison to healthy bone cells.

15.5.3 Redox-Responsive

Recent developments in pH- and redox-responsive polymer nanocomposites for site-specific drug delivery in cancer treatment have been reviewed by Gaddimath et al. (2024). The special ability of these nanocomposites to respond to two stimuli increases their accuracy and efficacy for delivering drugs in cancer treatment.

Redox-manipulating nanocarriers increase treatment effectiveness while reducing damage to healthy cells by taking advantage of tumour redox imbalances to facilitate targeted medication release. Designing tumour-specific, stimulus-responsive DDS requires the use of disulphide bonds, which are redox-sensitive connections. The production of redox-responsive nanocarriers that release drugs following glutathione (GSH)-triggered disulphide bond breaking is facilitated by the high glutathione (GSH) levels in the TME. Furthermore, these nanocarriers can deplete GSH, which raises oxidative stress and enhances the efficacy of cancer treatments based on ROS. Recent developments in redox-manipulating nanocarriers based on disulphides hold promise for theranostics and combination cancer treatments (Meng et al. 2024).

15.5.4 Enzyme-Responsive

The use of enzyme-based capping systems to MSNs is a peculiar approach. With this design, receptor-mediated cell uptake and regulated drug release are made possible by the capping system's response to enzyme activity.

This approach makes use of the pH-responsive interaction between the enzyme carbonic anhydrase (CA) and MSNs functionalized with aryl-sulphonamide. Sensitive targeting ligands like folic acid and anandamide can connect to CA site specifically by adding an artificial amino acid that contains norbornene. This design improves the specificity and effectiveness of drug administration by facilitating receptor-mediated absorption by cells and stem cells (Datz et al. 2016).

To accomplish controlled drugs release, researchers have developed nanoplatforms that use certain peptide sequences that are broken down by target enzymes. By ensuring that drugs are delivered mostly in the vicinity of tumour cells displaying overexpression of specific enzymes, this approach improves the specificity and effectiveness of therapeutic administration, especially in the treatment of cancer (Li et al. 2019a).

15.5.5 Hypoxia-Responsive

Hypoxia-activated prodrugs (HAPs) are an effective method for targeted cancer treatment since studies have shown that they may change from non-toxic to harmful molecules in hypoxic environments. It has been demonstrated that HAPs and photodynamic treatment (PDT) work better together to eliminate hypoxic tumour cells (Zhang et al. 2023).

2-Nitroimidazole has been used to create a novel hypoxia-responsive nanomicelle that alters the redox balance in cancer cells and improves drug release in hypoxic environments. This focused strategy lowers systemic toxicity while boosting the effectiveness of chemotherapy drugs and thereby highlighting the importance of targeting hypoxia as a therapeutic approach to advance oncological results (Zhu et al. 2024). NPs employing various HAPs for cancer treatment and fluorescence imaging have been the subject of research. By enabling accurate identification and focused therapy of hypoxic tumour areas, these NPs enhance therapeutic results and lessen adverse effects (Zhang et al. 2025) (Table 15.3).

15.6 External Stimuli Triggering Drug Delivery

External stimuli like light, temperature, magnetic fields and ultrasound provide precise therapeutic control.

Light-Responsive: The progresses in light-responsive MSNPs prove how particular light wavelengths may be used to deliver drugs in a regulated manner, opening stimulating opportunities for focused treatments with fewer adverse effects (Xu et al. 2023).

MSNPs that Respond to Ultraviolet Light: By adding photocleavable linkers, MSNPs have been designed to release medications when exposed to ultraviolet (UV) light. When exposed to UV radiation, these linkers experience bond cleavage, which causes the encapsulated medicinal chemicals to be released. By controlling drug distribution both spatially and temporally, this technique improves treatment effectiveness and reduces adverse effects (Linsley and Wu 2017). In 2021, Zhao et al., have synthesized biodegradable MSNPs that have been altered using the photosensitive chemical azobenzene. Azobenzene isomerizes when exposed to visible light, which causes the medication to be released. This technology shows promise for transdermal therapeutic applications by efficiently releasing medications after entering skin tissue.

MSNPs that Respond to Near-Infrared Light: Researchers have developed near-infrared (NIR) light-responsive MSNPs to increase tissue penetration and reduce harm to surrounding tissues. NIR light may be transformed into UV or visible light by combining upconversion NPs (UCNPs) with MSNPs. This activates the photosensitive molecules inside the MSNPs. The potential uses of light-responsive MSNPs in therapeutic contexts are increased by this technique, which permits regulated medication release in deeper tissues. The potential of MSNs to deliver drugs can be utilized in the development of therapeutic systems to accomplish multimodal synergistic treatment, chemotherapy, or NIR-mediated PDT. NaYF4:Yb/Er, an upconversion nanomaterial that emits both green and red light when excited by NIR light, enables dual photosensitizers to be activated at a only NIR light excitation, leading to increased $^1O_2$ production by encasing both zinc(II) phthalocyanine (ZnPc) and merocyanine 540 (MC540) photosensitizers into the MSN's surface layer (Idris et al. 2012).

Table 15.3 Internal stimuli-responsive MSNPs.

Stimuli	Mechanism	Example	Application	References
pH-Responsive	The slightly acidic pH of tumour surroundings, which results from anaerobic metabolism, makes them a good target for pH-sensitive systems because it causes the protonation or breakage of pH-sensitive bonds, allowing for regulated drug release.	Hydrazone linkers and polymers that are sensitive to pH, such as polyacrylic acid.	Tumour-targeted drug delivery.	Zhang et al. (2019); Li and Mooney (2016)
Redox-responsive	Elevated levels of reducing agents like glutathione in tumour cells break disulphide bonds.	Disulphide bonds or thiol-sensitive linkers.	Intracellular drug delivery in cancer cells.	Cheng et al. (2011)
Enzyme-responsive	Coatings or functional groups are broken down by enzyme activity.	Enzyme-degradable coatings or MMP-cleavable peptides.	Enzyme-rich tumour microenvironment allows for specific release.	Hu et al. (2014); Huo and Zhang (2016)
Hypoxia-responsive	Hypoxia-sensitive moieties are activated by low oxygen levels.	MSNPs functionalized with nitroimidazole or hypoxia-sensitive prodrugs.	Targeted administration to hypoxic tumour areas.	Brown et al. (2004); Wilson and Hay (2011)

In order to deal with the clinical issues of therapeutic drugs in thrombolytic therapy, another study developed a therapeutic drug delivery technology employing NIR-mediated UCNPs/MSNs. When excited by 808 nm NIR light, the UCNPs release UV/blue light, which may stimulate azobenzene to increase the release of urokinase (UK) and trigger the responsive release of NO, leading to a successful synergistic thrombolytic and anticoagulation treatment (Liu et al. 2022; Xu et al. 2023).

15.6.1 Thermo-Responsive

The potential of thermo-responsive mesoporous silica NPs (MSNPs) in regulated DDSs has drawn a lot attention recently. Temperature differences can cause these NPs to alter structurally or conformationally, giving precise control over drug release patterns. Several approaches have been investigated in recent research to improve MSNPs' thermo-responsive behaviour.

Thermo- and pH-responsive Polymer Coatings: Thirupathi et al. (2023) have synthesized MSNPs coated with copolymers that are sensitive to both pH and temperature, such as poly(*N*-isopropylacrylamide)-*co*-poly(acrylamide) (PNIPAm-PAAm). These dual-responsive NPs are attractive options for targeted cancer therapy because they can efficiently regulate medication release in response to physiological circumstances.

Biodegradable Thermo-Responsive Systems: To develop DDSs that respond to temperature changes, biodegradable materials with thermo-responsive qualities have been included into MSNPs. This study uses a twofold emulsion solvent evaporation technique via Eudragit RS100 (PNIPAM@RS100) and poly(*N*-isopropylacrylamide) to capture and produce porous biodegradable micro/NPs coupled with thermo-responsive gatekeepers. It was examined how different types of initiators affected the polymerization of NIPAM monomer/methylene-bis-acrylamide (MBA) crosslinker at ambient temperature for redox initiators and 60 °C for thermal initiators. The crosslinked PNIPAM acts as thermal gatekeepers, guaranteeing maximum loading efficiency and precise release of a model active chemical, Nile Blue A (Thananukul et al. 2022).

Furthermore, it has been investigated to combine MSNPs with thermo-responsive hydrogels to enable drug delivery through a variety of channels, including nasal delivery. By controlling the release of medications contained within MSNPs, this method makes use of the temperature-sensitive characteristics of hydrogels, which may enhance therapeutic results (Ribeiro et al. 2022).

15.6.2 Magnetic-Responsive

A recent research established a magnetic MSN platform for combination magnetothermal treatment and chemotherapy (Zhao et al. 2022).

An MSN-based nano system was shown to have a synergetic therapeutic effect of PTT and MTT in a study published by Yao et al. (2017). They coated magnetic Fe_3O_4 with MSN shells and used graphene quantum dots with photothermal conversion capabilities as the cap to regulate the release of DOX. Following application of external light and magnetic field stimulation and arrival in the tumour's acidic microenvironment, DOX was released under control, resulting in rapid tumour hyperthermia. In another study, Santhamoorthy, et al. 2023 developed dopamine-modified MSNs with magnetic iron oxide (FeNP) formed on their surface. They have superparamagnetic characteristics because of the magnetic iron oxide. Their drug loading capacity is excellent (~62%), and their drug release is sensitive to temperature and pH. Their safety is confirmed by biocompatibility testing (MTT assay), which makes them attractive options for regulated chemotherapeutic drug administration in targeted cancer treatment.

The mesoporous silica shells and Fe_3O_4 cores of the MSNs have been developed to accomplish magnetothermal heating in the presence of an alternating magnetic field and transport DOX. While the addition of disulfide-linked chitosan enabled the pH- and redox-responsive reaction of DOX released from the nanoparticle, the combination of folic acid ensured target-specific drug delivery to cancer cells that promote the folate receptor. In conjunction with AMF, the scientists found that the MSNs' redox/pH/AMF-responsive release of DOX demonstrated better cytotoxicity towards HeLa cells, indicating the successful synergy between chemotherapy and magnetothermal treatment. With implications for the future advancement of precision oncology, this study emphasizes the possibility of multimodal synergistic therapy (Dutta Gupta et al. 2024).

15.6.3 Ultrasound-Responsive

Using lipid-supported MSNPs, Amin et al. (2021) synthesized a stimuli-responsive DDS for ultrasound-triggered release. DOX and perfluoropentane were entrapped in these NPs' porous structure. The lipid coating served as a gatekeeper to stop the medication from releasing too soon and improved cellular absorption. Perfluoropentane changed from a liquid to a gas when exposed to ultrasonic radiation, which caused the lipid layer to rupture and caused the drugs to be released. This technology is a potential option for sophisticated chemotherapeutic applications since it showed stability, non-toxicity and enhanced cytotoxic effects upon ultrasonic activation.

Recently, F-MSN-DOX@RBC, biomimetic, superhydrophobic, ultrasound-responsive MSNs loaded with DOX and modified with red blood cell membranes for RM-1 tumour sonodynamic treatment. They found that the F-MSN-DOX group, these NPs, which had an average size of 232 ± 78.8 nm and a zeta potential of −35.57 ± 0.74 mV, demonstrated much greater tumour accumulation and decreased spleen uptake. The combination of ultrasound therapy with F-MSN-DOX@RBC led to improved tumour inhibition (71.5 ± 9.54%). DHE and CD31 staining were used to confirm ROS generation and tumour vascular damage. Further suggested the use of red blood cell membrane-modified NPs improves tumour treatment through sonodynamic therapy, anti-vascular therapy and chemotherapy by offering an efficient method for ultrasound-responsive drug release (Jin et al. 2023).

External stimuli-responsive MSNPs.

Stimuli	Mechanism	Example	Application	References
Light-responsive	When exposed to light, photo-labile bonds break down. Azobenzene undergo structural alterations upon irradiation.	UV or NIR-responsive molecules like azobenzene.	Drug release is precise in terms of both space and time.	Jiang et al. (2020); Agasti et al. (2011)
Thermo-responsive	Drugs are released by temperature-sensitive polymers at high temperatures.	Polymers such as poly (N-isopropyl-acrylamide) (PNIPAM).	Hyperthermia-induced drug delivery.	Zhang et al. (2016) and Chouhan et al. (2018)
Magnetic-responsive	Magnetic NPs yield heat when subjected to discontinuous magnetic fields.	Magnetic iron oxide NPs integrated into MSNPs.	Magnetic hyperthermia and selective administration of drugs.	Laurent et al. (2008); Lu and Salabas (2007)
Ultrasound-responsive	Ultrasound stimulates drug release or breaks coatings.	Coatings that are sensitive to ultrasonic waves or gas-filled nanobubbles	Combined therapy and imaging (theranostics).	Qian et al. (2019); Wang and Gao (2014)

15.7 Conclusion and Future Prospective

For the previous 10 years, the use of MSNPs has grown significantly, which has resulted in the development of several therapeutically active treatment options. A number of studies have been conducted to identify the material features of MSNPs and their usefulness for delivering drugs against cancer.

These mechanisms can prevent drugs from breaking down too quickly or from interacting with healthy tissues, which will increase their bioavailability and therapeutic efficacy. The benefits of MSNPs include improved solubility due to a quicker rate of drug dispersion, as well as superior systemic availability and stability in an in vivo setting.

With the ongoing development in research and clinical applications of silica nano-DDSs, the 'gated' stimuli-responsive MSN are making promising success in treating several existing illnesses and deserve the benefit of continued development. Even with the encouraging benefits of MSNPs as multimodal cancer therapy modalities, real-world success depends on significant clinical application to help patients.

Glossary

DDS	drug delivery system
MSNP	mesoporous silica nanoparticle
DOX	doxorubicin
CPT	camptothecin
TME	tumor microenvironment
NP	nanoparticle

References

Agasti, S.S., Chompoosor, A., and Rotello, V.M. (2011). Light-responsive nanocarriers in drug delivery. *Advanced Drug Delivery Reviews* 64 (11): 1037–1050. https://doi.org/10.1016/j.addr.2011.06.007.

Albanese, A., Tang, P.S., and Chan, W.C. (2012). The effect of nanoparticle size, shape, and surface chemistry on biological systems. *Annual Review of Biomedical Engineering* 14 (1): 1–16.

Amin, M.U., Ali, S., Tariq, I. et al. (2021). Ultrasound-responsive smart drug delivery system of lipid coated mesoporous silica NPs. *Pharmaceutics* 13 (9): 1396.

Austin, L.A., Kang, B., Yen, C.W., and El-Sayed, M.A. (2011). Nuclear targeted silver nanospheres perturb the cancer cell cycle differently than those of nanogold. *Bioconjugate Chemistry* 22 (11): 2324–2331.

Baeza, A., Colilla, M., and Vallet-Regi, M. (2015). Advances in mesoporous silica NPs for targeted stimuli-responsive drug delivery. *Expert Opinion on Drug Delivery* 12 (2): 319–337.

Barkat, A., Beg, S., Panda, S.K. et al. (2021). Functionalized mesoporous silica NPs in anticancer therapeutics. *Seminars in Cancer Biology* 69: 365–375.

Bhavsar, D.B., Patel, V., and Sawant, K.K. (2020). Design and characterization of dual responsive mesoporous silica NPs for breast cancer targeted therapy. *European Journal of Pharmaceutical Sciences* 152: 105428.

Bray, F., Laversanne, M., Sung, H. et al. (2024). Global cancer statistics 2022: GLOBOCAN estimates of incidence and mortality worldwide for 36 cancers in 185 countries. *CA: A Cancer Journal for Clinicians* 74 (3): 229–263.

Brown, J.M. and Wilson, W.R. (2004). Exploiting tumour hypoxia in cancer treatment. *Nature Reviews Cancer* 4 (6): 437–447.

Burrows, F.J. and Thorpe, P.E. (1993). Eradication of large solid tumours in mice with an immunotoxin directed against tumour vasculature. *Proceedings of the National Academy of Sciences of the United States of America* 90 (19): 8996–9000.

Cabañas, M.V., Lozano, D., Torres-Pardo, A. et al. (2018). Features of aminopropyl modified mesoporous silica NPs. Implications on the active targeting capability. *Materials Chemistry and Physics* 220: 260–269.

Cabral, H., Matsumoto, Y., Mizuno, K. et al. (2011). Accumulation of sub-100 nm polymeric micelles in poorly permeable tumours depends on size. *Nature Nanotechnology* 6: 815–823. https://doi.org/10.1038/nnano.2011.166.

Chan, H.-K. and Ismail, S. (2014). Side effects of chemotherapy among cancer patients in a Malaysian general hospital: experiences, perceptions and informational needs from clinical pharmacists. *Asian Pacific Journal of Cancer Prevention* 15 (13): 5305–5309. *Chemistry: An Asian Journal* 2008, 3 (3): 548–552.

Chen, F. and Gerion, D. (2004). Fluorescent CdSe/ZnS nanocrystal–peptide conjugates for long-term, nontoxic imaging and nuclear targeting in living cells. *Nano Letters* 4 (10): 1827–1832.

Chen, H., Zhen, Z., Tang, W. et al. (2013a). Label-free luminescent mesoporous silica nanoparticles for imaging and drug delivery. *Theranostics* 3 (9): 650–657.

Chen, Y., Xu, P.F., Chen, H.R. et al. (2013b). Colloidal HPMO NPs: silica-etching chemistry tailoring, topological transformation, and nano-biomedical applications. *Advanced Materials* 25 (22): 3100–3105.

Chen, F., Hong, H., Shi, S. et al. (2014a). Engineering of hollow mesoporous silica nanoparticles for remarkably enhanced tumor active targeting efficacy. *Scientific Reports* 4 (1): 1–10.

Chen, F., Nayak, T.R., Goel, S. et al. (2014b). In vivo tumor vasculature targeted PET/NIRF imaging with TRC105 (Fab)-conjugated, dual-labeled mesoporous silica nanoparticles. *Molecular Pharmaceutics* 11 (11): 4007–4014.

Chen, Z.A., Wu, C.H., Wu, S.H. et al. (2024). Receptor ligand-free mesoporous silica NPs: a streamlined strategy for targeted drug delivery across the blood–brain barrier. *ACS Nano* 18 (20): 12716–12736.

Cheng, R., Feng, F., Meng, F. et al. (2011). Glutathione-responsive nano-vehicles as a promising platform for targeted intracellular drug and gene delivery. *Journal of Controlled Release* 152 (1): 2–12.

Cheng, Z., Li, M., Dey, R., and Chen, Y. (2021). Nanomaterials for cancer therapy: current progress and perspectives. *Journal of Hematology & Oncology* 14 (1): 85.

Cheng, W., Chuang, K.H., Lo, Y.J. et al. (2022). Bispecific T-cell engagers non-covalently decorated drug-loaded PEGylated nanocarriers for cancer immunochemotherapy. *Journal of Controlled Release* 344: 235–248.

Choi, J.Y., Ramasamy, T., Kim, S.Y. et al. (2016). PEGylated lipid bilayer-supported mesoporous silica nanoparticle composite for synergistic co-delivery of axitinib and celastrol in multi-targeted cancer therapy. *Acta Biomaterialia* 39: 94–105.

Chouhan, R.K., Bajpai, A.K., and Bajpai, J. (2018). Thermo-responsive drug carriers in targeted cancer therapy. *Materials Science and Engineering: C* 93: 1191–1200. https://doi.org/10.1016/j.msec.2018.08.057.

Christian, D.A., Cai, S., Garbuzenko, O.B. et al. (2009). Flexible filaments for in vivo imaging and delivery: persistent circulation of filomicelles opens the dosage window for sustained tumour shrinkage. *Molecular Pharmaceutics* 6 (5): 1343–1352.

Daniel, R., Nitin, J., Wei, T. et al. (2018). Progress and challenges towards targeted delivery of cancer therapeutics. *Nature Communications* 9: 1410.

Datz, S., Argyo, C., Gattner, M. et al. (2016). Genetically designed biomolecular capping system for mesoporous silica NPs enables receptor-mediated cell uptake and controlled drug release. *Nanoscale* 8 (15): 8101–8110.

Davis, M.E., Chen, Z., and Shin, D.M. (2008). Nanoparticle therapeutics: an emerging treatment modality for cancer. *Nature Reviews. Drug Discovery* 7 (9): 771–782.

Decuzzi, P., Godin, B., Tanaka, T. et al. (2010). Size and shape effects in the biodistribution of intravascularly injected particles. *Journal of Controlled Release* 141 (3): 320–327.

Dreher, M.R., Liu, W., Michelich, C.R. et al. (2006). Tumour vascular permeability, accumulation, and penetration of macromolecular drug carriers. *Journal of the National Cancer Institute* 98: 335–344. https://doi.org/10.1093/jnci/djj070.

Duo, Y., Yang, M., Du, Z. et al. (2018). CX-5461-loaded nucleolus- targeting nanoplatform for cancer therapy through induction of pro-death autophagy. *Acta Biomaterialia* 79: 317–330.

Dutta Gupta, Y., Mackeyev, Y., Krishnan, S., and Bhandary, S. (2024). Mesoporous silica nanotechnology: promising advances in augmenting cancer theranostics. *Cancer Nanotechnology* 15 (1): 9.

Dutta, B., Nema, A., Shetake, N.G. et al. (2020). Glutamic acid-coated FeO NP for tumour targeted imaging and therapeutics. *Materials Science & Engineering. C, Materials for Biological Applications* 112: 110915.

Farokhzad, O.C. and Langer, R. (2009). Impact of nanotechnology on drug delivery. *ACS Nano* 3 (1): 16–20.

Gaddimath, S., Payamalle, S., Channabasavana Hundi Puttaningaiah, K.P., and Hur, J. (2024). Recent advances in pH and redox responsive polymer nanocomposites for cancer therapy. *Journal of Composites Science* 8 (1): 28. https://doi.org/10.3390/jcs8010028.

Gavas, S., Quazi, S., and Karpiński, T.M. (2021). NP for cancer therapy: current progress and challenges. *Nanoscale Research Letters* 16 (1): 173.

Ge, Z. and Liu, S. (2013). Functional block copolymer assemblies responsive to tumour and intracellular microenvironments for site-specific drug delivery and enhanced imaging performance. *Chemical Society Reviews* 42 (17): 7289–7325.

Geng, Y., Dalhaimer, P., Cai, S.S. et al. (2007). Shape effects of filaments versus spherical particles in flow and drug delivery. *Nature Nanotechnology* 2: 249–255.

Ghaferi, M., Zahra, W., Akbarzadeh, A. et al. (2022). Enhancing the efficacy of albendazole for liver cancer treatment using mesoporous silica NPs: an in vitro study. *EXCLI Journal* 21: 236.

Givarian, M., Moztarzadeh, F., Ghaffari, M. et al. (2024). Dual-trigger release of berberine chloride from the gelatin/perfluorohexane core–shell structure. *Bulletin of the National Research Centre* 48 (1): 65.

He, Q.J., Zhang, Z.W., Gao, F. et al. (2011). In vivo biodistribution and urinary excretion of mesoporous silica NPs: effects of particle size and PEGylation. *Small* 7 (2): 271–280.

He, Y., Liang, S., Long, M., and Xu, H. (2017). Mesoporous silica NPs as potential carriers for enhanced drug solubility of paclitaxel. *Materials Science & Engineering. C, Materials for Biological Applications* 78: 12–17.

Hillaireau, H. and Couvreur, P. (2009). Nanocarriers' entry into the cell: relevance to drug delivery. *Cellular and Molecular Life Sciences* 66: 2873–2896. https://doi.org/10.1007/s00018-009-0053-z.

Hoek, E.M. and Agarwal, G.K. (2006). Extended DLVO interactions between spherical particles and rough surfaces. *Journal of Colloid and Interface Science* 298 (1): 50–58.

Hu, X., Hu, J., Tian, J. et al. (2014). Matrix metalloproteinase-sensitive polymeric micelles for tumour-triggered drug release. *Biomaterials* 35 (25): 7118–7127.

Huo, M. and Zhang, W. (2016). Enzyme-responsive nanocarriers in cancer therapy. *Journal of Controlled Release* 234: 107–118.

Jabir, N.R., Tabrez, S., Ashraf, G.M. et al. (2012). Nanotechnology-based approaches in anticancer research. *International Journal of Nanomedicine* 4391–4408.

Jain, R.K. and Stylianopoulos, T. (2010). Delivering nanomedicine to solid tumours. *Nature Reviews. Clinical Oncology* 7 (11): 653–664.

Jia, G.W., Cao, Z.Q., Xue, H. et al. (2009). Novel zwitterionic-polymer-coated silica NPs. *Langmuir* 25 (5): 3196–3199.

Jiang, Z., Shao, L., and Wang, X. (2020). Light-triggered drug release from mesoporous silica NPs. *Chemical Society Reviews* 49 (16): 6850–6866. https://doi.org/10.1039/D0CS00051A.

Jin, Q., Chen, D., Song, Y. et al. (2023). Ultrasound-responsive biomimetic superhydrophobic drug-loaded mesoporous silica NPs for treating prostate tumour. *Pharmaceutics* 15 (4): 1155.

Jovčevska, I. and Muyldermans, S. (2020). The therapeutic potential of nanobodies. *BioDrugs: Clinical Immunotherapeutics, Biopharmaceticals and Gene Therapy* 34 (1): 11–26. https://doi.org/10.1007/s40259-019-00392-z.

Karimi, M., Ghasemi, A., Sahandi Zangabad, P. et al. (2016). Smart micro/NP in stimulus-responsive drug/gene delivery systems. *Chemical Society Reviews* 45 (5): 1457–1501.

Laurent, S., Dutz, S., Häfeli, U.O., and Mahmoudi, M. (2008). Magnetic hyperthermia: advances in design and applications of magnetic NPs. *Nano Today* 2 (3): 22–38. https://doi.org/10.1016/S1748-0132(08)70086-9.

Li, J. and Mooney, D.J. (2016). Designing hydrogels for controlled drug delivery. *Nature Reviews Materials* 1 (12): 1–17.

Li, X., Xing, L., Hu, Y. et al. (2017). An RGD-modified hollow silica@Au core/shell nanoplatform for tumour combination therapy. *Acta Biomaterialia* 62: 273–283.

Li, Y., Du, L., Wu, C. et al. (2019a). Peptide sequence-dominated enzyme-responsive nanoplatform for anticancer drug delivery. *Current Topics in Medicinal Chemistry* 19 (1): 74–97.

Li, Z., Zhang, Y., and Huang, Y. (2019b). pH-Sensitive nanocarriers for drug delivery in cancer therapy. *Biomaterials* 232: 119680.

Li, X., Yue, R., Guan, G. et al. (2023). Recent development of pH-responsive theranostic nanoplatforms for magnetic resonance imaging-guided cancer therapy. *Exploration (Beijing)* 3 (3): 20220002. https://doi.org/10.1002/EXP.20220002. PMID: 37933379; PMCID: PMC10624388.

Liechty, W.B. and Peppas, N.A. (2012). Expert opinion: responsive polymer NPs in cancer therapy. *European Journal of Pharmaceutics and Biopharmaceutics* 80: 241–246. https://doi.org/10.1016/j.ejpb.2011.08.004.

Ling, P., Yang, P., Zhang, Q. et al. (2023). pH-Responsive multifunctional nanoplatforms with reactive oxygen species-controlled release of CO for enhanced oncotherapy. *ACS Applied Bio Materials* 6 (12): 5708–5715.

Linsley, C.S. and Wu, B.M. (2017 Feb). Recent advances in light-responsive on-demand drug delivery systems. *Therapeutic Delivery* 8 (2): 89–107. https://doi .org/10.4155/tde-2016-0060. PMID: 28088880; PMCID: PMC5561969.

Liong, M., Lu, J., Kovochich, M. et al. (2008). Multifunctional inorganic NPs for imaging, targeting, and drug delivery. *Acs Nano* 2: 889–896.

Kwon, I.K., Lee, S.C., Han, B., and Park, K. (2012). Analysis on the current status of targeted drug delivery to tumours. *Journal of Controlled Release* 164 (2): 108–114.

Lombardo, D., Kiselev, M.A., and Caccamo, M.T. (2019). Smart NPs for drug delivery application: development of versatile nanocarrier platforms in biotechnology and nanomedicine. *Journal of Nanomaterials* 2019: 1–26.

Lu, A.H., Salabas, E.L., and Schüth, F. (2007a). Magnetic NPs as heat triggers. *Angewandte Chemie International Edition* 46 (8): 1222–1241. https://doi.org/10.1002/anie.200602866.

Lu, J., Liong, M., Zink, J.I., and Tamanoi, F. (2007b). Mesoporous silica NPs as a delivery system for hydrophobic anticancer drugs. *Small* 3: 1341–1346.

Lu, F., Wu, S.H., Hung, Y., and Mou, C.Y. (2009). Size effect on cell uptake in well-suspended, uniform mesoporous silica nanoparticles. *Small* 5 (12): 1408–1413.

Luo, Z., Cai, K., Hu, Y. et al. (2011). Mesoporous silica NPs end-capped with collagen: redox-responsive nanoreservoirs for targeted drug delivery. *Angewandte Chemie International Edition* 3 (50): 640–643.

Maeda, H., Seymour, L.W., and Miyamoto, Y. (1992). Conjugates of anticancer agents and polymers – advantages of macromolecular therapeutics in vivo. *Bioconjugate Chemistry* 3 (5): 351–362.

Maeda, H., Wu, J., Sawa, T. et al. (2000). Tumour vascular permeability and the EPR effect in macromolecular therapeutics: a review. *Journal of Controlled Release* 65 (1-2): 271–284.

Martínez-Carmona, M., Lozano, D., Colilla, M., and Vallet-Regí, M. (2018). Lectin-conjugated pH-responsive mesoporous silica NPs for targeted bone cancer treatment. *Acta Biomaterialia* 65: 393–404.

Maruyama, K. (2011). Intracellular targeting delivery of liposomal drugs to solid tumours based on EPR effects. *Advanced Drug Delivery Reviews* 63 (3): 161–169.

Matsumura, Y. and Maeda, H. (1986). A new concept for macromolecular therapeutics in cancer chemotherapy: mechanism of tumoritropic accumulation of proteins and the antitumor agent smancs. *Cancer Research* 46 (Pt 1): 6387–6392.

Meng, H., Xue, M., Xia, T. et al. (2011). Use of size and a copolymer design feature to improve the biodistribution and the enhanced permeability and retention effect of doxorubicin-loaded mesoporous silica NPs in a murine xenograft tumour model. *ACS Nano* 5: 4131–4144. https://doi.org/10.1021/nn200809t.

Meng, X., Shen, Y., Zhao, H. et al. (2024). Redox-manipulating nanocarriers for anticancer drug delivery: a systematic review. *Journal of Nanbiotechnology* 22: 587. https://doi.org/10.1186/s12951-024-02859-w.

Mizutani, H., Tada-Oikawa, S., Hiraku, Y. et al. (2005). Mechanism of apoptosis induced by doxorubicin through the generation of hydrogen peroxide. *Life Sciences* 76 (13): 1439–1453.

Mohanan, S., Sathish, C.I., Ramadass, K. et al. (2024). Design and synthesis of cabazitaxel loaded core-shell mesoporous silica NPs with different morphologies for prostate cancer therapy. *Small* 20 (39): 2303269.

Mu, S., Liu, Y., Wang, T. et al. (2017). Unsaturated nitrogen-rich polymer poly(L-histidine) gated reversibly switchable mesoporous silica NPs using "graft to" strategy for drug controlled release. *Acta Biomaterialia* 63: 150–162.

Mundžić, M., Ultimo, A., Mladenović, M. et al. (2025). Chlorotoxin-functionalized mesoporous silica NPs for pH-responsive paclitaxel delivery to Glioblastoma multiforme. *Heliyon* 11 (1): e41151.

Nel, A.E., Madler, L., Velegol, D. et al. (2009). Understanding biophysicochemical interactions at the nanobio interface. *Nature Materials* 8 (7): 543–557.

Neri, D. and Bicknell, R. (2005). Tumour vascular targeting. *Nature Reviews. Cancer* 5 (6): 436–446.

Nichols, J.W. and Bae, Y.H. (2013). Nanotechnology for cancer treatment: possibilities and limitations. In: *Cancer Targeted Drug Delivery: An Elusive Dream* (ed. Y.H. Bae, R.J. Mrsny, and K. Park), 37–56. New York, NY: Springer.

Niu, Y.T., Yu, M.H., Hartono, S.B. et al. (2013). NPs mimicking viral surface topography for enhanced cellular delivery. *Advanced Materials* 25 (43): 6233–6237.

O'Brien, N.A., Huang, H., McDermott, M.S.J. et al. (2022). Tucatinib has selective activity in HER2-positive cancers and significant combined activity with approved and novel breast cancer-targeted therapies. *Molecular Cancer Therapeutics* 21 (5): 751–761.

Pan, L.M., He, Q.J., Liu, J.N. et al. (2012). Nuclear-targeted drug delivery of TAT peptide-conjugated monodisperse mesoporous silica NPs. *Journal of the American Chemical Society* 134 (13): 5722–5725.

Pan, L.M., Liu, J.A., He, Q.J. et al. (2013). Overcoming multidrug resistance of cancer cells by direct intranuclear drug delivery using TAT-conjugated mesoporous silica nanoparticles. *Biomaterials* 34: 2719–2730.

Park, W., Heo, Y.J., and Han, D.K. (2018). New opportunities for NP in cancer immunotherapy. *Biomaterials Research* 22: 24.

Qian, X., Zheng, Y., Chen, Y., and Zhang, X. (2019). Ultrasound-mediated drug delivery systems: advances and future trends. *Frontiers in Pharmacology* 10: 1663. https://doi.org/10.3389/fphar.2019.01663.

Qureshi, M., Viegas, C., Duarte, S.O. et al. (2024). Camptothecin-loaded mesoporous silica NPs functionalized with CpG oligodeoxynucleotide as a new approach for skin cancer treatment. *International Journal of Pharmaceutics* 660: 124340.

Rastegari, E., Hsiao, Y.J., Lai, W.Y. et al. (2021). An update on mesoporous silica nanoparticle applications in nanomedicine. *Pharmaceutics* 13 (7): 1067.

Rejman, J., Oberle, V., Zuhorn, I., and Hoekstra, D. (2004). Size-dependent internalization of particles via the pathways of clathrin- and caveolae-mediated endocytosis. *The Biochemical Journal* 377: 159–169. https://doi.org/10.1042/bj20031253.

Ribeiro, T.D.C., Sábio, R.M., Luiz, M.T. et al. (2022). Curcumin-loaded mesoporous silica NPs dispersed in thermo-responsive hydrogel as potential Alzheimer disease therapy. *Pharmaceutics* 14 (9): 1976.

Rosen, J.E. and Gu, F.X. (2011). Surface functionalization of silica nanoparticles with cysteine: a low-fouling zwitterionic surface. *Langmuir* 27 (17): 10507–10513.

Ruggiero, A., Villa, C.H., Bander, E. et al. (2010). Paradoxical glomerular filtration of carbon nanotubes. *Proceedings of the National Academy of Sciences of the United States of America* 107: 12369–12374. https://doi.org/10.1073/pnas.0913667107.

Salvador-Morales, C., Zhang, L., Langer, R., and Farokhzad, O.C. (2009). Immunocompatibility properties of lipid-polymer hybrid NPs with heterogeneous sur-face functional groups. *Biomaterials* 30: 2231–2240. https://doi.org/10.1016/j.biomaterials.2009.01.005.

Sanchez-Salcedo, S., Vallet-Regí, M., Shahin, S.A. et al. (2018). Mesoporous core-shell silica NPs with anti-fouling properties for ovarian cancer therapy. *Chemical Engineering Journal* 340: 114–124.

Santhamoorthy, M., Thirupathi, K., Krishnan, S. et al. (2023). Preparation of magnetic iron oxide incorporated mesoporous silica hybrid composites for pH and temperature-sensitive drug delivery. *Magnetochemistry* 9 (3): 81.

Shao, D., Li, M., Wang, Z. et al. (2018). Bioinspired diselenide-bridged mesoporous silica NPs for dual-responsive protein delivery. *Advanced Materials* 30 (29): 1801198.

Slowing, I., Trewyn, B.G., and Lin, V.S.Y. (2006). Effect of surface functionalization of MCM-41-type mesoporous silica nanoparticles on the endocytosis by human cancer cells. *Journal of the American Chemical Society* 128 (46): 14792–14793.

Subhan, M.A., Yalamarty, S.S.K., Filipczak, N. et al. (2021 Jun 18). Recent advances in tumour targeting via EPR effect for cancer treatment. *Journal of Personalized Medicine* 11 (6): 571. https://doi.org/10.3390/jpm11060571. PMID: 34207137; PMCID: PMC8234032.

Sun, Q., You, Q., Wang, J. et al. (2018). Theranosticnanoplatform: triple-modal imaging-guided synergistic cancer therapy based on liposome-conjugated mesoporous silica NPs. *ACS Applied Materials & Interfaces* 10 (2): 1963–1975.

Susanne, K.G., Jan-Niklas, M., and Benjamin, T. (2018). Tumor targeting vis EPR: strategies to enhance patient responses. *Advanced Drug Delivery Reviews* 130 (17–38): 21.

Teng, Z.G., Wang, S.J., Su, X.D. et al. (2014). Facile synthesis of yolk-shell structured inorganic-organic hybrid spheres with ordered radial mesochannels. *Advanced Materials* 26 (22): 3741–3747.

Tham, H.P., Xu, K., Lim, W.Q. et al. (2018). Microneedle-assisted topical delivery of photodynamically active mesoporous formulation for combination therapy of deep-seated melanoma. *ACS Nano* 12 (12): 11936–11948.

Thananukul, K., Kaewsaneha, C., Opaprakasit, P. et al. (2022). Biodegradable porous micro/NPs with thermoresponsive gatekeepers for effective loading and precise delivery of active compounds at the body temperature. *Scientific Reports* 12: 10906. https://doi.org/10.1038/s41598-022-15069-x.

Thirupathi, K., Santhamoorthy, M., Radhakrishnan, S. et al. (2023 Feb 28). Thermosensitive polymer-modified mesoporous silica for pH and temperature-responsive drug delivery. *Pharmaceutics* 15 (3): 795. https://doi.org/10.3390/pharmaceutics15030795. PMID: 36986656; PMCID: PMC10051764.

Timin, A.S., Postovalova, A.S., Karpov, T.E. et al. (2022). Calcium carbonate carriers for combined chemo- and radionuclide therapy of metastatic lung cancer. *Journal of Controlled Release* 344: 1–11.

Tkachenko, A.G., Xie, H., Coleman, D. et al. (2003). Multifunctional gold nanoparticle–peptide complexes for nuclear targeting. *Journal of the American Chemical Society* 125 (16): 4700–4701.

Urata, C., Yamada, H., Wakabayashi, R. et al. (2011). Aqueous colloidal mesoporous NPs with ethenylene-bridged silsesquioxane frameworks. *Journal of the American Chemical Society* 133 (21): 8102–8105.

Vallet-Regí, M., Schüth, F., Lozano, D. et al. (2022). Engineering mesoporous silica NPs for drug delivery: where are we after two decades? *Chemical Society Reviews* 51 (13): 5365–5451.

Van Schooneveld, M.M., Vucic, E., Koole, R. et al. (2008). Improved biocompatibility and pharmacokinetics of silica nanoparticles by means of a lipid coating: a multimodality investigation. *Nano Letters* 8 (8): 2517–2525.

Verma, A. and Stellacci, F. (2010). Effect of surface properties on nanoparticle–cell interactions. *Small* 6 (1): 12–21.

Viswanathan, T.M., Chitradevi, K., Zochedh, A. et al. (2022). Guanidine–curcumin complex-loaded amine-functionalised hollow mesoporous silica NPs for breast cancer therapy. *Cancers* 14 (14): 3490.

Wang, X. and Gao, Y. (2014). Ultrasound-mediated nanoparticle systems. *Advanced Drug Delivery Reviews* 72: 23–41. https://doi.org/10.1016/j.addr.2014.02.006.

Wei, Q., Chen, Y., Ma, X. et al. (2018). High-efficient clearable NPs for multi-modal imaging and image-guided cancer therapy. *Advanced Functional Materials* 28 (2): 1704634.

Wilson, W.R. and Hay, M.P. (2011). Hypoxia-activated prodrugs in cancer therapy. *Nature Reviews Cancer* 11: 393–410.

Wu, M.Y., Meng, Q.S., Chen, Y. et al. (2015). Large-pore ultrasmall mesoporous organosilica NPs: micelle/precursor co-templating assembly and nuclear-targeted gene delivery. *Advanced Materials* 27 (2): 215–222.

Wu, M., Zhang, H., Tie, C. et al. (2018). MR imaging tracking of inflammation-activatable engineered neutrophils for targeted therapy of surgically treated glioma. *Nature Communications* 9: 4777.

Wu, H.M., Wang, S.L., Li, X.X., and Ai, K.X. (2025). Fabrication of gemcitabine and mitoxantrone loaded PLG/mesoporous silica nanofibers: investigation of in vitro drug release and anticancer activity in pancreatic cancer cells. *BioNanoScience* 15 (1): 1–16.

Xie, X., Yue, T., Gu, W. et al. (2023). Recent advances in mesoporous silica NPs delivering siRNA for cancer treatment. *Pharmaceutics* 15 (10): 2483.

Xu, C., Niu, Y., Popat, A. et al. (2014). Rod-like mesoporous silica nanoparticles with rough surfaces for enhanced cellular delivery. *Journal of Materials Chemistry B* 2 (3): 253–256.

Xu, B., Li, S., Shi, R. et al. (2023). Multifunctional mesoporous silica NPs for biomedical applications. *Signal Transduction and Targeted Therapy* 8: 435. https://doi.org/10.1038/s41392-023-01654-7.

Xu, C.J., Xie, J., Kohler, N. et al. (2008). Monodisperse magnetite nanoparticles coupled with nuclear localization signal peptide for cell-nucleus targeting. *Chemistry, an Asian Journal* 3 (3): 548–552.

Yang, Y. and Yu, C. (2015). Advances in silica-based NPs for targeted cancer therapy. *Nanomedicine: Nanotechnology, Biology, and Medicine*. https://doi.org/10.1016/j.nano.2015.10.018.

Yao, X., Niu, X., Ma, K. et al. (2017). Graphene quantum dots-capped magnetic mesoporous silica NPs as a multifunctional platform for controlled drug delivery, magnetic hyperthermia, and photothermal therapy. *Small* 13: 1602225.

Yu, M.H., Jambhrunkar, S., Thorn, P. et al. (2013). Hyaluronic acid modified mesoporous silica NPs for targeted drug delivery to CD44-overexpressing cancer cells. *Nanoscale* 5 (1): 178–183.

Yu, Z., Zhou, P., Pan, W. et al. (2018). A biomimetic nanoreactor for synergistic chemiexcited photodynamic therapy and starvation therapy against tumour metastasis. *Nature Communications* 9 (1): 5044.

Yuan, Y.Y., Mao, C.Q., Du, X.J. et al. (2012). Surface charge switchable NPs based on zwitterionic polymer for enhanced drug delivery to tumour. *Advanced Materials* 24 (40): 5476–5480.

Zhang, Z.P. and Feng, S.S. (2006). NPs of poly(lactide)/vitamin E TPGS copolymer for cancer chemotherapy: synthesis, formulation, characterization and in vitro drug release. *Biomaterials* 27 (2): 262–270.

Zhang, X., Zheng, C., Hou, Y. et al. (2016). Thermo-sensitive polymer-modified mesoporous silica NPs for controlled drug release. *Journal of Materials Chemistry B* 4 (44): 7113–7121. https://doi.org/10.1039/C6TB02010K.

Zhang, Z., Feng, J., Zhang, T. et al. (2023). Application of tumour pH/hypoxia-responsive NPs for combined photodynamic therapy and hypoxia-activated chemotherapy. *Frontiers in Bioengineering and Biotechnology* 11: 1197404. https://doi.org/10.3389/fbioe.2023.1197404.

Zhang, Y., Xing, J., Jiang, J. et al. (2025). Hypoxia-responsive NPs for fluorescence diagnosis and therapy of cancer. *Theranostics* 15 (4): 1353–1375. https://doi.org/10.7150/thno.104190. https://www.thno.org/v15p1353.htm.

Zhao, W., Wang, H., Wang, H. et al. (2021). Light-responsive dual-functional biodegradable mesoporous silica NPs with drug delivery and lubrication enhancement for the treatment of osteoarthritis. *Nanoscale* 13 (13): 6394–6399.

Zhao, Q., Xie, P., Li, X. et al. (2022). Magnetic mesoporous silica NPs mediated redox and pH dual-responsive target drug delivery for combined magnetothermal therapy and chemotherapy. *Colloids and Surfaces A: Physicochemical and Engineering Aspects* 648: 129359. https://doi.org/10.1016/J.COLSURFA.2022.129359.

Zhu, Y., Sundaram, H.S., Liu, S. et al. (2014). A robust graft-to strategy to form multifunctional and stealth zwitterionic polymer-coated mesoporous silica nanoparticles. *Biomacromolecules* 15 (5): 1845–1851.

Zhu, Y., Bai, Y., He, J., and Qiu, X. (2023). Advances in the stimuli-responsive mesoporous silica NPs as drug delivery system nanotechnology for controlled release and cancer therapy. *3 Biotech* 13 (8): 274.

Zhu, M., Ren, G., Guo, J. et al. (2024). Hypoxia-responsive nanomicelle based on 2-nitroimidazole for tumour treatment through chemotherapy and modulation of the tumour redox microenvironment. *ACS Applied Nano Materials* 7: 12452–12465.

Zitvogel, L., Apetoh, L., Ghiringhelli, F., and Kroemer, G. (2008). Immunological aspects of cancer chemotherapy. *Nature Reviews. Immunology* 8 (1): 59–73. https://doi.org/10.1038/nri2216.

16

Role of Silicon Nanoparticles (SiNPs) in the Treatment of Dermatitis

Aparna Sharma

Acharya Narendra Dev College, University of Delhi, New Delhi, Delhi, India

16.1 Introduction

Contact dermatitis remains among the most widespread skin inflammatory disorder that results from exposure to external substance (Scheinman et al. 2021). Estimated prevalence of contact dermatitis ranges from 1.7% to 6.35%, and it is known to go up during the hotter months or the places with higher temperature. Among the general population, elderly and females have higher manifestation of contact dermatitis (Statescu et al. 2011). The primary contributors to contact dermatitis include exposure to irritants, genetic makeup and other environmental influences. The irritants can be chemicals or allergens, and exposure arises from various occupational, social and household activities (Doi et al. 2017; Friedmann et al. 2015; Schnuch et al. 2011). Skin macerations also add on to the risk of dermatitis (Friedmann et al. 2015).

Traditional ways to manage dermatitis include prevention of irritant exposure, and steroid creams are used to soothe the lesions on skin (Kimyon and Warshaw 2019). However, recent research has shown that amorphous SiNPs can be used in dermatitis treatment (Palmer et al. 2019).

Amorphous SiNPs are prevalent and are presently utilized in various applications, including cosmetics, pharmaceuticals and food products (Contado 2015). Additionally, these nanoparticles (NPs) are being researched for their potential in drug delivery systems (Nafisi et al. 2018; Sapino et al. 2015, 2017) and bio-sensing technologies (Robinson et al. 2018). Being ubiquitous part of skin products, they are being tested on both normal healthy skin and inflammatory skin samples.

Silicon Nanoparticles: Synthesis, Uptake, and Applications, First Edition.
Edited by Archana Singh, Aryadeep Roychoudhury, and Indrakant K. Singh.

Although the impact of silicon nanoparticles (SiNPs) when tested on the healthy skin of mice is minimal, these NPs exhibit an immune-modulatory effect in the case of allergic contact dermatitis (ACD) (Matsuo et al. 2016; Nabeshi et al. 2011; Palmer et al. 2019; Ryu et al. 2014). This effect seems to be partially mediated by keratinocytes, leading to a reduction in epidermal hyperplasia (increase in the number of cells in a tissue), the release of inflammatory cytokines, immune cell invasion and a consequent decrease in skin swelling (Palmer et al. 2019). Further investigation is necessary in the understanding of the underlying mechanisms, especially when applied on humans. However, the immunomodulatory properties of SiNPs do hold immense potential for the preventive treatment of allergic skin disorders.

16.2 Dermatitis

Contact dermatitis is characterized as an inflammatory eczematous ailment of the skin. It arises after skin is exposed to chemicals such as metal ions that produce toxic effects.

When T-cell response is not triggered post-exposure, the chemical is known as contact irritants. On the other hand, if small reactive chemicals that alter proteins and provoke both innate and adaptive immune responses, it is referred to as contact allergens (Anderson et al. 2019; Basketter et al. 2019; Esser et al. 2019; Romita et al. 2018).

Contact dermatitis primary subtypes include (Nosbaum et al. 2009):

a) Irritant contact dermatitis (ICD)
b) Allergic contact dermatitis (ACD)
c) Combination of both

A chemical causing direct damage to the skin that leads to inflammatory response is a source of ICD. Alternatively, ACD is caused by allergens coming in contact with the skin and leading to type IV (delayed type) hypersensitive reactions (Gittler et al. 2013; Martin 2017). The immunological responses involved are a result of the collaboration between cytokines and T cells (Martin 2017). In cases of photoallergic contact dermatitis, lacerations are restricted to areas of the skin that came in contact with sunlight, despite the allergen being in contact with covered regions.

16.3 Contact Dermatitis: Symptoms

Symptoms associated with ICD can manifest as burning, itching, stinging, tenderness and pain. These symptoms are most prominent during the preliminary stages of the disorder (Scheinman et al. 2021). In contrast, pruritus is more frequently observed in the cases of ACD. Individuals with a prior history of skin issues have a heightened possibility for the development of nonspecific hand dermatitis and ICD (Scheinman et al. 2021).

The symptoms appear in three stages in the case of contact dermatitis. These stages consist of the acute phase, followed by the sub-acute phase and finally the chronic phase (Scheinman et al. 2021). ACD and ICD, both exhibit similar morphological characteristics in its acute and sub-acute stages. In the acute stage, both exhibit erythema, edema and vesiculation, with a positive patch test for ACD. The sub-acute stage includes lesions with crusts, scales and chronic stage that leads to lichenification (thickened/leathery skin) (Nosbaum et al. 2009). Additionally, in cases of ICD, blisters and erosions are present along the lesions.

The reaction towards the irritant escalates rapidly in the acute phase, peaking within minutes to sometime in a few hours following the contact, after which it begins to resolve. On the contrary, inception of ACD is influenced by the chemical properties of the sensitizing agent. The severity of allergic reaction also depends upon individual's sensitivity level. Lesions commonly manifest 24–72 hours post-exposure to the triggering substance, reaching their maximum severity around 72–96 hours. ACD tends to improve at a slower rate compared to ICD and is more prone to recurrence within a few days upon re-exposure (Nosbaum et al. 2009).

16.4 Contact Dermatitis: Treatment

In cases of contact dermatitis, adhering to avoidance strategies is crucial. The foundation of effective avoidance lies in the accurate assessment and identification of the allergenic triggers. It is advisable to wear a suitable attire to shield oneself from irritants both at residence and in the place of work (Nedorost 2018; Soltanipoor et al. 2019). To alleviate inflammation, topical application of high-efficacy corticosteroids (0.05% clobetasol propionate cream) is used (Vernon and Olsen 1990). Generally, these corticosteroids should be avoided on delicate skin areas, including the face and genitals, to mitigate the risk of skin atrophy. Itching is managed with the help of antihistamines. Cetirizine or hydroxyzine are the most recommended one for the same. The cases of severe reactions are managed with systemic steroids, but their use should be gradually decreased to prevent the recurrence of the reaction. It is also important to refrain from using soaps, application of perfumes and dyes. Ointments and creams assist in keeping the skin hydrated. Additionally, tacrolimus ointment and pimecrolimus cream are immunomodulatory agents. They inhibit calcineurin and can be effective in treating ACD (Vernon and Olsen 1990).

16.5 Contact Dermatitis: Immune Response

16.5.1 Immune Response in Irritant Contact Dermatitis (ICD)

ICD occurs when irritants/chemicals compromise the skin's barrier function (Scheinman et al. 2021). Skin breach leads to intercellular edema and possible necrosis. Necrosis triggers the release of damage-associated molecular patterns (DAMPs) (Scheinman et al. 2021).

Innate immune cells identify these DAMPs through pattern recognition receptors (PRRs). In mammals, toll-like receptors expressed on the cell surface of macrophages and dendritic cells interact with DAMPs. Subsequently, these cells generate a variety of cytokines, neuropeptides and chemokines. This includes factors such as tumor necrosis factor (TNF)-α, interferon (IFN)-α, substance P, and variety of interleukins such as IL-23, IL-18, IL-12, IL-6 and IL-1β (Gittler et al. 2013). Additionally, damaged keratinocytes contribute to this immune response by producing innate immune cytokines IL-23, IL-18, IL-12, IL-6 and IL-1β and factors like TNF-α, which further exacerbate damage of skin tissues (Gittler et al. 2013) (Figure 16.1).

16.5.2 Immune Response in Allergen Contact Dermatitis (ACD)

The pathogenesis of allergen contact dermatitis involves two distinct phases exhibiting delayed-type hypersensitivity (Type IV):

a) **The Sensitization Phase:** Immune system is exposed to the allergen for the first time and the response is mild (Gittler et al. 2013; Milam et al. 2019; Nosbaum et al. 2009; Rustemeyer 2022; Scheinman et al. 2021; Vocanson et al. 2009).
b) **The Elicitation Phase:** It refers to the immunological response triggered upon re-exposure to the allergen, which results in exaggerated response (Gittler et al. 2013; Milam et al. 2019; Nosbaum et al. 2009; Rustemeyer 2022; Scheinman et al. 2021; Vocanson et al. 2009).

Type IV hypersensitivity is a subtype of the allergic reaction, where the body responds several hours later after the exposure to the allergen. It is also the only cell-mediated hypersensitivity that involves tissue damage caused mainly by T cells and cytokines (Rustemeyer 2022; Vocanson et al. 2009).

16.5.2.1 Sensitization Phase

The sensitization phase is often denoted as the induction phase. During this phase, the allergen/hapten comes in contact with the skin for the very first time. This can occur either locally or systemically through the gastrointestinal tract or respiratory pathways.

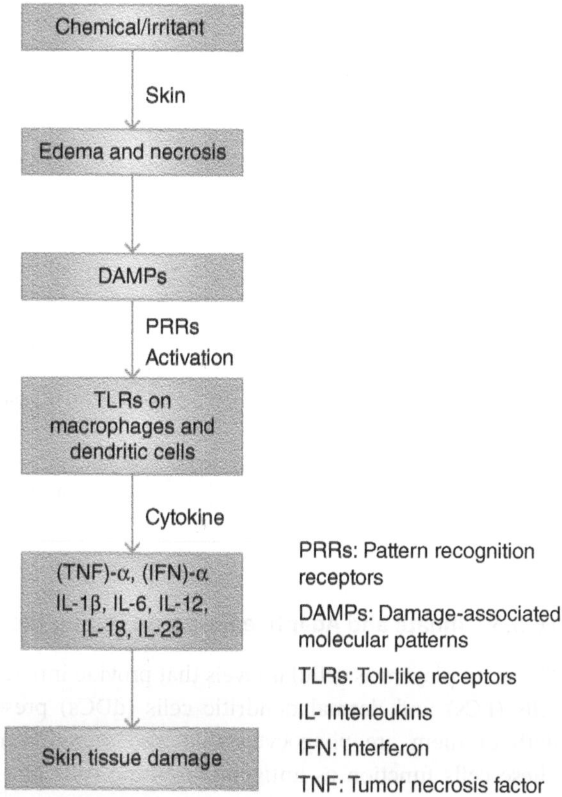

Figure 16.1 Flowchart illustrating the activation of irritant contact dermatitis after the skin is exposed to chemicals/irritants.

Haptens are antigens which on their own are unable to illicit an immune response or works as immunogens. They need to conjugate with either amino acids or protein molecules to be recognized by the immune system.

Haptens are frequently utilized as agents in patch testing on model animals and include substances listed in Table 16.1.

Since haptens are non-immunogenic antigens, they need to conjugate with self-proteins or amino acids. This process is termed as haptenization. In case of dermatitis, test model haptenization takes place in stratum corneum. It requires covalent bonding between proteins/amino acids and metal hapten molecules, resulting in a hapten–protein conjugate. This complex now acts as an immunogen and illicit immune response characterizing the sensitization phase. Some chemicals are termed as pre-haptens, and their activation is dependent on external agents like oxidation via air or UV rays. Some pre-haptens require activation by internal agents, for example, cytochrome P450 enzyme catalysis (Scheinman et al. 2021).

The following factors determine the extent and intensity of ACD (Scheinman et al. 2021; Yamaguchi et al. 2023) (Table 16.2).

Table 16.1 Haptens: subtypes and examples.

S. No.	Types of haptens	Examples and sources
1.	Chemicals (Schalock et al. 2020)	a) Nickel (nickel sulphate) b) Cobalt (cobalt chloride) c) Gold (gold sodium thio-sulphate) d) Fragrance mix I e) Fragrance mix II f) Limonene g) Linalool h) Lyral i) Peppermint j) Jasmine k) Latex (black rubber)
2.	Plant derived (Ito et al. 2022; Kim et al. 2019)	a) Urushiol (poison ivy) b) Urushiol (oak) and c) Urushiol (sumac)

16.5.3 Innate and Adaptive Response During the Sensitization Phase

The main players as cellular levels that provide innate immunity are Langerhans cells (LCs) and dermal dendritic cells (dDCs) present in the dermis region. Both of them are phagocytes that also secrete various chemotactic factors. These cells function as antigen-presenting cells and initiate adaptive immune responses involving T cells (Rustemeyer 2022). These cells utilize their extended dendrites to monitor the surrounding environment and are capable of engulfing allergen/hapten–protein conjugates.

As outlined in the pathogenesis of inflammatory skin conditions, LCs and dDCs identify DAMPs through PRRs. Subsequently, they release variety of cytokines and neuropeptides to complement phagocytosis (Scheinman et al. 2021). They also activate nearby keratinocytes to produce cytokines and chemokines. Under the influence of specific factors, such as IL-1β and TNF-α, LCs and dDCs travel from the skin to the nearby lymph nodes. In the lymph nodes, LCs and dDCs present antigens using major histocompatibility complex class 1 (MHC-I) and class 2 (MHC-II) molecules (Scheinman et al. 2021). This presentation activates naïve T cytotoxic cells (CD8+) and T helper cells (CD4+). As a result, naïve T cells, recognizing hapten-conjugates (TCRs), undergo clonal expansion. This leads to the formation of effector and memory T cells in lymph nodes, which can migrate to the skin via blood or lymphatic vessels (Pilkington et al. 2018; Yamaguchi et al. 2023). Effector T cells carry out sensitization response, whereas memory T cells play the major role in the elicitation phase.

Table 16.2 Factors contributing to allergic contact dermatitis (ACD).

S. No.	Factors
1.	Types of allergen/haptens involved i) Metal ii) Fragrance iii) Size iv) Lipophilic characteristics v) Electrophilic characteristics
2.	Concentrations of allergen/haptens
3.	The condition of the skin that is exposed to allergen/haptens i) Barrier function ii) Thickness iii) Maceration iv) Underlying skin diseases v) Comorbidities such as ICD, atopic dermatitis and stasis dermatitis
4.	Specific body areas that are exposed such as the face, dorsal hands and mucous membranes
5.	Environmental influences i) Occupational factors ii) Psychological well-being iii) Lifestyle choices, nutritional status iv) Medication use v) Exposure to pollutants vi) Climatic conditions
6.	Functionality of regulatory T cells
7.	Genetic predispositions i) Variations in detoxification enzymes ii) Ethnic background

16.5.3.1 Elicitation Phase

This phase often referred to as the efferent/challenge phase. It occurs when the same allergen interacts with the skin region it has previously infiltrated following the completion of the sensitization phase. This interaction results in the display of acute ACD clinical symptoms (Vocanson et al. 2009).

ACD is characterized by symptoms, such as erythema, population and vesiculation, typically emerging within 24–72 hours post hapten contact. Cytotoxic T cells (CTLs) are recognized as the principal allergen-specific effector T cells. They reside in both dermal and epidermal regions of the skin. These cells play a crucial role in the histological alterations observed, which include dermatitis at the epidermal–dermal interface, accompanied by spongiosis, i.e., fluid build-up in the skin, and an infiltration of inflammatory cells, predominantly lymphocytes and neutrophils (Saint-Mezard et al. 2004).

Effector T helper cells that are specific to haptens, generate sub-populations of helper T (Th) cells and regulatory T cells. They contribute significantly to the initiation of inflammatory cell migration from the bloodstream to local tissues, keratinocytes apoptosis and the stimulation and alteration of inflammatory cells response (Mark and Slavin 2006).

Various haptens and allergens were detected in skin biopsies from individuals with positive patch test results. These tests also indicated that various haptens present distinct immune signatures. Nickel was found to elicit the most significant immune activation, engaging pathways linked to innate immune cells for instance T helper cell (Th1, Th17 and Th22) responses (Dhingra et al. 2014). In contrast, fragrances and latex prompted a robust Th2 response along with some Th22 polarization, while exhibiting diminished Th1 and Th17 pathway activity (Pavel et al. 2023) (Figure 16.2).

This suggests that ACD pathogenesis is heterogeneous and varies according to the specific haptens/allergens involved. The immune signatures revealed that hapten diphencyprone provoked the most intense immune responses, as indicated by markers of innate immunity. It exhibits expression of various cytokines such as

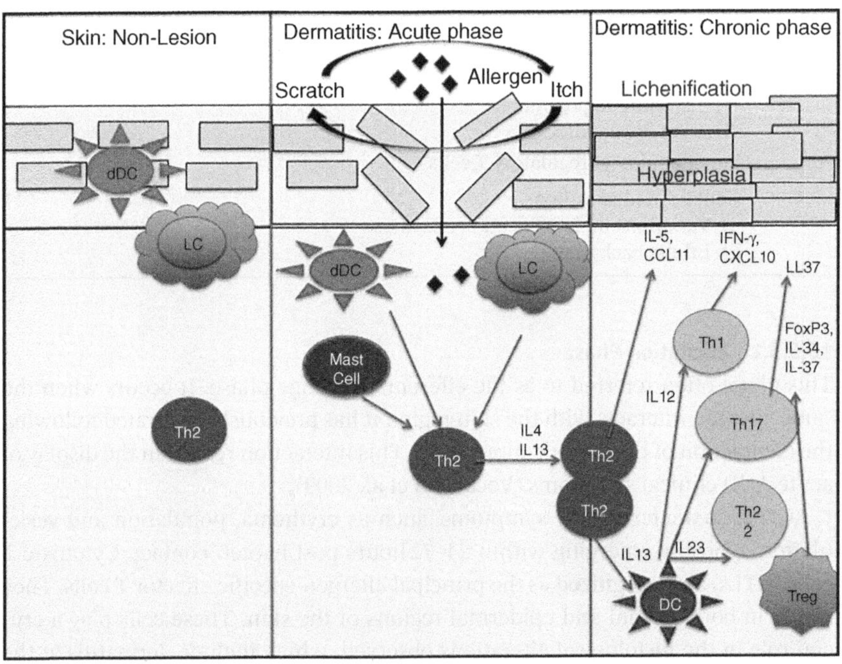

Figure 16.2 Immune response in allergic contact dermatitis (ACD). Three stages of ACD: non-lesion skin, acute phase response and Th2 activation and chronic phase response. dDCs, LCs, Th (T helper cells), Treg (T regulatory cells), IL (interleukins).

IL-1α and IL-8. It also displays T helper cells managing adaptive immune response with the help of Th1 (T helper 1) secreting CXCL10, IFN-γ, Th2 (T helper 2) cells secreting CCL11, IL-5, Th17 cells expressing LL37, and finally regulatory T cells expressing IL-34, IL-37, and FoxP3 to complete the response. Dust mites were shown to modulate the Th2 pathway, further underscoring the complexity of ACD pathogenesis (Gittler et al. 2013; Honda et al. 2013) (Figure 16.2).

Innate immune cells encompass a variety of types, including LCs, dDCs, monocytes, macrophages and mast cells(Honda et al., 2013). Additionally, granulocytes like neutrophils, basophils and eosinophils are part of this immune response in ACD.

This indicates that innate immune cells play a significant role in the progression of ACD during both the sensitization and elicitation phases along with adaptive immune responses (Correa da Rosa et al. 2015; Leonard and Guttman-Yassky 2019).

16.6 Contact Dermatitis: Treatment Utilizing Silicon Nanoparticles (SiNPs)

As interest in nanotechnology continues to expand across various industries, its use in the cosmetics sector has become the center of attention due to it being the pioneers in adopting nanomaterials. Among the various NPs widely used, silica is particularly notable due to its exceptional physicochemical properties and advantageous effects. The rising application of SiNPs in cosmetic products, coupled with emerging research that explores their potential as transdermal drug delivery systems, is expected to enhance skin's surface exposure to these NPs (Nafisi et al. 2018; Sapino et al. 2015, 2017).

There are numerous investigations that have focused on the in vivo toxicity studies using high-dose exposures to amorphous SiNPs in the lungs and gastrointestinal systems of rodents. However, only a few research works addressed the dermal toxicity of these NPs.

The initial research into the dermal toxicity of SiNPs indicated a dose-dependent cytotoxic effect on in vitro keratinocytes and LCs, which was linked to an increase in the reactive oxygen species (ROS) production (Jatana et al. 2017). These studies also correlate topical applications of SiNPs and its effect on the normal skin vs. dermatitis.

Various studies have shown somewhat conflicting results with regard to the effect of SiNPs on dermatitis. It has been demonstrated by Nabeshi et al. that 70 nm SiNPs can infiltrate the murine skin. It can also induce apoptosis in skin cells 28 days after exposure to high dose of SiNPs (250 μg/ear/day) (Ryu et al. 2014). In contrast to these findings, Ryu et al. reported no signs of either toxicity at the dermal level or systemic level in rats subjected to everyday applications (up to 2000 mg/kg) of 20 nm SiNPs over a 90-day period (Matsuo et al. 2016). More recently, Matsuo et al. conducted an experiment where mice were exposed to high doses (125 μg/ear/day) for around 28 days. The study demonstrated 70 nm SiNPs infiltrating the skin with negligible dermal toxicity (Luckhaupt et al. 2013).

Palmer et al. studied the topical administration of 20 nm SiNPs on ear swelling induced by DNFB (2,4-dinitrofluorobenzene) a potent allergen. Application of 20 nm SiNPs on the surface of the skin resulted in a reduction of ear inflammation and epidermal hyperplasia, which is associated with contact hypersensitivity (CHS) (Contado 2015; Yamaguchi et al. 2023). In cases of 20 nm SiNPs, it appears that a decrease in mast cell degranulation plays a noteworthy role in the immunosuppressive effects of SiNPs on the skin. However, it must be noted that mast cell degranulation at the 24-hour mark is also observed in treatments involving DNFB alone (control animals). This indicate that mast cell degranulation might not be the contributing factor in lowering the ear swelling/dermatitis resulting due to SiNP application.

Nevertheless, the in vitro study using cell line HaCaT (human keratinocytes) suggests a direct interaction between SiNPs and keratinocytes. The study considered that SiNPs significantly contribute to the observed reductions in epidermal tissue thickening and secretion of cytokine in vivo. The chemokines (MIP-2, KC) are responsible for recruiting neutrophils to the skin, and the notable decrease in MIP-2 resulted in a moderate reduction in the overall neutrophil count in the skin at 24 hours (Dilulio et al. 1999; Toda and Yoshino 2016). Neutrophils are known to produce IP-10 and are essential for the later activation of CTLs that are involved in the late-phase swelling response. At the 24-hour interval, there was a significant drop in the number of CTLs present in the dermal region accounts for the diminished late-phase swelling response that can be attributed to SiNP application (Askenase and Tsuji 2000; Askenase et al. 2015; Biedermann et al. 2000; Contado 2015).

There is a limited body of research examining the impact of lab-manufactured nanomaterials on immune response to the surface of the skin. Most studies highlight the immune-stimulating properties of SiNPs and their application as adjuvants in vaccine formulations, whereas dose-dependent use of SiNPs indicates an immunosuppressive effect, standing in contrast to the existing literature. This discrepancy may be attributed to the high concentrations of SiNPs (ranging from 30 to 300 μg) utilized in vaccine studies, coupled with the systemic exposure routes involved (Hirai et al. 2012; Yanagisawa et al. 2009). In a separate investigation, SiNPs were found to enhance symptoms of atopic dermatitis induced by dust mite allergens in a size-dependent manner following an intradermal injection of 250 μg of the particles (Hirai et al. 2012). Additionally, intradermal administration of titanium dioxide NPs has been shown to intensify both the sensitization and elicit inflammatory responses towards allergens (Ostrowski et al. 2014). A latest study revealed that the application of 250 μg of SiNPs on the skin surface that has been exposed to dust mite allergens increased the likelihood of atopic reactions in mice ears (Jatana et al. 2016; Yanagisawa et al. 2009). While the NPs examined in the aforementioned studies were negatively charged, positively charged functionalized SiNPs have not demonstrated any significant effects on ACD. This proves that both size and charge influence the immune-modulatory capabilities of NPs in the skin. The unchanged, amorphous SiNPs diminish allergen-induced skin immune responses potentially due to the topical administration route and the lower, less cytotoxic dosage used. Beside the use of 20 nm SiNPs in the reduction of ear swelling, the only skin-surface applied engineered nanomaterials that have been shown to reduce skin immune responses are silver nanocrystals and calcium phosphate NPs that bind nickel ions (Contado 2015).

There is ongoing discussion regarding the ability of NPs to penetrate healthy human skin. Nevertheless, evidence indicates that NPs can infiltrate mouse skin, which possesses a thinner stratum corneum and a greater density of hair follicles in comparison to human skin (Jatana et al. 2016). While mouse skin serves as a permeable barrier, there are also data suggesting that NPs may easily breach human skin that has experienced barrier disruption. Given that a significant segment of population suffers from inflammatory skin disorders, this raises concerns about the risk of unintended immunomodulation due to exposure to SiNPs. Consequently, it is crucial to evaluate the toxicity and potential immune-modulatory effects of chronic, low-dose SiNP exposure in animal models that mimic inflammatory skin disorders in humans. However, when administered in a controlled manner, low doses of SiNPs provide a reliable and effectual means of alleviating symptoms associated with inflammatory skin disorders, such as ACD. It is important to note that these findings are further needed to be replicated in human models with effective efficacy, prior to any therapeutic application.

The effects of 20 nm SiNP downstream mechanism are studied using the CHS model. The study finds keratinocytes as a probable upstream mediator of SiNP-induced immunosuppression pathways (Contado 2015). It is noteworthy that this effect is not specific to the particles themselves, as we have observed that several small negatively charged NPs also exhibit immunosuppressive properties in the CHS model. Given the diverse range of particle sizes and surface chemistries that can alleviate CHS symptoms, the underlying mechanism is likely associated with a common characteristic of many NPs, such as the generation of ROS (Contado 2015). For instance, exposure to cigarette smoke has been shown to diminish the ear inflammation in CHS.

Although there are a few studies that are investigating the potential preventive effects of SiNPs, the existing research indicates that these NPs can promote keratinocyte proliferation and immunomodulation. This makes them a good candidate for the treatment of dermatitis.

16.7 Conclusion

In conclusion, it has been observed that contact dermatitis is a complex condition arising from heterogeneous factors. Both innate and adaptive immunity contribute in its manifestation. It is the interplay of immune cells like LCs, dDCs, T-cells and secretion of numerous cytokines that regulate the symptoms of contact dermatitis. This chapter delves into the role that SiNPs can play to prevent dermatitis. SiNPs were tested at multiple doses on both healthy and inflamed skin of model animals.

The high doses of amorphous SiNPs exhibit pro-inflammatory and cytotoxic properties. However, the negatively charged 20 nm SiNPs at low doses applied on the skin surface possess immunosuppressive effects in a DNFB-induced model. These SiNPs reduce DNFB-induced epidermal thickening diminish skin cytokine expression and limit the infiltration of CTLs into the skin. It results in the reduction of swelling.

Since these studies were conducted on animal model systems, further investigation into this immune-modulatory effect is warranted in the human model/cell lines. SiNPs are already being tested for drug delivery and vaccination. If future studies prove effective immunomodulation by SiNPs, they can be considered as transdermal drug delivery systems for patients suffering from inflammatory skin disorders. SiNPs are already being used in various creams and cosmetic products. At an appropriate dosage in topical application, they can play an important role in the prevention of dermatitis. Granting additional studies are required, SiNPs holds a substantial role in the prevention and treatment of inflammatory skin disorders including dermatitis.

References

Anderson, L.E., Treat, J.R., Brod, B.A., and Yu, J. (2019). "Slime" contact dermatitis: case report and review of relevant allergens. *Pediatric Dermatology* 36 (3): 335–337.

Askenase, P.W. and Tsuji, R.F. (2000). B-1 B cell IgM antibody initiates T cell elicitation of contact sensitivity. *Current Topics in Microbiology and Immunology* 252: 171–177. https://doi.org/10.1007/978-3-642-57284-5_18.

Askenase, P.W., Bryniarski, K., Paliwal, V. et al. (2015). A subset of AID-dependent B-1a cells initiates hypersensitivity and pneumococcal pneumonia resistance. *Annals of the New York Academy of Sciences* 1362 (1): 200–214. https://doi.org/10.1111/nyas.12975.

Basketter, D.A., Huggard, J., and Kimber, I. (2019). Fragrance inhalation and adverse health effects: the question of causation. *Regulatory Toxicology and Pharmacology: RTP* 104: 151–156. https://doi.org/10.1016/j.yrtph.2019.03.011.

Biedermann, T., Kneilling, M., Mailhammer, R. et al. (2000). Mast cells control neutrophil recruitment during T cell-mediated delayed-type hypersensitivity reactions through tumor necrosis factor and macrophage inflammatory protein 2. *The Journal of Experimental Medicine* 192 (10): 1441–1452. https://doi.org/10.1084/jem.192.10.1441.

Contado, C. (2015). Nanomaterials in consumer products: a challenging analytical problem. *Frontiers in Chemistry* 3: 48. https://doi.org/10.3389/fchem.2015.00048.

Correa da Rosa, J., Malajian, D., Shemer, A. et al. (2015). Patients with atopic dermatitis have attenuated and distinct contact hypersensitivity responses to common allergens in skin. *The Journal of Allergy and Clinical Immunology* 135 (3): 712–720. https://doi.org/10.1016/j.jaci.2014.11.017.

Dhingra, N., Shemer, A., Correa da Rosa, J. et al. (2014). Molecular profiling of contact dermatitis skin identifies allergen-dependent differences in immune response. *The Journal of Allergy and Clinical Immunology* 134 (2): 362–372. https://doi.org/10.1016/j.jaci.2014.03.009.

Dilulio, N.A., Engeman, T., Armstrong, D. et al. (1999). Groα-mediated recruitment of neutrophils is required for elicitation of contact hypersensitivity. *European Journal of Immunology* 29 (11): 3485–3495. https://doi.org/10.1002/(sici)1521-4141(199911)29:11<3485::aid-immu3485>3.0.co;2-b.

Doi, T., Mizukawa, Y., Shimoda, Y. et al. (2017). Importance of water content of the stratum corneum in mouse models for contact hypersensitivity. *The Journal of Investigative Dermatology* 137 (1): 151–158. https://doi.org/10.1016/j.jid.2016.08.03.

Esser, P.R., Mueller, S., and Martin, S.F. (2019). Plant allergen-induced contact dermatitis. *Planta Medica* 85 (7): 528–534. https://doi.org/10.1055/a-0873-1494.

Friedmann, P.S., Sanchez-Elsner, T., and Schnuch, A. (2015). Genetic factors in susceptibility to contact sensitivity. *Contact Dermatitis* 72 (5): 263–274. https://doi.org/10.1111/cod.12362.

Gittler, J.K., Krueger, J.G., and Guttman-Yassky, E. (2013). Atopic dermatitis results in intrinsic barrier and immune abnormalities: implications for contact dermatitis. *The Journal of Allergy and Clinical Immunology* 131 (2): 300–313. https://doi.org/10.1016/j.jaci.2012.06.048.

Hirai, T., Yoshikawa, T., Nabeshi, H. et al. (2012). Amorphous silica nanoparticles size-dependently aggravate atopic dermatitis-like skin lesions following an intradermal injection. *Particle and Fibre Toxicology* 9: 3. https://doi.org/10.1186/1743-8977-9-3.

Honda, T., Egawa, G., Grabbe, S., and Kabashima, K. (2013). Update of immune events in the murine contact hypersensitivity model: toward the understanding of allergic contact dermatitis. *The Journal of Investigative Dermatology* 133 (2): 303–315. https://doi.org/10.1038/jid.2012.284.

Ito, A., Suzuki, K., Matsunaga, K. et al. (2022). Patch testing with the Japanese baseline series 2015: a 4-year experience. *Contact Dermatitis* 86 (3): 189–195. https://doi.org/10.1111/cod.14027.

Jatana, S., Callahan, L.M., Pentland, A.P., and DeLouise, L.A. (2016). Impact of cosmetic lotions on nanoparticle penetration through *ex vivo* C57BL/6 hairless mouse and human skin: a comparison study. *Cosmetics* 3 (1): 6. https://doi.org/10.3390/cosmetics3010006.

Jatana, S., Palmer, B.C., Phelan, S.J., and DeLouise, L.A. (2017). Immunomodulatory effects of nanoparticles on skin allergy. *Scientific Reports* 7 (1): 3979. https://doi.org/10.1038/s41598-017-03729-2.

Kim, Y., Flamm, A., ElSohly, M.A. et al. (2019). Poison ivy, oak, and sumac dermatitis: what is known and what is new? *Dermatitis: Contact, Atopic, Occupational, Drug: Official Journal of the American Contact Dermatitis Society, North American Contact Dermatitis Group* 30 (3): 183–190. https://doi.org/10.1097/der.0000000000000472.

Kimyon, R.S. and Warshaw, E.M. (2019). Airborne allergic contact dermatitis: management and responsible allergens on the American contact dermatitis society core series. *Dermatitis: Contact, Atopic, Occupational, Drug: Official Journal of the American Contact Dermatitis Society, North American Contact Dermatitis Group* 30 (2): 106–115. https://doi.org/10.1097/der.0000000000000439.

Leonard, A. and Guttman-Yassky, E. (2019). The unique molecular signatures of contact dermatitis and implications for treatment. *Clinical Reviews in Allergy & Immunology* 56 (1): 1–8. https://doi.org/10.1007/s12016-018-8685-0.

Luckhaupt, S.E., Dahlhamer, J.M., Ward, B.W. et al. (2013). Prevalence of dermatitis in the working population, United States, 2010 National Health Interview Survey. *American Journal of Industrial Medicine* 56 (6): 625–634. https://doi.org/10.1002/ajim.22080.

Mark, B.J. and Slavin, R.G. (2006). Allergic contact dermatitis. *The Medical Clinics of North America* 90 (1): 169–185. https://doi.org/10.1016/j.mcna.2005.08.008.

Martin, S.F. (2017). The role of the innate immune system in allergic contact dermatitis. *Allergologie Select* 1: 39–43.

Matsuo, K., Hirobe, S., Okada, N., and Nakagawa, S. (2016). Analysis of skin permeability and toxicological properties of amorphous silica particles. *Biological & Pharmaceutical Bulletin* 39 (7): 1201–1205. https://doi.org/10.1248/bpb.b16-00258.

Milam, E.C., Jacob, S.E., and Cohen, D.E. (2019). Contact dermatitis in the patient with atopic dermatitis. *The Journal of Allergy and Clinical Immunology in Practice* 7 (1): 18–26. https://doi.org/10.1016/j.jaip.2018.11.003.

Nabeshi, H., Yoshikawa, T., Matsuyama, K. et al. (2011). Systemic distribution, nuclear entry and cytotoxicity of amorphous nanosilica following topical application. *Biomaterials* 32 (11): 2713–2724. https://doi.org/10.1016/j.biomaterials.2010.12.042.

Nafisi, S., Samadi, N., Houshiar, M., and Haibach, I.M. (2018). Mesoporous silica nanoparticles for enhanced lidocaine skin delivery. *International Journal of Pharmaceutics*. https://doi.org/10.1016/j.ijpharm.2018.08.004.

Nedorost, S. (2018). A diagnostic checklist for generalized dermatitis. *Clinical, Cosmetic and Investigational Dermatology* 11: 545–549. https://doi.org/10.2147/CCID.S185357.

Nosbaum, A., Vocanson, M., Rozieres, A. et al. (2009). Allergic and irritant contact dermatitis. *European Journal of Dermatology* 19: 325–332.

Ostrowski, A., Nordmeyer, D., Mundhenk, L. et al. (2014). AHAPS-functionalized silica nanoparticles do not modulate allergic contact dermatitis in mice. *Nanoscale Research Letters* 9 (1): 524. https://doi.org/10.1186/1556-276X-9-524.

Palmer, B.C., Jatana, S., Phelan-Dickinson, S.J. et al. (2019). Amorphous silicon dioxide nanoparticles modulate immune responses in a model of allergic contact dermatitis. *Scientific Reports* 9: 5085. https://doi.org/10.1038/s41598-019-41493-7.

Pavel, A.B., Del Duca, E., Cheng, J. et al. (2023). Delayed type hypersensitivity reactions to various allergens may differently model inflammatory skin diseases. *Allergy* 78 (1): 178–191. https://doi.org/10.1111/all.15538.

Pilkington, S.M., Ogden, S., Eaton, L.H. et al. (2018). Lower levels of interleukin-1beta gene expression are associated with impaired Langerhans' cell migration in aged human skin. *Immunology* 153: 60–70.

Robinson, K.J., Huynh, G.T., Kouskousis, B.P. et al. (2018). Modified organosilica core-shell nanoparticles for stable pH sensing in biological solutions. *ACS Sensors* 3: 967–975. https://doi.org/10.1021/acssensors.8b00034.

Romita, P., Foti, C., Calogiuri, G. et al. (2018). Contact dermatitis due to transdermal therapeutic systems: a clinical update. *Acta Bio-Medica: Atenei Parmensis* 90 (1): 5–10. https://doi.org/10.23750/abm.v90i1.6563.

Rustemeyer, T. (2022). Immunological mechanisms in allergic contact dermatitis. *Current Treatment Options in Allergy* 9 (2): 67–75. https://doi.org/10.1007/s40521-022-00299-1.

Ryu, H.J., Seong, N.-W., So, B.J. et al. (2014). Evaluation of silica nanoparticle toxicity after topical exposure for 90 days. *International Journal of Nanomedicine* 9 (Suppl 2): 127–136. https://doi.org/10.2147/ijn.s57929.

Saint-Mezard, P., Rosieres, A., Krasteva, M. et al. (2004). Allergic contact dermatitis. *European Journal of Dermatology: EJD* 14 (5): 284–295.

Sapino, S., Ugazio, E., Gastaldi, L. et al. (2015). Mesoporous silica as topical nanocarriers for quercetin: characterization and in vitro studies. *European Journal of Pharmaceutics and Biopharmaceutics: Official Journal of Arbeitsgemeinschaft fur Pharmazeutische Verfahrenstechnik e.V.* 89: 116–125. https://doi.org/10.1016/j.ejpb.2014.11.022.

Sapino, S., Oliaro-Bosso, S., Zonari, D. et al. (2017). Mesoporous silica nanoparticles as a promising skin delivery system for methotrexate. *International Journal of Pharmaceutics* 530: 239–248. https://doi.org/10.1016/j.ijpharm.2017.07.058.

Schalock, P.C., Dunnick, C.A., Nedorost, S. et al. (2020). American Contact Dermatitis Society Core Allergen series: 2020 update. *Dermatitis: Contact, Atopic, Occupational, Drug: Official Journal of the American Contact Dermatitis Society, North American Contact Dermatitis Group* 31 (5): 279–282. https://doi.org/10.1097/DER.0000000000000621.

Scheinman, P.L., Vocanson, M., Thyssen, J.P. et al. (2021). Contact dermatitis. *Nature Reviews. Disease Primers* 7: 38.

Schnuch, A., Westphal, G., Mössner, R. et al. (2011). Genetic factors in contact allergy – review and future goals. *Contact Dermatitis* 64 (1): 2–23. https://doi.org/10.1111/j.1600-0536.2010.01800.x.

Soltanipoor, M., Kezic, S., Sluiter, J.K. et al. (2019). Effectiveness of a skin care programme for the prevention of contact dermatitis in healthcare workers (the Healthy Hands Project): a single-centre, cluster randomized controlled trial. *Contact Dermatitis* 80 (6): 365–373. https://doi.org/10.1111/cod.13214.

Statescu, L., Branisteanu, D., Dobre, C. et al. (2011). Contact dermatitis – epidemiological study. *Maedica* 6: 277–281.

Toda, T. and Yoshino, S. (2016). Enhancement of ovalbumin-specific Th1, Th2, and Th17 immune responses by amorphous silica nanoparticles. *International Journal of Immunopathology and Pharmacology* 29 (3): 408–420. https://doi.org/10.1177/0394632016656192.

Vernon, H.J. and Olsen, E.A. (1990). A controlled trial of clobetasol propionate ointment 0.05% in the treatment of experimentally induced Rhus dermatitis. *Journal of the American Academy of Dermatology* 23 (5): 829–832. https://doi.org/10.1016/0190-9622(90)70297-u.

Vocanson, M., Hennino, A., Rozières, A. et al. (2009). Effector and regulatory mechanisms in allergic contact dermatitis. *Allergy* 64 (12): 1699–1714. https://doi.org/10.1111/j.1398-9995.2009.02082.x.

Yamaguchi, H.L., Yamaguchi, Y., and Peeva, E. (2023). Role of innate immunity in allergic contact dermatitis: an update. *International Journal of Molecular Sciences* 24 (16): https://doi.org/10.3390/ijms241612975.

Yanagisawa, R., Takano, H., Inoue, K.-I. et al. (2009). Titanium dioxide nanoparticles aggravate atopic dermatitis-like skin lesions in NC/Nga mice. *Experimental Biology and Medicine (Maywood, N.J.)* 234 (3): 314–322. https://doi.org/10.3181/0810-RM-304.

17

Si/SiO_2 Nanoparticles as a Versatile Tool for the Treatment of Bacterial Infections in Human

Maya Madhavan[1], *Seena Thomas*[2], *R.S. Silpa*[3], *Ameen Pareesa Zidhiq*[4] and *Sabeena Mustafa*[5]

[1] *Department of Biochemistry, Government College Kariavattom, Thiruvananthapuram, Kerala, India*
[2] *Department of Biostatistics, King Abdullah International Medical Research Center (KAIMRC), King Saud Bin Abdulaziz University for Health Sciences, King Abdulaziz Medical City, Ministry of National Guard Health Affairs (MNGHA), Riyadh, Kingdom of Saudi Arabia*
[3] *Thejas, Neyyattinkara, Kerala, India*
[4] *M. Sc. Biotechnology, Vellore Institute of Technology (VIT), Vellore, Tamil Nadu, India*
[5] *Department of AI and Bioinformatics, King Abdullah International Medical Research Center (KAIMRC), King Saud Bin Abdulaziz University for Health Sciences, King Abdulaziz Medical City, Ministry of National Guard Health Affairs (MNGHA), Riyadh, Kingdom of Saudi Arabia*

17.1 Introduction

The emergence of antibiotic resistance represents a critical challenge in modern medicine, necessitating innovative approaches to antimicrobial therapy (Chinemerem Nwobodo et al. 2022). One of the potential breakthroughs in this sector is the use of silica nanoparticles, which has received substantial attention for their distinctive features and multiple uses (Nayl et al. 2022). These nanoparticles, distinguished by their extensive surface area and adjustable morphology, have the potential to augment the route of administration of antimicrobial agents and enhance their effectiveness against a variety of pathogens. By facilitating targeted delivery and controlled release, silica nanoparticles not only augment the activity of conventional antibiotics but also serve as platforms for novel therapeutic agents. Mesoporous silica materials, especially mesoporous silica nanoparticles (MSNs), exhibit significant potential for the regulation and sustainable administration of pharmaceuticals, attributable to their advantageous chemical characteristics, thermal stability and biocompatibility. MSNs

Silicon Nanoparticles: Synthesis, Uptake, and Applications, First Edition.
Edited by Archana Singh, Aryadeep Roychoudhury, and Indrakant K. Singh.

enable efficient drug loading and regulated release, and their characteristics could be modified through the incorporation of additives. They are widely used in diagnosis, target drug delivery, bio-sensing and cellular uptake in the biomedical field (Bharti et al. 2015). This chapter will delve into the diverse applications of silica nanoparticles in antimicrobial therapeutics, exploring their mechanisms of action, effectiveness against various micro-organisms and potential to revolutionize treatment strategies in an era plagued by increased resistance. The exploration of this technology emphasises the importance of multidisciplinary approaches in developing medicines, combating infectious diseases, and improving patient outcomes.

The development of novel and effective antimicrobial agents has become a pressing concern in the face of the growing problem of antimicrobial resistance (AMR). Antibiotics have saved millions of lives, but they have developed acquired AMR due to increased use and misuse. AMR spreads through human–human interactions, particularly in livestock feed. The prevalence of AMR is increasing worldwide, posing a silent pandemic threat to public health. Treatment options are limited, leading to significant morbidity and mortality. Immediate interventions include surveillance, minimizing antibiotic use, access to quality medicines, vaccines, etc. (Salam et al. 2023). Nanomaterials have emerged as a possible alternative because of their unique physical and chemical features (Altammar 2023). Nanoparticles have a high surface area relative to volume ratio, allowing close interactions with microbial membranes and the possibility of surface functionalization to boost their antibacterial activity (Díez-Pascual 2020). Silicon and silicon oxide nanoparticles have received substantial interest due to their flexibility in biological applications. Silicon-based nanoparticles, such as Si/SiO_2, have significant benefits, including their biocompatibility, simplicity of manufacturing and capacity to be functionalized with diverse therapeutic agents (Bruckmann et al. 2022). Si/SiO_2 nanoparticles' antibacterial action can be linked to a variety of processes, including the release of reactive oxygen species, disruption of bacterial cell membranes and the capacity to administer antimicrobial drugs in a targeted manner (Seil and Webster 2012). Furthermore, the combination of Si/SiO_2 nanoparticles with regular antibiotics might result in a synergistic impact, thereby solving the issue of antibiotic resistance (Mondal et al. 2024). Using Si/SiO_2 nanoparticles as a delivery mechanism for antimicrobial drugs can also assist to lower the dosage necessary and avoid unwanted side effects (Hemeg 2017). Their unique features, together with their capacity to improve the effectiveness of antimicrobial drugs, make them a prospective route for future study and development in antimicrobial treatment. Si/SiO_2 nanoparticles have potential antibacterial capabilities due to their capacity to break bacterial cell membranes, create reactive oxygen species and serve as excellent delivery routes for antimicrobial drugs (Ozdal and Gurkok 2022).

17.2 Different Types of Nanocomposites of Si/SiO_2 Nanoparticles and Their Biocompatibility and Biodistribution

Nanocomposites of silicon (Si) and silicon dioxide (SiO_2) nanoparticles provide unique methods to combat antibiotic resistance, which is a rising worldwide health problem. These nanocomposites can be engineered to enhance the antibacterial properties of traditional antibiotics while simultaneously addressing the limitations associated with antibiotic resistance. For instance, Si/SiO_2 nanoparticles can be combined with various antimicrobial agents, creating hybrid systems that not only improve drug solubility and stability but also facilitate controlled release mechanisms, ensuring prolonged therapeutic effects. Furthermore, the nanoparticles' distinct surface features may be adjusted to optimize their interaction with bacterial membranes, improving their efficiency in dissolving biofilms and suppressing bacterial growth. Furthermore, research has shown that Si/SiO_2-based nanocomposites can possess intrinsic antimicrobial properties, providing a dual-action approach that targets resistant strains (Bhattacharjee et al. 2023). By integrating these advanced materials into medical devices, coatings and drug formulations, researchers are paving the way for more effective treatments that could significantly reduce the impact of antibiotic resistance in clinical settings.

17.2.1 Biocompatibility

The biological compatibility of nanoparticles is a significant factor of consideration for their usage in biomedical applications. Studies have showed that silicon-based nanoparticles display relatively low toxicity and high biocompatibility (Yanmei Huang et al. 2022). These nanoparticles' surface qualities may be modified to improve their interaction with biological systems, hence increasing their overall biocompatibility. Extensive research has been conducted on the biocompatibility of silicon nanoparticles and silicon oxide nanoparticles (Bragato et al. 2024). These studies have shown that these nanomaterials can be readily integrated with biological systems without causing adverse immune responses or cytotoxic effects. The low toxicity of these nanoparticles renders them as an attractive candidate for a variety of medicinal applications, including biosensing, bioimaging and targeted medication administration.

17.2.2 Biodistribution

The biodistribution of nanoparticles is another crucial aspect to consider for their biomedical applications. Studies have demonstrated that silicon-based nanoparticles exhibit favorable biodistribution profiles, with the nanoparticles being cleared from the body via renal excretion (Dos Santos et al. 2014). This reduces the possibility of long-term build-up and possible toxicity, which is a significant advantage over other nanoparticle forms. Numerous *in vitro* and *in vivo* investigations have been conducted to examine the biodistribution of Si/SiO_2 nanocomposites. According to this research, there is less chance of long-term damage because the nanoparticles may efficiently target particular tissues or cells and are also quickly removed from the body. Through intravenous injection, the pharmacokinetics and *in vivo* biodistribution of silica nanoparticles (SiO_2) with systematically modified geometries, porosities and surface characteristics were investigated in immune-competent CD-1 mice. The liver and spleen extensively absorbed the nanoparticles (Yu et al. 2012).

17.3 Applications of Silica Nanoparticles in Antimicrobial Therapeutics

Innovations in nanotechnology, particularly involving silica nanoparticles, are revolutionizing medical applications, especially in the realm of antimicrobial therapeutics. As traditional antibiotics lose their effectiveness due to rising bacterial resistance, novel strategies are imperative. Silica-based materials demonstrate significant potential due to their unique surface properties that facilitate the targeted delivery of antimicrobial agents. These nanoparticles can enhance the efficacy of existing treatments, combating pathogenic bacteria that contribute to increasing rates of hospitalizations and mortality (Matthews et al. 2010). Furthermore, their versatility extends to diagnostics, providing rapid and accurate identification of infectious agents, which is crucial in clinical settings where timely intervention can drastically alter patient outcomes. Beyond bacteria, silica nanoparticles are also being explored in viral contexts, emphasizing their role in combating diseases like coronavirus disease 2019 (COVID-19) (Hildebrandt et al. 2022). Ultimately, the integration of silica nanoparticles into medical settings represents a significant stride towards effective, modern healthcare solutions.

17.3.1 Targeted Delivery of Antimicrobials

Targeted medication delivery is one of the many biological uses for silica nanoparticles that have shown promise (Abeer et al. 2020). Their unique characteristics that includes their large surface area and adjustable porosity enable the effective encapsulation of several antimicrobial substances, improving their stability and bioavailability (Guo et al. 2015). These nanoparticles have numerous benefits, including high stability, ease of functionalization and biocompatibility, which makes them ideal for creating targeted and efficient drug delivery systems (Hsu et al. 2023; Huang et al. 2024). These nanoparticles can selectively deliver antimicrobials to infected tissues by functionalizing the silica surface with particular targeting ligands. This maximizes therapeutic efficiency while reducing adverse effects on healthy cells. In addition to increasing the treatment's overall efficacy, this focused strategy lowers the risk of drug resistance, which is a major concern in antimicrobial therapy. Through surface modifications and functionalization, silica nanoparticles can be designed to selectively target and deliver antimicrobial compounds to specific sites of infection, thereby enhancing their efficacy and reducing potential side effects (Selvarajan et al. 2020). Furthermore, silica nanoparticles can be engineered to release their payload in response to environmental triggers, such as changes in pH or temperature, providing a controlled release mechanism that optimizes the therapeutic outcomes in combating infections (Guo et al. 2015). Due to its versatility and better functionality to bypass physiological barriers, nanoparticle-based drug delivery, particularly silica nanoparticle-based is getting an upper hand than all the other existing drug delivery systems. In the light of antibiotic resistance developed by microbes against traditional antibiotics, the need for a novel strategy is of utmost importance. High surface area, biocompatibility and stability in acidic tissues are characteristics of inorganic silica nanoparticles (SiNP), which enhance thermal and physiochemical characteristics and prevent other compounds from eroding and degrading thermally (Yujing Huang et al. 2024). These nanoparticles can enhance the efficacy of existing treatments, combating pathogenic bacteria that contribute to increased rate of hospitalization and mortality (Matthews et al. 2010). In addition to the actual antibacterial action of the cargo load, nanoparticles can attack pathogens through a variety of mechanisms, including as physical damage to cell membranes, ROS generation and endo lysosomal burden. The window for the emergence of antibiotic resistance is therefore quite limited. The Food and Drug Administration (FDA) has authorized silica. Its structure contains silanol groups (Si—OH), which facilitate covalent bonding and facile mixing with other substances including antibodies, nucleic acids and fluorescent compounds (Foglia et al. 2017). Furthermore, their versatility extends to diagnostics, providing rapid and accurate identification of infectious agents, which is crucial in clinical settings where timely intervention can drastically alter patient outcomes. Beyond bacteria, SiNPs are also being explored in viral contexts, emphasizing their role in combating diseases like COVID-19 through controlled drug delivery and diagnostic advancements. Owing of its high surface energy and potential for toxicity, silica is often used to coat magnetic nanoparticles (MNP), such as magnetite and ferrite-based transition metals (Sharma et al. 2016). The coating improves the biocompatibility and physiochemical stability as well as enables the synthesis of multifunctional particles (Kavkhani et al. 2022). MNPs can be used in biological applications because of their large surface area, relativity and responsiveness to an external magnetic field (Kubíčková et al. 2020). According to an earlier research, silica-based nanoparticles that are magnetic have a number of biological and medicinal properties, including antitumour, anti-protozoal, antibacterial and stroke treatment (Gao et al. 2018; Itatahine et al. 2018; Santha Moorthy et al. 2017).

Table 17.1 shows nanoparticle agents for drug delivery, their mechanisms, advantages and limitations. The FDA approved liposomes and micelle for the carriage and delivery of mRNA and siRNA drugs exhibited some limitations like low solubility, short shelf life and relatively high cost of equipment and raw materials for the production of lipids (Trucillo et al. 2020). Dendritic mesoporous silica nanoparticles (300 nm) have been used lately to deliver mRNA vaccinations, suggesting their potential utility as alternate carriers for nucleic acid-based therapeutics (Theobald 2020; Wang et al. 2018). Also, these SiNPs are of low production cost than lipid liposomes. Orally administered drugs and nanoformulations can undergo structural changes or complete breakdown and result in the premature release of drug and degradation due to the acidic environment of the stomach (Pasarin et al. 2023). Enzymes and efflux transporters in the gastrointestinal tract can also reduce drug stability interfering with the bioavailability of the drug (Gavhane and Yadav 2012). Because of their *in vivo* stability and biosafety, SiNPs are a flexible method of oral medication delivery. Surface coating of the SiNPs can increase their stability. After being stable in the stomach's acidic environment, the SiNP can be designed to be pH sensitive, allowing them to release the payload in the colon or small intestine's more neutral pH (Juère et al. 2020).

Sulfasalazine, an antirheumatic medication, was encapsulated in mesoporous SiNP and coated with succinylated soy protein isolate to prevent drug release at pH 1.2 in the stomach and pH 5.5 in the duodenum. The pancreatic enzyme hydrolyzes the succinylated soy protein isolate, releasing sulfasalazine from the mesoporous SiNP at pH 7.4 in the small intestine (Popat et al. 2014).

The approval for silica nanoparticle use depends on its intended purpose. Synthetic amorphous silica is generally recognized as a safe material by the FDA for food, drug and cosmetic applications (Kirla et al. 2023).

17.3.2 Intracellular Targeting for Anti-tuberculosis Therapy

The World Health Organization (WHO) identifies tuberculosis as a leading infectious killer, but conventional treatments face challenges like heavy pill regimens and prolonged treatment durations. Nanotechnology offers a novel approach for targeted drug delivery, reducing drug doses, reducing side effects, improving solubility, bioavailability and patient compliance. These nanocarriers can be linked with ligands for even more targeted treatment (Kumar et al. 2023). Silica nanoparticles are increasingly recognized for their potential in enhancing

Table 17.1 Agents for drug delivery with their mechanisms, advantages and limitation.

Agents	Mechanism	Advantage	Limitation
Cellular-based drug delivery • RBC • Stem cells	Living cells (e.g., red blood cells (RBCs), dendritic cells, stem cells) can be modified to carry drugs or therapeutic agents to target sites. These cells can either encapsulate the drug within themselves or act as carriers that deliver the drug directly to tissues or tumours.	High specificity, biocompatibility, natural ability to navigate biological systems.	Limited by cell viability, immune response and regulatory challenges.
Bacteria-based drug delivery • Salmonella • Lactobacillus	Genetically engineered bacteria can be used to transport and deliver drugs to specific sites in the body, particularly in the case of localized infections or cancer. The bacteria may target certain tissues or organs based on their ability to interact with specific biological markers.	Ability to target specific tissues or infections, potential for self-replication at the site of disease.	Potential for immune rejection, risk of uncontrolled bacterial growth and infection.
Viral-based drug delivery • Adenovirus • Oncolytic viruses	Viruses can be genetically modified to serve as vectors for delivering genes, drugs or vaccines. The virus delivers the therapeutic agents directly into cells, where they can express the desired therapeutic effects.	High specificity, ability to deliver complex genetic material, efficient cell targeting.	Risk of immune response, potential for viral replication outside target cells and regulatory concerns.
Protein and antibody-based drug delivery • Monoclonal antibodies • Fusion proteins	Proteins and antibodies can be modified to bind specifically to target cells, tissues, or pathogens, delivering the drug directly to the site of action. This system improves the targeting of therapeutic agents.	Highly targeted delivery, low systemic toxicity and the ability to treat specific conditions like cancer or autoimmune diseases.	High production cost, potential immune response and limited stability.

(Continued)

Table 17.1 (Continued)

Agents	Mechanism	Advantage	Limitation
Nanoparticles and nanocarriers • Liposomes • Polymeric nanoparticles • Gold nanoparticles	Nanoparticles are used to encapsulate drugs, which can then be released at controlled rates or in response to environmental stimuli (e.g., pH, temperature). Nanocarriers are typically made from lipids, polymers, or metals.	Improved bioavailability, controlled release, targeting of specific tissues or cells and reduced side effects.	Complex production, potential toxicity of certain nanomaterials and regulatory hurdles.
Microencapsulation • Polymeric microsphere • Hydrogels	In microencapsulation, drugs are enclosed in a microscopic shell made of natural or synthetic materials. This allows for controlled release of the drug over an extended period, protecting it from degradation and providing targeted release.	Long-term controlled release, protection of sensitive drugs and reduced dosing frequency.	Slow release, and certain types of microcapsules may have poor stability or require specialized manufacturing processes.
Lipid-based drug delivery • Liposomes • Solid lipid nanoparticles	Lipid-based carriers (liposomes or solid lipid nanoparticles) are used to encapsulate drugs. Lipid bilayers help in drug protection, and these carriers can fuse with cell membranes, allowing for intracellular drug delivery.	Biocompatible, can be designed to deliver drugs across biological barriers (like the blood-brain barrier), and can improve drug solubility.	Production costs, potential instability, and challenges in scaling up.
Polymer-based drug delivery • Polymeric micelles • PLGA	Polymers, particularly biodegradable ones, can be used to create drug delivery systems that release drugs in a regulated manner over time. These polymers may be in the form of nanoparticles, microspheres, or hydrogels.	Controlled release, biocompatibility, and flexibility in design.	Complex manufacturing processes, potential for incomplete biodegradation, and variability in drug release rates.
Inorganic nanoparticles and metals • Magnetic nanoparticles • Gold nanoparticles	Drug delivery can be accomplished with inorganic materials including magnetic, silica and gold nanoparticles. For targeted drug delivery, these materials can be functionalized with medications, antibodies, or targeting ligands.	Precise targeting, controlled release, ability to combine with imaging techniques for monitoring drug delivery.	Potential for toxicity, difficult to scale up for large-scale manufacturing.

anti-tuberculosis therapy through intracellular targeting. Tuberculosis, caused by *Mycobacterium tuberculosis*, often requires the delivery of antibiotics that can effectively penetrate host cells where the bacteria reside. Silica nanoparticles can be engineered to encapsulate anti-tuberculosis drugs, facilitating their uptake by macrophages, the primary immune cells that harbor the bacteria. By modifying the surface of these nanoparticles with specific ligands that bind to receptors on macrophages, researchers can achieve targeted delivery, ensuring that the therapeutic agents are released directly within the infected cells. This approach not only improves drug bioavailability and efficacy but also helps to reduce systemic side effects. Furthermore, silica nanoparticles can be designed for controlled release, allowing for sustained therapeutic levels of the drug over time, which is crucial for effective TB management and reducing the risk of resistance development. Overall, the application of silica nanoparticles in intracellular targeting represents a promising strategy to enhance the effectiveness of anti-tuberculosis therapies (Montalvo-Quirós et al. 2021).

Infectious diseases caused by multidrug-resistant pathogens are a significant threat to public health, with methicillin-resistant. *Staphylococcus aureus* being a prime example (Abebe and Birhanu 2023). In order to address this challenge, researchers have explored the potential of nanoparticles, particularly silver nanoparticles, as a viable alternative to conventional antimicrobial agents (Bruna et al. 2021). Silver nanoparticles have demonstrated broad-spectrum antibacterial activity against both Gram-negative and Gram-positive bacteria, as well as potent anti-biofilm properties. Furthermore, the combination of silver nanoparticles and antibiotics has shown promising results, suggesting that this approach could be a valuable strategy in the fight against drug-resistant pathogens (Alotaibi et al. 2022). One particular application of nanoparticles in antimicrobial therapeutics is the targeting of intracellular pathogens, such as Mycobacterium tuberculosis, the causative agent of tuberculosis. A key benefit of using silica nanoparticles for intracellular targeting in TB therapy is the ability to minimize systemic toxicity. By directing the therapeutic agents specifically to the site of infection—macrophages harbouring *Mtb*—the concentration of the drug at the target site can be greatly increased while minimizing exposure to non-target tissues. This targeted approach can enhance the bioavailability of the drug at the infection site and improve treatment outcomes, while reducing side effects typically associated with conventional systemic drug administration.

The ability to design silica nanoparticles for dual or multi-drug delivery is another notable advantage. This approach can combine several anti-tuberculosis drugs into a single nanoparticle formulation, improving the overall therapeutic efficacy. For example, combining drugs like isoniazid (INH) and rifampicin (RIF), which are commonly used in the treatment of TB, within a single silica nanoparticle carrier, can help achieve synergistic effects while minimizing the

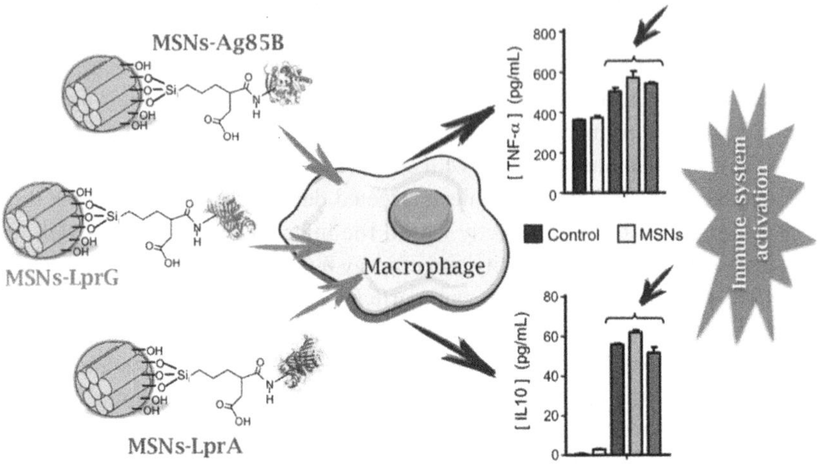

Figure 17.1 Interaction of mesoporous silica nanoparticles (MSNs) functionalized with specific antigens or molecules (such as Ag85B, LprG and LprA) with macrophages. *Source*: Montalvo-Quirós (2020) / MDPI / CC BY 4.0.

risk of drug resistance (Nair et al. 2023). Additionally, nanoparticles can be designed to include agents that improve the penetration of the drug into the macrophages, enhancing its ability to reach the site of infection (Figure 17.1).

Beyond silica nanoparticles, other types of nanoparticles, such as silver nanoparticles, have also been explored for intracellular targeting against *Mtb*. Silver nanoparticles have been shown to possess potent antibacterial activity against a wide range of pathogens, including *Mtb*, and have demonstrated strong anti-biofilm properties, which are critical in overcoming the protective biofilms formed by bacteria during chronic infections. Silver nanoparticles have the added advantage of being able to target both intracellular and extracellular forms of *Mtb*, which may contribute to a more comprehensive treatment strategy for TB. The application of nanoparticles in the treatment of tuberculosis is not limited to silica and silver nanoparticles. The emergence of multidrug-resistant mycobacterial strains is causing a crisis in antimycobacterial therapeutics, highlighting the global health concern of TB. Nanotechnology, particularly silver nanoparticles (AgNPs), has emerged as a promising approach to TB therapy. AgNPs have broad-spectrum antimicrobial activity, potentially enhancing the effectiveness of traditional antibiotics and improving treatment outcomes. However, the successful implementation requires addressing challenges like suboptimal delivery, variability in intra-macrophagic antimycobacterial effect and potential toxicity

(Barua and Buragohain 2024). Other nanomaterials such as liposomes, solid lipid nanoparticles (SLNs) and polymeric nanoparticles have also been studied for their ability to encapsulate anti-tuberculosis drugs and deliver them directly to macrophages or other relevant immune cells. These nanomaterials offer distinct advantages, including biodegradability, controlled release and improved bioavailability. For instance, liposomes, which are composed of lipid bilayers, have been used extensively in drug delivery due to their biocompatibility and ability to protect encapsulated drugs from degradation. Liposome-based systems have shown promising results in delivering drugs like rifampicin and pyrazinamide to macrophages, enhancing the effectiveness of TB therapy (Kumar et al. 2023).

17.4 Conclusion and Future Prospective

The rise of antibiotic resistance underscores an urgent need for innovative strategies in combating infectious diseases. Silica-based nanoparticles, such as Si and SiO_2, have demonstrated immense potential as antimicrobial agents and delivery platforms, offering promising solutions to this global challenge. Their unique properties, including a high surface area, biocompatibility and functionalization versatility, position them as highly effective tools for enhancing the efficacy of existing antibiotics and mitigating resistance. By leveraging mechanisms, such as reactive oxygen species generation, bacterial membrane disruption and targeted drug delivery, Si/SiO_2 nanoparticles not only improve therapeutic outcomes but also reduce the required dosages and associated side effects.

Future research should focus on optimizing the synthesis and functionalization of these nanoparticles to maximize their antimicrobial activity while ensuring safety and scalability. Investigating the synergistic effects of Si/SiO_2 nanoparticles with conventional antibiotics across a broader spectrum of pathogens will provide deeper insights into their therapeutic potential. Furthermore, the exploration of their applications in combating biofilm-associated infections and multidrug-resistant organisms can expand their utility in clinical settings.

The integration of silica nanoparticles into antimicrobial therapy also calls for interdisciplinary collaboration, bridging material science, microbiology and pharmacology to address existing limitations and uncover new possibilities. As we move forward, rigorous preclinical and clinical evaluations will be pivotal in translating the promising laboratory findings into real-world applications. In the era of rising antibiotic resistance, the innovative use of Si/SiO_2 nanoparticles represents a significant stride towards revolutionizing infection management and safeguarding global health.

References

Abebe, A.A. and Birhanu, A.G. (2023). Methicillin resistant *Staphylococcus aureus*: molecular mechanisms underlying drug resistance development and novel strategies to combat. *Infection and Drug Resistance* 16: 7641–7662. https://doi.org/10.2147/IDR.S428103.

Abeer, M.M., Rewatkar, P., Qu, Z. et al. (2020). Silica nanoparticles: a promising platform for enhanced oral delivery of macromolecules. *Journal of Controlled Release* 326: 544–555. https://doi.org/10.1016/j.jconrel.2020.07.021.

Alotaibi, A.M., Alsaleh, N.B., Aljasham, A.T. et al. (2022). Silver nanoparticle-based combinations with antimicrobial agents against antimicrobial-resistant clinical isolates. *Antibiotics (Basel, Switzerland)* 11 (9): https://doi.org/10.3390/antibiotics11091219.

Altammar, K.A. (2023). A review on nanoparticles: characteristics, synthesis, applications, and challenges. *Frontiers in Microbiology* 14: 1155622. https://doi.org/10.3389/fmicb.2023.1155622.

Barua, N. and Buragohain, A.K. (2024). Therapeutic potential of silver nanoparticles (AgNPs) as an antimycobacterial agent: a comprehensive review. *Antibiotics (Basel, Switzerland)* 13 (11): https://doi.org/10.3390/antibiotics13111106.

Bharti, C., Nagaich, U., Pal, A.K., and Gulati, N. (2015). Mesoporous silica nanoparticles in target drug delivery system: a review. *International Journal of Pharmaceutical Investigation* 5 (3): 124–133. https://doi.org/10.4103/2230-973X.160844.

Bhattacharjee, R., Negi, A., Bhattacharya, B. et al. (2023). Nanotheranostics to target antibiotic-resistant bacteria: strategies and applications. *OpenNano* 11: 100138. https://doi.org/10.1016/j.onano.2023.100138.

Bragato, C., Mazzotta, R., Persico, A. et al. (2024). Biocompatibility analysis of bio-based and synthetic silica nanoparticles during early zebrafish development. *International Journal of Molecular Sciences* 25 (10): https://doi.org/10.3390/ijms25105530.

Bruckmann, F.D., Nunes, F.B., Salles, T.D. et al. (2022). Biological applications of silica-based nanoparticles. *Magnetochemistry*. https://doi.org/10.3390/magnetochemistry8100131.

Bruna, T., Maldonado-Bravo, F., Jara, P., and Caro, N. (2021). Silver nanoparticles and their antibacterial applications. *International Journal of Molecular Sciences* 22 (13): https://doi.org/10.3390/ijms22137202.

Chinemerem Nwobodo, D., Ugwu, M.C., Oliseloke Anie, C. et al. (2022). Antibiotic resistance: the challenges and some emerging strategies for tackling a global menace. *Journal of Clinical Laboratory Analysis* 36 (9): e24655. https://doi.org/10.1002/jcla.24655.

Díez-Pascual, A.M. (2020). Antibacterial action of nanoparticle loaded nanocomposites based on graphene and its derivatives: a mini-review. *International Journal of Molecular Sciences* 21 (10): https://doi.org/10.3390/ijms21103563.

Dos Santos, C.A., Seckler, M.M., Ingle, A.P. et al. (2014). Silver nanoparticles: therapeutical uses, toxicity, and safety issues. *Journal of Pharmaceutical Sciences* 103 (7): 1931–1944. https://doi.org/10.1002/jps.24001.

Foglia, S., Ledda, M., Fioretti, D. et al. (2017). in vitro biocompatibility study of sub-5 nm silica-coated magnetic iron oxide fluorescent nanoparticles for potential biomedical application. *Scientific Reports* 7 (1): 46513. https://doi.org/10.1038/srep46513.

Gao, Q., Xie, W., Wang, Y. et al. (2018). A theranostic nanocomposite system based on radial mesoporous silica hybridized with Fe_3O_4 nanoparticles for targeted magnetic field responsive chemotherapy of breast cancer. *RSC Advances* 8 (8): 4321–4328. https://doi.org/10.1039/C7RA12446E.

Gavhane, Y.N. and Yadav, A.V. (2012). Loss of orally administered drugs in GI tract. *Saudi Pharmaceutical Journal* 20 (4): 331–344. https://doi.org/10.1016/j.jsps.2012.03.005.

Guo, Y., Andrew Davidson, R., Peck, K.A., and Guo, T. (2015). Encapsulation of multiple large spherical silica nanoparticles in hollow spherical silica shells. *Journal of Colloid and Interface Science* 445: 112–118. https://doi.org/10.1016/j.jcis.2014.12.055.

Hemeg, H.A. (2017). Nanomaterials for alternative antibacterial therapy. *International Journal of Nanomedicine* 12: 8211–8225. https://doi.org/10.2147/IJN.S132163.

Hildebrandt, R., Skubacz, K., Chmielewska, I. et al. (2022). Implementing silica nanoparticles in the study of the airborne transmission of SARS-CoV-2. *Molecules (Basel, Switzerland)* 27 (12): https://doi.org/10.3390/molecules27123896.

Hsu, C.-Y., Rheima, A.M., Kadhim, M.M. et al. (2023). An overview of nanoparticles in drug delivery: properties and applications. *South African Journal of Chemical Engineering* 46: 233–270. https://doi.org/10.1016/j.sajce.2023.08.009.

Huang, Y., Li, P., Zhao, R. et al. (2022). Silica nanoparticles: biomedical applications and toxicity. *Biomedicine & Pharmacotherapy* 151: 113053. https://doi.org/10.1016/j.biopha.2022.113053.

Huang, Y., Guo, X., Wu, Y. et al. (2024). Nanotechnology's frontier in combatting infectious and inflammatory diseases: prevention and treatment. *Signal Transduction and Targeted Therapy* 9 (1): 34. https://doi.org/10.1038/s41392-024-01745-z.

Itatahine, A., Ait Mehdi, Y., Fizir, M. et al. (2018). Multifunctional carbon nanomateriels for camptothecine low-water soluble anticancer drug delivery. *New Journal of Chemistry* 42 (2): 1326–1336. https://doi.org/10.1039/C7NJ04609J.

Juère, E., Del Favero, G., Masse, F. et al. (2020). Gastro-protective protein-silica nanoparticles formulation for oral drug delivery: in vitro release, cytotoxicity and mitochondrial activity. *European Journal of Pharmaceutics and Biopharmaceutics* 151: 171–180. https://doi.org/10.1016/j.ejpb.2020.03.015.

Kavkhani, R., Hajalilou, A., Abouzari-Lotf, E. et al. (2022). CTAB assisted synthesis of $MnFe_2O_4$@ SiO_2 nanoparticles for magnetic hyperthermia and MRI application. *Materials Today Communications* 31: 103412. https://doi.org/10.1016/j.mtcomm.2022.103412.

Kirla, H., Henry, D.J., Jansen, S. et al. (2023). Use of silica nanoparticles for drug delivery in cardiovascular disease. *Clinical Therapeutics* 45 (11): 1060–1068. https://doi.org/10.1016/j.clinthera.2023.08.017.

Kubíčková, L., Kaman, O., Veverka, P. et al. (2020). The ε-$AlxFe_2$-xO_3 nanomagnets as MRI contrast agents: factors influencing transverse relaxivity. *Colloids and Surfaces A: Physicochemical and Engineering Aspects* 589: 124423. https://doi.org/10.1016/j.colsurfa.2020.124423.

Kumar, M., Virmani, T., Kumar, G. et al. (2023). Nanocarriers in tuberculosis treatment: challenges and delivery strategies. *Pharmaceuticals (Basel, Switzerland)* 16 (10): https://doi.org/10.3390/ph16101360.

Matthews, L., Kanwar, R.K., Zhou, S. et al. (2010). Applications of nanomedicine in antibacterial medical therapeutics and diagnostics. *Open Tropical Medicine Journal* 3 (1): 1–9. https://doi.org/10.2174/1874315301003010001.

Mondal, S.K., Chakraborty, S., Manna, S., and Mandal, S.M. (2024). Antimicrobial nanoparticles: current landscape and future challenges. *RSC Pharmaceutics* 1 (3): 388–402. https://doi.org/10.1039/D4PM00032C.

Montalvo-Quirós, S., Vallet-Regí, M., Palacios, A. et al. (2020). Mesoporous silica nanoparticles as a potential platform for vaccine development against tuberculosis. *Pharmaceutics*. https://doi.org/10.3390/pharmaceutics12121218.

Montalvo-Quirós, S., Gómez-Graña, S., Vallet-Regí, M. et al. (2021). Mesoporous silica nanoparticles containing silver as novel antimycobacterial agents against *Mycobacterium tuberculosis. Colloids and Surfaces B: Biointerfaces* 197: 111405. https://doi.org/10.1016/j.colsurfb.2020.111405.

Nair, A., Greeny, A., Nandan, A. et al. (2023). Advanced drug delivery and therapeutic strategies for tuberculosis treatment. *Journal of Nanobiotechnology* 21 (1): 414. https://doi.org/10.1186/s12951-023-02156-y.

Nayl, A.A., Abd-Elhamid, A.I., Aly, A.A., and Bräse, S. (2022). Recent progress in the applications of silica-based nanoparticles. *RSC Advances* 12 (22): 13706–13726. https://doi.org/10.1039/D2RA01587K.

Ozdal, M. and Gurkok, S. (2022). Recent advances in nanoparticles as antibacterial agent. *ADMET & DMPK* 10 (2): 115–129. https://doi.org/10.5599/admet.1172.

Pasarin, D., Ghizdareanu, A.-I., Enascuta, C.E. et al. (2023). Coating materials to increase the stability of liposomes. *Polymers*. https://doi.org/10.3390/polym15030782.

Popat, A., Jambhrunkar, S., Zhang, J. et al. (2014). Programmable drug release using bioresponsive mesoporous silica nanoparticles for site-specific oral drug delivery. *Chemical Communications* 50 (42): 5547–5550. https://doi.org/10.1039/C4CC00620H.

Salam, M.A., Al-Amin, M.Y., Salam, M.T. et al. (2023). Antimicrobial resistance: a growing serious threat for global public health. *Healthcare (Basel, Switzerland)* 11 (13): https://doi.org/10.3390/healthcare11131946.

Santha Moorthy, M., Subramanian, B., Panchanathan, M. et al. (2017). Fucoidan-coated core–shell magnetic mesoporous silica nanoparticles for chemotherapy and magnetic hyperthermia-based thermal therapy applications. *New Journal of Chemistry* 41 (24): 15334–15346. https://doi.org/10.1039/C7NJ03211K.

Seil, J.T. and Webster, T.J. (2012). Antimicrobial applications of nanotechnology: methods and literature. *International Journal of Nanomedicine* 7: 2767–2781. https://doi.org/10.2147/IJN.S24805.

Selvarajan, V., Obuobi, S., and Ee, P.L.R. (2020). Silica nanoparticles-a versatile tool for the treatment of bacterial infections. *Frontiers in Chemistry* 8: 602. https://doi.org/10.3389/fchem.2020.00602.

Sharma, R.K., Yadav, M., and Gawande, M.B. (2016). Silica-coated magnetic nano-particles: application in catalysis. In: V. K. Sharma, R.-a. Doong, H.Kim, R. S. Varma, & D. D. Dionysiou (Eds.), *Ferrites and Ferrates: Chemistry and Applications in Sustainable Energy and Environmental Remediation*, vol. 1238, 1. American Chemical Society. https://doi.org/10.1021/bk-2016-1238.ch001.

Theobald, N. (2020). Emerging vaccine delivery systems for COVID-19: functionalised silica nanoparticles offer a potentially safe and effective alternative delivery system for DNA/RNA vaccines and may be useful in the hunt for a COVID-19 vaccine. *Drug Discovery Today* 25 (9): 1556–1558. https://doi.org/10.1016/j.drudis.2020.06.020.

Trucillo, P., Campardelli, R., Iuorio, S. et al. (2020). Economic analysis of a new business for liposome manufacturing using a high-pressure system. *Processes.* https://doi.org/10.3390/pr8121604.

Wang, Y., Song, H., Yu, M. et al. (2018). Room temperature synthesis of dendritic mesoporous silica nanoparticles with small sizes and enhanced mRNA delivery performance. *Journal of Materials Chemistry B* 6 (24): 4089–4095. https://doi.org/10.1039/C8TB00544C.

Yu, T., Hubbard, D., Ray, A., and Ghandehari, H. (2012). in vivo biodistribution and pharmacokinetics of silica nanoparticles as a function of geometry, porosity and surface characteristics. *Journal of Controlled Release1: Official Journal of the Controlled Release Society* 163 (1): 46–54. https://doi.org/10.1016/j.jconrel.2012.05.046.

18

Si/SiO$_2$ NPs-Based Therapies for Diagnostic Applications

Nancy Sanjay Gupta[1], *Shubham Kumar Shrivastav*[1], *Archana Singh*[2] and *Indrakant K. Singh*[1,3,4]

[1]*Molecular Biology Research Lab, Department of Zoology, Deshbandhu College, University of Delhi, New Delhi, Delhi, India*
[2]*Department of Plant Molecular Biology, University of Delhi South Campus, New Delhi, Delhi, India*
[3]*Delhi School of Public Health, Institute of Eminence, University of Delhi, New Delhi, Delhi, India*
[4]*Dr. B. R. Ambedkar Centre for Biomedical Research, & Delhi School of Public Health, Institute of Eminence, University of Delhi, Delhi, India*

18.1 Introduction

Nanoparticles are usually particulate substances whose preferably one-dimension ranges between 1 and 100 nm (Chen et al. 2016). They can take many different forms, including rods (Nikoobakht and El-Sayed 2003), multipods (Yong et al. 2006), wires (Cui et al. 2001), cages (Chen et al. 2005), stars, planes and spheres (Chen et al. 2016; Perrault and Chan 2009). Among the many special qualities that set nanoparticles apart from their bulk counterparts are their high surface energy, high surface-to-volume ratio and distinctive optical, mechanical, magnetic, thermal and electrical characteristics (Burda et al. 2005). Because of these characteristics, they may be used in a variety of fields, including electronics, communications, biology, medicine, energy harvesting and storage, and more. By adding desirable functional groups to their surface, their chemical reactivity and dispersibility in different solvents may be controlled to fit a specific application (Nam et al. 2013). In all branches of medicine, nanotechnology is at the forefront of the quickly evolving field of novel therapeutic and diagnostic ideas. The use of nanoparticles in medicine is known as nanomedicine (Moghimi et al. 2005). This area of advanced medicine aims to solve several issues and problems that traditional medicine faces, such as low target specificity, systemic and organ toxicity, and inadequate bioavailability (Boisselier and Astruc 2009). The capacity to integrate medications into nanosystems is a novel approach to pharmacotherapy that may be applied to the delivery of pharmaceuticals to specific cells.

Silicon Nanoparticles: Synthesis, Uptake, and Applications, First Edition.
Edited by Archana Singh, Aryadeep Roychoudhury, and Indrakant K. Singh.

18.1.1 Overview of Nanoparticle-Based Therapies

Drugs must overcome a number of biological obstacles in order to effectively reach the illness locations they are meant to treat. For oral medications to have a high systemic bioavailability, they must be very stable in the digestive tract and able to pass through intestinal epithelium (José 2004). Intravenous (IV) administration is still the most direct and effective way to administer medications that contain large molecules, proteins and polynucleotides, even though the majority of new drug development for small molecules is concentrated on oral delivery, with drug chemistry aimed at good oral absorption. To achieve their goals, drugs still in use must go past a number of obstacles. Drug delivery across the blood–brain barrier is being developed using a variety of nanoparticle-based techniques, such as cationic nanospheres, liposomes and albumin nanoparticles (Silva 2008).

The science of nanomedicine has grown and advanced tremendously during the past 25 years. Because it has the potential to transform medical therapy with more effective, less harmful, and intelligent treatments that might target disease regions like the elusive magic bullet, the field of nanoparticle-based medicine is given special attention (Desai 2012). A small number of nanomedicines have been effectively developed and authorized for clinical use, thanks to significant efforts by the biopharmaceutical sector and academic institutions. With a few and far between success stories, the area of nanomedicine is undoubtedly still in its infancy. Despite being fascinating and technically sophisticated, the majority of nanomedicine research is still in its infancy and is only gradually being applied to clinical trials along with everyday medical practice (Desai 2012).

18.1.2 Importance of Nanoprobe-Based Diagnostic Applications in Medicine

Traditional diagnostic techniques have been crucial in the detection of many diseases, but they have drawbacks, including being imprecise, slow, costly, as well as needing skilled personnel, especially in poor countries (Thwala et al. 2023). The increasing need for POC testing, high sensitivity and early identification necessitates the development of novel and creative diagnostic methods, such as nanodiagnostics, particularly for community medical centers in emerging economies. Due to a number of reasons, including location, population and socioeconomic situations, developing nations are more exposed to and vulnerable to a variety of issues (Thwala et al. 2023). In these nations, diseases are impeding human advancement by having a negative impact on life expectancy, income, education and other health indicators.

Nanodiagnostics is a new area that manipulates and analyzes single-molecule systems as well as channels for clinical diagnostics by using nanoscale characteristics (Wang et al. 2017). Nanomaterials' distinct physiochemical and optical properties have made a significant contribution to prompt and precise illness detection (Asdaq et al. 2021). It is becoming more and more important to create rapid, robust, user-friendly and economical diagnostic techniques based on nanotechnology for the detection of clinical samples since many illnesses have the potential to cause epidemics and high rates of morbidity and death (Thwala et al. 2023). Numerous nanomaterials for use in biomedicine have been created and documented in recent years. The introduction of paper-based sensors that could potentially be used for cancer screening, particularly in low-income nations, was claimed to have advanced by Dukle et al. (2022). Yang et al. described microfluidic POC equipment along with its uses in the detection of cancer, cardiovascular disease, infectious illnesses and chronic diseases in the same year. It was also explored how microfluidic POC devices may be used in the future in telemedicine and wearable technology (Yang et al. 2022). The state of sickness treatment in developing nations was also brought to light by this section. The electrochemiluminescence (ECL) signal amplification labelling for low-cost, portable devices that have supplanted conventional instrumentation and various light detecting methods used in paper ECL devices were evaluated by Chinnadayyala et al. (2019). Considerable attention has also been paid to how the analytical efficiency of the above ECL biosensors has been improved by downsized systems using immobilized chemicals on cellulose paper as well as nanoscale technologies, with potential applications in point-to-care diagnostics (Thwala et al. 2023).

18.1.3 Properties of Si/SiO_2 NPs

Among the most extensively prevalent materials in the world is silicon dioxide (SiO_2) or silica (Khan et al. 2020). Because of their varied physiochemical characteristics, silica nanoparticles (SNPs) have gained a lot of attention from the scientific community (Liou and Yang 2011). These particles can be classified as mesoporous or nanoporous based on the size of their pores. By changing the mix of the surfactants during synthesis, the particles' sizes may be changed (Liou and Yang 2011). Due to their hydrophobic nature, high surface area, pore volume and biocompatibility, SNPs are often affordable to produce on a large scale and have a wide range of uses (Akhter et al. 2022). Specific locations found in SNPs facilitate the particles' facile functionalization. For instance, the pore walls, pore entrance and interior/exterior of the surface of the particles are the three main functionalization sites found in mesoporous silica nanoparticles (MSNs) (Alsmaeil et al. 2021). SNPs are very desirable because of their promise for scale synthesis, simplicity of surface functionalization, colloidal stability, tuneable size of particles and biocompatibility. Researchers have recently effectively used silicon nanoparticles just like loading agents for a variety of payloads, including medicines and macromolecules including proteins, RNA and DNA (Mohamed Isa et al. 2021). Crucially, SNP has been shown to improve a variety of cargo's stability without compromising its distinctive physical and chemical characteristics (Janjua et al. 2023).

18.2 Diagnostic Applications of Si/SiO$_2$ NPs

One of the most important aspects of biomedical research is the study of medication delivery. One well-known aspect of the study is the use of tailored nanostructures to facilitate the targeted administration of medications within the patient's body (Manzano and Vallet-Regí 2020). These nanostructures have the ability to target particular biological tissues or organs and function as carriers (Dhinasekaran et al. 2020). Because of their enormous surface area as well as porous nature, MSNs have gained significant recognition in this respect. In recent years, these particles have been deployed as drug delivery nano-carriers (Manzano and Vallet-Regí 2020). Because of their small dimensions and wide surface area, SNPs have been used as biomarkers to identify certain proteins. Although quantum dots (QD) are widely used in this context, their toxicity from heavy metals and insoluble nature in water have prompted researchers to look for other alternatives (Fanizza et al. 2020). To this end, a number of researchers covalently bonded substances like tetramethyl rhodamine isothiocyanate (TRITC) dye as well as fuorophores to the silica nanoparticles (Ha et al. 2020). The produced materials were a good alternative to QDs since they demonstrated water solubility and less/nontoxic behaviour with encouraging outcomes (Lin et al. 2020). Table 18.1 discusses the types of silica nanoparticles involved in the clinical applications and in which phase they are in.

18.2.1 Bioimaging

SiO$_2$ NP have developed into potent medication delivery along with imaging contrast agent carriers that can provide details about the condition and course of diseases. Silica nanoparticles are used in a variety of imaging techniques (Baeza and Vallet-Regí 2020). Below is a discussion of several imaging methods.

Light Imaging: The application of functionalized fluorescent SiO$_2$ NP in imaging has been thoroughly investigated. α-L-fucose targeting lectin-labeled fluorescent as well as PEGylated SiO$_2$ NP can be used as endoscopic contrast chemicals to detect colorectal polyps early on, which may lead to colorectal cancer (Chen et al. 2017). PhENH2, which is an aggregation-induced emission (AIE) dye, was grafted onto SiO$_2$ NP by Wang et al. via chemical bonding with FA and

Table 18.1 Clinical applications, mechanism of action and clinical phases of SiO_2 NP.

Types of silica nanoparticles	Clinical application	Mechanism of action	Clinical stage	References
Fluorescent silica nanoparticles	Diagnostics and imaging	Tracking of lymph nodes and fluorescence imaging	Clinical trials	Hu et al. (2016)
Functionalized silica nanoparticles	Gene delivery	CRISPR/Cas9 delivery for genome editing	Preclinical	Bharali et al. (2005)
Mesoporous silica nanoparticles	Drug delivery systems	Drug solubility, bioavailability	Preclinical, clinical	Dutta Gupta et al. (2024)
Gold-coated silica nanoparticles	Photothermal therapy	Targets cancerous cells	Clinical	Kang et al. (2009)
Mesoporous silica nanoparticles	Antimicrobial and osteogenic	Bone regeneration, release of antimicrobial agents	Preclinical	Dutta Gupta et al. (2024)
Silica-based quantum dots	Detection of nucleic acids	Binds to nucleic acid	Diagnostic tools	Duan et al. (2017)
Silica-coated magnetic nanoparticles	Point-of-care diagnostics	Biomarker detection	Commercialization	Rong et al. (2019)

embedded molybdenum disulphide (MoS2) nanosheets (Wang et al. 2019). The resultant SiO_2 NP showed stable fluorescence for photothermal treatment of breast cancer. Functionalized SiO_2 NP with covalently attached and physically inserted rhodamine 101 were described by Prieto-Montero et al. for the purpose of visually targeting HeLa cells. In human breast cancer cells, FA-modified SiO_2 NP implanted with the photosensitizer produced targeted two-photon excitation fluorescence imaging (Prieto-Montero et al. 2020). Crucho et al. covalently glued a thermally triggered delayed fluorescent emitter inside SiO_2 NP, which mostly resided in the cytoplasm, to enable imaging within living cells (Crucho et al. 2020). Kirla et al. found that methylene amine-grafted, blue-doped, glucose-functionalized M-SNPs exhibited high fluorescence for bacterial imaging (Kirla et al. 2020).

Ultrasound Imaging: The low cost, favourable safety profile and superb temporal and spatial resolution of ultrasound make it a popular real-time imaging tool in operating rooms (Huang et al. 2022). Perfluoropentane gas-filled iron SiO_2 NP enriched using diethylenetriamine pentaacetic acid were described by Liberman et al. for the purpose of detecting the biodistributions of malignancies. Perfluorodecyltriethoxysilane, which has a low surface energy and also superhydrophobicity, was covalently bonded to SiO_2 NP that had been surface-capped with β-cyclodextrin in order to boost their hydrophobicity. The resultant particles were used in chemo-sonodynamic and antivascular treatment (Liberman et al. 2013).

Magnetic Resonance Imaging: Both in vitro and in vivo imaging may be accomplished with magnetic nanoparticles. M-SiO_2 NP with magnetic nanoparticle decoration and magnetic carbon multiwalled nanotube wrapping efficiently loaded siRNA and demonstrated strong cell uptake by magnetism. Mathieu et al. created ROS-responsive M-SiO_2 NP with PEG-conjugated chlorin e6 as a response element and a gadolinium (Gd)–DOTA complex as a barrier that demonstrated an increased T1-MRI signal (Mathieu et al. 2019). However, iron–iron oxide or Fe@FeOx continues to be investigated as a replacement for Gd-based contrast agents due to the possibility of their toxicity (Huang et al. 2022).

Radio Labeled Imaging: SiO_2 NP can be used in radiolabelled imaging as radiosensitizers. The in vivo distribution, decline, as well as clearance behaviours of SiO_2 NP may be identified using novel radioactive labels (Huang et al. 2022). After labelling the mesoporous shell and dense silica core of core-shell SiO_2 NP using zirconium-89, Bindini et al. discovered that the particles' biodistribution was constant for six hours (Bindini et al. 2021). Chen et al. used zirconium-89 to identify the rigid oxygen donors of several deprotonated groups of silanol on the outer layers of SiO_2 NP using a distinct synthetic technique. In order to detect melanoma, Portilho et al. synthesized technetium-99 m-labelled SiO_2 NP with cores that were magnetic and loaded them with dacarbazine (Chen et al. 2015).

18.2.2 Biosensing and Cell Tracking

There is an emerging new generation of biosensors that use MSNs. In addition to their enormous surface area, a typical characteristic of nanoparticles, MSNs have a number of special qualities that are very advantageous for developing biosensors. The potential for stable functionality of their membrane is one of these characteristics. In a one-step reaction with various alkoxysilanes, various functional groups, including thiol, amine, hydroxyl, halogen and others, can be readily grafted on the silica surface (Mladenović et al. 2021). This process is usually carried out in organic solvents and dry conditions, ideally at high temperatures to encourage the evaporation of alcohols. This characteristic makes it possible to use a variety of functionalization techniques to get the final functioning on the surface of nanoparticle that is needed (Mladenović et al. 2024; Yi et al. 2018).

Another special characteristic of MSNs that significantly encourages their usage in the development of biosensors is their organized porosity. To create electrochemical biosensors, the pores can be filled with signaling molecules like dyes or electroactive species (Cheng et al. 2019). More significantly, pore-blocking and surface-functionalization species can control the dispersal of the loaded cargo molecules. Therefore, it has been shown that big molecules or nanoparticles may successfully seal pore openings and ensnare cargo molecules (Cheng et al. 2019; Kim et al. 2019).

Furthermore, the potential of MSNs in cell tracking applications has been thoroughly investigated. These nanoparticles are perfect vectors for cell tracking using magnetic resonance imaging (MRI) because of their special qualities, which include higher internal efficiency, biocompatibility and pore architecture (Khoz et al. 2024). According to studies, mesenchymal stem cells or hMSCs may be effectively tracked by T1-enhancing MSNs functionalized to contain gadolinium (Gd) (Hsiao et al. 2008). The potential of MSNs in cell tracking applications has also been demonstrated by the development of silica-coated manganese oxide nanoparticles that serve as positive T1 contrast agents for the labeling as well as MRI tracking of adipose-derived mesenchymal stem cells (Khoz et al. 2024; Kim et al. 2011).

18.2.3 Targeted Diagnostics

The therapeutic effectiveness of medications with limited clinical application will be significantly impacted by MSNs' substantial drug load potential combined with targeted action that reduces contact with healthy tissue along with increase in cellular absorption of the pharmaceuticals in sick cells (Rosenholm et al. 2011). Given the diversity of carcinogenic activity, MSNs' high drug load capacity makes them effective instruments for multidrug delivery, an appealing feature of cancer treatment. Furthermore, since many of the currently marketed contrast agents have very short half-lives and are hazardous, using the particles as transporters for contrasting agents can greatly extend the imaging time frame since the particles are likely to be ingested by cells (Rosenholm et al. 2011).

Through the tailored administration of imaging agents for increased signal detection, this nanoplatform will provide diagnostic advantages in addition to better treatment. Additionally, the potential use of MSNs for the targeted distribution of biomarker/signaling sensors to improve the recognition of the often modest bioactivity is an intriguing application (Rosenholm et al. 2011). This drug delivery platform offers great instruments for multiple drug delivery and combination therapeutic and diagnostic activity also known as theragnostics since it may transport numerous entities (Girija and Balasubramanian 2019). The invention of MSNs for therapies as well as diagnostics is still in its early stages, and MSNs for targeted treatment of disease have not yet entered clinical trials, despite recent developments that highlight MSNs as intriguing drug delivery platforms (Rosenholm et al. 2011).

18.2.4 Drug Delivery System Using Silica Nanoparticles

Numerous causes contribute to the interest in nanoparticles and, more especially, MSNs for medication delivery. The first benefit, and arguably the most obvious one, is the size of MSNs' interactions with living systems. Human cells are known to vary widely in size; however, the majority typically fall within the micrometer range (Löffler 2017; Vallet-Regí et al. 2017). Because of their comparable scales, MSNs along with cells would interact intimately. The pharmacokinetic mechanism of the delivered medicinal substance may be precisely controlled with the help of MSNs. This is significant since the concentration of a free medication in the serum would reach its maximum when it is supplied, regardless of whether it is given orally or intravenously (Löffler 2017; Shi et al. 2017). Then, until the subsequent medication delivery, which would result in another concentration peak, that concentration would decrease. A 'roller coaster' concentration profile would result from this process; occasionally, the minima would fall below the ideal values and the maximum would exceed the toxicity thresholds (Etheridge et al. 2013). By maintaining the medication concentration at ideal levels for extended periods of time, MSNs can be used as tailored drug delivery systems, increasing therapeutic effectiveness and preventing possible toxicity. When used for medication delivery, MSNs can also shield the therapeutic chemicals as they move through the body. This would prevent any possible cargo degradation, which is especially important when administering soft therapeutic compounds like proteins or RNAs (Vallet-Regí et al. 2017). Because of MSNs' high loading capacity, two or more medications may be transported into the identical nanoparticles, enabling the development of combination treatments to treat malignancies that are resistant to several drugs.

Furthermore, this property makes it possible to use specific contrast compounds for biomedical image processing, which may be highly intriguing for tracking the course of treatment in real time (Etheridge et al. 2013; Vallet-Regí et al. 2017). The potential to give MSNs a responsive behaviour is an additional advantage of using them as medication delivery systems. This makes it possible to precisely regulate the release of cargo when a stimulus is provided, whether it be externally applied by the physician or internal and specific to the condition being treated (Vallet-Regí et al. 2017). In this way, the stimuli-responsive MSNs delivery system's regulation prevents the therapeutic substance from releasing too soon, which may be crucial when systemic toxicity is to be avoided (Vallet-Regí et al. 2017).

18.3 Advanced Diagnostic Techniques Using Si/SiO_2 NPs

Because of the level of sensitivity the detecting system, silica nanoparticles with varying methods provide varying detection limits. Numerous factors, including target analyte validation, signal amplification mode, sample separation and integration techniques, sensor interfaces as well as modification conditions, low interference and cost-effectiveness, might be blamed for the discrepancy. By incorporating new nanomaterials-assisted signal amplification technologies into the immune systems for target identification, these issues can be addressed (Bidram et al. 2021).

18.3.1 Point-of-Care Diagnostics

Due to its affordability and lack of toxicity, silicon is an exceptionally often utilized metalloid element in target detection aided by nanomaterials. The most widely used kind of silicon-based nanoparticles is spherical silica nanoparticles or SiO_2 NP. Mesoporous SiO_2 NP have been used as nanocarriers for medications, dyes and biomacromolecules due to their wide pore channels and easily accessible interior surfaces. To increase their usefulness, these particles have been doped with dyes or luminous rare earth complexes (Hou et al. 2023).

SiO_2 NP were usually created using a reverse microemulsion technique and then functionalized with amino or carboxyl groups. Fluorescent silica spheres were created by loading functionalized SiO_2 NP with fluorescent compounds or nanoparticles (Hou et al. 2023; Song et al. 2013). The fluorescent spheres were used to quantitatively identify targets after being covalently bonded with antibodies using either the enhanced periodate oxidation technique or carbodiimide cross-linking. To create colourful core-shell SiO_2 NP, for instance, the SiO_2 NP are capable of carrying dyes like Reactive Violet 5 before being encased in a silica shell. In order to identify β-agonists, rare-earth complexes, including Eu and Ru complexes, were covered with SiO_2 NP (Hou et al. 2023; Huang et al. 2018). To improve the robustness and sensitivity, QDs were additionally added to silica spheres and further functionalized and shielded by silica shells. The fluorescent spheres proved useful to detect CRP quantitatively after being covalently bound with antibodies (Hou et al. 2023). Procalcitonin (PCT) and C-reactive protein (CRP) were detected quantitatively and simultaneously in blood using an on-site detection method based on dual-color SiO_2 NP@QDs. The electrostatic adsorption of dense green or red CdSe/ZnS–COOH QDs on the surface of SiO_2 NP was used to create the dual-color SiO_2 NP@QDs nanotags, which have monodispersity and excellent luminescence (Hou et al. 2023; Zhou et al. 2020). These QDs nanotags were then conjugated with anti-PCT and anti-CRP monoclonal antibodies to create stable, fluorescence-enhanced QDs in the LFA system. The multiplex LFA system's sensitivity and specificity were impacted by the usage of SiO_2 NP@QDs with two distinct fluorescence signals (Zhou et al. 2020).

Methamphetamine was detected in urine samples by assembling AuNPs with dendritic silicon dioxide spheres via thiol-metal coordination. SiO_2 NP loaded with organic dyes exhibit adequate signal amplification and are significantly less expensive than Au (Hou et al. 2023). SiO_2 NP may also be used to load a wide range of materials, including big enzymes, tiny QDs and organic colours. To detect the rabies lyssavirus, for instance, pomegranate-shaped SiO_2 NP with thickness of 5 nm demonstrated a strong absorption of QDs and could be functionalized with antibodies (Hou et al. 2023).

18.3.2 Quantum Dot Analogues

Excellent optical characteristics, such as absorbance and photoluminescence, are displayed by colloidal semiconductor nanocrystals. QDs are common semiconductor nanocrystals made of two or more heavy metal components, such as CdTe, etc., having core-shell structures that are just a few nanometers in size. The NIR or UV regions of the electromagnetic range are where QDs release electromagnetic radiation when their high-energy electrons relax into holes (Hildebrandt et al. 2017; Hou et al. 2023). The electron bands in a quantum dot emerge as distinct energy levels because the diameters of QDs are of a comparable order equivalent to the exciton Bohr radius. Because of their special optical characteristics, QDs are the best particles among the many fluorescence markers. QDs' fluorescence emission wavelength may be changed by varying their chemical makeup and particle size (Hou et al. 2023).

The inherent features of a QD can be altered by adding more components. Core-shell QDs are composed of an inert inorganic core covered with a shell, which exhibits high photochemical stability. QDs have a huge molar extinction coefficient up to 106 l and a high fluorescence quantum yield (40–90%) (Michalet et al. 2005). Prolonged fluorescence durations, i.e. up to 20–50 ns, a high resolution threshold, comparatively small fluorescence emission, a wide excitation range and exceptional durability against photobleaching are just a few of the many benefits that QDs provide. The study of QDs in a variety of fields, from cellular and molecular biology to optical imaging including diagnostics in medicine, has garnered a lot of interest in recent decades. Because of these benefits, QDs have been used as labels in ICTSs to detect proteins, nucleic acids, tiny compounds and even microbes (Michalet et al. 2005).

Protein detection has made extensive use of such QD-labeled ICTS devices. In order to identify nitrated ceruloplasmin, nanocrystal QDs have been generated and used to ICA platforms. With a LOD of 8 ng/ml, the biosensor showed a broad range and a quick signal response for the targets with an amount of 1 ng/ml (Hou et al. 2023). Safety is still a crucial consideration even if several QDs have been recorded for ICAs. Because of their exceptional photostability, durability against bleaching, broad excitation range in the UV spectrum and narrow emissions, QDs have historically mostly been built on cadmium (Hildebrandt et al. 2017). Nonetheless, it is commonly recognized that certain heavy metals, such chromium and cadmium, are extremely harmful. Cadmium-free NPs are the best choice for labelling when the ASSURED standards are taken into account (Hou et al. 2023). Environmentally safe Cu:Zn–In–S/ZnS QDs have therefore been described using the ICA platform for tetanus antibody detection. This study offers an explanation for POC antibody determination, as the use of ICAs in antibody detection is often somewhat uncommon (Michalet et al. 2005).

It is well-established that ICST performance is significantly impacted by label size. Nevertheless, QDs are also tiny; therefore in order to maximize ICAs, the size must be increased. The intensity of fluorescence along with environmental endurance of QD nanobeads (QBs) can be greatly enhanced by encasing several individual QDs in a polymer matrix. Fluorescent sensors' stability and sensitivity were improved by this technique (Rong et al. 2019). To get over the drawbacks of low sensitivity and inadequate quantity, QBs were combined with ICTSs. According to the article, in order to achieve big QBs, the QDs were encapsulated with modified poly(*tert*-butyl acrylate-*co*-ethyl acrylate-*co*-methacrylic acid) (Michalet et al. 2005). The LOD was improved plus the prostate specific antigen

(PSA) was detected in 15 minutes using the QBs-based ICTS systems. To create fluorescent nanospheres larger than 200 nm, which were 380 times larger than QDs, Pang et al. used poly(styrene/acrylamide) polymeric nanospheres as carriers to insert CdSe/ZnS QDs. The detection of CRP using fluorescent nanospheres coupled with a monoclonal antibody was 257 times more precise compared with the colloidal gold-based ICA (Duan et al. 2017). Furthermore, both test lines were clearly visible under UV light. The International Agency for Research in Cancer has classified mycotoxins, which are harmful metabolic byproducts generated in mold-contaminated food, as a Group I carcinogen. Maize readily produces ochratoxin A and aflatoxin B1 (Duan et al. 2017). The findings of Xiong et al.'s synthesis of QBs with several sizes showed that 124 nm QBs were the best platform for ochratoxin A detection utilizing competitive ICA. Furthermore, QBs have been created to quantitatively detect carcinoembryonic antigen (CEA) and neuron-specific enolase (NSF) simultaneously. To put it briefly, large-sized QBs were more suited for the ICA than small-sized QBs or QDs since the latter exhibited lesser fluorescence intensities (Ren et al. 2014).

18.4 Combining Si/SiO_2 NPs with AI Machine Learning

Recently, many researchers are working on the development of a highly rigid machine learning model which can be used for the prediction of the SiO_2 NP efficiency in nanomedicine. A few studies have already depicted the usage of machine learning models to check the silica-based nanodiagnostics. In order to reduce and improve animal research, the Deep Learning model may be used as a platform to assist in the future creation of novel silica-based nanomedicines and assist scientists in making well-informed judgments about which SiO_2 NP should proceed to preclinical trials. Several domains, such as small molecule drug discovery, environmental health hazards assessments and animal-derived food safety evaluation, can benefit from the use of machine learning as well as deep learning models.

18.5 Si/SiO_2 NPs in Combination with Other Nanomaterials

Metal-Silica Nano Compositions: Since ancient times, people have used metals with natural antibacterial properties, such copper (Cu) and silver (Ag), to avoid getting sick. Nevertheless, they have poor pharmacokinetic characteristics and are highly cytotoxic to mammalian cells. Combining these metals with SiO_2 NP as codelivery payloads has therefore been successful in antibacterial applications. Copper is another typical heavy metal with bactericidal properties, similar to silver. Combining Cu with SiO_2 NP can aid in the long-term release of Cu, which has antibacterial properties. A scientist reported synthesizing core–shell copper-silica nanoparticles with antibacterial properties. The findings showed that compared to the insoluble Cu-OH particles, the core-shell nanoconstructs had superior antibacterial activity against *Bacillus subtilis* and *Escherichia coli* (Selvarajan et al. 2020).

AntiBiofilm Coating: According to Koo et al. (2017), microbial biofilm development is a major threat in nosocomial infections and an important aspect of virulence in localized chronic infections. Because of the low diffusion capability of traditional antibiotics, microorganisms within biofilms become resistant to them. It is extremely difficult to regulate biofilm production and eradicate premade biofilms since there are currently very few treatments in the pipeline (Koo et al. 2017). In addition, SiO_2 NP have demonstrated a strong inhibitory impact on wearable medical implants, making them invaluable in the fight against biofilms (Selvarajan et al. 2020).

Bone Implants and Dental Matrices: Drug-loaded SiO_2 NP can be utilized as an antibacterial component for dental matrices and bone cement, in addition to the many metallic biomedical implants covered above. One of the most widely used economically accessible orthopaedic cement matrices for orthodontic as well as bone implants is poly methyl methacrylate (PMMA). When antimicrobials are directly loaded onto this matrix, inadequate mixing or dispersion disparities make it challenging to achieve a sustained drug release. Therefore, in order to achieve regulated drug release, SiO_2 NP can serve as a reinforcing material (Selvarajan et al. 2020).

18.6 Biocompatibility and Toxicity

Determining the compatibility with life along with bio-distribution of MSNs that occur is necessary before incorporating the chemical molecule into an organism. According to a research by Hudson et al., mice who get intraperitoneal or intravenous injections die, while subcutaneous injections do not cause any harm. MSNs were directed towards distinct cells based on their properties (Hudson et al. 2008). Therefore, taking into account the size, shape and surface characteristics of MSNs, both variables rely on high dosage and big particles. The study of biomaterials' interactions with experimental animals' bodies to assess potential harmful consequences is known as biocompatibility (Shahbazi et al. 2012).

It is well-known that silica's permeability plus pore size affect its biocompatibility. Superficially porous silica as well as macroporous silicon are bioinert materials, but silica with porosity greater than 70% dissolves in all of the simulated bodily fluids except the stomach. PSi porosities below 70% are bioactive and slowly biodegradable (Hudson et al. 2008; Shahbazi et al. 2012). The reactive oxygen species (ROS), cytotoxicity and inflammatory responses in a number of cancer cell lines have been measured in order to evaluate the biocompatibility of silica-based products. Concentration of silica nanoparticles in the body along with its surface properties have a greater influence on the biocompatibility of the particles (Shahbazi et al. 2012).

Nearly all of the evaluated publications showed size-dependent toxicity caused by SiO_2 NP, whereas the majority of studies showed concentration and exposure time-dependent toxicity. Additionally, it was shown that the biological impacts of SiO_2 NP are influenced by their surface physicochemical characteristics (Mohammadi et al. 2018). In terms of hemolytic activity, thiol-added SiO_2 NP were less harmful than those containing aminopropyl. As a result, SiO_2 NP with thiol added may be a safe option for medication administration. SiO_2 NP' toxicity was influenced by their physical property, porosity. The toxicity of the particles was also influenced by their surface electrostatic properties; negatively charged SiO_2 NP were potentially hazardous than positively charged ones. SiO_2 NP that were delivered intravenously primarily accumulated in tissues that were enriched in the monocyte phagocyte network (Akhtar et al. 2010; Gong et al. 2012; Mohammadi et al. 2018).

There have been several attempts in recent years to address the drawbacks of traditional chemotherapy. One of the most promising is the creation of nano-drug delivery systems (nano-DDS) that may boost pharmacokinetic behaviour in vivo, enhance the bioavailability and solubility of drugs, increase their accumulation in tumour tissues, prolong their circulation time and improve their pharmacokinetic behaviour, all of which can improve the therapeutic effect while lowering side effects (Mu et al. 2020). MSNs are one type of materials among these nano-DDS that have attracted special attention from several researchers because of their enormous surface area, pore volume and easy large-scale manufacture. These characteristics allow MSNs to be prepared simply, such as with the sol-gel 'chimiedouce' procedures in aqueous solutions, and provide high drug packaging efficiency and delivery (Murugan et al. 2016). Dedicated attempts have been undertaken to use MSNs to create nanoplatforms for medication delivery and cancer treatment since the Food and Medication Administration deemed silica-based materials safe. The architecture of DDS has been incredibly flexible (Gao et al. 2020; Murugan et al. 2016). While some studies have used selective targeting groups to enhance chemotherapy selectivity and tumour targeting in MSNs, others have relied on the tumour microenvironment's features, such as its increased glutathione (GSH) concentration and lower pH compared to normal cells. Numerous MSNs that are sensitive to pH and/or redox have been developed to release chemotherapeutic medicines (Gao et al. 2020). These characteristics allow MSNs to be prepared simply, such as with the sol-gel 'chimiedouce' procedures in aqueous solutions, and provide high drug packaging efficiency and delivery. Dedicated attempts have been undertaken to use MSNs to create nanoplatforms for medication delivery and cancer treatment since the Food and Drugs Administration deemed silica-based materials safe (Gao et al. 2020). The architecture of DDS has been incredibly flexible. While some studies have used selective targeting groups to enhance chemotherapy selectivity and tumour targeting in MSNs, others have relied on the tumour microenvironment's features, such as its increased glutathione (GSH) concentration and lower pH compared to normal cells. Numerous MSNs that are sensitive to pH and/or redox have been developed to release chemotherapeutic medicines (Gao et al. 2020).

18.7 Current Clinical Applications

SiO_2 NP have been the subject of extensive investigation for a variety of biological applications, including illness diagnosis and control, genetic disorder identification and correction, and most importantly lifespan enhancement. As a result, SiO_2 NP have several benefits and have created new opportunities for biomedical research in a variety of cutting-edge fields.

18.7.1 Silica Nanoparticles in Gene Delivery Applications

For gene delivery, enriched silica nanoparticles (FSN) ought to be present in two distinct forms under two distinct circumstances. Gene loading is the first need, which calls for a high level of affinity among the genes as well as NPs. Under these circumstances, electrostatic contact is necessary for the NPs to absorb and store the gene molecules as well as for their interaction. Because DNA molecules have a negative charge, scientists attempt to create positively charging NPs, such amino groups. Gene release is the second need. To help release the genes through the NPs into the target, weak connections involving the FSN and gene are preferred in this situation. Bharali et al. documented the use of amino surface–enriched SiO_2 NP for gene transfer in vivo. Mouse brains were injected with free plasmid, plasmid DNA-binding SiO_2 NP and SiO_2 NP. Only plasmid DNA-binding SiO_2 NP, however, generated strong gene expression, indicating that the gene was well-shielded and transported into the cell nucleus (Bharali et al. 2005).

18.7.2 Pharmaceutical Applications

In edible as well as topical pharmaceutical preparations, colloidal dry silicon dioxide is often considered to be an excipient that is largely harmless and nonirritating.

However, because silica excipients are used for a number of purposes, such as viscocity-increasing agents, suspending agents, adsorbents, emulsion stabilizers, anticaking agents, glidants, tablet segregates and thermal stabilizers with a percent ratio ranging from 0.1 to 10, intraperitoneal and subcutaneous injections may result in local tissue reactions and/or granulomas, making parenteral administration impossible (Bitar et al. 2012).

Water adsorption benefits from silica's large surface area covered in polar silanol groups. These characteristics of silica were utilized by Gore and Banker to stabilize aspirin pills. For an analogous aspirin tablet, they investigated silica's moisture absorption capabilities and stabilizing impact. The experimental findings illustrated silica's water adsorption properties and how they rely on pore size and surface area (Bitar et al. 2012). Furthermore, the ideal concentration of silica for optimal stability was thought to be 3%. In a different study, magnesium stearate was mixed with silica. Although it was discovered to have adverse impacts on lubrication thus did not improve disintegration, it was observed that tablet strength had increased. Using amoxicillin as an example, the impact of using silica as a preservative on the drug's bioavailability was examined (Bitar et al. 2012). SiO_2 NP are frequently used as glidants in medicinal applications. The application of compacted hydrophobic and hydrophilic SiO_2 NP as pharmacological excipients (glidants) was examined by Jonat et al. Since hydrophobic silica only needs mild mixing conditions to obtain excellent flowability, they discovered that it is the most effective glidant. Hydrophilic silica, on the other hand, is highly dependent on mixing circumstances. Following equilibrating at high levels of humidity, all of the silica tested retained acceptable flowability at low glidant concentrations (Bitar et al. 2012; Jonat et al. 2004).

18.7.3 Application of SiO_2 NP in DNA Detection

In order to achieve variable fluorescence intensity, SiO_2 NP are also used to create DNA biosensors by functionalizing them with oligonucleotides and hybridizing them with complementary RNA, DNA or both probes. A paper described the use of disulphide-coupling chemistry to immobilize oligonucleotides onto SiO_2 NP. This was accomplished by first obtaining 60-nm silica particles, silanizing them using 3-mercaptopropyltrimethoxysilane (MPTS) and then immobilizing oligonucleotide probes on top of the SiO_2 NP before they were incubated for DNA hybridization (Zhou and Zhou 2004). To assess the effectiveness of hybridization, the signal from fluorescent molecules was examined at a radiation wavelength of 520 nm. In another study, researchers mounted oligonucleotide signaling probes and created novel core shell nanostructures. In microarray-based detection, these silica-coated gold nanoparticles emerged with dye were used for labelling. With photostable signals, sensitivity was increased (Jonat et al. 2004).

SiO_2 NP are also crucial in creating a reliable, quick and affordable isolation technique for DNA extraction, purification and analysis. The kinetics and structural variations of the plasmid DNA adsorption over silica in divalent as well as monovalent salts were investigated. They discovered that two types of electrostatic interactions need to be taken into account: the contact between the plasmid DNA plus the silica surface and the interaction between the plasmid DNA subunits that regulate the conformation of molecules (Bitar et al. 2012). Plasmid DNA adopts a noncompact structure with intramolecular electrostatic repulsion works well in a monovalent salt solution containing weak Na^+ ions. In contrast, the plasmid DNA molecule becomes more compact as the ionic strength increases, indicating that intramolecular repellent force is ineffective (Bitar et al. 2012). This connection greatly lowers the intramolecular repellent force of the DNA molecule, causing it to take on a very compact shape in a divalent solution containing Ca^{2+} ions, which create combinations with the atoms of oxygen of the phosphate groups of double-strand oligonucleotides. As a result, adsorption on silica is reversible, and the attachment coefficient measured in the presence of 300 mm NaCl is 0.01–0.03 (Bitar et al. 2012). In contrast, adsorption occurs quickly and irreversibly when 1 mm Ca^{2+} is present, and the attachment coefficient ranges from 0.75 to 0.87 (Bitar et al. 2012).

18.7.4 SiO_2 NP-Mediated DNA Extraction

DNA is frequently extracted from biological materials using silica-coated magnetic nanoparticles. Magnetic nanoparticles have emerged as preferred instruments for biological applications because of their quick and simple separation. While the operational shell of these structures has charges for interaction with the genetic material molecules, the magnetic core gives them separation qualities. Although mesoporous silica as well as magnetic particles were made to adsorb DNA molecules, their lack of a functional surface made them less selective, and the size of the pores affected the amount of DNA that could be adsorbed. To separate plasmid's DNA from genomic DNA and bacterial cell lysate from plant cells, nonporous silica-coated iron oxide nanoparticles (NPs) were made. This is dependent on plasmid DNA adhering to the silica surface at high salinity levels (Kang et al. 2009).

Moreover, functional groups were added to the particle surface to facilitate a more focused interaction among DNA and silica. Among the applications used in this sector is amino functionalization. The manufacture of silica-based magnetic nanoparticles with an average particle size of 25 nm was reported by researchers. In the presence of NaCl, these particles' adsorption effectiveness was four to five times greater than that of silica-coated magnetic nanoparticles. These efforts resulted in the effective and highly precise separation of human genomic

DNA from blood and saliva using amino-functionalized silica-coated magnetic nanoparticles (Kang et al. 2009).

To sum up, SiO_2 NP are continuously used in biomedical nanotechnology to create cutting-edge instruments and systems for both in vitro and in vivo uses. Table 18.2 depicts the different types of clinical application and their mechanism of actions, along with their features, etc. Proteins may be detected using the contact involving silica surfaces plus proteins, and proteins can be extracted from biologic media using the opposite interaction. Because silica surfaces provide a wide range of functionalization options, functionalized silica nanoparticles have been able to target proteins with excellent selectivity. SiO_2 NP' unique capacity to bind with nucleic acids has led to their widespread use in gene identification and purification. There are several benefits to using SiO_2 NP in this industry, including lower costs, quicker procedures and less usage of organic solvents. These days, DNA extraction and detection are helpful in gene therapy and diagnostics, which should result in the development of quick and innovative methods.

SiO_2 NP used as supplementary materials are also significant. Silica shells are a very effective way to preserve and enhance the effectiveness of imaging agents. SiO_2 NP have further use in pharmaceutical manufacturing, where they are added to powders to enhance their mechanical qualities. In conclusion, it is critical to emphasize that silica-based nanoparticles (NPs) are among the most crucial materials for comprehending and advancing biomedical applications that have the potential to significantly influence future health care because of their special qualities.

18.7.5 Silica Nanoparticles in Immunotherapy

Porous silica nanoparticles have garnered a lot of interest and are quickly emerging as a crucial class of compounds for immunotherapy due to their potent adjuvant action and other intriguing chemical and physical characteristics. More research is required to fully understand the mechanism behind MSNs' adjuvant activity. Thankfully, new research has shown the benefits of MSNs over traditional adjuvants and improved our knowledge of how MSNs interact with the immune system (depot impact, inflammasome activation, etc.). As a result, MSNs with logical designs are being created more frequently as stronger adjuvants. One example of such advancement is the work done by Kim and associates (Kim et al. 2015).

Apart from MSNs' inherent adjuvant effect, the extra-large pores of the nanoparticles offer a technique to efficiently transfer immunostimulatory molecules and high-molecular-weight antigens, enhancing antigen presentation on the immune cell surfaces via specific pathways. When complex or protein antigens must be included in vaccinations, this property is desirable. For instance, tumour extracts are occasionally used as an origin of antigen to detect the unknown alterations

Table 18.2 This table outlines different types of SiO_2 NP in the diagnostics of disease and also depicts the mechanisms of action involved, their specific features, challenges and clinical phases.

Disease	Diagnostics application	SiO_2 NP type	Mechanism of action	Features	Clinical stages	Challenges	References
Cardiovascular disease	Biomarker detection	Mesoporous silica nanoparticle	Cardiac marker detection	Device integration, detection	Early clinical phase	Efficacy of results	Bitar et al. (2012)
Genetic disorders	Diagnosis of nucleic acids	Quantum dots	Fluorescence tagging	Highly sensitive	Toolkits	Risk of nonspecific binding	Bharali et al. (2005)
Cancer	Cancer biomarkers	Functionalized SiO_2 NP	Conjugation with antibodies for detection of cancer via imaging signals	Specificity	Preclinical phase	Cross-reactivity possible	Gu (2017)
Viral disease	Disease detection	Si-coated florescent NP	Binding to viral RNA	Portable	Authorization for emergency use	Stability in storage	Florek et al. (2017)
Infectious disease	Detection of pathogen	Si-coated magnetic NPs	Capturing and isolation of pathogens	Very specific	Commercialization	Production cost	Bitar et al. (2012)
Diabetes	Blood glucose testing	MSN	Encapsulation of enzymes for glucose oxidation	Reusable	Preclinical	Sensor degradation	Bitar et al. (2012)

since antigens associated with tumours are not well-established (Gu 2018). A vaccination system that can include several kinds of protein antigens would be necessary due to the diverse makeup of tumour extracts. This kind of adaptability is also a crucial design factor to take into account when creating customized therapeutic vaccinations that use individual tumour specimens as the antigen source (Gu 2018). The development of vaccines will undoubtedly undergo a paradigm change if porous silica nanoparticles and other novel biomaterials-based adjuvants continue to advance. In addition to replacing the adjuvants now used in vaccinations, rationally created or even customized adjuvants may possibly open the door to novel forms of immunotherapy (Gu 2018).

18.7.6 SiO_2 NPs in Photodynamic Therapy

In recent decades, photodynamic therapy or PDT has drawn a lot of interest as a treatment for a variety of cancers, including those of the lung, bladder, skin, oesophagus and others. A contemporary, quickly evolving, noninvasive technique for identifying and treating cancer is photodynamic therapy. In PDT, a harmless dosage of a photosensitizer (PS) is applied either locally or systemically (Nady et al. 2023). Because photodynamic therapy is spatially and temporally selective – irradiation may be adjusted in terms of both place and time – it has garnered a lot of interest, primarily in the treatment of cancer. Furthermore, because it may be administered repeatedly to the same location, patients tolerate it well (Nady et al. 2023; Xu et al. 2022). Additionally, it has no effect on pregnancy or delivery and maintains fertility. PDT is said to have additional benefits, such as lowering systemic toxicity, inhibiting drug resistance mechanisms and being simple to combine with other treatment plans. PDT is a more promising option than traditional treatment modalities including chemotherapy, radiation and surgery because of all these benefits (Nady et al. 2023).

The level of light penetration and the drawbacks of PSs, including their hydrophobicity, cumulative toxicity, low cellular absorption and nonselective targeting of the targeted organs, restrict conventional PDT. In this context, a number of studies have documented the use of nanocarriers to overcome PSs' drawbacks and, as a result, modernize PDT in contrast to the traditional method (Nady et al. 2023; Xu et al. 2022). Compared to other kinds of organic and inorganic nanoparticles, MSNs are unique. This is explained by their optical transparency, biocompatibility, high loading capacities, ease of surface decorating with different functional groups and simplicity of production. Recent research has demonstrated that PSs and MSNs together may exhibit advantageous characteristics over free photosensitizers in a solution, including improving their internalization and selectivity in tumour cells (Nayl et al. 2022).

Furthermore, silica nanoparticles can reduce or completely eradicate the cytotoxic impact in dark environments and improve PSs' solubility and stability in physiological solutions. The use of mesoporous silica nanopartices in PDT for the treatment of cancer has been documented in recent research, which may give cancer patients new hope (Nayl et al. 2022).

There are still a number of issues that need to be resolved despite the efforts to create biocompatible MSNs that can act as carriers for different PSs to enhance cancer treatment. The lack of sufficient in-depth information on MSNs' immunological responses, bioelimination, cumulative toxic effects and in vivo biocompatibility along with biodistribution is one of these difficulties. Instead of concentrating solely on the synthesis, optimization and functionalization, of MSNs, future research should examine the biosafety issues associated with their use in cancer therapy (Nady et al. 2023).

18.8 Challenges Due to Application of SiO_2 NPs and Regulatory Considerations

Overcoming biological obstacles that impair nanomedicine's effectiveness and efficient delivery of therapeutic substances to the intended cells or tissues is one of its problems. Among these obstacles are

a) **Immune Clearance:** MSNs are susceptible to immune system recognition and elimination, particularly if they are not functioning or encased with biocompatible materials. This can result in inflammation, toxicity and decreased effectiveness of the medications loaded onto the MSNs (Dutta Gupta et al. 2024).
b) **The Blood–Brain Barrier (BBB):** This intricate structure regulates the movement of chemicals and cells to protect the brain from dangerous substances in the blood. However, because of the limited permeability caused by the size or shape of the nanoparticles, it presents a barrier to MSNs' capacity to penetrate and penetrate the brain tissue (Dutta Gupta et al. 2024).
c) **Endosomal Entrapment:** Mesoporous silica nanoparticles can enter cells by endocytosis, but because of their size and charge, they might not be able to get out of the endosomes. This may cause oxidative stress along with cell injury by causing the medicines to release too soon or by accumulating MSNs in the endosomes (Dutta Gupta et al. 2024).
d) **Lack of Specificity:** MSNs' therapeutic potential may be hampered by a number of variables, including tissue barrier, blood circulation, tumour heterogeneity, etc., that prevent them from reaching the tumour sites. MSNs' inability to target certain cancer cells as a result of their generic binding to receptors on cell surfaces or chemicals may also be the cause of their decreased effectiveness and selectivity (Dutta Gupta et al. 2024).

The high affinity of MSNs for blood proteins presents another difficulty since it may impact how well they work as drug delivery vehicles. Hundreds of proteins found in plasma have a tendency to group together on the nanoparticle's surface when it comes into contact with human blood. The existence of a protein-corona affects interactions like binding to specific cells, incorporation into cells, etc. The efficient operation of the MSN may be impacted by the formation of protein-corona, which also influences factors including the dispersity of the nanoparticle, changes in surface charge, size growth and solubility in the physiological milieu (Florek et al. 2017).

Numerous actions have already been made to mitigate the hazards posed by humans' inevitable exposure to silica nanomaterials. Around the world, several government agencies have put laws, rules and regulations into place to reduce or eliminate the dangers connected to exposure to silica nanomaterials (Kaur et al. 2020).

Nevertheless, there are no particular international regulations or legal guidelines for the synthesis, processing, or labelling of these kinds of nanomaterials, toxicity testing, or assessment of environmental exposure. There are a number of medical standards pertaining to biomedical professions that are pertinent to their morals and environmental safety. To address the concerns posed by these nanomaterials, it has been noted that the United States as well as the European Union (EU) have established robust regulatory agencies and guiding laws (Kaur et al. 2020).

18.9 Commercialization and Future Prospects

Because of their greater positive aspects over traditional nanoparticles, silica nanoparticles are considered as well as important nanomaterials with significant applications. Their market is expected to reach approximately $8.8 billion in 2020. Their use has increased significantly in recent years and has garnered a lot of attention (Nayl et al. 2022). Considering a compound annual growth rate (CAGR) of around 6.51% between 2022 and 2030, the worldwide nano-silica market is expected to reach a value of approximately USD 8.65 billion by 2030, up from its 2021 valuation of $4.65 billion (Anonymous 2023).

The market is anticipated to expand significantly as a result of the expanding uses of medication delivery in the medical field. P-Type was forecast to have the largest market share in 2021 based on product segmentation. Rubber was the most popular application in 2021, according to application segmentation. In 2021, the United States was the top revenue-generating area by region (Anonymous 2023).

As silica nanoparticles gain more attention worldwide, the medical community is under increasing pressure to fund comprehensive studies to gain a deeper understanding of the mineral and to develop systematic procedures to get rid of the harmful substances found in nano-silica, particularly when it comes to biomedical procedures. The size, shape, surface area, dosage and kind of cell used in the study are some of the variables that might significantly affect the cytotoxicity of silica nanoparticles (Nayl et al. 2022). A few of the frequent issues include haemolysis and caseation, which reduce the drug's efficacy and endanger patient safety. When it comes to demonstrating the value of investing in silica nanoparticles for use in the production of concrete buildings, the worldwide market participants confront difficulties. According to studies, although they have demonstrated a remarkable improvement in the durability of concrete, the material's mechanical qualities are mostly unaltered, making it only partially beneficial given the resources used (Nayl et al. 2022).

North America is expected to have the fastest growth in the worldwide nano silica market, with the United States holding the largest regional market share. The country's research departments' growing investments in nanotechnology are expected to be the main driver of the high income. Nanoscience as well as particles are key components of nanotechnology, for which the US administration has spent over 38 billion dollars in the past 20 years (Anonymous 2023; Nayl et al. 2022). More funding may be allocated in the upcoming year as a result of the increased focus on creating effective medications. Another contributing cause is the increasing number of players that produce quartz or silica dust. With its headquarters in the United States, NanoPore Incorporated is an established rival in the industry and had a 17% increase in sales in 2021 (Anonymous 2023).

18.10 Conclusion

MSNs are gaining widespread use as nanocarriers for conveying therapeutic, diagnostic and combined 'theranostic' cargos to cells and tissues because of their uniform and adjustable dimension, size of pores and shape, as well as their modular surface properties and biocompatibility. Despite extensive research, the FDA has only approved one MSN system for oral delivery. Not all MSNs are made equal in the lab or in the FDA's eyes, which might be due to the lack of a universally accepted, repeatable and acknowledged MSN synthesis procedure. Since the first SNP was authorized for use in human clinical trials, there is optimism for the future perspectives needed for the future use of these incredible nanocarriers in therapeutic applications. Even though the mechanisms by which functionalized silica nanoparticles function in a variety of fields are well understood and have been discussed, little is known about their side effects or degradation. Therefore, it is imperative that scientific research be continued in order to determine the fate of these nanoparticles, particularly in the fields of medicine and the environment.

Future applications are shaped by silica nanoparticles (SNPs), which are catalysts for revolutionary transformation in a variety of sectors. For its wider use, sustainable synthesis that is in line with green nanotechnology is essential. SNP synthesis may be made more environmentally friendly by using bioagents and natural extracts. Green synthesis standardization and improved industry-academia cooperation are essential. SNPs are excellent at medication delivery and diagnostics, which bodes well for future medical advancements. Cooperation in safety evaluations, long-term research and standardized testing is necessary for safe integration. Future industry-specific uses are hinted at by advanced coatings, which are advantageous for surfaces that are antimicrobial and resistant to corrosion. It is advised to do research on novel compatibility and capabilities for SNP-based coatings.

References

Akhtar, M.J., Ahamed, M., Kumar, S. et al. (2010). Nanotoxicity of pure silica mediated through oxidant generation rather than glutathione depletion in human lung epithelial cells. *Toxicology* 276 (2): 95–102.

Akhter, F., Rao, A.A., Abbasi, M.N. et al. (2022). A comprehensive review of synthesis, applications and future prospects for silica nanoparticles (SNPs). *Silicon* 14 (14): 8295–8310.

Alsmaeil, A.W., Hammami, M.A., Enotiadis, A. et al. (2021). Encapsulation of an anionic surfactant into hollow spherical nanosized capsules: size control, slow release, and potential use for enhanced oil recovery applications and environmental remediation. *ACS Omega* 6 (8): 5689–5697.

Anonymous (2023). Nano Silica Market Size, Share Global Analysis Report, 2022 – 2030, FAF-2191.

Asdaq, S.M.B., Ikbal, A.M.A., Sahu, R.K. et al. (2021). Nanotechnology integration for SARS-CoV-2 diagnosis and treatment: an approach to preventing pandemic. *Nanomaterials* 11 (7): 1841.

Baeza, A. and Vallet-Regí, M. (2020). Mesoporous silica nanoparticles as theranostic antitumoral nanomedicines. *Pharmaceutics* 12 (10): 957.

Bharali, D.J., Klejbor, I., Stachowiak, E.K. et al. (2005). Organically modified silica nanoparticles: a nonviral vector for *in vivo* gene delivery and expression in the brain. *Proceedings of the National Academy of Sciences of the United States of America* 102 (32): 11539–11544.

Bidram, E., Esmaeili, Y., Amini, A. et al. (2021). Nanobased platforms for diagnosis and treatment of COVID-19: from benchtop to bedside. *ACS Biomaterials Science & Engineering* 7 (6): 2150–2176.

Bindini, E., de Los Angeles Ramirez, M., Rios, X. et al. (2021). in vivo tracking of the degradation of mesoporous silica through ^{89}Zr radio-labeled core–shell nanoparticles. *Small* 17 (30).

Bitar, A., Ahmad, N.M., Fessi, H., and Elaissari, A. (2012). Silica-based nanoparticles for biomedical applications. *Drug Discovery Today* 17 (19–20): 1147–1154.

Boisselier, E. and Astruc, D. (2009). Gold nanoparticles in nanomedicine: preparations, imaging, diagnostics, therapies and toxicity. *Chemical Society Reviews* 38 (6): 1759.

Burda, C., Chen, X., Narayanan, R., and El-Sayed, M.A. (2005). Chemistry and properties of nanocrystals of different shapes. *Chemical Reviews* 105 (4): 1025–1102.

Chen, J., Saeki, F., Wiley, B.J. et al. (2005). Gold nanocages: bioconjugation and their potential use as optical imaging contrast agents. *Nano Letters* 5 (3): 473–477.

Chen, F., Goel, S., Valdovinos, H.F. et al. (2015). *in vivo* integrity and biological fate of chelator-free zirconium-89-labeled mesoporous silica nanoparticles. *ACS Nano* 9 (8): 7950–7959.

Chen, G., Roy, I., Yang, C., and Prasad, P.N. (2016). Nanochemistry and nanomedicine for nanoparticle-based diagnostics and therapy. *Chemical Reviews* 116 (5): 2826–2885.

Chen, N.T., Souris, J.S., Cheng, S.H. et al. (2017). Lectin-functionalized mesoporous silica nanoparticles for endoscopic detection of premalignant colonic lesions. *Nanomedicine* 13 (6): 1941–1952.

Cheng, C.A., Deng, T., Lin, F.C. et al. (2019). Supramolecular nanomachines as stimuli-responsive gatekeepers on mesoporous silica nanoparticles for antibiotic and cancer drug delivery. *Theranostics* 9 (11): 3341–3364.

Chinnadayyala, S.R., Park, J., Le, H.T.N. et al. (2019). Recent advances in microfluidic paper-based electrochemiluminescence analytical devices for point-of-care testing applications. *Biosensors & Bioelectronics* 126: 68–81.

Crucho, C.I.C., Avó, J., Nobuyasu, R. et al. (2020). Silica nanoparticles with thermally activated delayed fluorescence for live cell imaging. *Materials Science and Engineering: C* 109: 110528.

Cui, Y., Wei, Q., Park, H., and Lieber, C.M. (2001). Nanowire nanosensors for highly sensitive and selective detection of biological and chemical species. *Science (1979)* 293 (5533): 1289–1292.

Desai, N. (2012). Challenges in development of nanoparticle-based therapeutics. *The AAPS Journal* 14 (2): 282–295.

Dhinasekaran, D., Raj, R., Rajendran, A.R. et al. (2020). Chitosan mediated 5-Fluorouracil functionalized silica nanoparticle from rice husk for anticancer activity. *International Journal of Biological Macromolecules* 156: 969–980.

Duan, H., Huang, X., Shao, Y. et al. (2017). Size-dependent immunochromatographic assay with quantum dot nanobeads for sensitive and quantitative detection of ochratoxin A in corn. *Analytical Chemistry* 89 (13): 7062–7068.

Dukle, A., Nathanael, A.J., Panchapakesan, B., and Oh, T.H. (2022). Role of paper-based sensors in fight against cancer for the developing world. *Biosensors (Basel)* 12 (9): 737.

Dutta Gupta, Y., Mackeyev, Y., Krishnan, S., and Bhandary, S. (2024). Mesoporous silica nanotechnology: promising advances in augmenting cancer theranostics. *Cancer Nanotechnology* 15 (1): 9.

Etheridge, M.L., Campbell, S.A., Erdman, A.G. et al. (2013). The big picture on nanomedicine: the state of investigational and approved nanomedicine products. *Nanomedicine* 9 (1): 1–14.

Fanizza, E., Zhao, H., De Zio, S. et al. (2020). Encapsulation of dual emitting giant quantum dots in silica nanoparticles for optical ratiometric temperature nanosensors. *Applied Sciences* 10 (8): 2767.

Florek, J., Caillard, R., and Kleitz, F. (2017). Evaluation of mesoporous silica nanoparticles for oral drug delivery – current status and perspective of MSNs drug carriers. *Nanoscale* 9 (40): 15252–15277.

Gao, Y., Gao, D., Shen, J., and Wang, Q. (2020). A review of mesoporous silica nanoparticle delivery systems in chemo-based combination cancer therapies. *Frontiers in Chemistry* 8: https://doi.org/10.3389/fchem.2020.598722.

Girija, A.R. and Balasubramanian, S. (2019). Theragnostic potentials of core/shell mesoporous silica nanostructures. *Nanotheranostics* 3 (1): 1–40.

Gong, C., Tao, G., Yang, L. et al. (2012). The role of reactive oxygen species in silicon dioxide nanoparticle-induced cytotoxicity and DNA damage in HaCaT cells. *Molecular Biology Reports* 39 (4): 4915–4925.

Ha, C.V., Anh Tuan, C., Thi Bich Ngoc, N. et al. (2020). Synthesis and optical characterizations of the fluorescence silica nanoparticles containing quantum dots. *VNU Journal of Science: Mathematics – Physics* 36 (2).

Hildebrandt, N., Spillmann, C.M., Algar, W.R. et al. (2017). Energy transfer with semiconductor quantum dot bioconjugates: a versatile platform for biosensing, energy harvesting, and other developing applications. *Chemical Reviews* 117 (2): 536–711.

Hou, F., Sun, S., Abdullah, S.W. et al. (2023). The application of nanoparticles in point-of-care testing (POCT) immunoassays. *Analytical Methods* 15 (18): 2154–2180.

Hsiao, J., Tsai, C., Chung, T. et al. (2008). Mesoporous silica nanoparticles as a delivery system of gadolinium for effective human stem cell tracking. *Small* 4 (9): 1445–1452.

Hu, J., Zhang, Z.L., Wen, C.Y. et al. (2016). Sensitive and quantitative detection of C-reaction protein based on immunofluorescent nanospheres coupled with lateral flow test strip. *Analytical Chemistry* 88 (12): 6577–6584.

Huang, L., Liao, T., Wang, J. et al. (2018). Brilliant pitaya-type silica colloids with central–radial and high-density quantum dots incorporation for ultrasensitive fluorescence immunoassays. *Advanced Functional Materials* 28 (4): https://doi.org/10.1002/adfm.201705380.

Huang, Y., Li, P., Zhao, R. et al. (2022). Silica nanoparticles: biomedical applications and toxicity. *Biomedicine & Pharmacotherapy* 151: 113053.

Hudson, S.P., Padera, R.F., Langer, R., and Kohane, D.S. (2008). The biocompatibility of mesoporous silicates. *Biomaterials* 29 (30): 4045–4055.

Janjua, T.I., Cao, Y., Kleitz, F. et al. (2023). Silica nanoparticles: a review of their safety and current strategies to overcome biological barriers. *Advanced Drug Delivery Reviews* 203: 115115.

Jonat, S., Hasenzahl, S., Drechsler, M. et al. (2004). Investigation of compacted hydrophilic and hydrophobic colloidal silicon dioxides as glidants for pharmaceutical excipients. *Powder Technology* 141 (1–2): 31–43.

José, A.M. (2004). Nanomedicines for overcoming biological barriers. *Biomedicine & Pharmacotherapy* 58 (3): 168–172.

Kang, K., Choi, J., Nam, J.H. et al. (2009). Preparation and characterization of chemically functionalized silica-coated magnetic nanoparticles as a DNA separator. *The Journal of Physical Chemistry B* 113 (2): 536–543.

Kaur, J., Singh, H., and Khatri, M. (2020). Regulatory considerations for safety of nanomaterials. In: *Nanomedicine for Bioactives* (ed. M. Rahman, S. Beg, V. Kumar, and F.J. Ahmad), 431–450. Singapore: Springer Singapore.

Khan, W.A., Arain, M.B., and Soylak, M. (2020). Nanomaterials-based solid phase extraction and solid phase microextraction for heavy metals food toxicity. *Food and Chemical Toxicology* 145: 111704.

Khoz, R., Yazdian, F., Pourmadadi, M. et al. (2024). Current trends in silica based drug delivery systems. *European Journal of Medicinal Chemistry Reports* 12: 100206.

Kim, J., Li, W.A., Choi, Y. et al. (2015). Injectable, spontaneously assembling, inorganic scaffolds modulate immune cells in vivo and increase vaccine efficacy. *Nature Biotechnology* 33 (1): 64–72. https://doi.org/10.1038/nbt.3071.

Kim, T., Momin, E., Choi, J. et al. (2011). Mesoporous silica-coated hollow manganese oxide nanoparticles as positive T_1 contrast agents for labeling and MRI tracking of adipose-derived mesenchymal stem cells. *Journal of the American Chemical Society* 133 (9): 2955–2961.

Kim, C., Yoon, S., and Lee, J.H. (2019). Facile large-scale synthesis of mesoporous silica nanoparticles at room temperature in a monophasic system with fine size control. *Microporous and Mesoporous Materials* 288: 109595.

Kirla, H., Hughes, L., and Henry, D.J. (2020). Carbohydrate coated fluorescent mesoporous silica particles for bacterial imaging. *Colloids and Surfaces, B: Biointerfaces* 188: 110751.

Koo, H., Allan, R.N., Howlin, R.P. et al. (2017). Targeting microbial biofilms: current and prospective therapeutic strategies. *Nature Reviews Microbiology* 15 (12): 740–755. https://doi.org/10.1038/nrmicro.2017.99.

Liberman, A., Wu, Z., Barback, C.V. et al. (2013). Color doppler ultrasound and gamma imaging of intratumorally injected 500 nm iron–silica nanoshells. *ACS Nano* 7 (7): 6367–6377.

Lin, X., Kong, M., Wu, N. et al. (2020). Measurement of temperature distribution at the nanoscale with luminescent probes based on lanthanide nanoparticles and quantum dots. *ACS Applied Materials & Interfaces* 12 (47): 52393–52401.

Liou, T.H. and Yang, C.C. (2011). Synthesis and surface characteristics of nanosilica produced from alkali-extracted rice husk ash. *Materials Science and Engineering: B* 176 (7): 521–529.

Löffler, B. (2017). CLINAM summit May 7–10, 2017. *European Journal of Nanomedicine* 9 (2): 45. https://doi.org/10.1515/ejnm-2017-0006.

Manzano, M. and Vallet-Regí, M. (2020). Mesoporous silica nanoparticles for drug delivery. *Advanced Functional Materials* 30 (2): 1902634.

Mathieu, P., Coppel, Y., Respaud, M. et al. (2019). Silica coated iron/iron oxide nanoparticles as a nano-platform for T_2 weighted magnetic resonance imaging. *Molecules* 24 (24): 4629.

Michalet, X., Pinaud, F.F., Bentolila, L.A. et al. (2005). Quantum dots for live cells, in vivo imaging, and diagnostics. *Science (1979)* 307 (5709): 538–544.

Mladenović, M., Morgan, I., Ilić, N. et al. (2021). pH-Responsive release of ruthenium metallotherapeutics from mesoporous silica-based nanocarriers. *Pharmaceutics* 13 (4): 460.

Mladenović, M., Jarić, S., Mundžić, M. et al. (2024). Biosensors for cancer biomarkers based on mesoporous silica nanoparticles. *Biosensors (Basel)* 14 (7): 326.

Moghimi, S.M., Hunter, A.C., and Murray, J.C. (2005). Nanomedicine: current status and future prospects. *The FASEB Journal* 19 (3): 311–330.

Mohamed Isa, E.D., Ahmad, H., Abdul Rahman, M.B., and Gill, M.R. (2021). Progress in mesoporous silica nanoparticles as drug delivery agents for cancer treatment. *Pharmaceutics* 13 (2): 152.

Mohammadi, P., Abbasinia, M., Assari, M.J., and Oliaei, M. (2018). The toxicology of silica nanoparticles: a review. *Toxicological and Environmental Chemistry* 100 (3): 285–316.

Mu, W., Chu, Q., Liu, Y., and Zhang, N. (2020). A review on nano-based drug delivery system for cancer chemoimmunotherapy. *Nano-Micro Letters* 12 (1): 142.

Murugan, C., Rayappan, K., Thangam, R. et al. (2016). Combinatorial nanocarrier based drug delivery approach for amalgamation of anti-tumor agents in breast cancer cells: an improved nanomedicine strategy. *Scientific Reports* 6 (1): 34053.

Nady, D., Hassan, A., Amin, M. et al. (2023). Recent innovations of mesoporous silica nanoparticles combined with photodynamic therapy for improving cancer treatment. *Pharmaceutics* 16 (1): 14.

Nam, J., Won, N., Bang, J. et al. (2013). Surface engineering of inorganic nanoparticles for imaging and therapy. *Advanced Drug Delivery Reviews* 65 (5): 622–648.

Nayl, A.A., Abd-Elhamid, A.I., Aly, A.A., and Bräse, S. (2022). Recent progress in the applications of silica-based nanoparticles. *RSC Advances* 12 (22): 13706–13726.

Nikoobakht, B. and El-Sayed, M.A. (2003). Preparation and growth mechanism of gold nanorods (NRs) using seed-mediated growth method. *Chemistry of Materials* 15 (10): 1957–1962.

Perrault, S.D. and Chan, W.C.W. (2009). Synthesis and surface modification of highly monodispersed, spherical gold nanoparticles of 50–200 nm. *Journal of the American Chemical Society* 131 (47): 17042–17043.

Prieto-Montero, R., Katsumiti, A., Cajaraville, M.P. et al. (2020). Functionalized fluorescent silica nanoparticles for bioimaging of cancer cells. *Sensors* 20 (19): 5590.

Ren, M., Xu, H., Huang, X. et al. (2014). Immunochromatographic assay for ultrasensitive detection of aflatoxin B_1 in maize by highly luminescent quantum dot beads. *ACS Applied Materials & Interfaces* 6 (16): 14215–14222.

Rong, Z., Wang, Q., Sun, N. et al. (2019). Smartphone-based fluorescent lateral flow immunoassay platform for highly sensitive point-of-care detection of Zika virus nonstructural protein 1. *Analytica Chimica Acta* 1055: 140–147.

Rosenholm, J.M., Sahlgren, C., and Linden, M. (2011). Multifunctional mesoporous silica nanoparticles for combined therapeutic, diagnostic and targeted action in cancer treatment. *Current Drug Targets* 12 (8): 1166–1186.

Selvarajan, V., Obuobi, S., and Ee, P.L.R. (2020). Silica nanoparticles—a versatile tool for the treatment of bacterial infections. *Frontiers in Chemistry* 8: 602.

Shahbazi, M.A., Herranz, B., and Santos, H.A. (2012). Nanostructured porous Si-based nanoparticles for targeted drug delivery. *Biomatter* 2 (4): 296–312.

Shi, J., Kantoff, P.W., Wooster, R., and Farokhzad, O.C. (2017). Cancer nanomedicine: progress, challenges and opportunities. *Nature Reviews. Cancer* 17 (1): 20–37.

Silva, G.A. (2008). Nanotechnology approaches to crossing the blood-brain barrier and drug delivery to the CNS. *BMC Neuroscience* 9 (S3): S4.

Song, C., Zhi, A., Liu, Q. et al. (2013). Rapid and sensitive detection of β-agonists using a portable fluorescence biosensor based on fluorescent nanosilica and a lateral flow test strip. *Biosensors & Bioelectronics* 50: 62–65.

Thwala, L.N., Ndlovu, S.C., Mpofu, K.T. et al. (2023). Nanotechnology-based diagnostics for diseases prevalent in developing countries: current advances in point-of-care tests. *Nanomaterials* 13 (7): 1247.

Vallet-Regí, M., Colilla, M., Izquierdo-Barba, I., and Manzano, M. (2017). Mesoporous silica nanoparticles for drug delivery: current insights. *Molecules* 23 (1): 47.

Wang, Y., Yu, L., Kong, X., and Sun, L. (2017). Application of nanodiagnostics in point-of-care tests for infectious diseases. *International Journal of Nanomedicine* 12: 4789–4803.

Wang, J., Xu, M., Wang, K., and Chen, Z. (2019). Stable mesoporous silica nanoparticles incorporated with MoS2 and AIE for targeted fluorescence imaging and photothermal therapy of cancer cells. *Colloids and Surfaces, B: Biointerfaces* 174: 324–332.

Xu, Q., Yang, Y., Lu, J. et al. (2022). Recent trends of mesoporous silica-based nanoplatforms for nanodynamic therapies. *Coordination Chemistry Reviews* 469: 214687.

Yang, S.M., Lv, S., Zhang, W., and Cui, Y. (2022). Microfluidic point-of-care (POC) devices in early diagnosis: a review of opportunities and challenges. *Sensors* 22 (4): 1620.

Yi, S., Zheng, J., Lv, P. et al. (2018). Controlled drug release from cyclodextrin-gated mesoporous silica nanoparticles based on switchable host–guest interactions. *Bioconjugate Chemistry* 29 (9): 2884–2891.

Yong, K.-T., Sahoo, Y., Swihart, M.T., and Prasad, P.N. (2006). Growth of CdSe quantum rods and multipods seeded by noble-metal nanoparticles. *Advanced Materials* 18 (15): 1978–1982.

Zhou, X. and Zhou, J. (2004). Improving the signal sensitivity and photostability of DNA hybridizations on microarrays by using dye-doped core–shell silica nanoparticles. *Analytical Chemistry* 76 (18): 5302–5312.

Zhou, J., Ren, M., Wang, W. et al. (2020). Pomegranate-inspired silica nanotags enable sensitive dual-modal detection of rabies virus nucleoprotein. *Analytical Chemistry* 92 (13): 8802–8809.

19

Si/SiO_2 NPs as an Agent for Remediation of Contaminated Soil and Water

Upasana Medhi[1], *Sampa Paul*[1], *Bharat Engleng*[1], *Ramesh Chandra Deka*[1,2] *and Eeshan Kalita*[1]

[1] *Department of Molecular Biology and Biotechnology, Cotton University, Guwahati, Assam, India*
[2] *Department of Chemical Sciences, Tezpur University, Tezpur, Assam, India*

19.1 Introduction

The advent of industrialization, expansion in agriculture and urbanization have heightened the pollution level of natural resources, mainly soil and water. Contaminants like heavy metals (HMs), pesticides, herbicides, organic solvents and other hazardous materials pose a serious hazard to environmental and public health (Alengebawy et al. 2021). Conventional remediation methods for soil and water, including chemical treatments, bioremediation and physical removal, often encounter challenges related to their effectiveness, cost-effectiveness and eco-friendly sustainability (Zamora-Ledezma et al. 2021). As a result, researchers are investigating more innovative materials and methods for environmental cleanup, with nanotechnology showing great potential as a viable solution (Sathish et al. 2024).

As a step off the typical path, conventionally, one of the most innovative and promising forms in which nanotechnology could be applied to clean up the environment is through the use of silicon nanoparticles (SiNPs) in remediating polluted soil and water. This nanosystem shows an important contribution to solving environmental remediation (Mahawar et al. 2023). These silicon/silicon dioxide (Si/SiO_2) NPs are distinguished due to their vast surface area and compact stature, excellent mechanical properties and chemical composition, providing unique properties that enhance the efficiency of contaminant exclusion (Sharma et al. 2022). The inherent activity of SiNPs allows the efficient binding and immobilization of a variety of pollutants to restore ecological balance in contaminated

Silicon Nanoparticles: Synthesis, Uptake, and Applications, First Edition.
Edited by Archana Singh, Aryadeep Roychoudhury, and Indrakant K. Singh.

environments. This capability has attracted researchers to invest increasingly more time and effort into the deployment of these nanomaterials towards the dual aim of remediation and rehabilitation of ecosystems affected by human activities (Baragaño et al. 2021). The application of SiNPs spectroscopically will allow for effective and sustainable methods of environmental remediation, fostering better ecosystem health and public health outcomes. As recognition of NPs continues to seep in, these will employ a major advancement in the world of environmental remediation in environmental science and engineering (Chen et al. 2018). Its remedial and biocompatibility have led to its widespread application in the food and medical sectors (Seisenbaeva et al. 2021). These SiNPs' unique qualities qualify them to be customized for a wide range of uses, which include catalytic supports, advanced composites, sensing materials, adsorbents and polymer reinforcement (Kaya et al. 2020; Nassar et al. 2019; Shinde et al. 2021).

SiNPs have emerged as a crucial component in the remediation of the environment, predominantly addressing soil and aquatic contamination (Ahmed et al. 2023). SiNPs emerged as a very important player in remediation efforts, whereby soil and water pollution are tackled. Historically, SiNPs were made using physical methods such as laser ablation and grinding or through chemical vapor deposition (CVD) (Lumen et al. 2021). The production methods yield SiNPs with very high surface area, thus making them very efficient in pollutant adsorption. One of the first utilizations of SiNPs in environmental cleanup is their adsorption of HMs, namely cadmium (Cd), arsenic (As) and lead (Pb) from polluted water sources (Boora et al. 2023). The extensive surface area of SiNPs creates several binding sites for these metal ions. Furthermore, the use of SiNPs extends to catalytic processes, where they are primarily involved as carriers of catalytic agents. These agents could catalyze organic pollutants' catalytic degradation, including solvents and hydrocarbons. Regarding soil remediation, SiNPs enhance phytoremediation, the process wherein plants captivate and break down contaminated soil pollutants. SiNPs stimulate plant growth in the soil, thus significantly enhancing their ability to take up and detoxify harmful contaminants (Boora et al. 2023).

Recent breakthroughs in nanotechnology have resulted in significant progress towards using SiNPs for the remediation of the environment. One specific innovation is the functionalization of SiNPs, in which the surfaces of NPs are decorated with organic groups such as amines, carboxyl groups or thiols (Song et al. 2016). Such a function increases the specificity of NPs concerning target contaminants, such as HMs, organics and pathogens (Ji et al. 2023). Moreover, SiNPs are mostly combined with other materials (for example, activated carbon or metal oxides) to form hybrid nanocomposites. This combination enhances their adsorption capacity and catalytic performance (Li et al. 2020b; Mayakannan et al. 2023). For instance, combining SiNPs with zero-valent iron (FeO) forms composites that can effectively break down organic pollutants through reduction reactions (Li et al.

2021). Apart from that, in the current days, green synthesis from plant extracts, bacteria or fungi has taken a turn towards sustainable production of SiNPs. Such environmentally friendly production not only has lesser environmental consequences but also yields SiNPs with unique properties for remediation (Baran et al. 2023). Another development worth mentioning is the improvement of SiNPs' mobility in soil and water (Alsaeedi et al. 2019). Conventional methods were ineffective concerning the poor dispersion of SiNPs, mainly in the soil. They were hindered by simple aggregation from reaching deeper contaminants than just the soil surface (Okeke et al. 2023). The current plan is to modify the surface charge and size to ameliorate the stability of SiNP dispersion. Increasing the mobility of the NPs will help them to interact with the pollutants more effectively due to a more efficient penetration of delimited sites of contamination (Kadivar and Keshavarz 2022).

Various specific environmental circumstances, such as pH, temperature, or target chemical substances, trigger the activation of the 'smart' NPs, which have developed due to the advancement of SiNPs. This advancement in targeted remediation occurs at the appropriate time, when SiNPs are activated to release pollutant-absorbing agents or destruction catalysts, thereby increasing its efficiency in the remediation process and reducing unintended consequences to the uncontaminated site (Kazemi et al. 2023). In recent times, researchers have shifted attention from initial efficacy studies on SiNPs to their environmental and ecotoxicological implications, involving the addition of life-cycle assessments (LCAs) to ascertain the long-term implications of SiNPs on ecosystems and organisms present in soil and water (Martínez et al. 2020; Singh et al. 2024). Additionally, SiNPs are being utilized along with biological remediation processes like biosorption and biodegradation, where specific bacteria and fungi have been found to work synergistically with the NPs to enhance pollutant adsorption and degradation processes (Malik et al. 2022). In addition, SiNPs offer many additional benefits in bioremediation, such as treatment agents, including enzymes and microorganisms being highly effective carriers of treatment agents directly to the contaminated sites, thus improving remediation efficiency. The surface chemistry of SiNPs makes them a likely candidate for the exclusion of emerging pollutants, such as pharmaceuticals, personal care goods and microplastics; traditional efforts may have been neglected. Such adaptiveness brings forth the potential of SiNPs for combating modern environmental challenges (Anand et al. 2022; Chowdhury et al. 2021).

Therefore, this chapter delves into the novel field of nano-remediation based on silica. This cutting-edge technology provides new ways out of some persistent farmers' dilemmas and industrial pollutants in soils and waters. While using the singular properties of silica at a nanoscale, researchers are devising smart strategies to successfully tackle pollution and reinstate the balance of natural resources.

It also emphasizes the sources of these pollutants, whether caused by agricultural practices or industrial operations, as well as the negative effects on ecosystems and populations.

19.2 Pollutants and their Sources

The primary requirements for agriculture are soil and water resources, which provide a medium for crop production, plant growth and farm productivity. However, soil and water resource degradation are caused by various contaminants brought on by urbanization, fast industrialization and population increase. Food security, human health and the well-being of ecosystems are all directly impacted by the threat that this pollution of various chemicals poses to sustainable agriculture and environmental health (Lu et al. 2015). Below is a discussion of some of the potential soil and water contaminants.

19.2.1 Organic Pollutants

Natural contaminants have been increasingly prevalent in wastewater, biosolids and effluents. This is a result primarily of the incredible growth in the quantity of synthetic organic chemicals available for application by commerce over the last 50 years (Ghazaryan et al. 2024). The presence of organics varies depending on several interdependent variables involving the physicochemical characteristics of specific organic compounds, topographical aspects, practical parameters of the wastewater treatment plant and state conditions of the wastewater collection. However, organic pollutants are generally found in higher concentrations in industrial wastewater (Smith 2009).

In agriculture, agrochemicals are chemical substances to augment crop yields and shield plants against weeds, pests and diseases. These are broadly grouped into synthetic fertilizers and pesticides. The essential nutrients required by plants – nitrogen (N), phosphorus (P), potassium (K), sulphur (S) and numerous other elements, including vitamins and minerals – are provided by synthetic fertilizers, mostly inorganic, except for urea. Conversely, pesticides are naturally occurring organic compounds that kill, repel, or control plant and animal species seen as harmful to agriculture or domestic life. This wide group includes fungicides, herbicides, insecticides, plant growth regulators and other materials. Pesticides are sometimes classified according to their chemical structures, as organophosphorus, phenolic, phenols, metallo-organics, nitrogen benzenes, reactive alkanes, elementary or trisite acids, dinitrophenols, or other derivatives (Craswell and Karjalainen 1990; Romero and Nicewicz 2016).

While agrochemicals have increased crop yields, their capacity to be harmful has been the subject of debate for the last couple of decades. Agrarian use of pesticides and fertilizers escalated very fast in India due to the Green Revolution. Numerous industries exploited the utilization of these pesticides and fertilizers, profiting from them to the detriment of civilization and the ecosystem in a certain manner. The plants where these chemical manufacturing businesses operate may also have negative environmental consequences in addition to the chemicals themselves (Craswell and Karjalainen 1990). The improper usage of various agrochemicals has adversely influenced the beneficial bacteria in the soil (Meena et al. 2016).

19.2.2 Non-organic Pollutants

The term 'pollutants' includes a wide range of substances, particularly HMs, namely copper (Cu), lead (Pb), cadmium (Cd), zinc (Zn), chromium (Cr), arsenic (As), etc. HMs often enter the environment primarily through wastewater generated by residential and commercial activities, with the most prevalent elements being Cu, Pb and Zn. These HMs pose a significant risk since they accrue in edible crops. Through the food chain, it means that risks would be passed to livestock and human beings that consume the affected food (Cobb et al. 2000; Ghazaryan et al. 2024). The transmission of HMs to the aboveground parts of crops is considered unlikely; thus, under current regulatory frameworks, the consumption of agricultural products is not expected to pose significant health risks (Balali-Mood et al. 2021). While HMs may infiltrate ruminant tissues primarily through rapidly absorbing contaminated soil or biosolids, extensive buildup is normally improbable when pasture management practices are maintained suitably. However, emerging research indicates that HMs from biosolid applications may adversely affect soil microbial communities and biological functions (Li et al. 2020a). The full implications and potential applications of these findings remain largely unexamined.

The pollutants referenced earlier can originate from a varied selection of sources, each contributing to the total degradation of environmental quality. These sources can be primarily categorized as follows.

19.2.3 Industrial Activity

The products of industrial processes are different pollutants that are emptied into the environment, causing pollution of soil and water resources. Use of hazardous chemicals (Zou et al. 2023), bad waste disposal (Adeniran and Shakantu 2022) and industrial runoff (Yi et al. 2016) are some of the activities that aggravate the problem, hence causing long-term detriment to ecosystems and human health. Hence, degradation of these natural resources not only daunts the animals but also threatens agriculture and sources of drinking water, underscoring the need for more sustainable industrial practices. The toxic substances commonly used in the chemical industry's manufacturing processes pose a serious environmental threat. These processes may lead to hazardous waste generation, often composed of pesticides, various solvents, HMs such as Pb and mercury (Hg), and other toxic substances. If these chemicals are not handled properly, they may leak or be spilled accidentally, contaminating adjacent land and water supplies (Alengebawy et al. 2021). Besides, environmental disruption is also compounded by the chemical industry's well-known excessive use of water. Most of the wastewater is made up of a mixture of hazardous substances – such as acids, alkalis and solvents – that are then usually dumped into nearby lakes, rivers or other bodies of water (Singh et al. 2023). In addition to endangering human health and local ecosystems, tainted wastewater discharge pollutes the aquatic environment. Therefore, resolving these issues is essential to protecting our natural resources and ensuring sustainable practices.

To enhance and sustain the quality of fabrics, various dyes, bleaches and chemical treatments have been in use in the textile industry (Harsanto et al. 2023). However, wastewater generated in these processes is often a cocktail of toxic substances with potentially devastating effects on the environment. Discharge of this polluted water into the receiving rivers, lakes and soil may harm aquatic life and disrupt the ecosystems (Wilkinson et al. 2022). The extraction and transportation of gas and oil can also entail massive spills that destroy soils and lightly pollute the waters. Because oil hardly deteriorates in nature, these oil spills can exert horrendous ecological pressure (Shi et al. 2023).

Therefore, these environmental effects must be addressed to preserve natural resources and encourage sustainable business practices.

19.2.4 Agricultural Activities

Soil pollution remains a pivotal topic, the effects of which are primarily due to agricultural practices. Contributors to these are agrochemicals and organic waste from farms, such as manure and animal excrement (Azman et al. 2023). These materials can add different soil pollutants, with particular concern over some trace HMs, namely, Cd, Cu, Hg and Pb, which can cause irreversible disruption to plant metabolic processes, thereby reducing crop yield (Balali-Mood et al. 2021; Protesescu et al. 2015).

The continuous usage of pesticides, coupled with excessive manure avalanches, is a cause for worsening these problems, especially where fertilizers are not applied in an even or responsible manner, especially the essential nutrients of nitrogen (N) and phosphorus (P) (Ayilara et al. 2023; Gu et al. 2023; Khan et al. 2023). The reckless implementation of fertilizers is one of the chief causes of other ecological dilemmas associated with modern agriculture. Excessive use could lead to hazardous signs like accumulation of HMs, high nitrate concentration, soil salinization and many other things, including water pollution, all of which

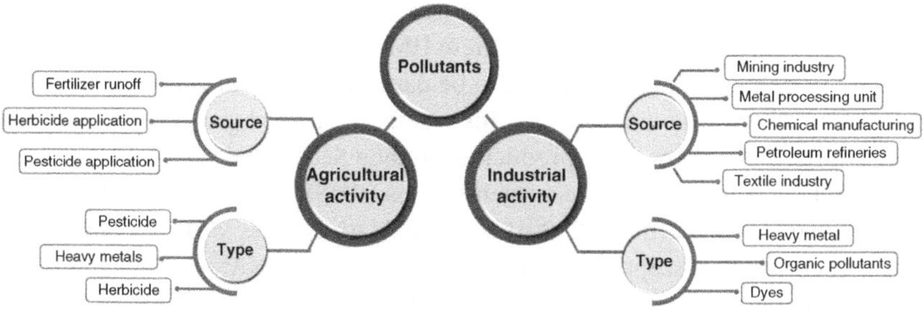

Figure 19.1 Different pollutants and their sources.

could indeed affect human and environmental health negatively (Jaishankar et al. 2014). Moreover, it has been suggested that naturally occurring radioactive elements and HMs come from the fertilizer business itself, further complicating the issue. Correct handling and storage of fertilizers would help in mitigating these challenges, ensuring the natural state of the soil. This would protect the soil and contribute to sound agricultural practice for future generations. Promoting sustainable farming, promoting the cautious management of soil and checking pollution would protect not only our agricultural ecosystem but also ensure the health of human beings (Khan et al. 2018).

Figure 19.1 summarizes different pollutants and their sources that can be observed in the environment.

19.3 Impact of Pollutants on Flora and Fauna

Agrochemicals, HMs and organic pollutants are among the pollutants that have a major effect on plants and animals, upsetting ecosystems and endangering biodiversity. Pesticides, industrial chemicals, HMs and polycyclic aromatic hydrocarbons (PAHs) are some examples of organic contaminants that can build up within organisms and cause bioaccumulation and biomagnification. As animals and plants ascend up the food chain, HM concentrations increase, called biomagnification, and accumulate within their tissues, called bioaccumulation, which has an impact on top-level predators, such as birds and mammals (Ray and Vashishth 2024). HMs like Pb, Cd and Hg can be harmful, even at small quantities, and alter an animal's or plant's physiological and biochemical functions. By interfering with cellular functions, HMs like Pb, Cd, Hg, As and Cr can prevent germination of seed, root elongation and general plant development (Saud et al. 2024). Deficits in vital nutrients including calcium, potassium and magnesium result from high metal concentrations can obstruct nutritional absorption. Consuming these contaminated plant items exposes humans to a variety of dangerous illnesses. The brain, kidneys, liver, intestines, lungs and other human body parts can all be harmed by the exposure to these HMs (Angon et al. 2024). However, in the case of plants, chlorophyll can also be damaged by HMs, which reduces the plant's capacity for food synthesis. This results in stunted growth, yellowing of leaves (chlorosis) and reduced productivity (Abou Seeda et al. 2024). Metals such as Cd and Cu generate reactive oxygen species (ROS) in green plants that result in oxidative stress (Berni et al. 2019). The oxidative stress weakens plants and makes them more susceptible to disease by destroying DNA, proteins and lipids (Balali-Mood et al. 2021). According to Gudadhe et al. (2024), Pb is neurotoxic and impairs an animal's motor and cognitive capacities (Gudadhe et al. 2024). As and Cd can harm the liver, kidneys and other important body parts (Balali-Mood et al. 2021). According to Badr and El-Habit (2018), reproductive toxicity from metals like Pb and Cr lowers fertility and offspring survival rates in males (Badr and El-Habit 2018).

Organic pollutants and dyes from industrial areas have significant adverse effects on both flora and fauna. These substances often persist in the ecology, are contaminated and can bioaccumulate, leading to widespread ecological harm. Organic contaminants, such as dyes, PAHs and polychlorinated biphenyls (PCBs), are mutagenic and immunotoxic to living beings, including animals, microbes, as well as humans (Bolden et al. 2017; Rengarajan et al. 2015). Dyes often block light penetration in water bodies, reducing the availability of sunlight for aquatic plants. This hampers photosynthesis and stunts growth (Farhan Hanafi and Sapawe 2020). Like HMs, many dyes and organic pollutants generate ROS in plants, damaging chlorophyll and other cellular components critical for photosynthesis (Sharma et al. 2021). Damage to cellular components, including proteins, nucleic acids, or cell membranes, is caused by reactive species and reactive intermediates that are created when PAHs absorb UV-A and visible light. When exposed to sunlight, human skin contaminated with PAHs can oxidize DNA nucleic acid, cleaving DNA single strands, and form covalent adducts, indicating that the toxicity of PAH when exposed to light is significantly higher than that in the dark (Borska et al. 2016).

Despite being designed to enhance agricultural production, agrochemicals like pesticides and fertilizers frequently leak into soil and water systems, eutrophicating them and endangering non-target species. The detrimental effects of the careless use of agriculture related chemicals on the environment and ecosystem are documented in a number of documents. Despite the fact that the green rebellion boosts farming output and productivity, the careless utilization of agrochemicals has long-term effects (John and Babu 2021). Residues of agrochemicals in the air, land, water and agricultural goods have been alarmingly stated worldwide (Evans et al. 2019; Zikankuba et al. 2019). There are currently about 1000 pesticides with almost 800 active ingredients that are utilized globally, and the countis constantly growing (Zhang et al. 2019a). The careless use of pesticides has a negative effect on soil microbial growth, soil fertility, HM concentrations and groundwater quality, all of which contribute to the fall in agricultural crop production (Ali et al. 2016). The use of fertilizers and pesticides in fields contaminates the soil first, followed by the subsurface water. Humans and animals are at risk from the tainted water, and the high nitrate can render haemoglobin immobile. Because of their fewer detrimental effects on society, organophosphate pesticides have recently supplanted organochlorine; nonetheless, when they are caught in vegetables, they can cause cancer in the human body in addition to symptoms including nausea, vomiting, dizziness and skin irritation (Miah et al. 2014). According to the Stockholm Convention, some pesticides are classified as persistent organic pollutants (POPs) (N.A. 2021). These include organic chemicals that are chlorinated, such as DDT, endosulfan, chlordane, endrin, aldrin, heptachlor, dieldrin and hexachlorobenzene, toxaphene, which have detrimental effects on the ecology and human health, are highly insistent in soil and can go long distances due to their prolonged residence (PMSLY 2017). The organophosphorus class of pesticides, which are hazardous to both people and wildlife, are the most prevalent in contaminated soils (Wang et al. 2024). Pesticide poisoning is most commonly caused by carbamates and organophosphates worldwide (Sharma et al. 2019). A number of studies have documented how pesticides affect the activity of soil enzymes that are known to preserve the soil's nutritional condition, including urease, oxidoreductases, nitrate reductase, dehydrogenase and nitrogenase (Meena et al. 2020).

Because it compromises the stability of entire ecosystems in addition to affecting the survival and reproduction of individual species, this environmental contamination presents a serious problem. A thorough grasp of pollution dynamics and the creation of long-term mitigation plans are necessary to protect ecosystems and biodiversity. SiNPs, one of the novel alternatives, have demonstrated great potential in environmental cleanup. They are very successful at reducing pollution because of their large surface area, adjustable qualities and ability to absorb a range of contaminants (Martins et al. 2020). By neutralizing pesticide residues, degrading organic contaminants and immobilizing HMs, NPs can lessen the toxicity and availability of these substances to both plants and animals (Alharbi et al. 2024). Furthermore, their biocompatibility ensures minimal secondary environmental harm (Namasivayam et al. 2023). A revolutionary strategy to maintain and restore ecological balance and create a better environment for all living things is to combine SiNP-based methods in mesoporous structures with other sustainable methods.

19.4 SiNPs-remediation of Agro-Pollutants

Various nutrients and contaminants, such as fertilizers, pesticides, HMs and halogenated and phenolic compounds, are now primarily known, in addition to fertilizers and pesticides, to play a detrimental role in agri-ecosystems as a result of human activities in agricultural production (El-Ramady et al. 2020; Ken and Sinha 2020). Agro-pollutants, which appear as a result of agricultural activities, are considered a class of pollutants that have drawn public attention worldwide, mainly because agro-pollutants do not directly or quickly overbalance any limiting effect upon the factors of the studied ecosystem (Cui et al. 2024). This results in the environmental protection control system not being engaged in their regulation. Continuous progress in food biotechnology and the application of a range of chemicals, such as herbicides, pesticides and other agrochemicals, can contribute to the production of modern high-quality products, but on the other hand, it causes soil degradation (Wang et al. 2023). One of the most exciting new technologies for the degradation of agro and other organic pollutants deposited in soil is nano-cultivation using SiNPs (Farooq et al. 2022; Luqueño et al. 2023). Nano-silica, with its unique biocompatibility, enhanced adsorption capacity and exceptional physicochemical properties, can be suggested for the dilapidation of an extensive range of different organic contaminants (Seaf El-Nasr et al. 2021; Singh et al. 2022). Due to the controlled size, shape and a certain charge compared with other sorbents, nano-silica can be an environmentally friendly alternative material applicable in pollution control over a long period (Yadav et al. 2023b). The application of SiNPs and materials infused with Si/SiO_2 NPs in agriculture has greatly benefited sustainability. The versatility of SiNPs has recently been documented since silicon is one of the most common elements in the universe (Dhakate et al. 2022).

Many research groups are involved in the burgeoning fields of research related to silica and its use in material science and various technologies to remediate organic, organometallic and inorganic metal pollutants worldwide due to their highly effective adsorption capacities (Perez et al. 2017). Following their participation in the deterioration process, they can be retrieved or turned into soil using a simple heating procedure (Wang and Wang 2018). Therefore, the information on the degradation of organometallic pollutants using SiNPs is important and welcomed, as these particles can contribute not only to an efficient degradation process but also to a more environmentally friendly response, which is crucial when pollution prevention is a primary task (Huang et al. 2022).

The increasing prevalence of soil and water contamination poses substantial coercions to ecosystems, human health and agricultural production (Zhang et al. 2023). Conventional remediation methods often fail to address the complexities associated with various pollutants, necessitating innovative approaches to environmental clean-up. Among these emerging technologies, SiO_2NPs have gathered attention for their unique properties and potential applications in the remediation of contaminated environments. Derived from various sources, including agricultural waste, Si/SiO_2NPs present a viable option not only for pollutant removal but also for enhancing soil quality and crop resilience (Okeke et al. 2023; Yadav et al. 2023b). SiO_2NPs are efficient agents for adsorbing a variety of pollutants from soil and water due to their large surface area, stability, reactivity and biocompatibility (Samani et al. 2024b). These NPs hold promise in alleviating the environmental impact of agricultural pollutants, such as pesticides, herbicides, HMs and fertilizers. Recent studies have highlighted their efficacy in removing HMs and organic pollutants through mechanisms that enhance pollutant immobilization and degradation. Furthermore, the biosynthesis of SiO_2NPs from sustainable sources such as rice husks demonstrates an environmentally friendly approach to nanomaterial production (El-Gazzar et al. 2021).

The unique characteristics of SiNPs help in the adsorption of a larger number of agro-pollutants such as pesticides, herbicides and HM ions like Pb, Cd, As, and so on. A significant factor influencing this capacity is the SiNP's large surface area (Okeke et al. 2023). This large surface area enables them to catch and neutralize these toxic compounds more effectively. Adsorption of these pollutants decreases their bioavailability, and SiNPs are essential for this process. This includes how much and how quickly these substances can be made available for uptake by living things. This is important because it reduces the bioavailability of these agro-pollutants and consequently the toxic effect of these compounds on humans and the environment (Maroušek et al. 2022).

Furthermore, SiNP's capacity to concentrate these contaminants will facilitate their removal from the environment and lead to the advancement of sustainable farming practices (Boora et al. 2023). Thus, adopting SiNPs in pollutant management practices will be a huge leap towards lowering agriculture's ecological footprint and safeguarding our water and soil resources (Ramos-Carrazco et al. 2023). In summary, the properties of SiNPs highlight their potential as effective agents for combating the pervasive issue of agro-pollution, ultimately contributing to a healthier and safer environment.

19.4.1 Mechanism of Action for Contaminant Removal

The mechanisms of this adsorption-based removal of HMs using Si/SiO_2 NPs or silica-based materials have become a focal point for increasing environmental concerns about metal pollution. This type of NP possesses unique physical and chemical properties that allow it to efficiently absorb, degrade and interact with various contaminants, including HMs and organic pollutants (Cannas et al. 2019). Furthermore, SiO_2-based nanomaterials are not only environmentally benign but also biocompatible, positioning them as a viable alternative to traditional remediation strategies (Islam et al. 2023). The adsorption mechanism consists of interactions between the absorbent SiNPs or silica-based materials and the adsorbate: HM ions, both chemically and physically, for removing pollutants from water or soil environments. Several parameters affecting the effectiveness of removal and adsorption capabilities include physicochemical properties of the silica material, specific types of HMs, environmental conditions and the interaction between metallic ions and the presence of functional groups on the surface of the silica particle (Huang et al. 2022).

SiNPs are generally produced by sol–gel synthesis, microemulsion methods and hydrothermal procedures. Techniques like these generate NPs characterized by large surface areas, small sizes and a high density of hydroxyl groups on their surfaces (Ashour et al. 2021). By facilitating ion exchange, co-precipitation, electrostatic attraction and complexation, these hydroxyl groups are essential for the adsorption of HMs (Foo and Hameed 2010). The silanol groups (Si—OH) present on the surface of the silica can engage with metal cations, forming surface compounds that render HMs immobile (Matos et al. 2017). This interaction is predominantly operative for divalent and trivalent metals, such as Pb, Cu and Cr. For example, the interaction between Pb^{2+} ions and surface Si—OH groups could form stable Pb—O—Si bonds, which significantly reduce the levels of free Pb ions present in the surrounding environment (Xiong et al. 2013). This is important with respect to the SiNPs' efficiency and effectiveness as potential agents for the HM pollutants remediation in various ecological settings (Samani et al. 2024b). In particular, the improved silica structure, in which the silica structure is modified with functional groups, highly increases its affinity to these HMs and allows selective binding to a range of HMs, and not only just protonated silica adsorption via electrostatic attraction (Kankala et al. 2020). These functional groups are often introduced either through surface modifications after synthesis or simply because of functionalization on the silica surface. Ion exchange of metal ions with Si/SiO_2 NPs modified with functional groups like amines or thiols is another technique to remove HMs from contaminated water. Ion exchange of metal ions with Si/SiO_2 NPs modified with functional groups like amines or thiols is another technique to eradicate HMs from polluted water (Youssif et al. 2024). For instance, with metal ions such as Cu^{2+}, Ni^{2+} and Zn^{2+}, an amine group can create stable complexes effectively, thus increasing the value of its adsorption capacity. Thiol-modified SiNPs exhibit a high level of attraction to Hg^{2+}, As^{3+} and Cd^{2+} ions, simply on the basis that sulphur has a high affinity for these metal cations (Godiya et al. 2022). Here mechanistic studies showed that the increase in surface area increases adsorption overall. Furthermore, silica-loaded metal-organic frameworks (MOFs) or those incorporated in polymer composites have demonstrated considerable potential in removing HMs, as they increase surface area and offer additional metal ion sorption sites (Arya et al. 2025). Furthermore, metal ions can be adsorbed and entrapped by these NPs in co-precipitation to give stable precipitates that can be readily removed from solution (Figure 19.2) (Aragaw et al. 2021).

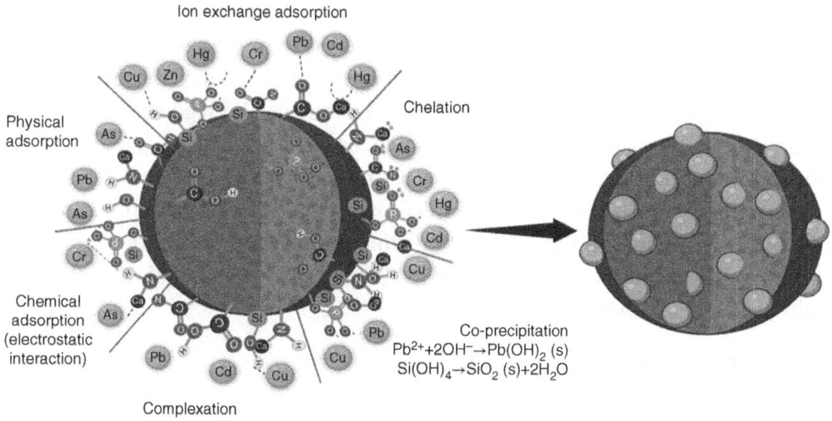

Figure 19.2 Action mechanism of Si/SiO_2 NPs.

The existence of competing ions in a solution can impact the adsorption of HMs, either enhancing or inhibiting it, depending on the type and charge of these ions. For example, divalent cations such as Ca^{2+} or Mg^{2+} might compete with Pb^{2+} ions for available sites for adsorption, which lowers the efficiency in removing Pb (Kankala et al. 2020). On the other hand, some ions can favour adsorption via synergistic reactions, either by forming complexes between metal ions and surface functional groups, among others. Temperature is another factor influencing adsorption, with temperature increase generally enhancing adsorption by promoting the diffusion of adsorbates. On the contrary, highly elevated temperatures may lead to the structural collapse of the silica and, hence, lose its adsorption capacity (Foo and Hameed 2010).

Silica compounds are critical to eliminate metallic contaminants from soil that frequently consists of toxic metals of elements like Pb, As and Hg can primarily have an adverse effect on human health as well as the environment. The use of SiNPs or composites embedded with silica allows for the cleaning of contaminated soils through techniques like immobilization or stabilization, where metal ions adhere to the silica surface, reducing their mobility and bioavailability (Samani et al. 2024b). The silica-based material can either be directly applied to the soil or can be mixed with other amendments, such as biochar, polymers or organic matter, to promote their adsorption efficiency. By adding SiNPs to the soil, HMs can be remedied, and the soil's structure, water retention and nutrient availability can be further improved (Yaashikaa et al. 2020).

Research displays that silica-based materials efficiently decrease the leachability and mobility of Pb in contaminated soils, lowering the risk of groundwater contamination (Pedro et al. 2023). Likewise, SiNPs modified with thiol or amine groups have been effective in adsorbing both Cd and Hg, effectively reducing their bioavailability and adverse effects on the soil ecosystem (Balali-Mood et al. 2021; Kankala et al. 2020). A key advantage of silica materials is that they can be recycled, which means that after they absorb HMs, they can be regenerated by using desorption methods like washing with acidic or chelating agents, such as EDTA (Andrunik and Bajda 2021). This not only enhances sustainability but also lowers remediation costs. Additionally, silica materials can be customized for specific HMs by modifying their surfaces or integrating them with other materials to improve adsorption efficiency and selectivity (Amaraweera et al. 2023; Ghaedi and Zhao 2022; Jaishankar et al. 2014).

19.4.2 Remediation of Agro-Pollutants

Si/SiO_2 NPs have surfaced with a flurry of interest in the remediation of the environment (Cannas et al. 2019). The capability of NPs to effectively adsorb HMs in contaminated soils caught this attention. These NPs express some very uniquely favourable qualities, such as having a high surface area, and are reactive; hence, they can be advertised to wreak environmental detoxifications induced by industrial activity or other pollution sources (Khan et al. 2019). Studies have coupled NPs with the treatment of HMs within a polluted soil environment, while the bioavailability and toxicity of those metals largely shrink. Thus far, Si/SiO_2 nanoparticles (NPs) have emerged as a promising material in the restoration of contaminated land for the environment, especially in the efforts to prevent the effects of HM pollution. Their applications continue to be investigated, and this is further revealing their possibilities as agents of environmental health and sustainability. An experiment by Samani et al. showed that nano-silica treatment influenced soil nutrition. Following the circumference of silica, concentrations of Cu, Ni, Cr, Pb and Zn dropped. With that addition, potassium and phosphorus nutrient uptake was increased among *Calendula officinalis* (Samani et al. 2024a). In another study, SiO_2 NPs were extracted from rice husk by Nguyen et al and were used to reduce Fe^{2+} (Nguyen et al. 2019). A study by Lian et al. exhibited that nano-silica effectively lessened the amounts of DTPA-extractable Cd in soil. The amounts of nano-silica applied in their studies, however, exceeded the threshold for general applicability, which is 6% (Lian et al. 2019). Stabilization of HM ions in the soil with the application of functionalized SiNPs has shown excellent outcomes in various studies, albeit at prohibitively high economic rates (Samani et al. 2024b). Zhang et al. synthesized thiol-functionalized nano-SiO_2 (SiO_2-SH) to passivate Cu^{2+}, Ni^{2+} and Zn^{2+} in soil (Zhang et al. 2019b). In today's world, Cd pollution in agricultural fields has turned out to be a raging issue worldwide. Hence, Wang and others assumed surface modified nano-silica or SMNS for the Cd-polluted soil remediation. The concentration decreased from 45.78% to 18.95% (Wang et al. 2020b).

An emerging field of environmental remediation is the elimination of agrochemicals from contaminated soil using Si/SiO_2 NPs and nanomaterials. Pesticides, herbicides and fertilizers are examples of agrochemicals that frequently linger in the soil, endangering human health and creating long-term environmental problems (Ali et al. 2023; Wahab et al. 2024). The pesticides that are widely used in the environment have to be detoxified since they have an adverse impact on ecosystems. Usually, detoxification methods for pesticides consist of chemical treatment, including hydrolysis via catalysis, incineration or adsorption on organic or inorganic sorbents, and the use of enzymes and biological systems (Al Hattab and Ghaly 2012; Bapat et al. 2016). Most chemical and physical methods tend to be quite expensive and sometimes ineffective (Ortiz-Hernández et al. 2013). Because they work efficiently under ambient circumstances, microbial and enzymatic techniques are beneficial. Furthermore, once immobilized on various inert supports, they can be used repeatedly. Desired characteristics of silica NPs are crucial for their successful application as a matrix for enzyme immobilization (Bapat et al. 2016). On silica nanostructures, a variety of enzymes, including organophosphate hydrolases, laccase and carboxyesterases, are immobilized along with bacteria expressing recombinant proteins for the detoxification of pesticides. The initial study of the specific hydrolase enzyme-catalogued under carboxyesterases, by dilapidation of the organophosphate Malathion and then carbamate Diethofencarb pesticide, was carried out on SiNPs immobilized with carboxyesterase from the insect *Spodoptera litura* (Diao et al. 2013). Esterases which are produced by insects help them metabolize organophosphates, carbamates and synthetic pyrethroids, contributing to their resistance against these pesticides. In this process, two kinds of mesoporous SiNPs, namely SBA-15 and MCM-41, were implemented as supports for the immobilization of the insect carboxyesterase (Diao et al. 2013).

Pesticide degradation can also be attained through whole cell immobilization. *Escherichia coli* cells encoding a recombinant enzyme named AtzA were enclosed on silanized SiNPs in a gel which is porous (Reátegui et al. 2012). In this set-up, silanization of SiNPs had been done by MTMS and cross-linked with polyethylene glycol (PEG). The stability of the system was tested at both 23 °C and 45 °C. The encapsulated cells performed similarly at both temperatures, but the gel was more structurally stiff at higher temperatures.

Advanced oxidation processes (AOPs) are very innovative ways in reducing environmental pollutants utilizing photocatalytic technology (Sorolla et al. 2012; Vanichvattanadecha et al. 2021). Such photocatalysts include titanium dioxide (TiO_2), which is a smart candidate for photocatalysts, is cheap, non-toxic and chemically stable and possesses good oxidizing capacity. Because of light irradiation, TiO_2 induces electron–hole pair formation (e^- and h^+), which combines with organic molecules, mineralizing them to water, carbon dioxide and other products (Vanichvattanadecha et al. 2021). High-porosity materials, including SiO_2 and mesoporous SiNPs (MSNs), are used as effective supports for TiO_2 NPs, enhancing photodegradation of organic compounds and minimizing agglomeration because of their large surface areas and diverse porous character. In 2021, Vanichvattanadecha et al. implemented the sol-gel method to develop highly mesoporous silica SBA-16, which they used to support TiO_2 NPs. They studied the degradation of commercial paraquat (PQ) herbicide under UV illumination conditions, and they observed a 70% reduction of PQ concentrations within 24 hours (Vanichvattanadecha et al. 2021). In a different investigation, the degradation of the herbicide Metsulfuron-methyl was attained with the implementation of TiO_2 coated with SiO_2 (Montañez et al. 2018).

Below in Table 19.1, different materials made up of Si/SiO_2 NPs and their activity in the agro-pollutant remediation process are described.

19.5 Nano-Silica-Mediated Remediation of Industrial Pollutants

Industrial wastewater pollution has become an urgent environmental issue as chemicals released into ecosystems become more complex and destructive. HMs, organic pollutants and other hazardous compounds produced by industrial operations pose substantial threats to human health, aquatic ecosystems and soil quality (Singh et al. 2023). Action procedures alone, such as chemical ion exchange, precipitation and biological treatment, frequently fail to completely remove these pollutants or produce secondary waste. Moreover, they have several boundaries, such as excess energy consumption, limited removal effectiveness and the production of contaminated sludge, which limits their broad deployment (Ahmed et al. 2022; Qasem et al. 2021). Therefore, nanotechnology presents an approach to environmental restoration with nano-silica emerging as a potential material due to its distinct physicochemical features. Nano-silica, with its high surface area, chemical durability and variable porosity, has shown great promise in adsorbing, decomposing or immobilizing a wide spectrum of industrial contaminants. In water remediation, silica can immobilize organic contaminants like dyes, insecticides and hydrocarbons in addition to adsorbing HMs like Cd, Pb and Hg. However, adding additional chemical functions to the nano-silica surface is the chemical process known as functionalization of nano-silica or silanization where silano groups that is OH functional group on the surface of nano-silica are replaced by one more function of improved adjustability in the chemical reaction (Martins et al. 2020). It is found that functionalizing nano-silica with certain chemical groups improves its selectivity and efficiency in targeting

Table 19.1 Different Si/SiO_2 NPs and nanomaterials in environmental remediation.

Sl. no	Materials	Target	Application in remediation	Method	Reference
1	Mesoporous SiNp /iron oxide nanocomposite	Lindane pesticide	Pesticide remediation	Adsorption	El-Said et al. (2018)
2	SiNp	Cd^{2+}	Cd-contaminated soil remediation and improved growth in *Triticum aestivum* L.	Adsorption	Ali et al. (2019)
3	SiNp	Pb and Cd	Elimination of Cd and Pb toxicity in *Oryza sativa* L.	Adsorption	Hussain et al. (2020)
4	SiO_2Np	Cr	Optimization in *Triticum aestivum* L. growth against Cr stress-induced phytotoxicity	Adsorption	Manzoor et al. (2022)
5	Ferrous sulphate-modified nano-silica	Cd, As and Pb	Remediation of HM-contaminated soils in the field	Adsorption	Cao et al. (2020)
6	Mercapto-functionalized nano-silica (MPTS/nano-silica)	Cd	Stabilization of Cd-contaminated soil	Adsorption	Wang et al. (2020a)
7	SiNp	Cr (VI)	Alleviation of Cr (VI) phytotoxicity in seedlings of *Pisum sativum* (L.)	Adsorption	Tripathi et al. (2015)
8	Mesoporous SiNPs MCM 41 and 48	Hg_2^+	Mitigating stress and increased growth of *Vigna radiata* and probiotic soil bacteria *Bacillus coagulans*	Adsorption	Mitra et al. (2023)

(Continued)

Table 19.1 (Continued)

Sl. no	Materials	Target	Application in remediation	Method	Reference
9	Alumina–silica nano-amendment (ASNA)	Pb and Zn	Immobilization of the HMs	In situ immobilization	Lei et al. (2022)
10	SBA-16/TiO_2 composites	Diuron	Helping in the reduction of diuron herbicide in dark conditions	Adsorption	Vanichvattanadecha et al. (2024)
11	SiNP	Cyprodinil, deltamethrin, difenoconazole, fipronil, pendimethalin, permethrin, oxadiazon and tetraconazole	Photolysis degradation of the pesticides	Adsorption	Samia et al. (2024)
12	Silicon fertilizer (SF)	Cd and Pb	Alleviation of Cd and Pb in soil aids tobacco growth	—	Ren et al. (2024)
13	Silicon (Si) fertilizer	Al, As, Cd, Cu, Zn and Cr		Foliar spray	Zhao et al. (2024)

specific pollutants, including dyes, HMs and organic pollutants (Ethaib et al. 2022; Grisolia et al. 2023).

19.5.1 Water Treatment

To improve the treated effluent's quality, a number of alternative action techniques have recently been studied. These include membrane separation, electro technologies, photocatalytic processes and adsorption with inexpensive materials. Wastewater with high concentrations of HMs is frequently treated using adsorption and membrane separation (Ahmed et al. 2022). Recently, new advantages for the removal of several pollutants have been demonstrated by nanoscale systems (NPs that are supported on various other inorganic adsorbents) (Trujillo-Reyes et al. 2014). Because of their unique qualities, nanomaterials derived from silica, another family of metal oxide nano-adsorbents, have a lot of potential for ion adsorption of metal. For the removal of HMs from water, these NPs have become a significant metal oxide class. Moreover, nano-silica is an eco-friendly, non-toxic adsorbent. These nano-adsorbents exhibit enhanced metal selectivity and adsorption capabilities upon treatment with amino and thiol groups. As shown in Figure 19.3, adsorbents and membranes made of silica can be used for the cleansing of wastewater from HMs and other hazardous contaminants. Some popular pollutants such as Cd^{2+}, Cu^{2+}, Cr^{6+} and Pb^{2+} cause heavy contamination of the water and are known to be caused by industrial effluents. To successfully eliminate Cd^{2+}, Cu^{2+} and Pb^{2+} from wastewater, Li et al. altered

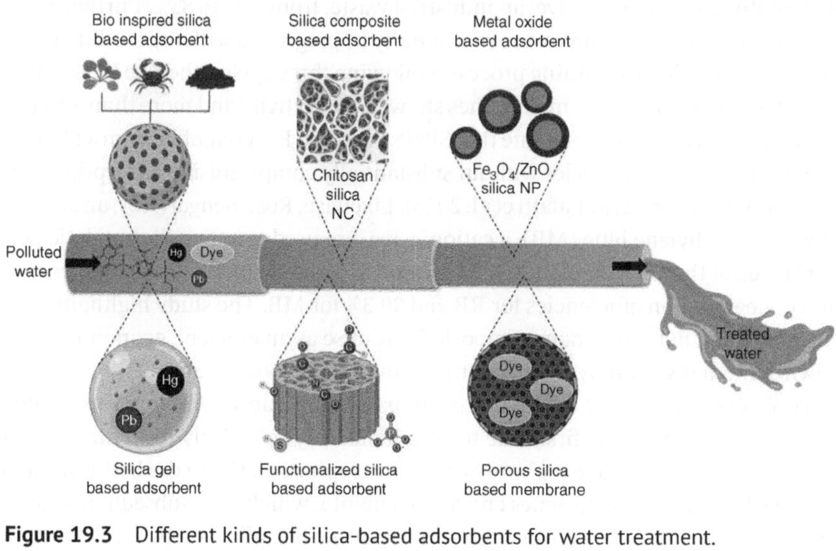

Figure 19.3 Different kinds of silica-based adsorbents for water treatment.

silica gel with nitrilotriacetic acid (NTA-silica gel) (Li et al. 2019). The produced material removed metal ions quickly, taking 2–20 minutes, with adsorptive values of 76.22, 63.5 and 53.14 mg/g for Pb^{2+}, Cu^{2+} and Cd^{2+}, respectively. Another study used amino-functionalized (3-aminopropyl and phenyl groups) nanospheres of silica to improve the removal of Cu^{2+} ions (Kotsyuda et al. 2017). The adsorption capacity rose in a linear fashion as the amino group content rose. Three distinct amino-functionalized silica nano-hollow spheres (NH_2-SNHS), amino-functionalized nano-silica gel (NH_2-NSG) and nonfunctionalized nanomaterials based on silica (i.e., nano-silica spheres and nonfunctionalized SNHS) were used to compare the removal of Cd^{2+}, Pb^{2+} and Cu^{2+} (NH_2-SG) (Najafi et al. 2012). Functionalized nano-silica had a higher adsorption capacity than SNHS; utilizing NH_2-SNHS, the maximal adsorption capacities for Cu^{2+}, Cd^{2+} and Pb^{2+} were 31.39 mg/g, 40.73 mg/g and 96.79 mg/g, respectively. In another study, silica waste produced while cutting with a diamond-wire saw was subjected to nano-silver-assisted chemical etching (Ag-ACE) to create hybrid materials of 3-aminopropylethoxysilane (APTES)-functionalized nanoporous silica. At room temperature, with a highest adsorption capability of 103.75 mg/g after 60 minutes, the APTES-NPSi generated showed a highly efficient adsorption capability for Cr^{6+} from the aqueous solution that was considerably pH-dependent (Yang et al. 2020). The method through which Cr^{6+} was adsorbed was the reduction of Cr^{6+} to Cr^{3+} over protonated amino groups over nano-silica (Ethaib et al. 2022).

Highly pure mesoporous SiO_2 NPs with surface areas of 245 m^2/g and diameters less than 20 nm can be extracted from rice husks (RH). The surface alteration of these SiO_2 NPs with trichloro (methyl) silane (TMCS) increases the adsorption of Rhodamine (RhB) dye, an industrial waste, from 49% to 83%. Furthermore, poly-lactic acid (PLA) fibres were successfully impregnated with the modified SiO_2 NPs using the electrospinning process. Following three cycles, the rice husk-based PLA:SiO_2 nanocomposite membranes showed 60% activity and more than 80% dye removal. In order to demonstrate that SiNPs based on PLA nanofibers provide simple reuse and removal cycles without substantially compromising adsorption, this study was performed (Malafatti et al. 2023). Likewise, Rose Bengal (RB), an anionic dye, and methylene blue (MB), a cationic dye, are used to assess the catalytic performance of the shell core ZnO@SiO_2 nanocomposite, which remarkably reaches 98.0% degradation efficiencies for RB and 99.3% for MB. The study highlights and suggests the ZnO@SiO_2 nanocomposite's promise as an efficient, economical and environmentally friendly water treatment method (Gharbi et al. 2023).

Fe_3O_4@SiO_2@UiO-67 is a unique MOF-based smart adsorbent that was created by Wang's group for the first time to simultaneously selectively recognize, detect and remove the organophosphorus herbicide glyphosate. Using the hydrothermal approach, Wang and colleagues created magnetite, which they subsequently used the Stöber technique to cover with a non-permeable silica shell. The ability of the produced $Fe_3O_4@SiO_2$ NPs, magnetite to remove the anionic dye Congo Red from aqueous solution was shown to be notably significant. Similarly, to allow organic contaminants to be continuously destroyed by photocatalyst, a granular medium based on silica was created for the removal of textile dyes known as MB (Yang et al. 2018).

An alternate perspective for the environmental remediation through SiNPs is to utilize the biological residue to remove industrial contaminants. A method to lessen waste from the industry that processes palm oil and boost additional value is to convert waste, like palm oil fuel ash (POFA), into SiNPs, which are used as filler material in the creation of filter membranes. This will help to realize an industry of waste-free and sustainable palm oil. Another option is to use this waste as a biomass-based (green synthesis) source of silica. The main goal is to alter the properties of composite filtration membranes to make them more efficient, ecologically friendly and effective for use in water purification or wastewater purification (Wardanu and Indrasti 2024). Recently, the adsorption of MB dye was investigated using a novel particle system known as raspberry-like supra-particles, which is composed of amorphous SiNPs and magnetite/maghemite (Oppmann et al. 2019). Li et al. used a two-step procedure to make silica-coated nano-zero valent iron (nZVI) NPs and polydopamine. When used to remove phenanthrene and anthracene from aqueous solutions, the resulting adsorbent ($Fe@SiO_2@PDA$) achieved maximum adsorption capabilities of 0.367 and 0.185 mg/g, respectively (Li et al. 2017). Likewise, different modifications of SiNPs, such as $Fe@SiO_2@PDA$, aid in the effective elimination of various types of industrial contaminants and results in the purification of wastewater.

19.5.2 Soil Treatment

An essential part of the ecosystem, soil significantly affects anthropological, animal and plant well-being. HM contamination of soil poses a significant danger to ecosystems and human well-being in the continuous fight against environmental pollution. Industrial effluents constitute a significant portion of contaminants that cause soil pollution. Effluents like HM must be moved, removed, or their impact reduced because they resist microbial and chemical breakdown and stay in the soil for prolonged periods of time, unlike organic contaminants. Because HM contamination in soils is complicated, traditional remediation techniques frequently fall short. However, new technologies are providing creative and long-lasting ways to clean up soils tainted by these pollutants (Samani et al. 2024b).

A number of processes have been documented earlier for the redeployment and immobilization of HMs within soils caused by the use of nano-silica, including adsorption, co-precipitation, complexation and modifications to soil characteristics, including surface area and pH. Since nano-silica has a large surface area, it can offer numerous active sites for the inner-sphere surface complexation-based adsorption of HMs as Zn, Pb, Ni, Cr, Cu and Cd (Lian et al. 2019; Morsy et al. 2023; Yadav et al. 2023a). For the soil remediation, immobilization has been found to be effective. Immobilization is an economical way to reduce HM contamination in soil. By combining polluted soil with appropriate substances, immobilization reduces the toxicity and mobility of contaminants by altering the specific surface area, ion exchange, pH, adsorption and stabilizing methods (Stegmann et al. 2013). HMs can be trapped in the lattice structures of many soil constituents, including silicate minerals, or they can form connections with them. The ability of soil to retain or release HMs is influenced by the strength of these connections (Samani et al. 2024b).

HMs can be successfully sequestered and their leaching or mobility prevented by sorption onto SiO_2 surfaces, which can lower their concentrations in the way more readily extractable portions (such as carbonated and exchangeable bound). As they become more difficult to extract. The remaining portion will also be linked to HMs that are directly adsorbed to the nano-silica surface (Lian et al. 2019). To look at how well nano-silica stabilizes HMs in calcareous soil that has been polluted by HMs, Samani and his group studied and found that the dispersion of HMs in Fe—Mn fractions bonded with oxides, residue rose and organic matter, and carbonate-bound and the exchangeable (EX) fractions of Cu, Pb, Ni, Cr, and Zn decreased as a result of the addition nano-silica adsorbent to the soil (Samani et al. 2024b). Additionally, SiO_2 and trace metals can co-precipitate, integrating them into the SiO_2 crystalline structure (Yadav et al. 2023a). Because the co-precipitated metals will be tied to the residual fraction, which signifies the most robust and structurally integrated forms of the metals, co-precipitation of HMs with SiO_2 can lower their concentration in the carbonated-bound and Fe—Mn oxide fractions. (Tessier et al. 1979). By changing the pH in the soil, nano-silica can affect the solubility and mobility of HMs. Metal hydroxides may precipitate as a result of a higher pH, and these may then combine with Fe/Mn oxyhydroxides. Because metals become immobile in lesser readily available forms, these changes can lead to a possible decline in exchangeable and carbonate-bound fractions (Lei et al. 2022). Metal ions and organic functional groups are connected via nano-silica and consequently can enhance metal-humic interactions and thus facilitate the binding of HMs with organic molecules in soil (Liang et al. 2011). Consequently, the organic matter-bounded percentage of HMs rises. A study reveals that non-colloidal silica (NCS) treated soil can better retain HMs than soil alone because of colloids in the phase of soil-metal solution. The sorption of metal ions is influenced by pore widening caused by carbonate dissolution. NCS has an increased immobilization capacity due to adsorption, trapping, encapsulating and binding with immobile salts. The sorptive process is heavily influenced by the pH effect. Nitric acid (HNO_3) was found to be as successful as a chelant, while hexavalent chromium (DTPA) was eventually shown to be the best chelator. The order in which metal ions desorb remained consistent even with NCS addition (Mohammed and Moghal 2021). In a study, active silicon/magnesia biochar (ALB) based on lignin showed effective elimination of Pb^{2+}, Cd^{2+}, Zn^{2+} and Cu^{2+} from a variety of actual contaminated water sources. In addition to producing more oxygen-containing functional groups and Mg_2SiO_4 and MgO NPs, the modified lignin-based biochar also improved the specific surface area. The adsorption capacity of ALB/Si—Mg for Pb^{2+}, Cd^{2+}, Zn^{2+} and Cu^{2+} was also 29.09–140.45 times that of ALB, with values of 665.07 mg/g, 848.17 mg/g, 245.78 mg/g and 151.84 mg/g, respectively. Incorporating ALB/Si—Mg into the remediation of soil improved the physicochemical characteristics of the soil (dissolved organic carbon, exchangeable magnesium and effective silicon), promoted food synthesis and biomass in plants, decreased the availability of HMs in soil and hindered plant uptake of HMs. This study clarified how functional groupings that contain oxygen (O=C—O, Si—O and Mg—O) contribute to the cleanup of composite HM contamination in the environment, providing a novel approach to fight against soil polluted by industrial pollutants (Li et al. 2024).

Apart from direct aids to remove industrial pollutants from soil, SiNPs give alternate perspectives such as enhancement of soil health and plant growth by remediating the contaminated soil due to industrial wastes. Hence, an excellent strategy to lessen or eliminate the hazardous pollutant loads from soil is to use direct NPs to improve the microbial community. SiNPs have been shown to enhance microbial biomass and colonization, including rhizospheric microorganisms that help better soil health (Gajić et al. 2018; Srivastava et al. 2021). A nano-silica with thiol functionalization (SiO_2-SH) was used to clean up Pb, Cd and Cu-contaminated soil in order to diminish the biotoxicity of HMs. The uptake of HMs into pakchoi and lettuce in a pot experiment was used to examine the effect of remediation of SiO_2-SH on polluted soil. The fraction distribution of Pb, Cu and Cd was used to assess the biotoxicity of the SiO_2-SH and suggest immobilization strategies. It was discovered that the SiO_2-SH may considerably lower the absorption of Pb, Cu and Cd into lettuce by 43.41%, 5.76% and 89.81%, respectively, and into pakchoi by 68.03%, 76.34% and 92.02% (Lian et al. 2021). Thus, for the treatment of damaged and contaminated soils, especially those impacted by industrial pollutants, nano-silica has shown itself to be an extremely effective material. HMs, hydrocarbons and other hazardous substances frequently found in industrial effluents can be absorbed and neutralized by it due to its large surface area and chemical reactivity. Nano-silica aids in restoring soil health and productivity by increasing the physicochemical characteristics of the soil and reducing the mobility and bioavailability of dangerous contaminants. The schematic diagram in Figure 19.4 below indicates that Si/SiO_2NPs will be used in the adsorption process to remediate a variety of contaminants produced from agricultural waste.

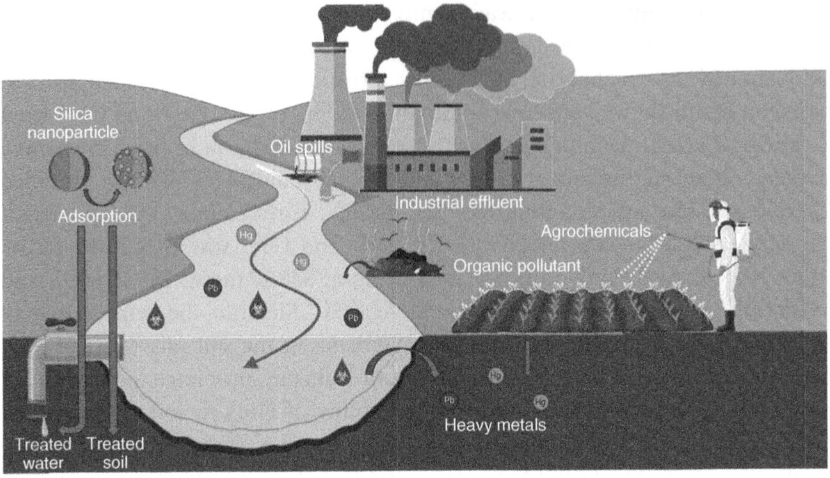

Figure 19.4 Schematic illustration of Si/SiO$_2$NPs as an agent for remediation of various contamination.

Below in Table 19.2, different materials made up of Si/SiO$_2$ NPs and their activity in the industrial pollutant remediation process are described.

19.6 Conclusion and Future Prospective

In recent decades, technological progress has spurred industrialization, urbanization and improvements in human living standards, even in the face of a declining population. Nonetheless, this progress brings with it significant risks, particularly the release of various pollutants into the atmosphere stemming from advanced scientific and industrial practices. The prevalence of potentially hazardous elements (PTEs) among the various contaminants that have been studied thus far is concerning because of their serious effects on environmental preservation and public health. Silicon and silicon dioxide NPs (Si/SiO$_2$ NPs) hold great potential for applications in environmental remediation, showing promise in addressing various ecological challenges. However, several critical issues and limitations must be thoroughly addressed before these materials can be adopted on a larger scale. One of the foremost concerns is the potential toxicity of Si/SiO$_2$NPs to non-target organisms within the ecosystem. While these NPs are predominantly regarded as non-harmful, their minute size and large surface reactivity can have significant risks if they are not managed appropriately.

This reactivity gives rise to further worries about its impact on a wide range of living things, including plants, aquatic species and soil microbes that are very important in maintaining ecological balance. Therefore, it is imperative to

Table 19.2 Various uses for different kinds of silicon nanomaterials in the remediation of wastewater and contaminated soil by eliminating industrial pollutants.

Sl. No.	Materials	Target pollutants	Methods	Reference
		Wastewater treatment		
1	Surface-functionalized magnetic silica-malachite composite (Fe—M—Si tricomposite)	Cypermethrin (CP) (pesticide)	Adsorption	Ul Ain et al. (2024)
2	Diphenylcarbazide immobilized silica adsorbent	Pb (II) (HM)	Sol-Gel process (adsorption)	Silviana et al. (2024)
3	Glycine/SiO_2 nanocomposite	Cr, Fe, Co, Cu, As, Ni, Cd and Pb (HM)	Adsorption	Sabry et al. (2024)
3	EDTA modified porous silica composite	Pb (II) (HM)	Hydrothermal process (adsorption)	Vu et al. (2024)
4	Mesoporous silica-grafted deep eutectic solvent-based mixed matrix membranes	As and Cr (HM)	—	Rashid et al. (2024)
5	$ZnFe_2O_4$@SiO_2 NPs	Methylene blue (organic pollutant)	Precipitation	Hashem et al. (2024)
6	Nano-SiO_2	Pb (II), Cr (VI)	Sol-Gel process (adsorption)	Meky et al. (2024)
7	Fe_3O_4@SiO_2 grafted with PVA	Cu (II), Cd (II) (HM)	Adsorption	Asgharinezhad et al. (2024)
8	Dithiol-modified mesoporous silica adsorbent	Ni (II), Zn (II), Hg (II) Pb (II) Ni (II), Fe (II) (HM)	Co-condensation (adsorption)	Santhamoorthy et al. (2023)
9	Fe_3O_4/silica NC	Petroleum oil (industrial effluent)	—	Zabermawi and Bestawy (2023)

(Continued)

Table 19.2 (Continued)

Sl. No.	Materials	Target pollutants	Methods	Reference
10	5-Chloro-8-quinolinol fabricated SiNP	Al (III) (HM)	Refluxing (adsorption)	Al-Wasidi et al. (2023)
11	Amine-functionalized SiNP	Cu (II) and NO_3^-) (HM)	Precipitation	Sharma et al. (2023)
12	Mesoporous SiNPs, large-pore mesoporous silica	Fe (III), Ni (II), Cu (II) (HM) Methylene blue and methyl green (Organic pollutant)	Refluxing (adsorption)	Flores et al. (2023)
13	Melamine functionalized mesoporous silica SBA-15	Cr (VI) (HM)	Hydrothermal process and refluxing (adsorption)	Purrostam et al. (2023)
14	Fe_3O_4 SBA-15	Pb (II), Cd (II) (HM)	Surface imprinting technique	sadat Hashami et al. (2022)
15	SiO_2@poly (VI-co-AA)	Ni (II) (HM)	Co-polymerization (adsorption)	Xing et al. (2022)
16	Silica-based zinc oxide nanocomposite (NC) (Nano-SZO)	Cd (II), Ni (II), Cu (II) (HM)	Chemisorption	Garg et al. (2022)
17	β-cyclodextrin/MCM-48 composite	Hg (II), As (V) (HM)	Ultrasonication (Adsorption)	Jumah et al. (2021)
18	Chitosan/mesoporous silica nanocomposite	U (VI), Sr (II) (HM)	Ultrasonication (adsorption)	Abukhadra et al. (2021)
19	SiO_2/chitosan/Fe nanocomposite	Vanadium (HM)	Ultrasonication (adsorption)	Sharififard and Rezvanpanah, (2021)

20	Mesoporous silica-based composite	Crystal violet dye (organic pollutant)	Adsorption	Baig et al. (2021)
		Soil treatment		
21	SiO_2NPs@BC	Cd^{2+} (HM)	Adsorption	Peng et al. (2024)
22	Si-NPs	Cd^{2+} (HM)	Foliar spray (immobilization)	Alharbi et al. (2024)
23	Active lignin-based silicon/magnesia biochar (ALB/Si—Mg)	Cd^{2+}, Pb^{2+}, Cu^{2+}, Zn^{2+} (HM)	Adsorption	X. Li et al. (2024)
24	Biosynthesized SiO_2-NPs	Ribazol black b and Crystal violet dyes (Organic pollutant) Co, Pb, Cd, Zn (HM)	Photocatalytic degradation	Kaabo et al. (2024)
25	Green silicon (Si) NPs	Pb (HM)	Foliar application	Maryam et al. (2024)
26	Green-synthesized SiO_2NPs	Arsenic (As) (HM)	Mobility Inhibition	Labulo et al. (2024)

perform extensive toxicological studies and comprehensive environmental impact assessments prior to their widespread application. Such research will ensure that the use of Si/SiO_2 NPs is effective in attaining desired remediation objectives, safeguarding the ecosystem as a whole and ensuring maximized advantages while reducing risks posed to the environment and its dwellers.

The other constraint regards the qualification of Si/SiO_2 NPs for large-scale remediation projects. While laboratory studies have proven effective at some small-scale levels, the application of those NPs in practical settings needs careful consideration of factors like their cost, simplicity of manufacture and long-term stability. Moreover, the likelihood of secondary contamination, whereby contaminants could still be leached from the surfaces of the Si/SiO_2 NPs, should be considered in remediation approaches. After the pollutants have been captured by these materials, several regeneration strategies exist involving thermal treatment, chemical washing, magnetic separation or desorption that allow one to reuse these SiO_2 materials many times and considerably reduce wastes and operational costs. Silica-based materials also help to mitigate the harmful environmental effects of remediation efforts by encouraging sustainable pollutant management practices. Furthermore, silica-based materials exhibit excellent recyclability properties in the removal and remediation of heavy metals and organic pollutants from agricultural and industrial effluents. Their high surface area, functionalization capabilities and chemical stability enable them to adsorb contaminants effectively. After adsorption, regeneration of these materials can be done through techniques such as thermal treatment, chemical washing, magnetic separation or desorption, allowing for repeated use which results in reducing waste and operational costs. Recyclability can enable long-term pollution management while reducing the environmental impact of the remediation procedure.

Nano-silica improves tolerance, seed germination and soil fertility while inhibiting pesticide resistance. Nano-silica is known to reduce soil permeability to water, and this translates to improved plant performance. This enhancement was due to enhanced physiological capacity in the plants, increased CO_2 assimilation and increased efficiency of biochemical metabolism in the plants. SiNPs embedded with catalysts, such as Fenton-type catalysts or photocatalysts like TiO_2, can reduce the organic pollutants into less toxic molecules. Therefore, these kinds of NPs would run hand in glove with reeds to diminish the incidence of environmental pollution and the conservation of natural resources with knowledge advancement in nanotechnology and environmental studies.

Glossary

Adsorbent Substance that adsorbs another substance
Adsorption Adhesions of ions and molecules to a surface

Agrochemical Chemicals generated from agricultural activity
Al Aluminum
ALB Active silicon/magnesia biochar
AOP Advanced oxidation process
APTES 3-aminopropylethoxysilane
As Arsenic
AtzA Atrazine-dechlorinating enzyme
Bioaccumulation Accumulation of contaminants in a living organism
Biocompatibility Ability to interact with the living system without producing harmful effects
Biotoxicity Accumulation of toxins in the body
Ca Calcium
Cd Cadmium
Co-precipitation Simultaneous precipitation of two or more compounds
Cr Chromium
Cu Copper
Desorption Process where adsorbed molecules are released from the host surface
DTPA Diethylenetriamine pentaacetate or Pentetic acid
Effluents Liquid-based discharge
Fe Iron
Functionalization Addition of functional groups on the surface of the materials
Herbicide Substances used to control weeds
Hg Mercury
Hydrothermal Method Technique for crystallizing substances in aqueous medium at high temperature and pressure
Immobilization To reduce or eliminate mobility
Ion Exchange Exchange of Ions with another compound
K Potassium
MB Methylene Blue
Microemulsion Method Preparation method that enables to control the particle properties such as size, surface area, etc.
MOF Metal-Organic Frame
MSN Mesoporous Silica Nanoparticle
N Nitrogen
Nanocomposite Multiphase solid material of at least one dimension in the nanoscale range
NCS Non-colloidal silica
Ni Nickle
NP Nanoparticle
NTA Nitrilotriacetic acid

Organic Contaminants/Pollutants Compounds that are resistant to degradation
P Phosphorus
PAH Polycyclic aromatic hydrocarbons
Pb Lead
PCB Polychlorinated biphenyls
PEG Polyethylene glycol
Pesticide Substances used to control pests
Photodegradation Degradation of contaminants using sunlight
PLA Poly-lactic acid
POPs Persistent organic pollutants
PTEs Potentially hazardous elements
RB Rose Bengal
Remediation Process of cleaning or removing pollutants from the environment
ROS Reactive oxygen species
Silanization Process that attaches an organosilyl group to a substance to modify its surface
SiNps Silicon Nanoparticles
SiO_2NPs Silica Nanoparticles
SMNS Surface-modified nano-silica
Sol-Gel Method Process for creating solid materials from small molecules
Stöber Technique Chemical process creating uniform silica particles of controllable size
Sustainability The ability of being able to maintain over a period of time
Wastewater Polluted water
Zn Zinc

References

(N.A.) (2021). Listing of POPs in the Stockholm convention. https://www.pops.int/TheConvention/ThePOPs/AllPOPs/tabid/2509/Default.aspx (accessed 28 December 2024).

Abou Seeda, M.A., Abou El-Nour, E.A.A., El-Bassiouny Hala, M.S. et al. (2024). Functions and toxicity of chromium on plants physiological and molecular aspects: a review. *Middle East Journal of Agriculture Research*. https://doi.org/10.36632/mejar/2024.13.2.19.

Abukhadra, M.R., Eid, M.H., El-Meligy, M.A. et al. (2021). Insight into chitosan/mesoporous silica nanocomposites as eco-friendly adsorbent for enhanced retention of U (VI) and Sr (II) from aqueous solutions and real water. *International Journal of Biological Macromolecules* 173: 435–444.

Adeniran, A.A. and Shakantu, W. (2022). The health and environmental impact of plastic waste disposal in South African townships: a review. *International Journal of Environmental Research and Public Health* 19 (2): 779. https://doi.org/10.3390/ijerph19020779.

Ahmed, M., Mavukkandy, M.O., Giwa, A. et al. (2022). Recent developments in hazardous pollutants removal from wastewater and water reuse within a circular economy. *npj Clean Water* 5 (1): 1–25.

Ahmed, T., Noman, M., Qi, Y. et al. (2023). Dynamic crosstalk between silicon nanomaterials and potentially toxic trace elements in plant-soil systems. *Ecotoxicology and Environmental Safety* 264: 115422. https://doi.org/10.1016/j.ecoenv.2023.115422.

Al Hattab, M.T. and Ghaly, A.E. (2012). Disposal and treatment methods for pesticide containing wastewaters: critical review and comparative analysis. *Journal of Environmental Protection* 3 (5): 431–453. https://doi.org/10.4236/jep.2012.35054.

Alengebawy, A., Abdelkhalek, S.T., Qureshi, S.R., and Wang, M.-Q. (2021). Heavy metals and pesticides toxicity in agricultural soil and plants: ecological risks and human health implications. *Toxics* 9 (3): 42. https://doi.org/10.3390/toxics9030042.

Alharbi, K., Hafez, E.M., Elhawat, N. et al. (2024). Revitalizing soybean plants in saline, Cd-polluted soil using Si-NPs, biochar, and PGPR. *Plants* 13 (24): 3550.

Ali, A.B., Hong, L., Haofang, Y., and Elshaikh, N. (2016). Obsolete pesticide contamination: a new type of water and soil pollution. *Current Science* 111 (6): 966. https://www.semanticscholar.org/paper/Obsolete-Pesticide-Contamination%3AA-New-Type-of-and-Ali-Hong/ebc33964340dea17b87862c8f8df1e39b74d532f.

Ali, S., Rizwan, M., Hussain, A., ur Rehman MZ, Ali, B., Yousaf, B., Wijaya, L., Alyemeni, M. N., & Ahmad, P. (2019). Silicon nanoparticles enhanced the growth and reduced the cadmium accumulation in grains of wheat (*Triticum aestivum* L.). *Plant Physiology and Biochemistry*, 140, 1–8. https://doi.org/10.1016/j.plaphy.2019.04.041

Ali, S., Ahmad, N., Dar, M.A. et al. (2023). Nano-agrochemicals as substitutes for pesticides: prospects and risks. *Plants* 13 (1): 109. https://doi.org/10.3390/plants13010109.

Alsaeedi, A., El-Ramady, H., Alshaal, T. et al. (2019). Silica nanoparticles boost growth and productivity of cucumber under water deficit and salinity stresses by balancing nutrients uptake. *Plant Physiology and Biochemistry* 139: 1–10. https://doi.org/10.1016/j.plaphy.2019.03.008.

Al-Wasidi, A.S., Katouah, H.A., Saad, F.A., and Abdelrahman, E.A. (2023). Functionalization of silica nanoparticles by 5-chloro-8-quinolinol as a new nanocomposite for the efficient removal and preconcentration of Al^{3+} ions from water samples. *ACS Omega* 8 (17): 15276–15287. https://doi.org/10.1021/acsomega.3c00413.

Amaraweera, S.M., Gunathilake, C.A., Gunawardene, O.H.P. et al. (2023). Carbon capture using porous silica materials. *Nanomaterials* 13 (14): 2050. https://doi.org/10.3390/nano13142050.

Anand, U., Carpena, M., Kowalska-Góralska, M. et al. (2022). Safer plant-based nanoparticles for combating antibiotic resistance in bacteria: a comprehensive review on its potential applications, recent advances, and future perspective. *Science of the Total Environment* 821: 153472. https://doi.org/10.1016/j.scitotenv.2022.153472.

Andrunik, M. and Bajda, T. (2021). Removal of pesticides from waters by adsorption: comparison between synthetic zeolites and mesoporous silica materials. A review. *Materials* 14 (13): 3532. https://doi.org/10.3390/ma14133532.

Angon, P.B., Islam, M.S., Kc, S. et al. (2024). Sources, effects and present perspectives of heavy metals contamination: soil, plants and human food chain. *Heliyon* 10 (7): e28357. https://doi.org/10.1016/j.heliyon.2024.e28357.

Aragaw, T.A., Bogale, F.M., and Aragaw, B.A. (2021). Iron-based nanoparticles in wastewater treatment: a review on synthesis methods, applications, and removal mechanisms. *Journal of Saudi Chemical Society* 25 (8): 101280. https://doi.org/10.1016/j.jscs.2021.101280.

Arya, K., Kumar, A., and Kataria, R. (2025). Recent advances in MOF-based composites for the detection and adsorptive removal of Pb(II) ions in aqueous phase. *Materials Today Sustainability* 29: 101057. https://doi.org/10.1016/j.mtsust.2024.101057.

Asgharinezhad, A.A., Esmaeilpour, M., and Afshar, M.G. (2024). Synthesis of magnetic Fe_3O_4@SiO_2 nanoparticles decorated with polyvinyl alcohol for Cu(II) and Cd(II) ions removal from aqueous solution. *Chemical Papers* 78 (6): 3799–3814. https://doi.org/10.1007/s11696-024-03350-4.

Ashour, M.M., Mabrouk, M., Soliman, I.E. et al. (2021). Mesoporous silica nanoparticles prepared by different methods for biomedical applications: comparative study. *IET Nanobiotechnology* 15 (3): 291–300. https://doi.org/10.1049/nbt2.12023.

Ayilara, M.S., Adeleke, B.S., Akinola, S.A. et al. (2023). Biopesticides as a promising alternative to synthetic pesticides: a case for microbial pesticides, phytopesticides, and nanobiopesticides. *Frontiers in Microbiology* 14: 1040901. https://doi.org/10.3389/fmicb.2023.1040901.

Azman, N.F., Katahira, T., Nakanishi, Y. et al. (2023). Sustainable oil palm biomass waste utilization in Southeast Asia: cascade recycling for mushroom growing, animal feedstock production, and composting animal excrement as fertilizer. *Cleaner and Circular Bioeconomy* 6: 100058. https://doi.org/10.1016/j.clcb.2023.100058.

Badr, F.M. and El-Habit, O. (2018). Heavy metal toxicity affecting fertility and reproduction of males. In: *Bioenvironmental Issues Affecting Men's Reproductive and Sexual Health* (ed. S.C. Sikka and W.J.G. Hellstrom), 293–304. Elsevier. https://doi.org/10.1016/B978-0-12-801299-4.00018-9.

Baig, M.M., Zulfiqar, S., Yousuf, M.A. et al. (2021). $Dy_xMnFe_{2-x}O_4$ nanoparticles decorated over mesoporous silica for environmental remediation applications. *Journal of Hazardous Materials* 402: 123526.

Balali-Mood, M., Naseri, K., Tahergorabi, Z. et al. (2021). Toxic mechanisms of five heavy metals: mercury, lead, chromium, cadmium, and arsenic. *Frontiers in Pharmacology* 12: 643972. https://doi.org/10.3389/fphar.2021.643972.

Bapat, G., Labade, C., Chaudhari, A., and Zinjarde, S. (2016). Silica nanoparticle based techniques for extraction, detection, and degradation of pesticides. *Advances in Colloid and Interface Science* 237: 1–14. https://doi.org/10.1016/j.cis.2016.06.001.

Baragaño, D., Forján, R., Sierra, C., and Gallego, J.L.R. (2021). Nanomaterials for soil remediation: pollutant immobilization and opportunities for hybrid technologies. In: *Sorbents Materials for Controlling Environmental Pollution* (ed. A. Núñez-Delgado), 701–723. Elsevier. https://doi.org/10.1016/B978-0-12-820042-1.00026-2.

Baran, M.F., Keskin, C., Baran, A. et al. (2023). Green synthesis of silver nanoparticles from Allium cepa L. peel extract, their antioxidant, antipathogenic, and anticholinesterase activity. *Molecules* 28 (5): 2310. https://doi.org/10.3390/molecules28052310.

Berni, R., Luyckx, M., Xu, X. et al. (2019). Reactive oxygen species and heavy metal stress in plants: impact on the cell wall and secondary metabolism. *Environmental and Experimental Botany* 161: 98–106. https://doi.org/10.1016/j.envexpbot.2018.10.017.

Bolden, A.L., Rochester, J.R., Schultz, K., and Kwiatkowski, C.F. (2017). Polycyclic aromatic hydrocarbons and female reproductive health: a scoping review. *Reproductive Toxicology* 73: 61–74. https://doi.org/10.1016/j.reprotox.2017.07.012.

Boora, R., Sheoran, P., Rani, N. et al. (2023). Biosynthesized silica nanoparticles (Si NPs) helps in mitigating drought stress in wheat through physiological changes and upregulation of stress genes. *Silicon* 15 (13): 5565–5577. https://doi.org/10.1007/s12633-023-02439-x.

Borska, L., Andrys, C., Krejsek, J. et al. (2016). Influence of dermal exposure to ultraviolet radiation and coal tar (polycyclic aromatic hydrocarbons) on the skin aging process. *Journal of Dermatological Science* 81 (3): 192–202. https://doi.org/10.1016/j.jdermsci.2015.12.010.

Cannas, M., Camarda, P., Vaccaro, L. et al. (2019). Luminescence efficiency of Si/SiO_2 nanoparticles produced by laser ablation. *Physica Status Solidi (a)* 216 (3): 1800565. https://doi.org/10.1002/pssa.201800565.

Cao, P., Qiu, K., Zou, X. et al. (2020). Mercapto propyltrimethoxysilane- and ferrous sulfate-modified nano-silica for immobilization of lead and cadmium as well as arsenic in heavy metal-contaminated soil. *Environmental Pollution* 266: 115152. https://doi.org/10.1016/j.envpol.2020.115152.

Chen, L., Liu, J., Zhang, Y. et al. (2018). The toxicity of silica nanoparticles to the immune system. *Nanomedicine* 13 (15): 1939–1962. https://doi.org/10.2217/nnm-2018-0076.

Chowdhury, N.N., Cox, A.R., and Wiesner, M.R. (2021). Nanoparticles as vectors for antibiotic resistance: the association of silica nanoparticles with environmentally relevant extracellular antibiotic resistance genes. *Science of the Total Environment* 761: 143261. https://doi.org/10.1016/j.scitotenv.2020.143261.

Cobb, G.P., Sands, K., Waters, M. et al. (2000). Accumulation of heavy metals by vegetables grown in mine wastes. *Environmental Toxicology and Chemistry* 19 (3): 600–607. https://doi.org/10.1002/etc.5620190311.

Craswell, E.T. and Karjalainen, U. (1990). Recent research on fertilizer problems in Asian agriculture. *Fertilizer Research* 26 (1–3): 243–248. https://doi.org/10.1007/BF01048762.

Cui, Z., Li, Y., Tsyusko, O.V. et al. (2024). Metal–organic framework-enabled sustainable agrotechnologies: an overview of fundamentals and agricultural applications. *Journal of Agricultural and Food Chemistry*. https://doi.org/10.1021/acs.jafc.4c00764.

Dhakate, P., Kandhol, N., Raturi, G. et al. (2022). Silicon nanoforms in crop improvement and stress management. *Chemosphere* 305: 135165. https://doi.org/10.1016/j.chemosphere.2022.135165.

Diao, J., Zhao, G., Li, Y. et al. (2013). Carboxylesterase from Spodoptera litura: immobilization and use for the degradation of pesticides. *Procedia Environmental Sciences* 18: 610–619. https://doi.org/10.1016/j.proenv.2013.04.084.

El-Gazzar, N., Almanaa, T.N., Reda, R.M. et al. (2021). Assessment the using of silica nanoparticles (SiO_2NPs) biosynthesized from rice husks by Trichoderma harzianum MF780864 as water lead adsorbent for immune status of Nile tilapia (Oreochromis niloticus). *Saudi Journal of Biological Sciences* 28 (9): 5119–5130. https://doi.org/10.1016/j.sjbs.2021.05.027.

El-Ramady, H., El-Henawy, A., Amer, M. et al. (2020). Agro-pollutants and their nano-remediation from soil and water: a mini-review. *Environment, Biodiversity and Soil Security*. https://doi.org/10.21608/jenvbs.2020.47751.1111.

El-Said, W.A., Fouad, D.M., Ali, M.H., and El-Gahami, M.A. (2018). Green synthesis of magnetic mesoporous silica nanocomposite and its adsorptive performance against organochlorine pesticides. *International Journal of Environmental Science and Technology* 15 (8): 1731–1744. https://doi.org/10.1007/s13762-017-1530-9.

Ethaib, S., Al-Qutaifia, S., Al-Ansari, N., and Zubaidi, S.L. (2022). Function of nanomaterials in removing heavy metals for water and wastewater remediation: a review. *Environments* 9 (10): 123.

Evans, A.E., Mateo-Sagasta, J., Qadir, M. et al. (2019). Agricultural water pollution: key knowledge gaps and research needs. *Current Opinion in Environmental Sustainability* 36: 20–27. https://doi.org/10.1016/j.cosust.2018.10.003.

Farhan Hanafi, M. and Sapawe, N. (2020). A review on the water problem associate with organic pollutants derived from phenol, methyl orange, and remazol brilliant blue dyes. *Materials Today: Proceedings* 31: A141–A150. https://doi.org/10.1016/j.matpr.2021.01.258.

Farooq, M.A., Hannan, F., Islam, F. et al. (2022). The potential of nanomaterials for sustainable modern agriculture: present findings and future perspectives. *Environmental Science: Nano* 9 (6): 1926–1951. https://doi.org/10.1039/D1EN01124C.

Flores, D., Almeida, C.M.R., Gomes, C.R. et al. (2023). Tailoring of mesoporous silica-based materials for enhanced water pollutants removal. *Molecules* 28 (10): 4038.

Foo, K.Y. and Hameed, B.H. (2010). Insights into the modeling of adsorption isotherm systems. *Chemical Engineering Journal* 156 (1): 2–10. https://doi.org/10.1016/j.cej.2009.09.013.

Gajić, G., Djurdjević, L., Kostić, O. et al. (2018). Ecological potential of plants for phytoremediation and ecorestoration of fly ash deposits and mine wastes. *Frontiers in Environmental Science* 6: 124.

Garg, R., Garg, R., Eddy, N.O. et al. (2022). Biosynthesized silica-based zinc oxide nanocomposites for the sequestration of heavy metal ions from aqueous solutions. *Journal of King Saud University-Science* 34 (4): 101996.

Ghaedi, H. and Zhao, M. (2022). Review on template removal techniques for synthesis of mesoporous silica materials. *Energy & Fuels* 36 (5): 2424–2446. https://doi.org/10.1021/acs.energyfuels.1c04435.

Gharbi, A.H., Hemmami, H., Laouini, S.E. et al. (2023). Green synthesis of ZnO@SiO_2 nanoparticles using Calligonum comosum L. extract: an efficient approach for organic pollutant degradation in wastewater. *Biomass Conversion and Biorefinery*. https://doi.org/10.1007/s13399-023-05063-2.

Ghazaryan, K., Agrawal, S., Margaryan, G. et al. (2024). Soil pollution: an agricultural and environmental problem with nanotechnological remediation opportunities and challenges. *Discover Sustainability* 5 (1): 453. https://doi.org/10.1007/s43621-024-00666-9.

Godiya, C.B., Revadekar, C., Kim, J., and Park, B.J. (2022). Amine-bilayer-functionalized cellulose-chitosan composite hydrogel for the efficient uptake of hazardous metal cations and catalysis in polluted water. *Journal of Hazardous Materials* 436: 129112. https://doi.org/10.1016/j.jhazmat.2022.129112.

Grisolia, A., Dell'Olio, G., Spadafora, A. et al. (2023). Hybrid polymer-silica nanostructured materials for environmental remediation. *Molecules* 28 (13): 5105.

Gu, B., Zhang, X., Lam, S.K. et al. (2023). Cost-effective mitigation of nitrogen pollution from global croplands. *Nature* 613 (7942): 77–84. https://doi.org/10.1038/s41586-022-05481-8.

Gudadhe, S., Singh, S.K., and Ahsan, J. (2024). Cellular and neurological effects of lead (Pb) toxicity. In: *Lead Toxicity Mitigation: Sustainable Nexus Approaches* (ed. N. Kumar and A.K. Jha), 125–145. Springer Nature. https://doi.org/10.1007/978-3-031-46146-0_6.

Harsanto, B., Primiana, I., Sarasi, V., and Satyakti, Y. (2023). Sustainability innovation in the textile industry: a systematic review. *Sustainability* 15 (2): 1549. https://doi.org/10.3390/su15021549.

Hashem, A.A., Abd El-Wahab, R.M., Selim, M.M.A., and Badawy, A.A. (2024). Nanozinc ferrites @ silica as efficient adsorbent for dye removal from wastewater: synthesis and adsorption studies. *International Journal of Environmental Science and Technology* 21 (14): 9157–9173. https://doi.org/10.1007/s13762-024-05564-1.

Huang, Y., Li, P., Zhao, R. et al. (2022). Silica nanoparticles: biomedical applications and toxicity. *Biomedicine & Pharmacotherapy* 151: 113053. https://doi.org/10.1016/j.biopha.2022.113053.

Hussain, B., Lin, Q., Hamid, Y. et al. (2020). Foliage application of selenium and silicon nanoparticles alleviates Cd and Pb toxicity in rice (Oryza sativa L.). *Science of the Total Environment* 712: 136497. https://doi.org/10.1016/j.scitotenv.2020.136497.

Islam, S., Rana, B.M.J., Parvez, M.S. et al. (2023). Electroosmotic flow in ternary (TiO_2-SiO_2-Al_2O_3) blood-based sutterby nanomaterials with bio-active mixers. *International Journal of Thermofluids* 18: 100363. https://doi.org/10.1016/j.ijft.2023.100363.

Jaishankar, M., Tseten, T., Anbalagan, N. et al. (2014). Toxicity, mechanism and health effects of some heavy metals. *Interdisciplinary Toxicology* 7 (2): 60–72. https://doi.org/10.2478/intox-2014-0009.

Ji, Z., Belfield, E.J., Zhang, S. et al. (2023). Evolution of a plant growth-regulatory protein interaction specificity. *Nature Plants* 9 (12): 2059–2070. https://doi.org/10.1038/s41477-023-01556-0.

John, D.A. and Babu, G.R. (2021). Lessons from the aftermaths of green revolution on food system and health. *Frontiers in Sustainable Food Systems* 5: 644559. https://doi.org/10.3389/fsufs.2021.644559.

Jumah, M.N.B., Eid, M.H., AL-Huqail, A.A. et al. (2021). Enhanced remediation of As (V) and Hg (II) ions from aqueous environments using β-cyclodextrin/MCM-48 composite: batch and column studies. *Journal of Water Process Engineering* 42: 102118.

Kaabo, H.E., Saied, E., Hassan, S.E.-D. et al. (2024). Penicillium oxalicum-mediated the green synthesis of silica nanoparticles: characterization and environmental applications. *Biomass Conversion and Biorefinery*. https://doi.org/10.1007/s13399-024-05350-6.

Kadivar, E. and Keshavarz, Z. (2022). The dynamics of magnetic nanoparticle motion in a straight microvessel. *Iranian Journal of Physics Research* 21 (4): https://doi.org/10.47176/ijpr.21.4.11182.

Kankala, R.K., Han, Y., Na, J. et al. (2020). Nanoarchitectured structure and surface biofunctionality of mesoporous silica nanoparticles. *Advanced Materials* 32 (23): 1907035. https://doi.org/10.1002/adma.201907035.

Kaya, G.G., Yilmaz, E., and Deveci, H. (2020). Synthesis of sustainable silica xerogels/aerogels using inexpensive steel slag and bean pod ash: a comparison study. *Advanced Powder Technology* 31 (3): 926–936. https://www.sciencedirect.com/science/article/pii/S0921883119304595.

Kazemi, S., Hosseingholian, A., Gohari, S.D. et al. (2023). Recent advances in green synthesized nanoparticles: from production to application. *Materials Today Sustainability* 24: 100500. https://doi.org/10.1016/j.mtsust.2023.100500.

Ken, D.S. and Sinha, A. (2020). Recent developments in surface modification of nano zero-valent iron (nZVI): remediation, toxicity and environmental impacts. *Environmental Nanotechnology, Monitoring & Management* 14: 100344. https://doi.org/10.1016/j.enmm.2020.100344.

Khan, M. N., Mobin, M., Abbas, Z. K., & Alamri, S. A. (2018). Fertilizers and their contaminants in soils, surface and groundwater. *Encyclopedia of the Anthropocene*, 5 (pp. 225–240). https://doi.org/10.1016/B978-0-12-809665-9.09888-8.

Khan, I., Saeed, K., and Khan, I. (2019). Nanoparticles: properties, applications and toxicities. *Arabian Journal of Chemistry* 12 (7): 908–931. https://doi.org/10.1016/j.arabjc.2017.05.011.

Khan, F., Siddique, A.B., Shabala, S. et al. (2023). Phosphorus plays key roles in regulating plants' physiological responses to abiotic stresses. *Plants* 12 (15): 2861. https://doi.org/10.3390/plants12152861.

Kotsyuda, S.S., Tomina, V.V., Zub, Y.L. et al. (2017). Bifunctional silica nanospheres with 3-aminopropyl and phenyl groups. Synthesis approach and prospects of their applications. *Applied Surface Science* 420: 782–791.

Labulo, A.H., David, O.A., Hassan, I. et al. (2024). Mobility inhibition of arsenic in the soil: the role of green synthesized silica nanoparticles. *International Journal of Phytoremediation* 26 (10): 1683–1690. https://doi.org/10.1080/15226514.2024.2348044.

Lei, C., Huang, H., Ye, H. et al. (2022). Immobilization of Pb and Zn in contaminated soil using alumina–silica nano-amendments synthesized from coal fly ash. *International Journal of Environmental Research and Public Health* 19 (23): 16204. https://doi.org/10.3390/ijerph192316204.

Li, J., Zhou, Q., Liu, Y., and Lei, M. (2017). Recyclable nanoscale zero-valent iron-based magnetic polydopamine coated nanomaterials for the adsorption and removal of phenanthrene and anthracene. *Science and Technology of Advanced Materials* 18 (1): 3–16. https://doi.org/10.1080/14686996.2016.1246941.

Li, Y., He, J., Zhang, K. et al. (2019). Super rapid removal of copper, cadmium and lead ions from water by NTA-silica gel. *RSC Advances* 9 (1): 397–407.

Li, S., Zhu, L., Li, J. et al. (2020a). Influence of long-term biosolid applications on communities of soil fauna and their metal accumulation: a field study. *Environmental Pollution* 260: 114017. https://doi.org/10.1016/j.envpol.2020.114017.

Li, X., Rong, H., Zhang, J. et al. (2020b). Modulating the local coordination environment of single-atom catalysts for enhanced catalytic performance. *Nano Research* 13 (7): 1842–1855. https://doi.org/10.1007/s12274-020-2755-3.

Li, Q., Chen, Z., Wang, H. et al. (2021). Removal of organic compounds by nanoscale zero-valent iron and its composites. *Science of the Total Environment* 792: 148546. https://doi.org/10.1016/j.scitotenv.2021.148546.

Li, X., Zhang, Y., Huang, W. et al. (2024). Silica–magnesium coupling in lignin–based biochar: a promising remediation for composite heavy metal pollution in environment. *Journal of Environmental Management* 363: 121392.

Lian, M., Feng, Q., Wang, L. et al. (2019). Highly effective immobilization of Pb and Cd in severely contaminated soils by environment-compatible, mercapto-functionalized reactive nanosilica. *Journal of Cleaner Production* 235: 583–589. https://doi.org/10.1016/j.jclepro.2019.07.015.

Lian, M., Wang, L., Feng, Q. et al. (2021). Thiol-functionalized nano-silica for in-situ remediation of Pb, Cd, Cu contaminated soils and improving soil environment. *Environmental Pollution* 280: 116879.

Liang, L., Luo, L., and Zhang, S. (2011). Adsorption and desorption of humic and fulvic acids on SiO_2 particles at nano-and micro-scales. *Colloids and Surfaces A: Physicochemical and Engineering Aspects* 384 (1–3): 126–130.

Lu, Y., Song, S., Wang, R. et al. (2015). Impacts of soil and water pollution on food safety and health risks in China. *Environment International* 77: 5–15. https://doi.org/10.1016/j.envint.2014.12.010.

Lumen, D., Wang, S., Mäkilä, E. et al. (2021). Investigation of silicon nanoparticles produced by centrifuge chemical vapor deposition for applications in therapy and diagnostics. *European Journal of Pharmaceutics and Biopharmaceutics* 158: 254–265. https://doi.org/10.1016/j.ejpb.2020.11.022.

Luqueño, F.F., López-Valdez, F., and Pérez, G.M. (2023). *Bio and Nanoremediation of Hazardous Environmental Pollutants*, 1e. CRC Press. https://doi.org/10.1201/9781003052982.

Mahawar, L., Ramasamy, K.P., Suhel, M. et al. (2023). Silicon nanoparticles: comprehensive review on biogenic synthesis and applications in agriculture. *Environmental Research* 232: 116292. https://doi.org/10.1016/j.envres.2023.116292.

Malafatti, J.O.D., Tavares, F.A., Neves, T.R. et al. (2023). Modified silica nanoparticles from rice husk supported on polylactic acid as adsorptive membranes for dye removal. *Materials* 16 (6): 2429.

Malik, S., Dhasmana, A., Preetam, S. et al. (2022). Exploring microbial-based green nanobiotechnology for wastewater remediation: a sustainable strategy. *Nanomaterials* 12 (23): 4187. https://doi.org/10.3390/nano12234187.

Manzoor, N., Ali, L., Ahmed, T. et al. (2022). Silicon oxide nanoparticles alleviate chromium toxicity in wheat (Triticum aestivum L.). *Environmental Pollution* 315: 120391. https://doi.org/10.1016/j.envpol.2022.120391.

Maroušek, J., Maroušková, A., Periakaruppan, R. et al. (2022). Silica nanoparticles from coir pith synthesized by acidic sol-gel method improve germination economics. *Polymers* 14 (2): 266. https://doi.org/10.3390/polym14020266.

Martínez, G., Merinero, M., Pérez-Aranda, M. et al. (2020). Environmental impact of nanoparticles' application as an emerging technology: a review. *Materials* 14 (1): 166. https://doi.org/10.3390/ma14010166.

Martins, G.L.O., Fraga, Y.S.B., Vasconcellos, J.S. et al. (2020). Synthesis and characterization of functionalized nanosilica for cementitious composites: review. *Journal of Nanoparticle Research* 22 (11): 338. https://doi.org/10.1007/s11051-020-05063-7.

Maryam, H., Abbasi, G.H., Waseem, M. et al. (2024). Preparation and characterization of green silicon nanoparticles and their effects on growth and lead (Pb) accumulation in maize (Zea mays L.). *Environmental Pollution* 346: 123691.

Matos, M.P.S.R., Correia, A.A.S., and Rasteiro, M.G. (2017). Application of carbon nanotubes to immobilize heavy metals in contaminated soils. *Journal of Nanoparticle Research* 19 (4): 126. https://doi.org/10.1007/s11051-017-3830-x.

Mayakannan, S., Raj, J.B., Raja, V.L., and Nagaraj, M. (2023). Effectiveness of silicon nanoparticles on the mechanical, wear, and physical characteristics of PALF/sisal fiber–based polymer hybrid nanocomposites. *Biomass Conversion and Biorefinery* 13 (14): 13291–13305. https://doi.org/10.1007/s13399-023-04654-3.

Meena, H., Meena, R.S., Rajput, B.S., and Kumar, S. (2016). Response of bio-regulators to morphology and yield of clusterbean [Cyamopsis tetragonoloba (L.) Taub.] under different sowing environments. *Journal of Applied and Natural Science* 8 (2), Article 2. https://doi.org/10.31018/jans.v8i2.863.

Meena, R., Kumar, S., Datta, R. et al. (2020). Impact of agrochemicals on soil microbiota and management: a review. *Land* 9 (2): 34. https://doi.org/10.3390/land9020034.

Meky, N., Salama, E., Soliman, M.F. et al. (2024). Synthesis of nano-silica oxide for heavy metal decontamination from aqueous solutions. *Water, Air, & Soil Pollution* 235 (2): 154. https://doi.org/10.1007/s11270-024-06944-6.

Miah, S.J., Hoque, A., Paul, A., and Rahman, A. (2014). Unsafe use of pesticide and its impact on health of farmers: a case study in Burichong Upazila, Bangladesh. *Cancer* 21 (3): 22–30.

Mitra, S., Mukherjee, S., Sil, M. et al. (2023). Role of mesoporous silica nanoparticles in combating mercury-induced stress in Vigna radiata (mung bean) and Bacillus coagulans (soil bacteria). *Environmental Science and Pollution Research* 30 (50): 109343–109353. https://doi.org/10.1007/s11356-023-30088-5.

Mohammed, S.A.S. and Moghal, A.A.B. (2021). Nanomaterials-based solidification/stabilization of metal-contaminated soils. In: *Nanomaterials for soil remediation* (ed. A. Amrane, D. Mohan, T.A. Nguyen, et al.), 385–407. Elsevier. https://www.sciencedirect.com/science/article/pii/B9780128228913000189.

Montañez, J.P., Heredia, C.L., Sham, E.L., and Farfán Torres, E.M. (2018). Photodegradation of herbicide metsulfuron-methyl with TiO_2 supported on magnetite particles coated with SiO_2. *Journal of Environmental Chemical Engineering* 6 (6): 7402–7410. https://doi.org/10.1016/j.jece.2018.07.006.

Morsy, H.Y., El-Etr, W.M., Youssef, G.H. et al. (2023). Nano-silica synthesis, characterization and their effects on rice production in the presence of calcium humate under salinity soil stress. *World Journal of Agricultural Sciences* 19 (5): 189–203.

Najafi, M., Yousefi, Y., and Rafati, A.A. (2012). Synthesis, characterization and adsorption studies of several heavy metal ions on amino-functionalized silica nano hollow sphere and silica gel. *Separation and Purification Technology* 85: 193–205.

Namasivayam, S.K., Francis, A.L., Kavisri, M. et al. (2023). Biocompatible nanoscale silica particles fabricated from aminopropyltriethoxysilane functionalized brick ash induced versatile pesticidal activity. *Environmental Research* 238: 117090. https://doi.org/10.1016/j.envres.2023.117090.

Nassar, M.Y., Ahmed, I.S., and Raya, M.A. (2019). A facile and tunable approach for synthesis of pure silica nanostructures from rice husk for the removal of ciprofloxacin drug from polluted aqueous solutions. *Journal of Molecular Liquids* 282: 251–263. https://www.sciencedirect.com/science/article/pii/S0167732218354667.

Nguyen, T.T., Ma, H.T., Avti, P. et al. (2019). Adsorptive Removal of Iron Using SiO_2 Nanoparticles Extracted from Rice Husk Ash. *Journal of Analytical Methods in Chemistry* 1–8. https://doi.org/10.1155/2019/6210240.

Okeke, E.S., Nweze, E.J., Ezike, T.C. et al. (2023). Silicon-based nanoparticles for mitigating the effect of potentially toxic elements and plant stress in agroecosystems: a sustainable pathway towards food security. *Science of the Total Environment* 898: 165446. https://doi.org/10.1016/j.scitotenv.2023.165446.

Oppmann, M., Wozar, M., Reichstein, J., and Mandel, K. (2019). Reusable superparamagnetic raspberry-like supraparticle adsorbers as instant cleaning agents for ultrafast dye removal from water. *ChemNanoMat* 5 (2): 230–240. https://doi.org/10.1002/cnma.201800490.

Ortiz-Hernández, M.L., Sanchez-Salinas, E., Godínez, M.L.C. et al. (2013). Mechanisms and strategies for pesticide biodegradation: opportunity for waste, soils and water cleaning. *Revista Internacional de Contaminación Ambiental* 29: 85–104.

Pedro, K.C.N.R., Da Silva, J.V.V., Cipolatti, E.P. et al. (2023). Adsorption of lipases on porous silica-based materials for esterification in a solvent-free system. *3 Biotech* 13 (11): 380. https://doi.org/10.1007/s13205-023-03801-x.

Peng, S., Liu, J., Pan, G. et al. (2024). Combining SiO_2 NPs with biochar: a novel composite for enhanced cadmium removal from wastewater and alleviation of soil cadmium stress. *Environmental Geochemistry and Health* 46 (11): 456. https://doi.org/10.1007/s10653-024-02243-5.

Perez, E.O., Toro, I.D., and Hernandez, B.Y. (2017). Measurement of potential absorption capacity in Colombia's innovative companies [Medición de la capacidad de absorción potencial en las empresas innovadoras de Colombia]. *Revista Espacios* 38 (26): 1307–1321. http://repository.udem.edu.co/handle/11407/4289.

PMSLY, H. (2017). Research progress of OCPs distribution characteristics and health risk assessment from soil. *Journal of Baotou Medical College* 33 (6): 130–135.

Protesescu, L., Yakunin, S., Bodnarchuk, M.I. et al. (2015). Nanocrystals of cesium lead halide perovskites ($CsPbX_3$, X = Cl, Br, and I): novel optoelectronic materials showing bright emission with wide color gamut. *Nano Letters* 15 (6): 3692–3696. https://doi.org/10.1021/nl5048779.

Purrostam, S., Rahimi-Ahar, Z., Babapoor, A. et al. (2023). Melamine functionalized mesoporous silica SBA-15 for separation of chromium (VI) from wastewater. *Materials Chemistry and Physics* 307: 128240.

Qasem, N.A., Mohammed, R.H., and Lawal, D.U. (2021). Removal of heavy metal ions from wastewater: a comprehensive and critical review. *npj Clean Water* 4 (1): 1–15.

Ramos-Carrazco, A., Berman-Mendoza, D., Ramirez-Espinoza, R. et al. (2023). Al_2O_3/Si NPs multilayered antireflective coating to enhance the photovoltaic performance of solar cells. *Journal of Materials Science: Materials in Electronics* 34 (36): 2328. https://doi.org/10.1007/s10854-023-11667-1.

Rashid, K., Aslam, M., Rácz, E. et al. (2024). Mesoporous silica-grafted deep eutectic solvent-based mixed matrix membranes for wastewater treatment: synthesis and emerging pollutant removal performance. *Nanotechnology Reviews* 13 (1): 20230213. https://doi.org/10.1515/ntrev-2023-0213.

Ray, S. and Vashishth, R. (2024). From water to plate: reviewing the bioaccumulation of heavy metals in fish and unraveling human health risks in the food chain. *Emerging Contaminants* 10 (4): 100358. https://doi.org/10.1016/j.emcon.2024.100358.

Reátegui, E., Reynolds, E., Kasinkas, L. et al. (2012). Silica gel-encapsulated AtzA biocatalyst for atrazine biodegradation. *Applied Microbiology and Biotechnology* 96 (1): 231–240. https://doi.org/10.1007/s00253-011-3821-2.

Ren, K., Yang, X., Li, J. et al. (2024). Alleviating the adverse effects of Cd–Pb contamination through the application of silicon fertilizer: enhancing soil microbial diversity and mitigating heavy metal contamination. *Chemosphere* 352: 141414.

Rengarajan, T., Rajendran, P., Nandakumar, N. et al. (2015). Exposure to polycyclic aromatic hydrocarbons with special focus on cancer. *Asian Pacific Journal of Tropical Biomedicine* 5 (3): 182–189. https://doi.org/10.1016/S2221-1691(15)30003-4.

Romero, N.A. and Nicewicz, D.A. (2016). Organic photoredox catalysis. *Chemical Reviews* 116 (17): 10075–10166. https://doi.org/10.1021/acs.chemrev.6b00057.

Sabry, N.M., Badry, R., Ibrahim, M.A., and Ezzat, H.A. (2024). Role of SiO_2, TiO_2, and Fe_3O_4 adsorbed on glycine for remediation of heavy metals and antibacterial activity in water. *Scientific Reports* 14 (1): 27617.

sadat Hashami, Z., Taheri, A., and Alikarami, M. (2022). Synthesis of a magnetic SBA-15-NH_2@ dual-template imprinted polymer for solid phase extraction and determination of Pb and Cd in vegetables; box behnken design. *Analytica Chimica Acta* 1204: 339262.

Samani, M., Ahlawat, Y.K., Golchin, A. et al. (2024a). Nano silica's role in regulating heavy metal uptake in Calendula officinalis. *BMC Plant Biology* 24 (1): 598. https://doi.org/10.1186/s12870-024-05311-1.

Samani, M., Ahlawat, Y.K., Golchin, A. et al. (2024b). Nano silica-mediated stabilization of heavy metals in contaminated soils. *Scientific Reports* 14 (1): 20496. https://doi.org/10.1038/s41598-024-69182-0.

Samia, B., Socorro, J., Durand, A. et al. (2024). Photolytic degradation of commonly used pesticides adsorbed on silica particles. *Science of the Total Environment* 949: 174964. https://doi.org/10.1016/j.scitotenv.2024.174964.

Santhamoorthy, M., Thirumalai, D., Thirupathi, K., and Kim, S.-C. (2023). Synthesis of dithiol-modified mesoporous silica adsorbent for selective adsorption of mercury ions from wastewater. *Applied Nanoscience* 13 (9): 6015–6024. https://doi.org/10.1007/s13204-022-02531-5.

Sathish, T., Ahalya, N., Thirunavukkarasu, M. et al. (2024). A comprehensive review on the novel approaches using nanomaterials for the remediation of soil and water pollution. *Alexandria Engineering Journal* 86: 373–385. https://doi.org/10.1016/j.aej.2023.10.038.

Saud, A., Gupta, S., Allal, A. et al. (2024). Progress in the sustainable development of biobased (nano) materials for application in water treatment technologies. *ACS Omega* 9 (27): 29088–29113. https://doi.org/10.1021/acsomega.3c08883.

Seaf El-Nasr, T.A., Gomaa, H., Emran, M.Y. et al. (2021). Recycling of nanosilica from agricultural, electronic, and industrial wastes for wastewater treatment. In: *Waste*

Recycling Technologies for Nanomaterials Manufacturing (ed. A.S.H. Makhlouf and G.A.M. Ali), 325–362. Springer International Publishing. https://doi.org/10.1007/978-3-030-68031-2_12.

Seisenbaeva, G.A., Ali, L.M., Vardanyan, A. et al. (2021). Mesoporous silica adsorbents modified with amino polycarboxylate ligands–functional characteristics, health and environmental effects. *Journal of Hazardous Materials* 406: 124698. https://www.sciencedirect.com/science/article/pii/S0304389420326881.

Sharififard, H. and Rezvanpanah, E. (2021). Ultrasonic-assisted synthesis of SiO_2 nanoparticles and SiO_2/chitosan/Fe nanocomposite and their application for vanadium adsorption from aqueous solution. *Environmental Science and Pollution Research* 28 (9): 11586–11597. https://doi.org/10.1007/s11356-020-11346-2.

Sharma, A., Kumar, V., Shahzad, B. et al. (2019). Worldwide pesticide usage and its impacts on ecosystem. *SN Applied Sciences* 1 (11): 1446. https://doi.org/10.1007/s42452-019-1485-1.

Sharma, J., Sharma, S., and Soni, V. (2021). Classification and impact of synthetic textile dyes on aquatic flora: a review. *Regional Studies in Marine Science* 45: 101802. https://doi.org/10.1016/j.rsma.2021.101802.

Sharma, P., Prakash, J., and Kaushal, R. (2022). An insight into the green synthesis of SiO_2 nanostructures as a novel adsorbent for removal of toxic water pollutants. *Environmental Research* 212: 113328. https://doi.org/10.1016/j.envres.2022.113328.

Sharma, P., Kherb, J., Prakash, J., and Kaushal, R. (2023). A novel and facile green synthesis of SiO_2 nanoparticles for removal of toxic water pollutants. *Applied Nanoscience* 13 (1): 735–747. https://doi.org/10.1007/s13204-021-01898-1.

Shi, C., Murshed, M., Alam, M.M. et al. (2023). Can minimizing risk exposures help in inhibiting carbon footprints? The environmental repercussions of international trade and clean energy. *Journal of Environmental Management* 347: 119195. https://doi.org/10.1016/j.jenvman.2023.119195.

Shinde, P.S., Suryawanshi, P.S., Patil, K.K. et al. (2021). A brief overview of recent progress in porous silica as catalyst supports. *Journal Of Composites Science* 5 (3): 75. https://www.mdpi.com/2504-477X/5/3/75.

Silviana, S., Saputra, B., Oktavian, H.D.A. et al. (2024). Rice husk as a potential source for the sol-gel synthesis of diphenylcarbazide immobilized silica adsorbent for the adsorption of lead. *AIP Conference Proceedings* 2710 (1): 040022. https://pubs.aip.org/aip/acp/article-abstract/2710/1/040022/3261264.

Singh, G., Dizaji, H.B., Puttuswamy, H., and Sharma, S. (2022). Biogenic nanosilica synthesis employing agro-waste rice straw and its application study in photocatalytic degradation of cationic dye. *Sustainability* 14 (1): 539. https://doi.org/10.3390/su14010539.

Singh, B.J., Chakraborty, A., and Sehgal, R. (2023). A systematic review of industrial wastewater management: evaluating challenges and enablers. *Journal of Environmental Management* 348: 119230.

Singh, G., Thakur, N., and Kumar, R. (2024). Nanoparticles in drinking water: assessing health risks and regulatory challenges. *Science of the Total Environment* 949: 174940. https://doi.org/10.1016/j.scitotenv.2024.174940.

Smith, S.R. (2009). Organic contaminants in sewage sludge (biosolids) and their significance for agricultural recycling. *Philosophical Transactions of the Royal Society A: Mathematical, Physical and Engineering Sciences* 367 (1904): 4005–4041. https://doi.org/10.1098/rsta.2009.0154.

Song, L.L., Chen, C., Luo, F. et al. (2016). Isoreticular MOFs functionalized in the pore wall by different organic groups for high-performance removal of uranyl ions. *Journal of Radioanalytical and Nuclear Chemistry* 310 (1): 317–327. https://doi.org/10.1007/s10967-016-4884-4.

Sorolla, M.G., Dalida, M.L., Khemthong, P., and Grisdanurak, N. (2012). Photocatalytic degradation of paraquat using nano-sized Cu-TiO_2/SBA-15 under UV and visible light. *Journal of Environmental Sciences* 24 (6): 1125–1132. https://doi.org/10.1016/S1001-0742(11)60874-7.

Srivastava, S., Shukla, A., Rajput, V.D. et al. (2021). Arsenic remediation through sustainable phytoremediation approaches. *Minerals* 11 (9): 936.

Stegmann, R., Brunner, G., Calmano, W., and Matz, G. (2013). *Treatment of Contaminated Soil: Fundamentals, Analysis, Applications*. Springer Science & Business Media. https://books.google.com/books?hl=en&lr=&id=w6rwCAAAQBAJ&oi=fnd&pg=PA3&dq=Stegmann,+R.,+et+al.,+Treatment+of+contaminated+soil:+Fundamentals,+analysis,+applications.+Springer+Science+%26+Business+Media,&ots=tLU20Jbxmw&sig=tuVfERwKC_5DYOea3j-RAjP8Nrk.

Tessier, A., Campbell, P.G.C., and Bisson, M. (1979). Sequential extraction procedure for the speciation of particulate trace metals. *Analytical Chemistry* 51 (7): 844–851. https://doi.org/10.1021/ac50043a017.

Tripathi, D.K., Singh, V.P., Prasad, S.M. et al. (2015). Silicon nanoparticles (SiNp) alleviate chromium (VI) phytotoxicity in Pisum sativum (L.) seedlings. *Plant Physiology and Biochemistry* 96: 189–198. https://doi.org/10.1016/j.plaphy.2015.07.026.

Trujillo-Reyes, J., Peralta-Videa, J.R., and Gardea-Torresdey, J.L. (2014). Supported and unsupported nanomaterials for water and soil remediation: are they a useful solution for worldwide pollution? *Journal of Hazardous Materials* 280: 487–503.

Ul Ain, S.N., Khan, M.S., Riaz, N. et al. (2024). Surface-functionalized magnetic silica-malachite tricomposite (Fe–M–Si tricomposite): a promising adsorbent for the removal of cypermethrin. *ACS Omega* 9 (12): 13803–13817. https://doi.org/10.1021/acsomega.3c08419.

Vanichvattanadecha, C., Jaroenworaluck, A., Henpraserttae, P. et al. (2021). Ordered mesoporous silica (SBA-16) supporting titania (TiO_2) nanoparticles for photodegradation of paraquat (PQ) herbicide. *Journal of Porous Materials* 28 (4): 1137–1153. https://doi.org/10.1007/s10934-021-01065-5.

Vanichvattanadecha, C., Jaroenworaluck, A., Henpraserttae, P., and Wimuktiwan, P. (2024). Synthesis of highly ordered mesoporous silica (SBA-16) to support titania (TiO_2) nanoparticles for enhancing the degradation of diuron herbicide. *Silicon* 16 (16): 5907–5919. https://doi.org/10.1007/s12633-024-03124-3.

Vu, A.-T., Nguyen, M.V., and Nguyen, T.H. (2024). Fabrication of ethylenediaminetetraacetic modified porous silica composite from rice husk for enhancing the remove of Pb^{2+} from aqueous solution. *Results in Materials* 21: 100525.

Wahab, A., Muhammad, M., Ullah, S. et al. (2024). Agriculture and environmental management through nanotechnology: eco-friendly nanomaterial synthesis for soil-plant systems, food safety, and sustainability. *Science of the Total Environment* 926: 171862. https://doi.org/10.1016/j.scitotenv.2024.171862.

Wang, J. and Wang, S. (2018). Activation of persulfate (PS) and peroxymonosulfate (PMS) and application for the degradation of emerging contaminants. *Chemical Engineering Journal* 334: 1502–1517. https://doi.org/10.1016/j.cej.2017.11.059.

Wang, Y., Liu, Y., Zhan, W. et al. (2020a). Long-term stabilization of Cd in agricultural soil using mercapto-functionalized nano-silica (MPTS/nano-silica): a three-year field study. *Ecotoxicology and Environmental Safety* 197: 110600. https://doi.org/10.1016/j.ecoenv.2020.110600.

Wang, Y., Zhan, W., Zheng, K. et al. (2020b). Effect of surface-modified nano-silica on the mobility and fraction of Cd in contaminated agricultural soils. *Soil and Sediment Contamination: An International Journal* 29 (1): 96–106. https://doi.org/10.1080/15320383.2019.1679086.

Wang, J., Zhen, J., Hu, W. et al. (2023). Remote sensing of soil degradation: progress and perspective. *International Soil and Water Conservation Research* 11 (3): 429–454. https://doi.org/10.1016/j.iswcr.2023.03.002.

Wang, F., Xiang, L., Sze-Yin Leung, K. et al. (2024). Emerging contaminants: a one health perspective. *The Innovation* 5 (4): 100612. https://doi.org/10.1016/j.xinn.2024.100612.

Wardanu, A.P. and Indrasti, N.S. (2024). Green synthesis of silica nanoparticles (Si-NPs) from palm oil fuel ash (POFA) and its application to purification water: a review. *IOP Conference Series: Earth and Environmental Science* 1358 (1): 012015. https://iopscience.iop.org/article/10.1088/1755-1315/1358/1/012015/meta.

Wilkinson, J.L., Boxall, A.B.A., Kolpin, D.W. et al. (2022). Pharmaceutical pollution of the world's rivers. *Proceedings of the National Academy of Sciences* 119 (8): e2113947119. https://doi.org/10.1073/pnas.2113947119.

Xing, Y., Li, Q., Chen, X. et al. (2022). Preparation of isoelectric point-switchable polymer brush-grafted mesoporous silica using RAFT polymerization with high performance for Ni (II) adsorption. *Powder Technology* 412: 117980.

Xiong, Y., Yao, S., and Driess, M. (2013). Chemical tricks to stabilize silanones and their heavier homologues with E=O bonds (E=Si–Pb): from elusive species to isolable building blocks. *Angewandte Chemie International Edition* 52 (16): 4302–4311. https://doi.org/10.1002/anie.201209766.

Yaashikaa, P.R., Kumar, P.S., Varjani, S., and Saravanan, A. (2020). A critical review on the biochar production techniques, characterization, stability and applications for circular bioeconomy. *Biotechnology Reports* 28: e00570. https://doi.org/10.1016/j.btre.2020.e00570.

Yadav, M., George, N., and Dwibedi, V. (2023a). Emergence of toxic trace elements in plant environments: insights into potential of silica nanoparticles for mitigation of metal toxicity in plants. *Environmental Pollution* 333: 122112.

Yadav, V.K., Amari, A., Wanale, S.G. et al. (2023b). Synthesis of floral-shaped nanosilica from coal fly ash and its application for the remediation of heavy metals from fly ash aqueous solutions. *Sustainability* 15 (3): 2612. https://doi.org/10.3390/su15032612.

Yang, Q., Wang, J., Chen, X. et al. (2018). The simultaneous detection and removal of organophosphorus pesticides by a novel Zr-MOF based smart adsorbent. *Journal of Materials Chemistry A* 6 (5): 2184–2192.

Yang, Z., Chen, X., Li, S. et al. (2020). Effective removal of Cr(VI) from aqueous solution based on APTES modified nanoporous silicon prepared from kerf loss silicon waste. *Environmental Science and Pollution Research* 27 (10): 10899–10909. https://doi.org/10.1007/s11356-019-07427-6.

Yi, X.-H., Jing, D.-D., Wan, J. et al. (2016). Temporal and spatial variations of contaminant removal, enzyme activities, and microbial community structure in a pilot horizontal subsurface flow constructed wetland purifying industrial runoff. *Environmental Science and Pollution Research* 23 (9): 8565–8576. https://doi.org/10.1007/s11356-016-6083-9.

Youssif, M.M., El-Attar, H.G., Hessel, V., and Wojnicki, M. (2024). Recent developments in the adsorption of heavy metal ions from aqueous solutions using various nanomaterials. *Materials* 17 (21): 5141. https://doi.org/10.3390/ma17215141.

Zahermawi, N.M. and Bestawy, E.E. (2023). Effective treatment of petroleum oil–contaminated wastewater using activated sludge modified with magnetite/silicon nanocomposite. *Environmental Science and Pollution Research* 31 (12): 17634–17650. https://doi.org/10.1007/s11356-023-26557-6.

Zamora-Ledezma, C., Negrete-Bolagay, D., Figueroa, F. et al. (2021). Heavy metal water pollution: a fresh look about hazards, novel and conventional remediation methods. *Environmental Technology & Innovation* 22: 101504. https://doi.org/10.1016/j.eti.2021.101504.

Zhang, H., West, D., Shi, H. et al. (2019a). Simultaneous determination of selected trace contaminants in drinking water using solid-phase extraction-high performance liquid chromatography-tandem mass spectrometry. *Water, Air, & Soil Pollution* 230 (2): 28. https://doi.org/10.1007/s11270-018-4066-9.

Zhang, L., Shang, Z., Guo, K. et al. (2019b). Speciation analysis and speciation transformation of heavy metal ions in passivation process with thiol-functionalized nano-silica. *Chemical Engineering Journal* 369: 979–987. https://doi.org/10.1016/j.cej.2019.03.077.

Zhang, Y., Lei, M., Li, K., and Ju, T. (2023). Spatial prediction of soil contamination based on machine learning: a review. *Frontiers of Environmental Science & Engineering* 17 (8): 93. https://doi.org/10.1007/s11783-023-1693-1.

Zhao, X., Yang, M., Shi, Y. et al. (2024). Multifunctional bacterial cellulose-bentonite@ polyethylenimine composite membranes for enhanced water treatment: sustainable dyes and metal ions adsorption and antibacterial properties. *Journal of Hazardous Materials* 477: 135267.

Zikankuba, V.L., Mwanyika, G., Ntwenya, J.E., and James, A. (2019). Pesticide regulations and their malpractice implications on food and environment safety. *Cogent Food & Agriculture* 5 (1): 1601544. https://doi.org/10.1080/23311932.2019.1601544.

Zou, H., Wang, T., Wang, Z.-L., and Wang, Z. (2023). Continuing large-scale global trade and illegal trade of highly hazardous chemicals. *Nature Sustainability* 6 (11): 1394–1405. https://doi.org/10.1038/s41893-023-01158-w.

Section E

Biosafety and Ethics

20

Environmental Impact, Risk Assessment and Biosafety Issues of Si/SiO_2 NPs

Himanshu, Arun Kumar, Vikanksha and Jatinder Singh

Department of Horticulture, School of Agriculture, Lovely Professional University, Phagwara, Punjab, India

20.1 Introduction

Nanotechnology, with its many uses across industries, has enormous potential for changing a variety of fields, including agriculture, health and electronics, among others. Silicon (Si) and silicon dioxide (Si/SiO_2) NPs have gained substantial interest since they have prospective uses in environmental remediation, biological research and nanotechnology. These NPs are among the varied spectrum of NPs that are now being investigated. However, along with their potentially useful uses come issues about the influence they have on the environment, the threat assessments they need and the consequences they have for biosafety (Tejedor et al. 2022). The incorporation of silicon and silicon dioxide NPs into a wide range of consumer and industrial goods raises concerns about the possible impact these NPs might have on human well-being, ecosystems and the environment. As such, NPs enter the soil, water bodies and atmosphere from production operations, product consumption and the removal of waste. It is therefore necessary to comprehend their destiny, transportation and biological relationships in order to conduct a full assessment of the influence they have on the environment. An evaluation of the possible dangers and exposure pathways posed by Si/SiO_2 NPs constitutes a component of the risk assessment process (Wang et al. 2022). This evaluation takes into account a variety of characteristics, including the size of particles, surface chemical composition, agglomeration conduct and longevity in the environment. In addition, the relationships of NPs with living species, including plants, animals and microbes, call for exhaustive toxicity studies

Silicon Nanoparticles: Synthesis, Uptake, and Applications, First Edition.
Edited by Archana Singh, Aryadeep Roychoudhury, and Indrakant K. Singh.

and environmental risk evaluations in order to protect the well-being of the environment and the biodiversity of the world. Not only do biosafety assessments take into account the direct effects that Si/SiO_2 NPs have on human health, but they also take into account the indirect effects that these NPs have via the food supply chain and environmental pollution. There are worries over the safety of food and the long-term health ramifications for consumers as a result of the possibility that plants would take up NPs and then accumulate them in edible crops. Furthermore, the introduction of NPs into aquatic habitats represents a potential threat to aquatic animals, as it has the potential to alter the dynamics of ecosystems and compromise the strength of ecological systems. It is necessary to have multidisciplinary cooperation, thorough scientific study and governmental control in order to tackle the environmental effects, risk assessment and biosafety concerns that are related to Si/SiO_2 NPs (Harrison et al. 2023). Stakeholders are able to design effective approaches for the secure and environmentally friendly implementation of nanotechnology by combining insights from nanoscience, environmental science, toxicology and regulatory frameworks. This allows them to minimize the possible adverse impacts that nanotechnology may have on human health and the environment. The latest advances in nanoscience have resulted in the creation of NPs made of silicon (Si) and silicon dioxide (SiO_2), each of which has distinctive characteristics and may be used in a variety of contexts. The outstanding durability, biological compatibility and programmable physicochemical features that these NPs possess make them attractive possibilities for a wide range of applications in the fields of industry, biomedicine and the environment. The adaptability of Si/SiO_2 NPs provides potential answers to urgent societal concerns across a wide range of applications, including the enhancement of medication delivery systems, the improvement of agricultural production and the development of remedial technology (Bhat et al. 2021).

The extensive usage of Si/SiO_2 NPs and the possibility that they may be released into the environment nevertheless raises issues regarding the environmental repercussions and hazards to the well-being of humans and the legislative consequences of these NPs. In addition, the one-of-a-kind characteristics of NPs, such as their substantial surface area-to-volume proportion and reaction time, have the potential to magnify the interactions between organisms and the environmental consequences of NPs, which calls for a comprehensive examination and the implementation of preventative measures. In spite of all these worries, academics, lawmakers and commercial participants have begun to concentrate their attention on comprehending the ecological fate and impact of Si/SiO_2 NPs, carrying out thorough risk analysis and creating strong legislation that guarantee the secure and environmentally friendly utilization of these small particles (Tiancheng 2023). A concerted effort is now being made to solve discrepancies in knowledge, reduce possible dangers and encourage ethical advancement in nanotechnology.

This is being accomplished via the integration of multidisciplinary expertise from domains that include nanoscience, the science of the environment, the field of toxicology and evaluation of risks. With regard to the environmental effects, risk assessment approaches and biosafety issues that are related to Si/SiO_2 NPs, this chapter presents an in-depth investigation (Jeong 2020).

20.2 Environmental Fate and Transport

When it comes to evaluating the possible dangers posed by silicon (Si) and silicon dioxide (SiO_2) NPs in the environment, it is essential to have an in-depth knowledge about their destiny and how they move through the environment. While transportation relates to the transportation and dispersion of Si/SiO_2 NPs between and within environmental spaces, the term 'fate' relates to the patterns of action and changes that occur within external divisions, including soil, water and air. The physical and chemical makeup of Si/SiO_2 NPs, the circumstances of the environment and their relationships with both biotic and abiotic elements are some of the variables that determine the environmental destiny and transit of these particles (Naidu et al. 2023).

- **Physical and Chemical Characteristics:** The dimension, form, charged surface and surface functionality of Si/SiO_2 NPs affect their environmental destiny and transit. Nanoparticle size affects movement and reaction time, with tiny particles being more mobile and of longer range. Surface functionalization changes NP durability and bioreactivity, whereas surface charge impacts accumulation, placement and environmental matrix contact (Ahmad et al. 2022).
- **Ecological Factors:** Environmental parameters, including pH, moisture, temperature and ion concentration, affect the conductivity and destiny of Si/SiO_2 NPs. Fluctuations in pH may impact the surface charges and agglomeration tendencies of NPs, resulting in alterations in their movement and distribution. Temperature and humidity affect the durability, aggregation dynamics and interactions of NPs at external interfaces. The clustering and depositing of NPs in aquatic settings are significantly affected by the strength of the ion (Naidu et al. 2023).
- **Transport Pathways:** Si/SiO_2 NPs may be transferred by many environmental channels, via advection, dispersion and deposition. Nanoparticles in a watery environment may be transported by water currents, diffuse over water-air boundaries and settle in bottom sediments. Nanoparticles are mostly transported in soil by diffusion and advection in pore waters, and their placement and absorption are affected by soil characteristics and organic matter concentration (Ghosh et al. 2022).

20.2.1 Dispersion and Stability of Si/SiO_2 NPs in Environmental Matrices

Various parameters such as temperature, pH, salinity and the presence of non-organic matter (NOM) affect the dispersion and stability of Si/SiO_2 NPs in the environmental matrix. Silica NPs dissolve in aquatic settings and eventually approach equilibrium in silica levels, whereupon they become unstable. High temperatures and alkaline pH levels enhance dissolution, while acidic pH, high salinity and high starting silica levels inhibit it. The durability of silica NPs is influenced by the amount of water retention, size and permeability (Khan et al. 2019).

20.2.1.1 Sedimentation, Aggregation and Transformation

The sedimentation, aggregation and modification of silicon (Si) and silicon dioxide (SiO_2) NPs are important factors that influence the destiny and behaviour of the environment, affecting their spread, movement and possible hazards. These activities take place in various environments, including water, soil and sediment, influenced by a mixture of physical, chemical and biological elements.

- **Sedimentation:** Sedimentation is the process by which NPs settle out of suspension due to gravity. Si/SiO_2 NPs in aqueous settings may settle on soil surfaces when gravity forces surpass inherent buoyant properties, causing sedimentation. Sedimentation pace is influenced by NP dimensions, density and fluid characteristics, causing bigger and heavier particles to settle faster. Sedimentation may lead to the buildup of NPs in bottom sediments, and these particles can remain for a lengthy time and affect benthic species and biogeochemical reactions (Little and Fernandes 2019).
- **Aggregation:** Aggregation is the process when NPs reversibly come together to form bigger clusters or aggregates. Si/SiO_2 NPs may assemble via Brownian motion, van der Waals forces, electrostatic contacts and hydrophobic reactions. Aggregation tends to rise as NP concentration increases, leading to the creation of bigger aggregates that are more prone to sedimentation or filtration. Aggregation may affect the movement, availability and transportation of NPs, which can influence their dispersal and outcome in water and soil settings (Givens et al. 2019).
- **Transformation:** Nanoparticle metamorphosis involves modifications in physical, chemical and biological aspects that impact their characteristics and actions in the surroundings. Physical alterations could entail modifications to NP dimensions, morphology, or exterior characteristics as a result of interactions with ambient substances or outside factors. Chemical changes may occur via dissolution or surface interactions, altering the chemical makeup and reactivity of NPs. Biological processes may result in NP biosynthesis or biological accumulation due to relationships with microbial populations.

These changes may affect the durability, ease of movement and harmfulness of NPs, eventually determining their environmental destiny and effect (Roma et al. 2020).

20.3 Ecotoxicity and Environmental Effects

The increasing usage of silicon (Si) and silicon dioxide (SiO_2) NPs has raised concerns about their ecotoxicity and environmental implications, which might have significant repercussions on ecosystems. Nanoparticles may be introduced into the environment by many pathways, such as emissions from industries, sewage treatment facilities and farming methods, potentially endangering aquatic and land-based creatures.

Aquatic Environment: Silicon/silicon dioxide NPs may build up in aquatic environments, including water bodies, sediments and organisms, causing harmful impacts on aquatic life. Research has shown that NPs may be hazardous to several aquatic creatures, including algae, zooplankton, fish and other aquatic species. Toxicity pathways may involve oxidative stress, cell destruction of membranes and genotoxicity, resulting in hindered development, reproduction and mortality (Elli et al. 2022).

Terrestrial Environment: Si/SiO_2 NPs can build up in terrestrial ecosystems, impacting soil microbes, plants and terrestrial creatures. Nanoparticles have the ability to impact soil microbial populations, the metabolism of nitrogen and relationships among plants and microbes. Research has shown that Si/SiO_2 NPs may be absorbed by plants and transported to various regions of the plant, which may impact plant growth and development (Roy 2023).

Bioaccumulation and Trophic Transfer: Si/SiO_2 NPs have the ability to build up in life forms, which may result in trophic transfer and biomagnification within food chains. This buildup may endanger creatures at higher levels of the food chain, such as predators and people that consume infected organisms (Suresh 2021).

Environmental Persistence and Transformation: Si/SiO_2 NPs may remain in the environment for prolonged durations, causing possible long-lasting impacts on ecosystems. Nanoparticles may experience chemical reactions such as aggregation, sedimentation and surface changes, which might change their surrounding destiny and toxicity (Sonia et al. 2023).

Risk Assessment and Regulation: Extensive risk evaluation and management are necessary due to the possible environmental hazards of Si/SiO_2 NPs. This involves analysing the toxicity of NPs in actual environmental settings, studying their environmental behaviour and movement and creating laws and standards for their secure utilization and destruction (Sokovnin and Ilves 2022).

20.3.1 Effects of Si/SiO_2 NPs on Aquatic Organisms (Fish, Algae and Invertebrates)

The impact of silicon (Si) and silicon dioxide (SiO_2) NPs on aquatic species such as fish, algae and invertebrates has attracted significant interest because of their possible ecological consequences. Scientists investigate the effects of Si/SiO_2 NPs on these important aquatic organisms:

Fish

- **Physiological Effects:** Si/SiO_2 NPs exposure may trigger physiological reactions related to stress in fish, leading to changes in the composition of the blood, enzymatic activity and hormonal control. Nanoparticles may interfere with ion control, osmoregulation and acid-base equilibrium in fish, causing abnormalities in metabolism and cell damage (Stanca et al. 2023).
- **Oxidative Stress:** Si/SiO_2 NPs may induce oxidative stress in fish by the production of reactive oxygen species (ROS), leading to the peroxidation of lipids, oxidation of proteins and destruction of DNA. Oxidative stress may weaken protective antioxidant systems, causing damage to cells and resulting in damage to tissues and organ failure (Taghavizadeh Yazdi et al. 2023).
- **Reproductive Effects:** Si/SiO_2 NPs have been shown to impact many fish developmental processes, such as the generation of gametes, implantation achievement and the growth of embryos. Nanoparticles might disturb sexual hormone production, the growth of gonadal cells and spawning, resulting in decreased fertility and population reduction (Lin et al. 2023).

Algae

- **Photosynthetic Inhibition:** Si/SiO_2 NPs may hinder the photosynthesis of algae by disrupting chlorophyll production, light absorption and electron transport chain functions. Nanoparticles might interfere with the activity of photosystem II (PSII), resulting in decreased photosynthesis effectiveness and the generation of biomass in algae (Jampílek and Kráľová 2019).
- **Cellular Damage:** Si/SiO_2 NPs contact may lead to damage to cells in algae, such as changes in membrane permeability, organelle destruction and cytoskeletal abnormalities. Nanoparticles may cause oxidative stress, disturb cellular balance and hinder metabolic functions, resulting in dead cells and a decrease in population (Jampílek and Kráľová 2019).

Invertebrates

- **Feeding Inhibition:** Si/SiO_2 NPs may deter feeding activity in aquatic invertebrates such as zooplankton and benthic species. Nanoparticles may hinder feed appendages, decrease feeding rates and interfere with food particle intake and preparation, resulting in energy shortage and deficiency in nutrition (Siddiqui et al. 2023).

- **Bioaccumulation and Trophic Transfer:** Si/SiO_2 NPs may collect in invertebrate tissues and increase in concentration as they travel along aquatic food chains, creating potential dangers for organisms at higher trophic levels, such as predators and humans. Nanoparticles may build up in lipid-rich organs such as the hepatopancreas and gonads, causing prolonged contact and possibly harmful consequences (Shi et al. 2023).

20.3.2 Impact on Terrestrial Organisms

Silicon (Si) and silicon dioxide (SiO_2) NPs have various impacts on terrestrial species, including soil microbes, plants and animals at higher trophic levels. Here is an in-depth analysis of the effects of Si/SiO_2 NPs on land-dwelling species.

Soil Microorganisms

- **Microbial Diversity:** Si/SiO_2 NPs may impact microbial populations in soil by changing the diversity of bacteria, population size and function. Nanoparticles may specifically impact expected microbiological groups, causing changes in the overall population, both structurally and functionally (Grün et al. 2019).
- **Nutrient Cycling:** Si/SiO_2 NPs may influence soil nutrient cycle activities such as nitrogen fixation, calcification and organic matter degradation. Nanoparticles may influence microbial enzyme activity, nutrient modifications and soil fertility, thus impacting the accessibility of nutrients to plants and production (Elemike et al. 2019).
- **Biodegradation:** Si/SiO_2 NPs might disrupt the breakdown of organic contaminants by soil microbes. Nanoparticles may hinder the activity of microbial enzymes that are crucial for breaking down pollutants, resulting in lower rates of elimination of pollutants and a reduced effectiveness in the restoration of soil (Rajput et al. 2022).

Plants

- **Nutrient Uptake:** Si/SiO_2 NPs may impact the absorption and use of nutrients by plants, such as nitrogen, phosphorus and micronutrients. Nanoparticles may influence nutrient carriers, channelization of ions and root transpiration structures, resulting in alterations in plant nutrition levels and production (Mathur and Roy 2020).
- **Photosynthesis:** Si/SiO_2 NPs might affect plant photosynthesis via influencing chlorophyll production, electron transfer and carbon assimilation. Nanoparticles may impact the efficacy of photosynthesis, the rate of stomatal activity and

the water usage effectiveness in plants, ultimately altering their development, production of biomass and carbon storage (Poddar et al. 2020).

- **Stress Responses:** Si/SiO_2 NPs may trigger stress reactions in plants, such as induction in antioxidant defence systems, osmotic regulation and plant hormone signaling. Nanoparticles may improve a plant's ability to withstand several environmental challenges, including drought, salt and heavy metals, by reducing oxidative damage and preserving cellular balance (Mathur and Roy 2020).
- Terrestrial Invertebrates and Vertebrates
  - **Invertebrates:** Si/SiO_2 NPs may impact soil-dwelling invertebrates such as earthworms, collembolans and mites. Nanoparticles may affect soil microhabitat circumstances, food availability and the utilization of assets by invertebrates, which can result in alterations in population patterns and ecological procedures (Salam et al. 2023).
  - **Vertebrates:** Si/SiO_2 NPs may affect land-dwelling species, including birds, mammals and reptiles, by ingestion or pollution of their environment. Nanoparticles may impact the reproductive success, immunological function and behaviour of vertebrates, thereby causing population decreases and biodiversity loss in impacted habitats (Devasena et al. 2022).

20.3.3 Bioaccumulation and Biomagnification Potential in Food Webs

- **Bioaccumulation:** Si/SiO_2 NPs may accumulate in organisms by many exposure pathways, such as ingestion, skin contact and breathing. Once taken in, these small particles may gather in tissues and organs, resulting in higher levels gradually. Si/SiO_2 NP bioaccumulation is affected by NP size, surface chemistry, ambient circumstances and organism-specific traits. Increased NP size, together with a high surface area-to-volume ratio and surface alterations, may lead to greater bioaccumulation potential. Aquatic creatures such as fish, invertebrates and algae may acquire silicon/silicon dioxide NPs from water and sand. Land-dwelling creatures, including plants, soil invertebrates and vertebrates, may accumulate NPs from soil, air and plants (Banu et al. 2021).
- **Biomagnification:** biomagnification is the process by which pollutants, such as NPs, become progressively concentrated at higher trophic levels of food webs. This process causes the concentration of toxins to increase. In both aquatic and terrestrial food chains, Si/SiO_2 NPs have the ability to biomagnify, which poses a threat to predators as well as humans who are at the top of the food web. Bioaccumulation of Si/SiO_2 NPs from water sources is a possibility for aquatic animals that are at lower trophic levels, which include algae and zooplankton (Shi et al. 2020).

20.4 Risk Assessment and Hazard Analysis

20.4.1 Identification of Potential Exposure Pathways for Si/SiO_2 NPs

It is essential to comprehend the potential entry and interaction of these NPs with various environmental compartments and the ensuing effects on ecosystems and human health. Comprehending these exposure pathways is crucial for evaluating risks and formulating efficacious risk mitigation tactics. Additionally, it serves as a reference for the adoption of monitoring programmes that track the concentration and presence of Si/SiO_2 NPs in various environmental compartments, assisting in the mitigation of potentially harmful impacts on ecosystems and human health (Figure 20.1) (Abbas et al. 2020).

20.4.2 Evaluation of Hazards Based on Physicochemical Properties and Toxicological Data

20.4.2.1 Physicochemical Properties

Nanoparticle size is an important component in determining toxicity. Smaller particles could be more reactive and capable of cellular absorption. The form of NPs can influence their interactions with biological systems. Increased surface area may enhance reactivity and relationships with biological tissues. Surface charge regulates relationships between cells and proteins. The existence of any surface coating can alter the endurance, reactivity and toxicity levels of

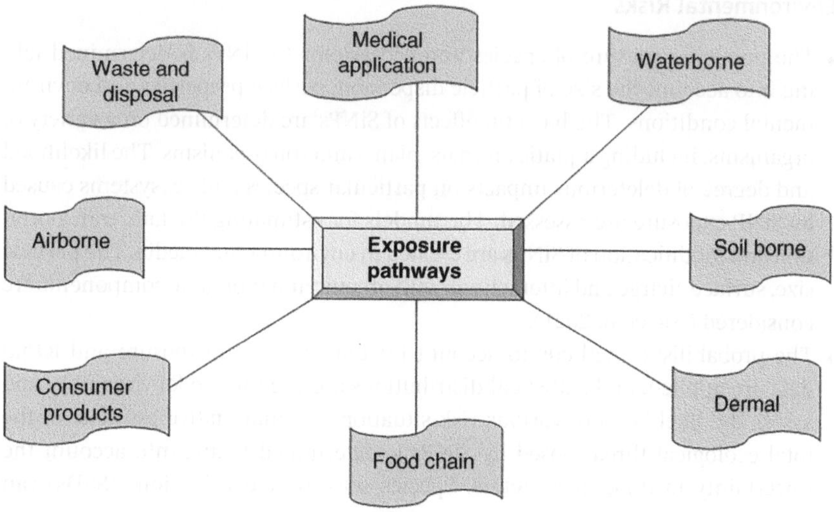

Figure 20.1 Different exposure pathways for NPs.

nanomaterials. The capacity of NPs to aggregate or scatter influences their actions in various situations. The solubility of Si/SiO_2 NPs in various fluids can affect their bioavailability and toxicity (Sabaraya 2021).

20.4.2.2 Toxicological Data

- Si/SiO_2 NPs were also observed to elicit oxidative stress in a dose-dependent way, as evidenced by glutathione depletion, ROS production and peroxidation of lipids. Furthermore, caspase-3 and caspase-9 enzyme activity increased dramatically in both types of cells when subjected to SiO_2 NPs. This work suggested that Si/SiO_2 NPs induce cytotoxicity and death in A431 and A549 cells, which is most likely caused by ROS production and oxidative stress (Rafieepour et al. 2021).
- Studies have looked into the consequences of inhaling SiNPs, with a specific focus on respiratory impacts. SiNPs accumulated in the lungs have been linked to inflammation, oxidative stress and tissue damage. In vivo studies show that SiNPs can bypass biological obstacles to move from their point of entrance (e.g., the lungs) to other organs. Dispersion patterns can vary depending on the size of the particles, surface charge and coating. SiNPs have been observed to enter the circulatory system, increasing concerns regarding their cardiovascular impacts. Animal studies have shown that SiNPs accumulate in the liver and kidneys. In some situations, SiNPs have been linked to liver and kidney damage, which could impair organ function (Lin et al. 2023).

20.4.3 Quantitative Risk Assessment Methodologies for Assessing Environmental Risks

- The possible exposure of species and ecosystems to SiNPs is determined taking into account the size of particle dispersion, surface properties and environmental conditions. The harmful effects of SiNPs are determined on a variety of organisms, including aquatic animals, plants and soil organisms. The likelihood and degree of deleterious impacts on particular species and ecosystems caused by SiNP exposure are assessed. The models for estimating the fate, transportation and modification of SiNPs are created in environmental media. The particle size, surface charge and interactions with organic and inorganic components are considered (Ale et al. 2021).
- The probabilistic methods to account for uncertainty in exposure and lethal data are applied, and statistical distributions are used to model variability and assess the likelihood of various risk situations. A quantitative estimate of the total ecological threat posed by SiNPs is determined taking into account the uncertainty in input parameters. Species sensitivity distributions (SSDs) can

be used to examine the progressive effects of SiNPs on different species in an ecosystem. The data on species sensitivity are included when estimating potential ecological threats to assess the possible effect of SiNPs on various trophic levels and evaluate the overall danger to biodiversity. Monte Carlo simulation can be used to simulate the variability and uncertainty of exposure and toxicity parameters. Several simulations are run to determine the range of possible results (Sanaei et al. 2021).

20.5 Fate and Transport Modelling

20.5.1 Modelling Approaches for Predicting the Environmental Fate and Transport of Si/SiO_2 NPs

Environmental fate models (EFMs) are exposure models that use a quantitative and mechanistic explanation of fate processes to forecast both the action and quantity of contaminants in the environment. The primary idea behind EFMs is to incorporate contaminant transport, transfer and breakdown mechanisms into mass balance equations for the pollutant. Multimedia compartmental models (MCMs) and spatial river/watershed models (SRWMs) are two categories into which engineered nanomaterials (ENM) EFMs fall. SRWMs are continually changing and globally resolved models that take into account variations in stream hydrology and morphology as well as sediment transport in river networks. MCMs include the technological explanation of the fate and transport methods of ENMs. They can also consider details of the water quality related to the watershed hydrology (Suhendra et al. 2020).

20.5.2 Uncertainty and Sensitivity Analysis in Fate and Transport Modelling

The model parameters (particle size, surface charge and environment) are determined in relation to the fate and transport of Si/SiO_2 NPs. The probability distributions or ranges based on studies and currently available data are used to describe the uncertainty in these parameters. The degree of uncertainty surrounding the choice of particular models or algorithms is examined for simulating the movement and fate of NPs made of SiO_2. The outcomes produced by applying various model structures or strategies are analysed. The sensitivity of the model to changes in boundary conditions is analysed, including starting levels, emission rates and background concentrations in the environment (Nguyen et al. 2020).

20.6 Environmental Monitoring and Surveillance

20.6.1 Strategies for Monitoring Si/SiO_2 NPs Release and Exposure in Environmental Compartments

- **Transmission Electron Microscopy (TEM):** Quantitative results have been obtained by examining nanomaterials smaller than 100 nm using TEM. The primary cause is that TEM devices require at least 100 keV of accelerated electron energy. There are several issues with the distant background observation of the samples while using TEM equipment that uses high-energy electrons. It is advised to utilize a scanning electron microscopy (SEM) device in this situation that can operate with an electron flow of up to 30–50 keV. TEM apparatus is applied, keeping in mind these characteristics, to fully observe nano SiO_2 samples (Verleysen et al. 2019).
- **Inductively Coupled Plasma Mass Spectrometry (ICP-MS):** Particle quantification in complicated aquatic matrices has been made easier with the development of single-particle inductively coupled mass spectrometry (spICPMS), one of the few techniques that can provide both the distribution of sizes and the number of suspended particles. Based on the mass-to-charge ratio, ICPMS equipment is utilized to quantify the majority of elements included in the periodic table of elements. The technique has become widely accepted due to a number of important advantages: high selectivity, which means that particles containing elements of interest can be easily quantified; high sensitivity, which enables the measurement of particles as small as 5 nm; high throughput, which enables the analysis of a significant number of samples in an appropriate amount of time; and easily accessible machinery (Patriarca et al. 2022).

20.6.2 Development of Analytical Methods for Detecting and Quantifying Nanoparticles in Complex Matrices

- NPs must be well characterized before creating analytical techniques. Their dimensions, form, surface charge and makeup should be assessed, as these attributes may affect the choice of suitable detection methods. Effective and dependable techniques should be provided for preparing samples so that NPs can be extracted from intricate matrices. Depending on the kind of material, techniques could include centrifugation, filtration and digestion (e.g., water, soil, biological tissues). To separate NPs from chemicals that interfere, separation procedures are to be adopted. To effectively separate NPs from bigger particles, size-based separation techniques like size exclusion chromatography and field-flow fractionation can be used. Sophisticated microscopic methods are used for direct imaging and characterization of NPs, such as TEM or SEM.

These techniques yield important details about the size distribution and shape of the particles (Mast et al. 2020).

- Spectroscopic methods are used for quick screening and qualitative investigation of NPs in intricate matrices, such as Raman, FTIR and UV-Vis spectroscopy. Mass spectrometry-based techniques are used to identify and measure NPs. The identification and quantification of NPs can be aided by elemental analysis by methods such as ICP-MS. Optical techniques are exploited to find the size distribution and concentration of NPs in solution, such as dynamic light scattering (DLS) and NP tracking analysis (NTA). These methods are useful for monitoring in real time. Biosensors or electrochemical sensors can detect NPs with sensitivity and selectivity. To improve selectivity, sensor surfaces are functionalized with certain ligands or antibodies. Biosensors are created to detect particular NPs. Sensitive and targeted detection is possible by functionalizing biosensor surfaces with receptors that bind to Si/SiO_2 NPs selectively. Real-time monitoring devices can be used to track the release of NPs in environmental compartments continuously (Yu et al. 2020).

20.7 Biosafety and Health Implications

The biosafety and health consequences of NPs made of silicon (Si) and silicon dioxide (SiO_2) are important to be taken into consideration because of the possible influence that these NPs might have on both human health and the environment (Table 20.1). In professional contexts, workers may come into contact with Si/SiO_2 NPs throughout the production, processing and disposal processes. Individuals may be exposed to Si/SiO_2 NPs via the use of items such as cosmetics, food packaging and electrical devices. The inhalation of Si/SiO_2 NPs may result in a variety of pulmonary consequences, such as inflammation of the lungs, fibrosis of the lungs and reduced lung function. It is possible for these NPs to go deep into the lungs, where they might cause inflammatory reactions, oxidative stress and damage to individual cells (Taghavizadeh Yazdi et al. 2023).

20.7.1 Potential Health Effects of Si/SiO_2 NPs on Human and Animal Populations

The potential health effects of silicon (Si) and silicon dioxide (SiO_2) NPs on human and animal populations are of significant concern due to their widespread use and potential for exposure. Here is an overview of the potential health effects of Si/SiO_2 NPs:

- **Respiratory Effects:** Inhalation of Si/SiO_2 NPs can lead to respiratory health issues in both humans and animals. These NPs have the potential to

Table 20.1 Addressing issues related to Si/SiO_2 NPs and their description.

Issues	Description	References
Environmental impact	Assessment of potential environmental hazards and risks associated with Si/SiO_2 NPs, considering their physicochemical properties and behaviour in environmental matrices such as air, water, soil and sediment.	El-Sayed, (2020)
	Evaluation of NP release pathways, including manufacturing processes, product use and end-of-life disposal, to identify potential sources of environmental contamination.	Zahra et al. (2022)
	Examination of NP fate and transport mechanisms, including dispersion, sedimentation and bioaccumulation, to understand their environmental distribution and persistence.	Lead et al. (2018)
	Identification of ecological receptors, such as aquatic organisms, terrestrial flora and fauna, and assessment of potential impacts on ecosystems, biodiversity and ecosystem services.	Peña et al. (2022)
Risk assessment	Conducting comprehensive life cycle assessments (LCAs) and hazard analyses to identify and evaluate potential hazards and risks associated with Si/SiO_2 NPs throughout their life cycle.	Earley (2022)
	Determination of exposure pathways, levels and potential health effects on human populations, including workers involved in NP production, handling and end-users of NP-containing products.	Thakur and Kumar (2023)
	Prioritization of mitigation measures based on risk assessment results, considering factors such as hazard severity, exposure likelihood and population vulnerability to ensure effective risk management strategies.	Fakhruddin et al. (2020)
Biosafety issues	Evaluation of NP toxicity, biocompatibility and immunogenicity through in vitro and in vivo studies to understand their potential health effects on biological systems.	Dobrovolskaia and McNeil (2013)
	Identification of biosafety guidelines and regulatory frameworks for NP use and handling, including exposure limits, protective equipment requirements and emergency response protocols to ensure safe and responsible NP utilization.	Alberto et al. (2022)

penetrate deep into the lungs, causing inflammation, oxidative stress and pulmonary fibrosis. Chronic exposure to airborne Si/SiO_2 NPs may increase the risk of respiratory diseases such as asthma, chronic obstructive pulmonary disease (COPD) and lung cancer in humans. Similarly, animals exposed to Si/SiO_2 NPs may experience respiratory distress, reduced lung function and histopathological changes in lung tissues (Taghavizadeh Yazdi et al. 2023).

- **Cardiovascular Effects:** For both people and animals, the presence of Si/SiO_2 NPs has been associated with heart-related health problems. Both inhalation and systemic exposure to these NPs have the potential to cause endothelial dysfunction, oxidative stress and inflammation, all of which are factors that contribute to the development of cardiovascular disorders. Studies conducted on animals have shown that NPs composed of silicon dioxide and silicon oxide may have a negative impact on vascular function, accelerate the advancement of atherosclerosis and raise the likelihood of cardiovascular diseases, including myocardial infarction (MI) and stroke. These microscopic particles may also interfere with the coagulation of blood and the function of platelets, which would make cardiovascular risks much more severe (Mostovenko et al. 2022).
- **Neurotoxic Effects:** Si/SiO_2 NPs may have neurotoxic consequences for the central nervous system (CNS) of both children and animals, according to an increasing body of research that suggests this possibility. It is possible for neuroinflammation, oxidative stress and neuronal injury to occur as a result of inhalation or systemic exposure to these NPs. According to research conducted on animals, NPs of silicon dioxide and silicon oxide have the ability to pass through the blood-brain barrier and accumulate in the brain. This accumulation has the potential to cause neurobehavioural impairments, cognitive impairment and neurodegenerative illnesses. There is a possibility that these NPs may cause neuronal death, disrupt neurotransmitter systems and change synaptic dynamics (Milani et al. 2020).
- **Immunotoxic Effects:** Nanoparticles composed of silicon dioxide and silicon dioxide oxide have the capability to regulate immunological responses and generate immunotoxic effects in both humans and animals. Hyperregulation of immune cell function, cytokine production and proinflammatory signaling networks may occur as a result of respiratory or chronic contact with these NPs. Studies conducted on animals have shown that NPs composed of silicon dioxide and silicon dioxide oxide have the ability to inhibit the growth of immune cells, hinder phagocytic activity and change the phenotype and function of immune cells. Additionally, these NPs have the potential to worsen allergic reactions, autoimmune illnesses and symptoms of hypersensitivity in persons who are vulnerable to them (Garcés et al. 2021).

20.7.2 Mechanisms of NP Toxicity and Biological Interactions

Different cellular and molecular pathways are involved in the processes of NP toxicity and the biological responses of silicon (Si) and silicon dioxide (SiO_2) NPs (Table 20.2). These mechanisms are complicated and multidimensional. Over the nano-bio interface, these NPs have the potential to interact with biological systems,

Table 20.2 Synthesis mechanisms, NP sizes, benefits and environmental impacts.

Mechanism/process	Nanoparticle size	Benefits	References
Chemical vapour deposition (CVD)	1–100 nm	Enhanced electrical conductivity and optical properties. Miniaturization of electronic devices.	Kumar et al. (2023)
Sol-Gel synthesis	5–200 nm	Versatile synthesis method suitable for various applications. Precise control over particle size and composition.	Navas et al. (2021)
Stöber method	10–500 nm	Monodisperse NPs with uniform size and shape. Simple and scalable synthesis process.	Jundale et al. (2024)
Microemulsion technique	5–100 nm	High stability and homogeneity of NPs. Controlled synthesis of small-sized particles.	Mousavi et al. (2022)
Hydrothermal synthesis	5–200 nm	Production of NPs with high crystallinity and purity. Tailoring of particle size and morphology.	Harish et al. (2023)
Template-assisted synthesis	1–100 nm	Precise control over NP size and shape. Production of complex nanostructures with tailored properties.	Kaur et al. (2022)
Solvothermal method	10–500 nm	Synthesis of NPs with controlled size and crystallinity. High yield and purity of synthesized products.	Bouremana et al. (2022)
Electrochemical deposition	1–1000 nm	Direct deposition of NPs onto substrates. Controllable synthesis parameters for precise NP fabrication.	Hao et al. (2020)
Green synthesis	5–200 nm	Environment-friendly approach using natural or renewable resources as reducing agents. Reduced environmental impact compared to conventional synthesis methods.	Madani et al. (2022)

which may result in a wide variety of negative impacts. The formation of reactive oxygen species (ROS) is one of the ways in which Si/SiO_2 NPs may cause oxidative stress at the cellular level. Nanoparticles composed of silicon dioxide and silicon dioxide oxide have the potential to disturb normal physiological activities and homeostasis by interfering with cell signaling pathways, receptor-mediated procedures and intracellular movement mechanisms (Liu and Tang 2020).

20.7.3 Occupational Health and Safety Considerations for Workers Handling Si/SiO_2 NPs

For the purpose of mitigating the possible dangers associated with NP exposure, it is essential for workers who handle silicon (Si) and silicon dioxide (SiO_2) NPs to take occupational health and safety precautions. When identifying possible dangers associated with the handling of NPs, it is important to do exhaustive risk assessments that take into consideration pathways of exposure, stages and possible adverse effects on health. The use of airflow systems, confinement barriers and local exhaust ventilation (LEV) to collect NPs in the air is an essential component of engineering controls, which play a significant part in reducing exposure levels (Brand et al. 2020).

- **Risk Assessment:** To identify the possible dangers that are linked with the handling of Si/SiO_2 NPs, it is necessary to conduct comprehensive risk assessments. The various modes of exposure (inhalation, cutaneous contact and ingestion), the degrees of exposure, and the possible health impacts should be taken into consideration. An assessment of the working circumstances, procedures and controls should be undertaken in order to determine the possibility of NP exposure and the extent of its impact (Staroń et al. 2020).
- **Personal Protective Equipment:** Proper personal protective equipment (PPE), which may include gloves, respirators and protective clothes, should be provided in accordance with the dangers and risk levels. The PPE is chosen, fitted and used appropriately in line with the instructions provided by the manufacturer and the required regulations. It is important to provide workers with training on how to properly use, maintain and dispose of PPE in order to guarantee efficient protection against NP exposure (Loibner et al. 2019).

20.8 Risk Mitigation and Management

For the purpose of ensuring the safe handling and use of these materials while simultaneously limiting possible dangers to human health and the environment, risk mitigation and management measures for silicon (Si) and silicon dioxide

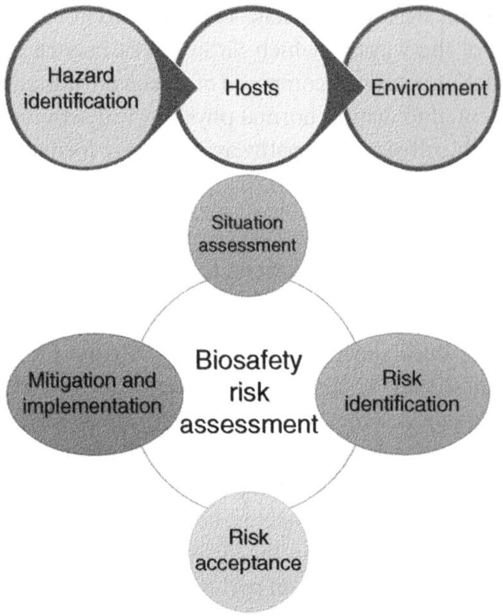

Figure 20.2 Strategies for biosafety risk assessment.

(SiO_2) NPs are of the utmost importance (Figure 20.2). The process of detecting, evaluating and mitigating the risks that are connected with NP exposure is an essential aspect of putting good risk management methods into use (Prajitha et al. 2019).

20.8.1 Strategies for Minimizing Environmental Exposure to Si/SiO_2 NPs

Green Synthesis Methods: The production of Si/SiO_2 NPs by the use of procedures that are both sustainable and kind to the environment is the primary focus of green synthesis methods. Reduction of the negative effects on the environment entails the use of non-toxic chemicals, renewable raw materials and environmentally friendly processes. During the manufacturing of NPs, green approaches strive to limit the amount of hazardous chemicals that are used and to minimize the amount of waste that is generated. Researchers and producers may reduce the environmental concerns that are connected with current NP synthesis techniques by adopting other ways that are more environmentally friendly (Naidu et al. 2023; Roychoudhury 2020).

Nanoparticle Stabilization: Enhancing the stability of Si/SiO_2 NPs and preventing their aggregation in environmental matrices like water and soil are the goals of the process of stabilizing these NPs. Surface modification approaches, which include coating NPs with polymers or surfactants, have the potential to enhance dispersibility and decrease the possibility of sedimentation or aggregation occurring. Nanoparticles that are stable are less likely to be transported and accumulated in the environment, which therefore lowers the likelihood that animals in the terrestrial or aquatic environment would be exposed to them (Naidu et al. 2023).

Environmental Monitoring: Environmental monitoring programmes are necessary in order to determine the existence of Si/SiO_2 NPs in the environment, as well as their distribution and ultimate destination. For this, it is necessary to perform routine sampling and analysis of the air, water, soil and sediment in order to identify microparticle pollution. The data collected via monitoring provides helpful insights into the amounts of environmental exposure, the modes of transport and the possible ecological concerns. Through the process of monitoring NP concentrations over a period of time, the stakeholders are able to recognize growing environmental problems and put relevant risk management measures into execution (Tomak et al. 2021).

20.8.1.1 Engineering Controls and Containment Measures in NP Manufacturing and Handling Processes

For the purpose of protecting both the atmosphere and staff health from the potential dangers that silicon (Si) and silicon dioxide (SiO_2) NPs may pose, it is essential that the production and handling procedures for NPs use tight technical controls and containment mechanisms. Due to the inherent difficulties that are linked with the tiny size and high reactivity of NPs, which may lead to major environmental and health problems if they are not adequately handled, these precautions are especially important. One of the most important engineering controls that are applied in NP facilities is the development of enclosed systems. Some examples of these systems are closed reactors and glove boxes. The purpose of these enclosed settings is to limit the processes of producing NPs, thereby preventing the NPs from dispersing into the air around them. There is a considerable reduction in the danger of NP exposure to both workers and the environment when these enclosed systems are maintained with a regulated atmosphere. In addition, the use of ventilation systems that are fitted with high-efficiency particulate air (HEPA) filters is of critical importance in the process of catching NPs that are in the air and preventing their release into the environment of workplaces. Local exhaust ventilation (LEV) systems are carefully positioned to collect NPs at their source. This successfully reduces the concentration of NPs that workers are exposed to via inhalation, hence reducing the likelihood of their experiencing respiratory health problems (Thakur and Kumar 2023).

20.9 Conclusion

Through rigorous risk assessment methodologies, such as life cycle assessments and hazard analyses, the potential hazards and risks posed by Si/SiO_2 NPs can be identified and evaluated across their entire life cycle, from production to disposal. Implementing robust engineering controls and containment measures, including enclosed systems, ventilation systems and secondary containment, is crucial to minimize environmental exposure and occupational health risks associated with NP manufacturing and handling processes. Additionally, green synthesis methods and NP stabilization techniques offer promising avenues for reducing environmental impact and enhancing NP stability. Effective environmental monitoring programmes are essential for assessing NP distribution, fate and ecological impacts, enabling stakeholders to identify emerging environmental concerns and implement appropriate risk management measures (Singh and Roychoudhury 2022). By addressing challenges and adopting proactive risk mitigation strategies, the environmental impact, risk assessment and biosafety issues associated with Si/SiO_2 NPs can be effectively managed, ensuring their safe and sustainable utilization in various applications.

References

Abbas, Q., Yousaf, B., Ali, M.U. et al. (2020). Transformation pathways and fate of engineered nanoparticles (ENPs) in distinct interactive environmental compartments: a review. *Environment International* 138: 105646.

Ahmad, F., Salem-Bekhit, M.M., Khan, F. et al. (2022). Unique properties of surface-functionalized nanoparticles for bio-application: functionalization mechanisms and importance in application. *Nanomaterials* 12 (8): 1333.

Alberto, A.R., Matos, C., Carmona-Aparicio, G., and Iten, M. (2022). Nanomaterials, a new challenge in the workplace. In: *Nanotoxicology in Safety Assessment of Nanomaterials*, vol. 1357 (ed. H. Louro and M.J. Silva), 379–402. Cham, Switzerland: Springer.

Ale, A., Gutierrez, M.F., Rossi, A.S. et al. (2021). Ecotoxicity of silica nanoparticles in aquatic organisms: an updated review. *Environmental Toxicology and Pharmacology* 87: 103689.

Banu, A.N., Kudesia, N., Raut, A.M. et al. (2021). Toxicity, bioaccumulation, and transformation of silver nanoparticles in aqua biota: a review. *Environmental Chemistry Letters* 19 (6): 4275–4296.

Bhat, J.A., Rajora, N., Raturi, G. et al. (2021). Silicon nanoparticles (SiNPs) in sustainable agriculture: major emphasis on the practicality, efficacy and concerns. *Nanoscale Advances* 3 (14): 4019–4028.

Bouremana, A., Mouaci, S., Berriah, A. et al. (2022). High yield solvothermal synthesis of Ni nanoparticles: structural, microstructural, and magnetic properties. *Journal of Nanoparticle Research* 24 (10): 204.

Brand, W., Van Kesteren, P.C.E., and Oomen, A.G. (2020). Potential Health Risks of Nanomaterials in Food: A Methodology to Identify Signals and Prioritise Risks (RIVM Letter Report 2019-0191). National Institute for Public Health and the Environment (RIVM), Bilthoven, The Netherlands.

Devasena, T., Iffath, B., Renjith Kumar, R. et al. (2022). Insights on the dynamics and toxicity of nanoparticles in environmental matrices. *Bioinorganic Chemistry and Applications* 2022 (1): 4348149.

Dobrovolskaia, M.A. and McNeil, S.E. (2013). Understanding the correlation between in vitro and in vivo immunotoxicity tests for nanomedicines. *Journal of Controlled Release* 172 (2): 456–466.

Earley, R. (2022). Sustainable textiles for fashion. *Journal of Physics: Materials* 5 (032001): 81–84.

Elemike, E.E., Uzoh, I.M., Onwudiwe, D.C., and Babalola, O.O. (2019). The role of nanotechnology in the fortification of plant nutrients and improvement of crop production. *Applied Sciences* 9 (3): 499.

Elli, G., Hamed, S., Petrelli, M. et al. (2022). Field-effect transistor-based biosensors for environmental and agricultural monitoring. *Sensors* 22 (11): 4178.

El-Sayed, M.E. (2020). Nanoadsorbents for water and wastewater remediation. *Science of the Total Environment* 739: 139903.

Fakhruddin, B.S., Boylan, K., Wild, A., and Robertson, R. (2020). Assessing vulnerability and risk of climate change. In: *Climate Extremes and Their Implications for Impact and Risk Assessment* (ed. J. Sillmann, S. Sippel, and S. Russo), 217–241. Elsevier.

Garcés, M., Cáceres, L., Chiappetta, D. et al. (2021). Current understanding of nanoparticle toxicity mechanisms and interactions with biological systems. *New Journal of Chemistry* 45 (32): 14328–14344.

Ghosh, D., Chakraborty, K., Pulimi, M. et al. (2022). The effects of pH, ionic strength, and natural organics on the transport properties of carbon nanotubes in saturated porous medium. *Colloids and Surfaces A: Physicochemical and Engineering Aspects* 647: 129025.

Givens, B.E., Wilson, E., and Fiegel, J. (2019). The effect of salts in aqueous media on the formation of the BSA corona on SiO_2 nanoparticles. *Colloids and Surfaces B: Biointerfaces* 179: 374–381.

Grün, A.L., Manz, W., Kohl, Y.L. et al. (2019). Impact of silver nanoparticles (AgNP) on soil microbial community depending on functionalization, concentration, exposure time, and soil texture. *Environmental Sciences Europe* 31 (1): 1–22.

Hao, N., Zhang, M., and Zhang, J.X. (2020). Microfluidics for ZnO micro-/nanomaterials development: rational design, controllable synthesis, and on-chip bioapplications. *Biomaterials Science* 8 (7): 1783–1801.

Harish, V., Ansari, M.M., Tewari, D. et al. (2023). Cutting-edge advances in tailoring size, shape, and functionality of nanoparticles and nanostructures: a review. *Journal of the Taiwan Institute of Chemical Engineers* 149: 105010.

Harrison, D.M., Briffa, S.M., Mazzonello, A., and Valsami-Jones, E. (2023). A review of the aquatic environmental transformations of engineered nanomaterials. *Nanomaterials* 13 (14): 2098.

Jampílek, J. and Kráľová, K. (2019). Impact of nanoparticles on photosynthesizing organisms and their use in hybrid structures with some components of photosynthetic apparatus. In: *Plant Nanobionics: Volume 1, Advances in the Understanding of Nanomaterials Research and Applications* (ed. R. Prasad), 255–332. Cham, Switzerland: Springer International Publishing.

Jeong, C.K. (2020). Toward bioimplantable and biocompatible flexible energy harvesters using piezoelectric ceramic materials. *MRS Communications* 10 (3): 365–378.

Jundale, R.B., Prasad, B.L., Devi, R.N., and Kulkarni, A.A. (2024). Continuous flow synthesis of mesoporous silica particles with tunable size and structure. *Industrial & Engineering Chemistry Research* 63 (4): 1843–1852.

Kaur, A., Bajaj, B., Kaushik, A. et al. (2022). A review on template assisted synthesis of multi-functional metal oxide nanostructures: status and prospects. *Materials Science and Engineering B* 286: 116005.

Khan, R., Inam, M.A., Park, D.R. et al. (2019). The removal of CuO nanoparticles from water by conventional treatment C/F/S: the effect of pH and natural organic matter. *Molecules* 24 (5): 914.

Kumar, R., Goel, N., Jarwal, D.K. et al. (2023). Strategic review on chemical vapor deposition technology-derived 2D material nanostructures for room-temperature gas sensors. *Journal of Materials Chemistry C* 11 (3): 774–801.

Lead, J.R., Batley, G.E., Alvarez, P.J. et al. (2018). Nanomaterials in the environment: behavior, fate, bioavailability, and effects—an updated review. *Environmental Toxicology and Chemistry* 37 (8): 2029–2063.

Lin, X., Wang, Y., Yang, X. et al. (2023). Endocrine disrupting effect and reproductive toxicity of the separate exposure and co-exposure of nano-polystyrene and diethylstilbestrol to zebrafish. *Science of the Total Environment* 865: 161100.

Little, S. and Fernandes, T.F. (2019). Nanomaterials in aquatic sediments. In: *Ecotoxicology of Nanoparticles in Aquatic Systems* (ed. J. Blasco and I. Corsi), 208–218. Boca Raton, FL: CRC Press/Taylor & Francis.

Liu, N. and Tang, M. (2020). Toxic effects and involved molecular pathways of nanoparticles on cells and subcellular organelles. *Journal of Applied Toxicology* 40 (1): 16–36.

Loibner, M., Hagauer, S., Schwantzer, G. et al. (2019). Limiting factors for wearing personal protective equipment (PPE) in a health care environment evaluated in a randomised study. *PLoS One* 14 (1): e0210775.

Madani, M., Hosny, S., Alshangiti, D.M. et al. (2022). Green synthesis of nanoparticles for varied applications: green renewable resources and energy-efficient synthetic routes. *Nanotechnology Reviews* 11 (1): 731–759.

Mast, J., Verleysen, E., Hodoroaba, V.D., and Kaegi, R. (2020). Characterization of nanomaterials by transmission electron microscopy: measurement procedures. In: *Micro and Nano Technologies: Characterization of Nanoparticles – Measurement Processes for Nanoparticles* (ed. V.D. Hodoroaba, W.E.S. Unger, and A.G. Shard), 29–48. Amsterdam: Elsevier.

Mathur, P. and Roy, S. (2020). Nanosilica facilitates silica uptake, growth and stress tolerance in plants. *Plant Physiology and Biochemistry* 157: 114–127.

Milani, C., Farina, F., Botto, L. et al. (2020). Systemic exposure to air pollution induces oxidative stress and inflammation in mouse brain, contributing to neurodegeneration onset. *International Journal of Molecular Sciences* 21 (10): 3699.

Mostovenko, E., Canal, C.G., Cho, M. et al. (2022). Indirect mediators of systemic health outcomes following nanoparticle inhalation exposure. *Pharmacology & Therapeutics* 235: 108120.

Mousavi, S.M., Behbudi, G., Gholami, A. et al. (2022). Shape-controlled synthesis of zinc nanostructures mediating macromolecules for biomedical applications. *Biomaterials Research* 26 (1): 4.

Naidu, S., Pandey, J., Mishra, L.C. et al. (2023). Silicon nanoparticles: synthesis, uptake and their role in mitigation of biotic stress. *Ecotoxicology and Environmental Safety* 255: 114783.

Navas, D., Fuentes, S., Castro-Alvarez, A., and Chavez-Angel, E. (2021). Review on sol-gel synthesis of perovskite and oxide nanomaterials. *Gels* 7 (4): 275.

Nguyen, Q., Rizvandi, R., Karimipour, A. et al. (2020). A novel correlation to calculate thermal conductivity of aqueous hybrid graphene oxide/silicon dioxide nanofluid: synthesis, characterizations, preparation, and artificial neural network modeling. *Arabian Journal for Science and Engineering* 45 (11): 9747–9758.

Patriarca, M., Barlow, N., Cross, A. et al. (2022). Atomic spectrometry update: review of advances in the analysis of clinical and biological materials, foods and beverages. *Journal of Analytical Atomic Spectrometry* 37 (3): 410–473.

Peña, L.V.D.L., Taelman, S.E., Préat, N. et al. (2022). Towards a comprehensive sustainability methodology to assess anthropogenic impacts on ecosystems: review of the integration of Life Cycle Assessment, Environmental Risk Assessment and Ecosystem Services Assessment. *Science of the Total Environment* 808: 152125.

Poddar, K., Sarkar, D., and Sarkar, A. (2020). Nanoparticles on photosynthesis of plants: effects and role. In: *Green Nanoparticles: Synthesis and Biomedical*

Applications (ed. J.K. Patra, L.F. Fraceto, G. Das, and E.V.R. Campos), 273–287. Cham, Switzerland: Springer International Publishing.

Prajitha, N., Athira, S.S., and Mohanan, P.V. (2019). Bio-interactions and risks of engineered nanoparticles. *Environmental Research* 172: 98–108.

Rafieepour, A., Azari, M.R., Khodagholi, F. et al. (2021). Interactive toxicity effect of combined exposure to hematite and amorphous silicon dioxide nanoparticles in human A549 cell line. *Toxicology and Industrial Health* 37 (5): 289–302.

Rajput, V.D., Minkina, T., Upadhyay, S.K. et al. (2022). Nanotechnology in the restoration of polluted soil. *Nanomaterials* 12 (5): 769.

Roma, J., Matos, A.R., Vinagre, C., and Duarte, B. (2020). Engineered metal nanoparticles in the marine environment: a review of the effects on marine fauna. *Marine Environmental Research* 161: 105110.

Roy, P. (2023). Nothing is 'waste' in agriculture: from nanotechnology and bioprocesses perspectives. In: *Agriculture Waste Management and Bioresource: The Circular Economy Perspective* (ed. S. Singh, P. Singh, A. Sharma, and M. Choudhury), 131–148. Hoboken, NJ: Wiley.

Roychoudhury, A. (2020). Yeast-mediated green synthesis of nanoparticles for biological applications. *Indian Journal of Pharmaceutical and Biological Research* 8 (3): 26–31.

Sabaraya, I.V. (2021). Nanomaterials in complex environments: interactions affected by matrix composition, pH, natural organic matter, and bio-or geo-colloids. Doctoral dissertation. University of Texas at Austin.

Salam, M., Zheng, H., Liu, Y. et al. (2023). Effects of micro (nano) plastics on soil nutrient cycling: state of the knowledge. *Journal of Environmental Management* 344: 118437.

Sanaei, F., Amin, M.M., Alavijeh, Z.P. et al. (2021). Health risk assessment of potentially toxic elements intake via food crops consumption: Monte Carlo simulation-based probabilistic and heavy metal pollution index. *Environmental Science and Pollution Research* 28: 1479–1490.

Shi, Q., Wang, C.L., Zhang, H. et al. (2020). Trophic transfer and biomagnification of fullerenol nanoparticles in an aquatic food chain. *Environmental Science: Nano* 7 (4): 1240–1251.

Shi, C., Liu, Z., Yu, B. et al. (2023). Emergence of nanoplastics in the aquatic environment and possible impacts on aquatic organisms. *Science of the Total Environment* 906: 167404.

Siddiqui, S.A., Khan, S., Tariq, T. et al. (2023). Potential risk assessment and toxicological impacts of nano/micro-plastics on human health through food products. *Advances in Food and Nutrition Research* 103: 361–395.

Singh, A. and Roychoudhury, A. (2022). Role of nanoparticles in remediation of environmental contaminants. In: *Hazardous and Trace Materials in Soil and Plants* (ed. M. Naeem, T. Aftab, A.A. Ansari, et al.), 327–340. Academic Press (Elsevier).

Sokovnin, S.Y.E. and Ilves, V.G. (2022). Nanopowders based on silicon oxide: production and properties. In: *Nanopowders of Metal Oxides and Fluorides* (ed. S.Y.E. Sokovnin and V.G. Ilves), 383–420. Singapore: Jenny Stanford Publishing.

Sonia, P., Kumar, A.S., Khan, I. et al. (2023). Sustainable manufacturing of high-performance composites from recycled materials. In: *Proceedings of the 15th International Conference on Materials Processing and Characterization (E3S Web of Conferences*, vol. 430 (ed. S.S. Kumar), 01105. Les Ulis, France: EDP Sciences.

Stanca, L., Geicu, O.I., Serban, A.I., and Dinischiotu, A. (2023). Interplay of oxidative stress, inflammation, and autophagy in RAW 264.7 murine macrophage cell line challenged with Si/SiO_2 quantum dots. *Materials* 16 (14): 5083.

Staroń, A., Długosz, O., Pulit-Prociak, J., and Banach, M. (2020). Analysis of the exposure of organisms to the action of nanomaterials. *Materials* 13 (2): 349.

Suhendra, E., Chang, C.H., Hou, W.C., and Hsieh, Y.C. (2020). A review on the environmental fate models for predicting the distribution of engineered nanomaterials in surface waters. *International Journal of Molecular Sciences* 21 (12): 4554.

Suresh, V.S.R. (2021). Toxicity induced by engineered nanoparticles in fresh water fish – a review of recent developments. *EC Pharmacology and Toxicology* 9: 2.

Taghavizadeh Yazdi, M.E., Qayoomian, M., Beigoli, S., and Boskabady, M.H. (2023). Recent advances in nanoparticle applications in respiratory disorders: a review. *Frontiers in Pharmacology* 14: 1059343.

Tejedor, J., Guerrero, V.H., Vizuete, K., and Debut, A. (2022). Environmentally friendly synthesis of silicon dioxide nanoparticles and their application for the removal of emerging contaminants in aqueous media. *Journal of Physics: Conference Series* 2238 (1): 012005.

Thakur, A. and Kumar, A. (2023). Ecotoxicity analysis and risk assessment of nanomaterials for the environmental remediation. *Macromolecular Symposia* 410 (1): 2100438.

Tiancheng, A. (2023). *Change in Structural and Potentiodynamic Characteristics of SiO_2/TiO_2 Nanocomposite Due to Changing Mechanosynthesis Duration*. Igor Sikorsky Kyiv Polytechnic Institute.

Tomak, A., Cesmeli, S., Hanoglu, B.D. et al. (2021). Nanoparticle-protein corona complex: understanding multiple interactions between environmental factors, corona formation, and biological activity. *Nanotoxicology* 15 (10): 1331–1357.

Verleysen, E., Wagner, T., Lipinski, H.G. et al. (2019). Evaluation of a TEM based approach for size measurement of particulate (nano) materials. *Materials* 12 (14): 2274.

Wang, X., He, K., Hu, Y., and Tang, M. (2022). A review of pulmonary toxicity of different types of quantum dots in environmental and biological systems. *Chemico-Biological Interactions* 110247.

Yu, X., He, M., Chen, B., and Hu, B. (2020). Recent advances in single-cell analysis by inductively coupled plasma-mass spectrometry: a review. *Analytica Chimica Acta* 1137: 191–207.

Zahra, Z., Habib, Z., Hyun, S., and Sajid, M. (2022). Nanowaste: another future waste, its sources, release mechanism, and removal strategies in the environment. *Sustainability* 14 (4): 2041.

21

Ethical and Societal Considerations and Challenges of Si/SiO_2 Nanoparticle Application

Prasann Kumar[1] *and Joginder Singh*[2]

[1] *Department of Agronomy, School of Agriculture, Lovely Professional University, Phagwara, Punjab, India*
[2] *Department of Botany, Nagaland University, Kohima, Nagaland, India*

21.1 Introduction

Nanoparticles (NPs) composed of silicon (Si) and silicon dioxide (SiO_2) exhibit significant potential in transforming diverse technological domains, encompassing electronics, biomedical engineering, energy and environmental science (Cani et al. 2021; Hassan et al. 2023; Zhang et al. 2023b). These NPs are leading the way in innovation due to their distinctive properties and wide range of applications. Nevertheless, extensive use of these technologies also gives rise to numerous ethical and societal ramifications that necessitate meticulous examination and resolution. The ethical considerations surrounding the application of Si/SiO_2 NPs in technology are diverse, encompassing various aspects, such as human health and safety, environmental sustainability, social equity, transparency and informed consent. Integrating Si/SiO_2 NPs into consumer products, medical devices and industrial applications has become a growing concern due to the potential health risks and safety hazards associated with their use. With the increasing prevalence of NPs in various aspects of daily life, it is crucial for researchers and manufacturers to adhere to ethical standards and place a high priority on ensuring the safety of these materials. It is imperative to conduct comprehensive studies to evaluate the compatibility of Si/SiO_2 NPs with living organisms, potential adverse effects and the accumulation of NPs in biological systems. Furthermore, it is morally essential to guarantee that the advantages of NP-enabled technologies surpass any accompanying hazards. Scholars must comply with stringent safety protocols and effectively disseminate their research outcomes to various stakeholders, such as regulatory entities, healthcare practitioners and the wider public. The promotion of public trust and confidence in NP technologies is contingent

Silicon Nanoparticles: Synthesis, Uptake, and Applications, First Edition.
Edited by Archana Singh, Aryadeep Roychoudhury, and Indrakant K. Singh.

upon the presence of transparency. Providing precise and reliable information regarding the existence of Si/SiO_2 NPs in consumer goods and their potential ramifications on human health and the environment is of utmost importance. Consumers are entitled to make well-informed decisions regarding their products, which requires implementing clear labelling and providing information regarding the NP content. In addition, it is crucial to prioritize disseminating knowledge and awareness regarding NP technologies to foster a comprehensive understanding among individuals regarding their potential advantages and disadvantages. The concept of informed consent assumes particular significance in medical applications involving Si/SiO_2 NPs, as patients must be sufficiently educated about the potential risks and benefits of the treatment before consent. The potential exacerbation of existing social and economic disparities may arise from the widespread adoption of technologies enabled by Si/SiO_2 NPs. There is a strong correlation between access to advanced technologies and socioeconomic status, which results in disparities in healthcare, education and economic opportunities. Hence, it is imperative to acknowledge and rectify these disparities to guarantee fair and equal access to NP innovations. This may entail implementing strategies to enhance the affordability and accessibility of NP-based products for marginalized communities, addressing the digital divide and fostering STEM education in underserved regions (Galarza-Acosta et al. 2023; Li et al. 2023b; Wang et al. 2021). Furthermore, it is imperative for policy makers and industry leaders to carefully evaluate the potential societal ramifications of NP technologies and adopt strategies to alleviate any detrimental consequences on susceptible populations. Incorporating Si/SiO_2 NPs in technology offers significant prospects for innovation and progress in diverse industries. Nevertheless, it presents intricate ethical and societal dilemmas that must not be disregarded. Adopting ethical principles, such as safety, transparency, equity and informed consent, is imperative for all stakeholders, including scientists, policy makers and industry leaders. By adopting this approach, we can effectively address these challenges and guarantee that NP-enabled technologies positively impact society while minimizing potential negative consequences. To fully harness the potential of Si/SiO_2 NPs in an ethically sound and socially responsible manner, it is imperative to adopt a comprehensive approach encompassing scientific research, regulatory oversight, ethical guidelines and stakeholder engagement (Table 21.1).

21.2 Health Risks due to use of Si/SiO_2 NPs: A Moral Issue

The potential applications of Si/SiO_2 NPs have attracted considerable interest across diverse domains, such as consumer products and medicine. Nevertheless, the growing utilization of Si/SiO_2 NPs gives rise to significant moral and ethical

Table 21.1 Role and Impact of Si/SiO_2 NPs.

Consideration	Description
Environmental impact	Potential effects on ecosystems and biodiversity due to NP release into the environment.
Human health	Health risks associated with exposure to Si/SiO_2 NPs, including respiratory and other health issues.
Nanotoxicity	Study of the toxic effects of NPs on living organisms and ecosystems.
Regulatory compliance	They ensure that Si/SiO_2 NP applications adhere to relevant regulations and standards.
Privacy and data security	Concerns about collecting and using data generated by Si/SiO_2 NP technologies.
Equity and access	Ensuring fair access to Si/SiO_2 NP technologies across different socioeconomic groups.
Intellectual property rights	Issues related to ownership and protection of intellectual property associated with Si/SiO_2 NPs.
Public perception and acceptance	Understanding public attitudes and concerns regarding Si/SiO_2 NP applications.
Transparency and disclosure	Clear communication of risks and benefits associated with Si/SiO_2 NP technologies.
Cultural sensitivity	Respecting cultural values and beliefs in developing and deploying Si/SiO_2 NP applications.
Economic impact	Adopting Si/SiO_2 NP technologies could potentially affect jobs, industries and economic systems.
Supply chain ethics	Ensuring ethical practices throughout the supply chain of Si/SiO_2 NP materials.
Resource management	Responsible use of resources such as silicon and other materials in Si/SiO_2 NP production.
Occupational health and safety	Ensuring the safety of workers involved in the production and application of Si/SiO_2 NPs.
Education and literacy	Promoting understanding of Si/SiO_2 NP technologies among the general public and stakeholders.
Dual-use concerns	Potential misuse of Si/SiO_2 NP technologies for harmful purposes.
International collaboration	Collaboration and coordination among countries in developing and regulating Si/SiO_2 NP technologies.
Ethical research practices	Researching Si/SiO_2 NPs ethically and responsibly.
Displacement and disruption	Potential displacement of existing technologies or disruption of established systems by Si/SiO_2 NP technologies.

(Continued)

Table 21.1 (Continued)

Consideration	Description
Long-term impacts	Consideration of the long-term effects and legacy of Si/SiO_2 NP applications.
Governance and oversight	Mechanisms for governance and oversight of Si/SiO_2 NP research and applications.
Social justice	Addressing inequalities and promoting social justice in developing and deploying Si/SiO_2 NP technologies.
Security concerns	Ensuring the security of Si/SiO_2 NP applications against cyber threats and other risks.
Environmental justice	Ensuring a fair distribution of environmental benefits and burdens associated with Si/SiO_2 NP technologies.
Education and training	Providing education and training opportunities for Si/SiO_2 NP application stakeholders.
Public engagement and participation	It involves the public in decision-making processes related to Si/SiO_2 NP technologies.
Cultural impacts	Potential impacts of Si/SiO_2 NP applications on cultural practices and identities.
Technological dependence	Risks associated with over-reliance on Si/SiO_2 NP technologies for critical functions.
Health equity	Addressing disparities in access to healthcare and health outcomes through Si/SiO_2 NP applications.
Environmental stewardship	Promoting responsible stewardship of the environment in developing and using Si/SiO_2 NP technologies.
Scientific integrity	Upholding principles of scientific integrity and rigor in research involving Si/SiO_2 NPs.

Source: Kuo et al. (2024); Subramanian et al. (2020); Takari et al. (2021).

apprehensions, specifically regarding their influence on human well-being. This section explores the ethical considerations related to the increased use of Si/SiO_2 NPs. It focuses on several important aspects, including the widespread use of these NPs in consumer products and medicine, the potential long-term health impacts and concerns, the need for thorough investigations into compatibility, adverse effects and accumulation, and the ethical responsibility of manufacturers and researchers to prioritize safety. Nanotechnology has rapidly progressed, resulting in the extensive integration of Si/SiO_2 NPs into various consumer products and medical devices. The NPs exhibit distinctive physical and chemical characteristics that render them valuable for various applications, such as drug delivery systems, biomedical imaging, personal care products and food packaging materials. Nevertheless, the extensive incorporation of Si/SiO_2 NPs into commonplace commodities raises ethical concerns about their safety and potential hazards to human well-being. Si/SiO_2 NPs are frequently incorporated into consumer goods, including cosmetics, sunscreen lotions, food additives and antimicrobial coatings, owing to their capacity to confer advantageous characteristics, such as UV protection, improved drug delivery and extended shelf life (Choi et al. 2021; Gharsallah et al. 2024; Khanna et al. 2023b; Passaro et al. 2022; Yang et al. 2023). Although NPs present notable technological advantages, their utilization in consumer goods gives rise to apprehensions regarding potential exposure pathways and unintended health ramifications. Furthermore, the absence of thorough regulation and supervision in numerous jurisdictions worsens uncertainties regarding the safety of NP products, emphasizing the necessity for strong ethical frameworks to direct their creation and utilization. The potential of Si/SiO_2 NPs in medicine is significant, as they can transform diagnostics, drug delivery and therapeutic interventions. The NPs can be deliberately designed to selectively target particular tissues or administer therapeutic agents accurately, presenting novel prospects for personalized medicine and enhanced patient results. Nevertheless, incorporating Si/SiO_2 NPs in medical environment presents ethical questions about the well-being of patients, obtaining informed consent and the possibility of unanticipated negative consequences. The ethical considerations surrounding NP-based interventions necessitate careful navigation by healthcare professionals and researchers, who must prioritize patient well-being while effectively utilizing their therapeutic potential. The potential for long-term health effects and unintended consequences is a significant ethical concern associated with using Si/SiO_2 NPs in consumer products and medicine. Although NPs provide various advantages in improved functionality and performance, their diminutive dimensions and distinctive physicochemical characteristics may also present potential hazards to human well-being, especially when subjected to extended or recurrent exposure. Research findings have indicated that Si/SiO_2 NPs have the potential to demonstrate cytotoxicity, genotoxicity and inflammatory reactions in biological systems, thereby giving rise to apprehensions regarding their suitability for human consumption or exposure. The inhalation or dermal exposure to products containing NPs may lead to the accumulation of NPs in essential organs, thereby causing negative health consequences, including pulmonary inflammation, cardiovascular effects, neurotoxicity and carcinogenicity. Moreover, the enduring presence of Si/SiO_2 NPs in the natural surroundings gives rise to apprehensions regarding their possible accumulation in food webs and ecological systems, which could have significant ramifications for human well-being and ecological equilibrium. Assessing the long-term health effects of Si/SiO_2 NPs in biological systems is challenging due to their dynamic nature. The conventional methods used for toxicity testing may not comprehensively capture the intricate interactions between Si/SiO_2 NPs and living organisms (Khanna et al. 2023a; Ningsih et al. 2022; Paul et al. 2022; Sun et al. 2020; Zhao et al. 2023). Therefore, it is imperative to use innovative approaches and foster interdisciplinary collaboration to address these gaps in knowledge effectively. In addition, it is crucial to emphasize the significance of long-term monitoring and surveillance endeavours to detect emerging risks and develop evidence-based risk management strategies, given the potential for health effects caused by NPs to become apparent over prolonged durations. In light of the ambiguities surrounding the health hazards linked to Si/SiO_2 NPs, it is imperative to conduct thorough investigations to assess their compatibility with biological systems, potential adverse impacts and mechanisms of buildup. To comprehend the potential hazards and advantages of NP exposure, these investigations must encompass various fields, such as nanotoxicology, pharmacology, environmental science and biomedical engineering. Investigating Si/SiO_2 NPs at the cellular and molecular levels through in vitro and in vivo studies is crucial to understand their biological interactions comprehensively. These studies offer valuable insights into the mechanisms of action and potential pathways of toxicity associated with these NPs. Furthermore, it is imperative to research the pharmacokinetics and biodistribution of Si/SiO_2 NPs within animal models to gain insights into their absorption, distribution, metabolism and excretion in living organisms. It is essential to conduct long-term, multi-generational studies to evaluate the cumulative impact of NP exposure over time and detect any potential negative health consequences that may occur in later stages of life. Furthermore, studies must not only assess the immediate effects of Si/SiO_2 NPs on human health but also evaluate their environmental destiny and ecological consequences. This encompasses examining the behaviour and alteration of NPs within environmental matrices, such as soil, water and air, and their potential for bioaccumulation in aquatic and terrestrial ecological systems. In addition, it is imperative to research the ecotoxicity of Si/SiO_2 NPs to gain insights into their potential impacts on non-human organisms and the dynamics of the environment. This knowledge is crucial for informing risk assessment and management strategies to protect environmental well-being and biodiversity. It is the moral and ethical responsibility of manufacturers and researchers to prioritize ensuring the safety of Si/SiO_2 NPs at every stage of the lifestyle of the product, including development, production, use and disposal. To optimize the advantages of NP-enabled technologies while mitigating potential hazards to human health and the environment, it is imperative to uphold ethical principles, such as beneficence, non-maleficence, justice and respect for autonomy (Paul et al. 2023; Roychoudhury and Bhowmik 2023). The selection of materials, synthesis methods and quality control measures in manufacturing should be guided by ethical considerations to minimize the production of hazardous by-products and ensure the safety and purity of NP formulations. Implementing quality assurance and certification procedures is crucial in ensuring adherence to safety standards and regulatory mandates, fostering trust in the safety and dependability of NP products. Researchers are crucial in progressing scientific knowledge and facilitating evidence-based decision-making about using Si/SiO_2 NPs safely. Ensuring the integrity and reliability of NP toxicity studies necessitates adherence to ethical research practices, which encompass rigorous experimental design, transparent reporting of findings and compliance with ethical guidelines and regulatory requirements (Choudhury et al. 2022; Huang et al. 2018; Li et al. 2023a; Park and Bae 2019; Zhu et al. 2021). Moreover, researchers must actively involve various stakeholders, such as regulatory bodies, industry collaborators, healthcare practitioners and the general public, to foster well-informed communication and decision-making processes concerning NP research and ethical and societal ramifications. Addressing the complex ethical challenges of health risks requires collaboration and knowledge sharing among stakeholders. Manufacturers and researchers can effectively address the moral challenges of nanoparticle technology and contribute to the development of safe, sustainable and socially responsible solutions that benefit society by promoting a culture of responsible innovation and ethical stewardship (Figure 21.1).

21.3 Moral Consideration for Environmental Sustainability

The employment of Si/SiO_2 NPs in diverse technological contexts has elicited noteworthy moral and ethical apprehensions regarding their potential ramifications on environmental sustainability. This section delves into the ethical considerations linked to the environmental consequences of Si/SiO_2 NPs. It examines crucial elements, including their introduction into ecological systems, impacts on ecological systems, wildlife and the environment, the necessity for sustainable production, utilization and disposal to mitigate environmental

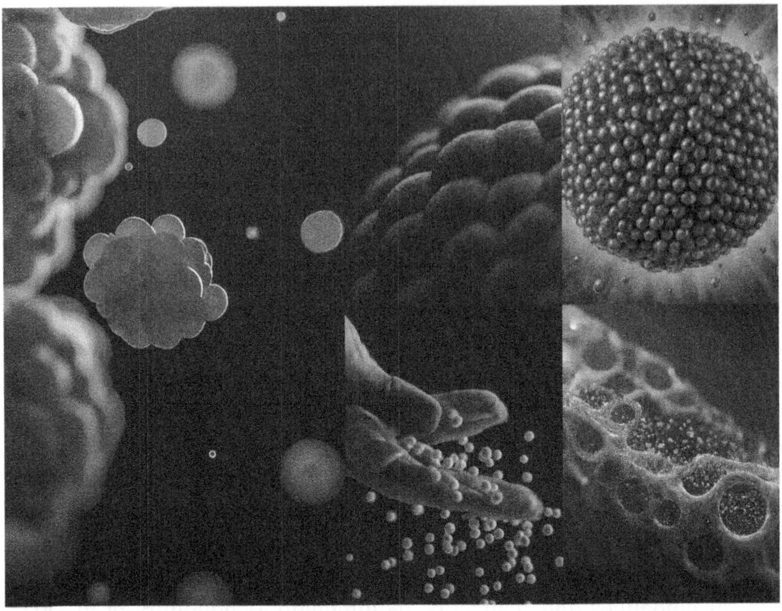

Figure 21.1 Si/SiO$_2$ NPs Ethical Use.

harm, and the significance of promoting environmental ethics and accountability (Chakraborty et al. 2022; Paul and Roychoudhury 2021). The widespread distribution of Si/SiO$_2$ NPs in consumer products, industrial processes and environmental remediation endeavours has resulted in their extensive presence in the environment, giving rise to apprehensions regarding their potential introduction into ecological systems. NPs have the potential to infiltrate ecological systems via multiple routes, encompassing atmospheric deposition, surface runoff, wastewater effluent and agricultural resources. When introduced into the environment, NPs can experience various physical, chemical and biological changes that impact their ability to move, absorb nutrients and cause harm. These changes can potentially jeopardize the ecological balance and biodiversity. The accumulation of Si/SiO$_2$ NPs in sediments, surface waters and aquatic organisms within aquatic environment can impact marine food webs and the functioning of ecological systems. The deposition of NPs and the subsequent uptake by soil organisms, plants and wildlife can also impact terrestrial ecosystems (Abduljabar et al. 2023; Hassan et al. 2022; Maniyazagan et al. 2022; Sanchez et al. 2011; Ulusal 2024; Zhang et al. 2023a). The introduction of Si/SiO$_2$ NPs into ecological systems underscores the necessity for proactive strategies to reduce their ecological consequences and protect the well-being and adaptability of the environment. The ecological, wildlife and environmental impacts of Si/SiO$_2$ NPs are intricate and diverse, involving intricate physical, chemical and biological interactions that can result in significant ecological ramifications. The presence of NPs can potentially modify soil properties, water quality and air quality, thereby inducing alterations in the structure and function of the environment. Furthermore, Si/SiO_2 NPs have the potential to cause direct harm to organisms, interfere with ecological processes and disturb the dynamics of the environment, thereby endangering biodiversity and ecological services. Research has demonstrated that Si/SiO_2 NPs can cause oxidative stress, inflammation, genotoxicity and other negative consequences in many organisms, such as bacteria, algae, aquatic invertebrates, fish and terrestrial plants. These effects have the potential to spread throughout food chains, impacting higher levels of the food chain and the overall resilience of the ecosystem. In addition, the enduring nature and capacity for bioaccumulation of Si/SiO_2 NPs give rise to apprehensions regarding their extended environmental destiny and the possibility of prolonged exposure to wildlife and ecological systems (Roychoudhury 2020). Si/SiO_2 NPs have the potential to interact with environmental contaminants, modify nutrient cycling and disrupt various ecological processes, including photosynthesis, nutrient uptake and decomposition, in addition to their direct impacts on biota. These secondary consequences can have a dominating effect on the structure and functioning of the ecosystem, potentially affecting its ability to withstand challenges, maintain stability and provide essential services. Hence, it is imperative to consider the wider ecological framework when assessing the environmental hazards linked to Si/SiO_2 NPs and to embrace a cautious strategy in their utilization and administration. Adopting a sustainable approach to the production, use and disposal of Si/SiO_2 NPs is crucial to address their environmental implications. Sustainable production practices aim to minimize the ecological impact of NP synthesis processes, decrease resource usage and address pollution and waste generation. Optimizing synthesis routes, using green chemistry principles and implementing pollution prevention measures may be necessary to mitigate the release of NPs and associated contaminants into the environment. Likewise, sustainable use practices involve reducing the ecological consequences of NP products at every stage, including production, distribution, utilization and disposal. This encompasses developing products that can be recycled and biodegraded, promoting environmentally friendly alternatives and implementing effective management strategies to reduce the release and exposure of NPs in both occupational and consumer environment (Arooj et al. 2022; Rochima et al. 2024; Sarac et al. 2022; Sarker et al. 2023; Sharma 2023; Yuan et al. 2022). Furthermore, sustainable disposal practices aim to mitigate the ecological impact of NP waste by promoting recycling, reuse and biodegradable disposal techniques. Ensuring responsible stewardship of Si/SiO_2 NPs and advancing ecological sustainability goals requires the promotion of environmental ethics and accountability. Environmental ethics emphasizes the inherent worth of the natural world, the interdependence of all living entities and the ethical obligation to safeguard and maintain the environment for both current and future cohorts. When policymakers, industry stakeholders and researchers integrate environmental ethics into their decision-making processes, they can give precedence to the preservation of the environment, advocate for sustainable practices and uphold principles of justice and intergenerational equity. Environmental accountability encompasses the principles of transparency, accountability and responsibility for the ecological consequences associated with NP-related activities. This encompasses revealing details regarding the utilization and release of NPs, evaluations of ecological hazards and executing strategies to avert or alleviate detrimental environmental effects. Moreover, ecological accountability involves the active involvement of various stakeholders, such as communities, indigenous people and civil society organizations, in decision-making procedures and promoting open communication, cooperation and collective efforts to tackle environmental issues effectively. The comprehensive and proactive integration of scientific knowledge, ethical principles and stakeholder engagement is necessary to address the moral considerations about the environmental sustainability of Si/SiO_2 NPs. Stakeholders can mitigate the environmental impact of Si/SiO_2 NPs and enhance the long-term health and resilience of ecosystems and the environment by prioritizing sustainable production, use and disposal practices, promoting environmental ethics and accountability and adopting a precautionary approach to environmental risk management (Figure 21.2).

21.4 Societal Accessibility and Fairness

The potential for innovation and improvement in quality of life exists through the incorporation of Si/SiO_2 NPs into diverse technological applications. Nevertheless, it is crucial to prioritize the fair allocation and availability of NP-based technologies to guarantee that all individuals in society have equal access to the advantages brought about by these advancements. This section examines the societal aspects of the accessibility and fairness of technological advancements enabled by Si/SiO_2 NPs (Chatterjee et al. 2021; Chen et al. 2024; Goswami et al. 2022; Kaur et al. 2023; Kumar et al. 2022; Rawat et al. 2018). It focuses on important factors, such as disparities in access, equitable distribution and pricing of products containing NPs and promoting inclusivity to mitigate social inequality. One of the key ethical obligations associated with adopting Si/SiO_2 NP technologies is to mitigate disparities in access, guaranteeing equitable availability of these advancements across all societal groups. Socioeconomic factors, such as income, education, geographic location and social status, often determine access to advanced technologies, leading to disparities in access and opportunity. Hence, it is imperative to implement focused approaches to address the digital divide and foster fair and equal availability of technologies enabled by Si/SiO_2 NPs. Efforts to mitigate

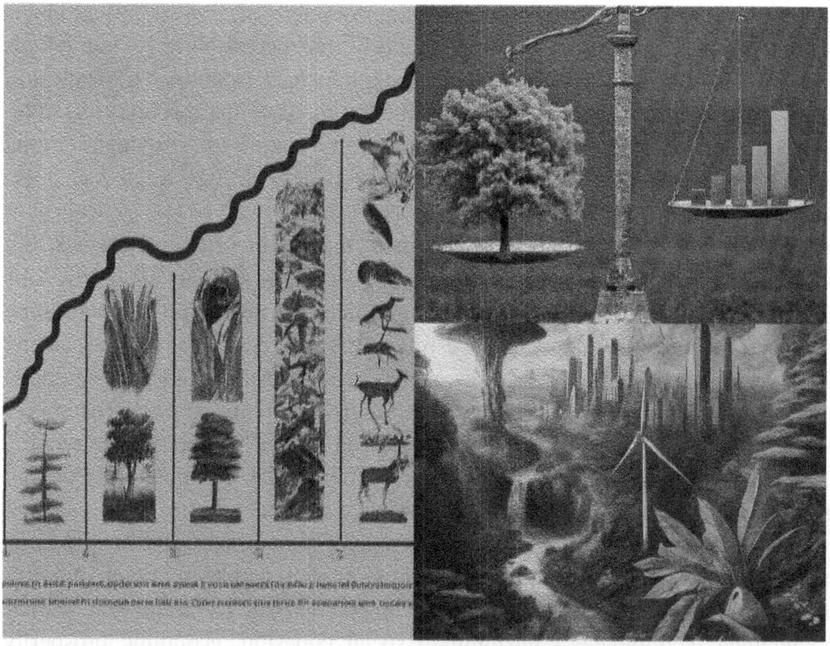

Figure 21.2 Moral consideration for environmental sustainability.

disparities in access may encompass various strategies, such as implementing initiatives to enhance infrastructure and connectivity in marginalized communities, providing subsidies or incentives to facilitate technology adoption and promoting programmes focused on digital literacy and skill training. Furthermore, collaborations among governmental entities, industrial sectors, academic institutions and non-profit organizations can play a crucial role in recognizing and resolving obstacles to entry, harnessing available resources and guaranteeing the dissemination of Si/SiO_2 NP-enabled technologies to underserved demographics, such as rural communities, minority communities and individuals with disabilities. Furthermore, resolving disparities in access necessitates acknowledging and resolving the interconnected manifestations of prejudice and disadvantages that contribute to the exclusion and marginalization of individuals within a society. This may entail implementing comprehensive and interconnected strategies that consider varied requirements of individuals and communities, viewpoints and advocates for fairness. To promote affordability and accessibility for all the members of the society, addressing inequalities in access and ensuring equitable distribution and pricing of NP-containing products is crucial. Fair distribution involves the equitable, transparent and inclusive provision of NP-enabled technologies, with resources allocated based on need and priority rather than wealth or privilege. Ensuring equitable pricing of NP products is imperative to guarantee their affordability and accessibility to individuals and communities across various socioeconomic backgrounds. The exclusion of low-income individuals and communities from accessing essential technologies and exacerbating existing inequalities can be attributed to high pricing barriers (Adegoke et al. 2023; Doğan et al. 2024; Ilves et al. 2024; Kalal et al. 2022; Kurumurthy et al. 2009; Machado et al. 2024; Zhao et al. 2024). Hence, manufacturers, retailers and policymakers must implement pricing strategies that strike a harmonious equilibrium between affordability and sustainability. This entails ensuring equitable distribution of production, distribution and marketing expenses without imposing excessive financial strain on consumers, especially those with limited financial resources. Moreover, endeavours to advance equitable allocation and pricing of products containing NPs should consider the social, economic and environmental consequences of pricing choices, encompassing their effects on public health, fairness and sustainability. One possible approach is to incorporate pricing models, considering external factors, such as environmental expenses, social advantages and long-term worth. Additionally, it is important to establish regulatory measures to prevent excessive pricing, monopolistic behaviours and unfair competition in the market. To ensure that Si/SiO$_2$ NP-enabled technologies contribute to social cohesion, economic prosperity and human flourishing, addressing access and pricing disparities, promoting inclusivity, and preventing social inequality is crucial. Inclusivity encompasses the establishment of policies and systems that actively accept and promote diversity, uphold human rights and encourage a sense of belonging and engagement for all individuals in society, irrespective of their race, ethnicity, gender, sexual orientation, age, ability or socioeconomic status. Initiatives that foster inclusivity and mitigate social inequality encompass a range of strategies. These strategies encompass augmenting workforce diversity and representation, eradicating prejudice and bias in recruitment, hiring and promotion procedures and establishing inclusive and accessible environment and services that cater to the requirements of diverse individuals and communities. Furthermore, cultivating an inclusive culture necessitates examining and questioning systemic obstacles and structural disparities that sustain social exclusion and marginalization. This entails advocating for policy changes that advance fairness and impartiality and empowering marginalized communities to engage in decision-making procedures that impact their existence. In addition, preventing social inequality necessitates examining and resolving underlying factors contributing to poverty, inequality and injustice. These factors encompass structural elements, such as disparities in educational access, healthcare provision, housing availability and employment prospects. Progressive social policies that involve the redistribution of resources, the expansion of social safety nets, and the promotion of economic empowerment and social

Figure 21.3 Societal accessibility and fairness.

mobility for marginalized populations may be necessary in addressing this issue. Moreover, promoting social unity and cohesion through community-driven initiatives, grassroot mobilization and collaborative efforts can develop resilience and cultivate a feeling of inclusion and common objectives among heterogeneous communities. To effectively tackle the societal aspects of accessibility and fairness in adopting Si/SiO_2 NP-enabled technologies, it is imperative to adopt a comprehensive and intersectional approach that considers diverse needs of individuals and communities, perspectives and experiences. Stakeholders can ensure that Si/SiO_2 NP-enabled technologies contribute to a more equitable, inclusive and sustainable society for all by addressing inequalities in access, promoting fair distribution and pricing, fostering inclusivity and preventing social inequality (Figure 21.3).

21.5 Ethical Concerns About Transparency and Informed Consent

The ethical implications surrounding transparency and informed consent have gained prominence as the incorporation of Si/SiO_2 NPs into various consumer products, medical devices and industrial applications continues to grow. This section explores the ethical considerations about transparency and informed consent within Si/SiO_2 NP technologies. It focuses on key elements such as establishing trust by providing public information regarding the NPs and their effects, advocating for transparent communication and comprehensive education, explicitly labelling Si/SiO_2 NP products and facilitating informed consumer decision-making based on values. Establishing trust and confidence in Si/SiO_2 NP technologies relies heavily on the fundamental principle of transparency. This technology enables consumers to make well-informed decisions regarding the products they consume while also promoting a sense of responsibility among manufacturers and regulators. It is crucial to offer easily understandable and precise information regarding the existence of NPs in consumer goods and their possible effects on health and the environment. This is necessary to foster public trust and confidence. The ethical responsibility of manufacturers, the main producers and distributors of Si/SiO_2 NP products, is of considerable importance (Choi et al. 2021; Gharsallah et al. 2024; Khanna et al. 2023a, b; Li et al. 2023b; Passaro et al. 2022; Takari et al. 2021; Yang et al. 2023). They must provide detailed information regarding the existence of NPs in their products, including any related hazards or safety factors. Providing comprehensive product labels, safety data sheets and informational materials is essential in effectively conveying the existence and characteristics of NPs, along with guidelines for properly handling and disposal of these particles. In addition, manufacturers must exhibit transparency regarding their production procedures, sourcing methodologies and quality assurance protocols to guarantee compliance with safety and regulatory benchmarks. In the Si/SiO_2 NP technologies domain, government agencies, regulatory bodies and independent organizations play a significant role rather than merely observing passively. They are vital in delivering unbiased and reliable information to the general public. This encompasses the performance of risk assessments, the surveillance of product safety and the distribution of educational materials and resources to enhance awareness of the risks and advantages of NPs. Facilitating communication and cooperation among various parties involved, such as industry, academia, healthcare professionals and civil society organizations, can contribute to the establishment of agreement and confidence in NP technologies. To empower individuals and communities to make informed decisions regarding Si/SiO_2 NP technologies, it is imperative to prioritize transparent communication and provide comprehensive education. This entails providing precise and reliable information regarding the potential advantages, drawbacks and uncertainties of using NPs and the prospects for substantial involvement and conversation. To foster a comprehension of NP technologies and their ramifications, educational endeavours must encompass a broad range of target audiences, such as consumers, healthcare professionals, educators, policymakers and the general public. The proposed approach entails the creation of educational resources, workshops, seminars and public outreach initiatives aimed at tackling prevalent misconceptions, debunking fallacies and fostering the development of critical thinking skills and informed decision-making. In addition, transparent communication necessitates continuous conversation and involvement of stakeholders to tackle the emerging problems, exchange optimal methods and cultivate a culture of accountability and responsibility. This may entail creating platforms, advisory groups and stakeholder meetings to promote transparent and productive conversations on NP-related matters and implementing mechanisms for receiving feedback and input from impacted communities. It is imperative to ensure that Si/SiO_2 NP products are clearly and explicitly labelled to empower consumers to make well-informed choices and exercise autonomy in purchasing decisions. Product labels must precisely indicate the existence of NPs and furnish details regarding their characteristics, intended applications and possible hazards or safety concerns. Manufacturers must conform to labelling standards and guidelines set forth by regulatory authorities and industry organizations to guarantee uniformity and precision in product labelling. Product labels, packaging and accompanying materials should provide detailed information about the NP content, concentration, size, surface modifications and other pertinent details. Moreover, it is imperative that labelling is crafted in a readily understandable manner and easily accessible to consumers with varying levels of literacy and diverse backgrounds. To effectively convey information and facilitate product comparison, it may be necessary to use clear and concise language, graphical symbols and standardized formats. The facilitation of informed decision-making among consumers regarding Si/SiO_2 NP products necessitates the provision of dependable information, the cultivation of critical thinking skills and the acknowledgement and consideration of individual preferences and priorities (Chatterjee et al. 2021; Chen et al. 2024; Doğan et al. 2024; Goswami et al. 2022; Ilves et al. 2024; Kalal et al. 2022; Kaur et al. 2023; Kumar et al. 2022; Machado et al. 2024; Rawat et al. 2018; Sarac et al. 2022). Consumers should be able to evaluate potential advantages and drawbacks of NP-enabled products based on their values, preferences and circumstances. Facilitating well-informed consumer decision-making may necessitate implementing clear labelling and transparent communication strategies, alongside providing independent and unbiased information and resources that enable thorough product comparison and informed choices. Possible sources of information for evaluating safety, efficacy and sustainability of NP-containing products encompass online databases, consumer guides and third-party certifications. In addition, promoting consumer empowerment necessitates resolving obstacles that impede access and engagement, including socioeconomic inequalities, restricted information availability and insufficient consumer consciousness. The empowerment of individuals and communities to advocate for their rights and preferences and hold manufacturers and regulators accountable for ethical and responsible practices can be facilitated by promoting digital literacy, consumer advocacy and community engagement. A comprehensive and cooperative strategy involving manufacturers, regulators, educators, healthcare professionals and consumers is necessary

Figure 21.4 Ethical concerns about transparency and informed consent.

to effectively tackle ethical considerations about transparency and informed consent in Si/SiO_2 NP technologies. Promoting transparent communication, comprehensive education, explicit labelling and informed value-based consumer choices can cultivate trust, accountability and responsible decision-making among stakeholders, developing and utilizing NP technologies (Figure 21.4).

21.6 Comprehensive Approach to Address Moral and Societal Impacts

The increasing adoption of Si/SiO_2 NP necessitates the implementation of a comprehensive strategy that effectively tackles the ethical and societal consequences linked to their utilization. This section examines key elements of this approach, such as the incorporation of scientific research, regulatory frameworks and ethical guidelines, the equilibrium between innovation and practical application, and the emphasis on ethical decision-making for sustainable and socially responsible technological progress. It is crucial to incorporate scientific research, regulatory frameworks and ethical guidelines to navigate the intricate realm of Si/SiO_2 NP technologies and tackle their moral and societal consequences. This comprehensive approach entails the cooperation and synchronization of various parties, such as scientists, policymakers, industry representatives, ethicists and civil society organizations, to create solutions grounded in evidence and prioritize safety, fairness and sustainability. Scientific inquiry advances knowledge regarding Si/SiO_2 NP technologies, encompassing their characteristics, dynamics and potential ramifications on human well-being and the natural surroundings. Collaborative research endeavours encompassing nanotechnology, toxicology, environmental science and bioethics are imperative to fill gaps in knowledge, detect emerging hazards and guide decision-making based on solid evidence. Regulatory frameworks establish the legal and institutional structures that govern the advancement, manufacturing and utilization of Si/SiO_2 nanoparticle technologies, guaranteeing adherence to safety protocols, ethical tenets and societal norms. Strong regulatory supervision is crucial for evaluating and controlling risks, setting safety limits and ensuring responsibility among nanoparticle product producers, distributors and consumers. Ethical guidelines in Si/SiO_2 nanoparticle technologies are normative frameworks that guide ethical decision-making and behaviour (Galarza-Acosta et al. 2023; Khanna et al. 2023b; Kuo et al. 2024; Li et al. 2023b; Passaro et al. 2022; Subramanian et al. 2020; Takari et al. 2021; Wang et al. 2021; Yang et al. 2023; Zhang et al. 2023b). All stages of the technology lifecycle, including research and development, commercialization and disposal, should incorporate ethical considerations such as safety, equity, autonomy and justice. Establishing ethical guidelines is crucial to guarantee the realization of the advantages offered by nanoparticle technologies while safeguarding human rights, environmental integrity and societal well-being. Ensuring a harmonious equilibrium between innovation and responsibility is crucial to fully exploit the revolutionary capabilities of Si/SiO_2 nanoparticle technologies while mitigating their adverse effects on society. Innovation propels advancement and economic expansion, cultivating ingenuity, competitiveness and affluence. Nevertheless, innovation must be accompanied by a sense of responsibility to guarantee that technological progress contributes to the collective welfare and effectively tackles urgent societal issues. Responsibility involves carefully examining the wider ethical, social and environmental consequences of Si/SiO_2 nanoparticle technologies and implementing strategies to minimize potential risks and optimize advantages. To address ethical concerns and societal needs proactively, it may be necessary to integrate principles of responsible innovation, such as anticipation, reflection, inclusion and responsiveness, into technology development processes.

In addition, responsible innovation requires active involvement with various stakeholders, such as consumers, communities, advocacy groups and regulatory agencies, to obtain input, collect feedback and integrate various perspectives into the decision-making procedures. Participatory methodologies, such as engaging stakeholders in consultations, implementing citizen science initiatives and organizing co-design workshops, can effectively promote transparency, accountability, and responsiveness in developing and deploying Si/SiO_2 nanoparticle technologies, aligning with societal values and preferences. Ensuring that Si/SiO_2 nanoparticle technologies contribute to sustainable and socially responsible advancements that benefit society requires prioritizing ethical decision-making. Ethical decision-making entails carefully evaluating conflicting interests, values and factors to arrive at well-informed choices that adhere to ethical principles and advance the collective welfare. When examining Si/SiO_2 nanoparticle technologies, ethical decision-making necessitates carefully evaluating the potential hazards and advantages associated with nanoparticle utilization and the wider societal ramifications of technological progress (Chatterjee et al. 2021; Chen et al. 2024; Goswami et al. 2022; Ilves et al. 2024; Kaur et al. 2023; Kumar et al. 2022; Rawat et al. 2018; Sarac et al. 2022). To address ethical dilemmas, conflicts of interest and unintended consequences, performing ethical risk assessments, engaging in stakeholder consultations and conducting impact analyses may be necessary. In addition, the ethical decision-making process necessitates transparency, accountability and integrity in developing and utilizing technology. This encompasses the act of revealing conflicts of interest, abiding by ethical codes of conduct and demonstrating respect for human rights, cultural values and environmental sustainability. To achieve sustainable and socially responsible technological advancements, it is essential for all stakeholders, such as researchers, policymakers, industry leaders and civil society organizations, to work together and prioritize ethical decision-making. By incorporating scientific research, regulatory frameworks and ethical guidelines, stakeholders can ensure that Si/SiO_2 nanoparticle technologies contribute to a more equitable, sustainable and ethical future for humanity and the planet. This involves striking a balance between innovation and responsibility and prioritizing ethical decision-making.

21.7 Conclusion

In summary, the extensive utilization of Si/SiO_2 NPs in diverse technological contexts has revealed a plethora of ethical and societal concerns that necessitate meticulous consideration and resolution. During this conversation, we examined the ethical implications related to the use of Si/SiO_2 NPs, such as potential health hazards, environmental sustainability, societal inclusivity and equity, transparency and obtaining informed consent. Concerns have been raised regarding the possible long-term health effects, environmental impact and equitable distribution of Si/SiO_2 NPs when incorporated into consumer products, medicine and industrial processes. Furthermore, the significance of implementing a comprehensive strategy to tackle these challenges effectively has been emphasized by concerns about transparency, informed consent and ethical decision-making. To progress, it is crucial to acknowledge the interdependence of moral and societal concerns about utilizing Si/SiO_2 NPs and to embrace a holistic strategy that incorporates scientific investigation, regulatory structures and ethical principles. The proposed methodology should emphasize safety, sustainability and social responsibility while promoting innovation and technological progress. Furthermore, achieving a harmonious equilibrium between incorporating Si/SiO_2 NPs into everyday existence and ethical deliberations is crucial. This equilibrium guarantees the optimization of advantages of these technologies while mitigating potential risks and adverse consequences. To attain this equilibrium, it is imperative to cooperate actively and involve various parties, such as researchers, policymakers, industry leaders and civil society organizations. They possess the capacity to formulate and execute strategies that foster ethical, sustainable and socially responsible progressions in technology. To unlock the transformative potential of nanotechnology, it is crucial to address the moral and societal issues associated with Si/SiO_2 NP applications through a comprehensive and collaborative approach. In addition to adhering to ethical principles and safeguarding human health and the environment, this approach also fosters the overall welfare of the society. This inherent capacity ought to motivate us to engage in proactive measures.

References

Abduljabar, R.S., Sajadi, S.M., and Al-Naqshabandi, M.A. (2023). Biosynthesis of novel *Hypercium sabrum* extract mediated functionalized Ag@Fe_3O_4@SiO_2 NCs for the efficient removal of poisonous pollutants: greener journey and bioactivity. *Molecular Catalysis* 550: 113590. https://doi.org/10.1016/j.mcat.2023.113590.

Adegoke, O., Oyinlola, K., Achadu, O.J., and Yang, Z. (2023). Blue-emitting SiO_2-coated Si-doped ZnSeS quantum dots conjugated aptamer-molecular beacon as an electrochemical and metal-enhanced fluorescence biosensor for SARS-CoV-2 spike protein. *Analytica Chimica Acta* 1281: 341926. https://doi.org/10.1016/j.aca.2023.341926.

Arooj, A., Tahir, K., Khan, A.U. et al. (2022). One-step fabrication of surfactant mediated Pd/SiO_2, a prospect toward therapeutic and photocatalytic applications. *Inorganic Chemistry Communications* 142: 109692. https://doi.org/10.1016/j.inoche.2022.109692.

Cani, D., van der Waal, J.C., and Pescarmona, P.P. (2021). Highly-accessible, doped TiO_2 nanoparticles embedded at the surface of SiO_2 as photocatalysts for the degradation of pollutants under visible and UV radiation. *Applied Catalysis A: General* 621: 118179. https://doi.org/10.1016/j.apcata.2021.118179.

Chakraborty, S., Singh, A., and Roychoudhury, A. (2022). Biogenic nanoparticles and generation of abiotic stress-resilient plants: a new approach for sustainable agriculture. *Plant Stress* 6: 100117.

Chatterjee, R., Lidin, S., Bhattacharyya, S., and Kuila, S.K. (2021). Structure, microstructure and photocatalytic properties of embedded spherical Cu nanoparticles on $Cu_2O–SiO_2$. *Materials Chemistry and Physics* 263: 124360. https://doi.org/10.1016/j.matchemphys.2021.124360.

Chen, Q., Liu, J., Tang, L. et al. (2024). A novel ex-situ method to fabricate pH-responsive material based on core-shell $Fe_3O_4@SiO_2$ nanoparticles for multi-functional oil-water separation and efficient recycling. *Journal of Environmental Chemical Engineering* 12 (2): 112422. https://doi.org/10.1016/j.jece.2024.112422.

Choi, H., Kim, E., Kwon, S. et al. (2021). Enhanced optical absorption in conformally grown MoS_2 Layers on SiO_2/Si substrates with SiO_2 nanopillars with a height of 50 nm. *Nanoscale Advances* 3 (3): 710–715. https://doi.org/10.1039/d0na00905a.

Choudhury, D., Rincon, C., Wei, R. et al. (2022). Polydopamine + SiO_2 nanoparticle underlayer for improving DLC coating adhesion and durability. *Surface and Coatings Technology* 429: 127964. https://doi.org/10.1016/j.surfcoat.2021.127964.

Doğan, B., Yeşilyurt, M.K., Yaman, H. et al. (2024). Green synthesis of SiO_2 and TiO_2 nanoparticles using safflower (*Carthamus tinctorius* L.) leaves and investigation of their usability as alternative fuel additives for diesel-safflower oil biodiesel blends. *Fuel* 367: 131498. https://doi.org/10.1016/j.fuel.2024.131498.

Galarza-Acosta, G.L., Parra, J.G., Hernández-Bravo, R. et al. (2023). A computational chemistry approach to the molecular design of SiO_2 nanoparticles coated with stearic acid and sodium stearate in ethanol solvent. *Colloids and Surfaces A: Physicochemical and Engineering Aspects* 679: 132527. https://doi.org/10.1016/j.colsurfa.2023.132527.

Gharsallah, F., Ulusoy, H.I., and Ghorbel, N. (2024). Highly selective analysis of glipizide as an antidiabetic drug in urine samples using HPLC-DAD technique via magnetic nanocomposites ($Fe_3O_4@SiO_2$@tryptophan). *Microchemical Journal* 197: 109861. https://doi.org/10.1016/j.microc.2023.109861.

Goswami, P., Sharma, M., Srivastava, N., and Mathur, J. (2022). Assessment of the fungicidal efficacy of biogenic SiO_2 NPs in *Eruca sativa* against fusarium wilt. *Journal of Natural Pesticide Research* 2: 100011. https://doi.org/10.1016/j.napere.2022.100011.

Hassan, Y.M., Guan, B.H., Chuan, L.K. et al. (2022). The synergistic effect of Fe_2O_3/SiO_2 nanoparticles concentration on rheology, wettability, and brine-oil interfacial tension. *Journal of Petroleum Science and Engineering* 210: 110059. https://doi.org/10.1016/j.petrol.2021.110059.

Hassan, S.S., Hubeatir, K.A., and Al-Haddad, R.M.S. (2023). Characterization and antibacterial activity of silica-coated bismuth ($Bi@SiO_2$) nanoparticles synthesized by pulsed laser ablation in liquid. *Optik* 273: 170453. https://doi.org/10.1016/j.ijleo.2022.170453.

Huang, H., Modanese, C., Sun, S. et al. (2018). Effective passivation of P+ and N+ emitters using SiO_2/Al_2O_3/SiNx stacks: surface passivation mechanisms and application to industrial p-PERT bifacial Si solar cells. *Solar Energy Materials and Solar Cells* 186: 356–364. https://doi.org/10.1016/j.solmat.2018.07.007.

Ilves, V.G., Zuev, M.G., Vasin, A.A. et al. (2024). Properties of an amorphous crystalline nanopowder Si–SiO_2 produced by pulsed electron beam evaporation. *Materials Chemistry and Physics* 316: 129026. https://doi.org/10.1016/j.matchemphys.2024.129026.

Kalal, P.R., Tomar, R.S., and Jajoo, A. (2022). SiO_2 nanopriming protects PS I and PSII complexes in wheat under drought stress. *Plant Nano Biology* 2: 100019. https://doi.org/10.1016/j.plana.2022.100019.

Kaur, M., Pandey, A., Yadav, R. et al. (2023). Growth and characterization of gold nanoislands & quantum dots on marker-based SiO_2/Si substrate for SEM resolution intercomparison studies. *Materials Today Communications* 35: 105741. https://doi.org/10.1016/j.mtcomm.2023.105741.

Khanna, S., Chidambaram, S., and Rathinam, M. (2023a). Preparation of urchin like structures of SiO_2/CuO by low temperature hydrothermal synthesis as adsorbent for dyes and efficient catalyst for 4-nitrophenol. *Surfaces and Interfaces* 41: 103174. https://doi.org/10.1016/j.surfin.2023.103174.

Khanna, S., Kiruthika, M., Chidambaram, S., and Rathinam, M. (2023b). Preparation of hierarchical self-assembled SiO_2/ZnO composite by low-temperature hydrothermal approach for enhanced adsorption of 4-nitrophenol and dyes. *Materials Science and Engineering B* 293: 116484. https://doi.org/10.1016/j.mseb.2023.116484.

Kumar, D., Ganat, T., Lashari, N. et al. (2022). Experimental investigation of GO-HPAM and SiO_2-HPAM composite for CEOR: rheology, interfacial tension reduction, and wettability alteration. *Colloids and Surfaces A: Physicochemical and Engineering Aspects* 637: 128189. https://doi.org/10.1016/j.colsurfa.2021.128189.

Kuo, Y.-H., Hsu, M.-C., Wang, W.-J. et al. (2024). Highly conductive riboflavin-based carbon quantum dot–embedded $SiO_2@MoS_2$ nanocomposite for enhancing bioelectricity generation through synergistic direct and indirect electron transport. *Nano Energy* 121: 109251. https://doi.org/10.1016/j.nanoen.2023.109251.

Kurumurthy, G., Shadak Alee, K., and Narayana Rao, D. (2009). Photoluminescence studies of Si/SiO_2 nanoparticles synthesized with different laser irradiation wavelengths of nanosecond pulse duration. *Optics Communications* 282 (17): 3509–3512. https://doi.org/10.1016/j.optcom.2009.06.003.

Li, X., Gao, D., Liu, M. et al. (2023a). Synthesis of amphiphilic Janus SiO_2/styrene butyl acrylate polymer microspheres and their application in oil recovery. *Colloids and Surfaces A: Physicochemical and Engineering Aspects* 675: 132076. https://doi.org/10.1016/j.colsurfa.2023.132076.

Li, Y., Li, H., Wang, X. et al. (2023b). Preparation, properties and application of multifunctional $CaWO_4$: Tb^{3+}/SiO_2/AgBr fluorescent composite. *Physics Letters A* 490: 129175. https://doi.org/10.1016/j.physleta.2023.129175.

Machado, Y.D.R., Germano, G.C.M., Pecoraro, E. et al. (2024). Impact of SiO_2 nanoparticle morphology on scattering efficiency for random lasers. *Optical Materials* 148: 114775. https://doi.org/10.1016/j.optmat.2023.114775.

Maniyazagan, M., Naveenkumar, P., Yang, H.-W. et al. (2022). Hierarchical SiO_2@$FeCo_2O_4$ core–shell nanoparticles for catalytic reduction of 4-nitrophenol and degradation of methylene blue. *Journal of Molecular Liquids* 365: 120123. https://doi.org/10.1016/j.molliq.2022.120123.

Ningsih, L.A., Yoshida, M., Sakai, A. et al. (2022). Ag-modified $TiO_2/SiO_2/Fe_3O_4$ sphere with core-shell structure for photo-assisted reduction of 4-nitrophenol. *Environmental Research* 214: 113690. https://doi.org/10.1016/j.envres.2022.113690.

Park, J. and Bae, S. (2019). Highly efficient and magnetically recyclable Pd catalyst supported by iron-rich fly ash@fly ash-derived SiO_2 for reduction of p-nitrophenol. *Journal of Hazardous Materials* 371: 72–82. https://doi.org/10.1016/j.jhazmat.2019.02.105.

Passaro, J., Imparato, C., Parida, D. et al. (2022). Electrospinning of PVP-based ternary composites containing SiO_2 nanoparticles and hybrid TiO_2 microparticles with adsorbed superoxide radicals. *Composites Part B: Engineering* 238: 109874. https://doi.org/10.1016/j.compositesb.2022.109874.

Paul, A. and Roychoudhury, A. (2021). Go green to protect plants: repurposing the antimicrobial activity of biosynthesized silver nanoparticles to combat phytopathogens. *Nanotechnology for Environmental Engineering* 6: 10.

Paul, R., Warkad, I.R., Arulkumar, S. et al. (2022). Facile synthesis of nanostructured TiO_2-SiO_2 powder for selective photocatalytic oxidation of alcohols to carbonyl compounds. *Molecular Catalysis* 530: 112566. https://doi.org/10.1016/j.mcat.2022.112566.

Paul, U., Singh, S., and Roychoudhury, A. (2023). Nanoparticles in developing plant tolerance against biotic stress. In: *Emerging Technologies to Combat Biotic Stress in Crop Plants and Food Security* (ed. A. Bhar and A. Roychoudhury), 291–309. New York, NY: Nova Science Publishers.

Rawat, R., Tiwari, A., Vendamani, V.S. et al. (2018). Synthesis of Si/SiO_2 nanoparticles using nanosecond laser ablation of silicate-rich garnet in water. *Optical Materials* 75: 350–356. https://doi.org/10.1016/j.optmat.2017.10.045.

Rochima, E., Octaviani, D., Azhary, S.Y. et al. (2024). Effect of fish gelatin as a structure forming agent to enhance transparency and speed up degradability of bionanocomposite semi-refined kappa carrageenan film reinforced with ZnO and SiO_2 filler. *Food Hydrocolloids* 152: 109965. https://doi.org/10.1016/j.foodhyd.2024.109965.

Roychoudhury, A. (2020). Silicon-nanoparticles in crop improvement and agriculture. *International Journal on Recent Advancement in Biotechnology & Nanotechnology* 3: 54–65.

Roychoudhury, A. and Bhowmik, R. (2023). Nanoremediation of heavy metals in agricultural soil. In: *Agricultural and Environmental Nanotechnology* (ed. F. Fernandez-Luqueno and J.K. Patra), 433–450. Springer Nature Singapore Pvt. Ltd.

Sanchez, D.F., Luce, F.P., Fabrim, Z.E. et al. (2011). Structural characterization of Pb nanoislands in SiO_2/Si interface synthesized by ion implantation through MEIS analysis. *Surface Science* 605 (7): 654–658. https://doi.org/10.1016/j.susc.2010.12.011.

Sarac, B., Ivanov, Y.P., Karazehir, T. et al. (2022). Multilayer crystal-amorphous Pd-based nanosheets on Si/SiO_2 with interface-controlled ion transport for efficient hydrogen storage. *International Journal of Hydrogen Energy* 47 (10): 6777–6788. https://doi.org/10.1016/j.ijhydene.2021.12.040.

Sarker, M.Z., Rahman, M.M., Minami, H. et al. (2023). A potential recyclable catalyst: in situ growth of bimetallic Cu-Ag nanoalloy on the magnetic SiO_2/Fe_3O_4-SiO_2-NH_2 nanocomposite support using a green approach. *Colloids and Surfaces A: Physicochemical and Engineering Aspects* 668: 131447. https://doi.org/10.1016/j.colsurfa.2023.131447.

Sharma, K.V., Sai, P.H.V.S.T., Sharma, P. et al. (2023). Prognostic modeling of polydisperse SiO_2/aqueous glycerol nanofluids' thermophysical profile using an explainable artificial intelligence (XAI) approach. *Engineering Applications of Artificial Intelligence* 126: 106967. https://doi.org/10.1016/j.engappai.2023.106967.

Subramanian, B., Theriault, G., Robichaud, J. et al. (2020). Large-area crack-free Au–SiO_2 2D inverse opal composite films: fabrication and SERS applications. *Materials Chemistry and Physics* 244: 122630. https://doi.org/10.1016/j.matchemphys.2020.122630.

Sun, H., Abdeta, A.B., Zelekew, O.A. et al. (2020). Spherical porous SiO_2 supported CuVOS catalyst with an efficient catalytic reduction of pollutants under dark condition. *Journal of Molecular Liquids* 313: 113567. https://doi.org/10.1016/j.molliq.2020.113567.

Takari, A., Ghasemi, A.R., Hamadanian, M. et al. (2021). Molecular dynamics simulation and thermo-mechanical characterization for optimization of three-phase epoxy/TiO_2/SiO_2 nano-composites. *Polymer Testing* 93: 106890. https://doi.org/10.1016/j.polymertesting.2020.106890.

Ulusal, F. (2024). The synthesis and cytotoxicity analysis of gold nanoparticles coated with mesoporous SiO_2 from gold treatment wastewater for medical applications. *Inorganic Chemistry Communications* 159: 111867. https://doi.org/10.1016/j.inoche.2023.111867.

Wang, K., Tan, Y., Li, P., and Wang, Y. (2021). Recycling Si waste cut from diamond wire into high performance porous Si@SiO_2@C anodes for Li-ion battery. *Journal of Hazardous Materials* 407: 124778. https://doi.org/10.1016/j.jhazmat.2020.124778.

Yang, Y., Ding, X., Li, J. et al. (2023). Highly stable core-shell structured SiO_2@C-Ag composites for organic contaminants degradation and antibacterial application. *Colloids and Surfaces A: Physicochemical and Engineering Aspects* 665: 131263. https://doi.org/10.1016/j.colsurfa.2023.131263.

Yuan, S., Zhang, H., Wang, X. et al. (2022). Molecular insights into the uptake of SiO_2 nanoparticles on phospholipid membrane: effect of surface properties and particle size. *Colloids and Surfaces B: Biointerfaces* 210: 112250. https://doi.org/10.1016/j.colsurfb.2021.112250.

Zhang, D.-S., Liu, Y., Ren, X. et al. (2023a). Rational design of zero-valence and well-dispersed copper nanocluster stabilized by carbon-coated SiO_2 for highly effective and ultrafast reduction of nitroarenes. *Colloid and Interface Science Communications* 56: 100730. https://doi.org/10.1016/j.colcom.2023.100730.

Zhang, X., Lv, X., Xie, T. et al. (2023b). Tailored hydrogel entrapped catalyst with highly active, controlled catalytic activity for the removal of 4-nitrophenol (4-NP). *European Polymer Journal* 200: 112533. https://doi.org/10.1016/j.eurpolymj.2023.112533.

Zhao, J., Tian, S., Li, P. et al. (2023). Molecular dynamics simulation and experimental research on the influence of SiO_2-H_2O nanofluids on wettability of low-rank coal. *Colloids and Surfaces A: Physicochemical and Engineering Aspects* 679: 132580. https://doi.org/10.1016/j.colsurfa.2023.132580.

Zhao, J., Tian, S., Jiang, Z. et al. (2024). Study on the mechanism of SiO_2-H_2O nanofluid enhanced water injection in coal seam. *Applied Surface Science* 658: 159843. https://doi.org/10.1016/j.apsusc.2024.159843.

Zhu, W., Sun, P., Ran, W. et al. (2021). Rational design and facile hydrothermal-thermal conversion synthesis of hierarchical porous urchin-like Cu_2–xSi_2O_5 $(OH)_3$·xH_2O and CuO/SiO_2 hollow microspheres for high efficiency catalytic reduction of nitroarenes and adsorption of organic dye. *Chemical Engineering Journal* 411: 128442. https://doi.org/10.1016/j.cej.2021.128442.

Index

Silicon Nanoparticles: Synthesis, Uptake, and Applications, First Edition.
Edited by Archana Singh, Aryadeep Roychoudhury, and Indrakant K. Singh.

g

h

i

k

l

m

Printed and bound by CPI Group (UK) Ltd, Croydon, CR0 4YY
17/09/2026
14964331-0007